Atlas of the Textural Patterns of Basic and Ultrabasic Rocks and their Genetic Significance

Atlas of the Textural Patterns of Basic and Ultrabasic Rocks and their Genetic Significance

S. S. Augustithis

Walter de Gruyter · Berlin · New York 1979

S. S. Augustithis

Professor of Mineralogy, Petrography, Geology
National Technical University of Athens
Athens, Greece

CIP-Kurztitelaufnahme der Deutschen Bibliothek

Augustithis, Stylianos Savvas:
Atlas of the textural patterns of basic and ultrabasic rocks und their genetic significance / S. S. Augustithis. – Berlin, New York:
de Gruyter, 1979.
ISBN 3-11-006571-1

Library of Congress Cataloging in Publication Data

Augustithis, S S
Atlas of the textual patterns of basic and ultrabasic rocks and their genetic significance.
Bibliography: p.
Includes indexes.
1. Rocks, Ultrabasic. 2. Rocks, Igneous.
3. Petrofabric analysis. I. Title.
QE462.U4A92 552'.1 78-27789
ISBN 3-11-006571-1

Typesetting: Otto Gutfreund & Sohn, Darmstadt.
Printing: Karl Gerike, Berlin.
Binding: Lüderitz & Bauer, Berlin.
Printed in Germany.

Dedicated to my former teacher and friend
Professor Dr. F.K. Drescher-Kaden

Preface

The recent progress in geodynamics has revolutionised our concept regarding the genesis of ultrabasic and basic rocks. In contrast to the explanation that the ultrabasics are differentiation products of a basaltic magma, mantle diapirism or oceanic crust-upper mantle obductions are put forward as explanations for the origin of ultrabasic complexes and rock types.

Parallel with the progress in geodynamics, progress in the petrofabric analysis of rocks has established the ultrametamorphic origin of rock types that were previously regarded to be magmatic (plutonic). Petrofabric analysis based on comparative microstructural studies provided a compounding amount of information in support of the ultrametamorphic origin of granites (see F.K. Drescher-Kaden, 1948, 1969, and S.S. Augustithis, 1973).

In a comparable and commensurable way present petrofabric analyses show evidence of an ultrametamorphic origin of a number of basic and ultrabasic rocks referred to as paragabbros and para norites, which were previously considered to be magmatic-plutonic.

However, realising that in addition to the mantle derivatives, and to the ultrametamorphic rock types, rock products of consolidation from melts exist, the present Atlas has as an aim to study and also present the magmatogenic textural patterns found in volcanic and subvolcanic basic and ultrabasic rocks, such as ophitic and spinifex textural patterns and the rocks in which they occur.

A most extensive petrofabric analysis of magmatogenic textural patterns of basic rocks is presented in the "Atlas of the textural patterns of basalts and their genetic significance" (Augustithis, 1978). There is therefore a limitation in the present Atlas as far as the presentation of basaltic textures is concerned.

Since the basic and ultrabasic rocks are the petrogenetic environment for the formation of deposits of Fe, Cr, Ni, Co, Pt in discussing the textural patterns of these rocks one is inevitably involved in the study of these elements and some of their important minerals and textures. However, there is a limitation imposed by the very nature of the present Atlas which is more concerned with the common and important textural patterns of the basic-ultrabasic rocks and on the petrogenetic significance of these textures.

The Atlas does not follow a classical scheme of classification of basic and ultrabasic rocks, since the petrofabric analysis based on textural patterns shows that rocks previously considered to be magmatic-plutonic are in reality mantle-derivatives or ultrametamorphic. The sequence of crystallization as established by petrofabric analysis and the recognition of textural patterns is considered genetically more important than a classification of these rocks on purely mineralogical or chemical criteria or even according to a combination of these.

The basic and ultrabasic rocks show most diverse and complex textures, the presentation of the most common and important textural patterns and their genetic interpretation is the main aim of the present volume, and is trying to fill a gap in the existing literature. Nevertheless an extensive consideration of the geological setting as well as the mineralogy and geochemistry of some of these rocks is taken in consideration and is discussed.

Another feature of the present volume is the study and the re-interpretation of the textures and rock types of well-known ultrabasic complexes and occurrences all over the world, admittedly though, with the result often to enhance existing controversies. Also as the list of contents shows, a great variety of metamorphic-ultrametamorphic and magmatic (volcanic) rock types are texturally analysed as well as mantle derived rock types. Realising the enormous possibilities of textural intergrowths which the rock-forming minerals and ores of these rocks could present the author feels that the present Atlas is only an introduction on the subject.

In the attempt to study the microstructure of basic and ultrabasic rocks, the author made use of material collected over a period of more than a quarter of a century and also collections kindly sent by colleagues. The author especially thanks the following for their kind help.

Prof. Alfredo San Miguel Arribas (Spain)
Mr. D.S. Boutsias (S.Africa)
Prof. D. Grigoriev (U.S.S.R)
Prof. Dr. F.K. Drescher-Kaden (Germany)
Mr. D. Minatidis (Canada)
Dr. E. Mposkos (Athens)
Prof. G. Panou (Belgium)
Dr. Th. Pantazis (Cyprus)
Dr. K. Spathi (Athens)
Dr. A. Vgenopoulos (Athens)
Prof. E.A. Vincent (England)

Acknowledgements

The author thanks the authors and publishers of the following figures, which have been used in the present Atlas:

Fig. 81, by F. K. Drescher-Kaden in »Granitprobleme« 1969 published by Academie Verlag GmbH, Berlin.

Fig. 461, by J. A. Wood et al, in »Lunar Anorthosites«, Science Vol. 167. No 3918. 30 January 1970, pubslished by the American Association for the Advancement of Science.

Fig. 539, 540 and 541 by E. Gübelin in the Journal of Gemmology published by the Gemmological Association of Great Britain, 1948, 1952.

Fig. 549 and 550 by A. Scharlau in »Petrographische und Petrologische Untersuchungen in der Hauptzone des östlichen Bushveldkomplex« 1972, Dr. Thesis, University of Frankfurt am Main, Germany.

Fig. 551, by Wager and Deer from Theoretical Petrology by T. W. Barth (1952). Published by John Wiley, New York.

Fig. 586, by J. Willemse in »The Geology of the Bushveld Igneous Complex...«, Magmatic Ore-deposits, 1969 published by Lancaster Press Inc. U.S.A.

Fig. 600 (a, b, c) by K. Spathi in »Microtectonic Studies of Chromites of Vourinon (Kozani), Greece« published by the Geological Survey of Greece 1966.

Fig. 609. Compiled by L. M. Bear, published by the Geological Survey of Cyprus.

Fig. 610 by J. G. Gass and D. Masson-Smith in »Principles of Physical Geology« by A. Holmes, (1965) published by Thomas Nelson and Sons Ltd.

Fig. 612 by M. E. Fleet in »The growth habits of olivine...«, Canadian Mineralogist Vol. 13, p. 293-297 published by Canadian Mineralogical Society.

Fig. 616 and 617a by Arndt et al in »Komatiitic and Iron rich Tholeiitic Lavas...«, Journal of Petrology, Volum. 18, 1977 published by Oxford University Press.

Fig. 663 by Dr. Fred. Seligmann.

Table I by F. K. Drescher Kaden, 1969. »Granitprobleme«, Akademie-Verlag, Berlin

Table III and IV by H. Rosenbusch 1898 (1923) Elemente der Gesteinslehre. E. Schweizerbart'sche Verlagsbuchhandlung, Stuttgart.

Table VI by L. J. Cabri, 1976. Glossary of Platinum-group minerals. Economic Geology. Vol. 71. No 7, pp. 1475–1480.

Contents

Errata

Page 25 (chapter 16 and its illustrations, Fig. 580, 581)
should be
(chapter 16 and its illustrations, Fig. 570, 571)

Page 73 *Ni and Cr are interrelated
should be
**Ni and Cr are interrelated

Page 187 Fig. 227 Spessartite crystalloblast (streptoplast)
should be
Fig. 227 Spessartite crystalloblast (streptoblast)

Page 238 (Fig. 376) Lekempti, W. Ethiopia
should be
Lekempti, W. Ethiopia.
With crossed nicols.

Page 238 Fig 387b
should be
Fig. 378b

(For Fig. 387 see page 239)
should be
(For Fig. 377 see page 239)

Page 306 (Lable on the Fig. 569)
mp-y
should be
m-py

Chapter 1 Olivine blastoids* in olivinefels with nematoblastic (velonblastic) tremolite. (Olivine megablasts along shearing planes of olivinefels)

The experimental work of Bowen and Schairer (1935) has shown that the synthetic olivines form a complete solid solution series, the pure magnesium mineral melting at 1890°C and the pure iron olivine at 1205°C.

In contrast to these high temperatures of synthetic olivines, olivines can crystallise under metamorphic-metasomatic conditions. Tilley (1947) has described forsterite porphyroblasts from dunite mylonites of St. Paul's Rocks, in which the (010) plane of the olivines is parallel to the foliation of the rock.

Harker (1950) describes forsterite formations by thermal metamorphism of impure magnesium limestone. Also Deer et al. (1962) mention that forsterite is an early product of the thermal metamorphism of impure dolomitic limestones, and is present in many marbles and ophicalcites. Furthermore, the thermo-chemical study of the equilibrium-phase relationships during the metamorphism of siliceous carbonate rocks by Weeks (1956) indicates that forsterite crystallises most probably at a temperature of 80°C above diopside in the "progressive metamorphism" of impure calcareous rocks.

A rather impressive olivine blastesis, with about 17.5 % iron is described by Drescher-Kaden (1961, 1969) from Prata South Chiavenna, Italy. The olivine blastesis is considered to be a post-kinematic topo-metasomatic growth. The following is quoted from Drescher-Kaden (1969).

"Allen diesen Talkschiefervorkommen gemeinsam ist wiederum die starke Durchbewegung und die weitgehende Vertalkung bei großer Dünnschiefrigkeit. Da die bisherigen Beobachtungen ergeben hatten, daß die Olivinkristallisation – wenn vorhanden – zeitlich vor der Serpentin- und Talkbildung lag, mußte die Feststellung einer späten Olivinblastese besonders auffallen. Hier handelt es sich um eine Olivingeneration, die sicher nach der Deformation der Schiefer und der Ausarbeitung ihrer Mineralgemeinschaft entstanden ist; denn sie ist auf Spalten angesiedelt, deren Bildung am Ende des Deformationszyklus der Schiefer liegt. Wenn mitunter noch letzte Bewegungen die Spalte und ihren Inhalt gerade noch mitergriffen haben, so ist die Hauptmasse der z. T. mehrere Zentimeter großen Olivine offenbar so spät in der Spaltenfüllung gewachsen, daß sie durch die letzten Deformationen des Gesteins nicht mehr beeinflußt werden konnten. Die Bildung dieser Olivinblasten ist höchst bedeutungsvoll und stellt ein Problem dar, das in seinen Einzelheiten noch nicht völlig gelöst ist. Zwar ist es keine Frage, daß bei der Kürze und geringen Reichweite der olivinerfüllten Spalten das Füllmaterial nicht von weither antransportiert worden sein kann, sondern dem unmittelbaren Nachbarbereich der Klüfte entstammen muß. Dann aber bleibt die Erklärung offen, weshalb zu Beginn der Mg-Metasomatose Olivin, sodann Serpentin und Talk – die ihn wieder abbauen – und zuletzt wiederum Olivin gebildet wird, wenn diese Vorgänge in erster Linie temperaturbedingt sind, wie häufig angenommen worden ist!

Bei der ersten Olivingeneration steht es fest, daß sie zur Zeit des ersten Bildungsintervalls der Mg-Metasomatose weit zahlreicher vertreten war als

* In support of a metamorphic-metasomatic origin of the ultrabasics is the enstatitisation process suggested by Carswell et al. (1974). Regarding the metasomatic origin of enstatite in ultrabasic bodies Carswell et al. have proposed that enstatite, anthophyllite, tremolite and chlorite are developed in a zoned sequence as a vein cross cutting a peridotite body located within the basal gneiss region of southern Norway. They are believed to have formed during the main regional metamorphic event at temperatures around 700° C and P total in excess of 6kb. Regarding the origin of enstatite Carswell et al. stated: "One interesting and perhaps important finding of the study was the undoubted presence of enstatite, a zonal mineral. Enstatitization therefore occurred as a metasomatic process on this very small scale and suggests the possibility of the same or a similar mechanism accounting for the development of the enstatite rich, ultrabasic masses which occur in some parts of this region of Southern Norway."

später. Sie wurde durch die nachfolgenden Serpentinisierungen und Vertalkungsvorgänge stark reduziert. Das ist auch der Grund, weshalb es so schwierig ist, in den Rahmengesteinen selbst noch reine Olivinfelse aufzufinden, welche dagegen dort noch anzutreffen sind, wo sie der differentiellen Durchbewegung im primären Schichtverband entzogen waren: in den Myloniten."

As a corollary to the olivine blastesis by Drescher-Kaden, Oliver and Nesbitt (1972) described olivine crystalloblasts in ultra-metamorphic rocks from Western Australia.

"The possibility that the megacrystic olivines are not relict phenocrysts but are in fact porphyroblasts of metamorphic origin was alluded to by Oliver and Ward (1971) and is indicated by the common inclusion in olivine of abundant tremolite prisms similar in size, shape and orientation to those constituting the matrix. These olivines appear to have "engulfed" the matrix tremolite prisms or to have crystallised synchronously with them."

As in the case of the Chiavenna olivine crystalloblasts, the Australian occurrence also shows a high fayalite content, from 13% to 25%. However, it should be mentioned that Oliver and Nesbitt consider the olivine crystalloblastesis as due to chlorite-olivine transformation, which implies considerable mobility of elements. Furthermore, according to them "Association in apparent equilibrium of olivine and tremolite, and the virtual absence of clinopyroxene, indicate temperatures (at 2Kbrs) between 420°C and 770°C, but the presence of anthophyllite in some rocks at Pioneer (i. e. the Western Australian occurrence) favours a figure near the top of this range."

Present studies (Fig. 1) show olivine crystalloblasts in a rupture zone of the tremolitic olivinefels of the Chiavenna. Microscopic observations Fig. 2 and Fig. 3 show the olivine crystalloblasts in contact with the olivinefels. Particularly Fig. 3 shows tremolite nematoblast transversing the olivine grains of the olivinefels and extending into the adjacent olivine crystalloblast. It is clear that the tremolite blastesis is post olivine and is penetrating the olivine due to the great crystalloblastic force of the tremolite nematoblasts, see Fig. 4. In contrast Fig. 5 shows tremolite nematoblasts between the olivinefels and the olivine crystalloblast.

The pattern of arrangement of the tremolite nematoblasts in the olivinefels (Fig. 6) clearly indicates a post olivine blastesis. The pattern of arrangement of the tremolite nematoblasts in the olivineblasts (see Figs. 7 and 8) clearly shows that the nematoblasts are preferentially orientated along two directions of crystal penetrability and are not at random as would be expected if they were engulfed and included by the later olivine crystalloblasts. Also, the shape and size of the nematoblasts (or velonoblasts), as straight elongated needles within the olivine and often transversing its boundaries and without corrosion appearances (see Figs. 9 and 10), support a post--olivine age of the tremolite nematoblasts, which do not only transverse both the olivine phases but are also later than the phlogopitic micaceous phase (Figs. 11 and 12) which is present within the olivinefels and can be associated with olivine crystalloblasts. Also, the post-phlogopite growth of the tremolite is indicated by a tremolite crystalloblast transversing the boundary of two differently orientated phlogopites.

The phlogopitic phase associated with the olivinefels and olivine crystalloblasts (Fig. 13) is pre-kinematic as indicated by the deformation effects exhibited and, in cases, is in intergrowth with olivine crystalloblasts. Fig. 14 shows tectonically deformed phlogopite, which supports a pre-kinematic growth. The intergrowth of olivine and deformed phlogopite (Fig. 15) is also of genetic interest, as are the olivine bodies with protuberances parallel to and extending into the cleavage of deformed phlogopite (Figs. 16 and 17).

In contradistinction, cases are exhibited where the olivine bodies follow the phlogopite cleavage (Fig. 18) and are delimited by its presence.

On the basis of the above observations it can be suggested that the olivineblasts (olivine megacrysts) are blastogenic growths in the olivinefels and can be attributed to topometasomatism within the olivinefels phase (a recrystallization of the olivine phase), the phlogopite is a pre-kinematic phase often indicating evidence that it is invaded by the later olivine recrystallization.

In contrast to the explanation of Oliver and Nesbitt, the tremolite in the case of Chiavenna occurrence is not a pre-olivine phase but as the orientation pattern of the tremolites, both in the olivinefels and in the olivine crystalloblast, indicates, we are dealing with blastogenic phase later than the olivine phases. Comparable and commensurable is the case of the olivine "breccia" from Chiavenna, in which case large olivine fragments consisting of olivine megacryst-mosaic are bound together by a tremolitic blastic phase. Fig. 19 shows the olivine "breccia" i. e. large olivine fragments bound together by a "matrix" consisting of radiating tremolitic nematoblasts and which extend into the olivine fragments.

Fig. 20 shows large olivine grains (of an olivine-fragment) marginally invaded by later tremolite-crystalloblasts. A good example of the later tremolite blastesis is shown in Fig. 21 where tremolite nematoblasts transverse the olivine grain and, in one case, transverse the boundary of two differently orientated olivines.

The blastic growth of the tremolite may exhibit "spinifex" textural patterns, often radiating tremolites in large olivine grains. Figs. 22 and 23 show tremolite-nematoblasts (velonoblasts) radiating and penetrating into the olivine, often extending from the boundary of the olivine fragments inwards due to their crystalloblastic force.

The "spinifex" tremolitic textures are definitely crystalloblastic and often in addition to the radiating patterns of the amphiboles, (Figs. 24 and 25) inter-penetration of the tremolites is indicated. (Comparable inter-penetration "spinifex" anthophyllite textural patterns are exhibited in the transition from leuchtenbergite/anthophyllite-rock types, see chap. 6). As Fig. 25 shows, the interpenetration of the tremolite-radiating growths, resulting in a blastogenic "spinifex" within large olivine grains, is additional evidence of their blastogenic growths.

As in the case of Chiavenna olivine-tremolite intergrowths, where no equilibrium of olivine/tremolite could be accepted, similarly in the case of the olivine fragments bound together by tremolite, there is no equilibrium of olivine-tremolite, in the sense of phases crystallising out of a liquid.

Chapter 2 Metasomatic transformations of marbles. (Olivine, pyroxene, amphibole, mica and plagioclase blastesis)

The growth of diopside and forsterite in dolomitic marbles has been discussed by a great number of workers e. g. Bowen (1940), Harker (1950), Deer et al. (1962), Weeks (1965).

Deer et al. (1962) in discussing the olivine growth in impure dolomitic marbles and regrowth of olivines in sediments, state the following:

"The (Mg, Fe) – olivines occur in a number of metamorphic rocks. Forsterite is an early product of the thermal metamorphism of impure dolomitic limestones, and is present in many marbles and ophicalcites. In the latter it has been mainly converted to serpentine. Forsterite also crystallises from regionally metamorphosed impure dolomitic limestones, in which environment its crystallization is preceded by a tremolitic or edenitic amphibole. From a study of the progressive metamorphism of siliceous limestone and dolomite, Bowen (1940) considered that during the progressive decarbonation of these rocks, forsterite crystallises earlier than diopside, but a more recent thermochemical study (Weeks, 1956) of the equilibrium phase relationships during the metamorphism of siliceous carbonate rocks indicates that, for a given pressure, the formation of forsterite occurs at a temperature some 80°C above that at which diopside crystallises. The formation of forsterite in an anhydrous environment is illustrated by the reaction:

$$\underset{\text{dolomite}}{2\,CaMg(CO_3)_2} + SiO_2 \longrightarrow \underset{\text{forsterite}}{Mg_2SiO_4} + 2\,CaCO_3 + 2CO_2$$

If water is available during metamorphism the formation of forsterite may result from the more complex reaction:

$$\underset{\text{tremolite}}{Ca_2Mg_5Si_8O_{22}(OH)_2} + \underset{\text{dolomite}}{11\,CaMg(CO_3)_2} \longrightarrow \underset{\text{forsterite}}{8\,Mg_2SiO_4} + 13\,CaCO_3 + 9\,CO_2 + H_2O$$

As pointed out, Harker (1950) also discusses the development of diopside and forsterite during the pure thermal and regional metamorphism of impure magnesian limestones. Harker, in discussing the thermal metamorphism of impure magnesian limestone, states:

"The metamorphism of dolomites and partly dolomitic limestones carrying various impurities presents some features of special interest. The salient fact that emerges is that silica reacts with the magnesian in preference to the calcic carbonate. It follows that, unless disposable silica is present in amount sufficient for complete decarbonation of the rock, one incident of the metamorphism is dedolomitization. This is an effect which we have already observed in some pure carbonate rocks as a consequence of the fact that the magnesian carbonate by itself is more easily dissociated than the calcic.

Consider first a dolomitic rock containing silica as its only impurity. Here are the materials for making a number of compounds, the simple lime metasilicate wollastonite and orthosilicate larnite; the doubled silicates diopside, tremolite and monticellite; and the purely magnesian silicates enstatite, anthophyllite, and forsterite. We find in fact that the first mineral to form and with a limited supply of silica, the only mineral is the magnesian forsterite."

Furthermore, Harker (1950) in discussing the metamorphism of impure magnesian limestone in his book "Metamorphism (Chap. XVI) Regional metamorphism of calcareous sediments" states the following:

"In limestones which were partly or completely dolomitized, metamorphism has followed a different course; for dolomite, with its relatively low dissociation-pressure as compared with calcite, has not normally been prevented from reacting with any free silica present. Here, however, the first product is not forsterite, as in simple thermal metamorphism, but the stress-mineral amphibole. It appears as numerous needles or slender crystals of tremolite, or less commonly a pale green actinolite, being at this early stage non-aluminous. With it may be associated more or less muscovite, biotite, and magnetite, representing micaceous, chloritic and ferruginous impurities in the original rock. By reaction with some of these minerals, tremolite gives place with advancing metamorphism to a green aluminous hornblende. If, however, the original magnesian limestone contained none but siliceous impurity, tremolite persists, in crystals of increasing dimensions, far into the garnet-zone (as determined in argillaceous sediments), being then replaced by diopside and forsterite."

Both Harker (1950) and Deer et al. (1962) attribute the formation of forsterite to the Mg present in the impure calcareous magnesian limestone (i. e. dolomites). In contrast Drescher-Kaden (1961, 1969), in discussing the metasomatic effects on the Trivena (Malga Trivena, Val di Breguzzo, Adamello) triassic marble, puts forward indisputable evidence of a Mg-SiO_2 metasomatism. Figs. 26 and 27 show a diffusion front of a metasomatic solution which has resulted in the formation of pyroxene, hornblende, olivine, chlorite and antigorite.

Particularly, the growth of forsterite and its textural patterns in the Trivena triassic marbles is presently discussed. The detailed phenomenology presents all transitions from "isolated" forsterite crystalloblasts to forsterite granoblastic textures (see Figs. 28 and Fig. 29).

Fig. 28 shows idioblastic olivine with the crystal-faces coinciding with the cleavage direction of the granoblastic calcite of the marble in which the olivine has been formed metasomatically. Most probably, the calcite-cleavage acted as a delimiting factor to the olivine growth.

In contradistinction granoblastic olivine partly following the calcite cleavage and partly grown independently is shown in Fig. 30; and as Fig. 31 illustrates granoblastic olivine transverses the perfect rhombohedral cleavage of the calcite.

A good example of granoblastic olivine in coarse grained marble and with an olivine individual the crystal faces of which are delimited by the rhombic cleavage of the calcite is shown in Figs. 32a and b.

In addition to the olivine crystalloblast, the shape of which can be delimited by the calcite cleavage, in the sense that olivine crystal-faces coincide with the calcite's cleavage, comparable relations of olivine crystalloblasts and the rhombic calcite twinning is shown in Figs. 33 and 34.

Often the granoblastic olivine transverses the boundary of two differently orientated calcite grains, Fig. 35. A most interesting case of spheroidal shapes of granoblastic olivine is shown in Fig. 36. All transitions of growth from calcite to olivine are shown in Fig. 36 in the sense that the "olivine drops" are produced by the metasomatic replacement of the calcite.

The replacement of calcite by olivine can be explained by the operation of the mechanism of advancing solutions or solution fronts as a result of intergranular and intracrystalline diffusion. Evidence of such metasomatic solution fronts is shown in Figs. 26 and 27.

An additional crystalloblastic phase in the triassic Trivena marbles is phlogopite (Fig. 37). In certain cases, the phlogopite blastesis has preceded the olivine growth, as Figs. 38 and 39 shows. The olivine crystalloblasts transverse the phlogopite and are often clearly delimited or following the mica's cleavage. Both the olivine granoblastic phase and the phlogopite are metasomatic growths in the Trivena marbles. Particularly the olivine crystalloblastesis can result in topo-concentration of the olivine which practically results in the transformation of the marble in an olivine rock (see Figs. 29, 40, 41, 42).

In the Trivena marbles in addition to the olivine, diopside and phlogopite, grossular crystalloblasts are quite abundant and in cases most characteristic poikiloblasts are shown.

Figs. 43 and 44 show a fine grained calcite-mafic-mineral granoblastic texture with a large grossular poikiloblast. The garnet has grown by the assimilation of the granoblastic calcite while the mafic granoblasts (forsterite, diopside, etc.) are preserved as relic inclusions in the poikiloblastic garnet. Similarly, a coarse-grained equigranular marble with a granular mafic phase indicates grossular crystalloblasts with the mafic grains as relics in the garnet (see Figs. 45 and 46).

A comparable case of poikiloblastic grossular blastically grown in a granular calcite with fine granoblastic mafic grains is illustrated in Figs. 47 and 48 from Val Palobia. The garnet has assimilated the calcite and maintains the mafic grains as relic inclusions in the garnet crystalloblast.

In addition to the poikiloblastic grossular, amphibole poikiloblasts may also exist enclosing granular mafic minerals. When a poikiloblastic grossular comes in contact with a poikiloblastic amphibole the garnet partly invades the amphibole and encloses relics of the granular mafic minerals (Figs. 49 and 50).

The relation of the blastic garnet to the granoblastic calcite of the marble may show variations. Figs. 51 and 52 show granoblastic equigranular calcite (marble) invaded by protuberances of a poikiloblastic grossular which maintains relic mafic grains. In contrast Figs. 53 and 54 show idioblastic grossular in a coarse grained calcite texture exhibiting rhombohedral twinning. In contradistinction to the xenoblastic and poikiloblastic grossulars (xenomorphic crystals), are poikiloblastic garnets, exhibiting idiomorphism (see Figs. 55 and 56).

Indeed, the granoblastic sequence in the metasomatically effected marbles of Trivena can show a wide spectrum of blastogenic phases and topologically olivine rich rocks or a garnet rich phase may result. In cases in a granoblastic phase consisting of calcite, mafic minerals, micaceous minerals (phlogopite), granoblastic Ca-rich plagioclases are blastically developed, Figs. 57 and 58.

In cases a Ca-rich plagioclase may grow in a coarse grained calcite in the near presence of a grossular crystalloblast (see Fig. 59). The plagioclase crystalloblastesis may result in a plagioclase calcite texture and if the Ca-rich plagioclase phase predominates topologically an anorthosite-like phase may result (see Fig. 60).

As a deuteric, in fact, diaphthoretic metasomatism (or even in the realm of atmospheric mineral disintegration) is considered the alteration of olivine crystalloblasts in marble, which shows all transition phases from olivines serpentinised along cracks and margins (Figs. 61 and 62) to completely serpentinised isolated olivines in calcite.

When a topological enrichment of granoblastic olivine occurs within a marble, the subsequent serpentinisation of the olivine may result in a serpentine or serpentine with calcite or an association of olivine-calcite-serpentine (i. e. an ophicalcite rocktype), see Figs. 63 and 64.

Chapter 3 Metasomatic serpentine in marbles

The association of serpentine and marble is genetically the result of a wide spectrum of possibilities.

Orthodox views consider the serpentine/calcite association as a result of different phases of igneous activity. We can consider, as a first phase, the intrusion of basic magma, which was subsequently serpentinised and at a concluding phase, due to "hydrothermal" activity, the calcite-phase was introduced, mainly as "hydrothermal" veins. This relation was later affected by tectonic disturbances as a result of which complex serpentine-calcite association could result. However, this hypothesis, despite its soundness, fails to explain the association of marbles with serpentine, that is, where a calcareous phase includes serpentine and eventually a transition from marble to serpentine takes place, e. g. Piz Lunghin, (by Maloja, Alps).

It should be noted that these associations (marbles with serpentine and transition of marbles to serpentines) have been explained by the orthodox views as the result of synsedimentogenic ejected or detached basic igneous materials which have been subsequently serpentinised. Another explanation is the intrusion of the basic phase in the marble and the subsequent tectonical intermingling due to mobilisation of the materials. Characteristic of the orthodox views is that despite what are the petrogenetic and geological relations, the serpentines are explained as a derivative of basic magmatic intrusions or effusions and the calcareous phase as initially sedimentogenic (if it is marbles) or hydrothermal vein fillings.

In contrast to these views the association serpentine – marbles seems to be more complex and Mg-metasomatism as the main mechanism can explain many of the strucutral and textural problems. The following transformations can be understood within the element mobilisation process – Mg-metasomatism being the main process:

a) Magnesium limestone → metamorphism (either under Mg mobilisation or Mg remobilisation) forsterite formation (olivine blastesis) → serpentisation.
b) Mg-SiO_2 metasomatism in marbles → olivine, diopside etc. blastesis → serpentinisation.
c) Mg-SiO_2 metasomatism in marbles → serpentine blastesis (tremolite blastesis) serpentinisation of marbles.

In the present chapter there is an attempt to present the phenomenology of serpentine blastesis in marbles and, in general, to discuss the association of metasomatic serpentine/marbles as well as the textures which may result in the tectonic mobilisation of these phases.

The blastogenic individuals of antigorite may occur either as isolated crystalloblasts or as aggregates which may result in the transition from isolated serpentine spots to complete transformation of the marble to serpentine (Fig. 65).

The textural phenomenology of serpentine in marbles supports a blastogenic origin of the antigorite. A series of photomicrographs (Figs. 66, 67, 68 and 69) shows gigantic antigorite crystalloblasts in recrystallised calcite (marble). As is indicated in Fig. 68, an antigorite tends to attain idioblastic form. In contradistinction, Figs. 66 and 67 show xenomorphic antigorites.

Most characteristic blastogenic textural patterns result when the antigorite crystalloblasts show an intersertal texture, clearly supporting evidence for a topological development (growth) of the antigorite, (see Figs. 70 and 71); similarly, blastogenic serpentine in marble shows radiating (in their orientation) antigorite crystalloblasts (Figs. 72 and 73). These textural patterns are comparable to the classical metamorphic, radiating crystalloblasts.

In addition to these radiating in their arrangement, crystalloblasts, comparable blastogenic antigorite textures are shown in Figs. 74 and 75 where clusters or bands of antigorite are exhibited. A very common textural pattern of metasomatic serpentine in marbles is the radiating or interpenetrating fine serpentine laths, see Figs. 76 and 77.

Also supporting the metasomatic origin of serpentine are the orientated antigorite crystalloblasts. Figs. 78 and 79 show gigantic antigorite crystalloblasts parallelly orientated within a recrystallised coarse calcite texture. The orientation of serpentine crystalloblasts and the fact that the fibrous antigorite is parallelly orientated supports a synkinematic origin of the metasomatic serpentine.

As a corollary to the synkinematic serpentine are structural patterns of serpentine "nodules" in marble indicating tectonic movements, fracturing and mobilisation of the serpentine in the marble (see Fig. 65).

Evidence for the blastogenic, metasomatic growth of antigorite is the development of blastogenic antigorite transversing the boundary of two calcite crystals, Fig. 80. In this case the crystalloblastic antigorite is post-calcitic in age and due to its crystalloblastic force has transversed the boundaries of the two calcites.

Similarly, Drescher-Kaden (1969) shows crystalloblastic laths of antigorite transversing the rhombic twinning of a coarse grained calcite (see Fig. 81).

In support of Drescher-Kaden metasomatic origin of the serpentine in marbles, a series of Figs. 82a, 82b, 83 and 84 shows metasomatic serpentine following and replacing the calcite along its rhombic twinning pattern.

In Fig. 82b the serpentisation is limited and restricted to the twinning and cleavage planes of the calcite, similarly Fig. 83 again shows a replacement of the calcite by metasomatic serpentine along its rhombic twinning and cleavage, however, in comparison to Fig. 82b the antigoritisation is more advanced and it is not limited to the cleavage and twin planes.

In contradistinction to the cases presented in Figs. 82b and 83, a more advanced case of replacement of the calcite along its rhombic cleavage and twinning is illustrated in Fig. 84. The replacement of calcite by antigorite and, in this case, it follows the rhombic internal pattern of the calcite (twinning and cleavage) but it is not restricted to the planes of twinning and cleavage, extensive replacement of the calcite by antigorite has taken place.

Comparable to Fig. 84 are the Figs. 85 and 86 which also show replacement of the calcite by metasomatic antigorite, following the initial internal pattern of the calcite (cleavage and twinning) which is rhombic. The metasomatism of the antigorite has resulted in the replacement of "rhombohedral" parts within the calcite. Furthermore, as Figs. 87 and 88 show, the replacement of the calcite by the antigorite can result in a complete transformation of calcite to antigorite, where "rhombohedral" calcite-relics are left in the serpentine.

As a corollary to the replacement of calcite by antigorite, where the metasomatic serpentine took advantage of the rhombic cleavage and twinning of the calcite, Fig. 89 shows gigantic antigorite crystalloblasts orientated parallel to the rhombic pattern of the calcite's cleavage and twinning, also in this case the serpentine is later and follows the rhombic cleavage and twinning of the calcite.

In certain cases the metasomatic antigoritisation is accompanied by tremolite blastesis. The series of Figs. 90, 91 and 92 shows tremolite crystalloblasts associated with metasomatic antigorite in marble. The topological development of tremolite crystalloblasts is in accordance with the hypothesis that, due to the diffusion of metasomatic solutions and progressive metamorphism-metasomatism, a series of mafic minerals may result.

In support of the possibility of metasomatic serpentine in marbles (despite the fact that both antigorite and calcite are mineralogical "mobile-mineral phases", i.e. deuteric mobilisation is possible), a series of Figs. 93, 94a, 94b, 95 and 96 shows vein-form metasomatic infiltrations of antigorite in marble. Fig. 93 particularly shows vein-form antigorite with the antigorite laths almost perpendicular to the walls of the veinlet and intersertal to one another. The veinform antigorite and the arrangement and intersecting of the laths clearly support a later infiltration of the serpentine in the marble.

Cases are shown where magnetite ore is metasomatically mobilised within the serpentinised marble and the antigorite laths are perpendicularly intergrown with it (see Fig. 97). Mobilisation of iron oxides is common in the serpentinised marbles, Fig. 98.

Geochemical studies and comparisons of metasomatic serpentine in marbles by Drescher-Kaden show that with the metasomatic mobilisation of Mg in addition to Fe, Ni and Cr are also metasomatically introduced in the marbles: Table (I) shows the analytical results of Ni and Cr in metasomatic serpentines.

So far, we have discussed cases of serpentinisation of fine grained and coarse grained mosaic marbles, however, most convincing pieces of evidence of serpentine metasomatism are obtained by considering coarse-grained granular marbles (often consisting of calcite individuals clearly showing zonal growths) with metasomatic serpentine clearly intergranular between the calcite crystal grains. As Fig. 99 indicates, in cases the granular calcite-serpentine textures show a tectonic fracturing and mobilisation.

Both the textural patterns introduced and field relations (the transition from spotted serpentine in marbles to a complete transformation of marble to serpentine, see Fig. 65) rather support a metasomatic and blastogenic origin of the serpentine in marbles. In contrast, textures may be observed that clearly show fragments of "serpentine" or green rocks within the calcite marble. Indeed, such textural patterns seem to contradict the metasomatic hypothesis, though they could be understood as fragmented pieces of the metasomatic serpentine in a tectonically remobilised calcite texture.

Fig. 100 shows a serpentine fragment (consisting of banded and folded serpentine) in tectonically remobilised calcite. Also Fig. 101 shows a rounded serpentine fragment with an idiomorphic spinel-magnetite in recrystallised marble. A recrystallised calcite band forms a margin around the serpentine fragment. In contrast to the serpentine veinlets transversing the marble, remobilised calcite veinlets may also transverse either serpentine veins (Fig. 95) or serpentine masses Fig. 102.

Table I (By F. K. Drescher-Kaden – 1969)
Die Mg-SiO_2-Metasomatose an Kalken

Ni-Cr-Bestimmungen an metasomatischen Serpentiniten und ihren Ausgangskalken
(Anal. Dipl.-Ing. Kipping)

Vorkommen	Gehalte in ppm Ni	Cr	Gestein
Westalpen			
1. Plaun da Lej, Silser See, Oberengadin	<10	<10	Kalk des Trias-Lias-Kalkzuges Piz Lunghin-Grevasalvas, ohne äußerlich erkennbare Serpentin-Einschlüsse
2. Plaun da Lej	120	240	Kalk, äußerlich wenig kristallin, mit ganz kleinen Serpentinenkörnchen
3. Plaun da Lej	425	730	Kalk, äußerlich stärker kristallin, mit kleinen Serpentinfetzen
4. Plaun da Lej	1175	1550	Kalk, deutlich marmorisiert, mit größeren Serpentineinschlüssen
5. Piz Lunghin, Oberengadin	1420	2400	Marmor, feinkörnig, mit stark geschieferten Serpentinflatschen
6. Piz Lunghin	1650	2900	Dunkler, dichter Serpentinit
7. Lanzada, östlich Chiesa Val Malenco	35	40	Trias-Marmor mit viel Serpentin-„Häcksel"
8. Lanzada	750	960	Marmor, sehr feinkörnig, mit viel Serpentin-„Häcksel", darunter pfenniggroßen Einschlüssen
9. Lanzada	1100	2000	Serpentinschiefer mit dünnen Calcitlagen
Ostalpen			
10. Geierspitze, Tarntaler Alpen	10	n. b.	Liaskalk, dicht
11. Geierspitze	2200	n. b.	Serpentinknollen im Kalk
12. Matrei am Brenner, Schloßberg	120	n. b.	Liaskalk, dicht
13. Matrei am Brenner	520	950	Liaskalk, mit wenigen kleinen Serpentinkörnern
14. Matrei am Brenner	3000	n. b.	Serpentinknollen aus Liaskalk
Zum Vergleich:			
Wunsiedler Bucht			
15. Stemmas, bei Holenbrunn, Fichtelgebirge	<10	<10	Feinkristalliner Marmor (algonkisch) mit hirsekorngroßen mitteldicht stehenden Serpentinschmitzen
16. „*Eozoon bavaricum*" im Marmor von Obernzell bei Passau	<10	<10	Mittelkristalliner, algonkischer Marmor mit granophyrischen Serpentin-Streptoblasten

Chapter 4 Basic and ultrabasic rocks related with granite and as inclusions in the granite (e.g. example provided by considering the Seriphos granite and skarns)

General concept of basic and ultrabasic phases related to granite bodies:

A wide range of basic rocks and even ultrabasics may occur as inclusions in granitic bodies. Drescher-Kaden (1969) describes olivinefels "spheroids" of Val Bondasca as being fragments of an olivinefels-mylonite engulfed and partly corroded by fine grained granite. Present studies show olivine xenoliths in fine grained granite (see Fig. 103).

In addition to the sedimentogenic in origin basic xenoliths in granites, consisting of biotite or biotite-hornblende schists, "igneous" in origin, even initially basaltic in composition xenoliths have been described by Sederholm (1910) and Pavlicova (1963, 1964).

Sederholm (1926) has noticed the disappearing of the basic rocks included in granitised-migmatitised regions, and has thus prepared theoretically the way for the Mg-front hypothesis of Wegmann (1935). The mobilisation of the basic material existing as a pre-granitic phase into which alkaline-metasomatism caused granitisation is schematically proposed as follows:

Alkaline solutions, through diffusion and metasomatism of a pre-granitic basic phase, result in the formation of agmatite, which transgresses to a phase of xenolithic relics interspersed into the granite → disappearing of the basic xenoliths → liberation of the basic front.

Petrographic studies of the intrusive granite of Seriphos suggest the existence of the following transition phases:

A hornblende-biotite quartzitic schist (country rocks, basic in appearance) often epitotitised, is invaded by alkaline solutions and changes into an agmatite zone, (granite rich in xenolithic inclusions).

The formation of agmatite is a well known granitisation process [Drescher-Kaden (1948, 1969), San Miguel Arribas (1956), Modesto Montoto San Miguel (1967), Augustithis (1973)].

The transition of agmatite to a phase of scattered xenoliths in granites is also a common phenomenon in granitisation and in granites in general. In cases though in addition to these processes, a mobilisation of the country rock, under granitisation, in the form of veins in the granite may occur. In particular in the Seriphos granite the following transitory phases exist:

A hornblende-biotite quartzitic schist (a country rock occurrence at the granite's periphery) changes to a zone of agmatite and eventually in a phase of interspersed xenoliths in granite. The hornblende-biotite quartzitic schist could be "mobilised" and due to plagioclase crystalloblastesis changes into a "veinform" hornblendite, consisting of hornblende-biotite quartzitic schist relics and new plagioclase and hornblende crystalloblast, some of the plagioclase phenocrysts show rounded and corroded outlines due to the mobilisation of the initial country rock into the veinform phase.

A series of Figs. 104 and 105 shows "transition" steps from hornblende-biotite quartzitic schist to agmatite and Figs. 106, 107 and 108 show the "transition" step from the country rock to the "veinform" hornblendite (actually a mobilised "xenolithic" phase).

In addition, the photomicrographs Figs. 109 and 110 show the sedimentogenic texture of the basic country rock consisting of hornblende, quartz and greenish biotite and as is particularly indicated in Fig. 110 epidote and plagioclase blastesis has occurred. In contradistinction, Fig. 111 shows the texture of the veinform hornblendite, consisting mainly of quartz and hornblende. Fig. 111 shows a quartzitic relic in the mobilised veinform hornblendite in the granite and Fig. 112 shows a quartzitic relic associated with a tectonically deformed hornblende. In the hornblendite veins in the granite in addition to idiomorphic hornblende crystalloblasts, (Fig. 113) plagioclases are also exhibited, corroded and affected within the hornblende-quartz groundmass of the veinform-hornblendite, Fig. 114.

In contrast to the previous magmatic hypothesis of these elongated bodies within the granite, the field association of the hornblende-quartzitic country rocks (Fig. 106), the transition of the metamorphic phase to an agmatite and the proximity of the veinform hornblendite to the agmatite and furthermore the textural evidence support the explanation that these veinform hornblendites are actually "mobilised country rocks" along fractural lines of the granite.

The metamorphogenic interpretation of the veinform basic bodies of the Seriphos granite, Augustithis (1973) was severely criticised by Pitcher (1974) and Koehnken (1976) who maintain that such bodies can only be basic magmatic intrusions (in fact volcanic dykes). Considering the critics it is necessary to discuss these bodies in greater detail than was discussed in the "Atlas of the textural pattern of Granites and associated rocks" by Augustithis (1973), in which only a field picture was presented. It is only through a detailed textural study that a clearer picture of these bodies can be obtained. Similar metamorphogenic basic veinform bodies have been studied by Drescher-Kaden and San Miguel Arribas (per.com).

In contradistinction to the hornblendite veins transversing the granite of Seriphos, a veinform phase described previously as lamprophyric* by Ktenas (1916) and Marinos (1951) transverses the Halara marbles. These basic veins transversing the Halara marbles produce a distinct Skarn zone at their contact with the marble (Fig. 115). The skarn zone consists clearly of epidote and garnet, in fact there is a succession towards the margin from epidote to garnet.

Microscopic studies and comparisons of the green veins transversing the Halara marbles with those veins following fracture-lines in the Seriphos granite at Halara show the following distinct differences.

Mineralogically the "veins" in the granite and the veins transversing the Halara marbles differ. The vein in the granite (Fig. 108) consists predominantly of fine and idioblastic hornblendes (Fig. 113), interlocking quartz mosaic grains and lens-shaped quartzitic "augen" structures composed of interlocking mosaic quartz grains exhibiting undulating extinction. The quartz clearly exhibits tectonic deformation syngenetic to the veins mobilisation otherwise it is difficult to explain the presence of "augen" quartzitic structures. As mentioned plagioclase laths and plagioclase phenocrysts exist either idioblastic or rounded (see Fig. 114). Zircons may be present in cases within the idioblastic hornblende.

Both the composition and texture of the green veins transversing the Halara marbles are different to the veins in the granite. Fine and idioblastic epidotes predominate (Fig. 116), idioblastic diopsides often zoned and twinned are also present (Fig. 117). The garnet (andradite-grossular) forms idioblasts or xenomorphic blastogenic masses. Often a "banding" of epidote-garnet exists in the margin of these "veins" in contact with the Halara marbles. The middle mass of the vein consists of fine epidote and in cases laths of plagioclase may exist exhibiting a "fluidal-like" orientation (Fig. 118). Also idiomorphic plagioclases often full of fine epidote are present in the veins transversing the Halara marbles. An additional metasomatic phase is scapolite** either as fillings of the interspaces between idioblastic epidote (Fig. 116) or as intergranular between the fine epidote. Poikiloblastic scapolites are often present. A secondary calcite mobilisation and chloritization are low temperature processes which have affected these veins.

As the mineralogical and textural study of the veins in the granite and the veins in the marbles showed, the veins in the granite are mobilised country rock material, whereas the green veins transversing the marbles are approaching in their mineralogical composition "skarn bodies" where blastogenic growths of epidote, garnet, diopside and scapolite played an important part. On the other hand the plagioclase laths and plagioclases are more difficult to be explained; perhaps the phenocrysts represent an idioblastic phase. In any case it is clear that the basic phases transversing the marbles have been effected and influenced by the basic front which produced the skarn in the granite*** and in the adjacent marbles (see Fig. 115). In fact these veins could be

* In contrast to the veinform hornblendite veins of the Seriphos granite and the "Skarn veins" in the Halara marbles, often ultramafic dykes occur in granitic massives (e. g. Adamello) which are lamprophyric in composition and differ in composition and texture (see Chapter 11).

** Scapolite was first determined in Halara by E. Mposkos (1977)

*** In addition to the derivation of the basic front by the assimilation of the pre-granitic basic phases during granitisation, mobilisation of material from a country rock source could perhaps better explain the basic front affecting the granite itself.

considered as mobilised skarn material, occupying fissures in the marble and reacting with it.

The hypothesis of basic front of Wegmann (1935) provides a satisfactory explanation for the formation of skarn bodies, particularly in already metamorphosed marble, in which the skarn bodies have intruded (Fig. 119). As Fig. 120 shows, a granitic apophysis invades marbles and causes a bending of the already metamorphosed marble. This indicates that the granitic apophysis was already solid (plastic) as it was pushed into the marble. Evidence supporting a diapiric intrusive character of the granites can be found in Augustithis (1973). The granitic apophysis (Fig. 120) in contact with marble indicates the formation of an epidote-grossular skarn. Fig. 121 apparently shows a marble-bending due to skarn formation, however, as microscopic evidence indicates, associated with the skarn body is also granitic material.

In contrast to the development of skarns, when granitic material comes in contact with marble, epidote-andradite-grossular-magnetite skarn* may develop within the marginal part of the granite and often clearly following cracks of the granite, indicating that the basic front mobilisation acted and after the granite was sufficiently solid to allow the development of cracks (which subsequently were followed by skarn formation within the granite).

In contradistinction to the development of skarns when the granite or apophyses of it come in contact with the marble, when the same granite comes in contact with a quartzite, no skarn or melting of the quartzite takes place. Fig. 124 shows a part of the Seriphos granite in contact with quartzite, no skarns are developed, however, extensive skarns are formed in the marbles (epidote-garnet bands), above the quartzite band, where the basic front has invaded the marbles (see Fig. 124).

The mineralogy and geochemistry of the skarns may be very complex and are discussed among others by Goldschmidt (1911), Sederholm (1926), Wegmann (1935), Reynolds (1947), Harker (1950), Oen (1968), Augustithis (1973).

As Fig. 125 indicates often marbles may be extensively altered to an epidote mass with relics of the initial marble indicated. In cases, specularite may be blastically formed within the relic marble pieces (Fig. 125).

In addition to the epidote replacing the granite itself, as well as the adjacent marbles, in a comparable way grossular may replace the granite or the adjacent to the granite marbles. Fig. 126 shows garnet forming solution (the basic front) invading the intergranular spaces of a granite texture (i. e. between granular quartz of the granite) also Fig. 127 shows garnet replacing the granite with feldspar relic in the grossular (actually grossular-andradite).

In addition to the replacement of the granite or the granitic components themselves most convincing evidence of marble replacement may be indicated. Fig. 128 shows idioblastic grossular with intergranular calcite relics.

Similarly, calcite is left as intergranular mass in a later-formed garnet mass, Fig. 129. In cases, the calcite may be secondarily mobilised in veinform within the skarn garnet.

The skarn bodies are not the exclusive "basic" phases that may develop due to the mobilisation of basic front around a granitic intrusion. As pointed out (in chapter 2) olivine crystalloblastesis may be formed due to Mg-SiO_2 introduction in marble, related to granitic intrusion.

Of particular interest is the dolomitisation of marble due to the migration of Mg within the metasomatic aureole round a granitic intrusion e. g. Halara Seriphos. In this case, the dolomitisation is considered to be a part of the basic front mobilisation.

Serpentinisation of marble may result as a consequence of the basic front mobilisation. Fig. 130 shows replacement of coarse (calcite) marble-texture by metasomatic solutions which resulted in the transformation of marble → to serpentine. Particularly as is shown in Fig. 131 the serpentinisation has invaded the calcite, replacing it.

In cases, within the complex process of basic front mobilisation the country rock marbles may be replaced by blastically formed magnetite either intergranularly developed or replacing the calcite (Figs. 132 and 133).

In other instances, the magnetite crystalloblastesis has followed a Mg-rich silicate crystalloblastis, e. g. diopside or olivine. Fig. 134 and Fig. 135 show diopside and forsterite crystalloblasts enclosed by a later magnetite crystalloblastesis. In cases a synkinematic mobilisation of the metasomatic serpentine and of magnetite blastesis may take place (see Fig. 136).

As mentioned, olivine crystalloblastesis (see chap-

---

* In addition to the andradite-grossular garnet of the Seriphos Skarn most impressive blastic zoned andradite-grossular garnets and epidotes occur in the skarns of Kato Vrontou (N.Greece). Fig. 122 shows blastic garnet (andradite-grossular) with tremolite needles orientated within the blastic garnet. Similarly Fig. 123 shows a finely zoned andradite-grossular crystalloblast.

ter 4), may develop in marbles in the Seriphos granite, as reported by Ktenas (1916) and Marinos (1951). Olivine may develop as (Fig. 137) olivine crystalloblasts in marble. Serpentinisation of the forsterite has also taken place.

Within the Seriphos granite aureole or in the zone of basic front, which is noticeable, as mentioned, around the Seriphos granite, occurs another type of skarn at Kouduros (Mavro Punti).

In contrast to the garnet-epidote magnetite skarns bodies at Vounies in the case of Kouduros, epidote-lievrite-hedenbergite skarns have been reported by Ktenas and Marinos, and recently studied by Mposkos and Vgenopoulos.

Field observation Fig. 125 shows that an epidotisation of the initial pre-granitic marbles has taken place. A subsequent fracturing of the epidote-zone followed, which resulted in the formation of a fragmental zone, a mega-breccia which was in turn invaded by "hydrothermal" solutions. Colloform structures consisting of alternating bands of radiating hedenbergite ($CaFe[Si_2O_6]$), and lievrite ($CaFe_2 Fe^{III} (OH/O/Si_2O_7)$ often surrounding the epidotised marbles, and fragments of granitic apophyses are shown (Fig. 138 and 139). These large colloform structures may consist of radially orientated alternating bands of hedenbergite and lievrite, often well developed crystals of both hedenbergite and lievrite may develop (Fig. 140).

As pointed out the banded colloform structure clearly representing the crystallization from colloidal solutions since unmistakable colloform structures are indicated, suggests that the basic front can attain a colloidal solution phase which has invaded a fracture zone of already epidotised marbles. Both the epidotisation ($Ca_2(Fe^{III}, Al)Al_2[O/OH/SiO_4/Si_20_7]$) and the hedenbergite-lievrite phase suggest the existence of "hydrothermal" metasomatism (skarns).

As a corollary to this hypothesis are the geodes and cavity space-fillings which exist between the colloform structures of hedenbergite/lievrite. Fig. 141 shows radiating bands of hedenbergite-lievrite forming large colloform structures between which exist geode spaces filled with calcite, quartz, agate, and often green calcite and green quartz is present.

These geodes represent the concluding phase of the "hydrothermal" skarns.

Chapter 5 Eclogites and eclogites

Yoder and Tilley (1962) consider the eclogites as the parental rocks of basalts and an eclogitic mantle* layer has been assumed mainly on geophysical hypotheses and on the evidence of eclogite xenoliths in kimberlites.

In contrast to the deep seated eclogites which consist mineralogically mainly of pyrope, phlogopite and pyroxenes (see chapter 13) eclogites may occur as regional metamorphic rocks** e. g. Table II by Vgenopoulos shows a comparison of the d-values of spessartite in a metamorphic-metasomatic eclogite from Galicia, North West Spain; pyrope in a quartzite in association with pegmatite from Evros, Greece and pyrope from an eclogite xenolith in kimberlite (S.Africa). Table II clearly shows that there is a marked mineralogical difference between the garnets in the metamorphic-ultrametamorphic eclogite and the garnets of the mantle (mantle-eclogites) which occur as xenoliths in kimberlites. In addition Table II shows that blastic pyrope may occur in the quartzitic background, also see Fig. 142.

From the above table it is clear that it is not the depth that determines the type of garnet to be formed but rather the topochemical conditions prevailing in a rock.

Considering both their mineralogical composition and textural patterns, these two eclogite types are essentially different, in other words there exist "eclogites and eclogites".

As a corollary to the explanation of eclogites and eclogites (i. e. eclogites of metamorphic-metasomatic origin within the crust and the mantle eclogites) Brueckner (1977) proposed the ratios of Sr^{87}/Sr^{86} as a criterion for distinguishing them: "Strontium isotopic data suggest that the classic eclogite-facies rocks of western Norway described by Eskola (1921) formed from several parental materials in a variety of environments. Mineral separates from essentially basic, bi-minerallic (clinopyroxene and garnet) eclogites that occur as lens-shaped masses within high grade gneisses (country rock eclogites) have Sr^{87}/Sr^{86} values that range from 0.704 for fine grained varieties to 0.716 for coarse grained, orthopyroxene-bearing varieties. These high, varied ratios contrast with the very low, restricted ratios (0.701 to 0.704) of similar minerals from ultrabasic, garnet-clinopyroxene-orthopyroxene-olivine assemblages (garnet peridotities) that occur as lenses within large peridotitic bodies. The eclogite-facies metamorphism that generated the garnet peridotites may have occurred in the mantle. However, the metamorphism that generated at least the more radiogenic country rock eclogites must have occurred in the crust."

The mineralogy and textural patterns of deep seated eclogites is considered under the heading of "Eclogite xenoliths in kimberlites" (Chapter 13). In contradistinction, the metamorphic-ultrametamorphic in origin eclogites are presently discussed.

An important and characteristic mineral component of the metamorphic-ultrametamorphic eclogites are spessartite garnets often "rounded" in shape and occasionally idiomorphic. Fig. 143 shows idioblastic garnet with quartz inclusions in a mylonitised and recrystallised quartz-hornblende mass. Such idioblasts are rare in comparison to the spessartite crystalloblasts which show rounded shapes.

---

* The garnet-clinopyroxene-disthene rocks (grospydites and associated Kyanite eclogites) from Zagadochnaya Kimberlitic pipe in Yakutia are considered by Sobolev et al. (1968) due to the presence of chromium-rich minerals (in the grospydite xenoliths) to be connected with the ultrabasic rocks. This represents another case of kimberlitic xenoliths of ultrabasic origin (most probably representing mantle derivation).

** There is a divergence of views regarding the origin of eclogites, magmatists consider them as cumulates and magmatic differentiates. Lappin (1974) explains the eclogites (an orthopyroxene eclogite containing clinopyroxenes with lamellae of garnet and orthopyroxene) from the Sunndal-Grubse Ultramafic mass, Almklovdalen, Norway as cumulates and differentiates of relatively high pressure (25–28 kb) melting of ultramafic rocks.

Most of the garnet-crystalloblasts show "rounded" forms due to mylonitisation and deformation. The garnet is a relatively hard and pressure resistant mineral-phase and will survive the crushing and fracturing which quartz and hornblende may suffer. Particularly quartz, under tectonic deformation, is easily mylonitised and plastically deformed.

Fig. 144 shows garnet tectonically deformed in an also deformed and recrystallised quartz mass. The shape of the spessartite crystalloblast is characterised by curving lines, these forms, the rounded shapes, could provide the maximum resistance to deformation. As microscopic observations reveal most of the garnet crystalloblasts, particularly in a quartz/hornblende mylonitised mass show rounded forms and "curving outlines" see Fig. 145.

In addition to quartz, hornblende inclusion may occur in spessartite crystalloblasts and occasionally calcite (see Figs. 146, 147, 148). In these cases, the enclosed quartz, hornblende, and calcite are most probably pre-garnet growths, enclosed by the crystalloblast.

In contrast quartz may be plastically mobilised under tectonic deformation and may be squeezed within cracks of the resistant spessartite.

Under comparable deformation conditions quartz and muscovite are plastically deformed and tectonically squeezed in the cracks of a grossular garnet in a pegmatite from Harrar, Ethiopia (see Fig. 597, Augustithis 1973). Similarly Figs. 149 and 150 show quartz and amphiboles tectonically mobilised and squeezed in the cracks of spessartite.

Fig. 151 shows a rounded spessartite crystalloblast in recrystallised quartzite with cracks of the garnet invaded by plastically mobilised quartz.

In addition to the pre-kinematic spessartite crystalloblasts a great part of the hornblende crystalloblast has been tectonically affected and has been mobilised with the quartz. Fig. 152 shows tectonically rounded garnet and hornblende crystalloblasts in a re-crystallised quartz-mylonite.

The hornblende crystalloblastesis can be later than the spessartite. Fig. 153 and Fig. 154 show spessartite crystalloblast enclosed in hornblendeblasts. Of particular interest is Fig. 155 which shows rounded garnets in hornblende, suggesting that the hornblende is formed after the roundening due to tectonic deformation of the spessartite. It is possible that amphibole crystalloblasts could be (in cases) post-kinematic, however, as is exhibited in most cases, the hornblende is pre-kinematic and has been clearly deformed and mobilised, see Figs. 152, 156, 157, 158.

The hornblende of the metamorphic-ultrametamorphic eclogites are often in symplectic intergrowth with quartz, Fig. 159(a and b). In cases, most complex "myrmekitic" like symplectic hornblende/quartz intergrowth may be exhibited, Fig. 160 and Fig. 161. Often these symplectic intergrowths may attain most complex structures, as is particularly indicated in Fig. 161.

A most complex textural pattern showing quartz/hornblende symplectic intergrowths, lenses of relic of quartzite (mylonitised fragments) and a resistent pre-kinematic apatite crystalloblast is illustrated in Fig. 162.

In addition to the symplectic hornblende/quartz intergrowth where the quartz is the later mobilised phase (Fig. 160), are metasomatic infiltrations of quartz invading the hornblende either as quartz "remobilisation", Fig. 163, or as quartz veinform infiltrations taking advantage of the amphibole interleptonic spaces of the cleavage, Fig. 164.

In the metamorphic-ultrametamorphic in origin eclogites, in addition to the blastic phases of spessartite and amphiboles, often apatite crystalloblasts are associated with the hornblende, Fig. 165 (a and b) and Fig. 166.

The association apatite/hornblende is common in granites (Augustithis 1973) and in basalts (Augustithis 1978). Also in the case of the metamorphic-ultrametamorphic eclogites apatite is associated with hornblende. Fig. 166 shows apatite crystalloblasts almost parallelly orientated in hornblende-blasts.

In addition to apatite, zircon may occur as an accessory phase in eclogites. Figs. 167 and 168 show quartz inclusions in the spessartite crystalloblasts. As is indicated in Fig. 168 prismatic zircons have grown in the quartz. Fig. 169 shows quartz included in the garnet crystalloblast with prismatic zircons in the quartz and transversing the boundary quartz/garnet extending and laying also in the spessartite.

As mentioned the spessartite crystalloblasts occur as "rounded" forms (in cases with curving outlines) in a mylonitised and recrystallised groundmass, consisting of quartz and hornblende, Fig. 170. As pointed out, garnets with rounded outlines occur in a recrystallised mylonitic groundmass (Figs. 145 and 171). Particularly Fig. 172 shows mylonitised groundmass mobilised between garnet and hornblende.

In cases the mylonitised quartz/hornblende groundmass can contain lenses of relic quartzite (Fig. 173). Fig. 174 shows lenses of relic quartz in the fine mylonitised groundmass. The quartzite fragments might be tectonically mobilised and may occur as intergrowths with hornblende/quartz symplectite (Figs. 175a and b).

In other instances, the quartzite lenses might be tectonically pushed and squeezed into cracks of the rounded spessartite crystalloblast. Fig. 176 (a and b) shows a spessartite crystalloblast in cracks of which quartzite and quartzite with hornblende are tectonically, under plastic deformation, mobilised.

The crystallization sequence of the metamorphic-ultrametamorphic eclogite may be outlined as follows:

Spessartite crystalloblasts have grown in a quartzite. The garnet crystalloblasts could be attributed to metasomatism, or topo-metasomatism of elements including Mn.

The spessartite crystalloblasts often enclose quartz, calcite and hornblende, phases associated with the initial quartzite.

Following the garnet crystalloblastesis there is an amphibole crystalloblastesis.

Apatite crystalloblastesis. Zircon crystalloblastesis (after the spessartite crystalloblastesis).

Mylonitisation is most probably the concluding event of the blastic sequences and processes in general that formed the metamorphic-ultrametamorphic eclogites of Galicia, North-West Spain.

The initial quartzite and the pre-kinematic blastic phases of garnet and amphibole have been involved and mobilised in the mylonitisation zones, resistant relics of spessartite and, to a lesser extent amphibole have survived crushing and mobilisation.

(Table II, Phyrope)

Eclogitic Xenoliths in Kimberlite		Blastic Pyrope in quartzite by Vgenopoulos		Literature
2ϑ	d	2ϑ	d	d
30.85	2.8955	30.85	2.896	2.88
34.6	2.590	34.63	2.587	2.58
				2.46
38.1	2.359	38.1	2.359	2.35
39.7	2.268	39.7	2.268	2.26
42.75	2.1125	42.8	2.111	2.10
48.55	1.873	44.3	2.042	2.03
		48.6	1.871	1.87
				1.82
54.9	1.670	54.95	1.669	1.66
57.5	1.601	57.35	1.605	1.60
59.85	1.5435	59.75	1.546	1.54
64.3	1.447	64.35	1.446	1.44
				1.42
64.5	1.4433	64.55	1.4423	
73.1	1.294	73.1	1.293	1.29
75.2	1.262	75.3	1.261	
75.3	1.261			
77.3	1.233	77.35	1.232	

(Table II, Spessartite, Galicia, Spain)

Blastic 2ϑ	Spessartite d	Literature d
30.75	2.9035	2.90
34.5	2.597	2.60
36.2	2.479	2.48
37.95	2.370	2.37
39.5	2.279	2.28
42.65	2.1175	2.12
45.00	2.012	2.05
48.35	1.8805	1.89
54.7	1.676	1.68
55.9	1.643	1.64
57.2	1.609	1.61
		1.58
57.3	1.606	
59.65	1.548	1.55
64.7	1.451	1.45
		1.30

Chapter 6 Blastic magnetite with ilmenite ex-solutions in epizonal chlorite with anthophyllite blastic growths

General

Blastic octahedra of magnetite with ex-solutions of ilmenite are present in a chlorite schist with anthophyllite. The interpretation of the blastic growths of the magnetite, i. e. iron concentration, necessitates the transportation and concentration of iron by intergranular solutions. A comparable origin is considered for the blastic amphiboles (anthophyllite) also present.

The microscopic observations show that the amphibole blastic growths follow tectonic cracks and microfaults (displacements of microscopic size); thus, it is suggested that there is a relationship between tectonic influence and the blastic growths.

Here the schistosity of the chlorite schist is considered to be the first influence (deformation I), and the cracks and microfaults the second (deformation II); the second being almost perpendicular to the first. The amphibole blastesis is obviously synkinematic to deformation II.

Microscopic observations show a postkinematic origin for the blastic magnetite in respect of deformation I; inclusions of chlorite in the magnetite follow the external schistosity, that is, the schistosity of the chlorite schist. Further observations show parts of the blastic magnetite following and extending into the cracks produced by the second deformation.

Considering the origin of the substance building up the blastic growths, the concept of topometasomatism seems to be a working explanation. Topometasomatism involves the transportation and mobilisation of elements to a certain spot (starting point of blastic crystallization within a rock under metamorphic processes) by intergranular solutions which have originated within the metamorphosed rock-mass itself; in fact, a type of autometasomatism.

Due to the fact that the magnetite is blastic, and that it occurs in chlorite schist (epizonal in origin) the temperature of formation of the ex-solutions of ilmenite is considered to be low. Most of the previous workers on ilmenite ex-solutions in magnetite consider their presence to be evidence of high temperature, that is, the magnetite formation has taken place between 400–700°C.

Growths Synkinematic to Deformation I, chlorite recrystallization (neocrystallization); Chlorite schistosity (Deformation I); "Augen Structures"

Under epizonal metamorphism an initial clay has been transformed into a chlorite schist. A relatively Al-rich and Fe-poor chlorite (leuchtenbergite) has been determined by X-ray studies carried out by Tatjana Wolbeck.

The schistosity can be considered as a product of deformation processes which have operated within the epizonal metamorphism, and is mainly caused by a flaky rearrangement and recrystallization of the chlorite crystal grains. These grains show an interlocking texture with their axes of elongation following the direction of schistosity (Fig. 177).

Within the epizonal metamorphism, and contemporaneous with schistosity, types of microscopic "augen structures" are formed (Fig. 178). These are products of the deformation responsible for the schistosity, that is, deformation I.

The chlorite crystallization processes are not only contemporaneous with deformation I, but often, as shown in Fig. 179, veinlets of chlorite with radiating structures exist within the chlorite schist and are clearly post-deformation I in age. In these structures, the chlorite has been formed by hot water solutions.

Also belonging to the initial phase of the metamorphism is ilmenite with hematite ex-solutions. These ore-minerals are discussed further.

Growths Postkinematic to Deformation I, but Synkinematic to Deformation II

Microdisplacement planes (Deformation II); recrystallization structures and iron mobilisation along these planes

Microscopic observations show cracks and microfaults, microtectonic displacements, often perpendicular to the schistosity of the chlorite phyllite, which are considered to be the result of deformation II. The differentiation between deformations I and II is of great importance in the understanding of the sequence of the processes of this chlorite-schist metamorphism.

Along the planes of weakness produced by deformation II, often iron mobilisation, recrystallization of chlorite and blastic growths have occurred. Fig. 180 shows chlorite schist with cracks and microdisplacements perpendicular to the direction of the schistosity, with iron-mobilisation having taken place along the microfissures.

Microdisplacements caused by deformation II, with a chlorite recrystallization are shown in Fig. 181. An "augen structure" consisting of a chlorite grain is present.

Anthophyllite blastesis

The series of photomicrographs of Figs. 182 to 184 shows that along the microdisplacements of deformation II blastogenic growths of amphibole (anthophyllite) have taken place. Figs. 182 and 183 show microfaults with iron mobilisation and with anthophyllite nematoblasts* (elongated blastic crystals).

A more detailed photomicrograph (Fig. 184) shows an anthophyllite, with alteration margins of talc, following a line of displacement, or which by the force of crystallization (blastogenic force) is "intruding" along weakness of the chlorite schist. It should be pointed out that the direction of the chlorite schistosity is different on both sides of the amphibole: this denotes a plane of tectonic disturbance prior to the growth of the amphibole.

The relation of anthophyllite blastesis and deformation II has been pointed out: it should, however, be seen as mobilisation of substances postkinematic to deformation I and synkinematic with deformation II. Further, the concentration of material building up the amphiboles necessitates the concept of material transportation, best understood by the operation of intergranular solutions (intracapillary solutions).

The sharp contrast between the blastic amphiboles and the chlorite schist shows that no metasomatic reactions have taken place between the blastic growths and the chlorite.

Often secondary alteration of the anthophyllite to talc has taken place, which sometimes can be so extensive and complete that no amphibole-rests can be seen (Fig. 185). In contrast, Fig. 186 shows anthophyllite with alteration margins of talc.

A most interesting metasomatic reaction between talc and leuchtenbergite is shown in Fig. 185 (see arrow). The interreaction margins of these two phases support the view that metasomatic reactions have taken place between the alteration product of the amphibole and the chlorite schist.

The observations introduced so far have shown amphibole blastesis along tectonic planes. A transition from chlorite schist with amphibole blastic growths to a rock essentially consisting of blastic anthophyllite has been observed in Hadabudussa/Gari-Boro area (Adola district, S. Ethiopia). Fig. 187 shows amphibolite with radiating anthophyllite blastic phases, a holoblastic rock, consisting entirely of blastic growths, with many radiating centres of crystal growth.

Comparison of experimental data on the synthesis of anthophyllite with conditions of formation inferred from studies of natural anthophyllite in a leuchtenbergite (chlorite) schist of Hadabudussa

Greenwood (1962) has shown the existence of a stability range of anthophyllite in the presence of excess H_2O. In contrast, a number of workers earlier than Greenwood agreed on theoretical and experimental bases that the mineral is not stable in the presence of excess H_2O.

Regarding the synthesis of anthophyllite Greenwood has said "The synthesis of anthophyllite is not

* Kalinin et al. (1976) in their study: Natural and synthetic amphibole asbestos (intergrowths, dislocation structures, kinetics and mechanism of fibre growth, genesis in nature), state that the "real structure of amphibole asbestos needles shows that the growth in three dimensional complex groupings is the most acceptable model for their formation. The result of electronic microscope studies is correlated with general physico-chemical conditions of amphibole asbestos synthesis, i. e. high ion activity of the groups X and Y, low temperature and high pressure. ... Natural amphibole asbestos are both growth and non-growth genesis. The growths are formed by the reaction of directed replacement, by growth in porous solutions on crack walls or by a disorderly crystallization in the solution volume" (Kalinin's et al. studies include also anthophyllite).

easy. Most of the hydrothermal experiments that have failed to produce the mineral have failed because of its extreme reluctance to nucleate, even well within its own field of stability... The only way in which anthophyllite could be produced in the absence of pre-existing nuclei was by the metastable decomposition of the talc at 1000 bars and 830° C for a period of 20 hours."

However, Greenwood (1963) succeeded to synthesise hydrothermally anthophyllite at a temperature range of 745°C to 667°C, which is still incompatible with the low temperature hydrothermal anthophyllite in the leuchtenbergite (chlorite of low temperature – about 250°C) of Hadabudussa, Gar-i-Boro.

By comparing this quoted experimental data with the conditions of anthophyllite formation inferred by studying natural material, the following are noted:

I. Observations on natural anthophyllite from Hadabudussa show that the mineral occupies cracks or microfaults in the leuchtenbergite schist. The concentration and transportation of the material to build this blastic phase is assumed to be by hydrothermal solutions working their way through intergranular spaces. From the relations that have been observed between the anthophyllite and the chlorite schist and with tectonic movements, a diagram showing hypothetical conditions of the anthophyllite formation can be drawn, as in Fig. 188. In this diagram arrows "A" mark the movement occurring along the microdisplacement and arrows "B" mark the hypothetical movement of the intergranular solutions which brought about the concentration of the material which built up the anthophyllite along the microdisplacement. On the basis of the observations introduced (Figs. 182 and 183) it can be pointed out that the anthophyllite growth has taken place by hydrothermal solutions but under conditions of tectonic movements (Schiebung).

II. The formation of anthophyllite in the epizonal phyllite allows us to infer that the temperature of anthophyllite formation must have been much lower than the temperature given by the metastable transformation of talc to anthophyllite [as mentioned in the work of Greenwood (1962) and the hydrothermal synthesis at 745–667°C, Greenwood (1963)]. Further, the transportation and concentration of material for the formation of natural anthophyllite has taken place by hydrothermal solutions; this is the only process that could explain the concentration of this material. The blastic (hydrothermal) origin of this anthophyllite is in accordance with the existence of a range of stability of anthophyllite in the presence of excess H_2O.

III. On the basis of microscopic observations it is clear that the relation between anthophyllite and talc is that of a later talc-formation by hydration alteration of the anthophyllite. Fig. 186 shows marginal talc, and talc following the cracks in the anthophyllite. These observations show that the anthophyllite changes to talc and that in the particular case it is not formed by the metastable decomposition of talc.

Blastic magnetite and its relation to blastic anthophyllite

The blastic magnetite is of particular importance in understanding the sequence of processes in the chlorite schist. Fig. 189a shows Fe_3O_4 with extensions along the schistosity of the chlorite. In addition, chlorite crystal grains are enclosed in the blastic magnetite following the direction of the schistosity. The magnetite is clearly postkinematic with respect to deformation I, which is responsible for the chlorite schistosity.

The inclusions of the chlorite in the blastic magnetite and the fact that it extends along the schistosity of the phyllite clearly support its postkinematic age (Fig. 189b). The additional and important fact that the magnetite builds idiomorphic octahedra in the chlorite schist also supports its blastogenic origin, that is, the magnetite is an idioblast. More detailed observations in Fig. 189c show blastic magnetite extending into the schistosity of the chlorite and, at the same time, enclosing chlorite flakes. Both photomicrographs of Figs. 189a and 189c show clearly that the magnetite is not a relic of clastic origin in the chlorite schist, but a later blastogenic growth which has extended parallel to the chlorite schistosity and partly enclosed the chlorite.

In Fig. 189d blastic magnetite is shown in contact with recrystallised chlorite. This is an interesting observation, although difficult to explain. It is uncertain whether the chlorite recrystallization is independent of the force of crystallization exercised by the blastic magnetite.

Additional observations in Fig. 190 show blastic magnetite, with a part extending along a crack partly occupied by blastic anthophyllite. It is clear that the blastic magnetite is synkinematic to deformation II. A consideration of the above observations shows evidence that the magnetite is postkinematic to deformation I (it enclosed flakes of chlorite and follows its schistosity) and is synkinematic to deformation II.

Since the origin of the Hadabudussa magnetite is considered to be a blastic growth in epizonal chlorite, it is surprising to find ex-solutions (intergrowth bodies) of ilmenite in the magnetite parallel to the (111) face of the host. In this case the magnetite formation with the ilmenite ex-solutions is a phase of crystal growth postkinematic to schistosity (deformation I) but synkinematic to deformation II. It is thus a blastic phase within the epizonal metamorphism and its temperature of formation should be within the limits of the epizonal metamorphism, which in this case is assumed to be about 300°C. In contrast, the majority of workers believe in a high temperature for the formation of magnetite with ilmenite ex-solutions, that is 400–700°C.

The relationship of the two blastic phases (magnetite and anthophyllite) is indicated in Fig. 191 which shows magnetite enclosing chlorite and being in contact with an amphibole. The blastic anthophyllite appears to be older than the magnetite. This is again shown in Fig. 190, where the magnetite follows the amphibole extending into a crack of deformation II.

Topogeochemically* (considering the formation of blastic growths, geochemically, in a particular spot within the rock) the two blastic phases can be regarded as synchronous, or approximately synchronous, element mobilisation phases. In the anthophyllite, blastesis mobilisation of Mg, Si, Al, Fe has taken place, however, in the case of blastic magnetite, the mobilisation has been mainly of Fe, and to a lesser extent Ti.

Ore-Microscopic Observations

Figs. 192 and 193 show ilmenite grains associated with the chlorite and being components formed during the initial stages of the epizonal metamorphism.

As shown in Fig. 192, ore-microscopically the ilmenites are rich in ex-solutions of haematite. The structural relation of the ilmenite (with haematite ex-solutions) to the blastic amphibole is clearly that of a pre-existing phase, often partly enclosed by the blastic anthophyllite, as indicated in Fig. 186. Furthermore, the ilmenite-haematite phase is prior to the blastic magnetite and, here again, the ilmenite-haematite phase is partly enclosed by the later blastic magnetite, as shown in Fig. 193.

It is interesting to note that the blastic magnetite encloses grains of ilmenite, some of which clearly contain ex-solutions of haematite.

Of particular importance are fine lamellae of ilmenite in the magnetite clearly following a geometrical pattern, that is, the (111) face of the host magnetite (Fig. 194). Also, larger lamellae or bodies of ilmenite are present (Fig. 195). However, these fine lamellae are of particular importance since they are "ex-solution" bodies of ilmenite in the magnetite. Often, fine spinel bodies marginal to the ilmenite are also observed, as in Fig. 196. Fig. 197 shows fine ilmenite and spinel bodies as orientated "ex-solutions" in the host-magnetite.

On the Origin of Ilmenite Intergrowths in Magnetite

The different opinions and important aspects of the origin of ilmenite ex-solutions in magnetite have been summarised by Basta (1960) and Lindsley (1961/2), and the following points are based on extracts from their work.

On the origin of ilmenite intergrowths the following opinions are mentioned:

I. Singewald (1913) and Warren (1918) due to an apparent proportionate amount of ilmenite in magnetite, advanced a eutectic origin.
II. Ramdohr (1960), Ödman (1932), Newhouse (1936), Edwards (1949) and Buddington and Lindsley (1964) explained these ilmenite bodies as ex-solutions in magnetite.
III. Foslie (1928), supported by explanations of Mogensen (1946). Vincent and Philips (1954), and Akimoto (1954), and lately by Verhoogen (1962) and Lindsley (1961/62), advanced that ilmenite "ex-solutions" are produced by unmixing of a $Fe_3O_4-Fe_2TiO_4$ solid solution followed by the decomposition (or oxidation) of the Fe_2TiO_4 (y-spinel = ulvospinel) components into $FeTiO_3$ + FeO.

The following facts shoud be considered in connection with the above mentioned opinions:

* Topogeochemical mobilisation = topometasomatism [Topos = place, location (Greek) and metasomatism], see Augustithis (1964)

I. Several workers, supporting the unmixing theory, have noted that ilmenite intergrowths cannot be homogenised by heating to 1000–1200°C., if bulk composition is maintained.
II. Verhoogen (1962), supporting the unmixing theory, points out that the ilmenite lamellae parallel to the (111) face of the host magnetite are formed by ex-solution. Ilmenites formed by oxidation of ulvospinel are parallel to the 100 faces of the magnetite, and this ilmenite has oblique extinction. Thus it differs from the natural occurrences of ilmenite intergrowths in the magnetite.

Considering both hypotheses of the origin of ilmenite intergrowths in magnetite (ex-solutions or oxidation from ulvospinel), it is assumed that the rock or ore which contains these magnetites must have attained high temperature at certain stages of its coming into existence, that is, it should have passed through a stage of a melt. (According to Verhoogen the ilmenite exists in solid solution with the magnetite at temperatures between 1200–1300°C. In addition, it is believed that at high temperatures y-spinels exist in solid solution with the magnetite).

In contrast to these explanations, the existence of ex-solution (or intergrowth) bodies of ilmenite in the blastic magnetite from Hadabudussa clearly shows that in nature there are other ways than those advanced on high temperatures for the formation of magnetite with ilmenite intergrowths. The petrographic observations introduced show that without any doubt the magnetite is blastic, that is, it was formed while the chlorite-schist existed, and without having been melted or partly melted and is formed contemporaneously with deformation II of the leuchtenbergite schist (chlorite-schist) of Hadabudussa/Gari-Boro (Adola) Sidamo Province, S. Ethiopia. Regarding the genesis of ilmenite intergrowths with magnetite, see also chapter 26.

Independent of the blastic magnetite with ilmenite ex-solutions metamorphic chromite occurs together with leuchtenbergite and anthophyllite in Hadabudussa area. The chromite occurrence together with chlorite and anthophyllite could be due to chrome remobilisation under metamorphism (see also chapter 23).

Chapter 7 Mafic crystalloblast

Crystalloblastesis and its significance in metamorphic-metasomatic basic and ultrabasic bodies.

In addition to the classical concept of crystalloblastesis in metamorphic rocks, Becke (1908), Harker (1950), crystalloblastesis in granites is considered to be the fundamental process of granitisation, Drescher-Kaden (1948, 1969) Erdmannsdörffer (1950) and Augustithis (1960a, 1973).

In understanding particularly the crystalloblastesis in granites of interest is the work of Erdmannsdörffer who considers endoblastesis as a definite metamorphic process. In contradistinction to the crystalloblastesis in metamorphic rocks and in the ultrametamorphics (as such can be considered granite), Augustithis (1960a, 1964a, 1973) has emphasised the growth of tecoblasts in volcanics where evidence is advanced of blastic plagioclases, phenocrysts less anorthitic, enclosing, corroding and assimilating groundmass. Blastic growths seem to be an important petrogenetic process in rock types, that up to now has not been considered as an important process. Drescher-Kaden (1969) emphasized the significance of olivine and serpentine blastesis in marbles and their transformation to olivine and serpentine rocks.

As has been shown in chapters 1, 2, 3, 4 and 5, cases have been presented where metamorphic metasomatic processes were responsible for the genesis of basic and ultrabasic rocks. In contradistinction to the importance of feldspar crystalloblastesis which is the process responsible for the granitisation in the basic and ultrabasic ultrametamorphics, the blastesis of mafic minerals has played the predominant role.

As mentioned in chapter 1, olivine crystalloblastesis has resulted in the formation of olivine megacrystalloblasts in olivinefels with tremolite and phlogopite crystalloblasts (see Fig. 3). In addition certain of the olivine megacryst described by Kenneth et al. (1975) from Nain Province, Labrador, are megacrystalloblasts. In addition, olivine crystalloblastesis is responsible for the transformation of marbles to olivine rich rock (see Figs. 29, 30, 32a). Mg-metasomatism in cases accompanied by SiO_2 metasomatic mobilisation may be considered a process responsible for certain olivinefels and olivine rock types, particularly in orogenic belts, see Drescher-Kaden (1969).

In addition to the olivine crystalloblasts described olivine poikiloblast enclosing olivinefels components and feldspars (plagioclases) are described from the picritic bands of the layered ultrabasic complex of Skaergaard, see Fig. 198.

Indeed, the olivine crystalloblastesis is a decisive piece of evidence for a metamorphic-metasomatic origin of the ultrabasic complex of Skaergaard. The microscopic evidence discussed in chapter 16 supports a crystalloblastic origin of the olivine in the olivine rich bands of the ultrabasic Skaergaard complex in contrast to the cumulate hypothesis of Wager and Deer (1939).

In support and agreement with the olivine crystalloblastesis is the pyroxene blastesis. The importance and significance of pyroxene crystalloblastesis in the layered complex of Skaergaard will be discussed in chapter 16. As Figs. 199 and 200 show, the pyroxene growths in the layered body of Skaergaard do not represent intracumulates due to fractional crystallization but poikiloblastic growth enclosing olivine and plagioclase as a pre-pyroxene metamorphogenic-metasomatic phase.

As a corollary to the pyroxene mega-poikiloblasts illustrated in Figs. 199 and 200 are blastogenic pyroxene metasomatic growths as interleptonic infiltrations between corroded plagioclases and crystalloblastic magnetites, see Fig. 201.

The amphibole group of minerals is primarily metamorphic metasomatic in origin, despite that amphibole crystalloblasts may exist in basalts, andesites and mantle-hornblendites. In volcanics the hornblende is often a crystalloblast (see Augustithis 1978) also the presence of hornblende megablasts in adamellites and granodiorites is regarded as a blastogenic phase, see Drescher-Kaden (1969) and Augustithis (1973).

In cases, hornblende crystalloblastesis may show topological differences and locally a hornblendite or hornblende-rich rock may develop within a tonalitic or adamellitic complex.

Figs. 202 and 203 show hornblende crystalloblast enclosing corroding and partly assimilating plagioclases. Similarly, Fig. 204 shows hornblende crystalloblast partly engulfing plagioclases as the xenomorphic hornblende crystalloblast invades a plagioclase texture Fig. 205.

Most convincing evidence of hornblende crystalloblastesis is shown in Fig. 206, where a hornblende crystalloblast poikilitically encloses corroded quartz and plagioclase crystal grains.

In contradistinction to the hornblende crystalloblasts which are full of inclusions, Fig. 207 and Fig. 208 show growths of hornblende crystalloblast that have cleaned themselves from inclusions.

Often the hornblende crystalloblasts include chloritised initial feldspar inclusions in the amphibole crystalloblasts Fig. 209, in other instances the hornblende metacrystals have been locally altered, see Figs. 205, 210 and 211.

In contradistinction to the allotriomorphic hornblende crystalloblast hornblende idioblast may exist. Full of corroded feldspar inclusions, and in turn enclosed by later plagioclase crystalloblast which in turn is full of hornblende inclusions, see Fig. 212.

Similarly, Fig. 213 shows idioblastic hornblende almost autocatharitically cleaned from inclusions surrounded by a plagioclase poikiloblast enclosing corroded and partly assimilated hornblendes.

In addition to the hornblende megablastic-poikiloblastic, a granoblastic idioblastic texture may, in cases, be included in plagioclase megablasts of clearly poikilitic character, see Figs. 212, 213 and 214.

In addition to the hornblende crystalloblasts a wide spectrum of amphiboles can exist as metacrystals clearly showing blastogenic textures (see chapters 6 and 11).

Co-existing hornblende/biotite crystalloblasts may exist, see Fig. 215. The biotite metacrystal encloses corroded pre-biotite plagioclases. Similarly, Fig. 216 shows biotite enclosing plagioclase which in addition to corroded outlines, exhibits reaction zones with inverted twin lamellae, see also Fig. 217. A most impressive biotite poikiloblast is shown in Fig. 218, here the mica enclosed and corroded a pre-biotitic texture of amphiboles.

In cases, most impressive biotite/hornblende intergrowths may exist whereby the mica, due to its crystalloblastic force of crystallization, penetrates through the hornblende crystalloblast often perpendicular to the elongation direction of the amphibole, traversing cleavage planes and often exhibiting interpenetration textures of the biotite crystalloblast within the hornblende host, Figs. 219, 220 and 221.

In contradistinction to these, we find biotite crystalloblasts clearly exhibiting evidence of blastogenic force. Fig. 222 shows biotite replacing the hornblende along direction of cleavage penetrability, see arrow "a" Fig. 222.

Another group of minerals of great petrogenetic significance in the metamorphic-metasomatic basic and ultrabasic rocks is garnet. Garnet occurs in a wide spectrum of rock types and in turn shows a wide spectrum of composition and textural patterns. Apart from the significance of the garnet in the classical metamorphic rock types, blastogenic garnet occurs in basic and ultrabasic rocks and signifies in these cases metamorphism-metasomatism.

Only a restricted number of cases will be presented here to show that garnet metacrysts are significant growths in rocks that up to recently have been considered by orthodox petrography to be of undisputable magmatogenic origin. Fig. 223 shows a rapakivi finish granite with a basic xenolith. Garnet crystalloblasts exist in the granite, in the basic xenolith and as Fig. 223 shows, garnet crystalloblast transverse the contact granite/basic inclusion. It is evident that this almandine garnet-phase is a crystalloblastic growth with an idioblastic tendency and which, due to its blastogenic force of growth, has transversed the margin basic xenolith/granite.

In banded ultramafics (Bushveld complex) blastogenic uvarovite is observed surrounding often rounded chromite crystal grains (see Fig. 224). In this case the Cr-garnet is blastogenic growth and represents second phase of Cr-mobilisation (see chapter 23).

More problematic is the pyrope which occurs with phlogopite in eclogitic xenoliths of mantle derivation in a number of rocks e. g. kimberlites, basalts. Fig. 225 shows idiomorphic pyrope in serpentine perhaps a mantle xenocryst in a serpentinised peridotite. In contradistinction to the mantle-pyrope, Vgenopoulos (1977) has determined crystalloblastic pyrope in a quartzitic texture, see Fig. 142.

In addition to these rather problematic garnet types a wide spectrum of garnets of metamorphic-metasomatic origin exist in metamorphogenic basic and ultrabasic rocks, such as the spessartite in eclogite (see chapter 5) and the most impressive garnet-andradite-grossular crystalloblasts in Skarns (see chapter 4).

A most impressive poikiloblastic, streptoblastic spessartite may also develop in a quartzitic texture with calcite, see Figs. 226 and 227. A wide spectrum of diverse garnet growths may exist in basic and ultrabasic rock some clearly metacrysts (crystalloblasts), others clearly mantle constituents.

As discussed in chapters 6, 7 and 16, magnetite may exist as a poikiloblastic and idioblastic phase in diverse metamorphic-metasomatic rocks.

Poikiloblastic magnetite with ilmenite and ulvite spinels may exist in ultramafics contradicting the magnetite cumulate hypothesis of fractional crystallization and the ex-solution hypothesis, namely that these ilmenite-ulvite bodies indicate relatively high magmatic temperatures of formation, see chapters 6 and 26.

On the basis of the textural patterns described blastogenic growths play a significant petrogenetic role in basic and ultrabasic rocks and can be taken as indisputable evidence that, when established in a rock, they signify that this rock, at a phase of its history of development, has passed through a phase of metamorphism-metasomatism.

Furthermore, the establishment of a blastic sequence in a mafic rock is equally important as in granites, for it is the history of the growth phases of a rock that shows its genesis.

Chapter 8 Gabbroic rocks – Essexites

In the group of gabbroic rocks are included sub-volcanic and plutonic rocks. In contrast to the sub-volcanics which are definitely "magmatic" in derivation, often all transitions from basalt → dolerite → microgabbro are exhibited, the origin of the plutonic gabbros is more problematic.

In contrast to the prevailing views that all gabbros are deep cooled magmatic derivatives, the present studies and particularly the analysis of their microstructure show traits definitely "metamorph", i. e. not deuteric metamorph but primary blastogenic growths.

The metamorphic-metasomatic growths and textural patterns observed are not post-consolidation tectonic influences, but blastogenic growths, re-crystallizations, synantetic and symplectic growths indicating an ultrametamorphic origin of these gabbroic rocks.

Commensurable to granitisation, the crystallization sequence depends on the blastogenic growths and their sequence may differ from the crystallization sequence of a consolidating magma.

In contradistinction to granitisation where, in addition to the neosoma (blastogenic growths), relics of the palaeosoma are also present in different extents of preservation; in the ultra-metamorphic gabbros the palaeosoma, if present, has undergone changes often beyond recognition. The ultrametamorphic processes in these rocks can be recognised by the blastic sequence, re-crystallizations, synantetic and symplectic intergrowths exhibited.

The relation of the paragabbros (ultrametamorphic) to the gabbroic protolytic layer is problematic. The gabbro-protolyte despite the geophysical evidences of existence and a possible occurrence of this rock, along the fault (volcanic lines of the great Rift Valley – Ethiopian part) – is itself problematic. The gabbroic protolyte is considered to be the parental rock, out of which, by fusion, basalts originate, however, its textural patterns differ from those of the basalts (see Augustithis 1978). Furthermore, considering the sub-volcanic gabbros particularly the micro-gabbroic rocks of the Karroo series, South Africa and the Palisade Sill, the ophitic textures are a characteristic textural pattern of these sub-volcanic basaltic intrusions, Augustithis (1978).

Comparing the ultrametamorphogenic (paragabbros) with the subvolcanic microgabbros, the question focuses on the ophitic intergrowths. Are the gabbroic "ophitic intergrowths" comparable and commensurable to the magmatogenic sub-volcanic ophitic intergrowths as shown in the Karroo and Palisade basalts – microgabbros?

As it is shown in the study of ophitic intergrowths from Karroo and Palisade by Augustithis (1978) in the sub-volcanic microgabbros the plagioclases represent a post-pyroxene crystallization phase often extending from its margins inwards, and in cases following penetrability direction (such as cleavage) of the pyroxene host, see Figs. 228 and 229. In contrast the pyroxene/plagioclase intergrowths in paragabbros depend on the blastogenic sequence of crystallization. As the study of gabbroic rocks from Skaergaard shows (chapter 16 and its illustrations, Figs. 580, 581), the pyroxene often extends as a post-plagioclase crystalloblastic phase between the plagioclase intergranular, enclosing and partly engulfing the plagioclases and attaining ophitic-like intergrowths.

In addition to these ophitic-like plagioclase/pyroxene intergrowths the paragabbros differ from the sub-volcanic microgabbros (dolerites) in their crystallization sequence and textural patterns. Microscopic studies of paragabbros show the following metamorph-metasomatic blastogenic traits.

Figs. 230 (a, b) and 231 show pyroxene invading a plagioclase and attaining graphic-like forms in the feldspar. The pyroxene partly follows the plagioclase twinning.

Often the crystallization of pyroxenes in paragabbros clearly follows the intergranular space between magnetite and plagioclase. In such cases, the pyroxene forming "solutions" have made use of the intergranular spaces.

However, often the contacts pyroxene/plagioclase are difficult to be interpreted and they are dubious (see Fig. 232); in such cases additional textural evidence showing the relation mineral to mineral is necessary in order to decipher the crystallization sequence of a rock. Fig. 233 shows plagioclases corroded, rounded and invaded by pyroxene. Similarly, Fig. 234 shows pyroxene invading the plagioclase and replacing the feldspar often along its twin planes. Comparable is the textural pattern shown in Fig. 235, the pyroxene extends between the intergranular of a plagioclase texture partly enclosing a plagioclase causing a reaction margin in the feldspar and invading an adjacent plagioclase, following its polysynthetic twinning.

Most typical pyroxene poikiloblasts are shown in Figs. 236 and 237. Fig. 236 shows plagioclase rounded and enclosed by the pyroxene poikiloblast. Similarly, Fig. 237 shows a pyroxene poikiloblast enclosing rounded and corroded plagioclases and olivines.

As a corollary to those later pyroxene growths is the intergranular pyroxene following the intergranular between an olivine (serpentinised) and plagioclase (Fig. 238).

The crystallization of pyroxenes in paragabbros is most complex; often more generations of pyroxenes may be present.

A series of photomicrographs, Figs. 239, 240 and 241 shows the relation of two pyroxene crystallization phases. The crystallization of phase-b (clinopyroxene) often follows interleptonic spaces of phase-a (orthopyroxene). As is clearly shown in Fig. 241 the phase-b extends along the cleavage interleptonic spaces of phase-a. Often the topometasomatic infiltration of phase-b may follow a pattern within the host phase-a (see Fig. 242). In cases the late pyroxene phase may follow two preference orientation directions within the pyroxene (see Fig. 243). As Fig. 244 shows not only phase-b may be invading phase-a, but also an adjacent plagioclase may send protuberances into the host pyroxene. Such textural patterns do not show "ex-solution" phases of pyroxenes, but later, often, topometasomatic pyroxene mobilisations, which in cases transverse the boundary of the pyroxene (phase-a) and also extend into the adjacent plagioclase, Fig. 245.

Often in gabbroic rocks most complex magnetite/pyroxene (hypersthene) symplectic intergrowths may be exhibited (see Fig. 246a). In cases, the symplectic magnetite transgresses to a magnetite phase (see Fig. 246b). Such symplectic intergrowths, often of mineral phases that do not belong to a crystallization-system (or isomorphous series) rather support complex crystallization sequences.

In contrast to the symplectic intergrowths synantetic reaction intergrowths may exist, between pyroxene, surrounding magnetite, and feldspar (see Fig. 247 (a and b)). Within the synantetic reaction, symplectic rhabdites of quartz may be present in pyroxene. Comparable are the quartz myrmekitic intergrowths in pyroxenes which surround magnetites and are in contact with plagioclases, see Fig. 248.

However, most complex textural patterns may result when magnetite with a complex pyroxene-corona is in contact with feldspar (see Fig. 249). Similarly complex pyroxene corona structures may develop around plagioclase. These represent complex reaction structures and belong to the palingenetic processes of the paragabbros, Fig. 250.

A most typical corona-structure (also common in norites), Fig. 251 (a and b) is often developed between olivine and plagioclase. This corona structure has been considered by Bowen (1928) as a result of the discontinuous crystallization series. In contrast to the explanation that these corona structures represent the consequence of the reaction of an early olivine reacting with the basaltic rest liquid, these corona structures represent rather a synantetic reaction of mineral phases, often complex in nature and they do not necessarily conform to the principle of the discontinuous reaction series of Bowen (see chapter 12).

The feldspar in the paragabbros may range from pre-pyroxene crystallizations (see Figs. 236 and 237) to late phase zonal growths. The composition of the gabbro plagioclases is considered to be intermediate in the series Ab-An. In contrast to an intermediate in composition plagioclase the recent research trend speaks rather of plagioclases showing sub-structure either of lamellar-twins or of ex-solved phases (ex-solutions). In most cases, the sub-structure of the plagioclases is in the range of Å units and is obvious with high voltage electron microscopy. In contrast comparable "substructures" but of different magnification may be seen with ordinary polarizing microscopy, see Figs. 252, 253 and 254.

In contradistinction to the hypothesis that the plagioclase sub-structure represents ex-solutions,

Fig. 253 shows that plagioclase phase-a invades and replaces phase-b and rather supports that topometasomatic processes are responsible for the "substructure" of this plagioclase crystalloblast.

In cases orientated inclusion of translucent minerals (?rutiles) are present in these plagioclases, see Fig. 254.

Often the plagioclases of gabbroic rocks show characteristics of tectonic deformation, bending of the polysynthetic twinning, undulating extinction and a type of lamellae that intersect the polysynthetic twinning (see Fig. 255). The phenomenology of the deformation of feldspars is extensively discussed in chapter 21.

In addition to the pyroxene the amphiboles and particularly hornblende are of petrogenetic significance in the gabbros. Often the hornblende exhibits a blastogenic character, enclosing corroding and reacting with the pre-blastic mineral phases. Fig. 256 shows crystalloblastic hornblende enclosing plagioclase and pre-hornblende mafic components in a texture, in which it has also grown blastically.

In cases the hornblende attains poikiloblastic character and is full of pre-hornblende, plagioclases and mafic components enclosed and corroded by the blastic amphibole.

In contradistinction, often the amphibole grows as prismatic growths replacing and invading plagioclases. As Fig. 257 shows the hornblende, due to its blastogenic force, has invaded extending from their margins inwards the plagioclases.

Comparable to the textural patterns where a blastic pyroxene encloses plagioclases, similarly crystalloblastic hornblende may engulf and corrode a pre-existing plagioclase. Fig. 258 shows plagioclases corroded and effected by a later hornblende crystalloblast which engulfs the plagioclase causing a reaction zoning in the feldspar. Also between the amphibole and the plagioclase a synantetic-symplectic reaction develops where a myrmekitic-like intergrowth of quartz and hornblende is formed (see Fig. 259).

Similarly, as Fig. 260 shows a reaction contact with hornblende and plagioclase a myrmekitic-like symplectic reaction margin of quartz and amphibole develops. In cases myrmekitic-like quartz bodies develop at the reaction margin of pyroxene in contact with a hornblende. The formation of quartz bodies can be interpreted either as a fore-runner silica-solution phase or as the liberation of a silica solution phase as a result of the complex synantetic reactions, Fig. 261.

These often complex symplectic-synantetic reactions as a whole support the blastogenic sequences within the ultrametamorphic processes of the gabbroic rocks.

Biotite often exhibits all traits of a blastogenic growth in gabbroic rocks. Fig. 262 shows interpenetrating biotite laths, exhibiting an intersertal blastoid textural pattern. As a corollary to the blastogenic growth biotite, Fig. 263 shows biotite enclosing and corroding a plagioclase.

The phenomenology of gabbros often shows synantetic and symplectic textural pattern that could hardly be expected in a basic rock often containing abundant olivine.

A series of Figs. 264, 265 and 266 shows the development of myrmekitic symplectic structures. As Fig. 264 shows a plagioclase in contact with other plagioclases shows a myrmekitisation margin (consisting of quartz worm-like bodies in intergrowth with plagioclase). The myrmekitic bodies are not strictly comparable to the typical myrmekites of the granitic and gneissic rocks where often myrmekites develop where K-feldspar reacts with plagioclase. In the cases illustrated here there is no K-feldspar present. However, in this case the quartz forming solutions have diffused through the intergranular of the plagioclase i. e. diffused through the interleptonic spaces of the intergranular spaces of the plagioclases and have, due to intergranular diffusion, invaded the plagioclase resulting in the myrmekitisation of the plagioclase.

In contradistinction to Fig. 264 where the myrmekite development was in a margin of a larger plagioclase, Fig. 265 shows a plagioclase myrmekitised almost entirely enclosed by another plagioclase.

Most characteristic myrmekitic patterns may develop in gabbros commensurable and comparable to myrmekites in granites (see Fig. 266). In addition, Fig. 267 shows idiomorphic plagioclase, in contact with pyroxene, again partly enclosed by another plagioclase and myrmekitised. In contradistinction, Fig. 268 shows a plagioclase surrounded by three other plagioclases "differently orientated" corroded and myrmekitised. A case also is exhibited where the myrmekitised plagioclase is adjacent to quartz (Fig. 269). Such cases support that the quartz growth and the myrmekitisation belong to a silica phase in the blastogenic sequence of gabbroitisation.

The gabbroic rocks, whether sub-volcanic or paragabbros often exhibit granophyric quartz (previously known as micropegmatitic quartz) in "graphic" intergrowth with plagioclases. A study of the granophyric quartz textural patterns in dolerites (microgabbros) of the Karroo and Palisade sub-volcanics shows that the quartz is a late phase (metasomatic phase) infiltrating and replacing not only the plagioclase but also the pyroxenes and other mafic components, see Augustithis (1978). Comparable textural patterns are exhibited in paragabbros. Fig. 270 shows a crystalloblastic plagioclase enclosing pyroxene, and being marginally invaded by metasomatic granophyric quartz. Similarly granophyric quartz invades plagioclase and is in intergrowth with the adjacent plagioclases (infiltration of the intergranular boundary), Fig. 271.

Most indicative textural traits of granophyric quartz replacing and invading the host plagioclase is shown in Fig. 272. The granophyric quartz has followed the plagioclase intergranular space and has replaced the feldspar, with plagioclase twin lamellae protruding into the quartz as non-replaced relics (see arrow in Fig. 272).

Another instance of late phase quartz corroding and invading plagioclases is shown in Fig. 273. The quartz, as intergranular phase, has corroded an adjacent plagioclase (indentation of the plagioclase, due to the quartz infiltration).

The quartz in gabbroic rocks is not restricted to the plagioclase phase, but symplectic quartz due to metasomatic replacements may be formed in biotite, see Fig. 274.

Another instance of colloform quartz invading the biotite along its cleavage is illustrated in Fig. 275. The elongated quartz colloform bodies (elongated and following the interleptonic cleavage spaces of the mica, consist of a fibre-chalcedony with a radiating arrangement of the fibres).

In the gabbroic rocks particularly in the paragabbros, blastogenic zircons and apatites may be present as shown in Figs. 276 and 277. Particularly, Fig. 277 shows an idioblastic zircon in a gabbroic plagioclase.

Magnetite is an important ore-mineral constituent in gabbros. As is shown in chapter 16 poikiloblastic magnetites are present in the Skaergaard layered gabbroic rocks. As is shown in Fig. 278, the magnetite has blastically crystallised enclosing and corroding not only mafic components but also plagioclases.

Essexites

Some essexites exhibit textural patterns typically magmatogenic, such as rhythmically zoned gigantic augites and feldspar phenocrysts Fig. 279 (a and b), see also Augustithis (1978).

Additional textural traits common in magmatogenic rocks of the doleritic-basaltic clan, such as hourglass structures of the pyroxenes and fracturing and rotation of the pyroxene phenocrysts (Fig. 280) are also abundant in magmatogenic essexites. The general appearance of some essexites is that of a hypabyssal rock (see Fig. 281), composed of early olivine phenocrysts, oscillatory zoned augite and plagioclase phenocrysts and of prismatic plagioclase groundmass. In cases the oscillatory zoned augites have plagioclase laths parallel to the augite fine zoning indicating that they have been incorporated as already crystallised bodies by the later grown tecoblastic augites (Figs. 282 and 283).

Often nepheline or analcime are present and are regarded as characteristic minerals of the essexite group.

Another textural pattern characteristic of the magmatogenic rock types, common in essexites, is the ophitic intergrowth of prismatic plagioclases and pyroxenes (see Figs. 284, 285, and 286). Often certain sections of the ophitic intergrowths show a "rectangular" section of the plagioclase lath included in the pyroxene, see Fig. 286.

In contrast to the magmatogenic essexites which could be comparable to the doleritic-microgabbroic rocks of magmatic origin, certain essexites are comparable to the paragabbros and show textural patterns comparable and commensurable to the paragabbroic rocks (see Chapter 8 "gabbros"). Blastogenic growths are characteristic of the paragabbros. Fig. 287 shows zoned augite, with interzonal fine apatite, in contact with orthoclase and biotite xenoblast. The para-essexites often show pyroxenes and magnetites or ilmenites in contact with the later crystalloblastic biotite, in groundmass of orthoclase or analcime. Often the ore-minerals show corroded outlines and the biotite follows as interleptonic fillings the space between the ore-minerals and the orthoclase, Fig. 288.

In addition to the biotite, hornblende (prismatic) and sphene are abundant idioblastic growths in essexites, see Figs. 289 and 290.

Apatite in addition to its occurrence as interzonal fine needles may occur as idioblastic growths which, due to their crystalloblastic force, may penetrate through the pre-existing mineral phases. Fig. 291 shows an apatite nematoblast transversing biotite and orthoclase. Often the apatites are large prismatic crystalloblasts (Fig. 292) and in cases they show feldspar groundmass enclosed in their central part Fig. 293.

Despite the presence of feldspathoids and of the alkaline nature of the essexite, quartz infiltrations, invading the mafic components, may be present (see Fig. 294) in such cases the quartz may be attributed to a metasomatic later phase.

Chapter 9 Norites – Troctolites – Shonkinites – Theralites – Jacubirangites

The norite group of rocks shows blastogenic traits comparable to the paragabbros. The crystallization sequence is rather dependent on the blastogenic sequence. However, in contradistinction to these meta-norites, cases exist where mantle traits are present and in such cases a mantle participation in the formation of the particular norite should be considered.

A study of these features shows, nevertheless, the preponderance of metamorphic-metasomatic (i. e. blastogenic) traits in a number of studied norites.

Pyroxene poikiloblastic growths where rounded and corroded olivines and plagioclases are enclosed in the pyroxene crystalloblasts, are common, see Figs. 295 and 296. Often the pyroxene crystallization follows the intergranular spaces between plagioclases. As Fig. 297 shows the pyroxene crystallization used the intergranular and interleptonic spaces between feldspars. In other instances the pyroxene crystallization shows protuberances occupying the spaces between feldspars, Fig. 298.

In cases the crystallization of a granular pyroxene may follow the twin lamellae of a plagioclase. Such cases of a granular replacement indicate the complexity of the pyroxene crystallization in norites.

The crystallization of amphiboles in norites invariably is crystalloblastic, where the blastogenic hornblendes enclose, corrode and produce reaction margins in enclosed plagioclases, see Fig. 299. A common textural pattern is where crystalloblastic hornblende encloses rounded hypersthenes, see Fig. 300.

Frequently in norites (in cases only in gabbros) a crystalloblastic sequence may be exhibited in a textural pattern. A series of photomicrographs, Figs. 301, 302, 303, 304, 305, 306 and 307 shows the following blastogenic sequence: pyroxene, biotite (or phlogopite), garnet, often crystalloblastic apatites either develop in the pyroxene or are included in the garnet crystalloblastic margins, Figs. 304, 305, 306 and 307.

The crystallization of the mica (biotite or phlogopite) often succeeds the pyroxene crystalloblasts and is followed by a margin of garnet crystalloblastesis, Figs. 301 and 302. In cases, though the succession of blastic phases results in the co-existence of mica and garnet margins of the pyroxene, see Fig. 303.

Comparable to the corona structures of gabbros are the corona structures of norites, where a reaction corona structure develops round olivines which are surrounded by plagioclases.

As the study of similar corona structures reveals, the question arises as to whether these reaction coronas are actually due to intergranular reaction-replacing processes, in which case complex synantetic-symplectic intergrowths take place.

Fig. 308 shows olivine with a complex reaction corona consisting of a pyroxene margin followed by a complex margin of symplectic-synantetic myrmekitic like intergrowths. Often these complex myrmekitic structures consist of "myrmekitic chlorite".

Also most complex and difficult to interpret corona reaction structures develop between feldspar and pyroxene. Fig. 309 shows a symplectic synantetic corona structure (consisting of "myrmekitic chlorite") between pyroxene and plagioclases.

Indeed, reaction symplectic textures in norites are abundant. As Figs. 310 and 311 show, symplectic-synantetic textures of plagioclase and pyroxene may result in great complexity of patterns. Particularly, as is shown in Fig. 312 the pyroxene invades and replaces the plagioclase causing the symplectic pattern of pyroxene/plagioclase.

Of particular importance are the metamorphic relic structures in the ultrametamorphic norites (para-norites). As Fig. 313 shows, initial sedimentogenic quartz granular texture, maintaining all the characteristics of a metamorphic texture, and with perthitic K-feldspar, is invaded by blastogenic hornblende. The hornblende-crystalloblasts often occupy the intergranular of the initial sedimentogenic texture. Similarly, Figs. 314 and 315 show initial sedimentogenic-metamorphic quartz-texture (often the quartz exhibits undulating extinction) with intergranular-poikiloblastic hornblende.

In cases, in the sedimentogenic-metamorphic texture consisting of mosaic quartz/plagioclase may be present zircons (Fig. 316).

The quartz (initial sedimentogenic) may show a recrystallization indicated by the maintenance of the traces of the initial granular texture and showing recrystallization (Fig. 317). In the metamorphic quartz texture, hornblende crystalloblastesis may often take place (Fig. 317) and with the granular quartz in cases engulfed by the amphibole (Fig. 318).

In contrast to the gabbros, the norites show a greater abundance of accessory minerals. Fig. 318 shows blastogenic apatites in a recrystallised quartz-amphibole texture.

Of significance in the norites is the abundance of zircon, which might represent "relic" phases (that is zircons of sedimentogenic derivation in the initial sedimentogenic-metamorphic phase of the meta-norites).

As Fig. 319 shows, zircons may show rounded and corroded boundaries and may exist as "scattered" remnants in the recrystallised sedimentogenic-metamorphic granular quartz. As a corollary to this, Fig. 320 shows corroded zircons with granular quartz recrystallizations. Similarly, Fig. 321 shows rounded zircons with granular quartz recrystallizations. Additional observations, Fig. 322 and Fig. 323 show rounded and corroded zircons associated with the metamorphic quartz-feldspar texture of a norite.

In contrast to these sedimentogenic zircons (zircons of the initial sedimentogenic-metamorphic phase) new growth of zircons, blastogenic in origin may be present in norites. Considering the origin in acid plutonic rocks and in particular the zircons in granitised granites, regrowth of zircons around initial sedimentogenic zircon-relic and new growths of zircons have been proposed by Wyatt (1954), Poldervaart (1955, 1956), Bader (1961), Hoppe (1962a, b, c, 1963, 1964, 1966), Augustithis (1967a, 1973) and Drescher-Kaden (1969). Comparable and commensurable is the origin of zircons in norites. In addition to the relic-zircons, new growths are common. Fig. 324 shows idioblastic zircon following the twinning in a blastogenic plagioclase. Similarly, idioblastic zircons are shown in granular quartz (Fig. 325) and in quartz, plagioclase-hornblende textures, Fig. 326.

Often there are present in the para-norites blastogenic biotites which, in cases, are invaded by later quartz which may invade the mica and produce indentation-margins with it. In contrast to the abundant myrmekitic and granophyric symplectic quartz of the gabbros in norites the quartz is often present as initial sedimentogenic-metamorphic quartz with recrystallization. The textural studies so far, consider the norites mainly as an ultrametamorphic rock, comparable to granite and gabbro, due to transformation and metamorphism of initial sedimentogenic material.

In contrast, certain textures exist in norites which are comparable and commensurable to mantle textures observed in olivine bombs in basalts and in ultrabasic diapirs mobilised in orogenic zones (see chapter 10). Fig. 327 shows graphic-like intergrowth of green spinel (pleonaste) with pyroxenes. Such textural patterns are comparable to mantle in origin ultrabasics or parts of them in basic rocks.

In support of the "xenocrystalline" or xenolithic origin of these mantle derived textures in norites and particularly in paranorites are the rounded and corroded boundaries of both pyroxene and spinel in contact with the plagioclase crystalloblast of the paranorite.

Similarly, Fig. 328 shows a mantle xenolith consisting of pyroxene, phlogopite and with pleonaste spinel in graphic intergrowth with them. The mantle xenocryst (association) is in contact with crystalloblastic plagioclase of the paragabbro.

Troctolites

Troctolites are basic to ultrabasic types with diverse proportions of olivine and with plagioclases ranging between labradorite-bytownite. Hatch et al. (1949), considering the idiomorphism of olivines and their relation to feldspars, state the following: "The degree of idiomorphism exhibited by the two chief minerals depends upon their relative proportions with olivine in excess of a certain amount, this mineral separates before the plagioclase, and is therefore euhedral towards it. With plagioclase in excess, the olivine crystallises late and is interstitial towards the earlier and therefore better formed plagioclase."

In contrast to this interpretation, present microscopic studies reveal that topologically within a troctolite (that is in parts of the troctolite where olivine predominates) the olivine is poikiloblastic towards the plagioclases, which are enclosed, rounded and corroded by the olivine crystalloblast. Fig. 329 shows plagioclase enclosed by the olivine crystalloblast. Similarly, Fig. 330 shows plagioclases rounded corroded and enclosed by the olivine crystalloblasts. In addition to the olivine poikiloblasts, idioblastic olivine may develop often free from feldspar inclusions. On the whole olivine tends to be free of inclusions due to "autocatharsis".

Often olivine crystalloblasts are surrounded by biotite or phlogopite crystalloblasts (Fig. 331), in case the biotite forms rims surrounding the olivine crystalloblasts (Fig. 332).

In the case of the typical troctolite reaction corona intergrowths are often observed. Fig. 333 shows a general view of a troctolitic olivine and the adjacent plagioclases. The olivine is partly surrounded by biotite and partly by a corona reaction structure which also surrounds the biotite.

Fig. 334 shows olivine (serpentinised) with pyroxene forming a "corona" between the olivine and the plagioclase. Such marginal growths of pyroxene are common not only in the troctolites but also in gabbros (see Fig. 251a, b). The corona structures surrounding the olivine may attain a most complex symplectic nature, Fig. 335 shows a pyroxene symplectic corona surrounding a serpentinised olivine. Another complex symplectic "corona" reaction-structure develops between serpentinised (antigoritised olivine) and pyroxene (see Fig. 336).

Considering the troctolites, of significance are the poikiloblastic and idioblastic tendencies of the olivine. The relation of troctolites in layered bodies with norites and the transition of troctolite through allivalite (essentially troctolite with anorthite) to anorthosites supports an ultrametamorphic origin of the troctolite due to ultrametamorphism and rock transformation.

Shonkinite

Pirsson and Weed (1895) first used the name to signify a coarse-grained basic rock* which forms the greatest part of the Shonkin Sag laccolith in the Bearpaw Mountains, Montana.

Microscopic studies of Shonkinites reveal two rock types, diverse in origin, with distinct textural patterns and genesis.

Fig. 337 shows rhythmically zoned augites magmatically rounded and corroded by the groundmass, which, while a melt, has affected the early formed pyroxene phenocrysts.

In contrast to the above described magmatic textures, shonkinite may exhibit textural patterns which suggest an ultra-metamorphic-palingenic** origin of the rock. Fig. 338 shows idioblastic biotite in a background of groundmass.

Similarly, Fig. 339 shows idioblastic augite enclosing feldspar parts. The co-existence of augite with orthoclase is understood on the basis of blastogenic growth of augites, independent of a rock equilibrium which would have been attained if the rock had crystallised out of a melt. Of particular interest is the presence of aegerine prismatic and radiating velonoblasts in cases transversing quartz, see Fig. 340. Similarly, aegerine prismatic velonoblasts are associated with biotite, and quartz is intergranularly associated with the pyroxene (see Fig. 341). Biotites, often exhibiting idioblastic growths are present in Shonkinite. In contrast to Fig. 338, Fig. 342 shows a hexagonal biotite idioblast enclosed in a larger, again hexagonal, biotite idioblast.

Apatite crystalloblastesis may attain significance in shonkinites. Fig. 342 shows accessory idioblastic apatite blastically penetrating into the idioblastic biotite. In contrast idioblastic apatite velonoblasts may transverse the boundaries augite-orthoclase, with the width of the apatite diminishing as it transverses the feldspar (Fig. 343).

In addition to the accessory apatite, associated mainly with the biotite and augites, apatite may attain an abundance of importance in the shonkinite. Figs. 344 and 355 show apatite mega-crystalloblasts exhibiting idiomorphism and comprising topologically the most abundant mineral constituent.

Apatite may often exhibit idiomorphism, perfect crystalline outlines may be exhibited, see Fig. 345. As is shown in Fig. 345 and as Fig. 346 and particularly Fig. 347 show magnetite may extend between the intergranular of the idioblastic apatite and by intracrystalline "diffusion" iron oxides may form "droplets" in the apatite. All transitions from intergranular iron-oxides to the development of intracrystalline iron oxide "droplets" in the apatite are shown in Fig. 347.

As a corollary of the crystalloblastic nature of apatite Fig. 348 shows regrowth zones of an apatite crys-

---

* orthoclase associated with clinopyroxene.

** In contradistinction to the blastogenic-metamorphic interpretation of the Kutzebuckel's shonkinite textures presented in the present Atlas and in harmony with the volcanic origin of the volcanic veins (basalt) occurring in the region, Frenzel (1953, 1954, 1954) considers the presence particularly of pseudobrookite in the volcanic "veins" as evidence of high temperature. Similarly certain textural patterns of the Kutzenbuckel shonkinite are definitely palingenetic (anatectic).

talloblast. Considering the main components of shonkinite, augite and orthoclase the presence of K-feldspar and in the case of the natron-shonkinite the presence of aegerine indicate a basic rock with an excess of K-Na.

Theralite

Table III Composition of Theralites by Rosenbusch

	1.	2.	3.	4.	5.	6.
SiO_2	46,53	46,47	44,42	43,68	45,06	51,26
TiO_2	2,99	1,21	1,63	1,10	1,72	1,66
Al_2O_3	14,31	18,77	13,33	20,15	16,56	23,78
Fe_2O_3	3,61	3,55	9,14	2,94	4,79	1,81
FeO	8,15	4,83	6,35	8,97	7,04	2,70
MnO	0,22	Sp.	–	0,23	0,11	0,10
MgO	6,56	3,90	5,74	4,38	5,11	1,96
CaO	12,13	7,28	10,60	7,24	9,27	8,00
Na_2O	4,95	3,73	5,60	7,00	5,29	6,72
K_2O	1,58	4,65	1,81	1,99	2,50	2,16
H_2O + ...	0,20[1])	4,93	1,75	0,89	1,93	0,55
H_2O	–	–	–	–	–	0,10
P_2O_5	–	0,14	0,35	1,38	0,62	n. best
Sa	101,23	99,80	100,35	100,22	100,00	100,00
Spez.Gew. .	2,96	–	3,008	–	2,98	2,77

[1]) Glühverlust.
1. Theralith. Umptck, Halbinsel Kola.
2. Theralith. Val dei Coccoletti, Mte. Mulatto, Pradazzo. Mit 0,34 CO_2. Reich an Mikroperthit.
3. Theralith, Flurhubl, Duppauer Gebirge, Bohmen, Mit 0,18 S.
4. Theralith-Canadit. Sagen b. Almunge, Schweden. Mit 0,07 Cl, 0,20 F.
5. Mittel dieser 4 Theralite (auf 100 berechnet).
6. Rouvillit. St. Hilaire, Monterigian Hills, Quebec.

Rosenbusch (1896/1923) proposed the name theralite for a coarse grained "igneous rock" equivalent to the fine grained volcanic nepheline tephrite, with corresponding mineralogical composition namely gabbroic plagioclases, clinopyroxene and nepheline.

As the chemical composition shows (see table III by Rosenbusch) the theralite is an alkali rich basic rock and according to Rosenbusch the plutonic equivalent of nepheline tephrite. A study of theralitic rocks from different known occurrences reveals however, mineralogical composition and textural patterns differing greatly from the initial concept of Rosenbusch.

Textural studies reveal that the theralite occurrence comprises coarse-grained volcanic rocks, with olivine phenocrysts (Fig. 349) and idiomorphic rhythmically zoned titano-augites comparable to the zoned augites found in basalts and essexites (see chapter 8). Fig. 350 shows the zonal growth and the form of the augite idiomorphic phenocrysts in the volcanic theralites.

In contrast to these volcanic theralites, other occurrences of theralites reveal an ultrametamorphic rock-type where crystalloblastic growths predominate. Figs. 351 and 352 show idioblastic hornblende enclosing magnetitic crystal grains. In contradistinction Fig. 353 shows crystalloblastic hornblende partly engulfing theralitic groundmass. Similarly, Figs. 354 and 355 show hornblende crystalloblast enclosing rounded and corroded calcite. Also Fig. 356 shows an idioblastic hornblende protruding into feldspar and enclosing a rounded and corroded calcite grain.

In the blastogenic sequence as is considered on the basis of textural patterns exhibited, the pyroxene is often surrounded by later hornblende crystalloblasts (see Fig. 357). In cases though, the pyroxene/hornblende crystallization sequence is dubious (see Fig. 358).

Apatites in theralites may show idioblastic forms, Figs. 359, 360, or may exhibit skeleton crystal forms due to incomplete crystallization. Particularly, Fig. 356 shows different developments of apatite crystal-skeleton forms. In contrast, in Fig. 359 both a crystal skeleton and a well developed apatite idioblast is illustrated.

As Fig. 357 and Fig. 361 show apatite crystalloblast may, due to blastogenic growth, transverse amphiboles. In contradistinction, Fig. 362 shows that an apatite crystalloblast has interfered in the development of an amphibole crystalloblast.

In cases, apatite idioblasts may be surrounded by biotite and with a reaction margin between the two (see Fig. 360) being formed. In contrast to these late phase apatite crystalloblasts large apatites showing a crystal corrosion and often associated with biotite may be present in these ultrametamorphic theralites, (see Figs. 360 and 363).

In addition to the presence of calcite in the groundmass of the theralites, also well developed calcite growths may exist, Figs. 364 and 365. Plagioclase enclosed in later biotite crystalloblast are common. Often magnetite pigments may follow the boundary plagioclase/biotite, however in cases the iron pigment may follow the biotite cleavage (see Fig. 366).

The biotite may blastically surround and corrode or follow corroded outlines of magnetite (see Figs. 367 and 368). The blastic biotite may be full of iron-oxide pigmentation and apatite crystalloblasts.

Jacubirangite an ultrametamorphic rock

Another ultrametamorphic ultrabasic rock is the Jacubirangite from Slada, Isle of Alnö, Sweden, which consists of crystalloblastic hornblende and mega-crystalloblast of apatite. Fig. 369 shows crystalloblastic hornblende penetrated by apatite mega-crystalloblast often the contact hornblende/apatite is indicating a mutual indentation of the two phases, in cases however, the apatite sends protuberances into the amphibole.

Similarly, Fig. 370 shows crystalloblastic amphiboles transversed by crystalloblastic later apatite.

An additional crystalloblastic-metasomatic phase is calcite often corroding and replacing the crystalloblastic apatite, see Figs. 371 and 372. As a corollary to the apatite replacement by the calcite, Fig. 372 shows calcite replacing the apatite and invading it, along one of the rhombohedral directions of the calcite.

The predominance of amphibole and apatite crystalloblastesis renders a unique textural pattern of blastogenic intergrowths in the case of Jacubirangite. Furthermore, the association hornblende/apatite signifies a petrogenetic and geochemical affinity between these mineral phases.

Chapter 10 Peridotites

On the origin of olivine rich ultrabasics (dunites, websterites, lherzolites, olivine-peridotites, olivine rich pyroxenites)

The dunitic bodies have been, till recently, believed to be ultrabasic magma crystallizations and in fact, according to Bowen's hypothesis, several workers e. g. Gwyneth Challis (1965), Miyashiro (1975), Pantazis (1973, 1977), have proposed that the dunitic magma is a basic differentiation of the basaltic magma. Indeed, the hypothesis of the dunitic magma has prevailed in petrological literature.

As an example of magmatic differentiation of ultrabasics from tholeiitic basaltic magma is proposed by Gwyneth Challis (1965) the Dun Mountain, Red Hills and Red Mountain associated with permian rocks of the South Island of New Zealand. According to Gwyneth Challis comparison of the textures and minerals chemistry of the ultramafic rocks with those of the ultramafic parts of layered intrusions suggest that the New Zealand rocks have been derived by gravitational differentiation from tholeiitic magma.

It is further proposed by him that the intrusions represent the subvolcanic differentiates of a chain of Permian volcanoes on the margin of the New Zealand geosyncline, and that the associated olivine poor tholeiitic Permian basalts are the extrusive differentiates.

In contrast to the ultrabasic-magmatic hypothesis Ross et al. (1954) have introduced the idea that dunitic and peridotitic bodies may represent mantle diapirs in orogenic zones. A number of workers support a mantle derivation hypothesis, e. g. Hess (1960), Loney et al. (1971), Moore (1973).

According to Green (1964) the Lizard peridotite produced a high-temperature metamorphic aureole during diapiric emplacement in a period of amphibolite facies regional metamorphism. It is further concluded by Green (1964) that the primary minerals of the Lizard peridotite have not crystallised and accumulated from a basaltic magma but have crystallised in a similar environment to that of peridotite nodules in basalts.

Furthermore regarding the emplacement of ultrabasic bodies Green states "The recognition of high-temperature peridotite intrusions accentuates the complexity becoming evident in the study of ultramafic rocks occurring in orogenic belts. Such ultramafic rocks include the large layered ultramafic complexes such as the Philippines, Cuba, and Papua characterised by movement and tectonic contacts against enclosing sediments or volcanics. These may preserve excellent accumulative textures, indicating tranquil accumulation from uncertain magma composition (Green, 1961) or such textures may be destroyed by solid flow as an inhomogeneous crystalline mush (Thayer, 1960). The suggestion by Hess (1960) that such large complexes (he instanced Cuba and Puerto Rico) are altered sub-oceanic mantle material exposed in orogenic belts may well be correct since their present environment and attitude seem unrelated to their environment of primary accumulation."

In addition, discussing the multiple hypothesis of ultramafic emplacement, Green comments "The genesis of the type of serpentinised peridotite sill or sheet in low-grade or unmetamorphosed sediment (Chesterman 1960) remains an outstanding problem but the answer is considered to lie in the deformation properties of serpentinised peridotite rather than in an appeal to low temperature peridotite magma. The reintrusion of serpentines as cold intrusions (Taliferro, 1943, Green, 1959) demonstrates the readiness of the rock to deform but the actual mechanism of deformation is not well understood."

Considering the data (the thermal contact of some ultrabasic inclusions, the mantle diapirism of the large ultrabasic intrusions and the "cold serpentine intrusions") Green suggests that the ultrabasic intrusions in orogenic belts (alpine type peridotites) could have a multiple origin.

Another example of diapir intrusion of ultrabasics is the alpine type peridotite at Burro Mountain, California discussed by Loney et al. (1971). The ultrabasic body is partly serpentinised harzburgite-dunite approximately 2 km in diameter. It lies in a chaotic mélange derived from the Franciscan Formation of the southern Coast Ranges of California. The peridotite appears to be one of a number of tectonic lenses, having a wide range in size, that make up the mélange. These lenses include metagrawacke, metachert, as well as ultramafic rocks, and represent a wide range of pressure-temperature environments.

The evidence according to Loney et al. (1971) indicates that the dunite and probably also the hartzburgite crystallised from an ultramafic magma, probably in the upper mantle. After the magmatic episode and crystallization, the peridotite was subjected to a deep seated plastic deformation and recrystallization. Following the deep seated deformation, which probably took place in the mantle, the peridotite mass was tectonically detached and moved upward to its present level in the crust.

Furthermore, regarding the multiple hypothesis explanation for the alpine-ultrabasic bodies, Moore (1973) states that "The Gosse Pile mafic-ultramafic intrusion is a layered igneous body, the upper part of which was involved in localised ductile deformation (low angle faulting) soon after crystallization.

A complete gradation between rocks showing typical igneous textures and layering and those showing typical tectonic, or metamorphic textures and layering can be observed."... "The metamorphic textures and layering of the deformed rocks are identical to "flow layering" which is used as an essential criterion for distinguishing "alpine-type" from "stratiform bodies"... "The Gosse Pile massif most probably indicates the transition mantle → block intrusion to tectonic mobilisation of mantle → into an alpine diapir."

On the basis of comparative textural patterns of mantle fragments in basalts (olivine bombs in basalts) and "intrusive" in character peridotitic bodies, identical patterns have been found both in the mantle fragments in basalts and in peridotite intrusives.

Figs. 373, 374, 375 and 376 show graphic spinel in intergrowth with olivine and pyroxene of the mantle fragments in basalts (i. e. olivine bombs in basalts). Identical textural patterns are found in a variety of olivine rich peridotite "intrusives" in orogenic belts. Figs. 377, 378 (a, b), 379, 380 and 381 show graphic spinel bodies in intergrowth with olivine and pyroxenes in ultrabasic intrusives from different parts of the world.*

On this occasion it should be pointed out that graphic spinel occurs in a variety of ultrabasic rocks, which despite their nomenclature variation (i. e. classified under different rock types such as olivine peridotites, dunite, websterites, lherzolites**, bronzitites) are in reality mantle diapirs in orogenic belts.

The mobilisation of the mantle, due to diapirs in orogenic belts*** of the crust, will result in different

---

* Fig. 382 shows orientated spinels in a pyroxene of "dunite" (actually olivine rich peridotite).

** In addition to the textural criteria comparisons which have been presented in this Atlas, the present trend of research [e. g. Mercier and Nicolas (1975), Schubert (1977), Maaløe and Aoki (1977)], considers the alpine-type peridotites as mainly "mobilised" spinel lherzolites, however, other ultramafics occur as mantle mobilisations in orogenic belts (see Chapter 18).
Maaløe and Aoki (1977) consider lherzolites as upper mantle and if the composition of the upper mantle may be estimated from that of lherzolites, the composition of spinel lherzolites should be the basis for this estimate.
In contradistinction a garnet lherzolite occurs below Africa in contrast to the spinel-lherzolites of continental and oceanic derivation.
Regarding the mantle origin of lherzolites Mercier and Nicolas (1975) state the following "A structural study of lherzolite xenoliths mainly from Western Europe and Hawaii leads to a classification of three representative groups with every traditional stage. Their origin in the upper mantle and their kinematic meaning can be understood by referring to Alpine type peridotite massifs in which the large scale structure can be observed."

*** In contrast to the diapiric intrusion of ultramafic bodies (alpine type of ultramafics), Jackobshagen (1977) and the French School of plate tectonics proposed various models of plate tectonic movements and subduction zones as an explanation of the Hellenides mountain belt. Furthermore, the ophiolites associated with the Hellenides have been interpreted as obducted oceanic crust-upper mantle.
Coleman (1977) considers the relation of the plate tectonics and ophiolites as follows:
"The development of the plate tectonic hypothesis over the last decade has led to some astounding changes in geologic dogma. The long-standing ophiolite controversy with its multiple theories was one of the first petrotectonic problems to yield new solutions within the framework of the plate tectonic theory (Gass, 1963; Dietz, 1963; Hess, 1965; Theyer, 1969b; Coleman, 1971; Dewey and Bird, 1971; Moores and Vine, 1971). The present-day paradigm considers that ophiolite forms as oceanic crust generated at midocean ridges, from whence it slowly migrates by ocean floor spreading toward continental margins – there to be subducted into the mantle. Under some circumstances at plate boundaries, slabs of oceanic lithosphere have become detached and override (obduction) continental margins (Coleman, 1971). The actual mechanism of ophiolite emplacement along continental margins

"ultrabasic intrusives" which despite an intrusive character and despite the lack of dynamic metamorphism in the area will exhibit all typical characteristics of mobilised mantle, such as deformation of olivines, graphic spinels etc.

In addition to the identical pattern of graphic spinel both in mantle fragments in basalts and in mantle diapirs in orogenes (i. e. ultrabasic intrusives), the typical granular olivine texture of some dunite, see Figs. 383 and 384, is encountered in olivine bombs in basalts, see Figs. 385 and 386. In both cases the granular olivine textural pattern is identical (i.e. the textures are comparable but the patterns identical) and clearly support a mantle mobilisation in tectonic mobile zones of the crust.

Another line of evidence that the intrusive peridotites in orogenic belts and the olivine "bomb" in basalts share a common origin from the mantle is the olivine bombs in basalts often show deformation lamellae and tectonic influences that can be attributed to mantle deformation, see Augustithis (1978) and also Figs. 387 and 388. This deformation evidence indicates that the olivine bombs are fragments of a rock in depth that exists in a state that can be deformed. It is evident that the deformation effects of the bombs have been produced before they have been engulfed by the basaltic melt that transported them to the surface as part of the effusive rock.

Comparable is the case of intrusive ultrabasics which lack the evidence of post-intrusion dynamic metamorphism and despite that, show deformation characteristics, such as deformation twin lamellae, undulating extinction etc. In such cases, the deformation effects indicate that the rock has been deformed either in its pre-intrusive state or due to deformation of a plastic solid on its actual intrusion by diaprism. In both cases, the pre- or syn-instrusive deformation evidence exhibited by many ultrabasic diapirs or ultrabasic intrusives supports a mantle derivation. Fig. 389 shows pre-intrusive deformation lamellae of a dunitic body.

Frequently identical textural patterns are exhibited both in olivine bombs in basalts (actually mantle fragments) and in ultrabasic intrusives which in turn are mantle diapirs in mobile crust zones. Fig. 390 (a, b) shows intergranular diopside between differently orientated bronzites, also extensions of the diopside follow the cleavage of the bronzite.

Identical textural patterns are exhibited in bronzitite which is believed to be a plutonic ultrabasic. Also in this case, in comparison to Fig. 390 (a, b) diopside intergranular between bronzite sends extensions which follow the cleavage interleptonic spaces of the bronzite, Fig. 391.

Comparing the textural pattern of mantle fragments in basalts with the textural pattern of a variety of ultrabasic peridotites such as dunites, websterites, lherzolites – the characteristic textural patterns are identical (in the sense that the textures are comparable but the patterns identical). It is therefore possible to believe that a great number of ultrabasic plutonic peridotites are mantle diapirs or mantle protuberances within the crust following-zones of weakness of the latter.

In contrast to the mantle diapirs, ultrabasics of diverse origin may exist with textural pattern and structural characteristics which may be in contrast to the mantle textural patterns (i. e. pre-intrusive deformation effect, graphic spinels, intergrowth with olivines and pyroxenes etc.). Often banded dunites may exist, Figs. 392 and 393, showing alternating bands of coarse grained olivine and finer grained olivine, with laths of tremolite parallel to the banding.

It should be pointed out that the banding is not deuteric due to deformation but to the olivine difference in size in the alternating bands, it should be emphasised that the fine grained olivine zone often shows the presence of tremolite crystalloblasts, see Fig. 393.

Comparable textural patterns with tremolite crystalloblasts following the intergranular between the olivine grain is shown in Fig. 394 (a, b). It should be pointed out that the granular olivine often shows undulating deformation. In contrast to the tremolite crystalloblast intergranular between the olivine, crystalloblastic tremolites often transversing the granular olivine textures are exhibited, see Fig. 395 (a, b).

As a corollary to the crystalloblastic tremolites,

---

is strongly debated, but most geologists agree that the slabs of ophiolite are allochthonous and that they originate in an environment distinctly different from where they occur today (Davies, 1971; Zimmermann, 1972; Gass and Smewing, 1973; Mesorian et al., 1973; Dewey, 1974; Coleman and Irwin, 1974). This modern view contrasts with the older debates regarding the ultramafic parts of ophiolites, which centered on the concept of their intrusion as magmas (Hess, 1938)."

In contradistinction, Augusthitis (1978) regards the Konitsa ultrabasics as mantle diapirds and similarly in the present Atlas the Vourinos ultrabasic complex of N. Greece and the Troodos complex of Cyprus are considered also of mantle derivation (see Chapter 18). In support of this suggestion are geophysical studies of Makris (1977 a, b) which point to a thinning of the crust over Epirus, and Euboea which is in harmony with mantle diapirism at least in north Greece (Epirus and Vourinus areas).

transversing granular olivine texture, Fig. 396(a) shows tremolite crystalloblast extending into the olivine due to the crystalloblastic force of the tremolite. Similarly Fig. 396(b) shows a developed elongated amphibole crystalloblast transversing the granular olivine texture of the Rödberg "Dunite", Norway.

In addition to the crystalloblastic amphibole, mica (phlogopite) crystalloblasts may also grow in "dunitic" rocks: "dunite" Sommergraben near Kranbath, Styria Austria, see Fig. 395.

These textural patterns, Figs. 393, 394 (a, b), 395 (a, b) and 396 (a, b) are comparable to the textural pattern exhibited by metamorphogenic olivinefels, see Figs. 4 and 397.

Fig. 4 shows a granular olivinefels texture transversed by tremolite crystalloblasts, similarly Fig. 397 shows tremolite crystalloblast in a granular olivinefels texture.

The above textural pattern comparisons show that "dunites" exist with different textural patterns i. e. dunites that are definitely mantle diapirs and "dunites" that are metamorphogenic olivinefels.

However, in certain peridotites, blastic mineral growths may be observed such as garnets and poikiloblastic pyroxenes.

Fig. 398 shows garnet crystalloblast in peridotite enclosing an olivine, between the garnet and "groundmass" of the peridotite a "corona" reaction structure is produced.

Similarly, a "corona" reaction structure is produced between garnet and the "groundmass" and also between an adjacent olivine and the peridotitic groundmass, see Fig. 399.

This textural pattern indicates that the corona structure is not necessarily a reaction structure of two mineral phases, i. e. garnet and the peridotitic groundmass, but rather a complex reaction structure of solution phases often taking advantage of the interleptonic-intergranular between mineral phases.

As a corollary to this, Fig. 400 shows a "corona" structure between garnet and pyroxene which for no obvious reason is thinning out.

In addition to the garnet crystalloblast, pyroxene crystalloblast may also develop in peridotites. Fig. 401 shows pyroxene poikiloblast enclosing olivines.

The formation of metamorphic peridotites with crystalloblastic phases is discussed in Chapter (1, 16).

The classification of basic and ultrabasic plutonics on the basis of their textural patterns.

On the basis of comparative microstructure the mineralogical-compositional classification of basic and ultrabasic rocks no longer holds.

The gabbro clan is subdivided into three, diverse in origin, group-rocks, i. e.

a) dolerites – microgabbros subvolcanic and with distinct magmatogenic textures, intersertal, ophitic etc.
b) The protolyte – gabbro: the gabbroic layer of the lower crust/upper mantle.
c) The paragabbros exhibiting blastogenic growth, i. e. with olivine and pyroxene poikiloblasts with symplectic and synantetic intergrowths due to metasomatism etc.

Also the norite clan of rocks should be differently classified, that is, the plutonic magmatogenic origin is in contradiction with the textural patterns which clearly are blastogenic and comparable to paragabbros. Particularly the presence of initial sedimentogenic metamorphic granular quartz is indicative of the paranoritic* derivation (origin).

Similarly, the troctolites are considered to represent a metamorphic group, with the poikiloblastic olivine, and with symplectic-synantetic corona-reaction structures.

In contrast to gabbros, norites, troctolites, the shonkinites and theralites are basic ultrabasic rocks that can be of diverse origin. Both shonkinites and theralites may exhibit sub-volcanic and volcanic textural patterns and are definitely magmatogenic.

In contrast, other shonkinites and particularly theralites may show beyond question metamorphic-metasomatic textural patterns and are products of ultrametamorphism and transformation. These para-shonkinites and para-theralites are not comparable to the sub-volcanic shonkinites and theralites. The misunderstanding and misclassification of rocks with entirely different origin and textural patterns was due to the fact that only compositional and mineralogical criteria were used and not the textural patterns and comparative analysis of the textural patterns.

On the other hand, rocks such as dunites, certain websterites, norites or parts of them and in general peridotites and pyroxenites show textural patterns comparable and commensurable to mantle derived xenoliths in basalts (Basaltic bombs). Indeed, these

* Ultrametamorphic norite.

plutonics, with deformed olivine, graphic spinels and granular olivines are mantle derived diapirs*, as presented in the discussed examples.

An exception is the banded dunites, often indicating alternating bands of equigranular olivine and tremolite rich bands, these could be olivinefels para-dunites, comparable might be the origin of certain other peridotites.

* or according to plate tectonics obducted oceanic crust-upper mantle.

Chapter 11 Lamprophyres

General

The lamprophyres (λαμπρός – Greek) are believed by magmatists to represent a special type of ultrabasic magma differentiate.

Seeliger (1975) in his paper "Was sind Lamprophyre?" compares the lamprophyres to the products of "metamorphosed" basic volcanic rocks by "natural autoclaves".

Despite the "magmatic origin" of these dyke-rocks certain textural and compositional peculiarities exhibited by some members of the group, render their explanation problematic. Particularly the lack of harmony between the phenocrysts and the groundmass has been emphasised by Hatch et al. (1949).

In addition present observations, particularly the granoblastic growths exhibited by the minette group (as well as their fragmented character) render their explanation as ultrabasic differentiate of granitic magma very doubtful. In addition considering the current granitisation and anatectic hypotheses for the origin of the granites, their origin (i. e. the minette group) as ultramafic (alkaline rich fraction) of a granitic differentiated magma is most improbable. The Vogesite is another lamprophyric group which again exhibits blastogenic growths and which is also associated with granitic bodies.

In contrast certain of the plagioclase and no-feldspar containing members of the lamprophyre group which occur as dyke-rocks (i. e. kersantites, camptonites and monchiquites) show textural pattern definitely magmatogenic (i. e. the rock has primarily crystallised out of a melt).

Based mainly on Rosenbusch (1896, 1923), and Hatch et al. (1949) we can classify the lamprophyres as follows: (Table IV).

In addition to their classification on the basis of the feldspars (type and presence) it is suggested that there is a relationship between the type of lamprophyre and the "parental rock type" (i. e. in the sense of the parental magma from which it was derived).

Table IV, Table showing Essential Mineral Composition of Some Types of Lamprophyres

	With Orthoclase	With Plagioclase	No Feldspar
Biotite	Minette (if with augite, augite-minette)	Kersantite (if with augite, augite-kersantite)	Alnöite (with melilite)
Common Hornblende	Vogesite	Spessartite Malchite (= aphyric Spessartite)	
Barkevikite and/or Augite		Camptonite (barkevikite in type-rock from Campton Falls, New Hampshire)	Monchiquite (with analcite)

"Thus the type of lamprophyre with dominant orthoclase-minette and vogesite – are associated with granites; the plagioclase bearing types, kersantites and spessartites are allied to diorites, and are often associated in the field with diorite-porphyries and microdiorites. Finally, camptonite and monchiquite are associated with highly alkaline deep-seated rocks such as foyaite and other sodic syenites."

The Minette group

In accordance with the orthodox views, the lamprophyric ultramafics are due to the crystallization of a special type of magma.

In contrast textural patterns in some lamprophyres are far from being convincing as magmatogenic textures, particularly the minette group.

As microscopic observations reveal there is a lack of harmony between the phenocrysts and the groundmass, which essentially consist of orthoclase.

Fig. 402 shows a biotite "phenocryst" actually a fragment of biotite with lenses of quartz in orthoclase predominating groundmass.

In contrast to the "magmatic textures" certain minettes show granoblastic texture consisting of granular orthoclase, plagioclase, micaceous minerals and quartz (see Fig. 403).

In contradistinction to the granoblastic textures described, unmistakable sedimentogenic in origin recrystallised quartz-granular textures are also exhibited (see Fig. 404). Occasionally crystalloblastic mica is associated with the granoblastic textures.

Often the orthoclase (a common and characteristic mineral component of the minette lamprophyres) may occur either as recrystallised granoblastic or as crystalloblastic growth corroding and assimilating the pre-existing biotitic texture (see Fig. 405). Similarly, intersertal crystalloblastic biotite is a common mineral constituent of the minette lamprophyres (Fig. 406). In addition as Fig. 407 shows idioblastic biotites may occur in minette lamprophyres.

Crystalloblastic quartz may also exist (see Fig. 408) with autocathartic margin and a central part with groundmass inclusions and also blastic quartz corroding biotites and feldspar, see Fig. 409. In addition granular interlocking carbonate-quartz aggregates may be present in the minette, see Fig. 410.

Apatite is an important mineral phase in the lamprophyre group often surpassing the accessory status. Fig. 411 shows nematoblastic (velonoblastic) apatite in orthoclase.

In contrast a gigantic prismatic apatite with carbonates rounded and included is shown in Fig. 412. Often the apatite crystalloblasts may exhibit subidioblastic forms in a background mass consisting of orthoclase and biotite fragments (see Fig. 413).

As is shown in Fig. 414 idioblastic prismatic apatite may grow in biotite, the elongated prismatic forms often following the biotitic cleavage. In other instances an aggregate of fine prismatic apatite crystalloblasts may grow in the minette orthoclase-groundmass associated with biotite (Fig. 415).

The amphibole crystalloblastesis is a common and predominant mineral phase in the minette lamprophyres. Fig. 416 shows idioblastic hornblendes with intergranular metasomatic quartz.

As pointed out the quartz in the minette lamprophyres may exist as granoblastic (sedimentogenic metamorphic in origin) see Fig. 404, as idioblastic (see Fig. 408) and as metamorphic-metasomatic, synantetic or symplectic intergrowths. Fig. 417 shows intergranular quartz between biotite and as synantetic "myrmekitic" intergrowth of the quartz in the biotite. Similarly, granophyric quartz may exist as symplectic intergrowth with plagioclase (Fig. 418) or as symplectic intergrowths in the outer zone of an orthoclase (see Fig. 419).

Vogesites

Another lamprophyre is the vogesite in which crystalloblastic hornblende may enclose plagioclases, see Fig. 420. In addition crystalloblastic biotite, often exhibiting undulating extinction, may exist with metasomatic quartz invading and replacing the mica often along cleavage directions, Fig. 421. Together with the minette the vogesite are the lamprophyres exhibiting predominantly blastogenitic textural patterns. As characteristic blastogenic growths are idioblastic hornblendes (Fig. 422) and recrystallised granular, initially sedimentogenic quartz (Fig. 423).

The Kersantite Group

In contrast to the minettes the kersantite lamprophyres are definitely "magmatogenic" in the sense that their predominant components crystallised out of a melt.

However, in this case "xenocrysts" exist indicating that there is a lack of harmony between the phenocrysts and the groundmass. Fig. 424 shows a quartz xenocryst rounded and corroded with undulating extinction supporting the hypothesis that the quartz is a "foreign mineral" component in the kersantite. It should be pointed out that a corona-reaction structure is exhibited between the quartz and the groundmass of the kersantite (see Fig. 424).

In addition to the quartz, plagioclase may exist as xenocryst again indicating a reaction corona structure with the groundmass, see Fig. 425. In contradistinction to the rounded and corroded plagioclase xenocrysts idiotecoblastic plagioclases (Fig. 426 and Fig. 427) may exist often with plagioclase margin autocathartically free of inclusions.

Despite the typical magmatogenic textural patterns of the kersantite, hornblende crystalloblasts may represent the most predominant mineral phase, see Fig. 428.

The Camptonite Group

In contrast to the minette, the camptonites are magmatogenic dyke-rocks, however, and in this case there is a lack of harmony between the groundmass and the phenocrysts. Fig. 429 shows idioblastic hornblende in a typical groundmass solidified from a melt. Similarly, Fig. 430 shows a twinned hornblende idioblast in a groundmass with calcite present.

Feldspars are characteristic mineral constituents of the camptonite group though they are often rounded and corroded by the camptonitic groundmass (see Fig. 431) in which fine hornblende is a predominant mineral phase.

Sodalite crystalloblast with rich zone of assimilated groundmass may co-exist with idioblastic hornblende (Fig. 432) in cases sodalite may be partly engulfed by a later hornblende crystalloblast, Fig. 433.

Augitic pyroxene may occur as sub-phenocrystalline in a groundmass consisting of feldspar and hornblendes. Fig. 434 shows an augite sub-phenocryst with sphenes included, orientated parallel to the augite's crystal faces.

Also in contrast to the minette the camptonites are dyke-rocks which have intruded as a "melt" (i. e. magmatogenic). This is evident by the textural patterns exhibited and by the "harmony between the phenocrysts and the groundmass". However, in this case the textural patterns exhibit tecoblastic growths (Fig. 432 and Fig. 433).

In cases a green hornblende with iron infiltrations may exist together with normal brown hornblende (see Fig. 435).

In addition to the idioblastic hornblende megacrystalloblasts, hornblende-blasts including groundmass may be exhibited in a feldspar groundmass background, in which apatite needles or sphene may be present (see Figs. 436 and 437).

Rarely together with the hornblende crystalloblasts in a background of feldspar (plagioclase) and later formed quartz "prismatic kyanitic crystalloblasts may grow" see Figs. 438 and 439.

Another blastogenic pattern is shown in Fig. 440, where mica crystalloblasts have grown perpendicular to blastic hornblende.

As a whole the camptonites, despite the abundant tecoblastic and crystalloblastic patterns exhibited, show characteristics of "magmatogenic gang rock".

*The Monchiquite Group**

In addition to the camptonites, monchiquites are "magmatogenic" lamprophyres exhibiting (in cases) a greater lack of harmony between the phenocrysts and their groundmass background.

Rhythmically zoned augites are in a groundmass composed of hornblende laths and analcime (see Figs. 441 and 442). In this case the pyroxene phenocryst could be considered to be in harmony with the groundmass. Similarly, twinned hornblende phenocrysts may exist in an analcime groundmass, Fig. 443. The amphiboles show "magmatic corrosion" appearances, and they are interpreted as true phenocrysts.

In contrast, hornblendes may exist in monchiquites with protuberances of the outer zones exhibiting idioblastic characteristics and enclosing parts of the groundmass (see Fig. 444 a, b).

A greater disharmony is exhibited between olivine phenocrysts and their groundmass. Fig. 445 shows rounded olivine xenocrysts in an isotropic groundmass (composed of analcime). Similarly, Fig. 446 shows rounded olivine xenocrysts with marginal antigorite serpentinisation, in an isotropic analcime groundmass. The disharmony between phenocrysts and groundmass is at last exhibited in the case of carbonate phenocrysts (calcite) which can be corroded and rounded by the analcime groundmass, Fig. 447 and Fig. 448.

Often in addition to the analcime, sodalite crystalloblast may be present with a free of inclusion margin and in their central part full of pigment-size inclusions, Fig. 449. In contrast to the "phenocrysts" and their disharmony to the groundmass, hornblende crystalloblastesis is a very common mineral phase in the monchiquites. Fig. 450 shows prismatic and idioblastic hornblende as a predominant mineral phase in the analcime. Similar is the behaviour of the sub-phenocrystalline hornblende crystalloblasts which might enclose groundmass components (see Fig. 451).

Sphene and in cases, pervoskite, may be present in the lamprophyres. Fig. 452 shows idioblastic sphene in analcime groundmass and penetrated perpendicularly by a crystalloblastic apatite.

---

* Ferguson and Currie (1971) suggest that evidence of liquid immiscibility exists in the alkaline ultrabasic dykes at Callander Bay, Ontario. Monchiquitic and camponitic dykes show inclusions displaying sharply rounded, ellipsoidal, coarse-grained segregations of predominantly leucocratic minerals, identical in both composition and zoning to minerals in the matrix, detailed mineralogy, and macro and micro chemical studies strongly suggest that they result from liquid immiscibility in ultrabasic alkaline magma.

Chapter 12 Feldspathisation (Anorthositisation)

Segregation of gabbroic anorthosite in dolerite; ultrametamorphic anorthosite, plagioclase crystalloblastesis; desilification due to $CaCo_3$ assimilation, and feldspathisation due to pyroxene assimilation.

In doleritic rocks from East Siberia veinlets and clots of anorthosite may separate and due to recrystallization polygonal anorthite may be formed, see Figs. 453, 454, 455, 456. As Figs. 456, 457 and 458 show the gabbroic-anorthosite consists of green diopside, iron-ore, recrystallised granular or polygonal anorthite and apatite.

Often within the granular polygonal anorthite, diopside granules may be enclosed though often, the diopside is inter-granular with the recrystallised granular anorthite (see Figs. 459 and 460). The textural pattern of this Siberian gabbroic anorthosite is comparable to lunar highland material, see Fig. 461.

In both cases the feldspathisation may be due to recrystallization, however, the restricted anorthosite clots and veinlets in the Siberian Dolerite are due to magmatic liquid segregation.

In contrast to the magmatogenic gabbroic anorthosites in the doleritic phase, anorthosite may be formed due to metamorphism-metasomatism.

As an example of anorthositisation by recrystallization and crystalloblastic growths is the spotted anorthosite of the "BIC" (Bushveld "Igneous" Complex) the granular anorthite texture is enclosed and corroded by a diopsidic poikiloblastic phase, see Fig. 462 (a, b).

Perhaps of significance in understanding the anorthosite problem (whether due to transformation or magmatic differentiation) is the textural patterns exhibited by the anorthosite of the Negra Massiva of Angola. Fig. 463 shows olivine (~ Fo 60 %) with rounded and corroded outlines in contact with anorthitic crystalloblasts.

The olivine represents relics often restricted in their distribution in the intergranular between plagioclase crystalloblasts. The corroded outlines of the olivine, the deformation twinning and the tectonic undulation exhibited by the olivine could suggest a mantle origin despite the fact that the olivine often shows relic parts between crystalloblastic plagioclases, see Fig. 464.

Furthermore, the composition of the olivine could be within the composition of the mantle hortonolitic olivines. A crystalloblastesic origin of the olivine is in contradistinction to its rounded and corroded outlines as it comes in contact with anorthite crystalloblasts (see Fig. 465).

Of significance is the diopside phase which in cases is crystalloblastic between the plagioclase crystalloblasts (see Figs. 466, 467 and 468), and often as celyphite* between the olivine and the plagioclase.

---

* Celyphitic structures in anorthosites:
Of particular petrogenetic significance is the formation of corona structures which are considered by Griffin (1971) to be symplectites due to the break down of the minerals due to pressure changes (in the sense that lower crust rocks have been pushed upwards). The other explanation, supported in the present Atlas (see Chapter 9) is that the corona structures are synantetic symplectites i. e. reaction structures).
Discussing the formation of coronas in anorthosites and gabbroic anorthosites, plagioclase (An 70 %) typically is accompanied by abundant disseminated green spinel. Such dark green clinopyroxene occurs as single grains and small lensoid pods, and appears to be primary. Olivine is also a primary phase, but wherever present it is surrounded by complex coronas. The original rocks appear to have been cumulates of plagioclase + spinel + clinopyroxene + olivine. There is no evidence for the presence of primary orthopyroxene, subsolidus corona forming reactions among these minerals have subsequently produced orthopyroxene, clinopyroxene, spinel, garnet and amphibole as secondary minerals.
The commonest type of corona in the Indre Sogn area, Norway, has a core of orthopyroxene and garnet. A smaller proportion of observed coronas are more complex, having olivine cores within the orthopyroxene, and in some cases an outermost rim of clino-pyroxene intergrown with spinel. Griffin (1971) furthermore suggests that the complex corona type represents a phase in the development of the more common simple corona.

Figs. 469, 470, 471 and 472 show the diopsidic celyphite often corroding the olivine and in cases consisting of granular diopsidic phase (see Figs. 473, 474). The crystalloblastic nature of the diopside is furthermore supported by the intergranular diopside which follows the interleptonic spaces between adjacent plagioclases, Fig. 475.

Of petrogenetic significance is the feldspathisation i. e. the crystalloblastesis of anorthite which results in the ultrametamorphic anorthosites.

The anorthosite of the Negra Massiva, Angola, is to be understood as a result of feldspathisation due to Na-Al (Ca) metasomatism as a counter mobilisation front to a Mg leaching of the mantle or mantle diapirs, relics of which are the intergranular olivines in the Negra Massiva anorthosite of Angola.

In contrast to the feldspathisation, which results in the formation of the anorthosite, a desilification by calcite assimilation may result in Ca-plagioclases within an acid rock phase. Fig. 476(a) shows a marble or initial limestone xenolith in a microcline granite (Harrar, Ethiopia). As a result of desilification, due to limestone assimilation, a granular plagioclase texture (with plagioclases showing An 30 %) emerges which supports the process, limestone assimilation → desilification → feldspathisation. Fig. 476(b) shows the feldspathisation phase in the Harrar granite due to the assimilation of the initial limestone xenolith. As a corollary to the feldspathisation of the mafic components and particularly of the "replacement" of olivines of the Negra Massiva, Angola, are the following textural patterns: Figs. 477 and 478 show crystalloblastic plagioclases in contact with olivine, as arrow "a" shows there is a reaction zoning of the plagioclase in contact with the olivine and the olivine itself shows rounded and corroded outline. Furthermore as a result of the replacement process, which involves Mg and Fe leaching and introduction of Ca, Al, Si an "iron margin" is formed in the reaction zone between the feldspar and the olivine. In contradistinction to it, it is believed that the Mg is more mobile (could lead to a Mg source).

Additional textural patterns observed in the Negra Massiva anorthosite of Angola is apatite crystalloblastesis (see Fig. 479) and a magnetite/diopside symplectite associated with phlogopite (see Fig. 480).

The sequence of the crystallization of the Negra Massiva anorthosite can be outlined as follows:

Olivine (Fo–50 %, in cases serpentinised), showing undulating extinction due to? mantle derivation under the influence of mantle tectonics; plagioclase (70–80 An %) megablastesis, as a result of the olivine/plagioclase reaction and independently diopside is formed (often celyphitic) and in cases symplectic with iron oxides. The apatite crystalloblastesis often preceded the diopside formation. The phlogopite formation is often symplectic and associated with the diopsidic phase.

In contradistinction to the blastogenic sequence of the Negra Massiva anorthosite, the anorthosite of Montreal, Canada, is exhibiting a different blastic sequence. The most important event is the blastogenic plagioclase (megablasts) and associated with it is a diopside crystalloblastesis.

The most important mafic component is hornblende often associated with it is biotite, and with the amphibole invading the plagioclase crystalloblasts often marginally. As Fig. 481 shows, the hornblende metasomatically replaces the plagioclase. In addition to the hornblende crystalloblast also garnet crystalloblast (grossular) is present, often including relics of engulfed hornblende, Fig. 482.

In contrast to these protogenic processes, deuteric metasomatic processes and reactions are observed in the Canadian anorthosite. Scapolitisation of the plagioclases is common (see Fig. 483), which is attributed to a deuteric metasomatism of the plagioclases. In addition the plagioclases show a spotting (spot-alteration of the plagioclases), which may eventually result in the entire disintegration of the anorthosite (see Fig. 484).

It is interesting to notice that each individual anorthosite so far described exhibits its own characteristic blastogenic sequences and textural patterns. The anorthosite occurrence from Rogland, Norway, is no exception, in the sense that it shows its own blastogenic peculiarities.

The plagioclase megablasts formation is the prevailing rock-forming process with a diopside blastesis often intergranular to the feldspars and in cases engulfing pre-existing apatite crystalloblasts (see Fig. 485). Most interesting recrystallization processes are exhibited by the plagioclases themselves. Fig. 486 shows a veinform plagioclase body trans-

Considering the anorthosite rocks of the Upper Jotum Nappe (Norway), Indre Sogn, Norway, Griffin (1971) suggests that these rocks are a part of a wedge-shaped mass extending down to the lower crust. This suggests a model in which the anorthosites and associated rocks represent an upthrust portion of the lower crust, without great lateral transport being involved. In either model the crystalline rocks could have been brought from the lower crust to shallow level conditions very rapidly, accounting for the breakdown symplectites.

versing the polysynthetic plagioclase twinning of an early plagioclase megablast. The twinning of the megablast is a relic structure (i. e. as a "ghost" twinning) in the veinform body (see Fig. 487). Also, as Fig. 488 shows within the same veinform metasomatic body and corresponding to the "ghost twin lamellae" of the replaced plagioclases, a polysynthetic fine twinning (lamination) is observed, in cases showing a micro-dislocation corresponding to the "ghost" twinning; see arrow "c" Fig. 488.

Furthermore in addition to these metasomatic veinform bodies, which may represent a more albitic phase (An 48 %), a type of antiperthitisation occurs as it is exhibited in Fig. 489. Within the wide range of feldspathisation processes is the assimilation of pyroxene by tecoblastic plagioclases, as a result of which the plagioclase assimilating the pyroxene indicates an increase in its An % content and pigment rest of the assimilated pyroxene (see Fig. 490 a, b and their descriptions).

In contrast to the transformist explanation of anorthosites, most of the anorthosites are regarded to be magmatogenic bodies and formed by magmatic differentiation e.g. the Rogland anorthosite of S.W. Norway discussed by Duchesne (1972).

The anorthosite-mangerite layered synkinematic lopolith of Egersund, South Rogland, S.W. Norway, is considered by Jean Clair Duchesne (1972) to be igneous layered body made up of cumulates of anorthosite-mangerite suite. The lower part of the massif presents a rhythmic structure. Intermittent supplies of undifferentiated magma are proposed as the geological mechanism controlling the chemical recurrences associated with rhythmic structure.

Chapter 13 Kimberlites

Kimberlite is the name given to a volcanic agglomerate or breccia from Kimberley, South Africa, and is in fact given to the diamondiferous volcanic pipe composed of fragmented materials of variable composition. It is now realised that "kimberlite" is a strange name given to a far more abundant rock type, namely to an explosion breccia.

Most of the "kimberlites" are in fact volcanic breccias (i. e. volcanic agglomerates or tuffitic types).

Supporting the tuffitic nature of kimberlites are the common presence of fragments of trachytes, serpentines, marble, quartzites, gneisses, mantlc eclogites (see Figs. 491, 492).

It should be noted, however, that regarding the origin of the kimberlites Mysen and Boettcher (1975) suggest "that the kimberlite can be related to partial melting of peridotite under conditions of X H_2O = 0.5–0.25 (X co_2 = 0.5–0.75). Such activities of H_2O result in melting at depths ranging between 125 and 175 km in the mantle. This range is within the minimum depth generally accepted for the formation of kimberlite".

The tuffitic kimberlites, despite the fact that they show a wide variation of mineralogical composition and textures, nevertheless share certain common characteristics, such as the possible presence of diamonds, pyrope, Mg-ilmenites*, biotites (red, brown) phlogopite, forsteritic olivines, iron-ores and, as common constituent, calcite.

Theoretically the kimberlite (a homogenised kimberlitic tuffite) is a volcanic breccia, the fragments of which may be derived from various mantle depths and from the crust. The materials that occupy the kimberlite-explosion pipe are derivatives of the entire traverse of rock types and "layers", that are crossed by the volcanic pipe.

The olivine in tuffitic kimberlites is mainly forsterite and often exhibits the typical deformation appearances, such as undulating extinction and deformation twinning.

Most often the forsteritic olivines, and particularly those exhibiting the deformation effects, are rounded due to "attrision" rather than corrosion since they can be regarded as being of fragmental derivation from the earth's mantle, i. e. mantle xenocrysts. Their composition 92% Fo, their deformation appearances and the rounded shapes support a mantle xenocryst origin.

Fig. 495 shows mantle-derived forsterite xenocryst indicating undulating extinction and "polysynthetic twinning". Furthermore, the rounded forms support an attrision of these olivine fragments, suffered during the transportation under explosion of the mantle pieces in the volcanic explosion pipe. The complex pattern of the deformation lamellae that can develop in an olivine xenocryst is shown in Fig. 496.

In cases ovoidal shapes of olivine xenocrysts are exhibited clearly indicating a margin due to the reaction of the olivine xenocryst with the binding "matrix" in which it occurs, see Fig. 497.

A corollary of the mantle derivation of olivine phenocrysts in kimberlites is the textural association of the olivine with spinels. Fig. 498 shows a rounded olivine xenocryst with "polysynthetic" lamellae exhibited and in intergrowth with a spinel.

As studies of mantle pieces in basalt indicate the graphic-like intergrowth of spinel with forsterite is indicative of mantle derivation, Augustithis (1972, 1978). The spinel/olivine intergrowth shown in

---

* Microprobe analysis of ilmenite from the Sloan kimberlite Larimer County, Colorado, by McCallum and Eggler (1971) shows the following composition in percentage contents: SiO_2 = 0.02; TiO_2 = 50.6; Al_2O_3 = 0.62; Cr_2O_3 = 3.17; FeO = 39.9; MnO = 0.24; CaO = 0.01; MgO = 13.4.
Similarly X-ray fluorescent Spectrochemical analysis of ilmenite xenocrysts (see Fig. 493), from the kimberlite of M'sipashi, plateau of Kundelungu, Zaire, shows Mg present (see Fig. 494).

Fig. 498 is comparable and suggests a mantle derivation.

In addition to those rather large size olivine fragments with rounded forms and exhibiting deformation twinning, cases exist where the fine grained olivine, due to mylonitisation and grinding of larger olivine xenocrysts may exist. Fig. 499 shows the transition from olivine fragment to fine olivine which is interspersed as fine grains in the matrix of the kimberlite agglomerate.

Supporting evidence that a great portion of olivine in tuffite kimberlites is a mantle derivative, is shown in Figs. 500 and 501, where the olivine xenocrysts are composed of an olivine granular rock comparable to the equigranular forsteritic mantle xenocryst in basalts (Fig. 386). Similarly, Fig. 502 shows a granular aggregate of olivine in a tuffitic kimberlite. In contradistinction, Fig. 503 shows an olivine xenocryst with an idiomorphic olivine inclusion.

The indisputable tuffitic character of some kimberlitic pipes is shown by the rounded olivine xenocrysts (see Figs. 495, 496, 497 and 498) and by the rounded apatites occurring together with biotite fragments (see Fig. 504).

In contrast to the rounded olivine xenocrysts, in cases, idiomorphic olivine phenocrysts may exist indicating a distinct reaction margin with "groundmass" of the tuffitic kimberlite (Fig. 505). In contrast a rounded and corroded olivine phenocryst exhibiting the reaction margin is also shown in Fig. 506. In both cases, the reaction margin is most probably due to Fe-oxides mobilisation.

In contradistinction, Figs. 507 and 508 show rounded, corroded olivines with a serpentinisation margin (antigorite formation). A distinct case of an idiomorphic olivine phenocryst with marginal antigoritisation and with extension of the serpentine along cracks, of the olivine is shown in Figs. 509 and 510.

Replacement of the olivine phenocrysts by the antigorite may assume a network within the orthosilicate (Fig. 511).

In most of the cases discussed the olivine xenocryst or phenocrysts in the kimberlitic tuffs or in the kimberlite proper were in a kimberlitic "groundmass". In contradistinction, Fig. 512 shows an idiomorphic olivine in a calcite rich portion of the kimberlitic "matrix" or groundmass. It should be noted that the calcite shows kinematic effects.

In addition to the olivine phenocryst, idiomorphic pyroxene phenocrysts (augite-diopsides) may rarely exist in kimberlites (Fig. 513).

One of the characteristic mineral components of the kimberlites and of the tuffitic kimberlite (volcanic breccias) is brown pleochroic biotite, and in cases, phlogopite.

As was the case with the olivine in the kimberlites, the biotite also shows comparable modes of occurrence. Fig. 514 shows biotite xenocryst showing a rounded and corroded appearance and with a reaction margin of the biotite with the "groundmass" or "matrix". In contrast, Fig. 515 shows a xenolith of biotite, interlocking granular mosaic, with intergranular iron-ore. The xenolith is rounded and corroded and is surrounded by the typical tuffitic kimberlitic "matrix".

Figs. 516 and 517 show parts of a xenolith consisting of interlocking granular biotite with iron oxides as later intergranular infiltrations. As is particularly indicated in Fig. 517, corroded biotite grains are enclosed in the iron-ore.

Comparable to the deformation structures exhibited by the olivine xenocrysts, deformation effects are shown by biotite xenocrysts in kimberlite. Fig. 518 shows deformation effects on biotite exhibiting an undulating effect. Similarly, Fig. 519 shows biotite showing "wedging" and bending of the biotite.

The individual biotite crystals may be phenocryst and in cases show a corrosion due to the reaction with the tuffitic "matrix"-groundmass.

Apatite crystalloblastesis may develop in the biotite phenocrysts, indicating crystalline prismatic form and often exhibiting intersertal growth within the mica (see Fig. 520).

In addition to the rounded and corroded biotite xenocrysts in kimberlite (Fig. 514), idiomorphic biotites may be corroded or show distinct reaction margins, Fig. 521.

A rather complex intergrowth pattern is shown in Fig. 522 where a biotite exhibiting undulating extinction due to tectonic effects is in juxtaposition to a calcite phenocryst exhibiting rhombohedral twinning. This biotite-calcite structure is in the typical tuffitic "matrix" of the kimberlite.

In contrast to the xenocryst and xenomorphic biotites in the tuffitic kimberlites are the idiomorphic biotite growths. Figs. 523 and 524 show a colourless mica with strongly pleochroic brown biotitic margins. Rarely a twinned biotite with an irregular intergrowth plane is exhibited (Fig. 525 a). More common are idioblastic biotite zonal growths, in an apatite iron-ore background of kimberlitic groundmass, Fig. 525(b).

As is shown in Fig. 522 calcite may exist in kimberlite exhibiting clearly a typical rhombohedral twinning. More common, though, are marble pieces composed of granular interlocking calcite, which exhibit rounded and corroded outlines and clearly support a xenolithic derivation of the marble pieces in the tuffitic kimberlites (see Fig. 526). Fig. 527 shows an angular fragment of marble consisting of granular calcite in tuffitic kimberlite.

In contrast to the marble xenoliths, Fig. 528 exhibits a corroded and rounded calcite in kimberlitic "matrix" and with iron-oxides mobilised. In contradistinction, idioblastic calcite phenocrysts may develop in kimberlites (Fig. 529).

In addition to the calcite idioblasts shown in Fig. 529, calcite growths may occur within the kimberlitic groundmass as microgeode-fillings, together with zeolites, see Fig. 530 (a, b).

One of the most important and characteristic mineral components is pyrope-garnet which may exist either as a fragment in tuffitic kimberlites of Ykutia or without any reaction effects or celyphites, or xenomorphic pyropes may exist exhibiting most impressive celyphites due to the reaction pyrope-kimberlitic groundmass, (Fig. 531a, b).

As was pointed out, the olivine xenocryst represents mantle fragments, similarly the pyrope in kimberlites may be regarded as fragments and xenocrysts derived from the eclogite mantle (see "Eclogite Xenoliths in Kimberlites"). The absence of idiomorphic pyrope and also the celyphites with the groundmass of the kimberlite support a xenocryst origin of the garnets in tuffitic kimberlites and in kimberlites in general, crystalloblastic pyrope is also a possibility.

Magnetite and particularly Mg-ilmenite (Geikilite) is an important mineral in tuffitic kimberlites. On disintegration of the kimberlites pyrope and Mg-ilmenites may be used as index mineral for kimberlite prospection. Fig. 493 shows a rounded ore-mineral xenocryst in kimberlite.

Among the amphibole group tremolite may develop crystalloblastically in tuffitic kimberlites, Fig. 532 (a, b). As the detail photomicrograph shows idioblastic tremolite crystalloblasts may be exhibited partly enclosing iron-ore minerals, Fig. 533.

As Fig. 534 shows, the tremolite crystalloblastesis may be later than the serpentinisation of the olivine, since the tremolite crystalloblast clearly develops within the serpentinisation margin of an altered olivine.

In addition to the mantle and lower crust xenoliths, in the tuffitic kimberlites quartzites and gneisses may also be present, see Fig. 535 and Fig. 536.

In contradistinction to these crust xenoliths also new growths of quartz perhaps a recrystallization from colloform silica (gels) may be present (see Fig. 537).

Another interesting growth in the tuffitic kimberlites are zeolites, see Fig. 538, again a crystallization phase from solutions.

From the textural studies of various tuffitic kimberlites and homogenised tuffites to kimberlite proper, it can be concluded that "kimberlites" are polymictic volcanic breccias or an explosion pipe, fragments of which may derive from the mantle (pyrope, and forsterite xenocrysts) or from the upper crust e. g. quartzites, and marbles. As already pointed out the feeding of the kimberlitic pipe was through the entire transverse of the rock-types crossed by the volcanic pipe. The mobilisation of calcite as a binding matrix or groundmass is a common characteristic particularly in the tuffitic kimberlites. Different estimations exist for the temperature and depth of origin of kimberlites. Ernst (1975) suggests on the basis of cubic F d 3 m high temperature modification form of diamond, a depth of 200 km. According to Rickwood and Mathias (1970) depths of 190–555 km are considered required for diamonds. Chaterjee (1969) concluded that kimberlite with phlogopite and olivine are stable phase at temperatures of 1150°C and 55 kb (i. e. at depth of 175 km).

In contradistinction Mitchell and Crocket (1971) consider the possibility of diamond growth in smaller depth, however, the nucleus formation should be within the stability field.

Ernst (1975), based on literature references, emphasised the transportation problem of the kimberlitic material from such depths often through pipes of relatively small diameter. Furthermore, Boyed and Nixon (1973) suggested that kimberlites are "a

* Considering the absence of celyphitic textures (structures) around diamonds included in kimberlites in contradistinction to the characteristic celyphites observed around enclosed pyropes in kimberlites, experimental studies by Harris and Vance (1974) showed that diamonds are attacked by degassed kimberlites at temperatures above 1100° C and pressure of 1 kb. At temperatures above about 1000° C etch features only developed, consistent with attack by wet CO_2.

mixed sample"* of the formations transversed either by the kimberlitic magma or gas current. As mentioned present observations support this explanation. In contradistinction, Sharp (1974) proposed "that diamond-bearing kimberlites are the result of activity at the deep end of a subducted oceanic plate. The source material consists of oceanic plate along with entrained hydrous and carbonaceous material that upon metamorphism or melting form diamond-bearing kimberlite. The emplacement should occur deep under continents and parallel to fold mountain ranges. The South African fields are cited as a specific example of the suggested process."

The mineral components of kimberlites may be either xenocrysts or new growths within the volcanic pipe

Considering the olivines, we have noted that in addition to the olivine xenocrysts (rounded forms and exhibiting deformation twinning, Figs. 495, 496, 497 and 498), idiomorphic olivine growths have been observed in the kimberlites, Fig. 510. Similarly, in addition to the biotite xenocrysts (Fig. 514) idioblastic biotites have been observed (Fig. 524). Comparable is the case of calcite, in addition to marble xenoliths, and calcite xenocrysts (see Figs. 526 and 528), calcite idioblasts are present (see Fig. 529).

In contrast the pyrope seems to be a deep seated xenocrystalline phase and so far in kimberlites occurs as xenocrysts.

The reported presence of diamonds in olivinefels, Murfreesboro in Arkansas, is problematic, however, growths of diamonds in milieu other than the kimberlites, where great pressures may be exercised should not be excluded.

The question now arises with the diamonds in kimberlites "Are the diamonds xenocrysts? (this means derivatives from a layer in depth as are the pyrope xenocrysts from the eclogite mantle) or are the diamonds idioblastic growths formed within the kimberlitic pipe under enormous pressures, created during a pipe explosion?"

Considering the growth of diamonds it is believed that, theoretically, in nature the conditions of 1400°C and pressures of 30,000 atm. are required. Observations of thin sections of diamonds by Lindley (in Ramdohr and Strunz, 1967) show "Spannungsdoppelbrechung und Zwillings-Lamellierung". As a corollary to the high pressure conditions of diamond growths are the often observed optical anomalies due to "Spannungen" which in the case of the smoky stones of Kimberley can result in the breaking down to diamond powder.

Considering the question whether the diamonds are xenocrysts or growths within the explosion pipe the following account by Ramdohr and Strunz (1967) seems of importance: –

"Die Mannigfaltigkeit der Formen und ihrer Ausbildung ist nur z. T. Folge ursprünglichen Wachstums, sehr wesentlich hat dazu nachträgliche Wiederauflösung (Anätzung durch die schmelzflüssige Mutterlauge) beigetragen. Auf sie ist die viele Diamant-XX charakterisierende Rundung der Formen, das häufige Auftreten von Ätzflächen und -hügeln, von Scheinflächen, die Streifung und Knikkung der Flächen, die Kerbung der Kanten zurückzuführen. Je stärker die Wiederauflösung gewirkt hat, um so mehr nähern sich die XX einem krummflächigen bis kugeligen Dodekaeder. – Auch verzerrte und ganz unregelmäßig gestaltete XX sind nicht selten."

In addition, the presence of inclusions in diamonds may help us to answer the question of the diamonds growth in the explosion pipes.

Inclusions and flakes are often present in diamonds and they consist of C-substances, haematite, ilmenite and also quartz and rutite, also small diamond crystals may occur as inclusions and microscopically, "steam-pores" have been observed. Gübelin (1948, 1952), has reported and presented illustrations showing the presence of diamond octahedron in a larger diamond, also distorted diamonds have been shown in diamond. In addition Gübelin has shown the presence of zircon, and garnet "pebbles" in diamonds (see Figs. 539 and 540) and a case of resorbed olivine crystal in diamond (see Fig. 541 and Gübelin et al 1978).

Considering both the overgrowths of diamonds and that small diamonds may exist as inclusions of larger idiomorphic ones and in addition the fact that

* Discussing the mineralogy and geochemistry of the kimberlitic pipe of Northern Larimer County, Colorado, McCallum and Eggler (1971) conclude that whereas the kimberlite is an intrusive breccia in which clasts consist of serpentine pseudomorphous after olivine and pyroxene, with variable magnesian ilmenite, pervoskite, pyrope, chrome-diopside, phlogopite, biotite, chromite, picotite and magnetite. The matrix of the breccia consists of finely crystalline serpentine, calcite, dolomite, phlogopite, haematite, chlorite and talc. Furthermore stable carbon and oxygen isotope analyses of carbonate inclusions suggest, according to McCallum and Eggler, a magmatic origin for the phlogopite and a few carbonates, which according to them represent a carbonatite liquid that was associated with the original kimberlitic magma.

garnet, ilmenite and quartz may exist as diamond inclusions (common compounds of kimberlitic tuffites), it is most probable that the diamonds, to a great extent, are idioblastic growths formed under great pressures in the explosion pipes.

In contradistinction the rounded and distorted crystal forms may suggest either a xenocrystalline derivation or an early crystallization within the explosion pipe. The presence of diamonds (and indeed the presence of distorted diamonds inclusion in larger idiomorphic diamonds and the mentioned diamond overgrowths) clearly support a multiple growth of diamonds, the last growth phase in the explosion pipe is definitely crystalloblastic, resulting in the idioblastic forms. The phenomenology of the diamond forms is excellently presented in the crystallographic atlas of Goldschmidt (1916), and in a monograph by Fersmann (1955), an example of diamond idioblast is shown in Fig. 542.

As the present textural studies show the kimberlitic mineral components may exist both as xenocrysts and as new growths (idioblasts), and most probably the diamond is not an exception.

Eclogite xenoliths in Kimberlites

As mentioned (see chapt. 5) the metamorphic ultra-metamorphic eclogites differ in their mineralogical composition and textures from the mantle eclogites.

Table II indicates that the garnet phase of the metamorphic metasomatic eclogites from Galicia, North-West Spain, is spessartite, in contrast the mantle eclogites (xenoliths in kimberlite) contain pyrope. Table II also shows that crystalloblastic pyrope may develop in quartzites associated with pegmatites (Vgenopoulos, 1977).

Most marked textural differences are also indicated by comparing the metamorphic ultra-metamorphic textures of the crystalloblastic spessartite with the coarse granular pyrope of the mantle-eclogite. As Figs. 144, 145 and 146 show, the spessartite crystalloblasts, in addition to their blastogenic characteristics show rounded shapes and curving outlines due to mylonitisation of the metamorphic-ultrametamorphic eclogite, in contrast, as Figs. 543 and 544 show, the pyrope of the mantle eclogites is coarse granular.

The following are the main intergrowth patterns indicated by mantle eclogites.

The pyrope garnets are in intergrowth with the pyroxene (diopside) see Fig. 543, mobilised kimberlitic groundmass may invade the intergrowth, however and replace the pyroxene. In addition to the pyrope-pyroxene intergrowth phlogopite may also exist intergranular between the coarse grained garnets (see Figs. 544 and 545). Furthermore, groundmass (mainly kimberlitic in derivation) may infiltrate into the intergranular spaces between the coarse grained garnets.

Microscopic studies of mantle eclogites, xenoliths in kimberlites often show mobilisation of the kimberlitic groundmass or matrix along cracks of the pyrope (see Figs. 546, 547) though cases are observed where olivine is mobilised and exists in cracks of the pyrope (see Fig. 548).

In discussing the mineral intergrowths of eclogite xenoliths in kimberlites, it is difficult to distinguish between intergrowths original and characteristic of the mantle-eclogite and intergrowths which may result when the mantle eclogite pieces have been infiltrated and replaced by the kimberlitic groundmass. It is possible that the pyroxene/pyrope intergrowth shown in Fig. 543 is most probably the only original mineral phase intergrowth in the mantle-eclogites.

Furthermore, the olivine mobilisation along cracks of the pyrope could also be original of the mantle eclogite. In contradistinction the cases where the kimberlitic groundmass invades the pyrope along cracks, and the mobilisation of phlogopite intergranular between the pyrope coarse grains seem to originate as a result of kimberlitic material invading the mantle eclogitic xenolith. In this connection it should be pointed out that phlogopite is a common mineral phase in kimberlites.

Chapter 14 Layered basic and ultrabasic complexes

Under the general heading "layered basic and ultrabasic complexes" petrogenetically diverse complexes are considered which either are banded or show in the field a "differentiation" from the centre outwards.

The following distinct groups are recognised: –

a) Layered basic and ultrabasic plutonic bodies, e. g. Bushveld Complex, Skaergaard Complex, see chapters 16, 17 and 18.
b) Banded dunites and peridotites (e. g. Vourinos, Troodos etc.), see chapter 18.
c) Hypabyssal (sub volcanic sills e. g. Karroo, Palisade Sill etc.), see Augustithis (1978).
d) Ultrabasic Archean Greenstone Peridotitic flows, see chapter 19.
e) Basic and ultrabasic ring intrusions, e. g. Yubdo, W. Ethiopia, etc. see chapter 20.

Layered basic and ultrabasic plutonic bodies

Often layered basic and ultrabasic rocks occur as petrological-structural distinct bodies, each one having its own peculiar petrological and structural characteristics.

However, certain of the layered basic-ultrabasic bodies may show comparable phenomenology, rock-types and genesis and these can be grouped together, e. g. the layered plutonic rocks complexes; the Duluth, Stillwater, Bushveld, Sierra Leone and Skaergaard "intrusive" bodies.

Considering the enormous diversity of basic-ultrabasic bodies that show "layered" characteristics, the variety of rock types and mineral assemblances that may be involved, justifies their treatment by a series of monographs.

Furthermore, the geology, petrology, geochemistry and metallogenesis of the basic-ultrabasic bodies are extensively discussed among others by Wager and Deer (1939), Jackson (1961) and Willemse (1969).

As pointed out, their common distinct characteristic is that they show a "layered structure". The understanding of these layered structures, from the textural point of view, is of particular concern in the present Atlas.

The textural studies of the layered structures of basic and ultrabasic bodies is the subject of many treatments.

Particularly, the layered structures have been considered from the point of view of the magmatic differentiation theory and the attempt was mainly to explain the microscopic textures observed in terms of the fractional crystallization concept of Bowen (1928). A rather teleological approach has been prevailing in the attempt so far to explain these textures, since their explanation had to clarify the structural layering which in turn was "a priori" accepted to be the result of fractional crystallization in accordance with the concept of magmatic differentiation which has been prevailing since 1928 (Bowen's publication "The evolution of igneous rocks").

It appears necessary to reconsider on the basis of textural patterns observed particularly in the "stratiform sheets of igneous rocks" our petrogenetic concepts (see chap. 7, 16).

On the layered plutonic complexes

The following are the main layered "plutonic" complexes: –

1. The Guillin gabbro complex of Tertiary age in Skye, studied by Geikie and Teall (1894), and Harker (1904).
2. The Duluth intrusion, Minnesota (1 to 9 miles thick) studied by Grout (1918).
3. The Bushveld igneous complex (2–3 1/2 miles thick), studied by Hall (1932).
4. The Bay of Islands complex, Newfoundland studied by Cooper (1936).
5. The Stillwater complex, Montana, (3 miles thick

with top not shown) studied by Peoples (1936), Buddington (1936) and Jackson (1961).
6. The Sierra Leone ultrabasic complex (about 4 miles thick) with neither base nor top exposed.
7. Isle of Rhum, Northwest Scotland, studied by Wager and Brown (1956).
8. Greenland. Skaergaard, Kap Edward Holm and Kaerven intrusions, by Wager and Deer (1939) and by many others.
9. Kiylapait intrusion, Canada.
10. Ultramafic intrusions of Duke Island, Alaska. Described by Irvine (1963).

Buddington (1943), by comparing several mafic stratified bodies, found a common systematic variation from bottom to top.

1. A gabbroic or noritic facies (a chill facies)
2. ultramafic rocks
3. Norite or olivine gabbro
4. Anorthosite with ilmenite-magnetite layers
5. Quartz, or granophyric bearing augite gabbro
6. Granophyre or granite.

It is interesting to mention that Buddington (1943) drew attention to parallelism between the layering of the earth's crust and the layering of the stratiformed sheets of igneous complexes (in the sense that the lower-crust/upper mantle is also considered).

Considering the rock types characteristic of the layered facies of the stratiform sheets of igneous rocks, particularly the rock types anorthorite, norite and granite, one has to be sceptical about their common derivation from a parental basaltic magma by the magmatic differentiation mechanism.

Considering the prevailing theories of the origin of granites in the last decades a tendency is becoming apparent, namely to explain the granitic rocks as derivatives of original sediments [either by grànitisation e. g. Drescher-Kaden (1948, 1969), Read (1957), Augustithis (1973), or by anatexis Sederholm (1910), Mehnert (1968)].

The explanation that the granite or granophyric phase is a separate event but belongs to the general differentiation of the parental "basaltic magma" seems to tend to be abandoned.

In addition the rock types anorthosites and norites are also of petrogenetic significance and the recent trend is to regard these rocks as products of metamorphism – metasomatism (see chapters 9, 12).

Regarding the origin of the layered plutonic bodies of particular interest is the following statement of Barth (1952). "For many decades geologists have pondered over these structures that seem to correspond to some universal tendency, repeatedly natural processes have operated in the same way, leading to the same conspicuous results. What are these processes? Among the numerous hypotheses, three major ones deserve special mention.

1. Fractional crystallization, with gravity sorting of crystals. Buddington adhered to this view, but thinks that the similarity to earth-crust structures suggests differentiation on liquefaction as an attendant process.
2. Successive intrusions of various mafic magmas, with or without the consequent basining of the sial under the large bodies of eruptive rocks.
3. Metasomatic replacement of a sedimentary complex. In the view the stratiform bodies are not magmatic, but the layers are relict structures indicating compositional differences in the pre-existing sediments.

The last hypothesis could not explain the systematic variations in the layered sequences as assumed by Buddington. Otherwise petrology today is not able to decide between the rivalling hypotheses: none of them is satisfactory. It seems that a promising field of research is here open for exploration."

After the elapse of more than a quarter of a century we are still facing the same question. The old hypotheses have been evolved and new criteria and evidences are introduced. However, the problem remains basically the same.* Despite that, in the present Atlas – in support of the metamorphic-metasomatic (ultrametamorphic) origin of these layered "plutonic complexes" a new criterion is introduced, the presentation of the textural patterns of these rocks under the principle of comparative anatomy.

As mentioned the genesis of the basic ultrabasic plutonic complexes has been a subject of controversy. The three alternative hypotheses, (a) magmatic differentiation by crystal fractionation, (b) multiple lava flows or hypo-abyssal lava injections, and (c) transformation of sediments by metamorphism-metasomatism, should be considered.**

---

* It has been recently suggested that the Bushveld Complex is a case of intercontinental rift deposit.

** See chapter 15 for the discussion of the multiple lava flows or hypo-abyssal lava injections hypothesis and chapters 16 and 17 for the metamorphic-metasomatic interpretation of layered plutonic complexes.

The layered plutonic complexes as a result of magma differentiation by crystal fractionation

The hypothesis that the layered basic and ultrabasic complexes are the result of basic magma differentiation by fractional crystallization (and gravity separation of minerals) has been introduced by several workers e. g. for the Skaergaard complex, Wager and Deer (1939) for the Stillwater complex, Jackson (1961) and for the Bushveld Complex, Boshoff (1942), Willemse (1964, 1969) also see Chap. 17.

The fundamental concepts of this hypothesis have been the principles of continuous and discontinuous crystallization series, the principle of fractional crystallization and the hypothesis of magma differentiation developed mainly by Bowen and Niggli. Bowen's book "The evolution of the igneous rocks" (1928) has formed the basis for further hypotheses and explanations, among which is the hypothesis of rhythmic and cryptic layering introduced by Wager and Deer (1939) for the Skaergaard intrusive complex.

Wager has written several papers on the Skaergaard plutonic complex e. g. Wager (1958, 1960, 1961, 1963) and Wager and Brown (1951).

Due to the continuation of his research over several years his basic thoughts have been evolved and a rather complex model has been finally proposed by Wager (1968) for the understanding of the Skaergaard complex.

Quotations from their work will be discussed with the aim to show the dependence of their hypothesis on the principle of fractional crystallization by Bowen and to comment on their explanation of rhythmic and cryptic layering. Furthermore, parts of their work, which bear particular reference to the textures of these layered rocks will be quoted and commented upon.

"By treating rhythmic layering and igneous lamination as sedimentation structures and textures, one can consider many mafic plutonic intrusions as a succession of sheets, resting one above the other, and a stratigraphic succession can be established. The minerals of such layered series, and their composition, often vary regularly with height. If the minerals are members of a solid solution series, the lowermost are the temperature solid solutions, and upward there is a change to lower temperature solid solutions, in the case of plagioclase the change is from anorthite rich toward albite rich plagioclase and in the ferromagnesian minerals the main change is from magnesium rich in the lower horizon toward iron rich in the higher.

The variation in composition of the minerals that are solid solutions is not readily appreciated in the field, and in describing the Skaergaard case the term cryptic layering was coined for the steady change in composition of the minerals with height, the implication being that the variation was "hidden" compared with the conspicuous rhythmic variation.

Besides the upward changes in the compositions of the solid solution minerals, it is found that at certain horizons new mineral phases may enter and then remain a constituent of the succeeding rocks, while in other cases a mineral phase may abruptly cease to be a constituent. The abrupt changes in the cumulus mineral phases present, as well as the changes in the solid-solution compositions, can be shown to be the result of crystal fractionation, and both sets of changes are included under the term cryptic variation. Thus there are two kinds of cryptic layering:

1. The steady upward change in composition of the minerals that belong to solid solution series;
2. The abrupt entrance or exit of cumulus minerals at particular horizons in the succession of layered rocks, called by Hess (1960), phase layering (cryptic layering)."

Wager and Brown and Wadsworth (1960) have thus introduced the concept of "cumulus crystals" and cumulate rocks, which is based on the concept of fractional crystallization by Bowen (1928).

"The mineral phases which by their variation give rise to rhythmic layering are believed to be capable of varying in abundance because at one time they were discrete crystals, suspended in the magma, and could be brought together in different proportions during precipitation. The discrete units that are assembled in different proportions are conveniently called cumulus crystals, and the igneous rocks formed from them are called cumulates (Wager, Brown and Wadsworth, 1960)."

Also, the related concept of gravity stratification introduced by Buddington (1936) is again based on the principle of fractional crystallization and crystal gravity separation by Bowen (1928). Quoting Wager (1968), Buddington understands gravity stratification as follows: – "The arrangement of heavy and light constituents immediately suggests to the observer the settling of particles under the influence of gravity to give what, among ordinary sediments, is called graded bedding. The heavier cumulus crystals apparently sank faster through the magma to the temporary floor, and they are most abundant in the lower part of the unit, while the lighter crystals, sinking more slowly, are more abundant in the upper part of the unit. To this type of layering Buddington (1936) gave the name gravity stratification. Sometimes density differences are the main factor in gravity stratification, and sometimes size variation."

The concept of "cumulus crystals" and cumulate rocks has been widely used and applied in addition to the Skaergaard. It is also used in the Bushveld complex, e. g. Scharlau (1972).

"Nach den Arbeiten von L.R. Wager, G.M. Brown und W.J. Wadsworth (1960) werden alle Gesteine, die durch Kristallakkumulation (N.L. Bowen, 1928) entstanden sind, als *kumulate* bezeichnet. Sie setzen sich zum einen Teil aus kumulaten Kristallen, die als Erstausscheidungen aus dem Magma gebildet wurden, zum anderen Teil aus intrakumulaten Bildungen zusammen, die aus der von kumulaten Kristallen eingeschlossenen Schmelze auskristallisierten. Ausgehend von dieser Vorstellung wurden von L. R. Wager et al. mehrere Typen von "cumulate textures" eingeführt, die je nach Ablauf des postkumulaten Bildungsprozesses zu unterscheiden sind."

Based on the interpretation of Wager et al. (1960) Scharlau (1972) presents some typical microscopic diagrams, Figs. 549 and 550, to illustrate the cumulate and intracumulate textures from the Bushveld complex. In contrast, present studies (see Chapters 16 and 17) explain identical textural patterns as metamorphic-blastogenic in origin.

In support of the magmatist's explanation Goode (1976) discussing the layering of the gabbroic intrusion of Kalka in Central Australia, considers discontinuous nucleation responsible for all the various types of small scale primary cumulus layerings, usually observed in the gabbroic inclusions, whereas continuous nucleation in such a model would result in massive unlayered rocks within gabbroic sequences.

Another magmatic explanation regarding the presence and absence of layering in ultrabasic rocks is proposed by Elsdon (1971) in discussing the upper Layered Series of the Kap Edward Holm Complex of Eastern Greenland, who indicates that consolidation occurred by accumulation of primocrysts near the floor of the magma chamber; nucleation occurred continuously near the roof of the Chamber and intermittently near the base.

Furthermore Elsdon, in order to explain the absence of detectable cryptic variation of the plagioclase over an exposed thickness of 4000 m, suggests that a combined liquid and crystal fractionation process was invoked.

Also the Dufek intrusion of Pensacola Mountains of Antarctica has been explained by Himmelberg and Ford (1976) as stratiform mafic body due to magmatic differentiation and fractional crystallization. According to them the textures, structures, magmatic stratigraphy and chemical variations indicate that layered gabbros and related rocks of the Dufek intrusion developed by accumulation of crystals that settled on the floor of a magma chamber. The major cumulus phase in the exposed part of the intrusion are plagioclase, pyroxene and iron-titanium oxides. This body is believed to be comparable to the Skaergaard and Bushveld intrusions.

Regarding the orientation of minerals in the ultrabasics and in particular that of olivine, Brothers (1964) puts forward the following magmatic interpretations.

"Petrofabric analyses of layered rocks from Rhum have revealed a preferred orientation for feldspar in the allivalites and for olivine in the peridotites; a regional petrofabric map of feldspar orientation contains a radial pattern which suggests the presence of convection currents during crystal settling. An orientated specimen of Skaergaard ferrogabbro from the margin of a trough band has allowed comparison to be made between a known magma current direction and the preferred orientation of feldspar, olivine, clinopyroxene, and apatite crystals in the rock."

Furthermore, Brothers discussing the orientation of minerals in the ultrabasic rocks of Skaergaard states "The preferred orientation displayed by plagioclase, olivine, clinopyroxene and apatite in the samples of Skaergaard ferrogabbro is in complete accordance with the role assigned by Wager and Deer (1939) to the activity of magma convection currents during crystal settling". Regarding the orientation of olivine he writes in particular "Olivine stands apart, for the crystals consistently have X=(OIO) most strongly orientated in a direction normal to the line of current flow and to igneous lamination, presumably by preferred planar coincidence of broad (OIO) faces with the laminar flow plane in the melt."

Chapter 15 Volcanic or subvolcanic hypothesis of the basic and ultrabasic intrusive plutonics

In contrast to the hypothesis of crystal fractionation and magmatic differentiation is the hypothesis that the layered plutonic bodies represent multiple lava flows or repeated hypabyssal injections.

As a corollary to the multiple ejection (flows) hypothesis are the recently recognised ultrabasic Archean Greenstone peridotites which on the presence of spinifex textures and other evidence are considered in cases to represent archean flows or subvolcanic injections. The Archean Greenstone peridotites (known also as Komatiites) and the petrogenetic significance of spinifex textures are discussed in Chapter 19.

Also, in the case of the layered basic and ultrabasic plutonic bodies, textural patterns are described which are, or simulate spinifex textures. The recognition of harrisitic textures (textures within the range of the spinifex-types) in layered basic and ultrabasic complexes goes back to Harker (1908) who described harrisitic textures in peridotites. Also Wager and Brown (1951) described spinifex textures in layered ultrabasic complexes called by them crescumulate textures. The following is a quotation from the work of Wager and Brown (1968), clearly showing that spinifex textures occur in ultrabasic rocks. "Another type of rhythmic layering first clearly recognised in the ultramafic rocks of the Isle of Rhum – northwest Scotland (Wager and Brown, 1951; Brown 1956), is due to an alternation of textures rather than of mineral proportions. Sheets of peridotites are found, in which large, elongated olivines are set roughly at right angles to the layering and are surrounded poikilitically by calcic plagioclase and augite. The sheets of coarse rock alternate with fine-grained, equigranular, peridotite layers. Harker (1908) thought the coarser rocks were ultramafic pegmatite injecting the fine-grained peridotites, but re-examination has shown that the elongated olivine crystals of the coarser sheets have apparently grown upward into the magma, and poikilitic plagioclase and augite have crystallized between them. After upward growth, to be measured in inches or feet, small olivines showered down forming each fine-grained peridotite layer. Harker gave the name harrisite to the rocks having elongated olivines, and the texture was called harrisitic by Wager and Brown (1951). The phenomenon is now known to occur with other minerals, and it is suggested that the general name crescumulate texture, in allusion to the upward or inward sprouting character of the crystals, should be used and that the rocks should be called crescumulates."

A somewhat similar, but even more remarkable type of crescumulate layering in norite and intermediate plutonic rocks, resulting from upward or inward growing crystals, has been described by Taubeneck and Poldervaart (1960) as Willow Lake type layering. This is a small-scale phenomenon; the maximum width of the examples so far described is 30 feet. Willow Lake type layering may occur near the boundaries of intrusions, and its origin is considered to be related to marginal cooling. The base of each layer is sharp, and from it crystals of plagioclase, orthopyroxene, augite or hornblende are seen to grow inward, usually at a constant small angle to the perpendicular to the layering. Above the base of the layer the proportion of ferromagnesian to plagioclase crystals usually decreases; then after a few inches, there is an abrupt beginning of a new unit of layering. The Willow Lake type of crescumulate layering, like the rather different kind from Rhum is ascribed by Taubeneck and Poldervaart to the inward growth of crystals into supercooled magma.

Considering the studies of Harker (1908), Wager and Brown (1951) and Taubeneck and Poldervaart (1960), the recognition of harrasitic, crescumulate, textures as well as the Willow Lake type layering which are textural patterns comparable and commensurable to the spinifex are of particular petrogenetic interest and significance. Harker interpreted the harrasitic textures as ultramafic pegmatite injected into the fine grained peridotites. In contradistinction the Willow Lake type layering was interpreted by Taubeneck and Poldervaart to be due to the inward growth of crystals into supercooled magma. As mentioned, we may find of particular interest the interpretation of crescumulate textures by Wager and Brown who pointed out the sprouting character of the crystals. The question arises whether the sprouting character of the olivines of the crescumulate textures signify a blastogenic growth?

Chapter 16 Metamorphic-metasomatic hypothesis (Metasomatic and blastogenic textural patterns in layered ultramafic plutonics)

In contrast to the interpretation that the layered ultrabasic body of Skaergaard represents a layered ultrabasic plutonic intrusion, (Fig. 551 shows a general sketch-diagram of the layered Skaergaard Complex); present textural studies distinguish the following metamorphic-metasomatic textural patterns and blastogenic sequences.

The gabbro-picrite (40 meters from the northern margin at Strömstedef), the peculiar eucrite (55 meters from the western margin on Mellemö) and the coarse olivine gabbro (30 meters from the western margin on Ivanarmuit) all show blastogenic olivine, and poikiloblastic pyroxene in groundmass consisting of olivine and prismatic plagioclase often intersertal in their arrangement.

The blastogenic nature of the "phenocrysts" is evident by the corrosion, assimilation and autocathartic tendencies shown by the crystalloblasts towards the included plagioclases and olivines.

Fig. 552(a) shows olivine megablast enclosing corroded and rounded plagioclase. The olivine crystalloblast is partly surrounded by pyroxene megacrystalloblast. Similarly, Fig. 552(b) shows olivine crystalloblast enclosing and corroding olivine and feldspars of the pre-blastic olivine-plagioclase texture.

Particularly, Fig. 552(b) shows that whereas the olivine megacrystalloblast is free of inclusion nevertheless, marginally it is full of rounded granular olivine and prismatic plagioclases. It is most probable that the olivine crystalloblast has grown into the intergranular of the pre-blastic granular texture. Fig. 553 shows (detail of Fig. 552(b)) corroded granular olivine and plagioclase corroded but still maintaining their prismatic form in the olivine crystalloblast. The part of the olivineblast that contains the pre-blastic inclusion definitely attains a poikiloblastic appearance. In cases the plagioclase inclusions in the olivineblasts, in addition to their corroded outlines occasionally exhibit reaction margins, see Fig. 554 (a, b). However, most of the plagioclase inclusions in the olivine crystalloblast show corroded outlines (Figs. 555, 556 and 557).

The plagioclase crystalloblastesis in the Skaergaard rocks (as mentioned) exhibits indisputable poikiloblastic appearances and character. Figs. 558 and 559 show olivine crystalloblast in turn surrounded by later pyroxene crystalloblasts which included both olivines (the granular pre-blastic) and blastic olivine, as well as plagioclase laths. A clinopyroxene megablast poikiloblastic in nature is shown in Figs. 560 and 561. In contrast Fig. 562 shows the poikiloblastic clinopyroxene extending as intergranular and enclosing the granular olivine and plagioclase pre-existing texture.

Comparable to the poikiloblastic growths exhibited in the Skaergaard ultrabasics, poikiloblastic pyroxene surrounding a rounded granular olivine phase is exhibited from Kaersut, Greenland (see Fig. 563 and Fig. 564).

The poikiloblastic pyroxene patterns observed in the Skaergaard Complex (see chapter 16) are also abundant in the Bushveld Complex (see chapter 17).

Figs. 565 and 566 show two generations of pyroxenes. Generation (I) surrounded by magnetite and generation (II) following the interleptonic space between pyroxene generation (I) and magnetite.

The second generation of pyroxene can be clearly seen as a blastogenic phase later than the plagioclase phase and the magnetite which surrounds it. Fig. 567 shows plagioclase rounded and corroded by later magnetite. The second generation pyroxene clearly takes advantage of the feldspar/magnetite contact and has grown interleptonically.

Fig. 568 shows second generation blastogenic pyroxene – sending its protuberances which, as extensions, invade the interleptonic spaces between the pyroxene (generation I) and the magnetite. Fig. 568 is comparable to Fig. 567 with the exception of the blastogenic, second generation, pyroxene which extends into the magnetite as well, Fig. 568.

In addition to the poikiloblastic pyroxene most of the clinopyroxene in the gabbroic rocks of Skaergaard show a post-plagioclase crystallization.

Despite the fact that the impression is often given that we have a coarse ophitic pyroxene/plagioclase intergrowth the detailed textural pattern exhibited rather supports a post-plagioclase pyroxene crystallization.

Fig. 569 shows a coarse ophitic gabbroic texture from Skaergaard. The indicated textural patterns are difficult to interpret despite that, prismatic plagioclase engulfed by the pyroxene shows corroded outline. Additional evidence is required for understanding the sequence of crystallization. Fig. 570 shows pyroxene enclosing rounded and corroded prismatic plagioclase and with the pyroxene extending into the intergranular spaces between the plagioclases. In this case clearly the clino-pyroxene has crystallised later than the plagioclase.

In cases, ophitic like prismatic plagioclase is enclosed by the pyroxene, with extensions of it clearly following the intertwinned plane (Fig. 571). It should be noted though that the plagioclase laths also show rounded outlines. As a corollary to these observations, Fig. 572 shows corroded plagioclase with an outline showing indentation due to reaction with the later pyroxene.

In cases, the plagioclase "laths" enclosed in the pyroxene show corrosion and replacement clearly selective and replacing the plagioclase along its direction of polysynthetic twinning, Fig. 573.

Additional evidence of the replacement of the plagioclase by the pyroxene is shown in Fig. 574 which shows a pyroxene prolongation extending into the plagioclase and across its twin lamellae.

An additional textural pattern supporting the later pyroxene growth is shown in Fig. 575 where a plagioclase grain and olivine are surrounded and enclosed by later pyroxene.

The crystallization of magnetite is of particular importance in the gabbroic rocks of Skaergaard. Often the magnetite corrodes and invades the plagioclases, clearly resulting in a plagioclase replacement (see Figs. 576 and 577).

In cases, in ferrogabbros the magnetite invades an olivine-plagioclase granular gabbroic texture through the intergranular spaces, replacing the preexisting mineral phases and resulting in poikiloblastic patterns (Fig. 578).

In contrast Vincent and Philips (1954) describe gabbro-rich magnetite as magmatic differentiates containing ex-solutions of ilmenite and ulvospinel (for a different interpretation of magnetite-ex-solutions see Chapters 6 and 26).

As Fig. 579 shows, accessory apatite may be replaced by magnetite along cracks.

The iron oxide replacement may invade and replace a pyroxene through magnetite forming solutions invading the pyroxene along interleptonic spaces, Fig. 580(a). Particularly Fig. 580(b) shows magnetite invading the pyroxene along its cleavage planes.

In addition to the mentioned blastic phases, metasomatic granophyric quartz in intergrowth with plagioclase may also be regarded as a replacement process. Fig. 581 shows granophyric quartz in intergrowth with plagioclase. A eutectic or co-eutectic quartz/plagioclase system is hardly possible in a consolidating gabbroic melt.

Fig. 582 shows that the quartz forming solutions have invaded the plagioclase along twin lamellae.

In cases the quartz metasomatic solution may invade a pyroxene and replace it, the quartz infiltrations in the pyroxene resemble granophyric intergrowths, see Fig. 583.

Cases exist where the granophyric quartz in intergrowth with the plagioclase may invade the plagioclase and take advantage of the pyroxene, or magnetite contact of an included or adjacent phase with the plagioclase, Fig. 584.

A special case of granophyric quartz in intergrowth with plagioclase has extension invading corroded inclusions of magnetite in the plagioclase (see Fig. 585).

The textural patterns shown clearly support a metasomatic origin of the quartz.

Chapter 17 Is the Bushveld complex igneous?

As mentioned, one of the most controversial layered complexes is the Bushveld "Igneous" complex in the Transvaal.

The majority of the workers consider the Complex to be igneous in origin and often is quoted as a classical example of magmatic differentiation.

Despite that the Bushveld is considered to be igneous in origin within the magmatists School scepticism is expressed concerning certain aspects of the complex. Characteristically Willemse (1969) states the following:

"The transgressive character of the layered Sequence to the encircling sedimentary rocks is emphasised and is considered to be due to funnel intrusion, perhaps at more than one center. The correspondence in the layering in areas some 200 miles apart presents a major problem. However, the effect of turbidity currents on a crystal mush accumulated at the margin of a gently dipping basin should receive serious consideration. To explain the very large volume of the Bushveld granite* one has to think in terms of anatexis of the epicrustal and sedimentary rocks."

The geology and petrology of the Bushveld complex is extensively discussed by many workers, the majority of whom consider it to be due to magmatic differentiation e. g. Boshoff (1942); Cameron (1963, 1964); Coertze (1958); Ferguson and Botha (1964); Hall (1932); Heckroodt (1958); Hiemstra and Biljon (1959); Kupferbürger, Lombaard, Wasserstein and Schwellnus (1937); Lombaard A.F. (1948); Lombaard B.V. (1934, 1956); Raal (1965); Schwellnus (1965); Van Zyl (1960); Wells (1952); Willemse (1964, 1969); Willemse and Bensch (1964).

In addition to the studies carried out in the period 1942–1969, which interpret the Bushveld Complex as a magmatic differentiation body the same trend is continued a decade later. Characteristically Vermaak (1976) attempts to explain the Bushveld anorthosite as follows:

"In the uppermost Critical Zone, anorthosites show remarkable persistency in thickness. Evidence is presented that plagioclase floated repeatedly to temperature-density-compositional inversions, thereby forming anorthosite mats which entrapped successive magmas and volatiles beneath them."

Similarly a magmatic interpretation is proposed for the Bushveld by Van der Merwe (1976) and Gruenwald (1976). Brynard et al. (1976) in discussing the formation of the Merensky Reef proposed a gravity concentration under magmatic conditions with a subsequent hydrothermal element redistribution:

"It is proposed that the Merensky Reef was formed by gravity concentration but that it was extensively modified by subsequent hydrothermal alteration, which gave rise to variations in the mineralogy and distribution of the platinum-group elements at different localities along the strike of the reef."

Considering the problems as outlined in the quotation of Willemse, as well as the problems involved with the intrusion of a body with the dimensions of Bushveld, the scepticism has led the author to reconsider the interpretation of textures observed in the rocks of the Bushveld Complex and which have been

* In contrast to the large volume of granophyre associated with the Bushveld Complex the leucocratic rocks are in relatively small proportions to their "associated ophiolites" as Coleman states:
"The close relation that leucocratic rocks have to the gabbroic parts of ophiolites and their compositional gradation from tonalite to albite granites have convinced many workers that these leucocratic rocks represent the end product of differentiation within ophiolite sequences (Wilson, 1959; Thayer, 1963; Coleman and Peterman, 1975). Where detailed mapping is available, the volume of the leucocratic rocks is small (2 %) in relation to the total exposure of the cumulate parts of the ophiolite (Wilson, 1959; Glennie et al., 1974). Clear-cut intrusive relations between these leucocratic rocks and associated mafic rocks are usually difficult to establish."

considered before to be "cumulate" and due to fractional crystallization under magmatitic differentiation.

In contrast to the prevailing views, the author presents a selection of textural patterns of the noritic, gabbroic and granitic rocks of the Bushveld Complex collected along the transverse A-A′ (Fig. 586), which are definitely metamorph (blastogenic growths). The blastogenic growths observed are definitely not deuteric; they are "protogenic" growths (rock forming growths), the results of the transformation of the initial sedimentogenic material under metamorphism-metasomatism and perhaps mantle involvement and mobilisation.

The blastogenic patterns observed in the Bushveld Complex are comparable to the blastogenic growths observed in gabbros, norites and anorthosites already described (see Chapters 8, 9 and 12).

It should be also emphasised that the Bushveld granites (granophyres) show graphic and micrographic quartz/feldspar intergrowths comparable to metasomatic quartz/feldspar intergrowths described by Drescher-Kaden (1948, 1969) and Augustithis (1962a, 1973).

The Bushveld granite, due to its huge quantities, is difficult to explain as an acid differentiate of the Bushveld complex (see Willemse 1969). The anatectic hypothesis is in contradiction to the metasomatic traits exhibited by the quartz in graphic intergrowth with the K-feldspar.

An attempt is made here to present only a selection of the blastogenic textural patterns observed and to emphasise that they are not rarely occurring cases but the prevailing textural patterns observed in the noritic, gabbroic and granophyric rocks observed, along the traverse A-A′ (Fig. 586).

Fig. 587 shows poikiloblastic pyroxene enclosing, corroding and following the intergranular of a plagioclase texture. Similarly Fig. 588 shows poikiloblastic pyroxene and biotite again enclosing, corroding and following the intergranular of a plagioclase texture. In both cases the pyroxene and the biotite are later crystallised than the plagioclase and are attributed to solutions following the plagioclase intergranular.

The Figs. 589 and 590 show the transition of a poikiloblastic pyroxene/plagioclase growth to an intergrowth simulating the ophitic pattern. These ophitic intergrowths are not due to simultaneous pyroxene/plagioclase crystallisation but are comparable to the textural patterns observed in the Skaergaard complex (see Figs. 570 and 571) and are due to later poikiloblastic pyroxene invading a plagioclase texture (i. e. enclosing laths of plagioclase).

In the complex and variable petrographical types that can be formed under the transformation processes and due to topo-metasomatism also the reverse case i. e. poikiloblastic plagioclase enclosing mafic components is also often exhibited (see Fig. 591 and 592).

In cases the noritic rocks of Bushveld show a two phase pyroxene crystalloblastesis. Fig. 593 shows hypersthene with a margin of diopside which often extends as intergranular in the plagioclase texture. As Figs. 239, 240, 245 show also in the Bushveld norites, and gabbros the phase "a" pyroxene (hypersthene) is invaded by phase "b" diopsidic pyroxene. Comparable textural patterns are also observed in the Skaergaard norites, see Fig. 566.

Samples from the Bushveld Complex show parablastic (blastic growths under tectonic influences) plagioclase/pyroxene textures exhibiting clearly plagioclase orientation and with the pyroxene partly delimited by the plagioclase and sending protuberances into the feldspar intergranular (see Fig. 594).

Within the Bushveld Complex and along the traverse A-A′ (Fig. 586) granoblastic olivinefels may occur which consist of olivine and pyroxene granoblasts (Fig. 595). This rock could attain dunitic patterns. However, dunitic patterns with spinel and chrome-spinels suggest mantle involvement (see Chapter 10 and Figs. 373–376).

In addition to the complex basic and ultrabasic rock types, granites and granophyres as mentioned occur in huge quantities in the central part of the complex (see Geological sketch map by Willemse, Fig. 586).

The granophyric intergrowth textures have been used as a criterion for a simultaneous crystallization of graphic quartz/K-feldspar under eutectic conditions of crystallization.

In contradistinction Figs. 596 and 597 show quartz veinlets transversing the K-feldspar and sending graphic shape protuberances (it should be noted that the quartz of the veinlets and the graphic in form protuberances are similarly orientated and form an entity). It is clear that the quartz veinlet and its graphic in form protuberance are later than the K-feldspar and are due to metasomatic solutions. Similarly Figs. 598 and 599 show fine quartz following interleptonic spaces of the K-feldspar and either attaining a granophyric form or are similarly orientated with quartz exhibiting a graphic shape. The origin of the graphic and granophyric quartz is extensively discussed in the work of Drescher-Kaden (1948, 1969) and Augustithis* (1962, 1973).

The Bushveld Complex remains a controversial subject, the present observations and their interpretation might not clarify the details of the geology and petrology of the complex, but I hope they will add more scepticism to an already complex problem or at least to the adjective "Igneous".

* The granitisation concept as presented in the "Atlas of the textural patterns of granites, gneisses and associated rock types" by Augustithis (1973) has been severely criticised by Anhaeusser (1975) who questioned whether such textures occur in the granites of South Africa. As the present textures of Bushveld granophyre show (Figs. 596 and 597), the textural patterns are identical to the metasomatic graphic quartz/feldspar intergrowths presented in the "Granite Atlas" (Figs. 259 and 260).

Chapter 18 Banded dunites with bands of chromite

Of particular petrogenetic significance are the banded dunites with bands of chromite. Often, as a result of this banding, fine bands of chromite alternate with bands of granular dunitic olivine which is subsequently serpentinised.

The "Schlieren" type of chromite ores (i. e. alternating chromite olivine banding) has been interpreted as magmatic banding due to repetition of fractional crystallization of chromite followed by olivine fractional crystallization of a crystallising magma.*

Often the olivine/chromite banding might be restricted or more abundant in the dunitic part of an ultrabasic complex composed of dunites and pyroxenites e. g. Vourinos ultrabasic complex, Greece.

Petrofabric analysis of the orientated olivine crystals (Fo 93 %) associated with the chromite bands by Spathi (1966), see Fig. 600 (a, b, c), shows that the ng axis of the olivine is parallelly orientated to the olivine banding (i. e. parallel to the chromite bands).

As Figs. 600 (a, b, c) show, there is a "Kornform Regelung", an orientation of the form of the olivine grains with the a-axis (ng) parallel to the chromite banding. Present microscopic studies indicate that most of the olivine relics in the serpentinised mass are exhibiting deformation effects such as undulating extinction and simple to complex pattern of deformation lamellae (deformation twinning), Figs. 601 and 602.

In addition to the orientation of the ("Kornform Regelung") forms of the olivines to the chromite banding, the plane of the deformation lamellae is apparently also orientated at an angle of 35°–45° to the chromite banding. It should be pointed out that in deformed olivines exhibiting deformation lamellae the translation plane is parallel to the (100) and the translation directions // (001) of the olivine.

Considering that deformation appearances are most abundant and almost the rule in the olivine grains associated with chromite bands; and the fact that in addition to the orientation of the forms of the olivine to the chromite bands there is also an orientation of the olivine translation plane to the chromite banding, it appears that the orientation of the olivines to the chromites is due to tectonic deformation.

Often the orientation of the olivine forms to the chromite banding has been used as a criterion to support the idea that the orientation of ng axis parallel to the chromite bands is due to the fractional crystallization of the olivines in their rhythmic precipitation with chromite bands, i. e. the a-axis of the settling olivine grains was parallel to the chromite bands.

In contrast to the fractional crystallization hypothesis (for the explanation of the orientation of ng axis of the olivine parallel to the chromite band-

* Coleman (1977) summarises as follows the magmatists' interpretation of the Vourinos Complex:
"Detailed petrologic studies of the layered sequences in ophiolites are just beginning and it has only been recently discovered that these cumulates are cyclic and that they may represent magmatic differentiation sequences (Mesorian et al., 1973; Hopson et al., 1975; Jackson et al., 1975). At Vourinos, a stratiform complex nearly 1500 m thick has been recognized and its cumulate stratigraphy has been described (Jackson et al., 1975). The lower unit consists primarily of olivine cumulates with thin olivine-clinopyroxenite zones near the base that may represent beheaded cyclic units. These units are overlain by a thick olivine cumulate (500 m) containing thin chromite cumulates. A 400m section above the predominantly olivine cumulates is characterised by cyclic unit changes that are thinner with each cyclic unit beginning with olivine cumulate and grading upward to clinopyroxene-olivine, then finally to clinopyroxene cumulates with some post cumulus plagioclase. These cyclic units are usually about 33 m in thickness and apparently the olivine cumulates decrease in thickness upward. Near the top of the Vourinos section, two-pyroxene cumulate layers form the bottom of cyclic units and grade upward into thick plagioclase two pyroxene cumulates. Jackson et al. (1975) interpret this cumulate section as a chamber slowly filling itself with magmatic sediments at or near a spreading ridge." (However it should be pointed out that according to the plate tectonic interpretation of eastern Europe the Vourinos ophiolite "series" and Troodos are considered to be obducted oceanic crust – upper mantle).

ing) as pointed out, abundant deformation appearances of the olivine and the ("internal texture") orientation (Feinbau Regelung) support the hypothesis of tectonic orientation.

In addition to the olivine the chromite also exhibits evidence of tectonic deformation. As Figs. 603, 604 and 605 show, most of the chromite of the "Schlieren" bands are tectonically rounded and show cataclastic phenomena. Also the chromite grains in most of the cases exhibit rounded outlines which are attributed to deformation causes (attrition deformation). In exceptional cases, though, idiomorphic chromite grains may be present particularly as isolated grains within the serpentinised dunite (see Fig. 606).

Taking into consideration that the olivines are tectonically orientated and that the chromite bands are also tectonically deformed, the question arises whether the fine banding of olivine/chromite is not only influenced by, but also caused by deformation and tectonic movements.

As a corollary to this tectonic hypothesis of chromite-olivine banding are the "leopard chromite ores" which are interpreted as deformation ellipsoid forms of the hard chromite in tectonically mobilised initial dunite (see chapter 23) and the formation of granular chromite by the cataclasis of compact chromite as is exhibited in the series of Figs. 607 and 608. Similarly, all transitions are exhibited in the formation of fine chromite bands and interspersed chromite "clots" and grains from compact chromite bands, which in turn may have been formed by the fracturing and tectonic mobilisation of more compact chromite masses in the initial dunitic diapir.*

It should be emphasised that the fine chromite bands and the interspersed chromites in the serpentinised dunite show ample evidence that they have been cataclastically fractured, rounded by tectonic "attrision" and detached from larger more compact chromite masses (see Figs. 607 and 608). In addition, the "Schlieren" chromite bands often exhibit ptygmatic folding which is again attributed to deformation effects.

The deformation evidences support a mantle-diapirism. The deformation and orientation of the olivine and chromite bands as well as the fracturing and mobilisation of the most compact initial chromite masses into bands and fine bands within the more plastic olivine masses could be interpreted on the basis of the hypothesis of mantle diapirism – of which the Vourinos complex may be considered as a model.

Comparable to the Vourinos, N.Greece, ultramafic complex, is the Troodos ultrabasic complex of Cyprus (see Fig. 609 and its description). The Troodos complex has been explained as magmatic differentiation mainly by Miyashiro (1975) and Pantazis (1973, 1977).

In contrast, as the sketch, Fig. 610, by Gass and Masson-Smith (1963) shows, the Troodos is regarded to be an upwards diapiric extension of mantle.

Pantazis (1977, per cum.) summarises the Troodos controversy as follows:

"The origin of the Troodos ophiolitic complex in Cyprus has recently been the subject of lenghty discussions by various authors. The Troodos complex has been regarded by several recent writers as typical ophiolites originally created in a mid-oceanic ridge (Gass, 1968; Moores and Vine, 1971; Dewey and Bird, 1971; Gass and Smewing, 1973). The advocates of this hypothesis endeavoured to show geological, tectonic and chemical similarities between the ophiolites and mid-oceanic ridges as evidence for their hypothesis. On the other hand Ewart and Bryan (1972) and then independently Miyashiro (1973) focused attention on the chemical resemblance of the Troodos rocks to some island arc volcanics, rather than to mid-oceanic ridge rocks, and proposed a new hypothesis that the Troodos complex was created in an island arc.

In subsequent discussions, Hynes (1975), Moores (1975) and Gass et al. (1975) strongly criticised Miyashiro's (1973) conclusion and claimed that the Troodos rocks were subjected to such intense metasomatic changes that Miyashiro's hypothesis of island arc or continental origin based on the bulk chemical analyses of rocks is not justified. Gass et al. (1975) claimed in addition that the known traditional demarcation between the Troodos Upper and Lower Pillow Lavas has been precisely defined by Moores and Vine (1971) and Gass and Smewing (1973), though this division had been strongly criticized by Govett and Pantazis (1971) and Pantazis (1973). However, Miyashiro (1975a, b) maintained his previous conclusions and considered that the presently observed compositional variation of the Troodos ophiolites resulted mainly from crystallization differentiation and that a considerable pro-

* The banded chromite-olivine dunite and the pyroxenites associated with it are "intruded" bodies due to mantle diapirism, or, according to the hypothesis of plate tectonics represent obducted oceanic crust and upper mantle.

portion of the volcanic rocks belong to the calc-alkalic series. Similar results for a fractional crystallization and calc-alkalic trends of Troodos rocks were found, independently, by Pantazis (1973, 1977).

Pantazis (1977, pers.comm.) based on the results obtained by the variation diagrams of the major oxides of representative rock samples of the Troodos Ophiolitic complex and in particular on the results of (a) the SI variation diagram, (b) the variation of Al_2O_3, Na_2O, K_2O and CaO in relation to total iron/MgO ratio and SiO_2 and (c) the FMA diagram as well as on the field relationships of these rocks concluded as follows: –

1. The continuous and smooth compositional variation of volcanic and plutonic rocks in the range SI 50 as well as the ways and extents of variation suggests that the main compositional variation of such Troodos rocks is controlled by crystallization differentiation.
2. Some intrusive rocks with SI = 50–60 plot in the SI variation diagram far from the extended trend indicating that intrusive rocks of this SI range were subjected to strong effect of crystal accumulations.
3. The quartz of the Troodos Sheeted Intrusive Complex and volcanic rocks is at least partly primary and cannot be attributed to a large-scale silica metasomatism. The assumption of silica metasomatism is not supported by any evidence.
4. The Troodos magma is subalkalic (non-alkalic). The complex belongs to class I of ophiolites as defined by Miyashiro (1975c), which is characterised by the presence of volcanic rocks of both calc-alkalic and tholeiitic series. This gives much support to the advocates of an island arc or continental origin of Troodos.
5. The Sheeted Intrusive Complex, the Pillow Lava Series and the plutonic rocks with basic to acidic compositions show similar differentiation trends. This is additional evidence that the Troodos Pillow Lava Series forms one stratigraphic unit closely related to the Sheeted Intrusive Complex, the latter constituting the feeders to the pillow lavas. Furthermore there are close field relationships and striking chemical similarities between the Sheeted Intrusive Complex and the Pillow Lava Series on one hand and the plutonic rocks on the other."

Both the Vourinos and Troodos massives show banded dunite and banded dunite-chromite alternations. Also in both cases leopard chromite-ores are present (see chap. 23). Similar to the Vourinos complex, harzburgite is present in the Troodos. However, the Cyprus ultramafic complex is characterised by the presence of lherzolites and gabbroic rocks. The association of the Troodos ultrabasic complex with the picritic and tholeiitic basalts has been suggested as evidence for magmatic origin.*

However, the dunite of the Troodos massif shows deformation effects (see Fig. 611) comparable and commensurable to deformation effects of the Vourinos ultrabasics, which suggest a plastic diapiric mobilisation of mantle, as also is the case of the Vourinos complex.

Furthermore the relation of harzburgite, due to recrystallization** changing to lherzolite, can be taken as evidence of the complex processes which take place within the mantle. It should be pointed out that spinel-lherzolites are widely considered, e. g. Mercier and Nicolas (1975), Maaløe and Aoki (1977) to be mantle diapirs in orogenic zones.

---

* Gwyneth Challis (1965) suggests that the Dun Mountain, Red Hill and Red Mountain of New Zealand are magmatic differentiates from tholeiitic basaltic magma and that they are genetically related to a chain of Permian Volcanoes on the margin of the New Zealand geosyncline, and that the associated olivine-poor tholeiitic Permian basalts are the extrusive differentiates.

** Mercier and Nicolas (1975) discussing the texture of lherzolites suggest that the progranular lherzolites might have first developed in a harzburgite paragenesis with high Al-Ca enstatite or alternatively in a high Al pigeonite bearing peridotite. It is mainly due to a recrystallization which is regarded as occurring during partial melting.

Chapter 19 The ultrabasic archean greenstone peridotitic volcanic flows and the significance of spinifex textures

General

In contradistinction to the "alpine type of ultramafic bodies" (see chapter 10) are the Archean greenstone ultramafic belts, e. g. the Abitibi orogenic belt, Ontario, Canada; the Pilbara block of Western Australia and the Berberton Mountainland of South Australia. Eskstrand (1973) outlines the petrographical studies made on the Archean Greenstone ultramafic belts and the significance of the spinifex textures for understanding the mode of emplacement of these ultramafics. "Early descriptions of spinifex occurrences in the Abitibi belt, were given by Bruce (1926), Berry (1940), Tremblay (1950), Prest (1950, 1951), Satterly (1951) and Abraham (1953). In many of these, the spinifex was attributed to metamorphic crystallization. Satterly believed the ultramafic rocks to be flows, but not on the basis of spinifex texture. Later, Drever and Johnston (1957) in an excellent study of skeletal textures of olivine, popularised the fact that fayalitic olivines in smelter slags and artificial melts form spinifex textures by rapid growth in a supercooled silicate liquid. Lewis (1970) further documented spinifex in slags. Subsequent accounts of natural spinifex (Naldrett and Mason, 1968, Pyke, 1970) inferred rapid quenching conditions to yield these textures, and postulated a volcanic or sub-volcanic origin for the rocks. A similiar interpretation was proposed by Nesbitt (1971) for some Australian occurrences. Viljoen and Viljoen (1969) suggested an extrusive origin for South African spinifex-bearing ultramafic units, but principally because they are conformable with, and co-extensive with mafic volcanic rocks."

The most convincing evidence to date for a volcanic origin of spinifex-bearing ultramafic rocks is the exceptionally well-exposed Munro township occurrence described by Pyke et al. (1973) – "The similarity between slags textures and spinifex textures observed in the Archean Greenstone peridotites and the fact that "spinifex" textures in slags can be formed due to rapid growth in a supercooled silicate liquid allowed Naldrett and Mason (1968) and Pyke (1970) to infer rapid quenching conditions to yield these textures and postulated a volcanic or subvolcanic origin for the rocks."

The spinifex textures could be understood as a result of the growth habits of olivine and clinopyroxene. Fleet (1975) in discussing the growth habits of olivines states "Analysis of the crystal structure of olivine, in terms of isolated SiO_4 tetrahedra and M-site cations, provides a qualitative explanation for the relative importance of the forms (010), (021), (110), (210), (101) and (001), for the unit cell with b c a. Factors contributing to the dominance of (010) in skeletal olivine are: (i) The uneven distribution of the compositional units in planes parallel to this form results in faces of low surface energy, (ii) the stereochemistry of the co-ordination polyhedra of the M (2) protosites is unfavourable for the nucleation of growth layers and (iii) the faces of (010) are parallel to directions of easy growth".

Fleet presents an idealised sketch (Fig. 612) of a skeletal olivine crystal indicating a lantern and chain habit – and points out that this chain or blade-like habit reflects a particular degree of undercooling or supercooling although the extensions usually appear to have grown later than the lanterns and may be associated also with a lower concentration of crystal-forming constituents in liquid.

The phenomenology and genetic interpretation of spinifex or spinifex-like textures, "chicken-track", "herringbone", "feather" or "quench" textures has been widely studied.

The presence of the spinifex textures in Archean Greenstone peridotites has been used as evidence of their volcanic and sub-volcanic origin. Pyke et al. (1973) have pointed to the conditions of spinifex textures developments in the Archean Greenstone peridotites; they state the following:

"The conclusions of Nesbitt (1971) on the physical conditions accompanying the formation of spinifex are largely accepted; in situ, rapid crystallization of crystal-free liquid. The absolute rate of crystallization, however, remains open to conjecture. The facts that (1) fine rather than coarse spinifex formed at the chilled upper surface of the flow units, (2) the size of the olivine blades continually increases down into the flow and away from the upper contact, and (3) the composition of the olivine blades is zoned, indicate that the conditions of formation probably did not involve an instantaneous quench as suggested by Viljoen and Viljoen (1969). Instead, they developed in a rapidly cooled environment in which heat was removed from the upper surface and the blades of olivine grew down into the underlying liquid, attached to nucleii in the chilled zone of the top and growing roughly perpendicular to isothermal lines in the magma. Analogy with slags suggests that this process might have lasted from a few tens of minutes to several days."

The following are some of the petrogenetic points raised by Nesbitt regarding the spinifex textures:

a. The spinifex textures are believed to be the in situ crystallization of a silicate liquid (a prerequisite for the formation of spinifex textures).
b. The morphology of the olivine crystal depends on the physical and chemical environment at the point of growth "thus elongate, parallel growth of skeletal olivine at the margins of unmetamorphosed Tertiary Sills is proof of the igneous origin of the textures observed in Archean greenstone rocks."
c. Fast growth is the mechanism by which large skeletal olivine crystals are produced.
d. In order to crystallise large single individuals of olivine, rather than a mass of small crystals (as in a chilled margin), it seems that nucleation was centred on very few sites. This suggests that initially the liquid was essentially free of nuclei and to achieve this the liquid would be superheated. We thus arrive at the postulate that fast-cooling of a super-heated liquid is responsible for a fast growth rate once nucleation commences.

The growth rate can be further increased if supersaturation (i. e. supercooling) can be produced prior to nucleation. This seems to be possible if cooling is fast enough.

In contradistinction to the volcanic interpretation of spinifex textures, Lewis (1970) points out the similarity between iron/silicate slag and the spinifex textures found in the mafic and ultramafic rocks. Also Lewis suggests that the spinifex can occur both in volcanics and plutonics and cannot be a criterion for determining the mode of emplacement.

"Textures found in an iron/silicate slag bear a remarkable resemblance to "spinifex texture" as found in the Archean mafic and ultramafic rocks of Western Australia, and provide evidence on the origin of the texture. Fayalite in the slag has crystallised as complex dendritic crystals or, more commonly, as thin plates elongated along the a and c axes.

The arrangement of the olivine plates gives rise to a variety of textures which are described and compared with the natural texture. By analogy with the slag textures, it is shown that spinifex texture is an original igneous texture formed by rapid chilling of the magma. Preservation of the delicate structure indicates a lack of convection currents and mechanical disturbance during crystallization, and recent work on ceramics suggests that crystallization may have been from an initially formed glass. Spinifex texture can be found in both intrusive and extrusive bodies and cannot, by itself, be used as a criterion for determining the mode of emplacement."

As already pointed out most of the early workers have attributed the spinifex to metamorphic crystallization. Similarly Collerson et al. (1976) interpreted the growth of large inter-locking olivine crystals as relic cumulate or harrisitic growths, mimicked by olivine regrowth during metamorphism. In contradistinction to spinifex they refer to the olivine growths as "bladed olivines in ultramafic rocks".

"Exceptionally coarse grained, interlocking olivine crystals up to 1 m long occur in small lenticular ultramafic bodies within the Nain Province of Labrador. The ultramafic rocks occur in association with Archean gneisses that vary in metamorphic grade from amphibole facies to granulite facies. Some of the interlocking textures (i. e. near Hopedale) are interpreted as relic cumulate or haristic growths, mimicked by olivine regrowth during metamorphism and preserved in structurally isotropic enclaves in their host rocks. Others (e. g. near Saglek) are regarded as solely of metamorphic origin. None of the olivine growths resemble spinifex textures reported in Archean ultramafic rocks elsewhere."

Present experimental studies by G. Kostakis and S.S. Augustithis produced "spinifex" forsterite (Fo 95–97 %) when mantle (fragments of olivine bombs in basalt from Gerona, Spain) have been crushed and melted at a temperature of about 1450°C.

The mineralogical composition of the mantle

fragments used is: forsterite (about 93 % Fo), enstatite and spinels.

The melted mantle fragments have been subsequently super-cooled and as a result a glassy mass, due to rapid cooling, has been produced at the contact with the crucible. In contrast, the melt in the central part of the crucible cooled relatively slowly and a granular texture consisting predominantly of olivine has been produced. As Fig. 613(a) shows, granular olivine sends extensions into the glassy chilled margin attaining spinifex forms. Fig. 613(b) also shows spinifex forsterite.

The present experimental studies show that superheated mantle when rapidly cooled (as would be the case of submarine ultrabasic effusions) would produce spinifex forsterite. The present experimental studies are in harmony with the prevailing explanations regarding the origin of the Archean Greenstones and the olivine spinifex textures. Most impressive skeletal "spinifex" hercynite textures are observed by E. Mposkos in the Larymna (lateritic Ni-Fe ores) slags, where super-heating and subsequent fast cooling has taken place (see Fig. 614).

In addition to forsterite, pyroxene and magnetite, spinifex plagioclases are common (see Fig. 615a, b, c). Fig. 615(a) shows a characteristic pattern of spinifex plagioclase in glassy basaltic groundmass from sub-marine basaltic effusions of the Mid-Atlantic ridge.

In contradistinction to the typical spinifex textures of Fig. 615(a), spinifex plagioclase growths may "initiate" from an early crystallised pyroxene nucleus (see Fig. 615(b)). The spinifex plagioclase may in this case exhibit a radiating orientation starting from the pyroxene nuclei and extending into the glassy basaltic groundmass. Often the plagioclase spinifex growths may exhibit "crystal skeleton" forms (see Fig. 615(c)) and also show interpenetration intergrowths due to the crystallization force (see Fig. 615(a)).

As mentioned natural olivine spinifex textures are a characteristic textural pattern of Komatiitic and Archean Greenstone usually ancient effusions, see Figs. 616, 617 a, b, c, d, and e and their descriptions.

Chapter 20 Intrusive ring complexes

The ring complexes consisting of a central nucleus of dunite followed by shells of pyroxenite, gabbro and hornblende "plutonics" have been among others explained as magmatic differentiations of a basic-ultrabasic magma. The sequence dunite → pyroxenite → gabbro has been considered as an indisputable piece of evidence of magmatic differentiation.

Another piece of evidence in support of a magmatogenic origin is their size, usually a few km in diameter and as a consequence of their "cylindrical shape" have been considered to be ultrabasic plutonic pipes.

The ring intrusion complexes have been explained as magmatic differentiations. James (1971) discussing the origin and emplacement of the ultramafic rocks of the Emigrant Gap area, California, compares the "zoned complexes" (intrusive ring bodies) with the stratiform intrusive bodies and with the "alpine intrusives". According to James "the ultramafic bodies of the Emigrant Gap area are part of a mafic complex within a large composite pluton of the northern Sierra Nevada. The pluton was magmatically emplaced and is surrounded by an aureole of hornblende-hornfels facies and appears to have partly melted. Gravity studies indicate that the ultramafic bodies have near vertical contacts extending to depths of at least 1 1/2 to 2 1/2 km."

As James states, "The mafic complex shows rough concentric zoning of rock types: ultramafic bodies occur at the core; gabbro forms a discontinuous intermediate unit; and diorite, tonalite and granodiorite occur at the margins.

Within the ultramafic bodies, unserpentinised wehrlitic peridotite is dominant; dunite and olivine clinopyroxenite are present but greatly subordinate.

The structural and chemical relations within the mafic complex suggest that all the rocks are derived by mechanical accumulation of early crystallised mafic minerals and the two pyroxene bearing granodiorite crystallised from felsic differentiate. It is likely that flowage differentiation was the dominant process in crystal segregation.... In most major respects, the mafic complex of the Emigrant Gap area is similar to zoned ultramafic complexes, such as those of the Urals and South Eastern Alaska (Ruckmick and Noble, 1959; Taylor and Noble, 1960; Noble and Taylor, 1960; Irvine, 1959, 1963, 1965, 1967; Taylor, 1967) and is unlike alpine-type ultramafic bodies (see Chapter 10), also Hess, 1955; Thayer, 1960, 1967; Ragan, 1963, 1967; Raleigh, 1965) or stratiform complexes (Hess, 1960, Jackson, 1961, 1967; Wager and Brown, 1968)."

The similarities of the "zoned intrusive complexes" or the zoned intrusive rings are outlined by James (1971) as follows:

1. Rock types tend to be distributed in concentric zones.
2. Rocks in the interior of the larger ultramafic bodies are dunite, wehrlitic peridotite, and olivine clinopyroxenite and are composed almost entirely of olivine and diopside, whereas gabbro surrounding these bodies contains both ortho and clinopyroxene.
3. Peridotite and pyroxenite have mosaic textures, gabbro has typical igneous texture, and all three are essentially undeformed (from comparison with Union Bay samples).
4. Magmatic structures are dominant (although at Duke Island these are clearly the result of crystal settling rather than magmatic flow, Irvine (1959, 1963, 1965, 1967).
5. Dunite replaces peridotite (Irvine 1959);
6. Late Hornblende gabbro intrudes ultramafic rocks;
7. Serpentine is rare or absent.
8. Country rocks adjacent to and included by ultramafic rocks are hornblende-hornfels and pyroxene-hornfels facies.

James points out that the Emigrant Gap complex

is atypical in comparison with the Alaskan bodies.

In contrast to these views, present studies suggest that the ring complexes are mantle diapirs which may exhibit certain particularities.

Mineralogical and petrographical study of the Yubdo-Wollaga Ring Complex of Western Ethiopia reveals the following special features which rather support a diapiric origin for that particular complex.

a. As the geological sketch map shows there is a dunitic nucleus followed by a pyroxenitic "shell", which in turn is followed by amphibole schist (hornblende schist) – see Fig. 618, i. e. the gabbro shell is missing.

b. There is an extensive area of alteration of the dunite into birbirite (see chapter 30).

c. There is a basaltic cover partly covering the complex and containing nodules of olivine and pyroxene, Augustithis (1965).

In contrast to the above characteristics which illustrate the particular geology of the area certain other mineralogical-petrographical characteristics are of greater petrogenetic significance. The following are features of petrogenetic importance:

1. The central dunitic nucleus consists of almost pure forsterite with Fo % of more than 98. The forsterite which is marginally antigoritised often shows orientated bodies of iron oxides, see Fig. 619.

2. The forsterite exhibits undulating extinction and deformation twin-lamellae which are considered to be pre-intrusive or syn-intrusive deformation features (see Fig. 620).

3. The dunite is an almost pure olivine ultrabasic with idiomorphic chromite often exhibiting perfect octahedra with a margin of magnetite (often martitised). Of significance is that the rock in its unaltered phase is predominantly composed of forsterite.

4. The complex is platiniferous and sperrylite occurs in the chromite and also in its magnetitic margin (see Chapter 24).

5. The pyroxenite next to the ultrabasic (dunitic nucleus) consists of diopside, grammatite, leuchtenbergite (also hornblende is microscopically determined). Ore microscopically magnetite with ilmenite and spinel and ilmenite grains are determined.

Considering the above geological, mineralogical and petrographical features of the Yubdo Ring Complex the following genesis is tentatively proposed.

The Yubdo complex particularly the dunitic nucleus is a mantle diapir that has also resulted in the diapiric mobilisation of the pyroxenitic layer. Particularly, the almost pure monomineralic composition of the unaltered dunite, with almost pure, more than 98 % Fo, renders it most "plastically mobile".

The abundance of pre- or syn-intrusive deformation features such as undulating extinction and deformation lamellae in the olivine support diapiric origin of the dunitic nucleus of the complex.

The fact that in the basalt adjacent to the complex, which partly covers the complex, olivine and pyroxene xenoliths, most probably of mantle origin have been found, supports that the earth mantle is not far below the earth's surface.

As a corollary to this the Moho map by Makris et al. (1975) shows the Moho-discontinuity (the upper part of the mantle) to be 30 km below the surface in the Yubdo region.

The mineralogical and petrographical characteristics suggest that the Yubdo dunitic pipe is a mantle diapir mobilised from a depth of the mantle*, on its way up it has caused the diapiric mobilisation of the pyroxenite as well.

The fact that the Yubdo dunite is almost monomineralic consisting of almost pure forsterite, furthermore suggests that a forsterite part of the mantle (most plastic indeed) has been mobilised and intruded in the crust or earth's upper part as a diapir.

In addition it should be pointed out that whereas the olivine in the basalt nodules is forsterite with a composition of Fo 92 %, the dunitic olivine is almost pure forsterite and contains "segregation Fe-bodies" – see Fig. 619, thus the dunite "diapir" should be contrasted with the mantle xenoliths in the basalts and should be regarded as perhaps a deeper mantle material. This could perhaps explain why it is platiniferous.

---

* By diapiric mobilisation of mantle is meant the diapiric intrusion of parts of it under plastic mobilisation and not necessarily the diapirism of hot mantle material as is suggested by Maxwell (1970, 1973, 1974).
The "Kristalloplastese" of the olivines of the mantle bodies as well as the relatively high elasticity modulus of the ultrabasics (dunites, actually mantle bodies), Niggli (1948) render a diapiric mobilisation of the mantle possible.

Chapter 21 Tectonic deformation textures in basic and ultrabasic rocks

In the acid plutonics quartz, despite its brittle nature, under tectonic deformation behaves as plastic-solid (undulating extinction, plastic deformation etc.). Often quartz shows mylonitisation and rupture zones, with mylonitisation and recrystallization of the quartz resulting in an interlocking tectonically produced mosaic is not seldom.

The plagioclases, in cases, exhibit bending of the twin-lamellae and seldom undulating extinction and rupturing. Mylonitisation may affect the granitic feldspars, and augen structures of resistant feldspars in mylonitisation zones are not rare. The other main component of the acid rocks, the micas, may show deformation effects exhibited by crystal-bending, undulating extinction and occasionally rupturing and mobilisation. However, considering the tectonic deformation effects on acid plutonics, quartz yields more easily to deformation effects while the feldspars are unaffected.

In the ultrabasic rocks the olivine is the most easily deformed component exhibiting undulating extinction, deformation lamellae and rarely rupturing and recrystallization. Also in the case of the ultrabasic and basic rocks, in addition to the olivine the pyroxenes, feldspars and micas may show deformation effect. However, it should be emphasised that olivine is the most susceptible to deformation mineral. This susceptibility of olivine to deformation is comparable to that of the quartz in the acid plutonics. Also the olivine may show deformation effects (undulating extinction) while the other components may be intact.

The deformation textures of basic and ultrabasics are not only interesting as evidence of rock metamorphism, i. e. as a phenomenology of deformation, but are of great petrogenetic significance. Augustithis (1978) has studied the deformation phenomenology of mantle xenoliths (olivine bombs) in basalts. Summarising, the following are the main deformation effects in the mantle xenoliths:

a. Undulation and deformation twinning of the forsterites.
b. The development of crushed zones surrounding the spinels. The crushed zones may consist of pyroxenes and olivines "marginal" to the spinels which remain as pressure resistant minerals.
c. The development of a cleavage pattern mainly in the bronzites (see Fig. 374).

The deformation effects in the mantle xenoliths are mainly attributed to the overall pressure which is exercised on the mantle by the kilometres thick crust and lower crust. The deformation of the olivine and particularly the development of the crush-zone around the pressure resistant spinels, as well as the pressure produced cleavage pattern in the pyroxenes, are to be understood rather by the effect of overall pressure. Directed stress, whereas possible, is difficult to be connected with a certain phenomenology in the mantle derived xenoliths.

In constrast the deformation mantle diapirs and particularly the petrofabric study of olivine-orientation support that directed stresses have resulted in the olivine grain orientation (see Paulitch (1963) and Spathi* (1966)).

Present studies of mantle diapirs in orogenic zones show that peridotites of mantle derivation may show deformation phenomenology which is clearly supporting the operation of directed forces (directed stresses). In contrast to the phenomenology of mantle deformation (see Figs. 374–376), a series of photomicrographs (Figs. 621, 622, 623 and 624) shows a deformation phenomenology due to directed stresses in mantle diapirs from the orogenic belt of Greece (see Chapters 10, 18).

Fig. 621 shows an "augen" structure of pyroxene

* Spathi's original interpretation of olivine orientation parallel to the chromite bands was due to magmatic settling of the olivines; however, in the present Atlas the results are differently interpreted (see chapter 18).

in a mylonitised zone consisting of crushed and recrystallised mafic minerals. Similarly, Fig. 622 shows a curved and rounded outline of a pyroxene due to tectonic attrision, in a mylonitised fine mass of mafic components.

Whereas in the case of mantle deformation (i. e. mantle xenoliths in basalts) the spinels as pressure resistant minerals are unaffected by the overall pressure, in the case of the mantle diapirs* which are affected by directed stress and in which zones of strain are produced (such as mylonitisation zones) the spinels are conspicuously deformed. Figs. 623 and 624 show "augen" structures of deformed and mobilised spinel in a mylonitised zone. In addition to the spinel also the pyroxenes show deformation as components of the mylonitised zones.

In addition to the diapiric mantle which is intruded tectonically within orogenic zones (see chapters 10, 18 and 21) most impressive tectonic deformation structures may develop in dunites (see chapter 10) where mylonitisation of the olivine has resulted in fine grained olivine interlocking mosaic (see Figs. 625, 626, 627, 628 and 629).

In the case of mantle diapirs in addition to the deformation effects described, often the pyroxenes exhibit a bending of their polysynthetic twinning (Fig. 630).

In peridotites and pyroxenites deformation effects may be pronounced particularly if phlogopite is present. If phlogopite is present in peridotites and pyroxenites, in addition to the olivine which will display undulating extinction and deformation lamellae (twinning) the phlogopite is most susceptible to deformation effects (see Fig. 631 and Fig. 632).

Fig. 633 illustrates bending of the cleavage pattern of the phlogopite in a tectonically deformed mica-peridotite. Similarly, Fig. 634 shows a complex pattern of tectonic lamination produced in phlogopite. In gabbroic rocks both the plagioclase and the pyroxenes are tectonically deformed and exhibit a complex deformation phenomenology.

Figs. 635, 636 and 637 show rupture of the plagioclases due to tectonic deformation and the formation feldspar-mylonitisation (fragmentation) along the crystal ruptures. As a result of fragmentation along the rupture zones and the accompanied recrystallization – the rupture zone occurs in zones of plastic deformation and in which the plastic limit** is surpassed and rupture occurred – an interlocking mosaic of granular plagioclases is developed. Comparable is the deformation of quartz in which a granular-mylonitisation due to recrystallization may develop in rupture zones of plastically deformed quartz (see Figs. 543–546, Augustithis, 1973). As Figs. 635 and 637 show, displacement of the plagioclase has taken place along the fracture zones (which in reality are microfaults).

Of particular interest is Fig. 638 which shows plagioclase indicating undulating extinction i. e. the feldspar has been plastically deformed and with ruptures along which mylonitisation of the plagioclase has been produced, resulting in a granular recrystallization of the plagioclase along the ruptures.

Also Fig. 639 shows deformed plagioclase (undulating extinction and deformation twinning) with a rupture along which feldspar mylonitisation (granular recrystallization) has taken place. The deformed plagioclase can be seen as remnants in a mass of mylonitised plagioclase (granular plagioclase recrystallization). Similarly, Fig. 640 shows deformed remnant-plagioclase in a mylonitised zone with the mylonitised plagioclase as granular recrystallization extending along a crack of the remnant plagioclase.

In the gabbroic mylonitised zones deformed plagioclase and pyroxenes may often form "augen" structures in a mylonitised plagioclase mass of plagioclase-recrystallization (the mylonitised plagioclase recrystallization shows an interlocking mosaic texture), see Figs. 641 and 642.

In deformed gabbros not only the plagioclase is mylonitised and recrystallised, comparable and commensurable is the mylonitisation of pyroxene. Fig. 643 shows a deformed pyroxene (i. e with deformation lamellae) with mylonitisation in the periphery of the deformed pyroxene which is as "augen" structure in a mylonitised zone of pyroxenes and plagioclases.

Often, along rupture lines of the pyroxene, mylonitisation and recrystallization may take place resulting in a granular pyroxene recrystallization with interlocking mosaic texture, see Fig. 644.

It should be pointed out that usually the ruptured and mylonitised zones within a crystal are occupied by topo-autochthonous material (i. e. material derived by the fragmentation of the crystal itself). As a corollary to this hypothesis, ruptured and recrystallised plagioclase and pyroxene are shown in Fig. 645. It is interesting to notice that the rupture zones of the

* or plastically obducted oceanic crust-mantle according to the plate-tectonic hypothesis.

** evidence that the plagioclase is plastically deformed is the undulating extinction which is often exhibited.

plagioclase are occupied by granular plagioclase and the rupture zones of pyroxene by recrystallised fine granular pyroxene.

Though cases showing the opposite, that is mobilisation of material along rupture zones, are common. Fig. 646 shows a mobilised and recrystallised pyroxene zone composed of fine granular pyroxenes and with biotite neocrystallizations. In addition Fig. 647 shows a tectonically affected pyroxene (with rounded margins due to tectonic deformation) sending as an extension fine granules of pyroxene, which extend along a deformation line of an adjacent plagioclase.

Frequently in strongly mylonitised gabbros, zones of "mobilised" plagioclases and pyroxenes may exist next to each other consisting of fine interlocking grains of plagioclase and pyroxene respectively, see Fig. 648.

Considering the tectonic deformation of gabbros, the deformation phenomenology of plagioclases assumes particular significance. An attempt is made to present "stages" of development of the deformation phenomenology in plagioclases. Fig. 649 shows plagioclase tectonically deformed and with the appearance of undulating extinction, however, in the same plagioclase and interrelated to undulating extinction the beginning of deformation lamellae is to be observed.

Similarly, deformation twinning may develop interrelated to a rupture zone along which plagioclase mylonitisation has taken place, see Fig. 650.

Fig. 651 shows a tectonically deformed plagioclase in a plagioclase mylonitised zone. The plagioclase shows undulating extinction and a pattern of bent deformation lamellae is produced. Indeed, often plagioclases may show deformation lamellae which are parallel to zones of mylonitisation (Figs. 652 and 653). Particularly, Fig. 653 shows all transitions from a "lineation" structure in the central and not so affected plagioclase part, to a deformation lamella next to the mylonitised peripheral part.

In addition to these deformation lamellae, often in deformed plagioclases (particularly when remnants of plagioclase occur in mylonitised zones) a pattern of cross-lamellar twinning may be produced, see Figs. 654, 655 and 656 (a, b). In contradistinction to these cross-lamellar twinning a pattern of deformation lamellae may result, in which case the twin lamellae may either intersect one another at an angle (see Fig. 657) or may show a bending (Fig. 658).

In cases, as a result of tectonic deformation and as a special case of cross-lamellar twinning development, plagioclase lamellae may be produced cutting across the fine deformation lamellae, see Fig. 659.

Chapter 22 The mineralogy, geochemistry, ore-microscopy of ore minerals and bodies associated with basic and ultrabasic bodies

General

Associated with basic and ultrabasic bodies are certain more or less characteristic ore-mineral parageneses and ore deposits.

Of particular importance is the presence of minerals of the spinel group. The paragenetic association of spinel groups (of a certain range of the spinel composition) with types of the ultrabasic rocks is of petrogenetic importance.

Augustithis (1978) has made a study of the mineralogy and textures of "olivine bombs" in basalts. These bombs are interpreted as mantle pieces. Spinels of the range spinel-chromite-magnetite are in graphic intergrowth with olivines and pyroxenes. The graphic shaped spinels represent skeleton crystals incompletely developed, however, transition and cases of more idiomorphic developed spinel have been observed.

The mineralogical and petrographical studies of the mantle pieces clearly show that whereas the spinel range (Al-spinels) predominates, nevertheless, chromite, or chrome-spinels are important mineral constituents of the mantle.

As a corollary of the importance and significance of chrome in the mantle is the abundance of chrome as "trace-element" in the different mineral constituents of mantle pieces (olivine bombs in basalt).

As pointed out in chapter 18, the tectonic (diapiric) mobilisation* of mantle, rich in olivine (forsterite) and chrome-spinels could provide a satisfactory explanation of dunitic ultrabasic complexes such as Vourinos and Troodos.

Of particular importance for the mantle-diapirism is a forsterite rich portion of the mantle (in the sense that forsterite prevails over the pyroxenes in a mantle portion, dunitic in composition). The relative abundance of forsterite will give to the mantle part a greater plastic mobility as is demonstrated by the deformation appearance and petrofabric analysis of dunites of such complexes (see chapter 20).

Another important factor for understanding the paragenetic association of chromite-dunite i. e. the association of chrome spinels with forsterite-rich portions of the mantle could be broadly understood by considering the geochemical interrelations of Mg in the olivines and Ni**-Cr (i. e. size of atomic radii $Mg^{+2} = 0.78$ Å and $Ni^{+2} = 0.78$ Å). It is therefore probable that the paragenetic association of the different types of spinels with the ultrabasics depend, among other factors, on MgO/FeO ratios.

Considering the paragenetic concept: spinel-type/ultrabasic type, it is clear that a distinguishing of the spinel paragenesis is possible, namely that chrome-spinels and chromites in the forsterite rich dunites and magnetite-titanospinels prevailing in ultrabasic and basic rocks where the FeO is above a certain marginal value in the MgO/Fe ratio.

Furthermore, these rather empirically proposed trends of preferential association of Cr (and to some extent Ni) in primary distribution, with ultrabasics in which the MgO is prevailing; and magnetite spinel, Al-spinels and Ti-spinels in the wide range of ultrabasic in which the FeO is above a certain value range, can provide a tentative explanation for the preference of Ti in pyroxenites and gabbros rather than dunites.

As the geochemical comparison of trace element distribution (see Augustithis, 1978) shows, Ti is virtually limited in abundance in the mantle fragment (olivine bomb) whereas relatively high values are noticeable in the basalt enclosing the olivine bomb.

The distribution of Ti in basic and ultrabasic rocks is also of interest particularly in an attempt to explain

* or, in accordance to its alternative hypothesis, based on plate tectonics, oceanic crust-upper mantle obduction.
* Ni and Cr are interrelated in accordance the periodic table.

the virtual absence of titano-magnetites, Ti-spinels in dunites, which, by contrast, are very abundant in the gabbroic rocks (see Chapters, 16, 26).

Considering the distribution of Ti in basalts the following should be noted:

a. The range of titania content in terrestrial basalts varies within the following limits, 1.9–6.43% TiO_2.

b. The terrestrial basalts are relatively poorer than the equivalent lunar basalts.

c. There is a difference in the titania content between inter-oceanic and circum-oceanic basalt, i.e. Ti values decrease with supposed depth of basalt derivation.

The above geochemical consideration which synoptically reviews the Ti in basic and ultrabasics shows that Ti compared with Cr and Ni, is more abundant in those basic and ultrabasics where FeO is above a value range and in which also Al_2O_3 is more abundant; in contradistinction Cr and Ni tend to be more abundant in ultrabasics with high ratio of MgO.

In this connection the following quotation from the work of Hartmut Kern (1968) is interesting:

"The FeO and the MgO contents (in chromites) are strictly antipathetic as well as the Fe_2O_3 and Al_2O_3 resp. the Cr_2O_3 and (Al_2O_3 + Fe_2O_3) values. The end members of the spinel group show similar relations. Ferrite-chromite decreases on account of increasing values of picrochromite. The spinel and magnetite diagrams are contrary in the same way."

Nevertheless, in dunitic complexes and intrusions, chromites may be present which show the presence of alumino-spinel and rutile ex-solutions. Fig. 660 (a, b) shows the presence of rutile ex-solution lamellae orientated parallel to the (111) of the chromite. In addition to the ex-solution lamellae rounded rutile and twinned rutiles may be as grain crystallizations, included in the chromite from Rodiani, Greece.

The presence of Ti and Cr together in the ultrabasic rocks of Rodiani, Greece (Augustithis, 1960) is of geochemical significance. The Ti and chrome are related as subgroup elements of the periodic system which belong to the same horizontal line (Sc, Ti, V, Cr, Mn, Fe, Co, Ni), as it has been pointed out, Ti is not abundant in the dunite (and in the dunitic mantle) and Ti-spinels are virtually absent in the dunitic types.

In this connection, it is interesting to note the presence of Al-spinels as orientated ex-solutions lamellae in the same Rodiani chromite (Augustithis, 1960). Fig. 661 shows the co-existence of Al-spinel ex-solution lamellae and rutile lamellae both following the (111) of the host chromite (Mg rich chrome-spinel picro-chromite). Similarly, Fig. 660(b) shows abundant Al-spinels in the Rodiani chromite grains.

Considering the statement made by Kern (1968), it is clear that despite the antipathetic tendency of Cr_2O_3 and (Al_2O_3 + Fe_2O_3) values, Al-spinel ex-solutions may be present in chromites.

Seldom does chromite and in fact idiomorphic chromites of dunitic "pipe" intrusions show margins of magnetite (see Fig. 662), so far the magnetite associated with chromite is mostly free of ilmenite ex-solutions.

Chapter 23 Chromite ores in ultrabasics (The Controversy of the Genesis of Chromite-Ores in Dunites)

Considering the controversy of the layered ultrabasic bodies and particularly the controversy of the layered dunitic complexes, the origin of the chromite-ores associated with the dunitic complexes is subject to the same controversy (see chapt. 18).

Bowen's magmatic differentiation theory has put-forward the basis for fractional crystallization hypothesis of the dunitic complexes, namely that the dunitic complexes and the chromite bodies associated with them are fractional crystallization products of a parental basaltic magma. Particularly the chromite layers have been interpreted as being due to crystal settling under fractional crystallization and gravitative separation of the heavier chromite crystal grains.

Among the many supporters of this explanation, Kern (1968), summarises as follows the main points of the fractional crystallization hypothesis of the chromite ore-bodies formation:

"Summarising the microscopical studies and the physico-chemical and petro-chemical considerations it is concluded that the genesis of chromite ore deposits must be attributed to the fractional crystallization and gravitative differentiation of initial-basaltic magmas in situ.

The development of the different types of chromite deposits and the different textural features – disseminated ores, banded ores, ovoid ores and lumpy ores – is ascribed to the dynamic interaction of the following factors:

1. dip of the primary magmatic layering (i. e. of the rhythmically arranged partial melts),

2. activity of magmatic tectonics (prototectonics) related to gravitational settling and rolling processes,

3. enrichment of volatiles during the course of fractionate crystallization.

Finally the importance of the postmagmatic serpentinisation during dynamo-metamorphic processes and the most significant phenomena of secondary tectonics and disturbances of the primary ore bodies are discussed. Distinct relationships between the tectonic elements and the primary magmatic layering are important for practical purposes and prospecting.

The statistical treatment of the tectonic elements also supports the theory that the chromite ore deposits originate by situ differentiation of a more or less homogeneously intruded basaltic magma."

In contrast to the fractional crystallization hypothesis of chromite ore-deposits formation, the progress in geotectonophysics, mantle-petrography and experimental mineralogy has created new trends of thinking which regard the dunitic complexes more or less as mantle-diapirs in orogenic and geotectonic lines (see chapt. 18, 20), or as obducted oceanic crust-upper mantle.

In support of the mantle-diapirism of the dunitic complexes and contrary to the fractional crystallization hypothesis is the following geotectonic, geological, petrographical and mineralogical evidence (see Chapters 10, 18, 20) which are quoted here as evidence against the fractional crystallization hypothesis of the chromite ore-deposits in dunites and ultrabasics.

1. Ultramafic complexes comprising dunitic massifs exist without proportional basic, intermediate and acid representatives as would be postulated on the basis of magmatic differentiation hypothesis, e. g. the Vourinos and Troodos massifs and complexes do not have proportional amounts of basic and acid differentiates as would be expected by the enormous masses of dunites which predominate in these complexes.

2. Ultramafic and particularly dunitic complexes and intrusions lack thermal contacts (see Chapter 10).

3. The olivine of the dunites associated with the chromites is almost forsterite pure (Fo – 92%–93%). The crystallization of olivine from a

basaltic magma is normally poorer in Fo. Forsterite (i.e. 82–85 Fo %) would crystallise out from a basaltic magma, as it is proposed if chromite-ores are fractional crystallization products of a basaltic magma (in situ).

4. Also, according to the fractional crystallization hypothesis, the crystallization of chrome-spinel must have preceded an enormous amount of olivine crystallization. If this is true the olivine that crystallised with chromite should be Fe rich, however, that is not the case.

5. Often intrusive in character, dunitic bodies show olivine indicating undulating extinction and deformation twin lamellae indicating that the deformation effects have been produced at a mantle-state of the material or during its plastic or diapiric mobilisation (see Chapters 10, 18, 20).

6. In cases, ultramafic diapirs in the alpine orogenic belt show graphic spinel in intergrowth with forsterite and pyroxene and are comparable and commensurable to graphic spinels observed in olivine bombs (mantle fragments in basalts), see Figs. 373, 374 and 375.

7. Trace element studies of ultramafic samples from the Konitsa region of Greece (diapiric intrusions in the alpine orogeny) show more than 10 ppb of Ir indicating a mantle derivation.

8. Geophysical gravity studies of both the Vourinos massif, Greece and Troodos, Cyprus, support that the intrusive complexes are continuations of the mantle and are "plastic" mobilisations of it (see Fig. 610, also chapter 18).

9. In intrusive dunitic complexes (diapiric in nature but not recently tectonically disturbed), the banding shows folding, cross bedding, ptygmatic folding and unmistakable deformation evidence. Fig. 663 shows sharp folding of banded ultrabasics from Ivrea. Most impressive are the complex folding of "Schlieren" chromite bands in Xerolivado, Vourinos (see Fig. 664). The tectonic phenomena mentioned and the deformation effects discussed in chapter 18 show that these phenomena are not strictly prototectonic (during the consolidation of the ultramafic magma).

10. Bowen's differentiation hypothesis and fractional crystallization cannot provide a satisfactory explanation of the origin of ultrabasic and granitic rocks as postulated in Bowen's hypothesis, Augustithis (1978). The present trend of thinking is to regard granites either as a granitisation or anatectic products and not basaltic magma differentiations. In contradistinction the ultramafic complexes are regarded as mantle diapirs or obducted oceanic crust-upper mantle.

11. Crystallization of basaltic magma, Augustithis (1978), as well as its derivation by fusion of the lower crust's protolytic layer (and with the participation of the ultrabasic mantle, i. e. olivine bomb, olivine xenocrysts) precludes the crystallization of dunitic complexes as a result of alkali basalt crystallization.

Primary Chromite Distribution in Dunites and Serpentinised Dunites

The chromite and general chrome minerals can be understood on the basis of: primary distribution, 1st phase redistribution (meta-tectonic mobilisation and metasomatic chrome mobilisation of chromites and chrome) and 2nd phase redistribution: (alteration and weathering-mobilisation).

As primary distribution is the presence of chrome-spinels and chromite in mantle (or mantle-diapirs or mantle fragments in basalts, i.e. olivine bombs). Fig. 376 shows spinels in olivine bombs, in basalt. As already mentioned, the Cr and Ni occur as trace elements in the mantle mineral components and their relative amounts are shown in Table V.

Table V, Trace-Element content (ppm) in forsterites, pyroxenes and spinels on mantle fragments in basalts

Trace element determination by A. Vgenopoulos

Locality	Mineral	Ni	Cr	Co	Cu	Zn	Mn	Ti	Zr	Sr	Ba
Lekempti,	forsterite	2410	145	54	52	202	2020	215	45	–	100
W. Ethiopia	pyroxene (bronzite)	265	2610	7	17	115	1100	800	9	–	850
	Spinel	80	27935	–	32	180	915	620	180	–	–
Canary	forsterite	2820	95	42	55	220	960	180	25	40	85
Islands,	pyroxene	1250	3400	85	56	225	450	345	–	–	175
Lanzarote	spinel (chromite)	2200	9880	99	58	220	140	465	250	565	312

We can also consider the presence of crystal-chromite grains either forming "Schlieren" chromite or as compact chromite in serpentinised dunitic mantle-diapirs. Fig. 603 shows rounded chromite crystal grains as bands in the serpentinised dunite. Similarly, Fig. 604 shows both rounded chromite and idiomorphic shapes in serpentinised dunite. As can be seen in Fig. 605 the crystal chromite grains are rounded and show cataclasis.

The alternating olivine/chromite banding has been considered by magmatists to represent fractional crystallization in the sense that the chromite bands have been formed by crystal settling, due to gravity.

Dunitic complexes free from metatectonic effect (tectonic effects after the dunite intrusion) do show an olivine grain orientation, with the olivines showing undulating extinction and twin lamellae (see chapt. 18). In addition to the deformation effects exhibited by the olivine, the chromite grains show a roundening and cataclasis which can be attributed to *prototectonic effects** within the complex processes of the mantle crystallization (magmatic attrision and deformation effects).

The undulating extinction and deformation lamellae of olivines and geophysical studies support the idea of a plastic mantle. Prototectonics (tectonics during the crystallization of the mantle) and metatectonic effects can be considered as being responsible for the "ptygmatic folding", "cross-bedding structures" and "disconformities" observed in the banding of "Schlieren" chromite in ultrabasics.

Often compact chromite ores, e.g. Rodiani, may show chromite enclosing idiomorphic olivines representing a pre-chromitic crystallization phase (see Fig. 665).

The compact chromite shows most impressive cataclastic effects (see Figs. 666, 667 and 668) and often serpentinisation as "re-binding" holds the fragments together. In cases most impressive mylonitisation of compact chromite is exhibited often along mylonitised planes. In addition there are transitions exhibited showing all phases from compact chromite which marginally, due to tectonic deformation, changes into a chromite granular phase (see Figs. 607 and 608).

Indeed, the chromite lens bodies, and the chromite "potatoes" (Fig. 669) which are showing tectonically polished faces, are chromite bodies within a serpentinised dunite which have been tectonically fractured and moved within the dunitic body.

The impressive leopard-ore, which has been considered to be a primary crystallization phase can be attributed to tectonic deformation, often the chromite "spheroids" are perfect strain deformation ellipsoids in yellow serpentine (see Fig. 670 a, b) which is tectonically mobilised. The above mentioned distribution patterns, structures, and texture of chromite ore-bodies within dunitic complexes are regarded as partly primary and partly due to tectonic deformation. This can be attributed to the proto-tectonic (i.e. on the tectonic influences operating within the mantle and during the mantle's solidification), and should be distinguished** from the distribution pattern caused by meta-tectonics, i.e. the tectonic deformation associated with serpentinisation and the resultant structure and textures should not be considered strictly speaking as primary.

Re-distribution of chromites and remobilisation of chrome metasomatically

The presence of chrome and particularly of chromite in metasediments (chloritic schists – leuchtenbergite schist with anthophyllite (see chapter 6) Hadabudussa Gari-Boro, Adola, Ethiopia) can be considered as a metasomatic chromitic mobilisation with a blastogenic chromite formation.

Indeed, Hutton (1942), see Geijer (1963), describes a most remarkable metasomatic chromite formation. The following is a quotation from Geijer.

"Another example of special interest is the fuchsite occurrence described by Hutton (1942) from Dead Horse Creek, Otago, New Zealand. There the chrome mica is found in schists consisting largely of quartz and feldspar. A remarkable feature is that the fuchsite contains small crystals of chromite. Hutton's interpretation is that "the fuchsite-schists are the results of chromium metasomatism brought about by the penetration and soaking of narrow zones of quartzo-feldspathic schists by solutions at high temperature, or aqueous chromium-bearing vapours, ... emanating from deep-seated intrusions of an ultrabasic nature" (Hutton, p.cit., p.64). An argument for a derivation from a source of this nature is the fact that fahlbands with nickeliferous pyrrhotite are also developed. In any case, the presence of chromite makes this occurrence unique."

---

* As prototectonic effects, tectonic processes within the solidification of the mantle, re-melting of mantle by fusion and plastic deformation as well as mantle diapirism.

** Of course it is difficult to draw a demarcation line between proto- and meta-tectonic influences in ultrabasics.

Also the "metamorphic" chromite mobilisations and deposits described by Petrascheck* (1938, 1957, 1958, 1959, 1966) should be considered as instances of chromite mobilisation, a redistribution of chromite from a source of primary distribution". In addition the association of chromite with sulphide parageneses as is described by Ramdohr (1961) from Outokumpu should be again regarded as redistribution from the primary distribution phase.

Perhaps a most interesting case of the co-existence of chromite of primary distribution and chromite redistribution is from the Bushveld Complex. Figs. 671, 672 and 673 show chromite spheroids (rounded and ellipsoid forms) surrounded by uvarovite garnet blastically formed (metasomatically). It is possible that the chromite spheroids represent clastic-sedimentary distribution-phase and the garnet crystalloblasts indicate chromium redistribution.

Another case of chromite re-distribution is the "mechanical" (i. e. tectonical) mobilisation of compact chromite-ore in marble. Figs. 674, 675 show fragmented chromite in marble from Drepanon, Kozani, Greece.

In contrast to the re-distribution of chromite and chrome described, chromite minerals may be subject to weathering and alteration processes.

Chrome-spinels, when subject to alteration process and differential leaching have as a result the formation of decoloration margins, Augustithis (1960b), see Fig. 676.

The preferential leaching out of Mg and Al results in the decoloration margin of the chromite which, in turn, shows a relatively greater concentration of Fe and Cr in the decoloration margin compared with the unaffected central part of the chromite.

However, alteration and weathering of chromite grains may result in more pronounced changes of the chromite with unmistakable evidence of chromite solution and replacement. Often the alteration may result in the formation of chrome-oxides and chrome-hydro-oxides. A series of illustrations shows the phenomenology of chromite alteration (see chapter 33).

The chemical mobility and redistribution of Cr in the Ni-Cr Iron ore deposits of Larymna/Greece due to diagenetic changes has been intensively discussed by Augustithis (1962b) and Kurzweil (1966).

The second phase Cr redistribution is extensively discussed in chapter 23. However, it can be concluded here that chrome and chromite, in addition to their primary distribution, can be subject to a redistribution due to hydrothermal activity and metasomatism, in addition chrome and chromite are subject to mobilisation processes, secondary in nature which, however, can result in pronounced redistributions.

* One of the prominent supporters of the origin of chromite deposits by basaltic magma differentiation.

Chapter 24 The distribution of Pt-group elements and the Pt-group minerals in the basic and ultrabasic rocks

The distribution of the Pt-group elements (respectively the Pt-group minerals) is in a comparable way to the Cr; and can be distinguished as a primary phase of distribution in mantle, as a remobilisation phase due to hydrothermal and topo-metasomatic solutions and as a third phase, of low-temperature Pt-group element mobilisation and Pt-group mineral formation (often nuggets may result). The low temperature Pt-group element mobilisation is interrelated with the lateritisation and weathering of ultrabasic rocks and as a result Pt-group element mobilisation, Pt-nuggets and a complex mineral paragenesis may result (see chapt. 31).

In the primary distribution Pt is often associated with the spinels in the mantle, usually values of Ir more than 10 ppb indicate mantle. Neutron activation analysis of diapiric mantle samples from a peridotite intrusion from Konitsa region, N.Greece have shown more than 10 ppb Ir. In this case the Ir is taken as an index for the Pt-group elements. Geochemically the Pt-group elements are related to the Fe of the spinels of the mantle, as elements of the same subgroup of the 8th family of the periodic table.

Most probably ferroplatin as early crystallization within the chromite, could also be regarded as a primary distribution phase in the basic and ultrabasic rocks. A probable example is the presence of P.G.E. in the chromite ore from the Island of Skyros*, Greece.

Of particular geochemical and mineralogical interest is the redistribution of the Pt-group elements in basic and ultrabasic rocks. In contrast to Wagner (1929) and Schneiderhöhn (1958), Genkin (1959), Ramdohr (1960) and Stumpfl (1962) have produced microscopic evidence of the existence of independent Pt-minerals as later hydrothermal veinlets often invading the sulphides in basic and ultrabasic rocks.

Whereas Stumpfl accepts the transportation and location of P.G.E. in the magmatic stage, he also recognises a redistribution of the P.G.E. in the pneumatolytic or hydrothermal phase (i. e. the second phase of Pt-elements distribution). In this connection it is interesting to quote some of the discussion by Stumpfl (1974):

"Platinum Deposits Associated with Basic and Ultrabasic Igneous Rocks

Views on the genesis and geochemistry of PGE-deposits associated with basic igneous rocks have, until very recently, been strongly influenced by Wagner's fundamental work on the platinum deposits of South Africa, to which Schneiderhöhn had contributed a chapter on the mineragraphy of the ores. In particular, the following statements are of interest: "There can be no reasonable doubt in the mind of anyone who has studied the various occurrences that both olivine-dunite and hortonolite-dunite are products of the consolidation of liquid dunitic magma fractions". The presence of phlogopite, black hornblende, and fluorapatite in some dunites is attributed to "mineralisers such as HO, F and H_2S". "If Professor Schneiderhöhn is correct in his conclusions as to the form assumed by platinum in the different ores of the Bushveld Complex, the fact that platinum crystallized in the dunites mainly in the metallic state is due to the poverty of the dunitic rest-magmas in sulphur, arsenic and antimony." Ore-microscopic and spectrographic investigations led Schneiderhöhn to conclude that, in the Merensky Reef, there are no independent PGM, but only PGE in solid solutions in the sulphides.

These views and their genetic implications have proved to be very long-lived and have achieved al-

* Agiorgitis (1978) has determined exceptionally high contents of Os, Ru and Ir for the Skyros and Vourinos chromite, which rather supports a mantle derivation for these ultrabasic bodies.

most classical significance by their acceptance into recent textbooks. Schneiderhöhn emphasizes that all PGE in nickel-copper deposits (Sudbury, Merensky Reef) are located in the lattice of pyrrhotite, pentlandite, and chalcopyrite. "Independent platinum minerals have never been found in the fresh rocks" (of the Merensky Reef). He also states that native platinum, alloyed with iron, is the only PGM in the Bushveld dunites – without, understandably, sounding the cautious note that Wagner has added to the same statement in 1929, i. e. "if Professor Schneiderhöhn is correct in his conclusions".

In 1962, the author presented evidence against the supposedly dominant role of solid solutions of PGE in the lattice of sulphides at Sudbury and in the Merensky Reef, and showed the complex mineralogical composition of the Driekop dunite ores. These observations, Genkin's initial work at Noril'sk, and Ramdohr's findings were taken to suggest that "many Pt-deposits may have formed at temperatures other than previously supposed and that processes other than pure magmatic differentiation of silicate and sulfide melts may sometimes have been involved in their formation".

In contradistinction to the hydrothermal and metasomatic Pt-group element mobilisation, is the low temperature mobilisation and transportation of the Pt-group elements in eluvial lateritic covers of platiniferous ultrabasics (i. e. the third phase of Pt-group elements distribution), Ottemann and Augustithis (1967), see also Chapter 31.

This low temperature Pt-group element mobilisation known as the "accretion hypothesis" has been proposed to explain also alluvial Pt-deposits by Cousin (1973a, b) and Stumpfl (1974) as is stated by Cabri and Harris, D. C. (1975).

"As the proposals of Cousins and of Stumpfl regarding the accretion hypothesis are based on the papers by Augustithis (1967b) and Ottemann and Augustithis (1967), close examination of the original papers is necessary to determine whether accretion originally proposed for a lateritic and eluvial environment, is applicable to Witwatersrand, Tasmanian, and other alluvial deposits" (see again chapter 31).

It is interesting to note that the Pt-group elements and the formation of the Pt-group minerals can be present in a wide range of distribution and genetic possibilities, e.g. mantle, magmatic rocks, pneumatolytic-hydrothermal occurrences and finally as lateritic paragenetic associations*.

Considering most of the known Pt group minerals from a wide range of parageneses Table VI it can be seen that the mineral building elements can be grouped into two independent groups. Groups A, consisting of the Fe, Ni, Co, Ru, Rh, Pd, Os, Ir, Pt (sub-group elements of the 8th family of the periodic system) and Cu and Au**; and group B consisting of the elements As, Sb, Bi, Sn, Te, and Pb, Se***. The elements As, Sb, Bi are related to one another as elements of the main group, homologue elements of the V family of the periodic system. Also the elements Sn, Sb, Te are related to one another as main group elements, next to each other. It is therefore to be seen that all the elements of the B group are interrelated in accordance with the empirical laws of the periodic system (see Augustithis, 1964c, 1967b).

Considering the element-geochemistry and particularly the abundance of the elements comprising groups A and B, which build-up most of the Pt-group minerals (see Table VII), it is clear that the joint segregation of very rare elements in the earth's outer part† cannot be accidental but is governed by "laws" which enable their joint segregation in Pt-minerals and parageneses††. As those "laws" are proposed the empirical "laws" of the periodic system which govern the relation of elements among themselves.

---

* As a corollary of the solubility of Pt elements in low temperature solutions is the presence of Pt in manganese nodules found in oceanic floors (Agiorgitis and Grundlach "Platinum content in manganese nodules"). Contents of 6–22 ppm of Pt have been found in manganese nodules.

** Cu and Au are related as sub-group elements of the I family and Hg to Au as adjacent elements.

*** Pb is related to Bi as next to it and Se is related to Te as homologues.

† Upper mantle and crust.

†† "In addition to the broad relations in accordance to the periodic system which are exhibited by the elements building the main platinum group-minerals, other specific interrelations exist. Keays and Davison (1976) state that palladium and iridium correlate strongly with the nickel sulfide content of the rocks. Another interrelation is stated by Naldrett and Cabri (1976):

"In the sulfide ores of the Merensky, Pechenga, Sudbury, and Noril'sk deposits, the Pt/(Pt+Pd) ratio decreases systematically with an increasing Cu/(Cu+Ni) ratio. Since the latter ratio is thought to increase with progressive differentiation of the host silicate magma, it is suggested that the trend of the decreasing Pt/(Pt+Pd) ratio is also related to differentiation. Ores associated with komatiites are major exceptions to the trend exhibited by the tholeiites, the komatiite ores having both a low Cu/(Cu+Ni) and a low Pt/(Pt+Pd) ratio. It is suggested that this is due to the fact that komatiites originate at unusually great depths in the mantle, depths at which mantle sulfides (rich in Pd) have accumulated and are hence incorporated in the magmas."

In addition Chyi and Crocket (1976) found that the "Individual noble metals are strongly associated with specific sulfide minerals. Palladium and gold are concentrated with

Table VI (based on the glossary of platinum-group minerals by Cabri, 1976).

Mineral	General composition
Arsenopalladinite	$Pd_5(As, Sb)_2$
Atheneite	$(Pd, Hg, Au, Cu)_3$ (As, Sb) Pd, Hg, As major
Atokite	$(Pd, Pt)_3$ Sn Pd > Pt
Biteplapalladite	(Pt, Pd) $(Te, Bi)_2$ Pt > Pd; Te > Bi
Biteplatinite	(Pd, Pt, Ni) $(Te, Bi)_2$
Borishanskiite	Pd_{1+x} $(As, Pb)_2$ As ≃ Pb; x < 0.2
Borovskite	$(Pd, Pt, Ni)_{15}$ $(Sb, Bi)_{05}Te_{200}$ Pd > Pt, Ni; Sb > Bi
Braggite	(Pt, Pd, Ni)S Pt > Pd > Ni
Cooperite	(Pt, Pd, Ni)S Pt > Pd, Ni
Daomanite	$(Cu, Pt)_2AsS_2$ Cu ≥ Pt
Erlichmanite	$(Os, Ir, Rh, Ru, Pd)S_2$ Os > other Pt group
Froodite	(Pd, Pt) $(Bi, Te)_2$ Pd > Pt; Bi > Te
Geversite	$Pt(Sb, As, Bi)_2$ Sb > As, Bi
Guanglinite	Pd_3As
Hexatestibiopa-nickelite	$(Ni, Pd)_2(Sb, Bi)$ Te Ni > Pd; Sb > Bi
Hollingworthite	(Rh, Ru, Pt, Pd, Co, Ni)AsS Rh > other Pt group
Hongshiite	PtCuAs
Insizwaite	(Pt, Pd, Ni) $(Bi, Te, Sb, Sn)_2$ Pt > Pd, Ni; Bi > Te, Sb, Sn
Irarsite	(Ir, Ru, Rh, Pt, Pd, Os, Ni, Co)AsS Ir > other metals
Iridarsenite	(Ir, Ru, Os, Rh, Pt, Cu) $(As, S)_2$ Ir > other metals
Iridium	Ir, Os, Ru, Pt, Pd, Rh, Fe, Ni Ir major
Iridosmine	Os, Ir, Ru, Pt, Pd, Rh, Fe, Cu, Ni Os, Ir major
Isoferroplatinum	~ $(Pt, Ir, Os, Ru, Rh)_3$ (Fe, Ni, Cu, Sb) Pt > other Pt group; Fe > other metals
Isomertieite	$(Pd, Cu)_5$ $(Sb, As)_2$ Sb ≃ As
Kotulskite	(Pd, Ni) (Te, Bi, Sb, Pb) Pd > Ni; Te > Bi, Sb, Pb
	$(Ru, Ir, Os)S_2$ Ru > Ir, Os
Malanite	$(Cu, Pt, Ir, Pd, Fe, Ni)S_2$ Cu, Pt, Ir major
Merenskyite	(Pd, Pt, Ni) $(Te, Bi, Sb)_2$ Pd > Pt, Ni; Te > Bi, Sb
Mertieite	~ $(Pd, Cu)_5$ $(Sb, As)_2$ Pd > Cu; Sb ≃ As = Group I Sb > As = Group II
Michenerite	(Pd, Pt, Ni) (Bi, Sb) Te Pd > Pt, Ni; Bi > Sb
Moncheite	(Pt, Pd, Ni) $(Te, Bi, Sb)_2$ Pt > Pd, Ni; Te > Bi, Sb
Niggliite	(Pt, Pd) (Sn, Bi, Te) Pt > Pd; Sn > Bi, Te
Oosterboschite	$(Pd, Cu)_7Se_5$ Pd > Cu
Osarsite	(Os, Ru, Ni, Ir, Pd, Pt, Rh)AsS Os > other metals
Osmiridium	Ir, Os, Ru, Pt, Pd, Rh, Fe, Cu, Ni Ir, Os major
Osmium	Os, Ir, Ru, Pt, Pd, Rh, Fe, Cu, Ni Os major
Palladium	Pd, Pb, Rh, Pt, Os, Ir Pd > other metals
Palladoarsenide	$(Pd, Pt, Au, Cu)_2$ (As, Sb, Te) Pd > Pt, Au, Cu; As > Sb, Te
Palladobismuthar-senide	$Pd_2As_{0.8}Bi_{0.2}$
Paolovite	$(Pd, Pt)_2Sn$ Pd > Pt
Platiniridium	Ir, Pt, Os, Ru, Fe, Cu, Ni Ir, Pt > other metals
Platinum	Pt, Pd, Ir etc. Pt > other metals
Plumbopalladinite	$(Pd, Ag)_3$ $(Pb, Bi, Sn, Cu, Sb)_2$ Pd > Ag; Pb > other elements
Polarite	Pd(Bi, Pb) Bi > Pb
Potarite	PdHg
Rhodium	Rh, Pt Rh > Pt
Rustenburgite	$(Pt, Pd)_3Sn$ Pt > Pd
Ruthenarsenite	(Ru, Ni, Rh, Ir, Pd, Os)As

pentlandite, and platinum and iridium are associated with chalcopyrite. Relative to the most abundant sulfide, pyrrhotite, palladium and gold are concentrated in pentlandite by a factor of approximately 14, and platinum and iridium are concentrated in chalcopyrite by a factor of approximately 6. Except for iridium, magnetite is depleted in noble metals relative to all sulfide minerals."

Mineral	General composition
Rutheniridosmine	Ru, Os, Ir, Pt, Pd, Rh, Fe, Cu, Ni Ru, Os, Ir major
Ruthenium	Ru, Ir, Rh, Pt, Os, Pd, Fe Ru major
Ruthenosmiridium	Ir, Os, Ru, Pt, Pd, Rh, Fe, Ni Ir, Os, Ru major
Sobolevskite	$Pd_{1.07}Bi$ minor Pt, Pb, Sb, Te reported
Sperrylite	$(Pt, Rh, Ir)(As, Sb, S)_2$ Pt > Rh, Ir; As > Sb, S
Stannopalladinite	$(Pd, Cu)_3Sn_2$ Pd > Cu
Stibiopalladinite	$(Pd, Cu)_{5+x}(Sb, As, Sn)_{2-x}$ Pd ≫ Cu; Sb ≫ As, Sn
Stillwaterite	$Pd_8(As, Sb, Te, Sn, Bi)_3$ As > Sb, Te, Sn, Bi
Stumpflite	Pt(Sb, Bi) Sb > Bi
Sudburyite	(Pd, Ni) (Sb, Bi, Te, As) Pd > Ni; Sb > Bi, Te, As
Telargpalite	$(Pd, Ag, Pb, Bi)_{4+x}(Te, Se)$ Pd, Ag > Pb, Bi; Te > Se
Temagamite	$Pd_2Hg(Te, Bi, Sb)_3$ Te > Bi, Sb
Testibiopalladite	(Pd, Ni) (Sb, Bi)Te Pd > Ni; Sb > Bi
Tetraferroplatinum	Pt, Fe, Ir, Cu, Ni, Sb Pt ⋝ Fe > Ir, Cu, Ni, Sb
Tulameenite	$(Pt, Ir)_2(Fe, Cu, Ni, Sb)_2$ Pt > Ir; Fe ⋝ Cu > Ni, Sb
Vincentite	$(Pd, Pt)_3(As, Sb, Te)$ Pd > Pt; As: (Sb + Te) ≃ 1:1
Vysotskite	(Pd, Ni, Pt)S Pd > Ni, Pt
Xingzhongite	(Ir, Cu, Rh, Pb, Os, Pt, Fe)S Ir, Cu, Rh major
Yixunite	$Pt_{1.0}In_{0.86}$
Zvyagintsevite	$(Pd, Pt, Au)_3(Pb, Sn)$ Pd > Pt, Au; Pb > Sn

Table VII, Precious metals in the upper lithosphere. Compiled from V.M. Goldschmidt, Mason, and Wedepohl.

A Group				B Group			
Ru	=	0.4	ppb	As	=	5	ppm
Rh	=	0.4	ppb	Sb	=	21	ppm
Pd	=	4	ppb	Bi	=	0.2	ppm
Os	=	0.4	ppb	Sn	=	40	ppm
Ir	=	0.4	ppb	Te	=	0.002	ppm
Pt	=	2	ppb	Pb	=	16	ppm
Fe	=	50,000	ppm				
Ni	=	80	ppm				
Co	=	23	ppm				
Cu	=	70	ppm				
Au	=	0.001	ppm				

Chapter 25 The mineralogical and geochemical distribution of sulphides in basic and ultrabasic bodies

General (magmatists' views based on literature)

The elements Fe, Ni, Co, Cu, Zn occur in the mineral components of the olivine bombs in basalts (i. e. mantle fragment brought up by basaltic effusions). Fe is a main constituent of the mantle minerals, in contradistinction Ni, Co, Cu, Zn occur as trace-elements in the mantle mineral components. The existence of these elements in the mantle could well form the source of all subsequent mobilisations and redistributions of these elements in rocks in general and in basics and ultrabasics in particular.

Of particular significance is the remobilisation and the redistribution of the elements Fe, Ni, Co, Cu, Zn in terms of ore bodies (sulphides*) in basic and ultrabasic intrusive bodies within the crust.**

Wilson et al. (1969), in discussing the geochemistry of some Canadian Nickeliferous ultrabasic intrusions, point out that in the case of an ultrabasic sill, which was intruded as a liquid-crystal mixture, cumulus silicate minerals accumulated at the centre of a flow system. More dense cumulus sulphides liquid accumulated near the bottom of the sill and collected in rifle shaped traps to form ore bodies. Wilson et al. (1969) regard also the McWatters, Ontario deposit (in a volcanic environment) as an intrusion of a liquid-crystal mush with enough liquid to allow segregation of cumulus olivine toward the centre of a dyke-flow-systcm, and gravity settling of cumulus sulphides liquid.

Considering the process of sulphide separation in the "magmatic rocks", Wilson et al. state the following:

"The ultrabasic rocks associated with these nickel deposits in Canada are intruded as magmas composed of a mixture of liquid and olivine, accompanied by a minor quantity of pyroxene, and ultramafic inclusions. The proportion of liquid to crystals is variable but always low. Magmas with a high proportion of liquid form lamellar flow systems with olivine accumulating at the centre and liquid at the edges. Magmas with small proportions of liquid are intruded as a crystal mush with little apparent segrcgation of cumulate crystals and liquid. The olivine-pyroxene textures indicate that "solid intrusion" is not an important process in the intrusions studied.

Sulphide occurs as immiscible droplets suspended in the liquid silicate phase and in solution in the silicate liquid. The immiscible sulphide droplets are heavy and, with appropriate flow velocity, tend to concentrate by gravity toward the bottom of relatively liquid intrusions during flowage although a considerable portion may remain throughout the magma, particularly in lamellar flow streaks. Basins or troughs in the bottom of the intrusion may cause turbulence and act as traps for settling and accumulation of heavy sulphide liquid during the flow stage. Thus accumulation may occur even in steeply dipping intrusions and the size of the orebody is dependent on the volume of magma passing the trap rather than the volume relations at any particular place in the frozen intrusion.

Nickel occurs in solid solution in the silicate minerals, in fine dust-size sulphide particles that have crystallized from sulphide in solution as the magma froze, as suspended accumulate sulphide blebs generally concentrated toward the bottom of the intrusion, and as massive sulphide accumulations at or near the base of the intrusion or in fractures in footwall rocks. The sulphur: nickel ratio is relatively constant in one intrusion but is affected by the relative amounts of nickel in each of the above forms and also by the relative chalcophile natures of copper, nickel, cobalt, and probably iron."

---

* pyrite, chalcopyrite, pentlandite, milerite, sphalerite, etc.

** The intercontinental rift deposits: Continental rifting basaltic effusions, leaching of elements by sea water from the basalts → generation of "hydrothermal deposits". An alternative explanation is that the elements are mobilised from the basic effusions and from the crust, e. g. Pb, Zn and others.

A further example of magmatic sulphide ore minerals is discussed by Skinner and Peck (1969) from Hawaii. The following is an excerpt from their summary:

"The basaltic lava trapped in Alae pit crater from the August 1963 eruption of Kilauea volcano, Hawaii, produced an immiscible sulphide melt during a late stage of the cooling of the lava. A differentiated siliceous liquid confined within interstices of the upper crust of the lava lake became saturated with respect to sulphide/sulphur at a temperature of 1,065°C and a sulphur content of 0.038% S by weight. Precipitation of two sulphide-rich phases, one an immiscible sulphide-rich liquid, the other a copper-rich pyrrhotite solid solution followed.

The sulphur-rich liquid, which quenched to a mixture of pyrrhotite, chalcopyrite, and magnetite, is estimated to have a composition of approximately 61% Fe, 4% Cu, 31% S and 4% O by weight. The copper-rich solid sulphide that coexisted with the sulphide liquid has an estimated composition of 53% Fe, 9% Cu, 3% Ni and 35% S by weight.

The basaltic liquid from which the sulphide-rich phases precipitated does not contain abnormally high contents of either Cu or Ni."

In contrast to the volcanic (lavas and sills) sulphide ore concentration, the orogeny category of ultrabasic rocks, predominantly peridotites and pyroxenites disseminated sulphides occur, irregularly distributed throughout the ultrabasic inclusion.

Kilburn et al. (1969) state the following, comparing the distribution of sulphide in orogenic and volcanic basic and ultrabasic rocks.

"Petrographic studies reveal a pattern of progressive recrystallization with serpentinisation. Increased serpentine grain size with recrystallization leads to displacement of sulphide minerals and finally complete destruction of primary silicate-sulphide textures.

In the orogenic category, ultrabasic rocks are predominantly peridotite and pyroxenite; disseminated sulphide is the most common mode of occurrence, irregularly distributed throughout the ultrabasic intrusion; contact deposits are encountered less frequently and deposits in the wall rocks are rare."

In contrast to high temperature sulphide ores in the volcanic types (sills and lavas) the experimental work of Kullerud et al. (1969) and Craig and Kullerud (1969) supports the possibility of "re-equilibration" of sulphide ores in ultrabasics at lower temperatures.

The following are excerpts from these publications Kullerud et al. (1969) and Craig and Kullerud (1969).

"The phase relations in the Cu-Fe-S, Fe-Ni-S systems were investigated by silica-tube quenching, differential thermal analysis, and high temperature X-ray powder diffraction experiments. In addition, portions of the Cu-Fe-S and Fe-Ni-S systems were studied by gold tube quenching and differential thermal analysis experiments under high confining pressures.

At elevated temperatures extensive liquid immiscibility fields span the sulphur-rich region of each of the three systems, whereas homogenous liquid fields dominate phase relations in their central portions. The average composition of the Sudbury Cu-Fe-Ni sulphide ore, when projected onto the Cu-Fe-S plane, is accounted for above 860°C by a mixture of copper containing hexagonal pyrrhotite and copper-rich sulphide liquid. Thus at high temperatures a mechanism exists that may be responsible for certain copper-rich segregations observed in this type of ore. The minerals of the Cu-Ni-S system, with the rare exception of millerite, do not occur in Sudbury-type ores. Knowledge of the phase relations in this system is prerequisite, however, for systematic investigations of the complex Cu-Fe-Ni-S system. Applications of the phase relations in the Cu-Fe-S and Fe-Ni-S systems to typical ore assemblages show that extensive re-equilibrium took place among the sulphides after their initial deposition. The sulphides in Sudbury-type ores commonly have compositions and crystal structures that can be produced in the laboratory only at low temperatures.

The phase relations in the Cu-Fe-Ni-S system are discussed in the light of experimental results from liquidus temperatures to below 550°C. Although liquidus temperatures vary greatly throughout the system, beginning of melting of a typical ore composition (1.5% Cu, 4.5% S, 55.5% Fe) was found to occur at 1,105°C. Magnetite lowers the beginning of melting temperature by less than 20°C.

Under conditions encountered in typical massive Ni-Cu sulphides, fractionation leading to formation of a Cu-enriched liquid is possible only above 850°C. Ni enrichment in a liquid phase is not encountered below 1,000°C except under sulphur pressure conditions much lower than those encountered in typical Ni-Cu ores.

The bulk compositions of Ni-Cu sulphide ores commonly lie within the confines of a quaternary monosulphide solid solution at temperatures above 500°C. This suggests that such ores formed a homogeneous phase at the time of their emplacement. Cooling below 600°C may result in exsolutions of chalcopyrite and/or pyrite or chalcopyrite and/or pentlandite depending on sulphur pressure. The commonly observed pentlandite + pyrite assemblage is stable only well below 300°C and cannot crystallise directly from a Cu-Fe-Ni sulphide melt.

The absence of natural analogues even of a ternary (Fe, Ni) $1\text{-}x^s$ monosulphide solution and the frequent occurrence of monoclinic pyrrhotite and the pyrite + pentlandite assemblage indicate low-temperature re-equilibration of mineral assemblages and mineral compositions subsequent to original emplacement which may have occurred at or near liquidus temperatures. Such re-equilibration negates usage of the pyrrhotite-pyrite geothermometer in magmatic Ni-Cu ores.

The present study offers a new view of the paragenesis of Ni-Cu ores and demonstrates that the observed mineral assemblages may be explained in terms of phase changes occurring in a system of isochemical bulk composition in response to decreasing temperature."

As was the case of Cr and the Pt minerals, the Fe, Ni, Co, Cu, Zn minerals may exhibit the same pattern of distribution, namely:

The elements Ni, Co, Cu, Zn may occur as trace elements in the mineral components of the mantle. This might be referred to as the primary distribution of these elements.

In contrast the high temperature sulphide cumulus, particularly in volcanic rocks, could again represent a primary distribution as mobilised derivatives from the mantle.

In contrast the possibility that in orogeny ultrabasic bodies a progressive recrystallization of the sulphides takes place with serpentinisation could be interpreted that serpentinisation results in a remobilisation of the elements Fe, Ni, Co, Cu, Zn from primary distribution in mantle [the orogeny ultrabasic represent mantle diapirs (see chapt. 10)] to sulphides within the serpentinised ultrabasics.

The "re-equilibration" of sulphide ores in ultrabasics under low temperatures (Kullerud et al., 1969, and Craig and Kullerud, 1969) could be understood as a case of remobilisation of these elements (Fe, Ni, Co, Cu, Zn).

The re-mobilisation particularly of Ni under lateritisation is discussed by Augustithis (1962b) and Agiorgitis (1972).

The element geochemistry of the group Fe, Ni, Co, Cu, Zn can be understood on the basis of their relationship in accordance with the empirical "laws" of the periodic system.

The elements Fe, Co, Ni are related as sub-group elements of the 8th family, the element Cu is related to Ni as next to it and similarly Zn with Cu.

The joint segregation of the elements in the mantle (primary distribution) and in the magmatic (lava and sills sulphides) as well as a re-equilibration phase of Fe-Ni-Cu-Co-Zn sulphides in ultrabasics is explicable by their relationship in the periodic system.

In this connection it should be pointed out that the intercontinental ore-deposits are recently considered to represent leached out elements from basaltic flows or from the crust due to percolating brines (sea water). They are thus considered to represent a redistribution of elements.

In contradistinction the Ni-deposits associated with the Archean Greenstones, again represent a redistribution of elements originally present in the mantle (particularly in the mantle forsterites); the mechanism, however, being remelting of material.

Chapter 26 The significance of magnetite and magnetite with titaniferous "ex-solution" in basic and ultrabasic rocks

The presence of magnetite in basic and ultrabasic rocks has been almost invariably explained as a product of fractional crystallization of a differentiating magma.

The presence of titaniferous "ex-solutions" and spinels (Al-spinels and ulvite) has been used as evidence to support the magmatic intracumulate crystallization of magnetite.

The "ex-solution" phases in magnetites have been used as a "reliable geothermometer" the presence of which has been taken as evidence that the magnetite and the rock containing these intergrowths has reached a certain temperature range, or that it has been formed within a temperature range of about 700°C to 1000°C.

Taylor and Noble (1969) in discussing the origin of magnetite in the zoned ultramafic complexes of Southeastern Alaska state the following:

"In several of the complexes (including Union Bay and Duke Island) a consistent zoning of ultramafic rock types occurs, with dunite, peridotite, olivine pyroxenite, and hornblende pyroxenite occurring in sequence from the centre outward to the margin. Only the hornblende pyroxenite zone is exposed at Klukwan and Snettisham, but the mineralogy of this zone is identical in all the complexes. The minerals are diopsidic augite (Di_{85}/Hed_{15}/to Di_{70}/Hed_{30}) hornblende (abnormally high in Al_2O_3 and low in SiO_2 relative to most igneous hornblendes), magnetite and accessory ilmenite and hercynitic spinel. Accessory amounts of olivine may occur in the interior parts of the magnetite pyroxenite zones of the complexes that contain olivine pyroxenite and/or dunite, but the magnetite concentration drops abruptly as the amount of olivine increases; magnetite is largely absent from the olivine pyroxenite zones".

It is interesting to note that the preferential paragenetic association of magnetite with ultrabasics "gabbroic in composition", is in accordance with the tendency of magnetite to disappear in olivine-pyroxenites and dunites. This, as already explained, could depend on the MgO/FeO ratio in an ultrabasic (see Chapter 22).

In this connection, one should attempt to understand the preferential association of titaniferous minerals with the magnetite paragenesis in ultrabasics rather than with the chromite (see Chapter 22).

Considering the geochemistry of the magnetite paragenesis in ultrabasics another antipathy should be considered, namely that of Ti and V.

Willemse (1969) in his study of the plug-like bodies and seams of vanadiferous magnetite in the Bushveld Complex states the following:

"Many seams have a sharp basal contact against anorthosite, and a transitional hanging wall composed of massive ore grading upwards into magnetite-anorthite and hyperite. A thin, highly puckered veneer of olivine and an intergrowth of pyroxene and plagioclase are developed at the sharp basal contact. The intergrowth could be due to diffusion of iron into the plagioclase. Irregularities at the contact consist of vein-like offshoots of magnetite into the foot wall and even brecciation of the foot wall is encountered.

The grain size of the ore in the Main Seam is variable. The closely packed nature of the grains is best explained by enlargement after concentration.

The magnetite started to crystallise earlier or simultaneously with the plagioclase and the seams could have formed in the same manner as any monomineralic rock in a layered sequence. In some of the country rocks magnetite continued to crystallise interstitially to the plagioclase. In these examples hydroxyl minerals such as biotite and amphibole are generally present.

Titanium minerals in the ore consist mainly of (1) small ilmenite granules that probably formed earlier or simultaneously with the magnetite, (2) rather rare, broad lamellae of ilmenite which exsolved from magnetite at an early stage, (3) ulvite, which is ubiquitously present and forms the characteristic cloth-pattern, and (4) a wide range of intergrowths of dispersed ilmenite ("proto-ilmenite") with magnetite, maghemite and martite. These intergrowths are considered to have formed from the ulvite cloth-pattern by surface or near surface oxidation.

The cloth pattern of ulvite seems to have had a profound effect on the formation of maghemite and martite – it promoted maghemitization and retarded martization.

An antipathetic relationship prevails between the V_2O_5 and the TiO_2 content of the ore. The lowermost seam contains about 2 % V_2O_5 and 14 % TiO_2, the uppermost one about 0.3 % V_2O_5 and 18–20 % TiO_2. As the titanium is mostly contained in ulvite it is considered that a decrease in oxygen fugacity in the magma determined the increase in the titanium content upward in the layered sequence.

The V_2O_5 content of the magnetite plugs seems to correspond approximately to that of the seams in their vicinity."

The understanding of the antipathy of Ti to V as mentioned by Willemse and also the mentioned "antipathy" of Ti to Cr are even so more difficult, since the three elements Ti-V-Cr are between themselves interrelated in accordance with the empirical laws of the periodic system, Remy (1961).

"Ähnlichkeiten im Verhalten der im Periodensystem nebeneinanderstehenden Elemente nehmen von der III bis zur VIII Nebengruppe von dort über die I bis zur III Nebengruppe wieder ab."

Their "antipathy" should be rather understood considering their paragenesis (i. e. paragenetic association of the minerals in which they tend to occur) than from the point of view of element to element interrelation.

In contrast to the so far discussed literature, Augustithis (1964b) has provided evidence of a low temperature magnetite crystalloblastesis (perfect magnetite octahedra) in leuchtenbergite phyllite (in which also grow anthophyllite-nematoblasts following tectonic fissures, see chapter 6). It should be emphasised that orientated ilmenite inclusion and Al-spinels following the (111) and (100) face of the magnetite respectively have been determined, see Fig. 194 and Fig. 197.

Also in support of a blastogenic poikiloblastic growth are the so-called magnetite "intracumulus" in the ferro-gabbro bands of Skaergaard which have been explained by Vincent and Phillips (1954) as products of magmatic crystallizations.

In contrast to their interpretation and particularly on the poikiloblastic phenomenology of the magnetites, the intracumulate status of these poikiloblastic magnetites is contradicted by the fact that they are later crystalloblasts enclosing, corroding and following twin planes and intergranular spaces of the plagioclases (see Figs. 576, 577 and 578) which in terms of fractional crystallization should have preceded the light feldspar fractions.

Additional evidence, in support of a blastogenic growth of the magnetites in ultrabasic, is the existence of the ilmenite, spinel and ulvite intergrowths in the magnetite.

As already mentioned the existence of idioblastic octahedral magnetite in leuchtenbergitic phyllite, and which magnetite contains ex-solutions of ilmenite and Al-spinels, is evidence that the ex-solution intergrowths in magnetite can take place under considerably lower temperature.

The intergrowths of ilmenite, spinel and ulvite in magnetite which follow a preferential orientation within the magnetite host have until now been interpreted as orientated ex-solution lamellae, in the sense that they represent an ex-solved phase of a previous mixed crystal or a crystal containing in solid solution another crystalline phase or something approximately that (see chapt. 6).

In contrast to the orthodox views an attempt is made to interpret spinel, ilmenite and ulvite intergrowths in magnetites, other than the classical approach, despite the fact that these textures occur in magnetites which were considered to represent unquestionable liquimagmatic crystallizations.

Fig. 677 shows three mineral intergrowths with the magnetite. Arrow (a) indicates an ilmenite, lamellar in form, most probably following the penetrability direction of magnetite cleavage after (111). Arrow (b) shows Al-spinels also orientated and following the probable penetrability direction of the cleavage set (100). In contrast arrow (c) again shows Al-spinels, which clearly follow the penetrability directions of the interleptonic spaces between the magnatite and the ilmenite. It is thus clear that the Al-spinel can grow at random within the magnetite since it can form a sheath round the ilmenite. The third intergrowth phase is a table-cloth ulvite.

In contrast to the ideal octahedral orientation of ilmenite lamellae almost following perfectly the octahedral cleavage pattern (Fig. 678 which shows perfect and diffuse lamellae, in case inter-penetrating one another) is the presence of lens shaped ilmenite bodies following white lines (indicating differences in reflectivity within the magnetite) and being often at random distribution within the magnetite (Fig. 679).

Fig. 680 shows ilmenite-spinel intergrowths, within magnetite, that are deviating entirely from an orientated pattern.

Fig. 681 shows spinel grains surrounded by ilmenite which also follows cracks within the magnetite; also the Al-spinel follows a crack pattern within the magnetite and in this case the ilmenite follows the margin of the Al-spinel and the magnetite.

Indeed, such patterns deviate from the "ideal" ex-solved pattern of orientated intergrowths and show that the ilmenite-Al-spinel-magnetite intergrowths are more difficult to understand than solid solutions, ex-solved phases and processes.

In addition to Figs. 679, 680 and 681 which show that the Al-spinels intergrowths can follow random orientation within the magnetite, Fig. 682 shows magnetite with lamellae of ilmenite and Al-spinels (following the 111 face of magnetite) also Al-spinels follow the intergranular between magnetite and an adjacent ilmenite free crystal grain. Clearly the spinel follows the intergranular between the magnetite and the ilmenite grains.

In cases, the Al-spinels seem to be perfectly orientated (Fig. 683) as lamellae following the cubic face of the magnetite, however, as Fig. 684 shows the Al-spinel is not an orientated lamella but a filling of a cleavage crack of the (100) cleavage of the magnetite.

Fig. 683 clearly shows that ulvospinel is not following the (100) as orientated lamellae but is forming a diffused pattern. Also the table cloth textures of ulvite rather indicate diffused patterns of this intergrowth (see Fig. 685).

The above intergrowth patterns of ilmenite, Al-spinel and ulvospinel should be considered at least problematic if they do not throw shadows of doubt on the hypothesis of "ex-solutions".

Chapter 27 On the alteration and weathering of basic and ultrabasic rocks

General

In order to understand the alteration of the basic and ultrabasic rocks it is necessary to consider the main alteration of the rock forming minerals characteristic of this group. The alteration of olivines, pyroxenes and feldspars is a subject widely discussed by many authors, e.g. Yoder (1952), Roy-Roy (1954), Gillery (1959), Veniale (1962) and Coleman (1977). Augustithis (1978) presents a series of microphotographs showing the textural pattern and transitions of the alteration of olivines, pyroxenes, mainly in basalts; comparable and commensurable is the alteration of these minerals in the other basic und ultrabasic rocks.

In the present Atlas we are more concerned with the rock changes and geochemical mobilisations involved by the alteration of rocks.

The basic and ultrabasic rocks are susceptible to alteration and weathering processes and this is due to:

a. gradoseparation of the femisialic minerals which give rise to residual sialic components by the leaching out (separation) of the mafic components. Mg is particularly mobile under low temperature hydration conditions and to a lesser extent Fe, Pieruccini (1962).
b. Another reason is that the ultrabasic rocks and particularly the mantle diapirics are rock types formed under different pressure-temperature environment on exposure on the exogenic atmospheric process, the mineral disintegration is relatively rapid and the process of rock weathering can be pronounced and can reach relatively great depths. Appearances of these alteration processes may be traced in depths of hundreds of metres. The exposure of the basic and ultrabasic rock to the atmospheric environment should be seen as a "reaction zone", in cases, extending hundreds of metres in depth.

In addition to the meteoric water penetration and its resultant rock alteration the hydrothermal influences should be considered as an additional serious factor contributing to the alteration of ultrabasics. However, and in this case the hydrothermal processes should be seen as "heated water" solution and in this case ultimately a part of the broad reaction atmosphere-lithosphere.

As atmospheric sciences show, and particularly the petrography of the rock weathering, all minerals to a greater or lesser extent are susceptible to complete alteration and solution by the sum total of chemical weathering processes. Forsterite and chromite are two minerals of different resistance to weathering as the phenomenology of their alteration process shows, both can be completely altered or solved.

Most interesting information regarding the formation of serpentine group minerals by serpentinisation of olivines is given in the work of Veniale (1962). Veniale, considering the composition, crystal structure and form of the serpentine minerals formed concludes that whereas septechlorite, serpentine and antigorite are low temperatures, lizardite is a mineral of the serpentine group and is the product of weathering and tectonic transformation of the ultrabasic rocks from the ophiolitic formation of Apennines. Giuseppetti et al. (1962) have suggested that lizardite originated from Mg-chrysotile as a consequence of the introduction of Al and Fe into the lattice; the parental fibrous serpentine is, in this case, the early mineral formed during the serpentinisation process.

In addition the work of Veniale, quoting Yoder (1952), Roy-Roy (1954) and Gillery (1959) gives a theoretical picture of crystal structure, composition and range of the temperature formation* of the serpentine group minerals. The following are excerpts from Veniale's publication.

"Gli studi di Yoder (1952) e di Roy-Roy (1954) hanno mostrato che clinocloro e cloriti con distanza basale 14 Å si formano ad alte temperature, mentre i tipi con struttura tipo caolinite a distanza reticolare basale 7 Å (septecloriti, serpentini ed antigoriti) si formano a temperature piu basse.

Gillery (1959) ha ottenuto sinteticamente i minerali del gruppo del serpentino a temperature $<500°C$ ed ha mostrato che la temperatura sembra avere maggiore effetto sulla struttura, mentre l'abito dipende piu dalla composizione. Nella preparazione di serpentini di Mg – Al sintetici Gillery e partito da miscele: $(6-x)\,MgO$; $x\,Al_2O_3$; $(4-x)\,SiO_2$; $4\,H_2O$, corrispondenti alla formula: $(Si_{4-x}Al_x)$. $(Mg_{6-x}Al_x)\,O_{10}\,(OH)_8$. Il serpentino fibroso e stabile per valori fra x = o e x = 0,25.

La varieta lizardite con cella ortoesagonale contenente I unita strutturale si forma, nel rango di stabilita delle forme piatte e cioe per $x > 0{,}25$, al piu basso contenuto di Al e mostra piu disordine nella direzione cristallografica b, mentre la varieta prodotta partendo da una miscela con piu alto contenuto di Al si puo indicizzare sulla base di una cella ortoesagonale contenente 6 unita strutturali. Al crescere della pressione aumenta la possibilita di formazione del minerale con 6 unita strutturali nella cella elementare a causa del suo minore volume specifico.

Il minerale di S. Margherita si puo interpretare come un ter nine di passaggio fra crisotilo tubulare, unicamente o quasi magnesiano, e la lizardite piatta; esso infatti ha un valore di x = 0,218 (distribuendo ugualmente l'Al nei due pacchetti ottaedrico e tetraedrico) che non raggiunge ancora il limite minimo del campo di stabilita delle forme piatte (x = 0,25). Si potrebbe cosi spiegare la forma inconsueta dei tubuli, non molto sviluppati in lunghezza e spesso fessurati o anche parzialmente svolti.

La natura dei sostituenti isomorfi ha, come gia detto, effetto sul raggio di curvatura delle fibre ed i due tipi a diversa morfologia possono convergere, a dare una varieta con caratteristiche intermedie, come quella di Caracas, Venezuela (Hess-Smith-Dengo, 1962), Zussman (1954), Whittaker-Zussman (1956), Zussman-Brindley (1957), Zussman-Brindley-Comer (1957) hanno gia mostrato come possono esistere vari "stati strutturali" nella regione di transizione fra forme tubulari e piatte.

La genesi piu probabile del minerale studiato sembra quella da originarie fibre di crisoltilo per cambiamento del l'ambiente chimico-fisico (facies); dapprima in un ambiente particolarmente ricco di magnesio si sono formate fibre ben sviluppate di crisotilo, poi, in seguito ad un relativo aumento del tenore di Al, Fe^{3+} e altri elementi, probabilmente per apporto da parte di soluzioni idrotermali, le fibre di crisotilo sono state modificate e si e originato il minerale con caratteristiche della varieta lizardite. Quest'ultima, per essere piu ricca in Al, Fe^{3+}, ecc. e quindi con tensione strutturale minore non presenta piu fibre ben sviluppate come quelle di crisotilo, bensi piu corte, fessurate e srotolate; l'eccesso di H^+ determina il conservarsi e persistere di alcune forme ancora chiaramente tubulari, nonostante le sostituzioni nei pacchetti ottaedrici e tetraedrici.

Potrebbe anche avere avuto luogo una contemporanea ricristallizzazione in seguito a fenomeni dinamometamorfici; la formazione ofiolitica e tettonicamente molto disturbata ed alcune anomalie ottiche (estinzioni ondulate, piegamenti e fratturazioni di alcuni costituenti) riscontrate all'esame microscopico nella roccia serpentinitica avvalorano questa assunzione. Gio avrebbe sicuramente comunque favorito i processi di trasformazione sopra detti".

In contradistinction to the element mobilisation for the formation of lizardite more pronounced geochemical mobilisation takes place in the lateritization of serpentine. Schellmann (1964) summarised the process as follows:

"A well preserved laterite profile of 7,5 m thickness in Kalimantan (Borneo), permits a deeper insight to the nature of lateritic weathering. This profile, complete without any gaps from the weathered top layer to the serpentinite lying below, was studied adopting X-ray, differential-thermal, microscopic, electronmicroscopic, chemical and soil chemical (cation-sorption and pH) methods. The chemical transport was calculated quantitatively by using Cr_2O_3 as internal standard.

A fresh formation of geothite, gibbsite, montmorillonite, chlorite and kaolinite is noticed. The pH value moves from alkalic on the upper surface of serpentinite to acidic on the surface of laterite. This change is in concordance with the formation of montmorillonite in the initial phase and the kaolinite formation in the end phase of lateritisation.

The percentage of Fe_2O_3 falls with increasing depth from 70 % to 10 % while SiO_2 and MgO increase from a small quantity to as much as 40 %. The chemical displacements are caused by the removal of components by weathering and this is demonstrated by the quantitative calculations made for chemical

* Coleman (1977) considers serpentinisation in addition to low temperature formation also due to internal metamorphism of the ophilite assemblage.

transport. Further, Fe_2O_3 is transported from the surface to the deeper parts of the profile by descending water.

The principal weathering and intensive removal of SiO_2 and MgO are confined to a small zone of 0,5 m bordering the serpentinite. At least 45 m of serpentinite was weathered to form a 7.5 m thick layer of laterite.

The fine needle like structure of geothite, as seen in electronmicrographs, explains the larger volume of the pores and the low density of the investigated high-grade lateritic iron ores."

The geochemistry, mineralogy and physicochemistry of the lateritisation of serpentine is a rather complex and special subject and it is not included as a topic of the present Atlas. Though it should be mentioned that within the complex of the geochemical mobilisations and new mineral formations garnierite and other secondary nickeliferous minerals may be formed, Siegel (1954), Augustithis (1962b), Kurzweil (1966) and Albandakis (1974). Also within the complex lateritisation element mobilisation processes in the sense of differential leaching of elements from the weathered ultrabasics, a material source of bauxite may be formed.* Aronis (1952/53) in discussing the derivation of Greek Bauxites has produced geological and mineralogical evidence (i. e. residual chromite grains indicating alterations and occurring as granules in banded bauxites). In addition Erhart (1973) and Thodoropulos (1971) similarly support a bauxite origin from an initial ultramafic source.

Considering Pieruccini's gradoseparation hypothesis, silica and alumina will be among the residual compounds of gradoseparation of initial ultrabasic material. The gradoseparation hypothesis in the sense of differential leaching of elements could provide a theoretical interpretation of the bauxite derivation.

* Makŝimovic and Crnković Bozìca (1968) consider altered ultramafics as a partial source for the origin of bauxite in addition to schists.

Chapter 28 Textures of olivine serpentinisation

One of the most significant alteration processes of olivine and olivine rocks is the serpentinisation of the olivine and the transformation of olivine rich rocks to serpentine.

The alteration of forsterite → antigorite involves a considerable volume increase. The serpentinisation of olivine can be seen as a process which brings expansion of volume, consequently when olivine is enclosed by pyroxene and when alteration of the olivine takes place to serpentine, cracks may develop transversing both the olivine and the surrounding pyroxene. It should be pointed out that these cracks are occupied by serpentine, see Fig. 686.

In certain cases, olivine may be serpentinised (transversed by antigorite veinlets) and may be surrounded by marginal pyroxene which in turn may be enclosed in plagioclase, Fig. 687. It should also be pointed out that not only the olivine but also the plagioclase is transversed by antigorite veinlets as is shown in Fig. 687.

In contrast, cases are observed where serpentinised olivine may be surrounded by pyroxene. In contradistinction to Fig. 686, in this case no serpentine veinlets extend from the olivine into and transversing the pyroxene.

Rarely serpentinised forsterites transversed by antigoritic veinlets may be surrounded by a thin margin of pyroxene which is an extension of a marginal pyroxene, see Fig. 688.

The serpentinisation of olivine in cases follows cracks in tectonically deformed olivine, Fig. 689. In addition to the antigorite, iron oxides are mobilised in the olivine crack partly representing iron oxides as residuals of the olivine serpentine "alteration". The amount of iron pigments mobilised within the serpentine cannot be accounted for by the amount of serpentinisation present within the olivine cracks and has to be explained by a topo-metasomatic mobilisation.

In contradistinction to the serpentinisation which is following cracks at random, the serpentinisation may follow a pattern of cracks traversing several of the olivine crystal grains of the dunite, the end result will be a parallelism of the outlines of the residual olivine grains (see Fig. 690(a)).

Similarly, antigoritisation may follow a set of "cleavage" of two perpendicularly intersected directions, see Fig. 690(b). A geometrical pattern of antigorite veinlets in residual olivine may thus be developed.

All transition phases may be seen from olivine with only a few antigorite veinlets to textures where only a few residual olivine grains may be left, see Figs. 691(a) and 691(b). In contradistinction to Fig. 691(a), where the olivines are scattered at random in a predominant mass of serpentine, Fig. 691(b) shows scattered olivine grains transversed and altered by a pattern of serpentine veinlets.

As already pointed out, the serpentinisation of the olivine rock also depends on the tectonic influences which the olivine rock has been subjected to prior to the serpentinisation. Fig. 692 shows dunite with ore-mineral (magnetite) crushed and invaded by serpentine veinlets. As is shown by arrows (a), Fig. 692, the serpentine veinlets extend between the intergranular spaces of the olivine crushed-granular aggregate. The co-existence of "crushed" olivine mass and lens shaped olivines in antigorite is interpreted as tectonically cataclastic dunite antigoritised in which case the lens shaped olivines are relics of cataclastically affected olivine as well as the cataclastic masses of olivine in the serpentine.

In contradistinction to the geometrical pattern of antigorite veinlets transversing an equigranular olivine texture, coarse grained olivines may show a parallel pattern of cracks along which antigoritisation takes place. The antigoritisation of coarse grained olivine may follow a randomly orientated crack pattern.

Considering serpentinisation, the alteration of the olivine to give rise to antigorite is also accompanied by topo-metasomatic iron oxide mobilisation. As a balancing of the "reaction" olivine → antigorite shows, the Fe which is not accommodated in the lattice of the antigorite is topo-metasomatically mobilised and forms iron oxides granules or pigments which are often following (in cases restricted, (Fig. 693) in the initial olivine grain boundary). A chemical autocatharsis of a sort may take place, whereby the iron-pigment-formation and granules and their enrichment may show all transitions within an olivine grain under alteration, see Fig. 693.

The topo-metasomatism of iron oxides and the concentration of the iron-pigments and granules is not restricted only to the margins of olivine. Mobilisation in the sense of topometasomatism of iron oxides may result in local concentrations.

The serpentinisation of olivine often is as serpentine fibres and fine serpentine, in contradistinction Fig. 694 shows olivine serpentinisation whereby large serpentine laths are formed.

Fig. 695 shows olivine-alteration with a margin of iron oxides-pigmentation at the front of olivine alteration to serpentine (see arrow 'a' of Fig. 695). The serpentine fibres are perpendicular to the initial boundaries of the olivine in which iron granules have been segregated.

The serpentinisation of the wehrlite often exhibits most intricate textures and patterns. Fig. 696(a) illustrates "cell-structures" of serpentinised olivine whereby the initial olivine grains are replaced by antigorite fibres which radiate from the centre of the "serpentine cell" structure. As Fig. 696(a) shows, iron granules indicate the initial boundaries of the olivine. Another instance of a cell serpentine structure again consisting of fibrous serpentine radiating from a centre, a sectorial-optical effect is produced due to the orientation of the antigorite fibres (see Fig. 696(b)).

The serpentinisation of wehrlite may result in the most complex antigorite replacement structures which may be pseudomorphs of the initial olivine grains or may be entirely recrystallised (see Fig. 697).

In addition to the typical structures of olivine replaced by serpentine, Fig. 698 shows parallelly developed antigorite "bands" in serpentinised dunite.

Most interesting replacement of original dunitic texture may take place by antigorite, in which case the cell structure of the dunite is preserved and is partly substituted by the relic iron-oxides (hydroxides) of the original iron in the olivine, after partial mobilisation of Mg, see Fig. 699.

Another instance of dunite serpentinisation is exhibited in Fig. 700(a) where the original cell-structure of the olivine is replaced by antigorite which in turn is transversed by an antigorite veinlet banded and "showing a pattern of folding" (comparable is the textural pattern shown in Fig. 700(b)). Of particular significance are the "veinlets" of chrysotile asbestos common in serpentinised dunite, see Fig. 701. Comparable is the network of antigoritic veinlets in a serpentine mass, see Fig. 702.

Serpentinisation may develop also in pyroxenes particularly orthopyroxenes whereby the antigoritisation may follow a crack pattern within the pyroxene and also replace it along cleavage directions (Fig. 703). In cases, the serpentinisation within pyroxenes may form veinlets with side extensions, Fig. 704.

In addition to olivine serpentinised, antigoritisation of enstatite may take place, often following a cleavage direction of the host (Fig. 705). Furthermore, Fig. 706 shows an initial pyroxene completely serpentinised and transversed by an antigoritic veinlet.

Often in serpentinised rocks relics of garnet may be present which could represent xenocrysts in the original rock prior to serpentinisation. Fig. 707(a) shows a rounded pyrope xenocryst in serpentine. Similarly, Fig. 707(b) shows a garnet with a corona reaction margin in serpentine with veinlets of serpentine extending into the garnet.

Chapter 29 The reaction chromite-serpentine

We have seen so far the serpentinisation of olivines and pyroxenes, whereby these minerals have been replaced by serpentine. In contradistinction the chromite grains in the dunites may show a phenomenology which indicates a preexisting chromite attacked by the later-formed serpentine.

"Intergrowth" structures of serpentine-chromite, decoloration rim of the chromite minerals.

Ore-microscopic observations show chromite grains in serpentine. Often the outer part or patches of the chromite show distinct decoloration and usually with it are present fine worm-like intergrowth structures of serpentine. These structures are not due to simultaneous crystallization, but later serpentine penetrating into the chromite. Besides this type of intergrowth, oil-immersion observations show that serpentine exists as binding material of fine grained chromite, the aggregate gives the impression of an individual with fine branching intergrowth structures, which in reality are intergranular space fillings. Another type is the re-binding "intergrowths" in which the serpentine acts as the binding material of cataclastic chromite.

The weight of observations establishes that there is a transition and interrelation between the serpentine existing as fine intergranular and the re-binding serpentine of cataclastic chromite. Also interrelation and transition exist between intergranular space filling serpentine and myrmekite-like structures. These relations and transitions of the one type to the other are illustrated in the photomicrograph, Fig. 708, also see Augustithis (1960).

In understanding the relation of the chromite with the serpentine, the nature of the latter is of importance. Oil-immersion observations distinguish a brownish-serpentine, in continuity with the main serpentine mass, which is more greenish with stronger internal-reflection. The brown type serpentine differs in no way from the main mass, it is only to be distinguished by its different reflection (Reflexionsvermögen). It must be pointed out that the recognition of this mass as a distinct unit is more on structural basis than on mineralogical differences. It is mainly this brown serpentine that occurs as intergrowth structures.

The serpentine "myrmekitic" intergrowths in spite of their resemblance to the true myrmekite observed in granitic and gneissic rocks, are different. In the proper myrmekite we have younger orthoclase, coming in contact with pre-existing albite-plagioclase and as a result of the synantetic reaction of these two minerals we have a third formed, quartz; which in minute worm-like structures penetrates into the plagioclase. In the case of the chromite serpentine, worm-like bodies occur in the chromite and clearly they are not a synantetic reaction product of two different minerals. Observations distinctly show that the serpentine is later than the chromite, surrounding the pre-existing grain and penetrating from its margin inwards, extending inside as far as there is a decoloration zone in the chromite. Also indentation of chromite margins due to corrosion of later serpentine have been microscopically observed. Photomicrograph, Fig. 709, shows later serpentine penetrating into chromite grain.

It can be seen that where there is "myrmekitic-serpentine" there is decoloration of the chromite, whereas if there is decoloration it is not necessary that there is present "myrmekitic intergrowths" of serpentine. It is observed that marginal areas of chromite without serpentine intergrowth structures show decoloration, photomicrograph, Fig. 710(a) indicating that the change of colour of the chromite is independent of the presence of "myrmekitic" serpentine. Often chromite fine aggregate with only intergranular serpentine, no true intergrowths present, shows distinct decoloration compared with larger mineral grains. Further, the same serpentine mass within the same chromite grain, passing from re-binding material to myrmekitic structures, shows only decoloration in the areas where worm-like bodies of serpentine are present, see photomicrograph, Fig. 710(b). It seems that in most cases, the change of colour is due to contact-reaction of the chromite mineral grains as a whole with the surrounding serpentine, a product of later serpentinisation.

The above observations suggest that the chromite can be affected by serpentinisation, but whether the decoloration is caused by synantetic reaction of the grain as a whole, or by the penetration of the minute serpentine worm-like bodies, is difficult to be sure of in every case. Most careful ore-microscopic observations with crossed nicols show that the areas indicating decoloration are slightly anisotropic.* Both the change in colour and slight anisotropic effect of the chromite can be understood as a loosening (Auflokkerung) of the lattice or intracrystalline space enlargement. Allowing thus to suggest, that because of the loosening of the lattice, an accommodation of the serpentine intergrowths in the chromite could take place, and thus to explain the observation that, wherever there is myrmekitic serpentine present, there is decoloration.

* See chapter 30.

Chapter 30 Differential leaching of elements from ultrabasic rocks and birbiritisation of dunites

Generally, element-leaching by "hydration" is considered to be a secondary process often involving mineral and rock disintegration. In constrast, examples are given in which no rock destruction has taken place, in the sense that the altered phase is a hard and compact rock, e. g. birbirite.

The release of elements from a crystal lattice gives rise to crystal changes which may be noticeable as changes in the physical properties and composition. With progressive element-leaching, more pronounced alterations take place and, as a result, new minerals and textures may develop.

Element leaching is essentially a chemical process, but only a postulation of the reactions is possible, based on mineralogical and chemical comparisons of the mineral phases involved. In addition, it is a chemical differentiation, with preferential mobilisation of some elements over others from a mineral or rock. No "rules" or "laws" seem so far to explain the overall differential leaching of elements – the experiences gained are mainly empirical. At this point, only a few cases of element leaching are selected for consideration: Mg-leaching in the transformation of dunite to birbirite, alteration and decoloration of chromites due to element leaching (decoloration rims).

In addition to chemical comparisons of the unaffected and leached phases, a study of the phenomenology is included since element leaching involves mineralogical and textural changes. Detailed microscopic observations on mineral alterations, relic structures of mineral transformations and replacement textures are considered basic for an understanding of the process.

Another process – which is, however, geochemically the reverse to differential leaching – is the element agglutination (accretion) which takes place in special geochemical environments. In a paper by Ottemann and Augustithis (1967) an example of element agglutination is considered in which platinoid elements agglutinate in an environment of lateritisation to form nuggets. In chapter 24 an attempt is made to explain the process of element agglutination (accretion) on the basis of the chemical affinities and relations which exist between the platinoid and other elements. These affinities are outlined by the "empirical laws" of the periodic system, see also chapter 31.

Mg-leaching (birbiritisation)

The significance of relic gel-structures in birbiritisation.

The leaching of Mg from dunites results in a rock which consists essentially of fine-grained quartz, chalcedony, limonite, and partly preserved chromite grains. Microscopic comparison by Augustithis (1965) of the dunite and birbirite of Yubdo, Ethiopia, has provided evidence of this transformation. In addition, a net-number of counts comparison of the same rocks by Ottemann and Augustithis (1967) has similarly shown that Mg-leaching is associated with the transformation of the ultrabasic to birbirite (Table VIII).

Experimental olivine alteration by Dittmann (1959) has shown that Fe^{3}-hydroxide, $Mg(OH)_2$ and colloform silica are produced. Microscopic observations on a wide range of birbiritic rocks from Yubdo/Birbir, Daleti/Wollaga, Mobassi-Asossa/Beni-Shangul and Chalalaka/Chambe (Adola) – all Ethiopian occurrences – show that colloform or relic gel-structures are common. On the basis of textural evidence it is postulated that the birbirite has passed through a silica gel-phase, subsequent crystallization of which has resulted in fine-grained quartz.

The phenomenology of the transformation of dunite to birbirite is difficult to interpret. Fig. 711(a) shows preserved olivine relics in serpentine, though other olivines have been altered to colloform fine chalcedony and to a semi-translucent mass stained by rich iron pigmentation (see arrow 'a',

Fig. 711(a)). Though it should be pointed out that despite the complex processes involved, the granular textural pattern of the initial olivine is preserved.

In contradistinction to Fig. 711(a) which actually shows the first and end phase of the birbiritisation, Fig. 711(b) shows an initial olivine granular texture in serpentine, whereby the physical properties of the "olivine" have been greatly changed.

Due to element leaching the olivine has been "synisotropised" and has changed to a grey mass almost "isotropic in optical behaviour". The synistropisation should be interpreted as a state of the break down of the olivine lattice due to leaching out of Mg.

Table VIII, Net number of counts comparison under equal conditions.

	Ni	Fe	Mn	Cr	Ca	K	Si	Al	Mg
Dunite	160	> 220	45	16	16	8	155	–	150
Birbirite	50	> 220	15	14	10	11	230	–	–

A series of photomicrographs illustrates the transformation of silica-gel to quartz. Fig. 712 shows a "veinlet" of chalcedony with quartz grains in the central part. Relic gel-structures are preserved simultaneously in several quartz grains. Both the chalcedony and the quartz have been formed from colloform solutions with subsequent crystallization to quartz.

Arrows (a) in Fig. 713(a) and 713(b) show a quartz individual with relic gel-structure simultaneously exhibiting a texture resembling crystal zoning. In this particular case it is difficult to differentiate between gel-relics and zonal growth. It is perhaps possible that the central part of the grain has grown as a crystal nucleus (a crystalline outline is shown). However, in the outer part of the same quartz, relic gel-structures are present (Fig. 714). Similarly, Fig. 713(a) shows structures in quartz grains which despite a geometrical outline might represent an original gel-structure. Additional observations (Fig. 713(b) arrow b) show relic gel-structures traversing different quartz individuals. (The gel structure is a relic of a colloform phase and is preserved in the individual quartz grains after the transformation of silica-gel to quartz).

The transformation from a gel-phase to a crystal is difficult to understand. It has been proposed by the author (Augustithis 1964) that the transformation of gel to crystalline uraninite involves an intermediate solution-phase which facilitates the rearrangement of the atoms from an amorphous state to a crystal lattice. Whether a gel-phase can change to a crystal without an intermediate solution-phase is difficult to answer. Levicki (1955) has shown that silica gel can change to quartz, leaving the original gel-structures preserved as gaseous inclusions in the crystal. Commensurably, Fig. 713(b) shows original colloform textures preserved as relics in quartz aggregates, whereby the quartz individuals show different extinction positions. The transformation of dunite to birbirite consisting essentially of fine quartz, limonite and chromite (see Figs. 715(a) and 715(b)) is generally understood on the basis of the hypothesis of gradoseparation* as put forward by Pieruccini (1962).

In femisialic compounds (e. g. olivines) the femic components are dissolved and the sialic are residual; that is, in the case of olivine alteration, silica is the residual phase. In the case of birbiritisation Mg is preferentially mobilised compared to Fe.

Chromite grains in birbirite subject to alteration and leaching processes.

As pointed out above, partly altered chromite grains are preserved in the birbirite, representing relics of the transformation of dunite to birbirite. Whereas chromite grains are relatively resistant to alteration processes, certain alterations (under low temperature hydration conditions) are nevertheless known to take place (Augustithis, 1960, 1962, 1965).

Fig. 716 shows a chromite grain in the birbirite, with corroded and leached margins. Also decoloration patches are present which are closely associated with the corrosion margins of the grain. Comparable decoloration margins and patches have been described by Augustithis (1960) in chromites from Rodiani, Greece, as being magnetite-like in reflection. The transitions from intergranular to intragranular

* Gradoseparation = gradual separation of elements.

serpentine (myrmekitic in intergrowth), the association of the decoloration margins with these intergrowths and, in addition, an extensive study of the phenomenology, lead to the conclusion that the decoloration margins and patches are due to alteration of the chromite. In contrast to this mode of origin, magnetite margins can be formed around chromite xenocrysts in basalts due to magmatic corrosion (Augustithis, 1965). The latter case, however, shows a diverse phenomenology.

Panagos and Ottemann (1966) on the basis of microprobe analysis (electron scanning comparisons) of the Rodiani chromite and its decolorised margins have determined that the chromite cores are relatively rich in Mg and Al compared to the margins, which in turn are richer in Fe and Cr. Furthermore, these authors explained the chromite decoloration as being either connected with a later magmatic stage or as penecontemporaneous with the serpentinisation.

The most conclusive evidence that decoloration can be caused by leaching out of elements is provided by comparing unaffected chromite grains of the Yubdo dunite with the partly altered chromite grains present in the birbirite.

The chromite grains in the dunite show the following characteristics:

1. They are free from decoloration margins or patches and do not show solution canals or a corrosive appearance.
2. They are often surrounded by martitised margins of magnetite; (Fig. 717(a)). These should not however, be confused with the decoloration margins and patches of the chromite.*

In contrast the same chromite grains, when in the birbirite, show the following changes:

1. The chromite grains are partly corroded and dissolved or leached (decolorised).
2. The martitised magnetitic margins are changed to limonite and as such remain marginal to the chromite (Fig. 717(b) arrow a.).

* Some of these chromite grains are euhedral, and the martitised magnetite margins are distinctly a separate mineral phase.

Chapter 31 Low-temperature mobilisation of the Pt-group elements in lateritic covers (Synoptical discussion)

Lateritisation of platiniferous dunite and the low temperature mobilisation of the Pt group elements to form nuggets (consisting of a complex platinum-mineral paragenesis) has been described by Ottemann and Augustithis (1967) in lateritic covers of Yubdo, W. Ethiopia dunitic intrusion.

In contrast to this hypothesis of "element agglutination" proposed by Ottemann and Augustithis (1967) and Augustithis (1967b) for the formation of Pt-nuggets in the lateritic eluvial cover of the Yubdo (W. Ethiopia) dunite, Cabri and Harris (1975) suggest that the nuggets are primary growths in the dunite and that they are liberated in the eluvial lateritic cover after the disintegration of the dunite. According to Cabri and Harris the absence of ferroplatin in the size of nuggets (see Fig. 718) in the unaltered dunite is attributed to the platinum grains being far more dispersed in the unaltered dunite.

Studies by Duparc et al. (1927) and Augustithis (1965, 1967b) have failed to show the presence of primary Pt-group minerals of the size of the nuggets discovered in the eluvial cover. In fact Duparc believed Pt to be in the lattice of the olivine. Fire assay of the dunite, despite the absence of nuggets in the crushed rock, showed the presence of Pt (more than 1 ppm), which suggests that Pt-group minerals (perhaps more than sperrylite which was identified, Augustithis 1965, are present in the unaltered dunite of Yubdo). In the period of the early seventies the U.N. special fund* made an extensive investigation by drilling for primary Pt in the unaltered dunite of Yubdo. *No ferroplatin of the size of the nuggets* was found in the primary unaltered dunite. Pt-nuggets of the size and appearance illustrated in Fig. 718 are abundant only in the eluvial cover of the dunite.

Of course the mobilisation of Pt-group of elements under low temperature and the possibility of Pt-group mineral formation under low temperatures needs additional work. The chemical accretion explanation for the growth of nuggets in alluvial Pt-deposits by Cousins (1973, a and b) and Stumpfl (1974) seems to support the low temperature growth of the ferroplatin nuggets in the lateritic cover of the Yubdo dunite. Also the veinform occurrence of the Pt-group minerals in the sulphides of ultrabasic rocks supports the low-temperature mobilisation of Pt-elements (see Chapter 24) also in the case of the eluvial covers.

Also Cabri and Harris (1975) discussing our papers regarding the presence of altered and unaltered chromite grain in the ferroplatin nuggets, suggest that the Pt could grow around the chromite grains during its crystallization in the dunite. However, it should be noted, and this is emphasised in the present Atlas, that the ferroplatin nuggets contain altered chromite grains i. e. chromite grains showing distinctly that they have been chemically altered during birbiritisation or lateritisation (see Figs. 719 a, b, c) prior to their engulfment by the later ferroplatin. Many chromite grains, in addition to other less altered ones, are included in the ferroplatin nuggets. Chromite grains showing unmistakable evidence of alteration effects are completely enclosed by the ferroplatin (Fig. 719 a, b, c).

Also chromite with margins of martitised magnetite (the martitisation being due to oxidation) are entirely enclosed by ferroplatin nuggets. Also in this case the martitisation of the magnetite margins indicating oxidation-alteration of the magnetite are prior to the ferroplatin formation, see Figs. 720a, b.

As Duparc reported and as Fig. 719 (a,b) shows, the Pt nuggets are often surrounded by a dark limonitic cover, formed during lateritisation. Such limonitic coatings may surround chromite grains; as Fig. 721 shows, chromite grains with limonitic coat-

* United Nations – Ethiopian Mineral Survey Preliminary Report on Magmatic and Geochemical Surveys over the Yubdo Ultrabasic Complex, Yubdo Wallega (1969).

ing are entirely enclosed in the ferroplatin, which again supports a growth of the nuggets in an environment of lateritisation.* The limonitic coating may also be mobilised along cracks from the external part into the ferroplatin of the nugget. Also of interest is the presence of elongated silicates at random enclosed in the ferroplatin nuggets (see Fig. 722).

If the Pt actually occurs as ferroplatin grains of the size of the observed nuggets; ferroplatin grains of the size of the nuggets (some weighing more than a gram) should occur on the surface of the exposed dunite just below the lateritic cover, since this lateritic cover is the disintegration of the upper part of the exposed dunite. However, as mentioned this is not the case. In contrast platin in small quantities (as the fire assay showed) in abundance of 1 ppm occurs in the dunite below the lateritic cover in the form of sperrylite (Augustithis 1965) and other relatively small P-G-minerals. X-ray fluorescence spectroanalysis of the metallic minerals components (after the separation of chromite) of the opaque minerals of the dunite showed the abundance of Ni, Co and As which also occur in the paragenesis of the nuggets.

The shape of the nuggets, the fact that in cases rounded and altered chromite grains are enclosed in the nuggets, the absence of ferroplatin grains in the ultrabasic that would compare with the nuggets and in general the "intergranular shape" of some of the nuggets (in the sense that they surround corroded chromite grains), see Fig. 719a, b, c, supports a growth under lateritic conditions.

The Pt-elements present in the ultrabasics as sperrylite (and possible other minerals) disintegrated under lateritisation (though it persisted birbiritisation) and by low temperatures "hydration" solution agglutinated and formed the nuggets including the other platinum group minerals present in paragenetic association with the "ferroplatin".

Considering the element-geochemistry of the minerals building up the nuggets it can be seen that the Pt-minerals paragenesis consists of elements of the subgroup of the 8th family of the periodic system (see chapter 24).

* In addition to the present argument in support of the accretion hypothesis Cousins and Kinloch (1976) state the following: "Physical evidence for the stage of accretion ("element agglutination" of Ottemann and Augustithis, 1967) is, in the opinion of the authors, evident in the rounded and sculpted form of alluvial and especially eluvial platinoids, in the large size of the grains (six or seven times larger than Merensky Reef platinoids), and in the zoning evident in microscopic and electron probe studies of grains (Stumpfl, 1974; Cabri and Harris, 1975; Barrass, 1974; and this paper)."
Also Stumpfl and Tarkian (1976) consider that the formation of platinum deposited is dominated by polyphase processes which also include transportation of platinum group elements in aqueous solutions. In addition McCallum et al. (1976) discuss the mobility of the precious metals Pt, Pd, Rh, Au and Ag in supergene processes. Synoptically they state the following: "A comparative study of the distribution and geochemical behaviour of the precious metals Pt, Pd, Rh, Au, and Ag in the weathered portions of the deposit has shown that Pt and Rh are substantially enriched in the strongly oxidized ore horizons, possibly due to supergene processes. Pd and Ag have been extensively mobilised from the upper levels during weathering. Ag has undergone dramatic enrichment in the supergene sulfide zone, but Pd apparently has been removed from the system."

Chapter 32 Alteration of dunite, magnesite formation

In contradistinction to the alteration of olivine to produce serpentine, weathering-alteration of olivine may give rise to magnesite. Fig. 723 shows forsterite changing marginally and along cracks into magnesite.

Often the alteration (magnesite) of different olivine grains segregates and forms "veinlets" within the dunite rock (Fig. 724).

Augustithis (1965) schematically outlines the alteration of forsterite to form magnesite as follows:

Weathering of olivines – due to hydration – with the participation of carbonic acid (CO_2) of the air, results in the formation of magnesite as pseudomorphs after olivine. The origin of magnesite in stock-work and in veinlets in altered dunite has long ago been recognised as a product of the alteration and weathering of olivine. However, two alternative hypotheses have explained the origin of magnesites in ultrabasics in general. In contradistinction to the afore-mentioned hypothesis of weathering alteration process, veinform magnesite occurrences have been regarded as the product of hydrothermal magmatic solutions, at a concluding phase of the basic and ultrabasic magma.

For the understanding of the origin of magnesite stock-work in altered dunite and for the understanding in general of the presence and origin of magnesite, particularly with dunites and "serpentinised dunites" (altered dunites), a geochemical understanding of the alteration processes of olivine is necessary, see Augustithis (1965, 1967, 1978).

Experimental work by De Noske Saltverker Av Bergen and Dittmann (1959/60) has shown that olivine was altered by weak acids if concentrated $MgSO_4$ solution was added. Furthermore, by heating at temperatures of 80–100°C the olivines were completely solved.

Additional experimental work in this line by Dittmann has shown that by the application of different concentrations of $MgSO_4$ solutions at a temperature range of 100–300°C i.e. within the range of hydrothermal solution, the solubility of olivines can be achieved without any addition of acids and at lower concentrations of $Mg^{++}SO_4$. In these experiments the role of $MgSO_4$ has been considered to be catalytic.

According to Dittmann, in addition to Fe-III-hydroxide, brucite Mg(OH) and colloidal silica are produced. These products are the result of olivine alteration, whereby the $MgSO_4$ has acted as a catalyst.

Considering the schematic reaction:

a) "Alteration of olivine due to hydration (weathering of dunites) with the participation of carbonic acid (CO_2) of the air, results in the formation of magnesite as pseudomorphs after olivine", and,

b) The experimental work of Dittmann (1960), an understanding can be reached of the observed mineral phases, textures and structures associated with the presence of magnesite and altered dunitic rocks.

Often magnesite veinlets transverse serpentinised dunite and an intimate association of the veinform magnesite with the antigorite is shown (Fig. 725). In other instances veins of magnesite intersecting one another are observed (Fig. 726).

The mechanism: olivine alteration → generation of magnesite and subsequent mobilisation of magnesite explains satisfactorily the stock-work veinlets (Fig. 727(a)) and the formation of magnesite "potatoes" (Fig. 727(b)) and rounded large magnesite concentrations in altered dunitic rocks.

In both cases the magnesite occurs associated with the so-called "yellow compact serpentine" which in reality is a birbiritic rock [see Duparc, et al. (1927) and Augustithis (1965), also chapter on birbiritisation (chapter 30)].

Fig. 728(a) shows most of the phases involved in the process: dunite → birbirite → Mg liberation and reaction with carbonic acid of the air to produce magnesite and the presence of residual silica which from an initial colloform phase is now represented by granular quartz.

Furthermore, the magnesite is "banded colloform" next to the birbiritised (dunite) and as colloform structures associated with the granular quartz of the birbirite. The colloform texture of the magnesite in the birbirite is also illustrated in Fig. 728(b).

Birbiritisation of dunite and atmospheric CO_2 are factors that explain the origin of magnesite and its common association with "the yellow variety of serpentine". In other instances, as pointed out, direct alteration of olivine may give rise to magnesite (see Fig. 723).

Chapter 33 Lateritisation processes of serpentine* and altered dunitic rocks

The lateritic covers of ultrabasic rocks may be either as lateritic eluvials (in situ lateritisation of dunites or serpentinites) or may be transported and redeposited, Siegl (1954), Augustithis (1962b), Kurzweil (1966), Agiorgitis (1973) and Albandakis (1974).

Both types of laterites are of economic importance for Ni (third phase of Ni-distribution – see chapter 22).

As mentioned the "supergene" mobilisation of the Pt-elements has resulted in the Pt-nugget mineral paragenesis (see chapter 31). In addition to the Pt-group elements mobilisation, Cr in addition to its mechanical distribution within the laterite particularly in the re-deposited lateritic types – see Siegl (1954) – may be geochemically mobilised as the studies on the chromite minerals and their alteration processes show during diagenetic processes in the lateritic iron Ni-Cr oolitic deposits of Larymna, Greece, see Augustithis (1962b) and Kurzweil (1966).

The mobilisation of Mg, Ni, Cr and Pt minerals in laterites is complex and despite similarities the complexities of geochemical conditions may be influenced by the climatological and in general physiographical conditions prevailing in a region during and after lateritisation.

Of particular interest is the case of lateritic covers of the Yubdo ultrabasics in western Ethiopia. As mentioned the Yubdo dunites have partly changed into birbirites. Ottemann and Augustithis (1967) have made a geochemical comparison between birbiritisation and lateritisation, the following is an excerpt from their work:

"Textural studies of the ultrabasics and birbirite of Yubdo/Birbir, W. Ethiopia (Augustithis, 1965) suggest that the latter is a derivative product of the dunite, mainly by leaching out of Mg. X-ray fluorescence spectroanalysis, using a relative net-number of counts comparison of dunite and birbirite, shows that the leaching out of Mg was the important process in the alteration and transformation of dunite into birbirite".

The process of "birbiritisation" (i. e. the alteration and transformation of dunite into birbirite) in its broad lines conforms to the hypothetical process of "gradoseparation" as put forward by Pieruccini (1962). Under the concept of gradoseparation the silica in the birbirite is thought as having been derived from the olivines through leaching, by which the femic components are removed from the minerals (separation) and with the sialic components remaining. New exploratory studies have revealed several new occurrences of "birbirite-like rocks" in southern and western Ethiopia, comparable in texture and composition to the original birbirite described by Duparc et al. in 1927 from Yubdo/Birbir.

As a corollary to the textural studies, the above described relative net-number of counts comparison of birbirite and dunite supports the derivation of the former from the dunite (Table VIII). However, it should be mentioned that the most decisive evidence of this transformation is obtained on the basis of detailed textural studies. The derivation of birbirite from dunite is supported and substantiated by the presence of chromite grains, often with martitised magnetitic margins, in both the dunite and birbirite; however, in the case of birbirite the chromite grains show alteration effects and the magnetitic margins can be changed into limonite (Augustithis, 1965). Furthermore, field and petrographic studies suggest that the lateritic covers of the ultrabasics and birbirites have been derived from these rocks through alteration (whereby a comparable Mg-leaching out has taken place as in the case of the transformation of the dunite into birbirite). This is supported by comparison of the net number of counts (Table VIII). This

* See also chapter 27.

comparison with dunite, pyroxenite, birbirite and with several samples of the lateritic cover has been carried out and produced the following results (see Table IX).

It should be pointed out here, that the platinum nuggets described (see chapt. 31) occur in these lateritic eluvial covers and are structures formed during the process of early alteration stages or lateritisation. Thus in contrast to any high temperature conditions, the formation of the "ferroplatin" osmiridium, "roseite" and other platinoid minerals has taken place by "element agglutination"* under a phase of hydration (low temperature) in the environment of early alteration-stages of the ultrabasic rocks (see chapt. 31).

Studies of the in situ lateritisation of serpentines by Schellmann (1964) provide a satisfactory explanation of the differential leaching of elements and compounds under the lateritisation of serpentines and serpentinised dunites.

"The lateritisation of the serpentines is quickly accomplished and new weathering minerals are formed, the most important are geothite, meghemite, hydrargillite and the clay minerals montmirillonite, chlorite and kaolinite. In the boundary zone between serpentine and the lateritic cover a slightly alkaline pH-value dominates due to hydrolysis. In this zone leaching of compounds takes place, particularly MgO, also SiO_2 and to a lesser degree Al_2O_3 and Fe_2O_3. This sequence can be understood in

Table IX, Net number of counts* comparison under equal conditions

	Zr	Sr	Ni	Fe	Mn	Cr	Ti	Ca	K	Si	Al	Mg
Dunite	–	–	160	> 220	45	16	–	16	8	155	–	150
Birbirite	–	–	50	> 220	15	14	–	10	11	230	–	–
Pyroxenite	–	15	12	> 220	15	10	25	230	30	230	8	90
Laterite (1)	–	–	75	> 220	85	110	40	36	26	45	11	–
Laterite (2)	Traces		120	> 220	125	85	22	35	16	25	7	–
Laterite (3)	Traces		65	> 220	66	55	50	48	40	82	20	–

* Net number of counts = ratio of counts after background correction.

terms of the solubility of the compounds. It is possible that the leaching out of Mg is as Mg^{2+} ions in a lateritisation environment about pH = 8. In contrast SiO_2 is soluble in a wide range of pH values and under these lateritisation conditions would be leached out.

The behaviour of Fe and Al is different, Fe^3 and Al^3 in a neutral environment precipitate as hydroxides and in this environment in slightly increased concentrations form Fe and Al soles which are easily coagulated. The enrichment of Fe and Al in the laterite is thus understandable since colloid particles are less transportable in solution compared with ions."

As X-ray fluorescence comparisons by Ottemann and Augustithis (1967) see table IX, between the unaltered dunite and pyroxenite of the ultrabasic complex of Yubdo and their lateritic cover show, in the lateritisation of the Yubdo complex, Mg is almost completely leached out and to a lesser extent Si is also leached out in constrast Al is slightly increased. These results are in accordance with the hypothesis of Schellmann and with the hypothesis of gradoseparation of Pieruccini (1962) and they rather support that in the differential leaching of elements the solubility of the compounds under the prevailing milieu of pH values of lateritisation may play a fundamental role.

The following is a tentative list of the mineral .phases, so far determined which constitute the Yubdo Pt-nuggets (see chapt. 33).

1. Chromite nuclei of the nuggets (partly altered remnants of the dunite) essentially: $FeCr_2O_4$.
2. Limonite, $2FeO_3 . 3H_2O$ (as coatings of the nuggets; some Mn may be present).
3. "Ferroplatin", Pt, Fe, Au and a little Co.
4. Osmiridium: Os, Ir.
5. Roseite, Os, Ir (S).
6. Mineral (a)** consisting of Ni, S, Pd, Rh and Fe.
7. Mineral (b), consisting of Ni, Co, Fe and little Pd.
8. Mineral (c), consisting of Rh, Pd and Pt.
9. Mineral (d), consisting of Ru, some Rh and Pd.

* (accretion)

** The minerals (a, b, c and d) have been determined in the nuggets by microprobe analysis (electron-scanning), however no other data are given.

A geochemical consideration of the elements constituting the minerals which form the nuggets shows that these elements are interrelated and show affinities in accordance with the empirical laws of the periodic system. On the basis of the composition of the minerals forming the nuggets, the following metallic mineral-building elements are recognised: Pt, Ir, Os, Pd, Rh, Ru, Fe, Co, Ni (i. e. elements of the subgroup of the VIII family of the periodic system) and, in addition, Mn, Cr and Au.

Considering the periodic table (Remy, 1961) where the elements are grouped on the basis of the "inert" gases ("Gruppierung der Elemente um die Edelgase") in Table X it can be seen that the elements Pt, Ir and Os are interrelated, in that they are next to each other and belong to the same subgroup of the VIII family, and similarly the elements of the groups Ru, Rh, Pd and Fe, Co, Ni. All these elements are interrelated because of an increase of the horizontal and vertical affinities of the elements in the sub-group of the VIII family (as stated in the "empirical laws" of the periodic system). The elements Mn and Cr are related, being that they are next to each other in Table X and are similarly related to the group Fe, Co, Ni. Also, Au is related to Pt, being adjacent in the periodic table.

Considering the relative abundance in the earth's crust of the mineral building elements present in the nuggets (Table VII), it can be seen that very rare elements have segregated jointly to form these nuggets in a lateritic milieu.

The lateritisation of the ultrabasics often results in nickel and iron deposits which can be of economic importance. In addition to the re-deposited lateritic Ni-Fe deposits of Larymna and Euboea, Greece, Siegl (1954), Augustithis (1962), Kurzweil (1966), Agiorgitis (1972/3), and Albandakis (1974); Coleman (1977) summarises the main types of laterite Ni-Fe deposits as follows:

"Nickel and iron laterite deposits are found in Cuba (Kemp, 1916; De Vletter, 1955), Guatemala, Colombia, Oregon (Hotz, 1964) New Caledonia (Chetelat, 1947; Routhier, 1952). Indonesia (Reynolds et al., 1973), Oman, Philippine Islands, New Hebrides (Coleman, 1970) and the Solomon Islands (Coleman, 1970). The laterites are almost invariably developed on partially serpentinized peridotites or on unsheared completely serpentinized peridotites. Laterites have not been reported on highly sheared and tectonized serpentinites because serpentine is much more stable than olivine in the weathering environment. The laterites can be placed into two broad types (Hotz, 1964).

1. Those deposits developed on serpentinized peridotite such as those in Cuba and the Philippines are called nickeliferous ferruginous laterites and contain approximately 40% Fe and average about 1% Ni. These deposits represent large reserves of low grade iron-ore and the nickel is highly dispersed.

2. The second type of deposit is typically developed on peridotites that are weakly serpentinized such as those from New Caledonia and Oregon and are referred to as nickel silicate deposits (Hotz, 1964). The nickel silicate deposits are relatively low in iron (35% Fe) and contain garnierite (nickel silicate 15% NiO.) as the main ore mineral in the lower part of the weathered zone."

Table X

Subgroups of the Periodic System									
IV	V	VI	VII		VIII		I	II	III
easily changeable Valencies									
+IV	+V	+VI	+VII	+VII...			(+I)	+II	+III
Neither volatile nor salt-type hydrogen compounds									–
$_{22}$Ti	$_{23}$V	$_{24}$Cr	$_{25}$Mn	$_{26}$Fe	$_{27}$Co	$_{28}$Ni	$_{29}$Cu	$_{30}$Zn	$_{31}$Ga
$_{40}$Zr	$_{41}$Nb	$_{42}$Mo	$_{43}$Tc	$_{44}$Ru	$_{45}$Rh	$_{46}$Pd	$_{47}$Ag	$_{48}$Cd	$_{49}$In
$_{72}$Hf	$_{73}$Ta	$_{74}$W	$_{75}$Re	$_{76}$Os	$_{77}$Ir	$_{78}$Pt	$_{79}$Au	$_{80}$Hg	$_{81}$Tl
$_{90}$Th	$_{91}$Pa	$_{92}$U	Trans-Uranium Elements			$_{93}$Np	$_{94}$Pu	$_{95}$Am	$_{96}$Cm
Elementary ions partly paramagnetic									
Forming mainly coloured elementary ions									

Chapter 34 Metasomatic alterations of ultrabasics (e.g. Rodingites)

In contradistinction to serpentinisation, magnesite-formation and birbiritisation (which are hydration processes) and which result in the disintegration of the ultramafic minerals, mainly olivines, more complex metasomatic processes may occur.

In zones of tectonic mobilisation – hydrothermal metasomatism may result in late formation (talc Mg_3 $[OH)_2/Si_4O_{10}]$). The talc formation is attributed to Mg-metasomatism under hydration solutions.

Augustithis (1960b) has described the formation of chromite margins of alteration where talc has affected chromite grains. In contradistinction to the decoloration margins described (see Fig. 710a and b Augustithis (1960), Panagos and Ottemann (1966) and Augustithis (1967b)), which are due to element leaching – serpentinisation, in the case of chromite alteration due to talc metasomatism more pro -nounced alteration effects are produced, i. e. changes in reflectivity and increase in the internal reflection, see Fig. 729 and its description.

In contrast to the Mg metasomatism (talc formation) often within the ultrabasics metasomatic hydrothermal process may result in the formation of rodingites, metasomatic alteration of ultrabasics in which hydrogrossular, diopside and vesuvian may be present.

Suzuki (1954), Bilgrami and Howie (1960), Majer (1960), Crnčević et al. (1962), Ducloz (1962), Müller (1962), Bilgrami (1963), Vuagnat and Pusztaszeri (1964), Cogulu and Vuagnat (1965), Korzhinskii (1965), Vuagnat (1965), Heflik and Zabiński (1969), Paraskevopoulos (1970), Quiser et al. (1970) and Demou (1972), have described rodingite occurrences.

Müller (1962) distinguishes between grossular-diopside and vesuvian rodingites, however, usually grossular-diopside-vesuvian rodingite are common.

Demou (1972) summarising the studies on rodingite and based on original observations, makes the following conclusions:

1. Rodingites are usually vein bodies, mineralogically mainly composed of hydro-grossular diopside and vesuvian.
2. The hydro-grossular may be alteration derivative of initial feldspars, without precluding a derivation from diopside.
3. Diopside and vesuvian may be new growths in rodingite (diopside, vesuvian metacrystals).
4. The CaO of the rodingites could derive due to the disintegration of Ca rich minerals such as diopside or due to Ca rich solutions.
5. Rodingites are related to serpentinisation and are formed by hydrothermal solutions.

Present studies of rodingite occurrences from the region of Pagoda (Euboea) show veinform or diffused rodingite bodies, often in tectonic zones or tectonically affected in ultramafics, mainly pyroxenitic and dunitic rocks.

A series of photomicrographs (Figs. 730a and 730b) shows later metasomatic hydro-grossular invading and replacing a pre-existing pyroxenitic or partly serpentinised mass. Diopsides or vesuvians may persist as relic in the metasomatically formed hydro-grossular (a debyegram was obtained only after heating the hydro grossular at temperatures of more than 1200°C – the regenerated lattice as determined by the debyegram was that of grossular), (per com. A. Vgenopoulos, 1977).

Particularly as shown in Fig. 731, diopside shows corroded outlines clearly indicating a remnant phase in the metasomatic and later formed hydrogarnet which represent a hydrothermal alteration phase in cases later than the serpentinisation, since serpentinised olivines may represent enclosed and partly corroded pre-hydrogrossular phases.

In cases the metasomatic hydrothermal solutions responsible for the rodingite formation may result in more complex alterations and neo-crystallizations.

Augustithis (1960) describes most complex alteration and neo-crystallizations in the hydrothermal metasomatic alteration of talc zones and metasomatism in contact with the Rodiani (Greece) chromite. Present X-ray studies show that hydrogrossular is an important silicate phase associated with the chromite alteration.

In cases due to metasomatism, hydrogrossular, carbonates and talc may be formed as a "rodingitic" phase which attacks the Rodiani chromite (the Rodiani chromite contains orientated rutile and spinel ex-solutions, Augustithis (1960) – see also chapter 23).

As a result of the metasomatic reactions chromite-rodingite all phases of chromite disintegration are noticeable (Augustithis, 1960) as well as perwoskite (suspected by Ramdohr, 1957) and actually determined later by microprobe studies by Skunakis (1977). In addition to ilmenite present, rutiles as disassociation products also occur.

In contrast to the cases of hydrogarnet metasomatism so far described, cases may exist where the hydrogarnets (determined by X-rays to be hibschite) form "augen" structures in tectonically mobilised and recrystallised antigorite.

Figs. 732a and 732b show tectonically affected hydro-garnet with rounded forms and outlines in which tectonically mobilised antigorite has been "intruded". A rather more distinct case of tectonic mobilisation is shown in Figs. 733a and 733b where the relatively hard hydro-garnets form "augen" structures with rounded outlines due to "attrision" and which are enclosed by tectonically mobilised and recrystallised antigorite.

In contradistinction to the veinform or diffuse rodingite and the cases of the hydro-garnet, talc, serpentine, carbonate mobilisation, the described case the "augen" structures of hydro-garnet in antigorite is a tectonically mobilised texture with a later serpentine recrystallization.

Illustrations

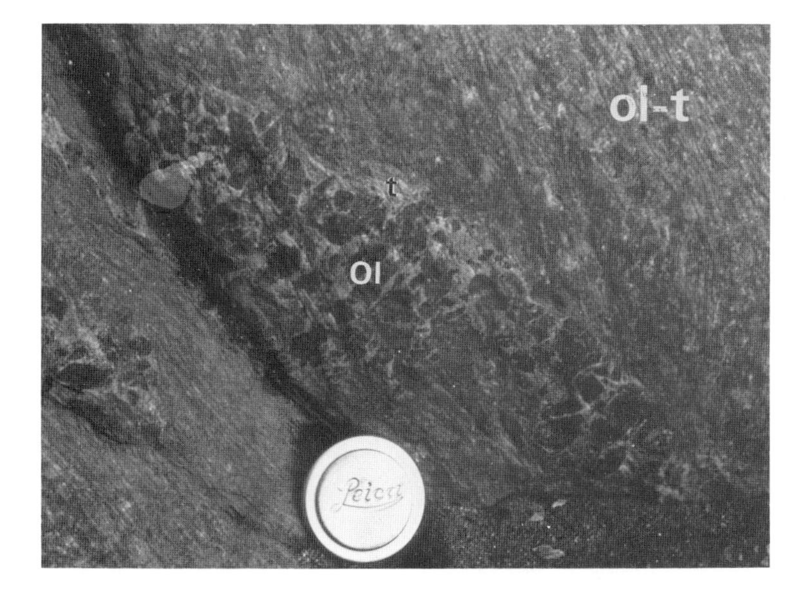

Fig. 1 Olivine megablast in a shear talc zone of an olivine-tremolitefels.
ol = olivine crystalloblasts
t = talc zone
ol-t = olivine-tremolitefels
Olivinefels, Prata S. Chiavenna, Italy.

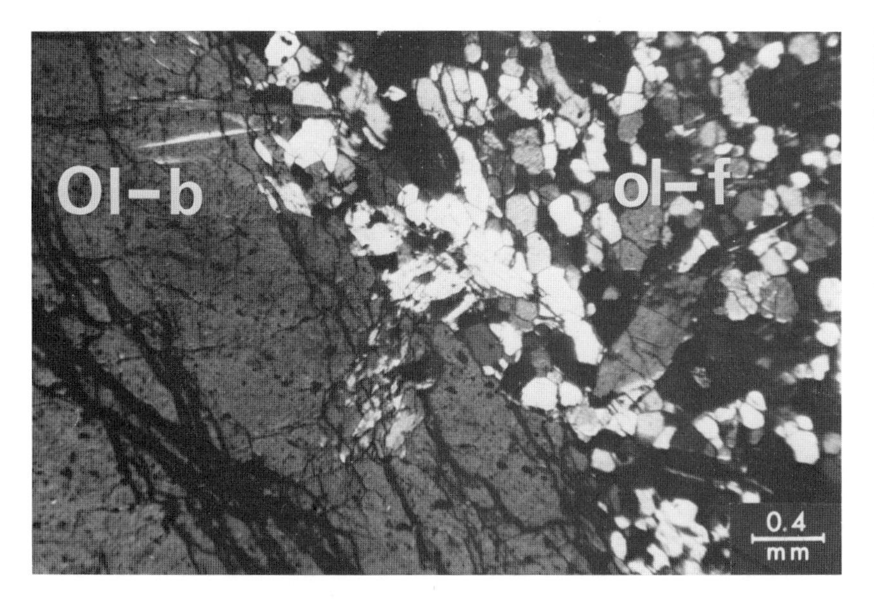

Figs. 2 and 3 (detail) Olivine megablast in an olivinefels (partly including granular olivines of the olivinefels). A later phase tremolite crystalloblastesis transverses both the olivines of the olivinefels and the megablast.
Ol-b = olivine megablast
ol-f = olivinefels
t = tremolite crystalloblast
Olivinefels, Prata S. Chiavenna, Italy.
With crossed nicols.

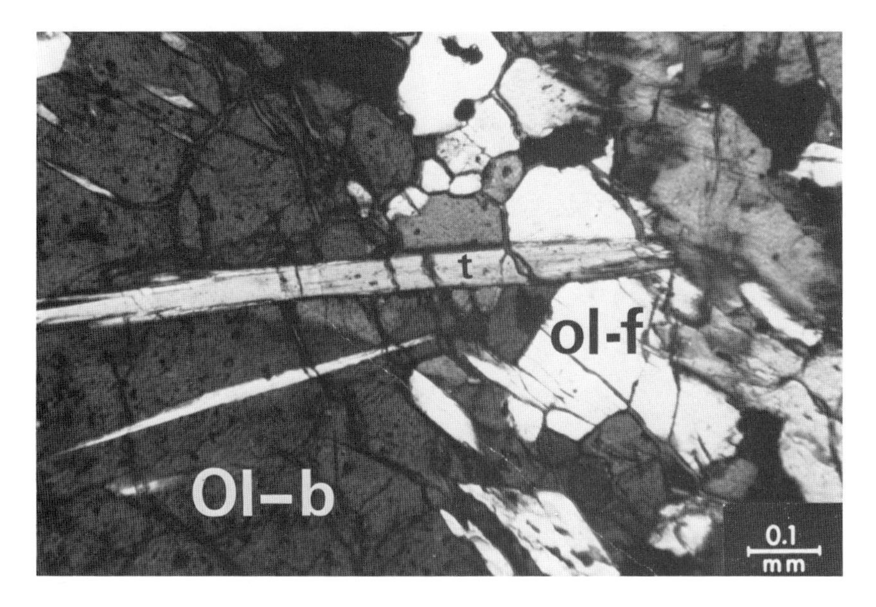

Fig. 3 (detail)

Fig. 4 Tremolite nematoblast transversing a granular olivinefels texture.
ol-g = granular olivine
t = tremolite nematoblast
Olivinefels, Prata S. Chiavenna, Italy.
With crossed nicols.

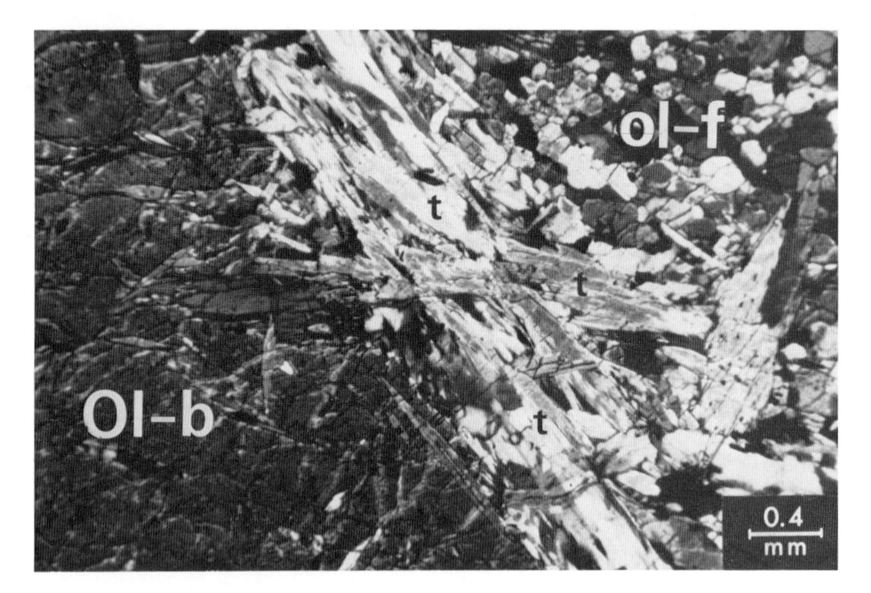

Fig. 5 Olivinefels and olivine megablast both transversed by a tremolite crystalloblastic phase.
ol-b = olivine megablast
ol-f = olivinefels
t = tremolite crystalloblasts
Olivinefels, Prata S. Chiavenna, Italy.
With crossed nicols.

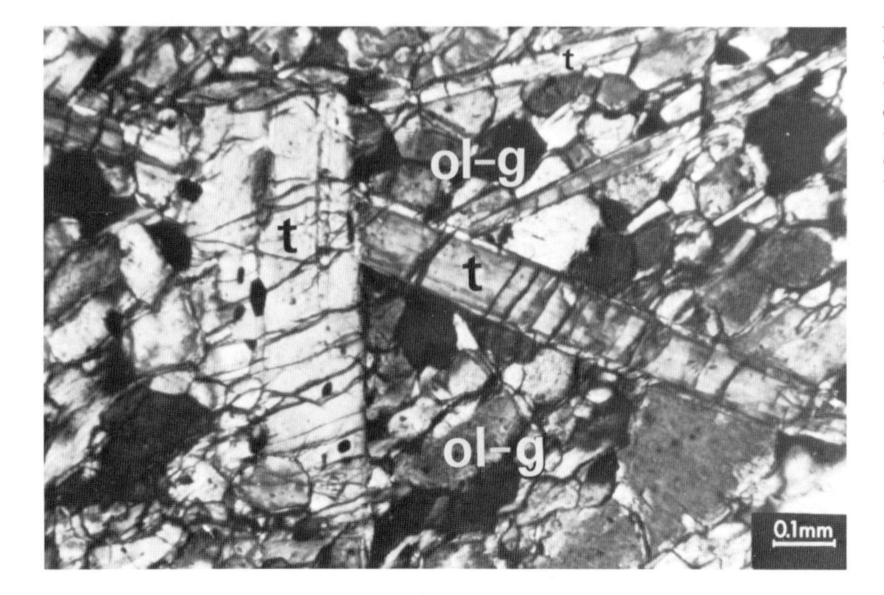

Fig. 6 Tremolite nematoblasts transversing the olivinefels and indicating interpenetration textures.
ol-g = granular olivine
t = tremolite nematoblasts
Olivinefels, Prata S. Chiavenna, Italy.
With crossed nicols.

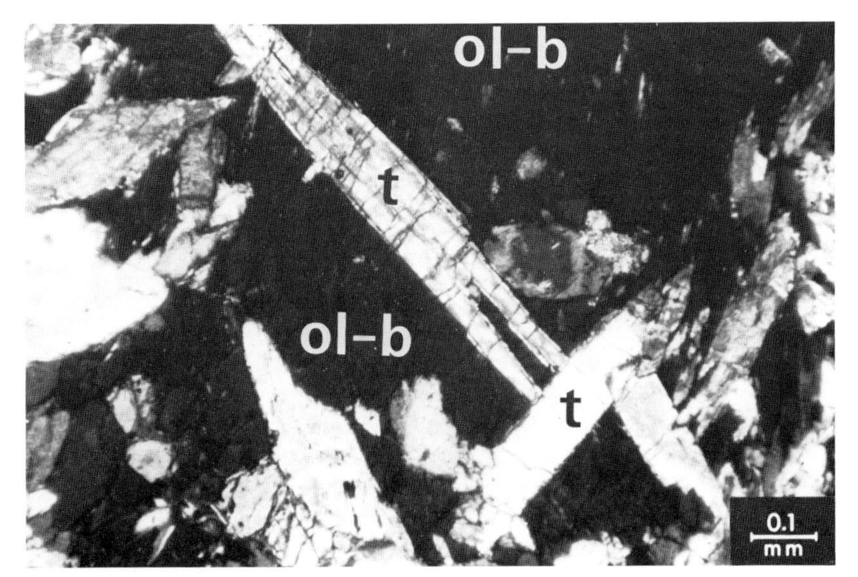

Fig. 7 Olivine megablast transversed by tremolite crystalloblasts following different directions within the olivine.
ol-b = olivine megablast
t = tremolite crystalloblasts
Olivinefels, Prata S. Chiavenna, Italy.
With crossed nicols.

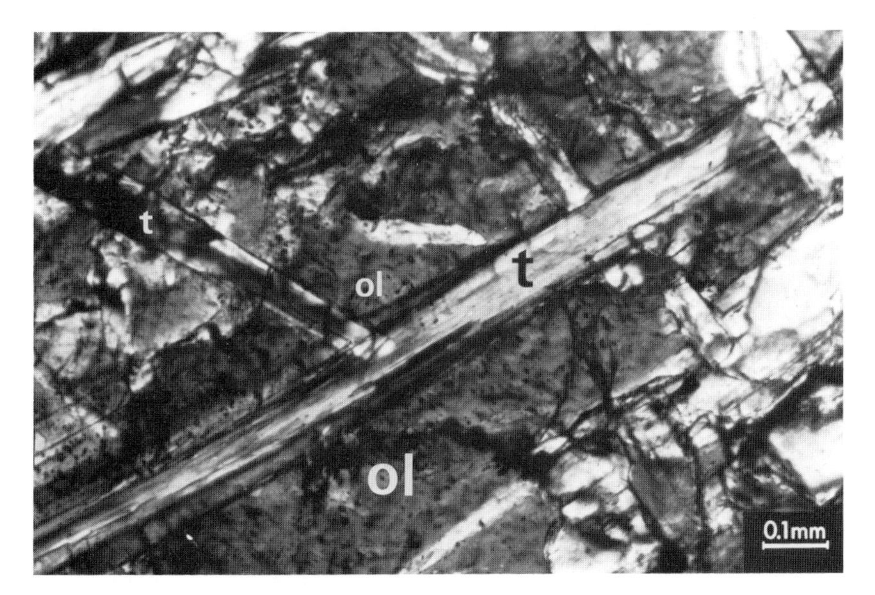

Fig. 8 Tremolite nematoblasts orientated and intersecting one another grown within an olivine host.
t = tremolite nematoblasts
ol = olivine grain
Olivinefels, Prata S. Chiavenna, Italy.
With crossed nicols.

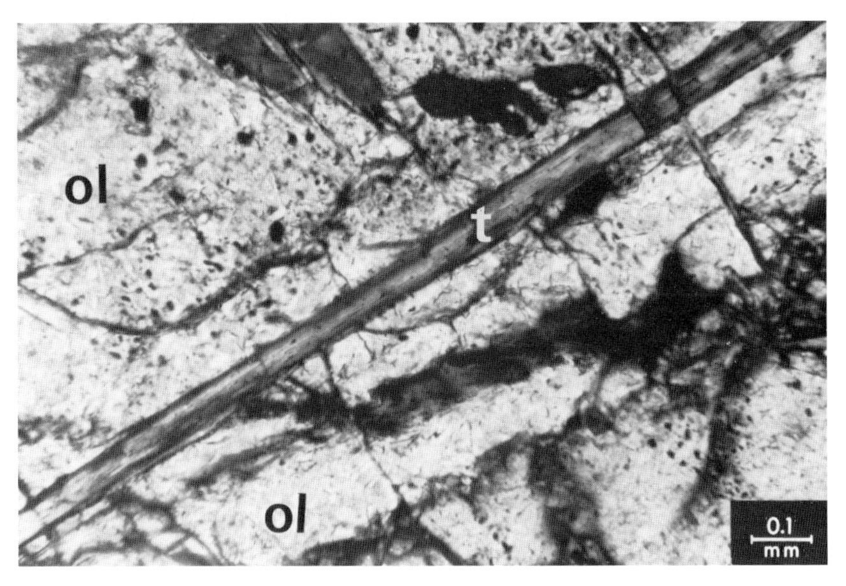

Fig. 9 Tremolite nematoblast transversing olivine.
t = tremolite nematoblast
ol = olivine
Olivinefels, Prata S. Chiavenna, Italy.
With crossed nicols.

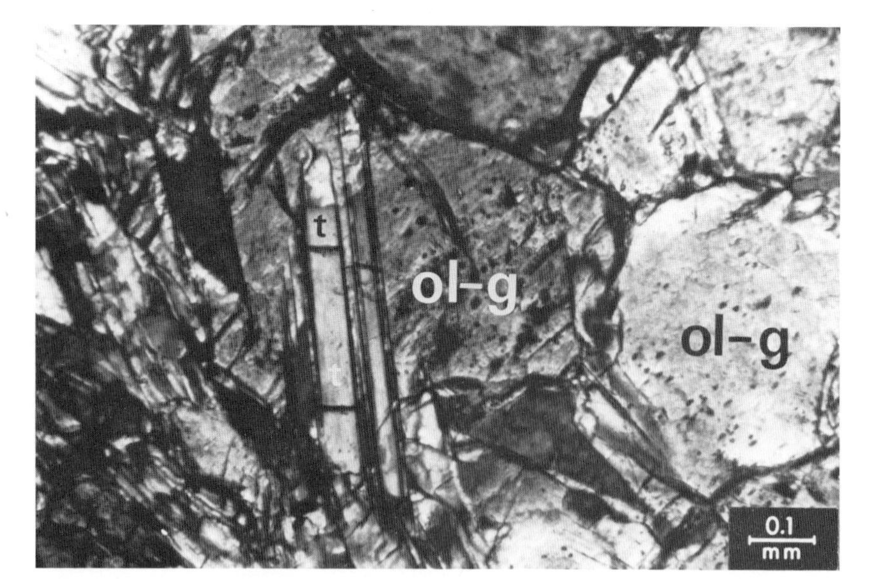

Fig. 10 Prismatic tremoliteblasts parallelly grown within an olivine grain of the granular olivinefels texture.
ol-g = granular olivine
t = tremolite
Olivinefels, Prata S. Chiavenna, Italy.
With crossed nicols.

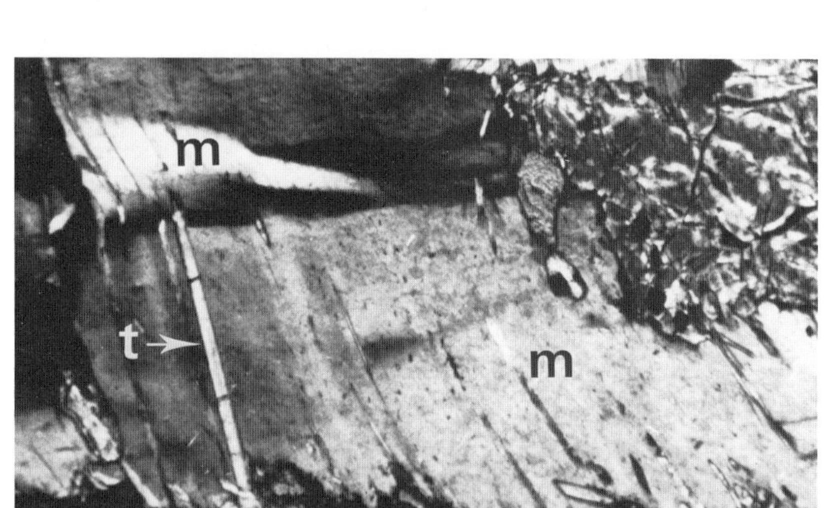

Fig. 11 and detail Fig. 12 Micaceous crystalloblast, phlogopite, tectonically deformed and penetrated by tremolite nematoblasts.
m = micaceous crystalloblast in granular olivinefels (undulating extinction due to deformation).
t = tremolite nematoblast.
Olivinefels, Prata S. Chiavenna, Italy.
With crossed nicols.

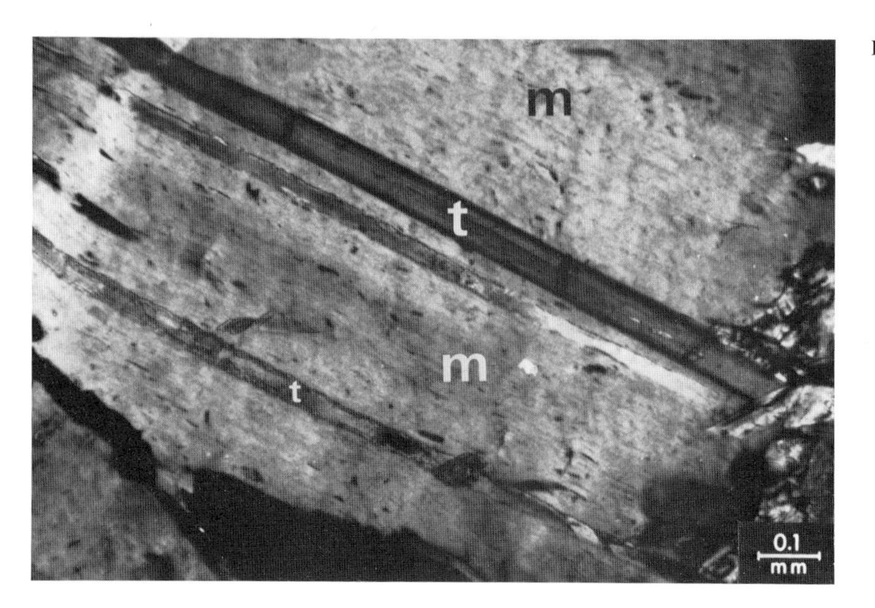

Fig. 12

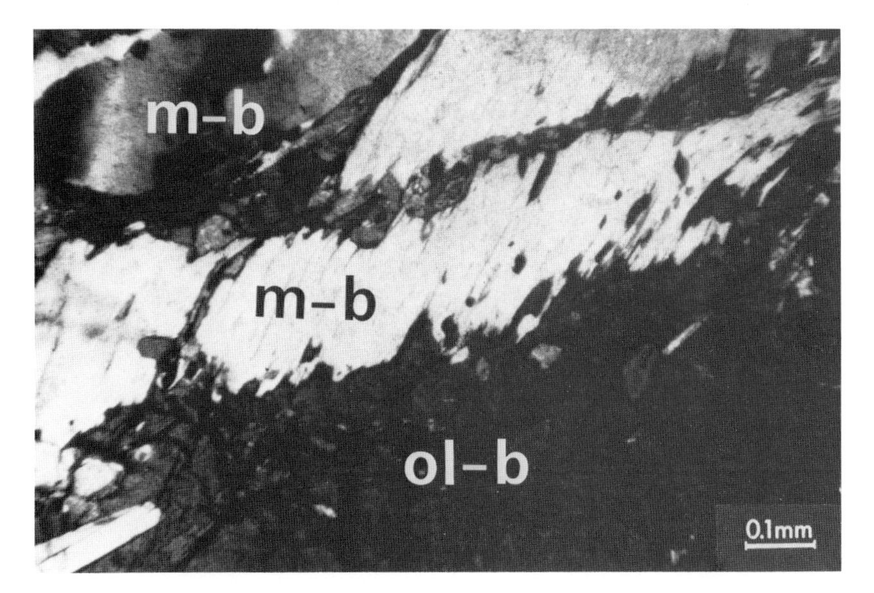

Fig. 13 Deformed and mobilised micaceous crystalloblast in intergrowth with the olivine megablast.
m-b = micaceous crystalloblast
ol-b = olivine megablast
Olivinefels, Prata S. Chiavenna, Italy. With crossed nicols.

Fig. 14 Tectonically deformed micaceous crystalloblast in the olivinefels. Olivinefels, Prata S. Chiavenna, Italy. With crossed nicols.

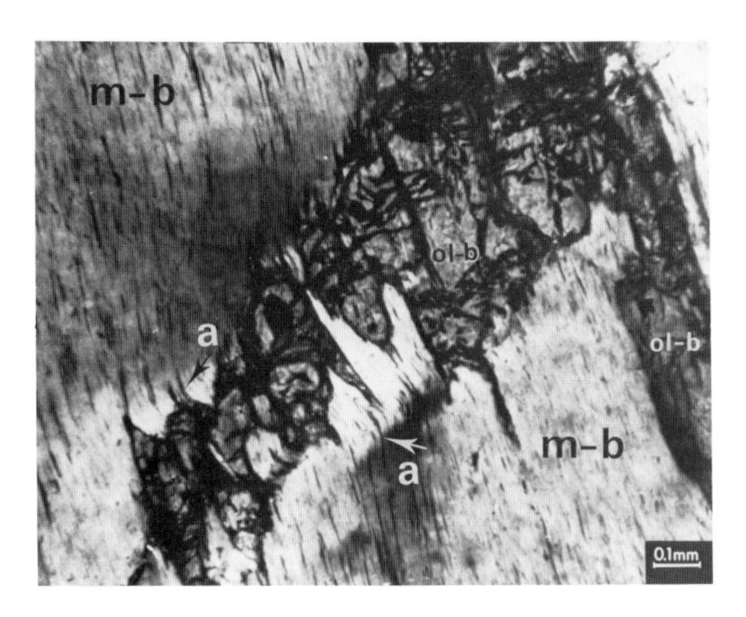

Fig. 15 An olivine late phase crystalloblastesis bursting through an already deformed micaceous crystalloblast. As arrows "a" show, the micaceous crystalloblast is further deformed by the crystalloblastic growth of the olivine.
m-b = micaceous crystalloblast
ol-b = olivine crystalloblast
Olivinefels, Prata S. Chiavenna, Italy. With crossed nicols.

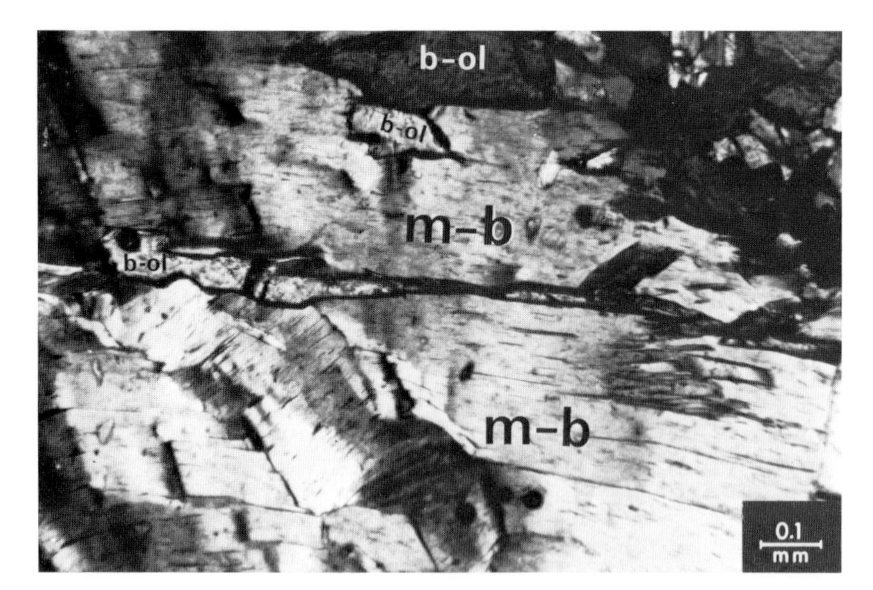

Fig. 16 Tectonically deformed micaceous crystalloblast with a new phase of olivine crystalloblastesis partly following the cleavage of the mica.
m-b = micaceous crystalloblast tectonically deformed.
b-ol = late phase of olivine blastesis following the interleptonic spaces of the micaceous crystalloblast.
Olivinefels, Prata S. Chiavenna, Italy.
With crossed nicols.

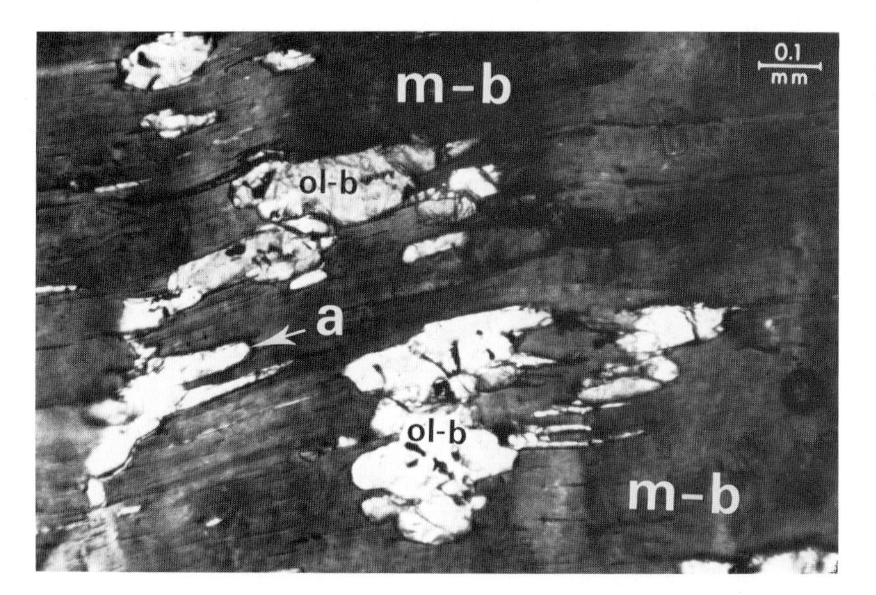

Fig. 17 Tectonically deformed micaceous crystalloblast with a new phase of olivine crystalloblastesis partly with rounded outlines (arrow "a") and partly sending protuberances following the cleavage of the micaceous crystalloblast.
m-b = micaceous crystalloblast
ol-b = late phase olivine crystalloblastesis
Olivinefels, Prata S. Chiavenna, Italy.
With crossed nicols.

Fig. 18 Micaceous crystalloblast with late phase olivine crystalloblastesis following the interleptonic cleavage space of the micaceous host.
m-b = micaceous crystalloblast
ol-b = olivine, late phase crystalloblastesis
Olivinefels, Prata S. Chiavenna, Italy.
With crossed nicols.

Fig. 19 Olivine breccia (dunitic fragments) in a tremolite binding mass.
Ol = olivine fragments (dunitic fragments)
Tr = tremolite binding mass, holding the dunitic fragments together.
Olivine breccia.
Prata South Chiavenna, Italy.
Handspecimen about natural size.

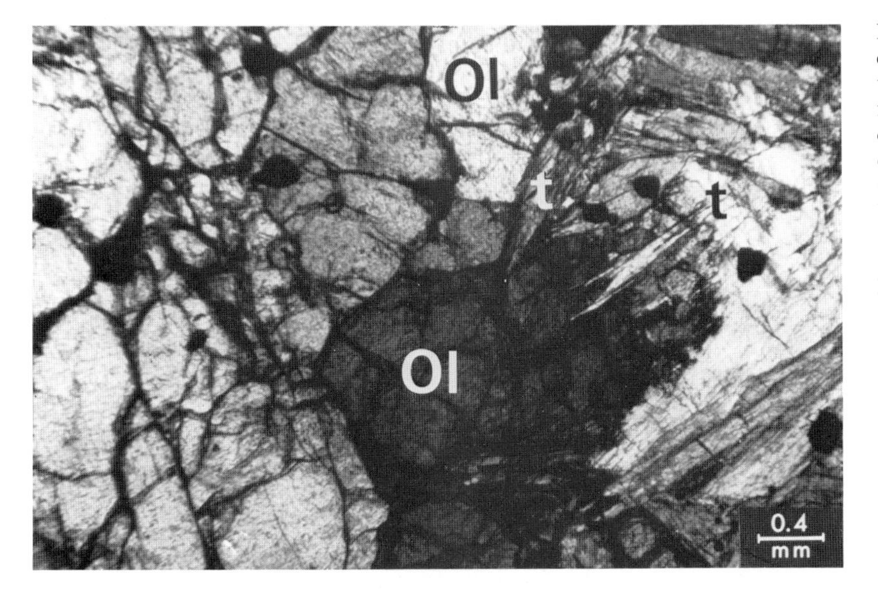

Fig. 20 Olivine breccia consisting of coarse grained olivine with metasomatic tremolite "cementing together the olivine fragments" and due to blastogenic force of crystallization, penetrating through the olivine fragments.
Ol = coarse grained olivine of the olivine breccia.
t = tremolite
Olivine breccia.
Prata S. Chiavenna, Italy.
With crossed nicols.

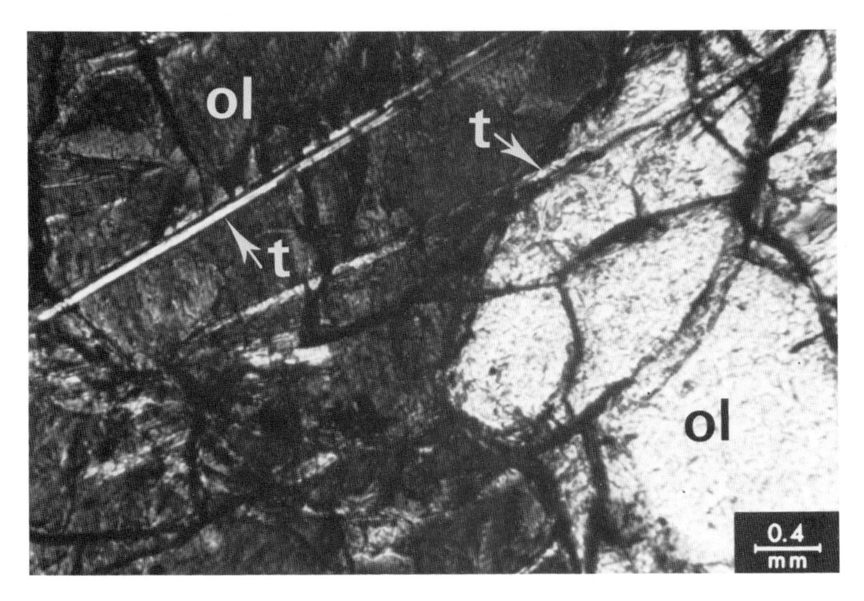

Fig. 21 Differently orientated coarse grained olivines transversed by tremolite nematoblasts.
ol = coarse grained olivine of the olivine breccia.
t = tremolite nematoblasts.
Olivine breccia.
Prata S. Chiavenna, Italy.
With crossed nicols.

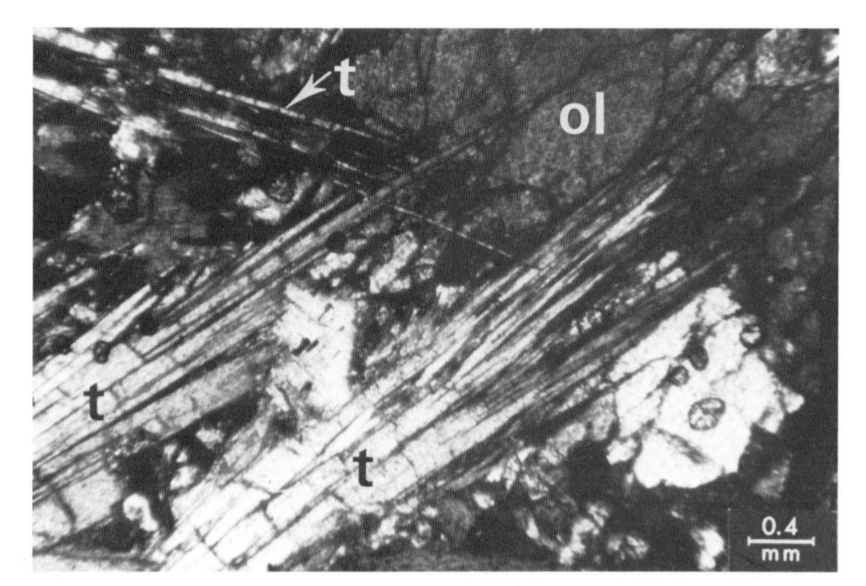

Fig. 22 Radiating tremolite "nematoblasts" due to blastogenic crystallization-force penetrating into the coarse grained olivine of the olivine breccia. Tremolite crystalloblasts intersecting one another can be seen.
ol = coarse grained olivine of the olivine breccia.
t = tremolite crystalloblast (nematoblasts).
Olivine breccia. Prata S. Chiavenna, Italy.
With crossed nicols.

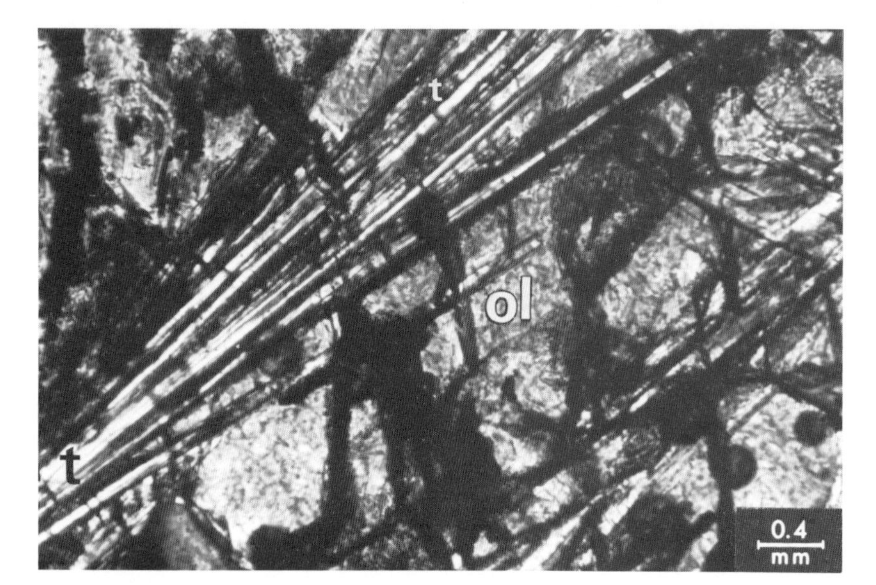

Fig. 23 Radiating tremolite "nematoblasts". Due to blastogenic crystallization-force the tremolite penetrates the olivine.
ol = olivine (olivine breccia)
t = crystalloblastic tremolite nematoblasts.
Olivine breccia. Prata S. Chiavenna, Italy.
With crossed nicols.

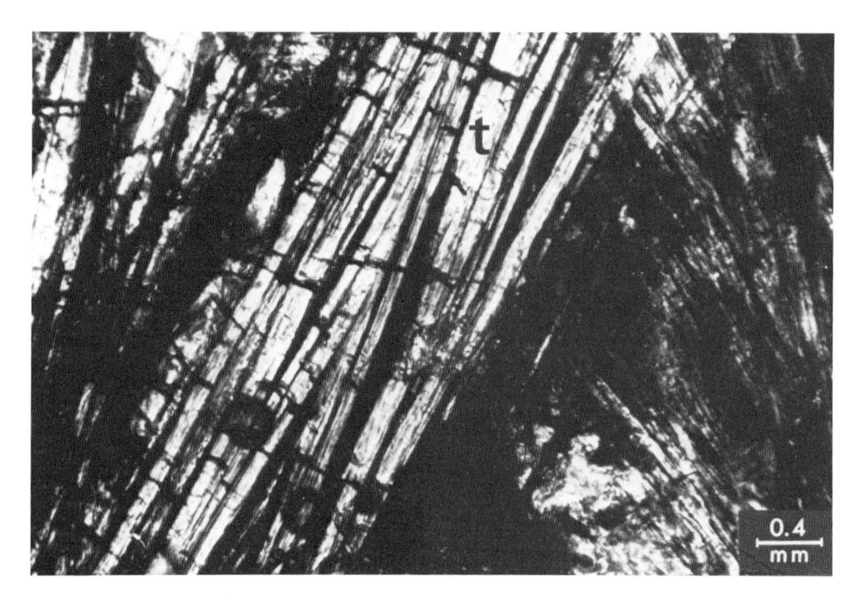

Fig. 24 Bands of parallelly orientated tremolite nematoblasts and in cases intersecting one another.
t = tremolite nematoblasts
Olivine breccia. Prata S. Chiavenna, Italy.
With crossed nicols.

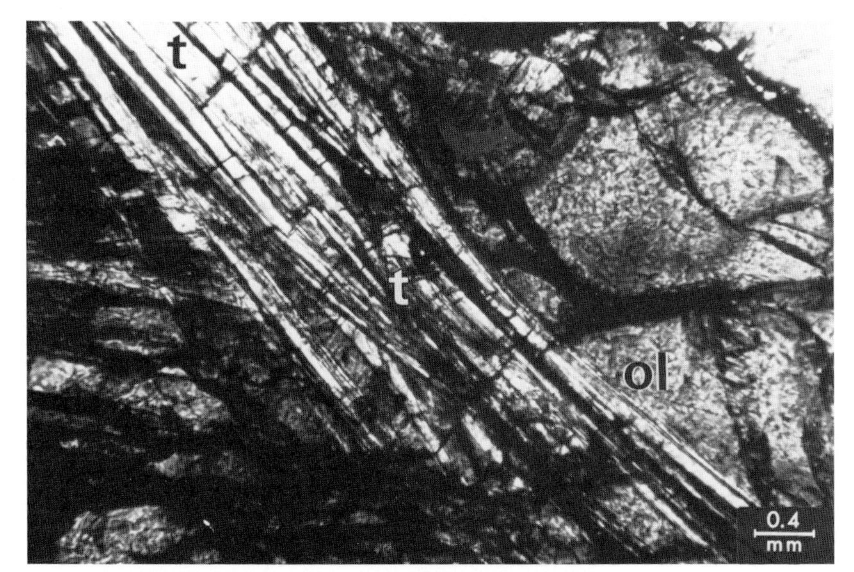

Fig. 25 Tremolite nematoblasts transversing, due to crystalloblastic force, coarse grained olivine. The tremolite crystalloblasts are also curved.
ol = olivine crystalloblast
t = tremolite nematoblast
Olivine breccia. Prata S. Chiavenna, Italy.
With crossed nicols.

Fig. 26 Shows folded triassic marbles effected by metasomatic solutions. As a result of metasomatic processes and reactions, pyroxene, hornblende, olivine, chlorite and antigorite may be formed (also grossular is formed).
m = marble.
b-s = basic solution front invading the marbles.
Malga Trivena, Val di Breguzzo, Adamello, Alps.
Photograph from an excursion with Prof. Dr. F. K. Drescher-Kaden (1961).

Fig. 27 (Similar to Fig. 26) Folded triassic marbles invaded by basic metasomatic solutions which resulted in the formation of pyroxene, hornblende, olivine, chlorite, antigorite and grossular.
m = marble.
b-s = basic solutions following an anticlinal fold of the marbles Malga Trivena, Val di Breguzzo, Adamello, Alps.
Photograph from an excursion with Prof. Dr. F. K. Drescher-Kaden (1961).

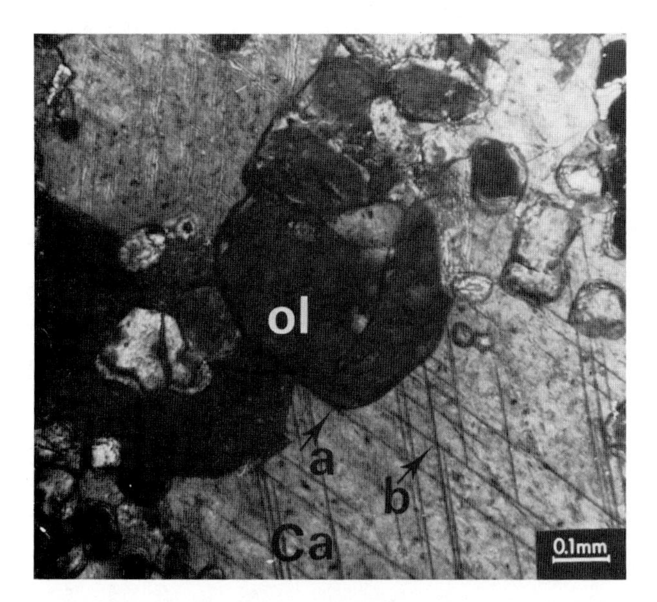

Fig. 28 Blastogenic olivine in coarse grained marble. The olivine crystalloblast has crystal faces coinciding with the rhombohedral cleavage pattern of the calcite host (see arrow "a").
Ca = coarse grained calcite.
ol = olivine crystalloblast; arrow "b" shows the rhombohedral cleavage of the calcite.
Metasomatically affected marbles.
Malga Trivena, Val di Breguzzo, Adamello, Italy.
With crossed nicols.

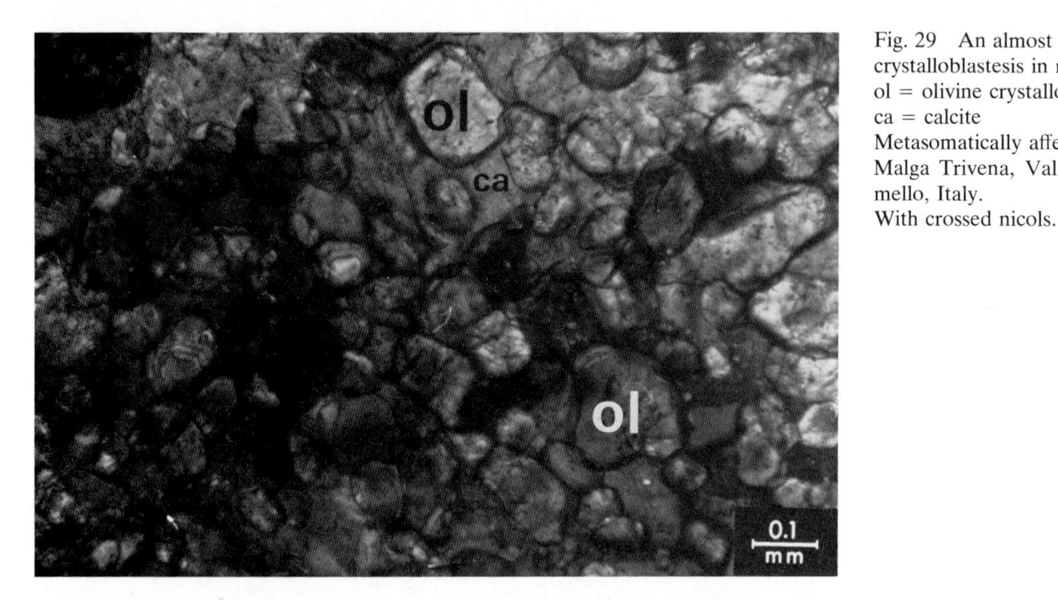

Fig. 29 An almost equigranular olivine crystalloblastesis in marble.
ol = olivine crystalloblast
ca = calcite
Metasomatically affected marbles.
Malga Trivena, Val di Breguzzo, Adamello, Italy.
With crossed nicols.

Fig. 30 Granular olivine crystalloblasts (partly serpentinised) in a coarse grained calcite (marble).
ca = calcite
ol = olivine crystalloblast
s = serpentinised olivine
Metasomatically affected marbles.
Malga Trivena, Val di Breguzzo, Adamello, Italy.
With crossed nicols.

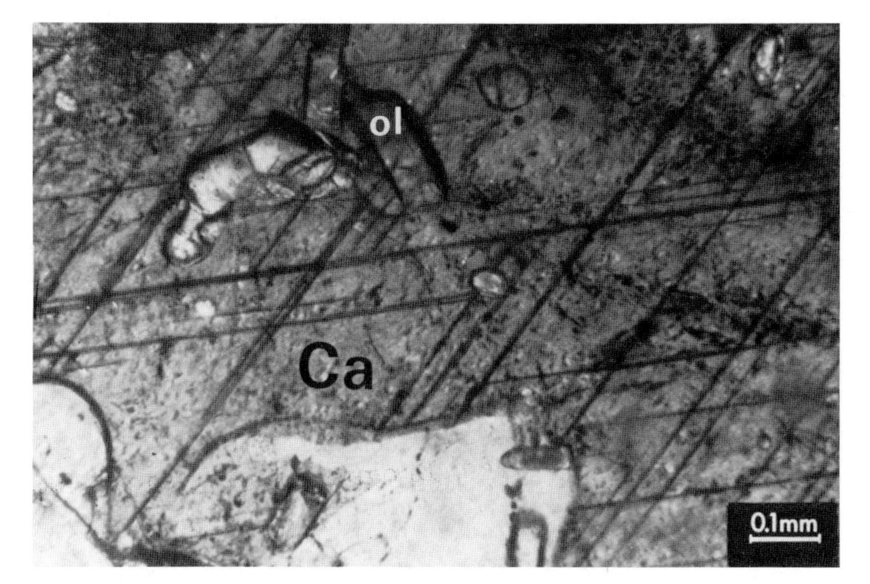

Fig. 31 Coarse grained calcite (marble) with rhombohedral calcite cleavage transversed by "granules" of olivine crystalloblasts.
Ca = calcite
ol = granules of olivine crystalloblasts
Metasomatically affected marbles
Malga Trivena, Val di Breguzzo, Adamello, Italy.
With crossed nicols.

Fig. 32a and detail 32b Coarse grained calcite (marble) with granular olivine crystalloblasts. The olivine crystalloblasts often show outlines dependant on the rhombohedral calcite cleavage.
Ca = calcite (coarse grained marble)
ol = olivine crystalloblasts
Arrows "a" show the coincidence between olivine outline and calcite's cleavage
Metasomatically affected marbles.
Malga Trivena, Val di Breguzzo, Adamello, Italy.
With crossed nicols.

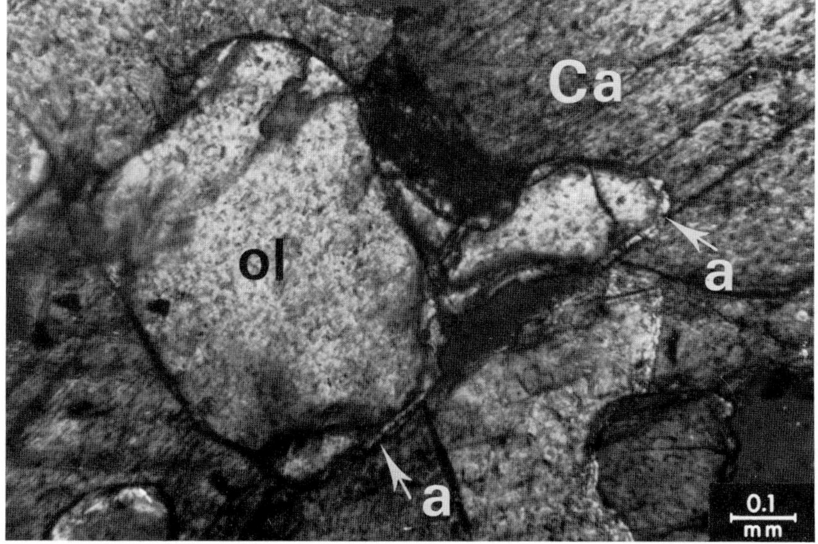

Fig. 32b

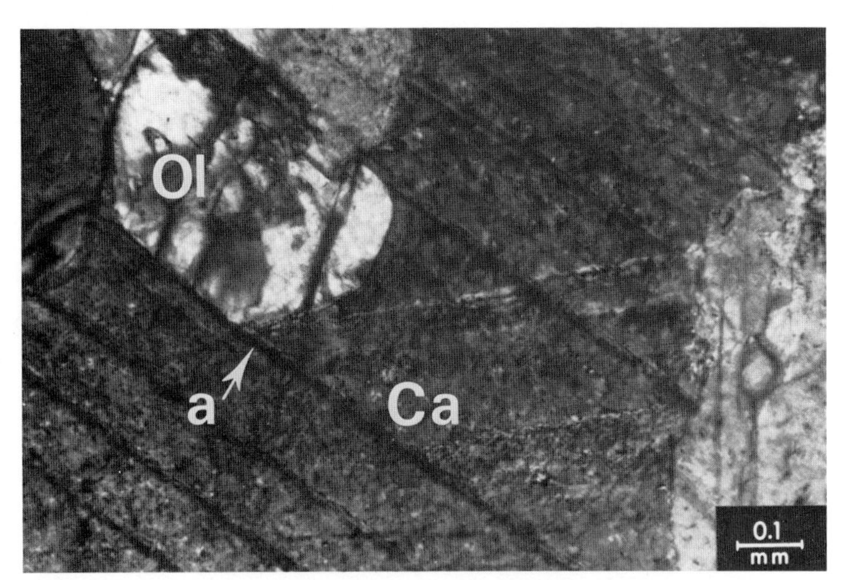

Fig. 33 Olivine crytalloblast with an outline delimited by the rhombohedral calcite twinning.
Ol = olivine crystalloblast (partly serpentinised)
Ca = calcite, coarse grained marble.
Arrow "a" shows the rhombohedral calcite twinning.
Metasomatically affected marbles.
Malga Trivena, Val di Breguzzo, Adamello, Italy.
With crossed nicols.

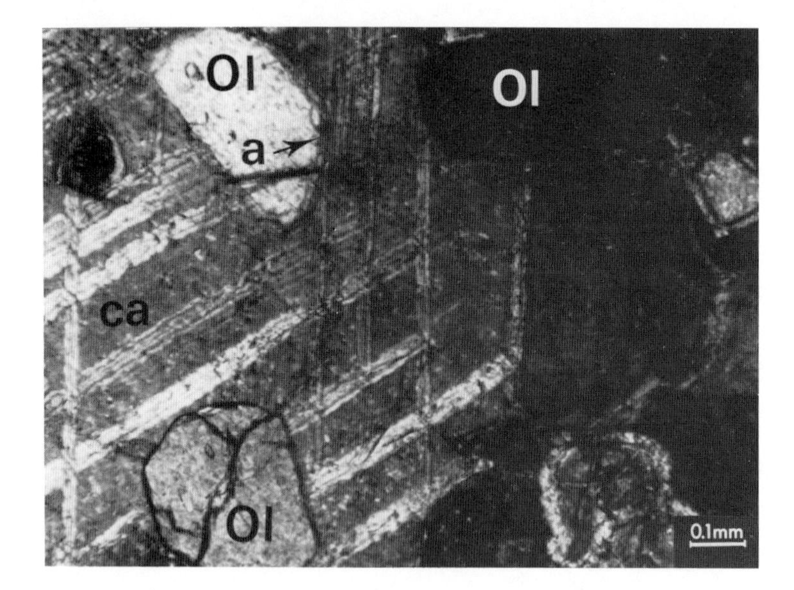

Fig. 34 Olivine crystalloblasts transversing the rhombohedral calcite twinning. In cases an olivine crystalloblast partly follows one of the calcite twinning directions (arrow "a").
Ol = olivine crystalloblast
ca = coarse grained calcite, with twinning.
Metasomatically affected marbles.
Malga Trivena, Val di Breguzzo, Adamello, Italy.
With crossed nicols.

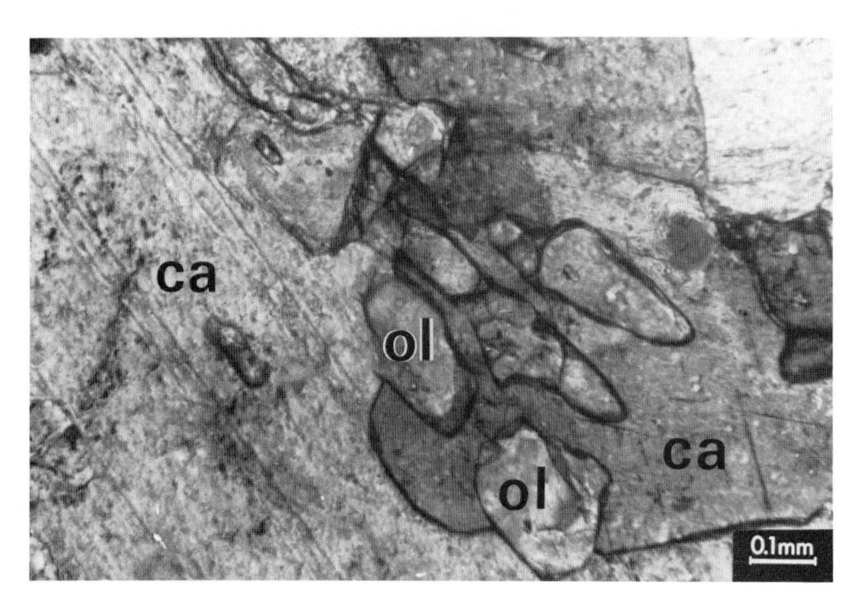

Fig. 35 Granular olivine crystalloblasts transversing the boundary of two differently orientated calcite grains.
ol = olivine crystalloblast
ca = calcite, coarse grained marble
Metasomatically affected marbles.
Malga Trivena, Val di Breguzzo, Adamello, Italy.
With crossed nicols.

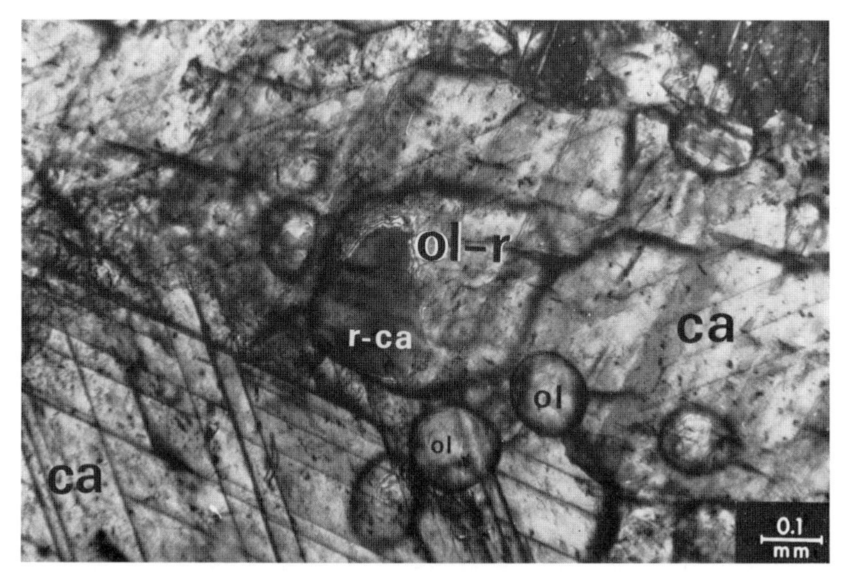

Fig. 36 Rounded olivine crystalloblast "crystal drops" in coarse grained marbles. Also a larger olivine crystalloblast with calcite "relics" and calcite-cleavage relics preserved within the olivine.
ca = calcite
r-ca = calcite relics within the olivine crystalloblast.
ol = rounded olivine crystalloblast
ol-r = olivine crystalloblast (anchiblast) with calcite relic and calcite cleavage preserved.
Metasomatically affected marbles.
Malga Trivena, Val di Breguzzo, Adamello, Italy.
With crossed nicols.

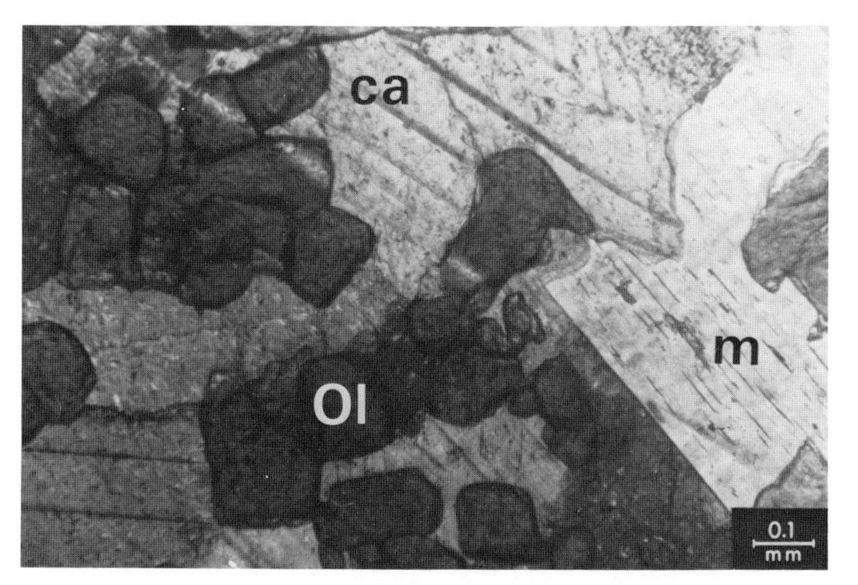

Fig. 37 Coarse grained marble with mica and olivine crystalloblasts.
Ol = olivine crystalloblasts
ca = calcite
m = micaceous mineral, crystalloblast in marble.
Metasomatically affected marbles.
Malga Trivena, Val di Breguzzo, Adamello, Italy.
With crossed nicols.

Fig. 38 Olivine crystalloblast cutting through a micaceous mineral in the Trivena marble. The olivine partly follows the cleavage of the mica.
ol = olivine crystalloblast
ca = calcite
m = micaceous mineral in the Trivena marble.
Metasomatically affected marbles.
Malga Trivena, Val di Breguzzo, Adamello, Italy.
With crossed nicols.

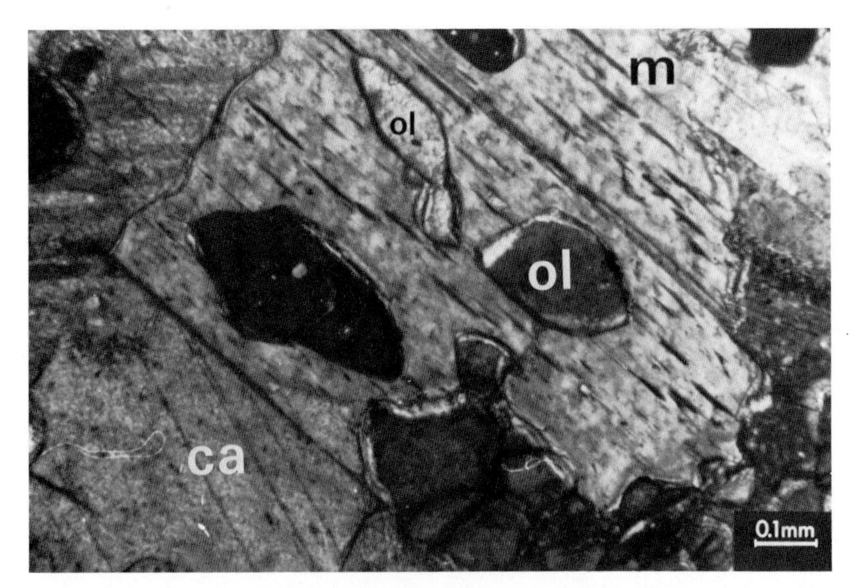

Fig. 39 Olivine crystalloblasts partly following and delimited by the micaceous mineral cleavage.
ol = olivine crystalloblast
ca = calcite
m = micaceous mineral in the Trivena marble.
Metasomatically affected marbles.
Malga Trivena, Val di Breguzzo, Adamello, Italy.
With crossed nicols.

Fig. 40 Coarse grained marble with granular olivine blastesis being enriched in a band within the marble.
Metasomatically affected marbles.
Malga Trivena, Val di Breguzzo, Adamello, Italy.
With crossed nicols.

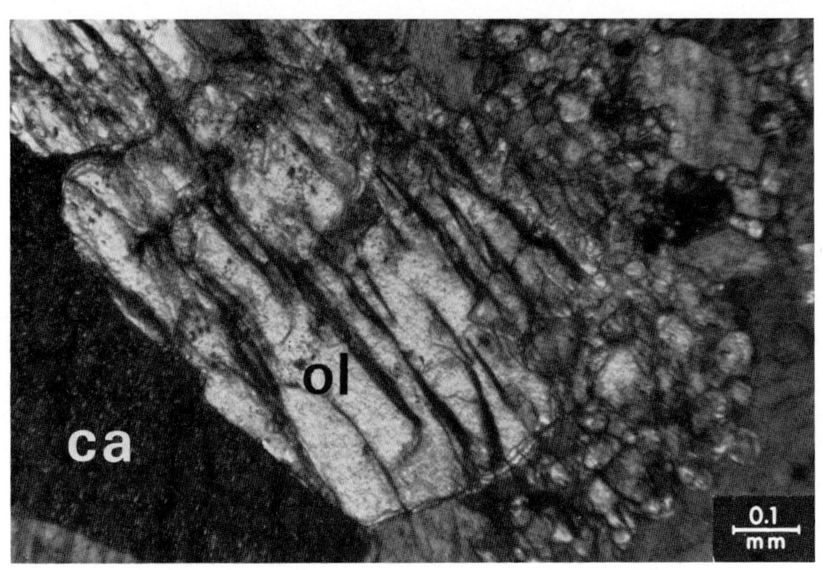

Fig. 41 A megablast and fine olivine crystalloblasts in marble.
ol = olivine crystalloblast
ca = calcite
Metasomatically affected marbles.
Malga Trivena, Val di Breguzzo, Adamello, Italy.
With crossed nicols.

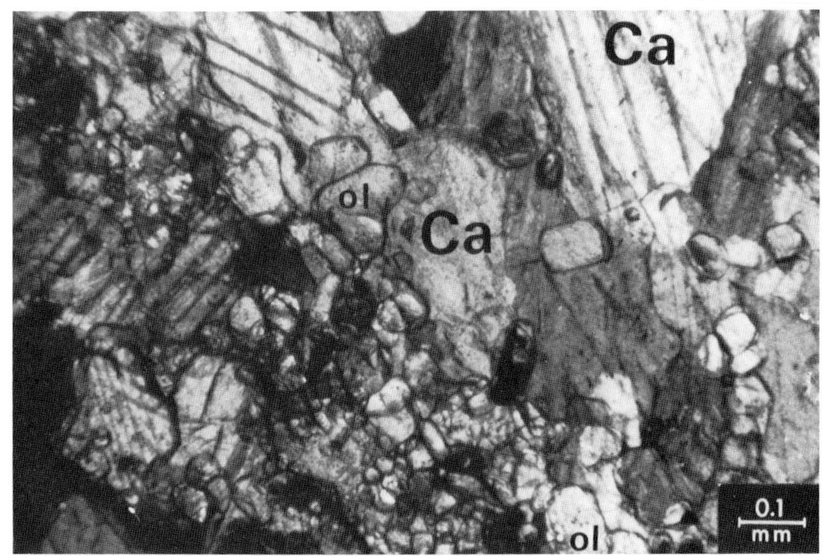

Fig. 42 Coarse grained marble with granular olivine crystalloblasts, top-concentration of the olivine crystalloblasts has taken place.
Ca = calcite, coarse grained marble.
ol = olivine crystalloblast.
Metasomatically affected marbles.
Malga Trivena, Val di Breguzzo, Adamello, Italy.
With crossed nicols.

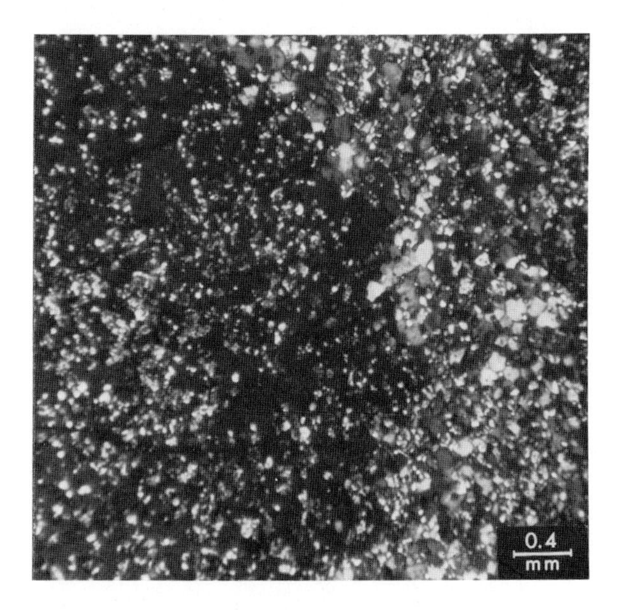

Fig. 43 Equigranular marble with mafic minerals in which a poikiloblastic garnet (grossular) has grown enclosing and assimilating the calcite and the mafic grains.
Garnet (isotropic, black).
Metasomatically affected marbles.
Malga Trivena, Val di Breguzzo, Adamello, Italy.
With crossed nicols.

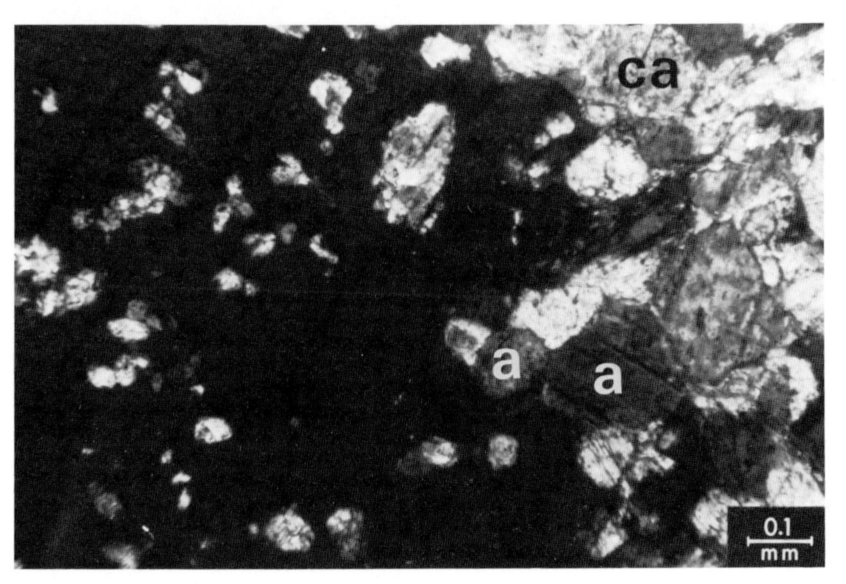

Fig. 44 Garnet poikiloblast enclosing and partly assimilating calcite and amphiboles (mafic grains) of the equigranular coarse grained marble, with mafic minerals; see Fig. 43.
Garnet (black)
a = amphiboles
ca = calcite
Metasomatically affected marbles.
Malga Trivena, Val di Breguzzo, Adamello, Italy.
With crossed nicols.

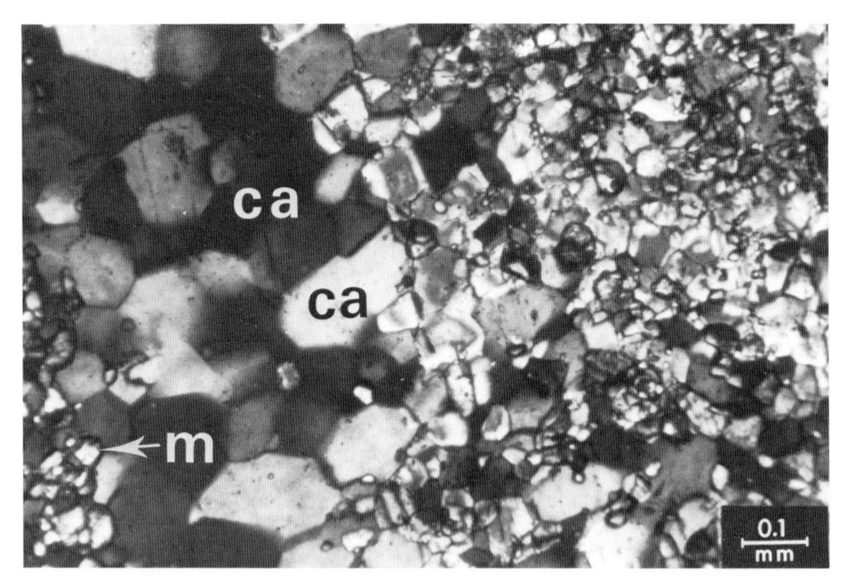

Fig. 45 Coarse grained marble, with equigranular calcite and fine mafic mineral grains.
ca = calcite
m = mafic mineral grains
Metasomatically affected marbles.
Malga Trivena, Val di Breguzzo, Adamello, Italy.
With crossed nicols.

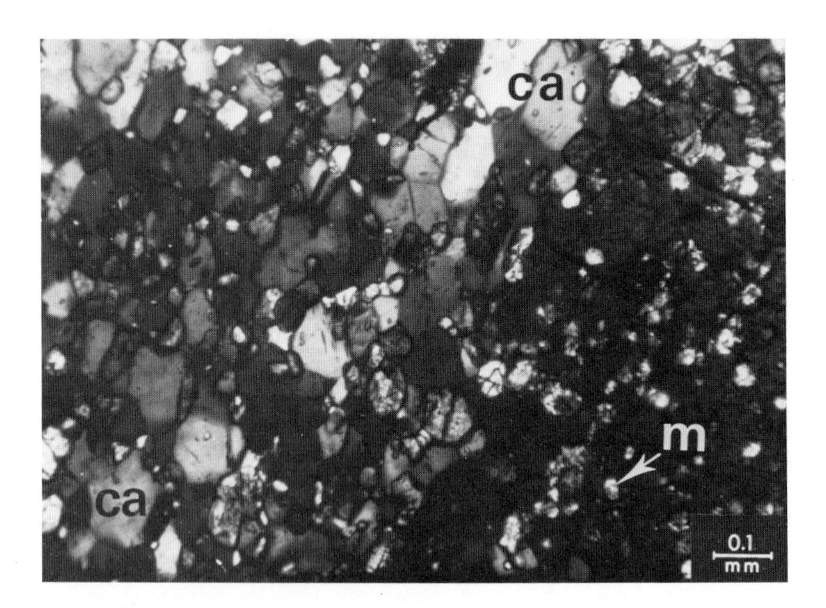

Fig. 46 Coarse grained marble with equigranular calcite and fine mafic minerals and a poikiloblastic garnet (isotropic-black) with relics of calcite and mafic minerals present.
ca = calcite (coarse grained marble)
m = fine mafic minerals
garnet (grossular) = black
Metasomatically affected marbles.
Malga Trivena, Val di Breguzzo, Adamello, Italy.
With crossed nicols.

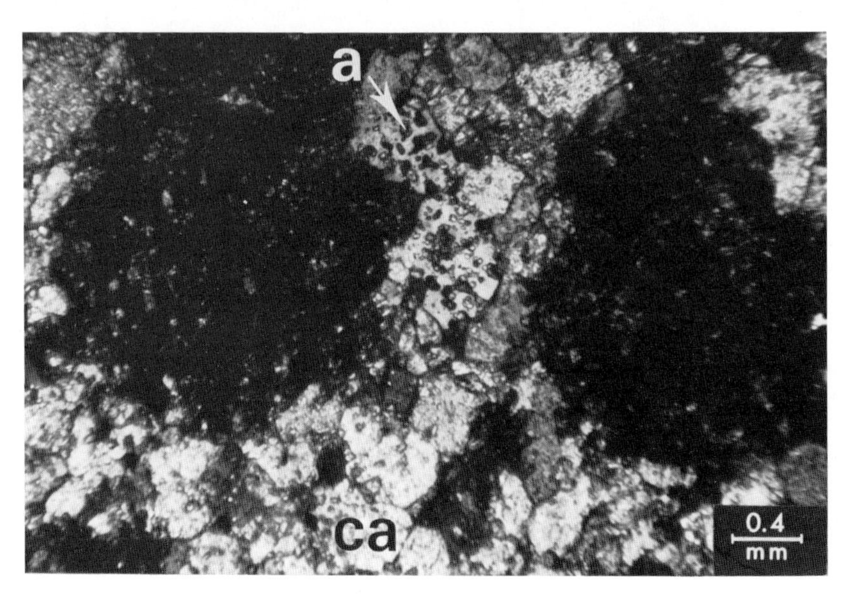

Fig. 47 Garnet poikiloblast in equigranular calcite texture (marble).
Arrow "a" shows fine mafic crystalloblasts within the calcite.
garnet (isotropic)
ca = calcite
Val Palobia, N. Italy.
With crossed nicols.

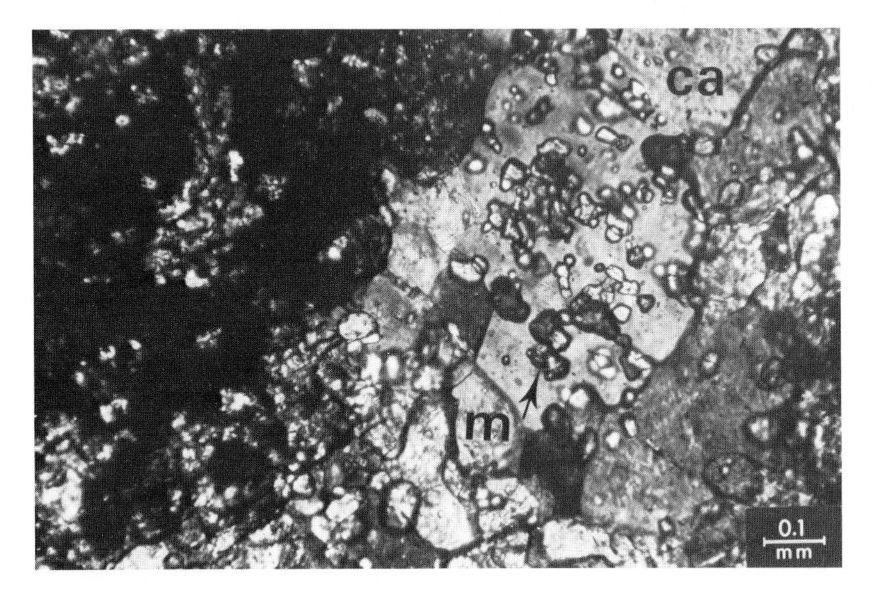

Fig. 48 Detail of Fig. 47 Garnet poikiloblast enclosing and assimilating calcite and mafic minerals of the calcite-mafic minerals of the marble.
garnet (isotropic)
ca = calcite (coarse grained marble)
m = mafic crystalloblasts within the calcite.
Olivine is, in cases, a component of the mafic minerals in the marble.
Val Palobia, N. Italy.
With crossed nicols.

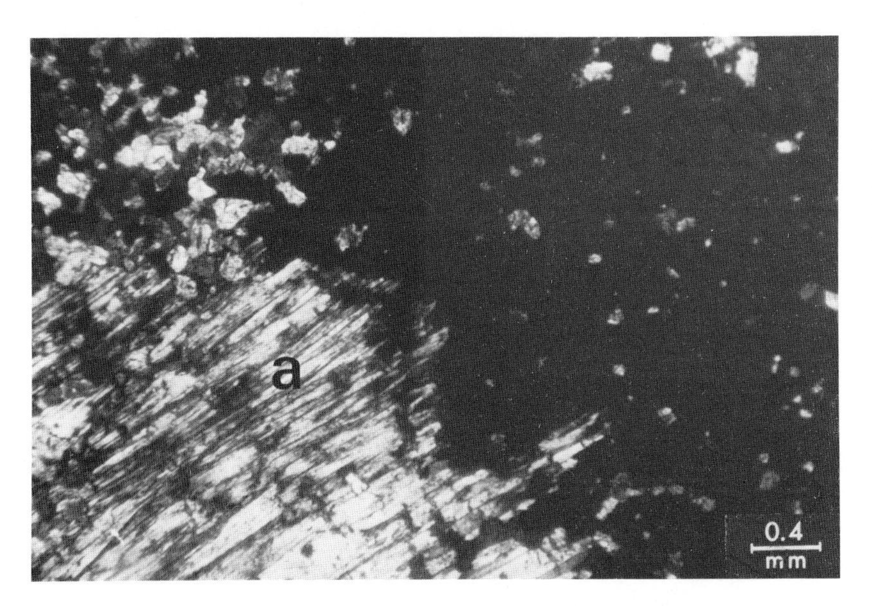

Fig. 49 A garnet poikiloblast (black) partly enclosing an amphibole crystalloblast, both grown blastically within an equigranular calcite texture.
a = amphibole blastesis.
Metasomatically affected marbles.
Malga Trivena, Val di Breguzzo, Adamello, Italy.
With crossed nicols.

Fig. 50 The garnet poikiloblast enclosing and extending into the intergranular of amphiboles which also represent a blastogenic phase within an equigranular marble.
a = amphibole crystalloblastesis.
garnet (isotropic, black).
Metasomatically affected marbles.
Malga Trivena, Val di Breguzzo, Adamello, Italy.
With crossed nicols.

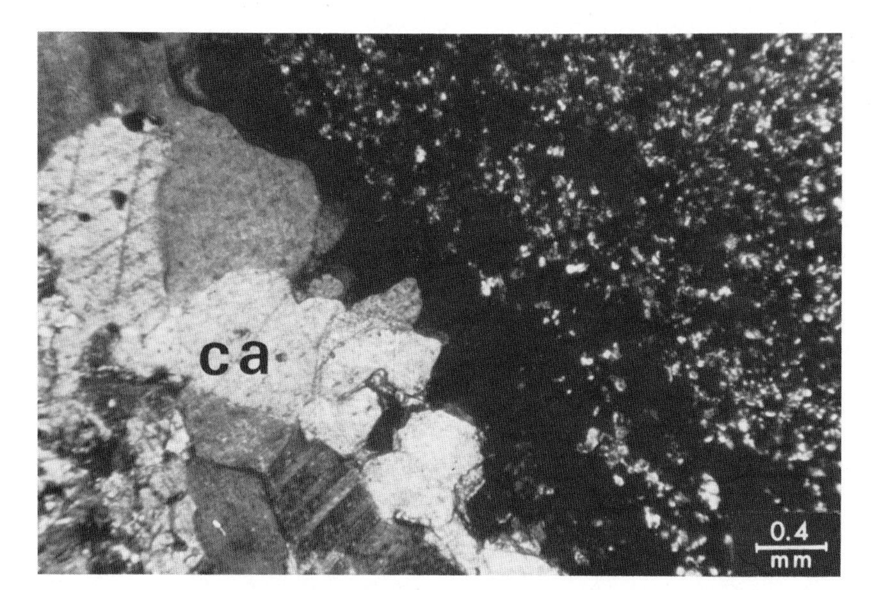

Fig. 51 Coarse grained marble with a garnet poikiloblast enclosing relics of calcite.
ca = calcite
garnet (isotropic)
Metasomatically affected marbles.
Malga Trivena, Val di Breguzzo, Adamello, Italy.
With crossed nicols.

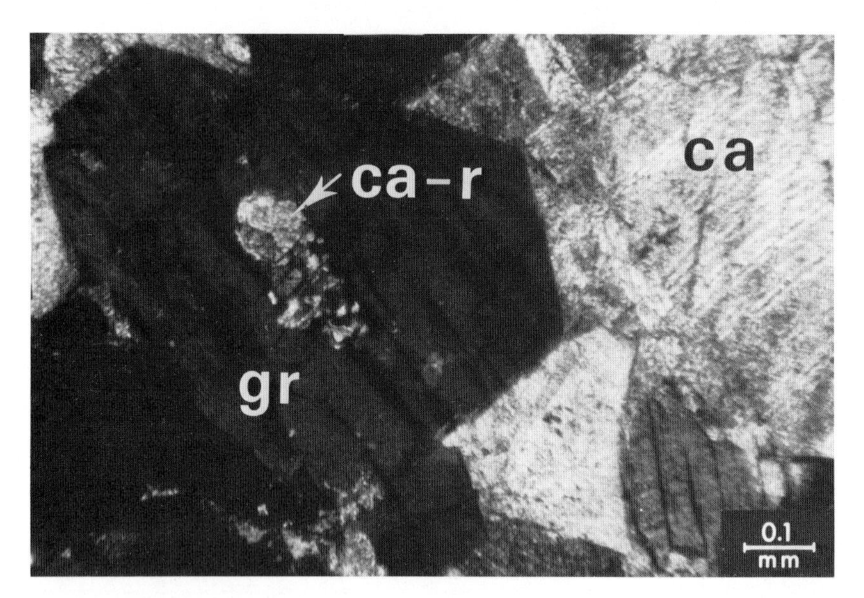

Fig. 53 and Fig. 54 Coarse grained calcite (marble) with idioblastic garnet enclosing calcite relics.
ca = calcite
ca-r = calcite relics
gr = idioblastic garnet
Metasomatically affected marbles.
Malga Trivena, Val di Breguzzo, Adamello, Italy.
With crossed nicols.

(For Fig. 52 see page 129)

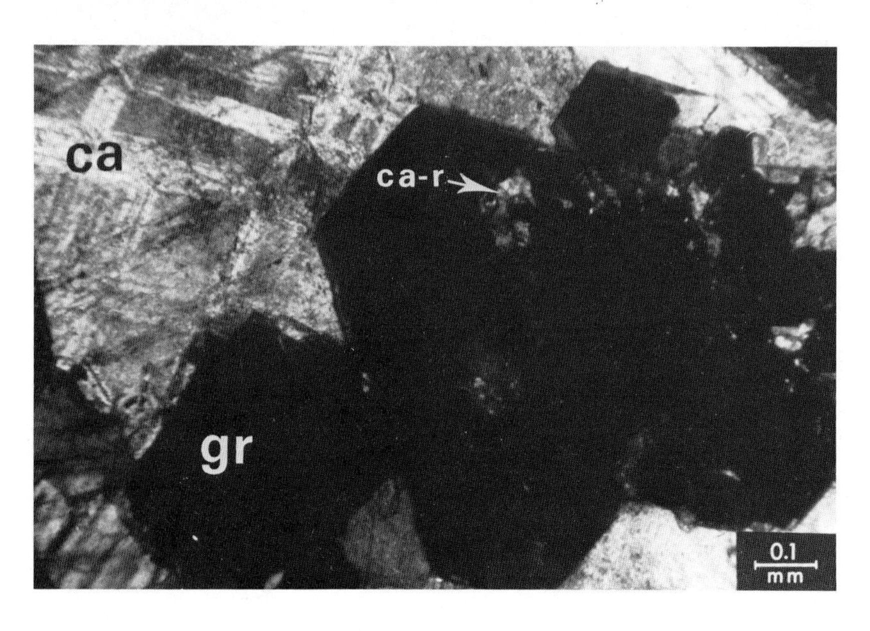

Fig. 54

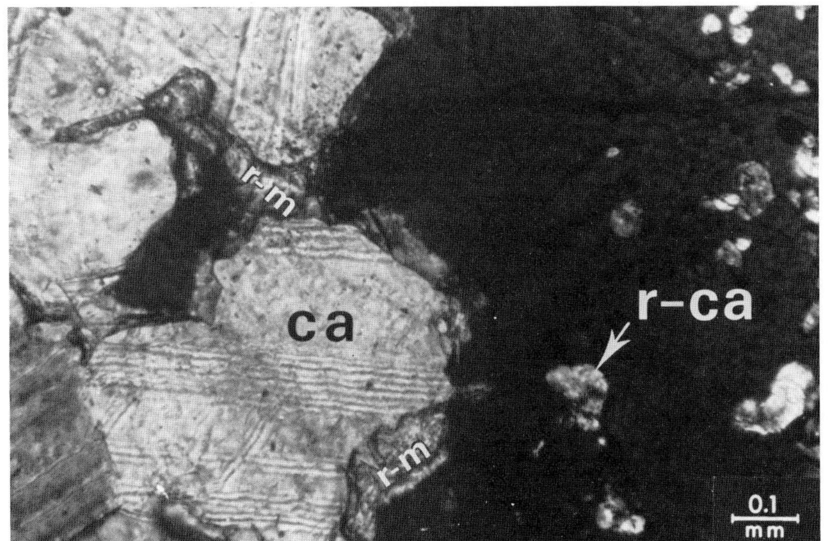

Fig. 52 Poikiloblastic garnet with protuberances invading the coarse grained marble along the intergranular and producing a reaction margin with calcite (r-m).
garnet = isotropic
ca = calcite
r-m = reaction margin of poikiloblast with the calcite.
r-ca = relic of calcite in the garnet poikiloblast.
Metasomatically affected marbles.
Malga Trivena, Val di Breguzzo, Adamello, Italy.
With crossed nicols.

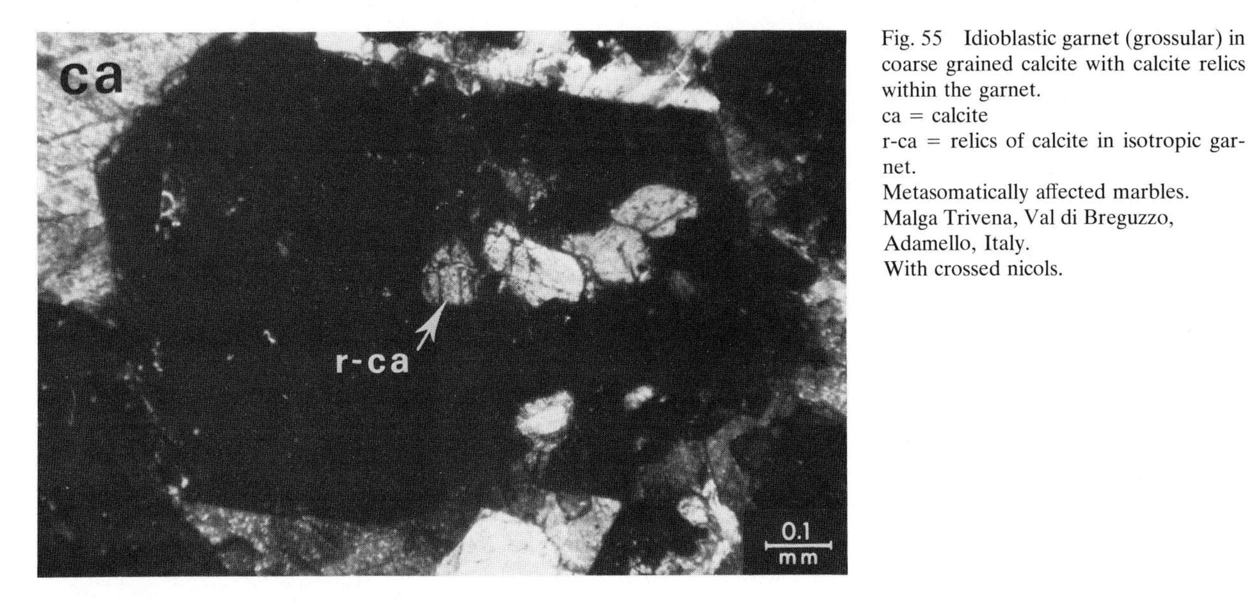

Fig. 55 Idioblastic garnet (grossular) in coarse grained calcite with calcite relics within the garnet.
ca = calcite
r-ca = relics of calcite in isotropic garnet.
Metasomatically affected marbles.
Malga Trivena, Val di Breguzzo, Adamello, Italy.
With crossed nicols.

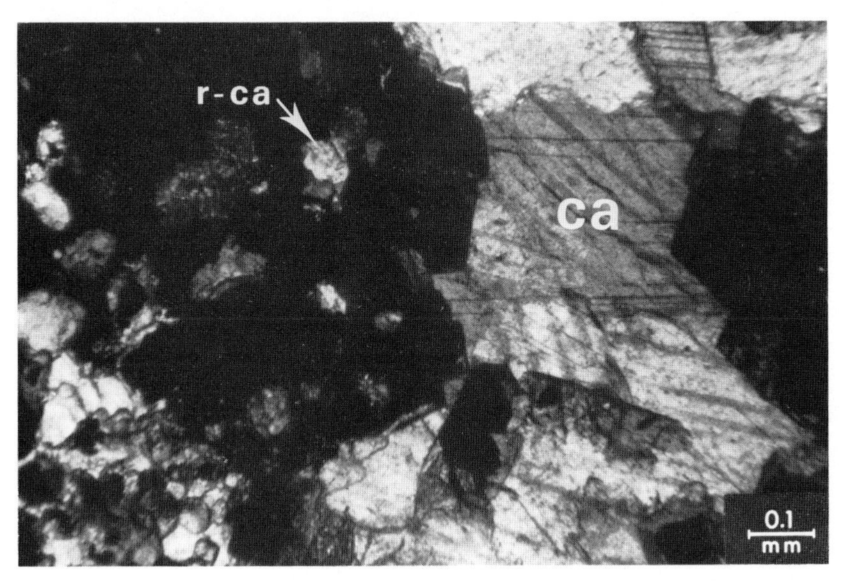

Fig. 56 Idioblastic garnet in coarse grained calcite (marble), enclosing calcite relics.
garnet = isotropic
ca = calcite
r-ca = relics of calcite in garnet
Metasomatically affected marbles.
Malga Trivena, Val di Breguzzo, Adamello, Italy.
With crossed nicols.

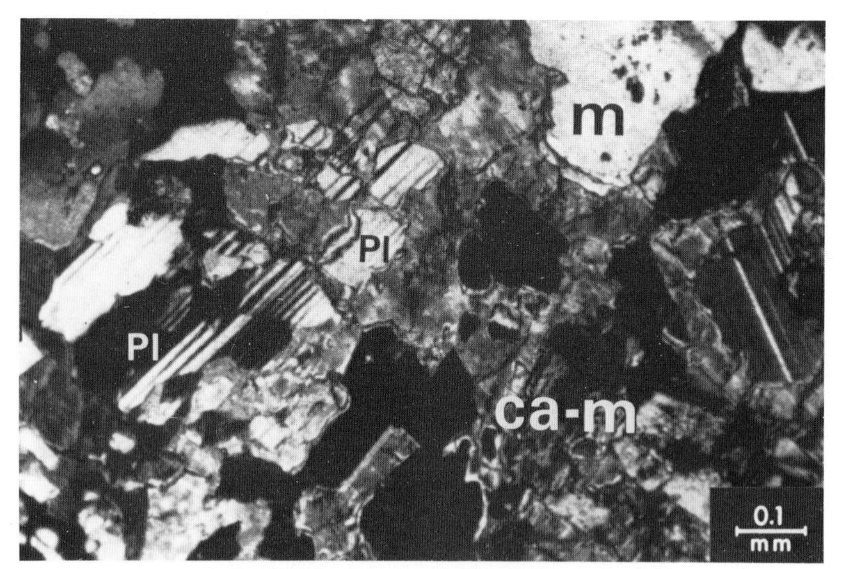

Fig. 57 Ca-rich plagioclase crystalloblasts and micaceous crystalloblasts in a calcite-mica granular texture.
Pl = plagioclase
m = mica crystalloblastesis
ca-m = calcite-micaceous mineral granular texture.
Metasomatically affected marbles.
Malga Trivena, Val di Breguzzo, Adamello, Italy.
With crossed nicols.

Fig. 58 Ca-rich plagioclase crystalloblasts in a calcite-micaceous-minerals texture.
Pl = plagioclase crystalloblast
ca-m = calcite micaceous texture
Metasomatically affected marbles.
Malga Trivena, Val di Breguzzo, Adamello, Italy.
With crossed nicols.

Fig. 59 Crystalloblastic ca-rich plagioclase and garnet in a calcite micaceous granular texture.
Pl = plagioclase crystalloblasts
gr = garnet crystalloblast
ca-m = calcite-micaceous-minerals
Metasomatically affected marbles.
Malga Trivena, Val di Breguzzo, Adamello, Italy.
With crossed nicols.

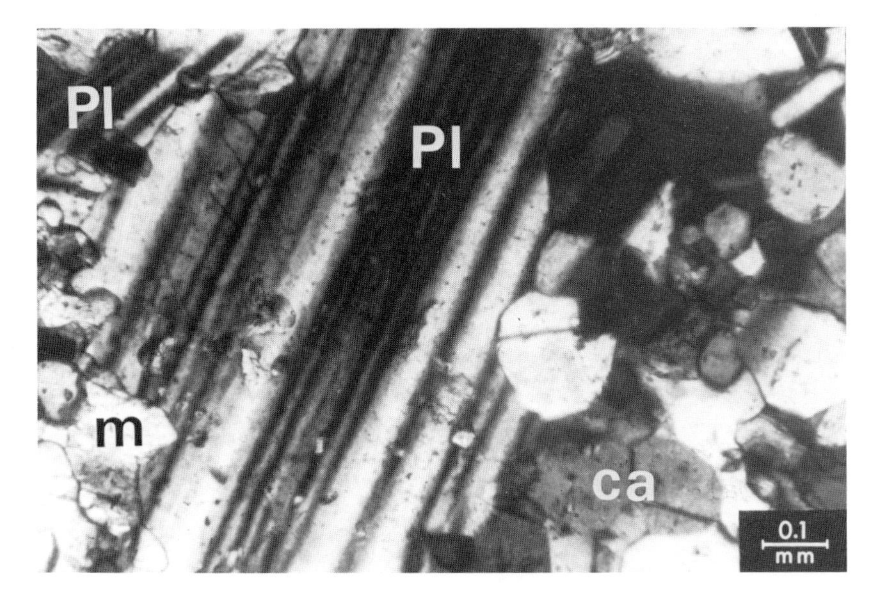

Fig. 60 Ca-rich plagioclase crystalloblasts in an equigranular calcite mica texture.
Pl = plagioclase crystalloblast
ca = equigranular calcite texture (marble)
m = mica
Metasomatically affected marbles.
Malga Trivena, Val di Breguzzo, Adamello, Italy.
With crossed nicols.

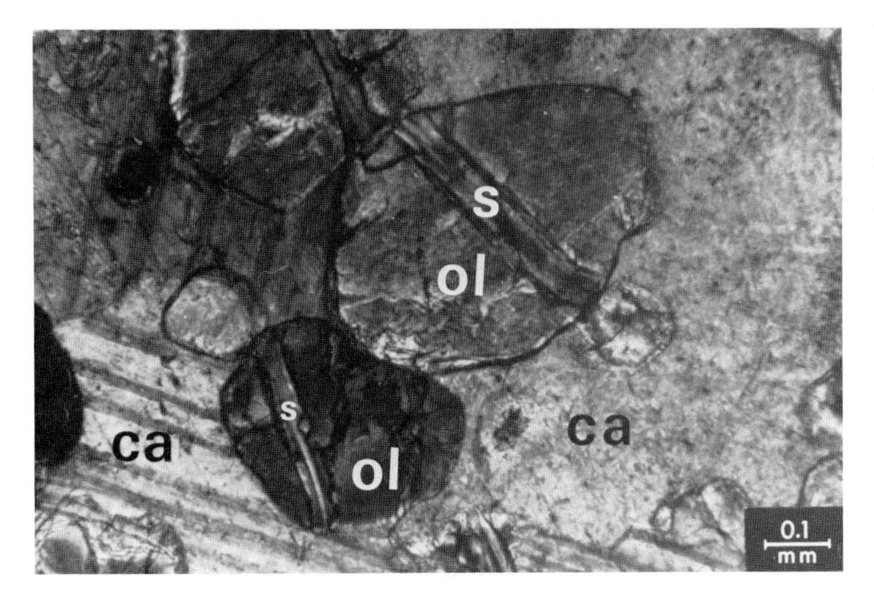

Fig. 61 Crystalloblastic and serpentinised olivines in calcite (coarse grained marble)
ol = olivine crystalloblasts
ca = calcite, coarse grained marble
s = serpentinisation of the olivine crystalloblast
Metasomatically affected marbles.
Malga Trivena, Val di Breguzzo, Adamello, Italy.
With crossed nicols.

Fig. 62 Crystalloblastic olivine serpentinised in coarse grained marble.
ol-b = olivine crystalloblast
s = serpentinisation of olivines
ca = calcite crystalloblast
Metasomatically affected marbles.
Malga Trivena, Val di Breguzzo, Adamello, Italy.
With crossed nicols.

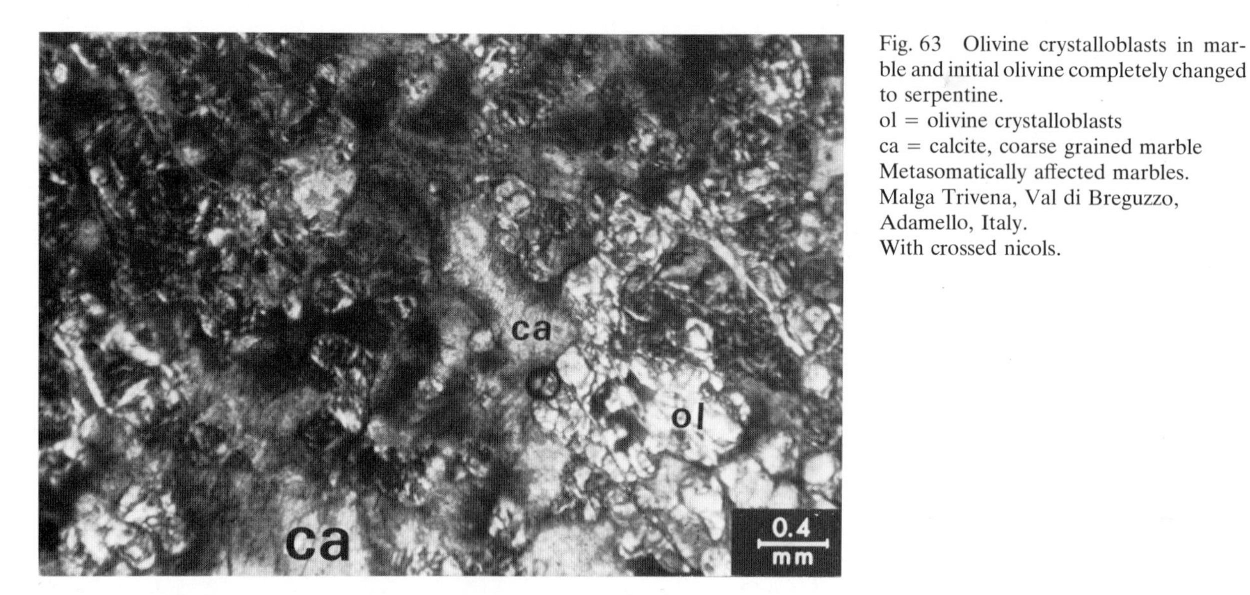

Fig. 63 Olivine crystalloblasts in marble and initial olivine completely changed to serpentine.
ol = olivine crystalloblasts
ca = calcite, coarse grained marble
Metasomatically affected marbles.
Malga Trivena, Val di Breguzzo, Adamello, Italy.
With crossed nicols.

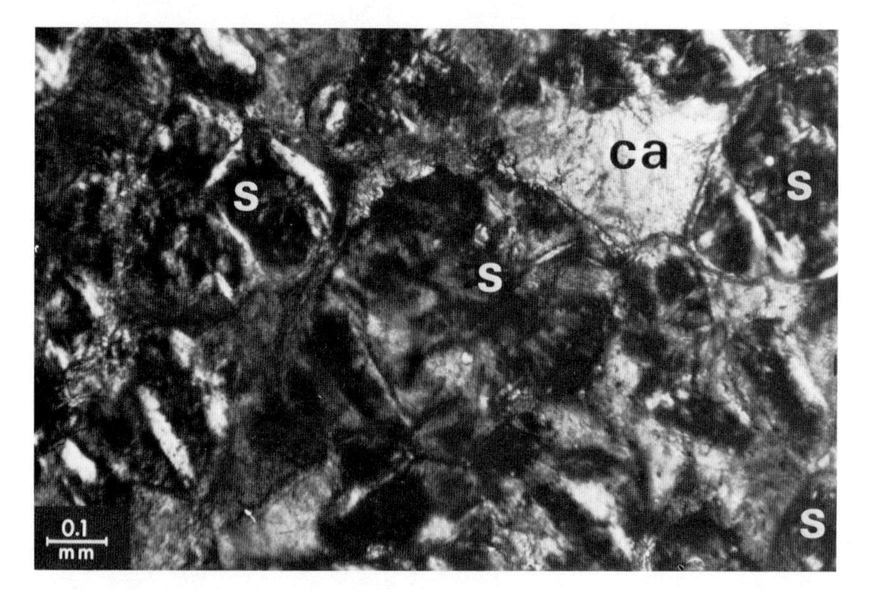

Fig. 64 Detail of Fig. 63 Initial olivine crystalloblasts in marble, coarse grained calcite.
s = serpentine, initial olivine grains completely changed into serpentine
ca = calcite, coarse grained marble
Metasomatically affected marbles.
Malga Trivena, Val di Breguzzo, Adamello, Italy.
With crossed nicols.

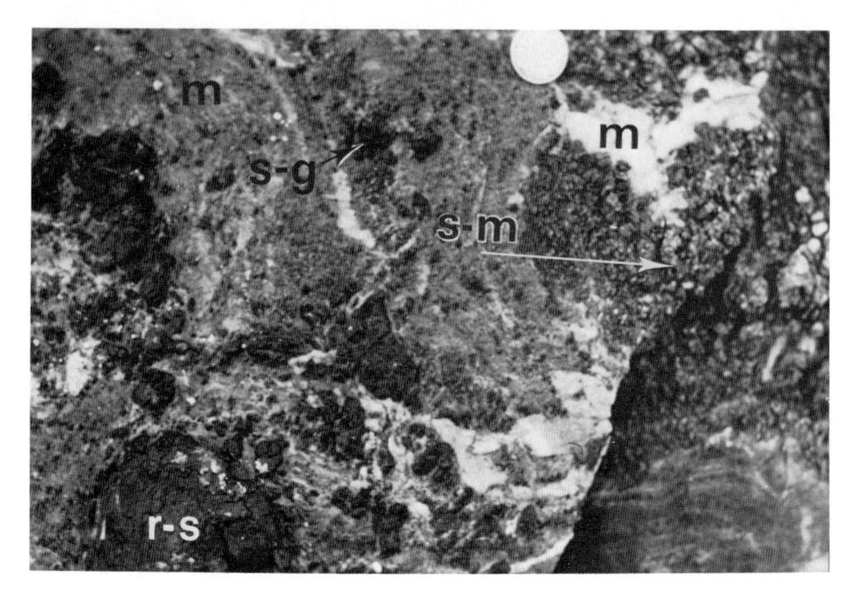

Fig. 65 Transition zone of serpentinised marble to metasomatic serpentine mass.
In the marble are included rounded bodies of serpentine (the roundening is synchronous or subsequent to serpentinisation and is due to tectonic effects) and scattered serpentine growths.
r-s = tectonically rounded serpentine bodies in marble.
m = marble
s-g = serpentine growths in marble
s-m = transition of scattered serpentine growths to massive metasomatic serpentine.
Piz Lunghin, Oberengadin, Alps.
Photograph from an excursion with Prof. Dr. F. K. Drescher-Kaden (1961).

Fig. 66 Gigantic antigorite crystalloblast with a banding, in recrystallised calcite.
An = antigorite
ca = calcite
Serpentinised marbles.
Piz Lunghin, Oberengadin, Alps.
With crossed nicols.

Fig. 67 Antigorite (serpentine) crystalloblast in coarse grained recrystallised calcite (marble).
An = antigorite
ca = calcite
Serpentinised marbles.
Piz Lunghin, Oberengadin, Alps.
With crossed nicols.

Fig. 68 Coarse grained recrystallised calcite with an antigorite idioblast.
ca = calcite
an = antigorite
Serpentinised marbles.
Piz Lunghin, Oberengadin, Alps.
With crossed nicols.

Fig. 69 An antigorite crystalloblast in a tectonically affected and recrystallised calcite.
ca = calcite
an = antigorite
Serpentinised marbles.
Piz Lunghin, Oberengadin, Alps.
With crossed nicols.

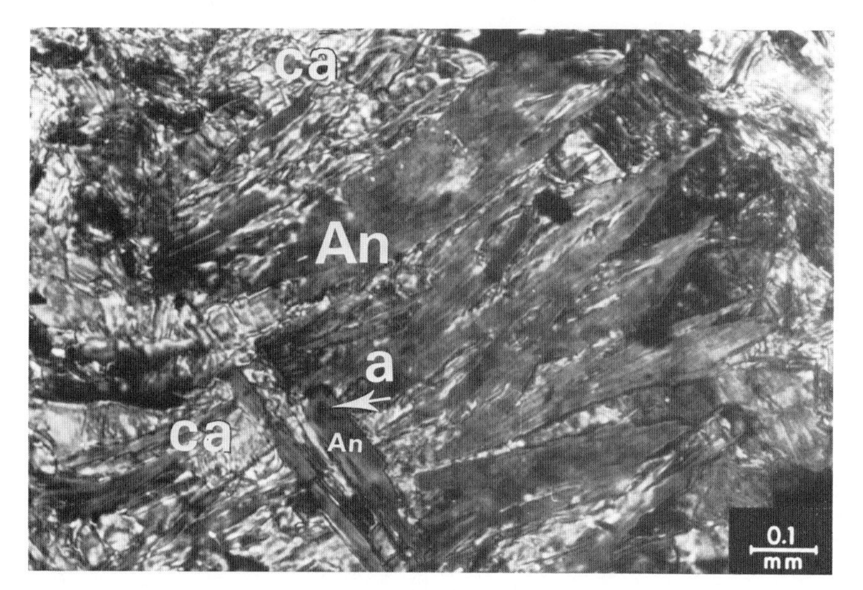

Fig. 70 Intersertal antigorite crystalloblast in calcite.
An = antigorite
ca = calcite. Arrow "a" shows interpenetration of the serpentine crystalloblasts.
Serpentinised marbles.
Piz Lunghin, Oberengadin, Alps.
With crossed nicols.

Fig. 71 Blastogenic often radiating and interpenetrating antigorite-laths in recrystallised calcite.
An = antigorite
ca = calcite
Serpentinised marbles.
Piz Lunghin, Oberengadin, Alps.
With crossed nicols.

Fig. 72 A blastogenic pattern of antigorite crystalloblast in recrystallised calcite.
An = antigorite
ca = calcite
Serpentinised marbles
Piz Lunghin, Oberengadin, Alps.
With crossed nicols.

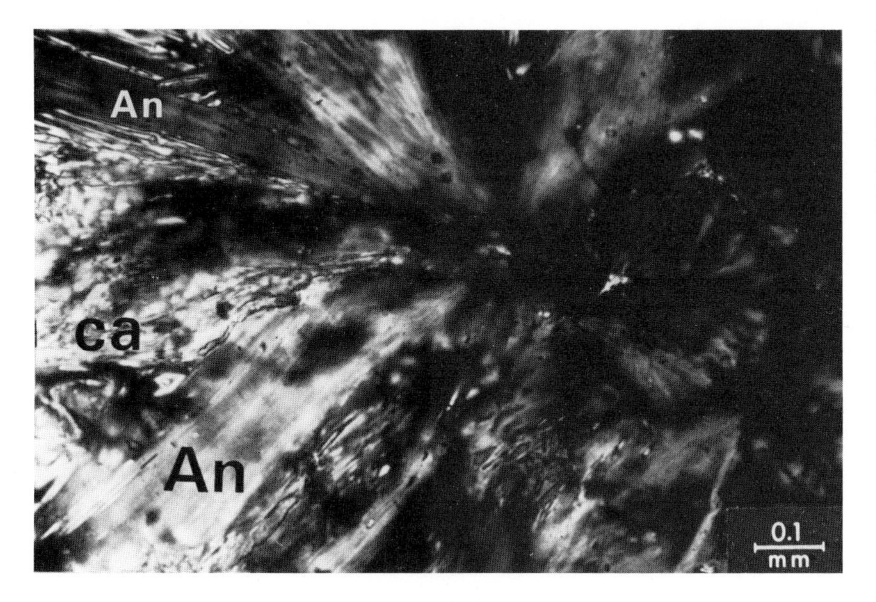

Fig. 73 A radiating aggregate of antigorite crystalloblasts in recrystallised calcite.
An = antigorite
ca = calcite (recrystallised)
Serpentinised marbles.
Piz Lunghin, Oberengadin, Alps.
With crossed nicols.

Fig. 74 Calcite (marble) with radiating serpentine crystalloblasts.
An = antigorite (serpentine)
ca = calcite
Serpentinised marbles.
Piz Lunghin, Oberengadin, Alps.
With crossed nicols.

Fig. 75 Calcite (marble) with blastogenic serpentine. The serpentineblasts form a band within the marble.
An = antigorite (serpentine)
ca = calcite (marble)
Serpentinised marbles.
Piz Lunghin, Oberengadin, Alps.
With crossed nicols.

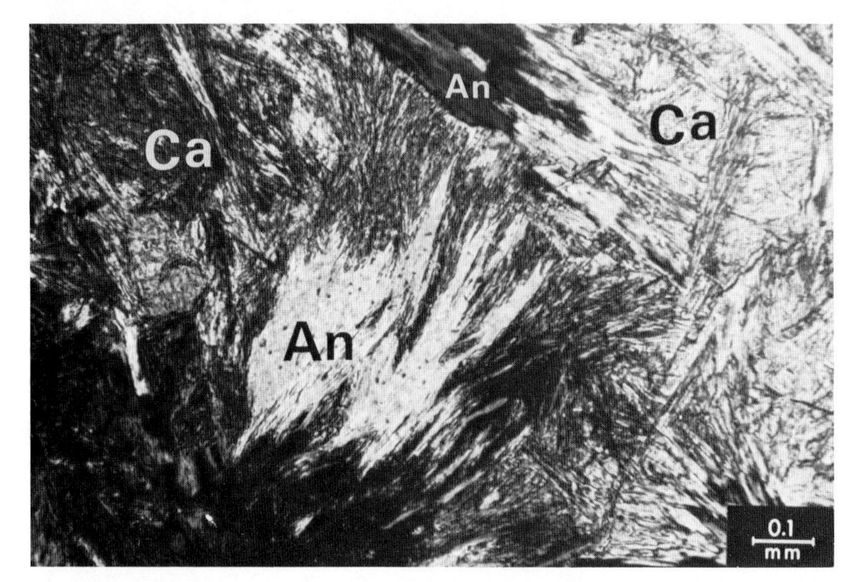

Fig. 76 Radiating serpentine laths within calcite (marble).
An = antigorite (serpentine)
Ca = calcite
Serpentinised marble. Phytia, Veria. N. Greece.
With crossed nicols.

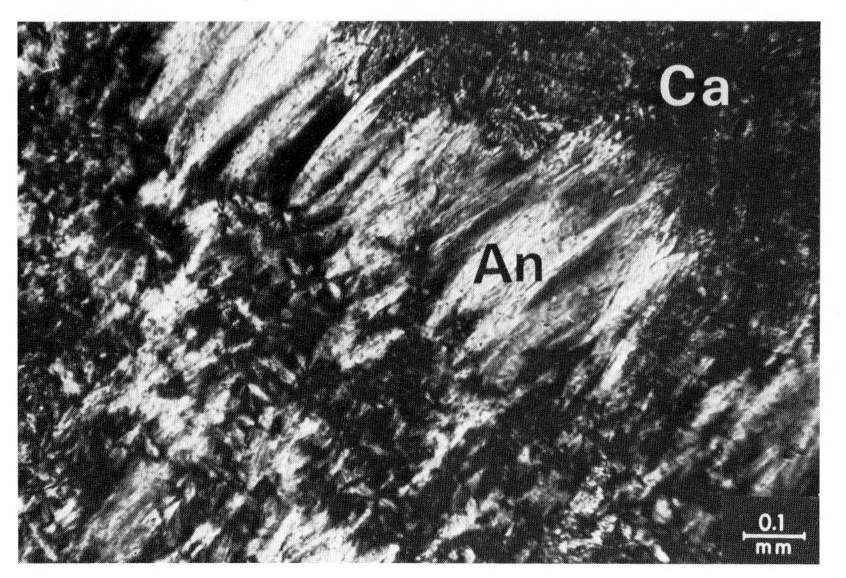

Fig. 77 Radiating serpentine crystalloblasts (antigorite-mass) within calcite (marble).
An = antigorite
Ca = calcite
Serpentinised marbles. Phytia, Veria, N. Greece.
With crossed nicols.

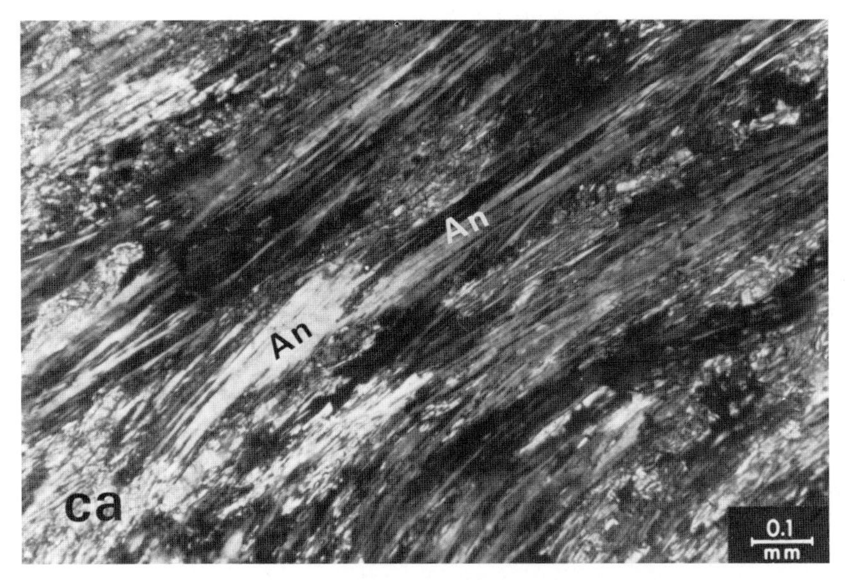

Fig. 78 Orientated antigorite crystalloblast within recrystallised calcite (marble).
An = antigorite (serpentine)
ca = calcite
Serpentinised marbles.
Piz Lunghin, Oberengadin, Alps.
With crossed nicols.

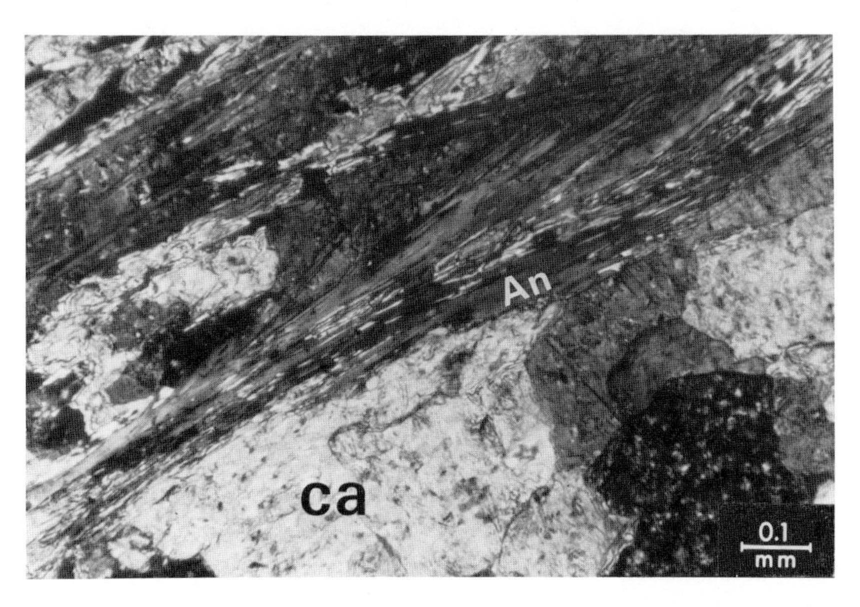

Fig. 79 Antigorite (serpentine) crystalloblast within recrystallised calcite (marble).
An = antigorite
ca = calcite
Serpentinised marbles.
Piz Lunghin, Oberengadin, Alps.
With crossed nicols.

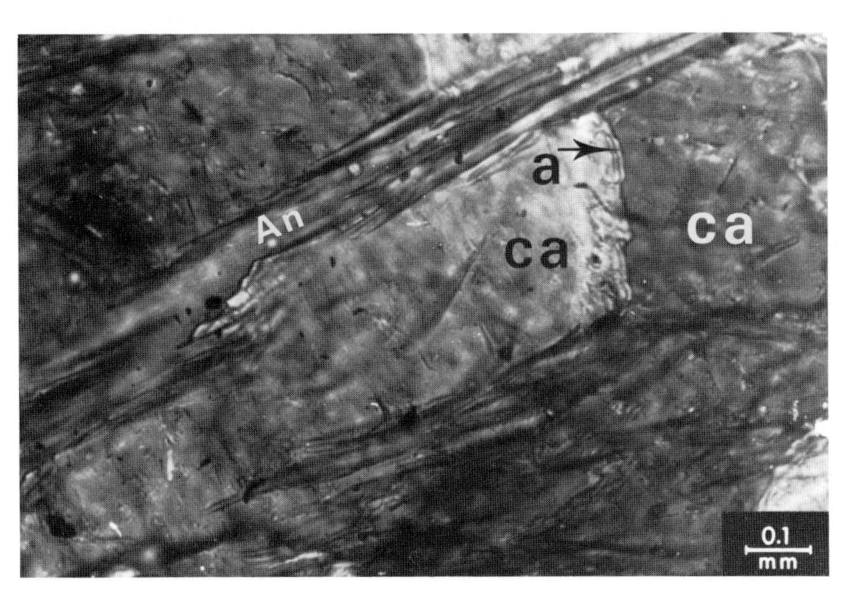

Fig. 80 Elongated antigorite crystalloblasts transversing the boundary of two coarse grained calcites.
An = antigorite crystalloblast
ca = calcite
Arrow "a" indicates the boundary between two adjacent coarse calcite grains.
Serpentinised marbles.
Piz Lunghin, Oberengadin, Alps.
With crossed nicols.

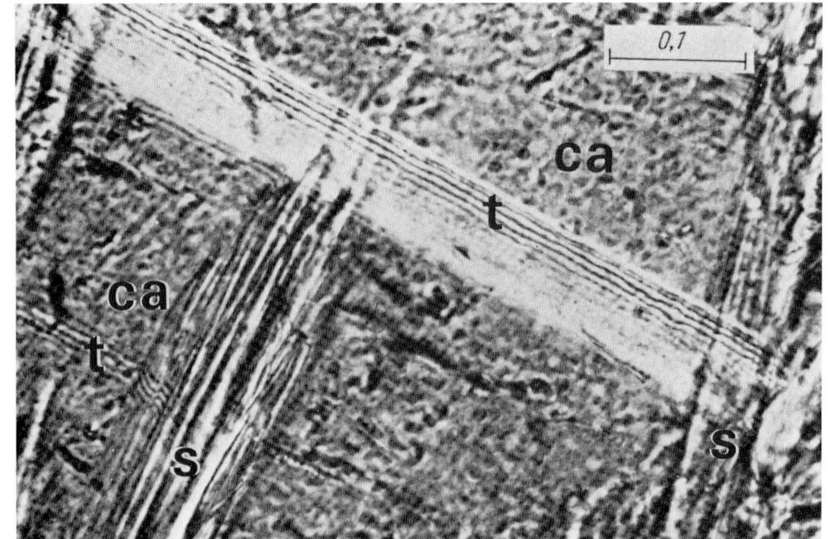

Fig. 81 Serpentine crystalloblasts transversing calcite twinning planes.
ca = calcite
s = serpentine
t = twinning plane
Malga Trivena, Val di Breguzzo, Adamello.
With crossed nicols.
Photomicrograph by Prof. F.K. Drescher Kaden (1969).

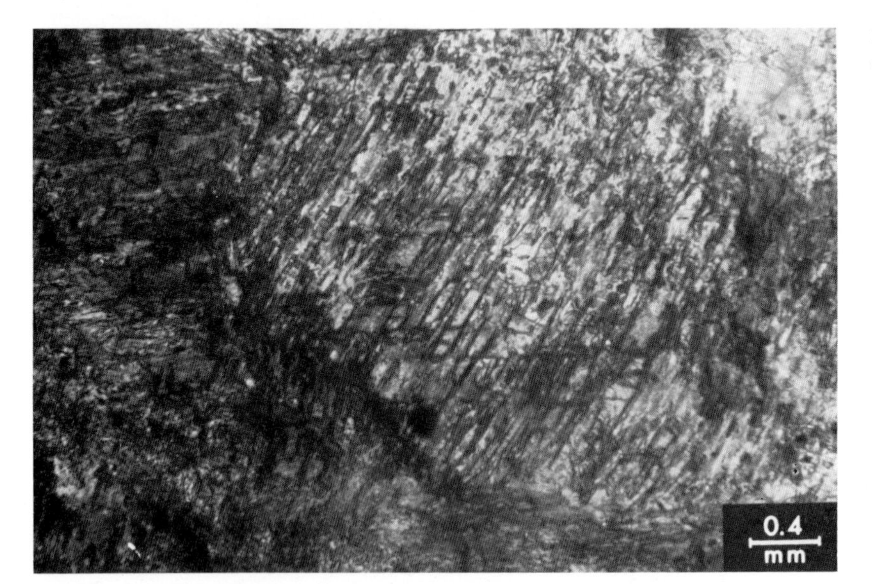

Figs. 82a and 82b Coarse grained recrystallised calcite with antigoritic serpentine orientated and following the rhombohedral calcite twinning.
An = antigorite (serpentine)
ca = calcite
Serpentinised marbles.
Piz Lunghin, Oberengadin, Alps.
With crossed nicols.

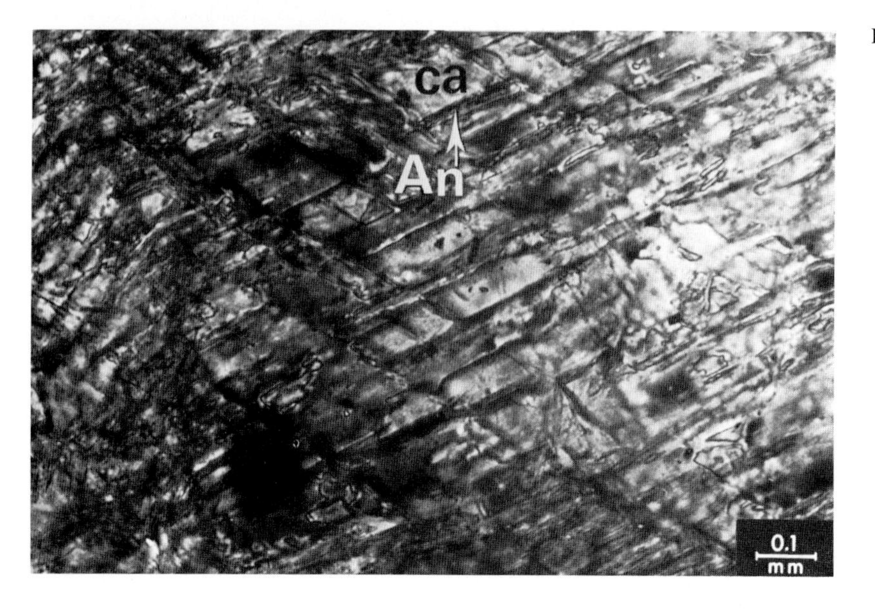

Fig. 82b

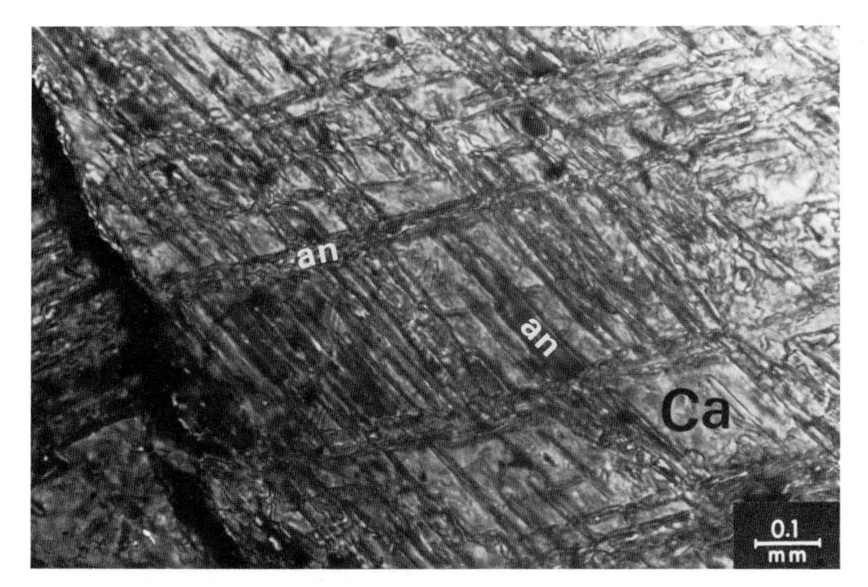

Fig. 83 Antigorite serpentine following the calcite twinning (coarse grained marble).
Ca = calcite
an = antigorite (serpentine)
Serpentinised marbles.
Piz Lunghin, Oberengadin, Alps.
With crossed nicols.

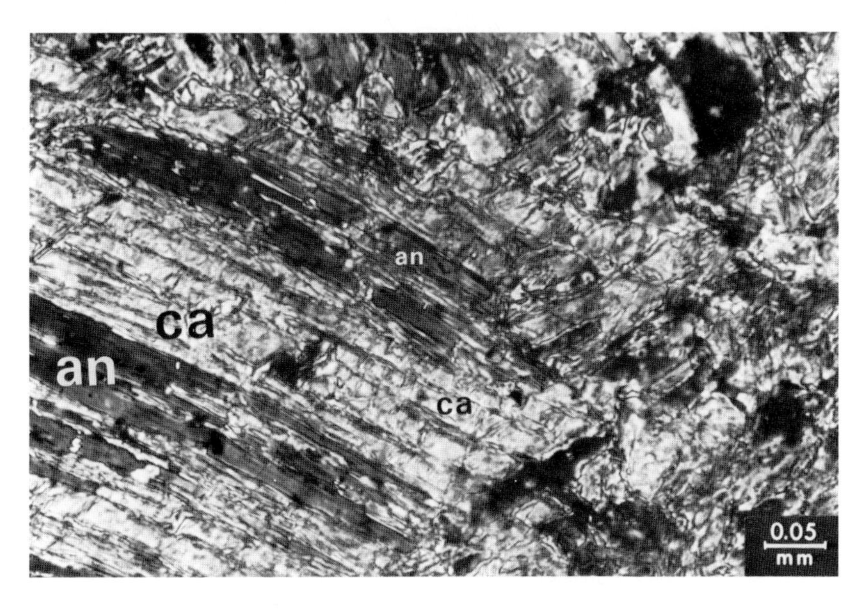

Figs. 85 and 86 Coarse grained calcite metasomatically replaced by later serpentine (antigorite), which has replaced the calcite not only along the twinning pattern but also in parts delimited by the calcite twinning.
ca = calcite
an = antigorite (serpentine). Arrows "a" show rhombohedral parts of calcite replaced by antigorite.
Serpentinised marbles.
Piz Lunghin, Oberengadin, Alps.
With crossed nicols.

(For Fig. 84 see page 140)

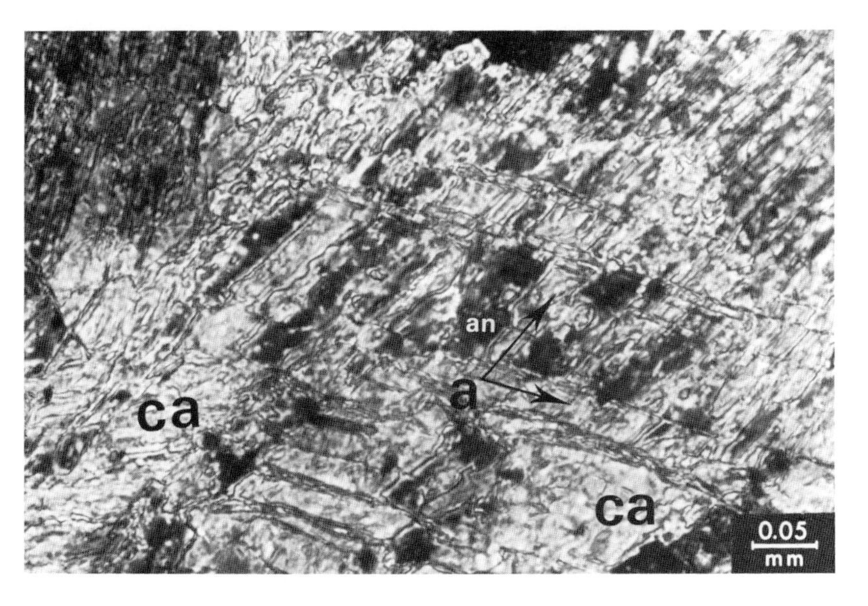

Fig. 86

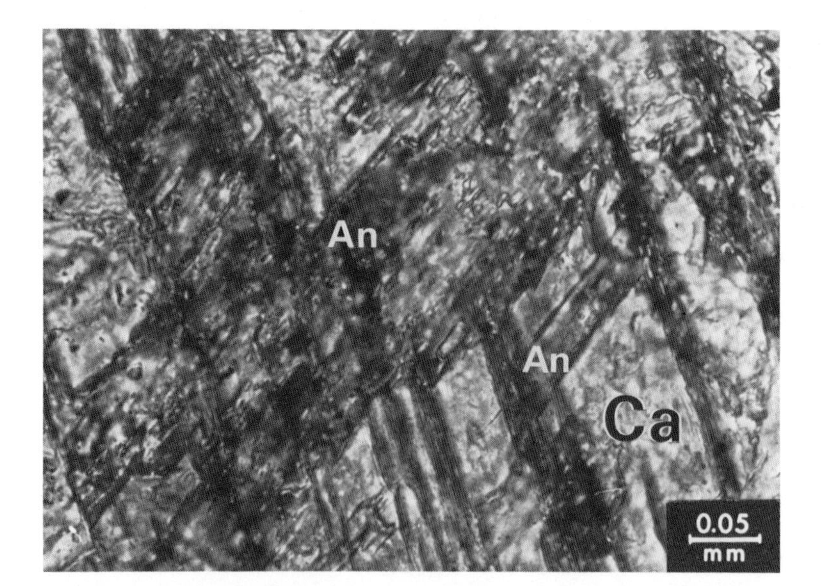

Fig. 84 Coarse grained calcite replaced along its rhombohedral twinning by later metasomatic serpentine (antigorite).
An = antigorite
Ca = calcite
Serpentinised marbles.
Piz Lunghin, Oberengadin, Alps.
With crossed nicols.

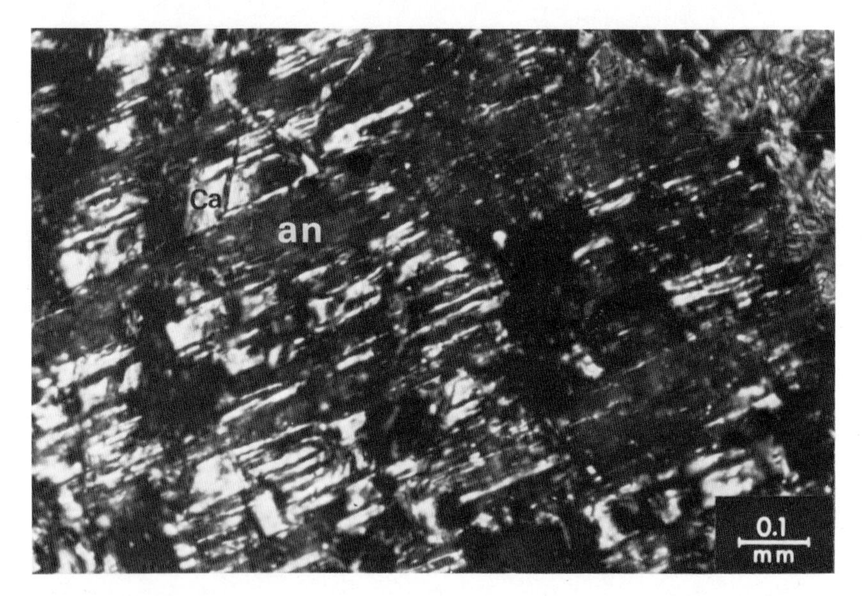

Fig. 87 Coarse grained calcite replaced by antigorite (serpentine). The antigorite has replaced the calcite along its rhombohedral twinning pattern.
Ca = calcite
an = antigorite
Serpentinised marbles.
Piz Lunghin, Oberengadin, Alps.
With crossed nicols.

Fig. 88 Coarse grained calcite metasomatically replaced by antigoritic serpentine which follows the rhombohedral twinning pattern of the calcite.
An = antigorite
ca = calcite
Arrows "a" show the rhombohedral pattern of the calcite.
Serpentinised marbles.
Piz Lunghin, Oberengadin, Alps.
With crossed nicols.

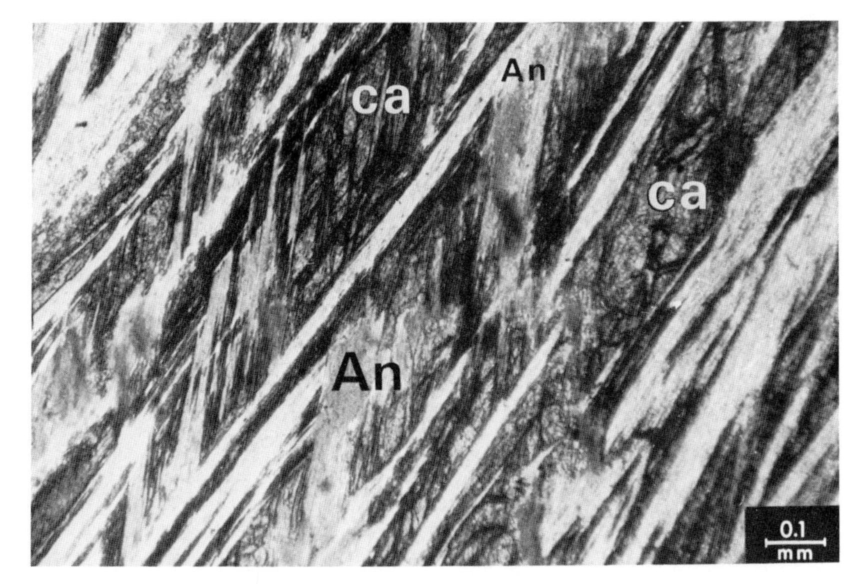

Fig. 89 Megablastic serpentine (antigorite) replacing the coarse grained calcite along its rhombohedral cleavage pattern.
ca = calcite
An = antigorite (serpentine)
Serpentinised marble. Phytia, Veria, N. Greece.
With crossed nicols.

Fig. 90 Metasomatic antigorite associated with calcite and with blastically grown tremolite.
ca = calcite
an = antigorite (serpentine)
tr = tremolite
Serpentinised marbles.
Piz Lunghin, Oberengadin, Alps.
With crossed nicols.

Fig. 91 Metasomatic antigorite associated with calcite and with blastically grown tremolite.
ca = calcite
an = antigorite (serpentine)
tr = tremolite
Serpentinised marbles.
Piz Lunghin, Oberengadin, Alps.
With crossed nicols.

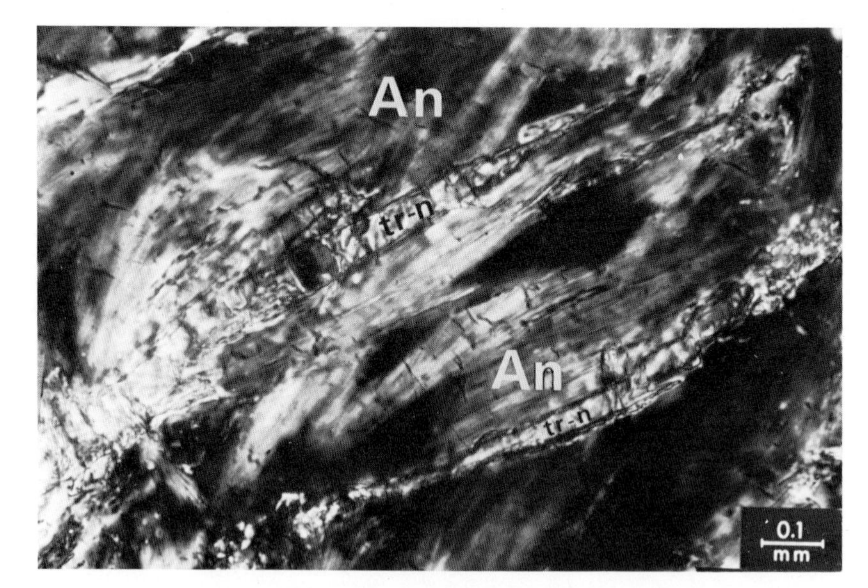

Fig. 92 Tremolite nematoblast in antigoritic (serpentine) which has metasomatically replaced the recrystallised calcite texture.
An = antigorite
tr-n = tremolitic nematoblast
Serpentinised marbles.
Piz Lunghin, Oberengadin, Alps.
With crossed nicols.

Figs. 94a and 94b Veinlets of antigorite transversing recrystallised calcite (marble).
An = antigorite (serpentine)
ca = calcite
Serpentinised marbles.
Piz Lunghin, Oberengadin, Alps.
With crossed nicols.

(For Fig. 93 see page 143)

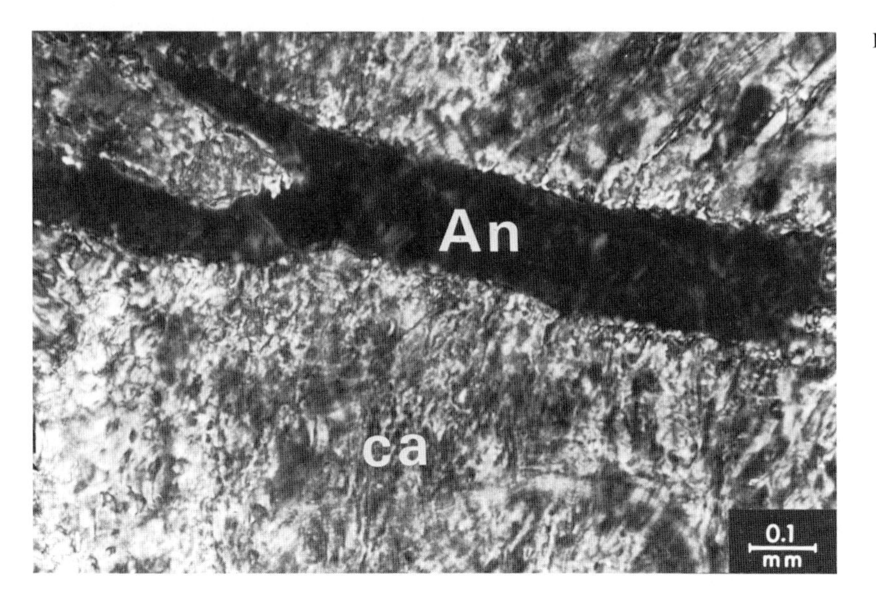

Fig. 94b

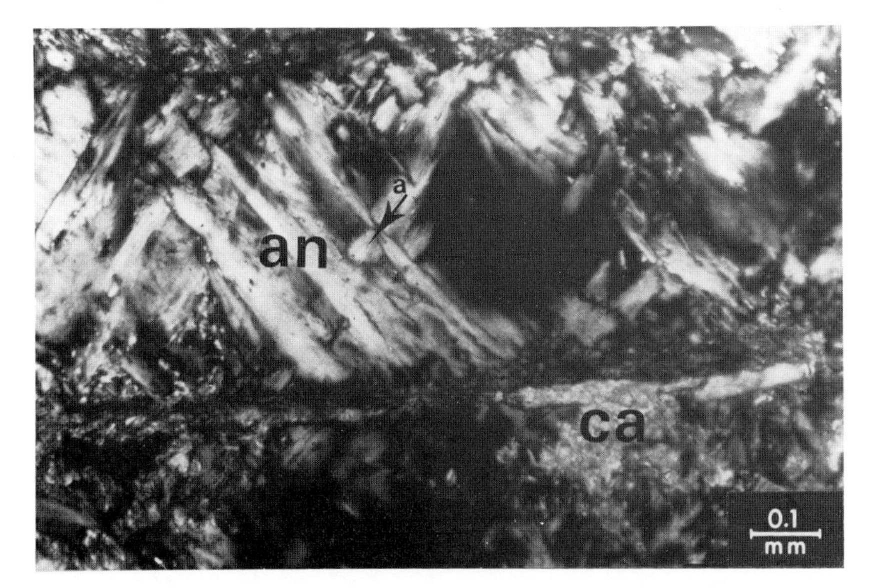

Fig. 93 An antigorite veinlet transversing a re-crystallised calcite texture. The serpentine veinlet consists of interpenetrating antigorite laths.
ca = calcite
an = antigorite
Arrow "a" shows antigorite interpenetration.
Serpentinised marbles.
Piz Lunghin, Oberengadin, Alps.
With crossed nicols.

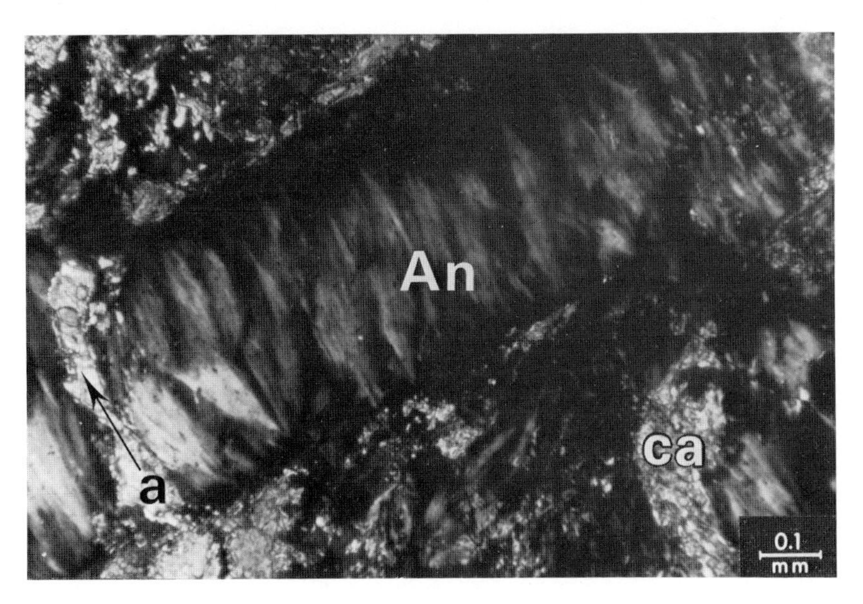

Fig. 95 Antigorite (serpentine) veinlet with antigorite cross-fibres, transversing recrystallised calcite (marble).
Also remobilised calcite transversing the serpentine veinlet.
ca = calcite
an = antigoritic veinlet. Arrow "a" remobilised calcite transversing the antigoritic veinlet.
Serpentinised marbles.
Piz Lunghin, Oberengadin, Alps.
With crossed nicols.

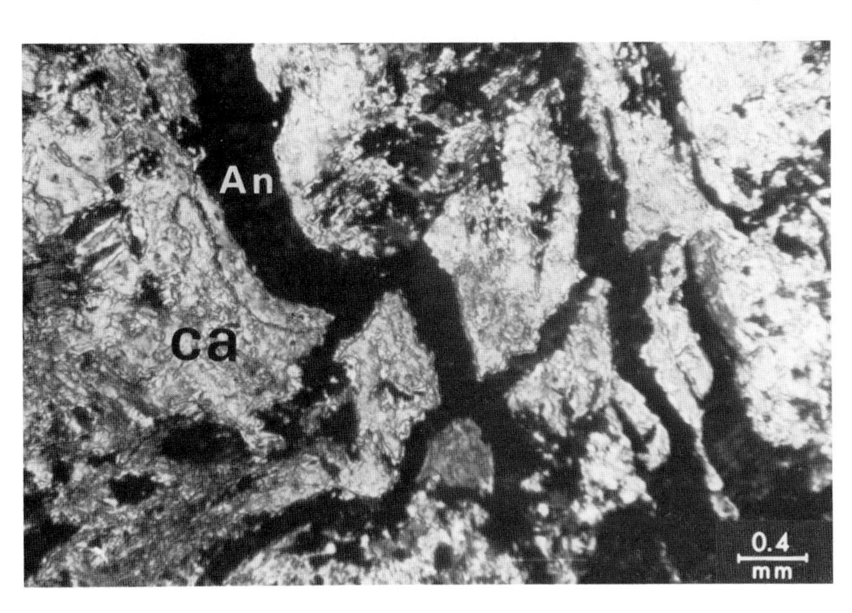

Fig. 96 Branching and anastomosing antigorite veinlets transversing a recrystallised calcite (marble).
ca = calcite
an = antigorite (serpentine)
Serpentinised marbles.
Piz Lunghin, Oberengadin, Alps.
With crossed nicols.

Fig. 97 Calcite (marble) with magnetite veinlet and blastogenic serpentine.
an = antigorite (serpentine), magnetite (black)
ca = calcite
Serpentinised marbles.
Piz Lunghin, Oberengadin, Alps.
With crossed nicols.

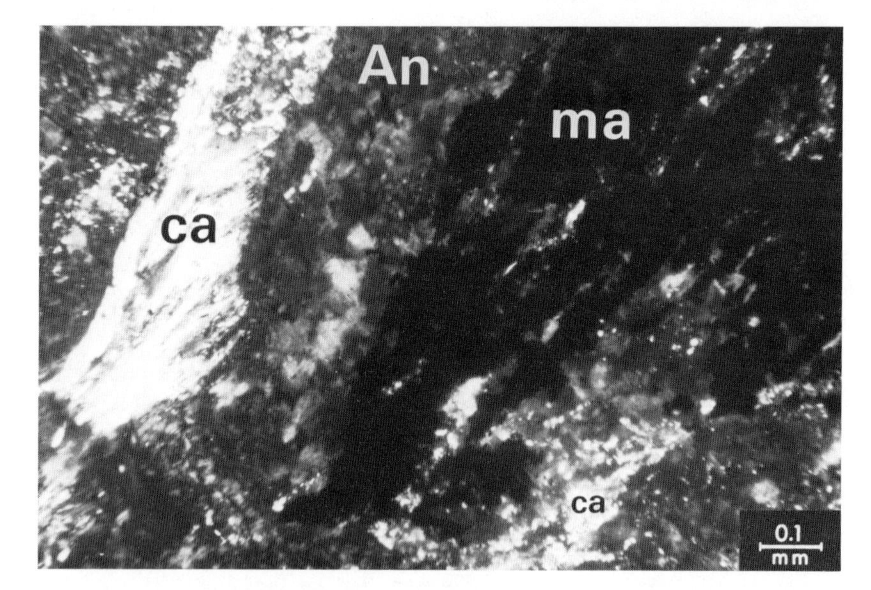

Fig. 98 Antigorite with magnetite aggregates associated with recrystallised calcite.
An = antigorite
ma = magnetite
ca = calcite
Serpentinised marbles.
Piz Lunghin, Oberengadin, Alps.
With crossed nicols.

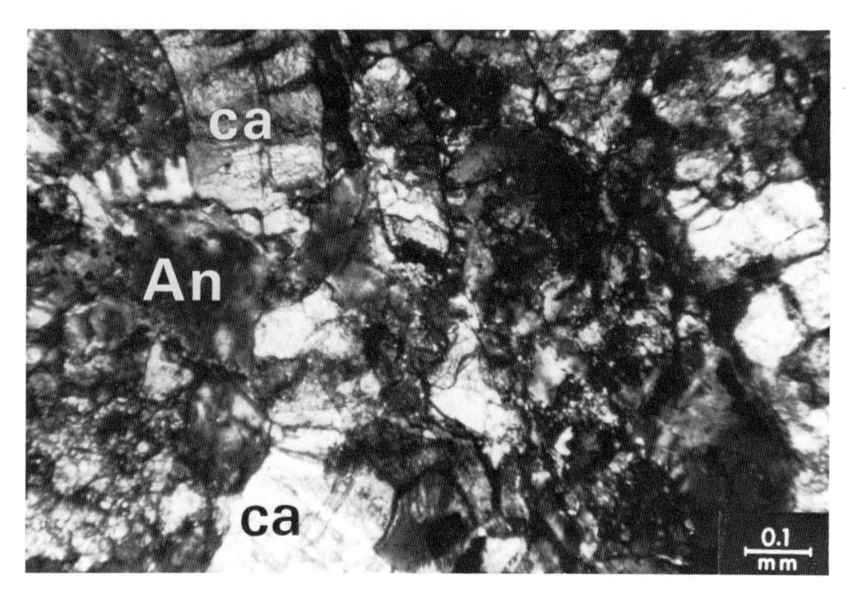

Fig. 99 Coarse grained granular marble with metasomatic antigorite (serpentine).
ca = calcite
An = antigorite (serpentine)
Serpentinised marble. Phytia, Veria, N. Greece.
With crossed nicols.

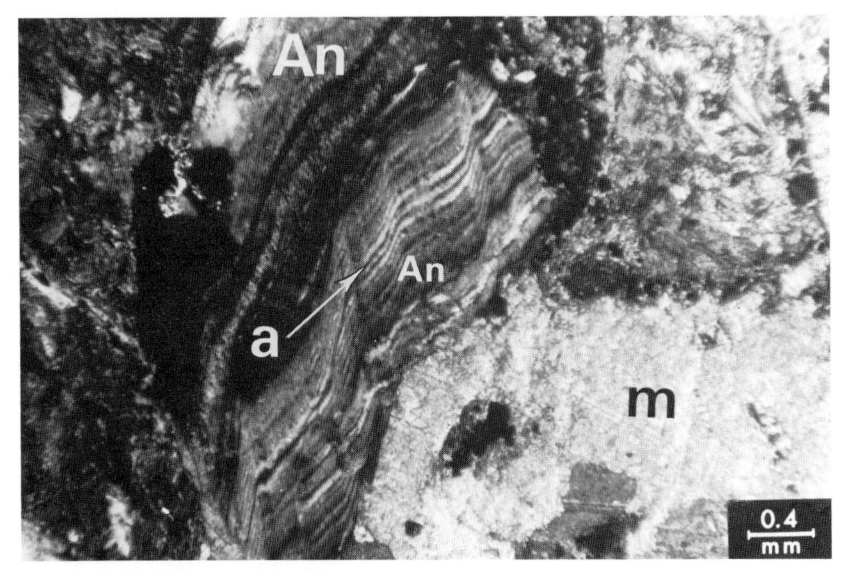

Fig. 100 Banded and tectonically deformed antigorite-serpentine fragment in a crystalline calcite texture.
An = antigorite serpentine fragment. Arrow "a" shows tectonic deformation of the banded serpentine fragment.
m = recrystallised calcite (marble), black (magnetite associated with the serpentine).
Serpentinised marbles.
Piz Lunghin, Oberengadin, Alps.
With crossed nicols.

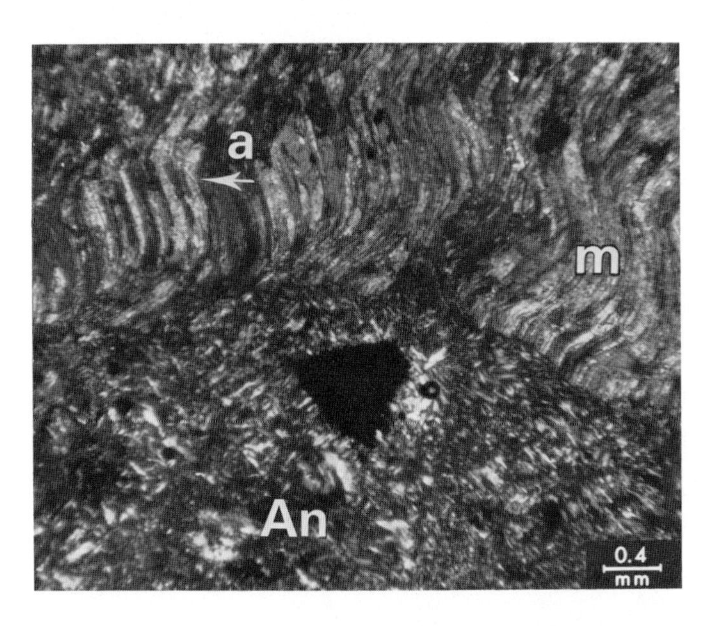

Fig. 101 An aggregate of antigoritic serpentine (a fragment?) with magnetite in marble. The marble shows an orientation of the calcite with a crystal bending indicated, adjacent to the serpentine aggregate.
An = antigorite (serpentine), magnetite (black)
m = marble. Arrow "a" shows calcite crystal bending.
Serpentinised marbles.
Piz Lunghin, Oberengadin, Alps.
With crossed nicols.

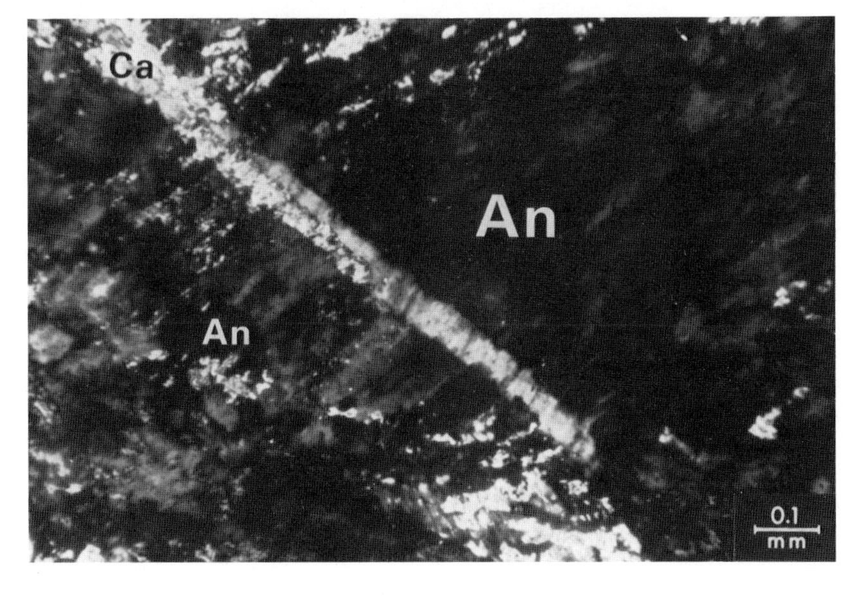

Fig. 102 Remobilised calcite veinlet transversing antigorite (serpentine) mass.
An = antigorite (serpentine)
ca = remobilised calcite veinlet
Serpentinised marbles.
Piz Lunghin, Oberengadin, Alps.
With crossed nicols.

Fig. 103 Olivinefels xenoliths of an olivinefels mylonite in a fine-grained granite. Corrosion and talc formation took place as a reaction of the olivinefels xenolithic fragments with the fine grained granite.
Val Bondasca.
Photograph from an excursion with Prof. Dr. F. K. Drescher-Kaden (1961).

Fig. 104 Shows the contact of the Seriphos Granite with the country-rock (hornblende quartzitic schist). Xenoliths (single bodies) and an agmatite "zone" exist in the transition zone country-rock → granite.
G = granite
C-R = country-rock (hornblende quartzitic schist)
ag = agmatite "zone" in the granite
x = larger xenolith
Halara, Seriphos, Greece.

Fig. 105 Agmatite zone as a transition phase between the contact of country rock (hornblende-biotite quartzitic schist) and a granitic phase with scattered xenoliths. The agmatite is a granitic phase rich in basic xenoliths.
x = xenoliths
g = granite
Agmatite, Halara, Seriphos, Greece.

Fig. 106 Shows the transition contacts, hornblende-biotite quartzitic schist (country rock) / agmatite granite. Also mobilised initial country rock as "veinform xenoliths" are shown.
g = granite
c-r = country rock (mainly a hornblende-biotite quartzitic schist).
ag = agmatite (transition zone between the country rock and the granite).
v-x = "veinform" xenoliths, mobilised intitial country rock material.
Halara, Seriphos, Greece.

Fig. 107 The Seriphos granite with "veinform xenoliths" (mobilised country rock).
v-x = "Veinform xenolith"
Halara, Seriphos, Greece.

Fig. 108 Veinform "xenolith" consisting of hornblendite, of comparable composition to the country rock, see Fig. 110. The veinform "xenolith" follows a joint-system of the granite and is telescoping within the granitic mass (see Fig. 107).
Halara, Seriphos, Greece.

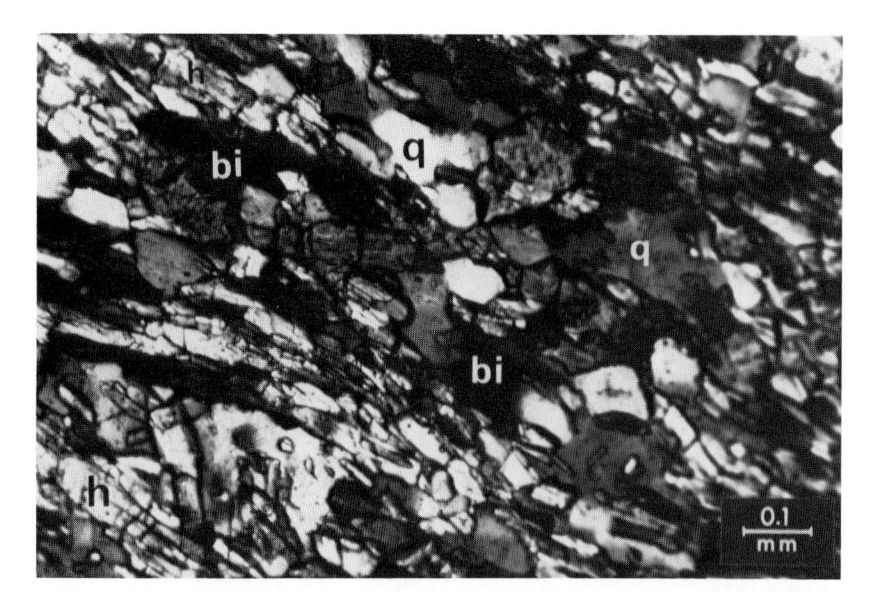

Fig. 109 Shows the texture and composition of the metamorphic country rock (see Fig. 106), consisting of hornblende, green biotite, quartzite.
h = hornblende
bi = biotite
q = quartz
Halara, Seriphos, Greece.
With crossed nicols.

Fig. 110 Metamorphosed country rock (hornblende quartzitic schist), composed mainly of hornblende, green biotite, quartz and epidote. Also a plagioclase crystalloblast is present.
pl =plagioclase crystalloblast
q = quartz
h = hornblende
bi = biotite (green)
e = epidote
Halara, Seriphos, Greece.
With crossed nicols.

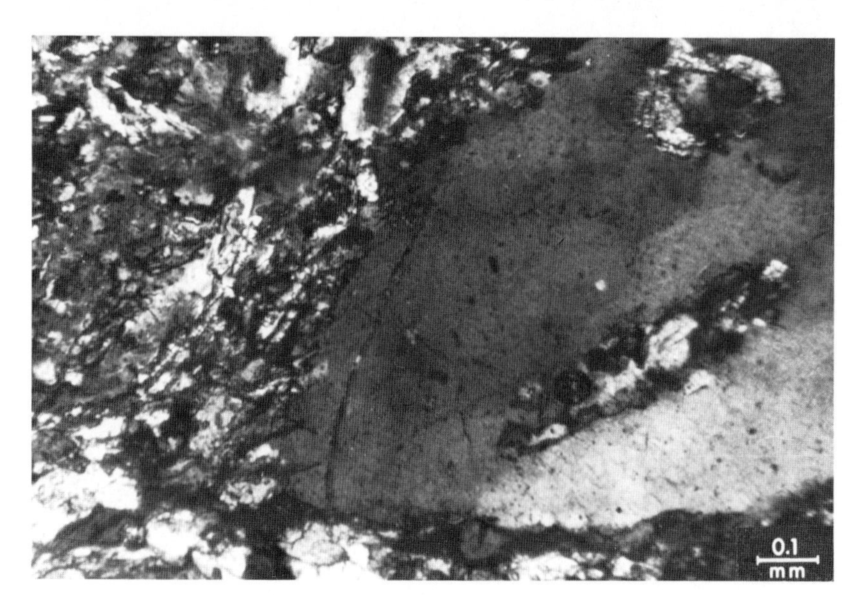

Fig. 111 Shows quartzitic relic, consisting of interlocking and undulating quartz grains in the hornblende-quartzite mass of the veinform "xenolith" (see Fig. 108).
The quartzite relic clearly shows a metamorphic origin (comparable to the country rock) for the veinform "xenoliths".
Halara, Seriphos, Greece.
With crossed nicols.

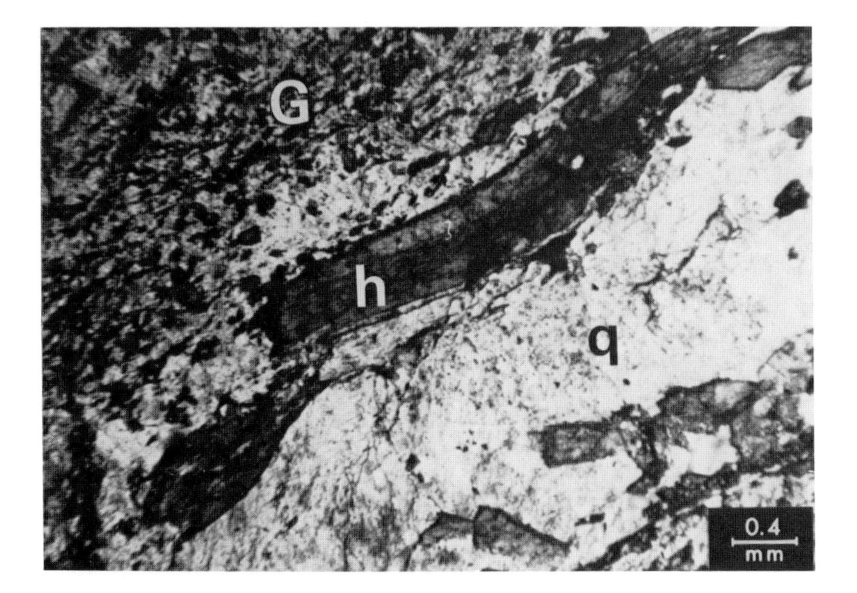

Fig. 112 Tectonically bent hornblende (paracrystalline) adjacent to a quartz which is actually also tectonically deformed (see Fig. 111).
h = hornblende (bent)
q = tectonically deformed quartz
G = hornblende-quartz groundmass
Hornblendite vein transversing the Seriphos granite.
Halara, Seriphos, Greece.
Without crossed nicols.

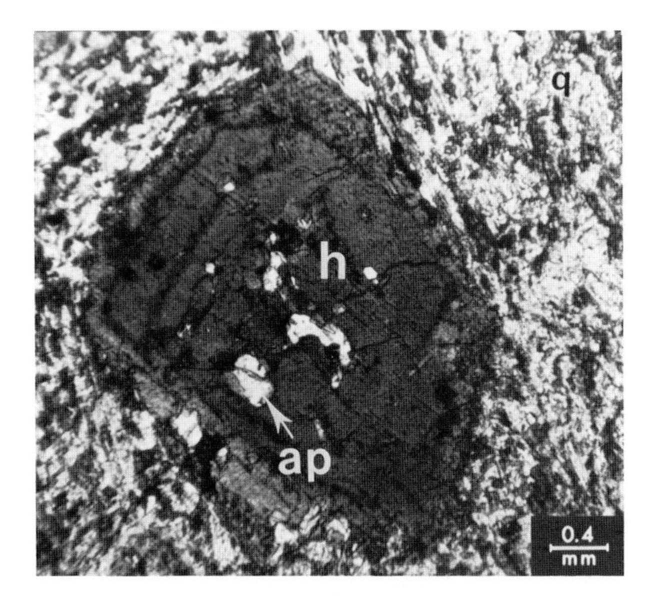

Fig. 113 Idioblastic hornblende in the quartzitic hornblendite "veins" (mobilised country rock) following granite fissures in a hornblende-quartzitic groundmass.
H = hornblende idioblast.
ap = apatite included in the hornblende crystalloblast
G = groundmass consisting of hornblende quartzite.
Hornblendite "vein" following fissures of the Seriphos granite.
Halara, Seriphos, Greece.
Without crossed nicols.

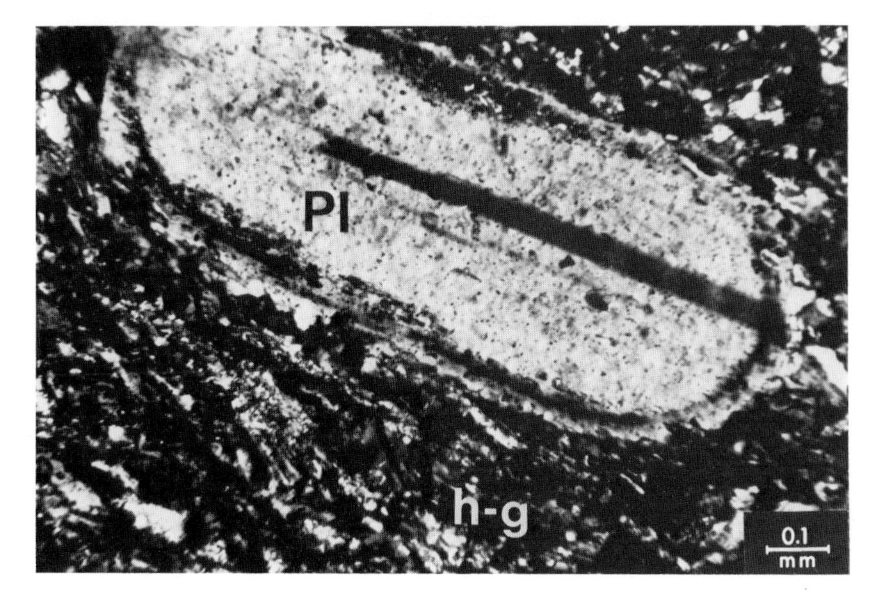

Fig. 114 Plagioclase crystalloblasts, corroded and affected by the hornblende-quartz groundmass of the veinform hornblendite (xenolith).
Pl = plagioclase
h-g = hornblende-quartz groundmass
Halara, Seriphos, Greece.
With crossed nicols.

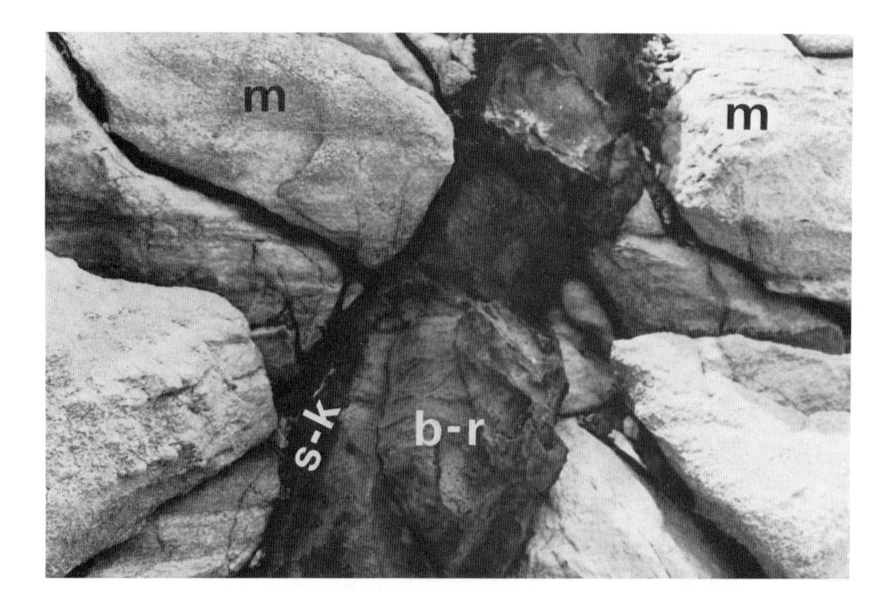

Fig. 115 "Basic veins" transversing the Halara marbles. A distinct skarn contact is produced in the contact of the "basic vein" with the marble.
m = Halara marble
b-v = basic veins transversing the Halara marbles
s-k = epidote-garnet skarns at the contact of the "basic vein" with the marble.
Halara, Seriphos, Greece.

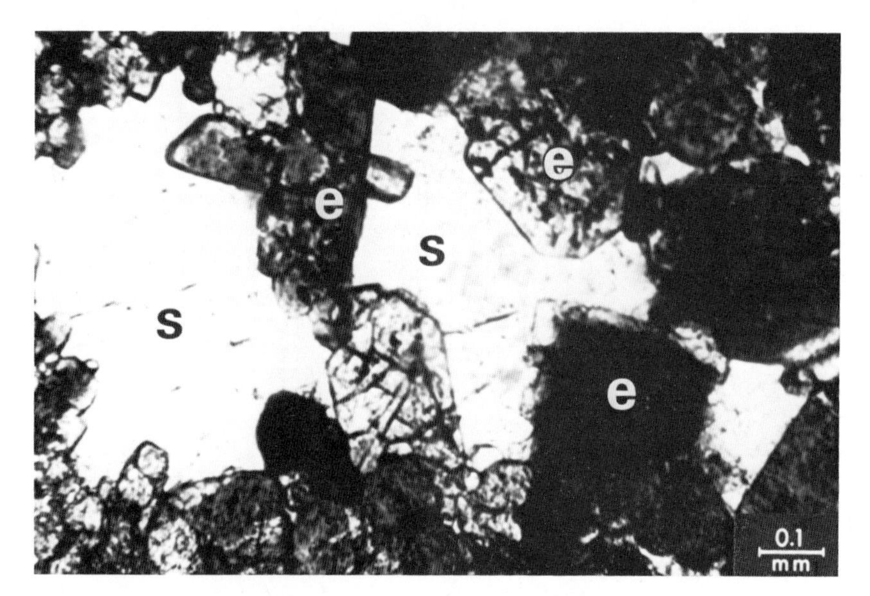

Fig. 116 Idioblastic epidote with interstitial scapolite.
e = epidote
s = scapolite
Veins transversing the Halara marbles.
Halara, Seriphos, Greece.
With crossed nicols.

Fig. 117 Idioblastic diopside in a groundmass consisting of interlocking quartz grains and epidote, also plagioclases are present.
d = diopside
e = epidote
q = quartz
f = feldspar
Veins transversing the Halara marbles.
Halara, Seriphos, Greece.

Fig. 118 Shows the middle mass of the green veins transversing the Halara marbles, consisting of fine epidote and laths of plagioclase showing a "fluidal like" orientation.
ep = fine epidote mass
f = plagioclase laths
Green veins transversing the Halara marbles (see Fig. 115).
Halara, Seriphos, Greece.

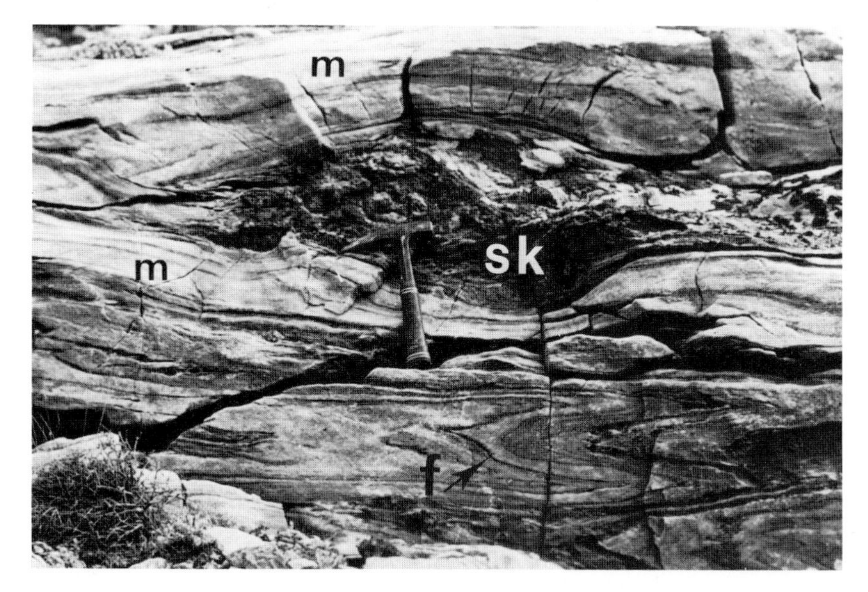

Fig. 119 Folded marbles (pre-granitic metamorphism) into which a skarn body has been emplaced.
m = marble
f = pre-granitic folding of marbles
sk = skarn emplaced in the folded marbles
Granite Skarns
Halara, Seriphos, Greece.

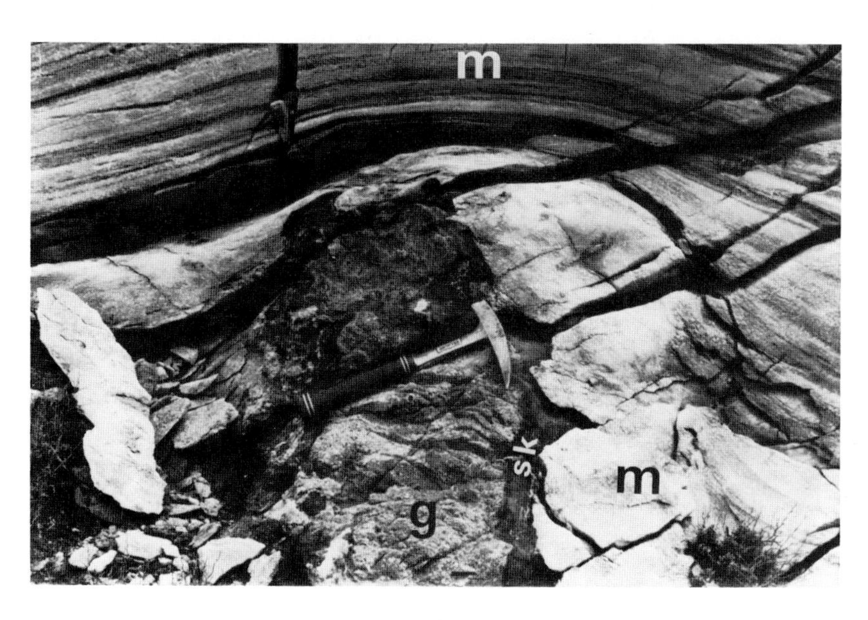

Fig. 120 A granitic apophysis intruding and arching marbles into which it has been pushed as a solid-plastic mass. Epidote-garnet skarns are produced between the marble and the granite.
m = marbles
sk = epidote-garnet skarns
g = granitic apophysis.
Halara, Seriphos, Greece.

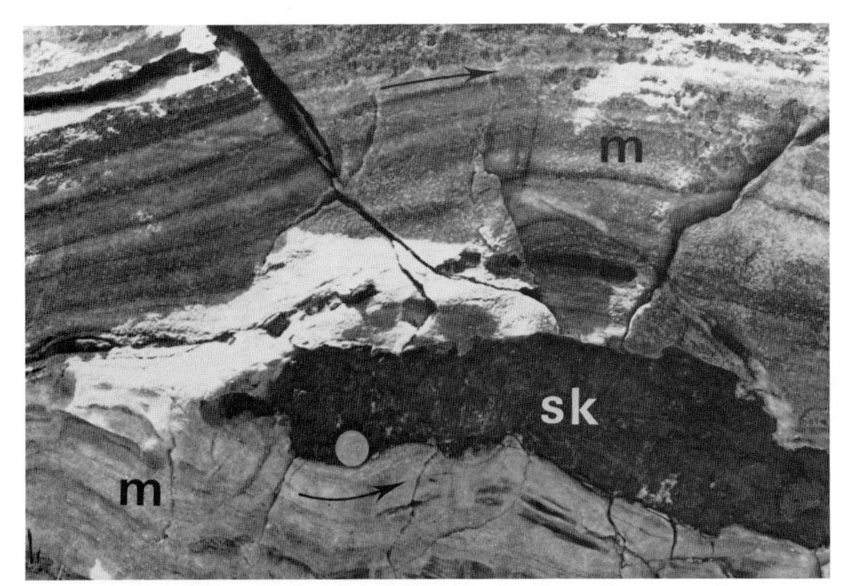

Fig. 121 A bending of the marbles is indicated associated with intrusion of a skarn body. Microscopic examination reveals that associated with skarn is also granitic material (probably a granitic apophysis).
m = marbles
sk = skarn
Arrows indicate the bending of the marbles.
Halara, Seriphos, Greece.

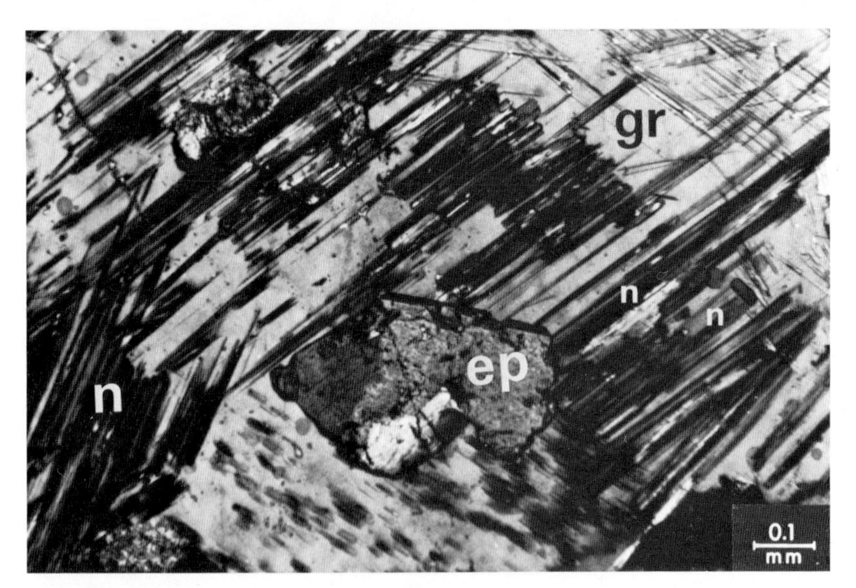

Fig. 122 Blastic garnet (andradite-grossular) with tremolite needles oriented within the blastic garnet.
gr = blastic garnet
n = needles of tremolite
ep = epidote
Granite Skarn.
Kato Vrondou, N. Greece.
(Sample courtesy of K. Spathi)
With crossed nicols.

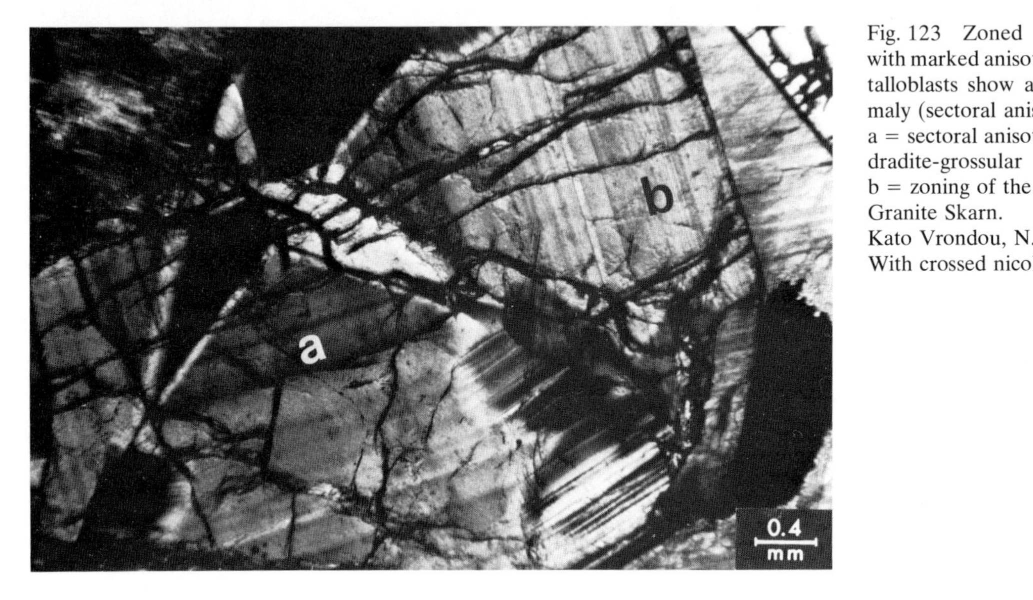

Fig. 123 Zoned andradite-grossular with marked anisotropy. The garnet crystalloblasts show a sectoral optical anomaly (sectoral anisotropy).
a = sectoral anisotropy of the zoned andradite-grossular
b = zoning of the garnet.
Granite Skarn.
Kato Vrondou, N. Greece.
With crossed nicols.

Fig. 124 Granite in contact with quartzite. The contact granite/quartzite is not characterised by skarn bodies or pronounced thermal effects, in contrast marbles above the quartzite band are metasomatically affected as is indicated by the development of skarns.
m = marbles
sk = skarn (epidote-grossular skarn)
q = quartzitic band
g = granite
Halara, Seriphos, Greece.

Fig. 125 Marble relics in epidote skarn. Specularite is present in the marble relic.
e = epidote
m = marble relics
s = specularite developed within the marble relics
Halara, Seriphos, Greece.

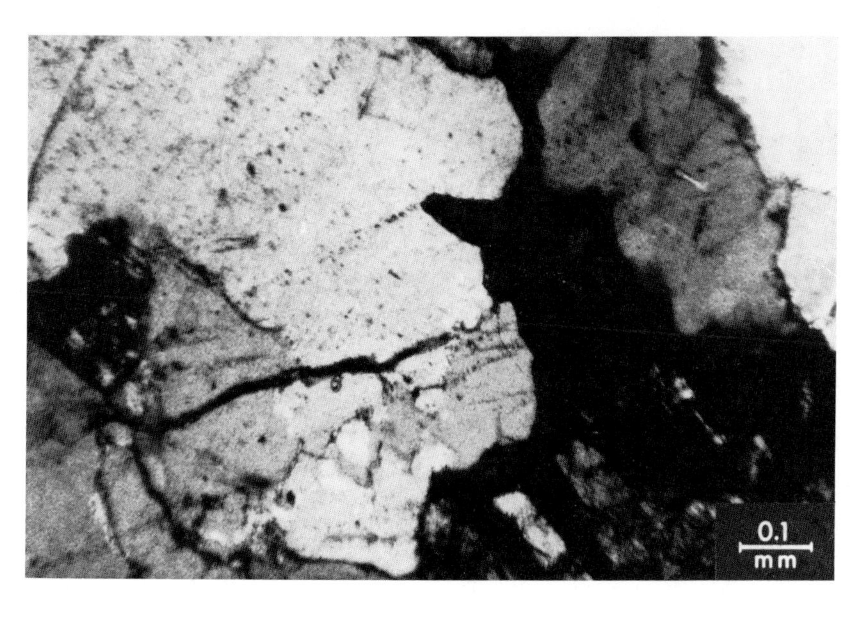

Fig. 126 Andradite-grossular garnet forming-"solutions" invading the granitic textures and following the intergranular spaces between quartz.
Granite skarn.
Halara, Seriphos, Greece.
With crossed nicols.

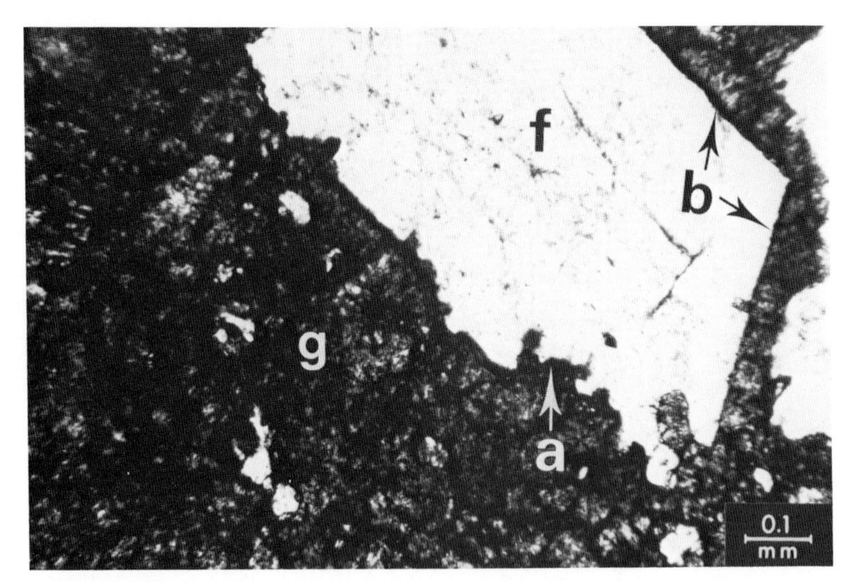

Fig. 127 Andradite-grossular garnet of a skarn body developed in the granite corroding and invading a feldspar of the granite.
Whereas the feldspar is corroded and replaced by the garnet, other faces of the feldspar are intact.
g = andradite grossular replacing and invading the feldspar (see arrow "a").
f = feldspar
Arrow "b" shows the unaffected crystal faces of the feldspar.
Halara, Seriphos, Greece.
With crossed nicols.

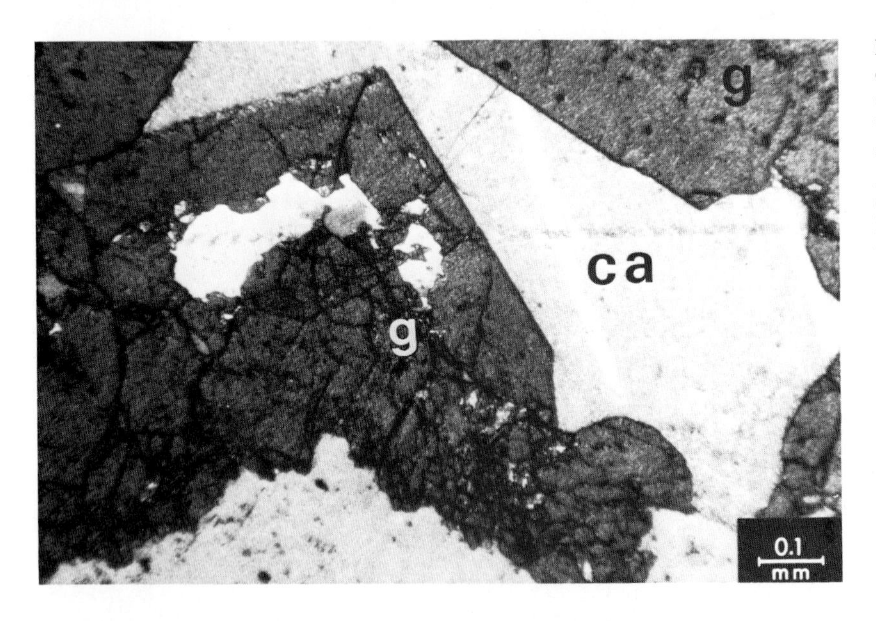

Fig. 128 Idioblastic grossular, partly enclosing recrystallised calcite (marble) and partly with calcite relics, is occupying the intergranular spaces between the garnet crystalloblasts.
g = grossular
ca = calcite
Halara, Seriphos, Greece.
With crossed nicols.

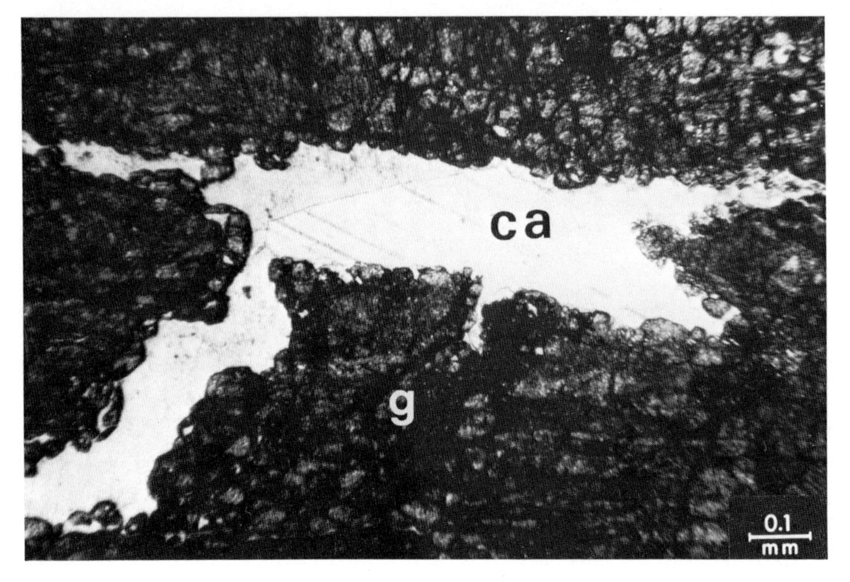

Fig. 129 Calcite as relics (intergranular) between and within grossular crystalloblasts.
g = grossular
ca = calcite
Halara, Seriphos, Greece.
With crossed nicols.

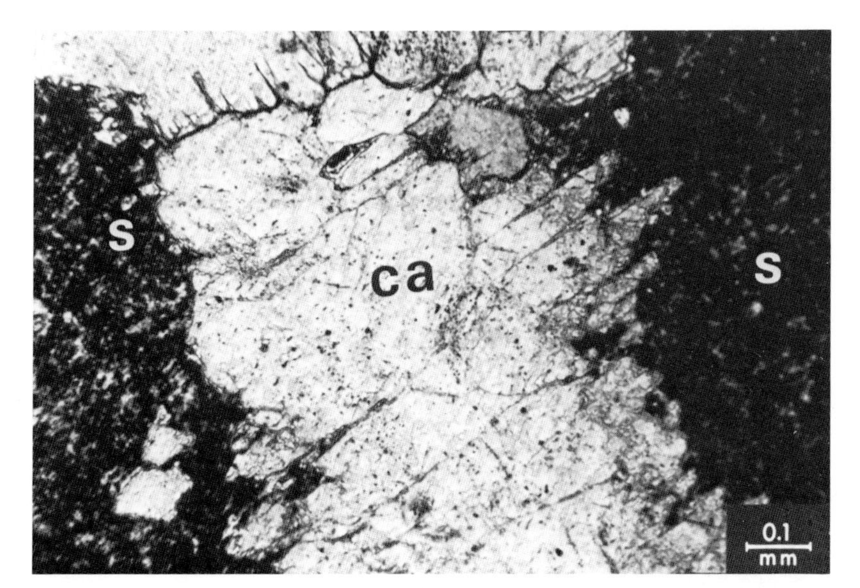

Fig. 130 $Mg\text{-}SiO_2$ metasomatic mobilization related to the basic front of the Seriphos granite.
Coarse grained calcite metasomatically replaced by serpentine.
ca = calcite
s = serpentine
Vounies, Seriphos, Greece.
With crossed nicols.

Fig. 131 A more advanced phase of metasomatic serpentinisation of marbles, due to the basic front ($Mg\text{-}SiO_2$ solutions related with the granitic basic front).
Vounies, Seriphos, Greece.
With crossed nicols.

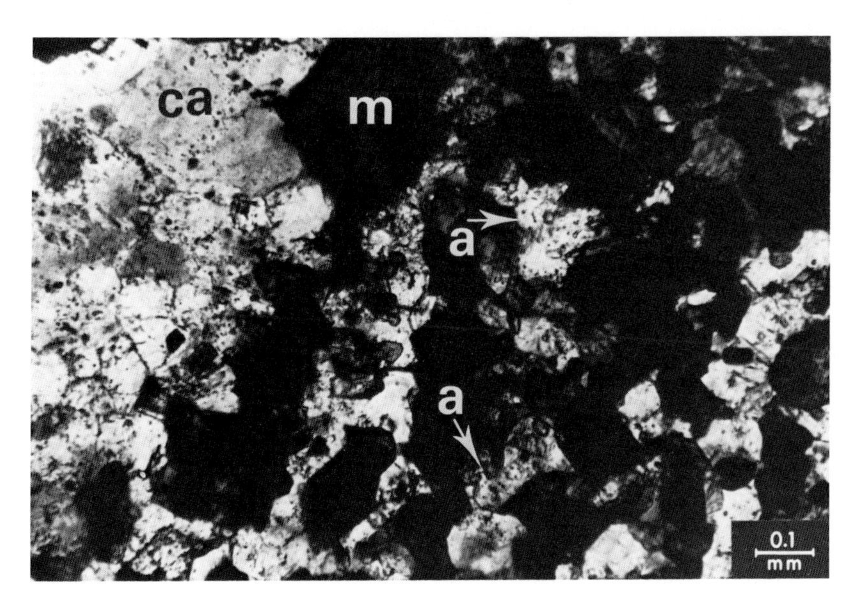

Fig. 132 Magnetite crystalloblastically developed and replacing the calcite texture (marble) mainly taking advantage of the intergranular spaces.
ca = calcite
m = magnetite (black)
Arrows "a" show magnetite engulfing calcite grain.
Vounies, Seriphos, Greece.
With crossed nicols.

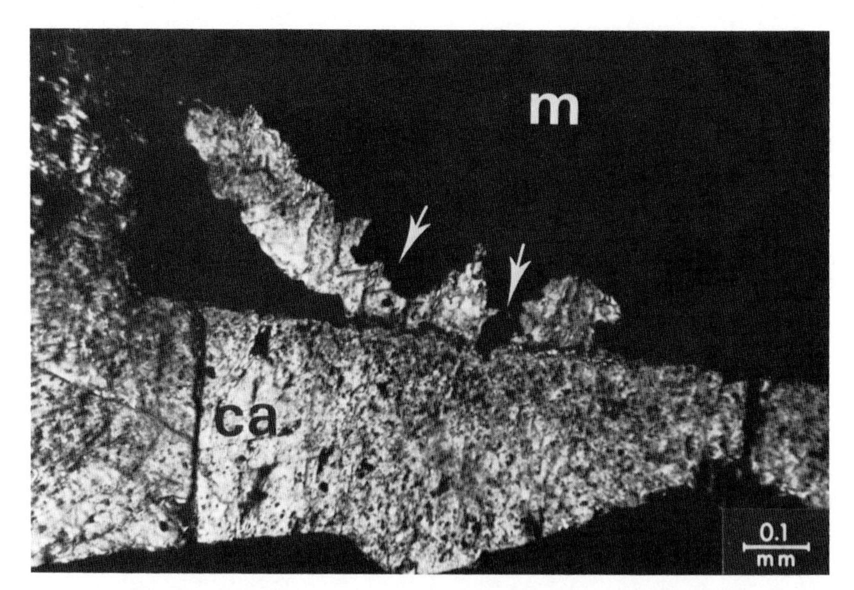

Fig. 133 Magnetite (with an idioblastic tendency) replacing and enclosing calcite.
m = magnetite
ca = calcite
Arrows show the replacement-front of magnetite replacing the calcite of the coarse grained marble.
Vounies, Seriphos, Greece.
With crossed nicols.

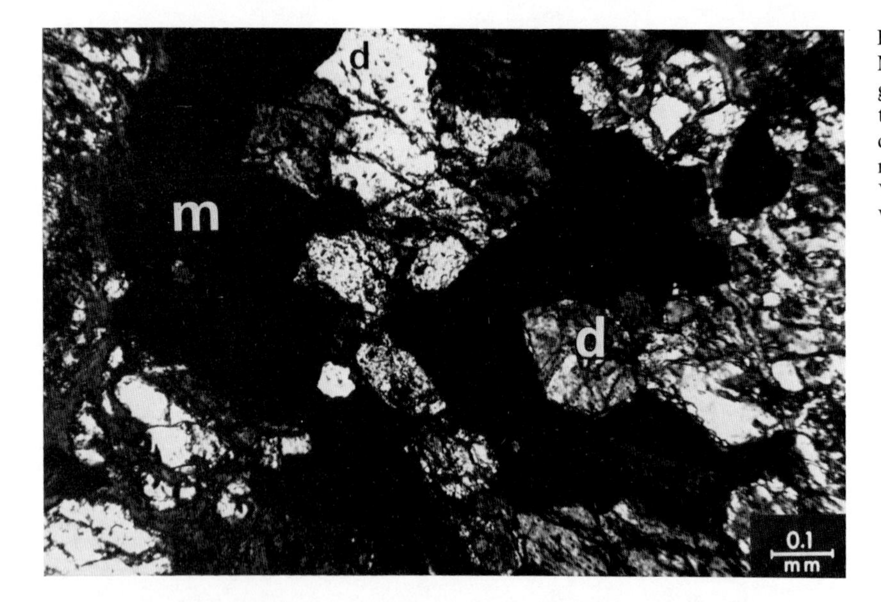

Fig. 134 Diopside crystalloblast (a Mg-SiO_2 phase of the basic front) engulfed and enclosed by a magnetite crystalloblastesis, also of the basic front.
d = diopside
m = magnetite crystalloblastesis
Vounies, Seriphos, Greece.
With crossed nicols.

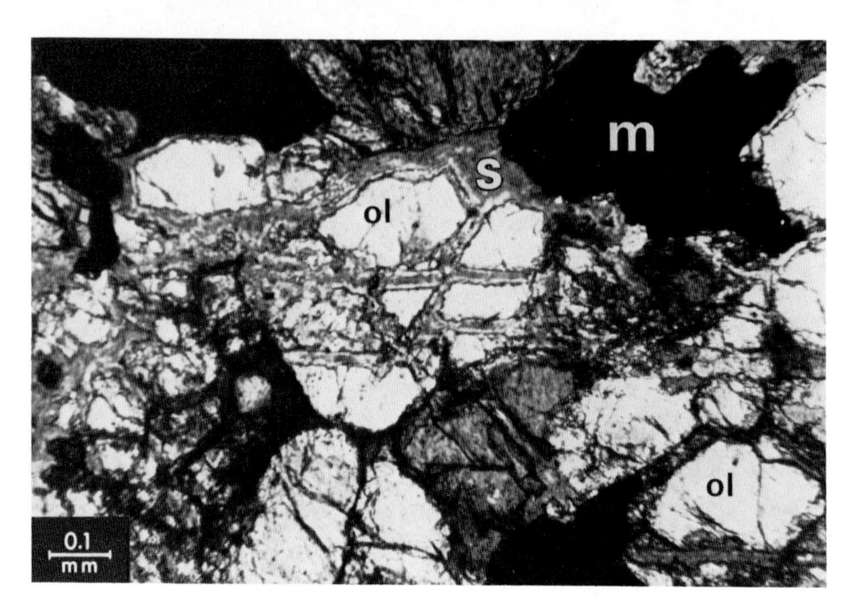

Fig. 135 Forsterite crystalloblasts, (partly serpentinised) partly engulfed by later crystalloblastic magnetite.
ol = olivine
s = serpentine
m = magnetite
Vounies, Seriphos, Greece.
With crossed nicols.

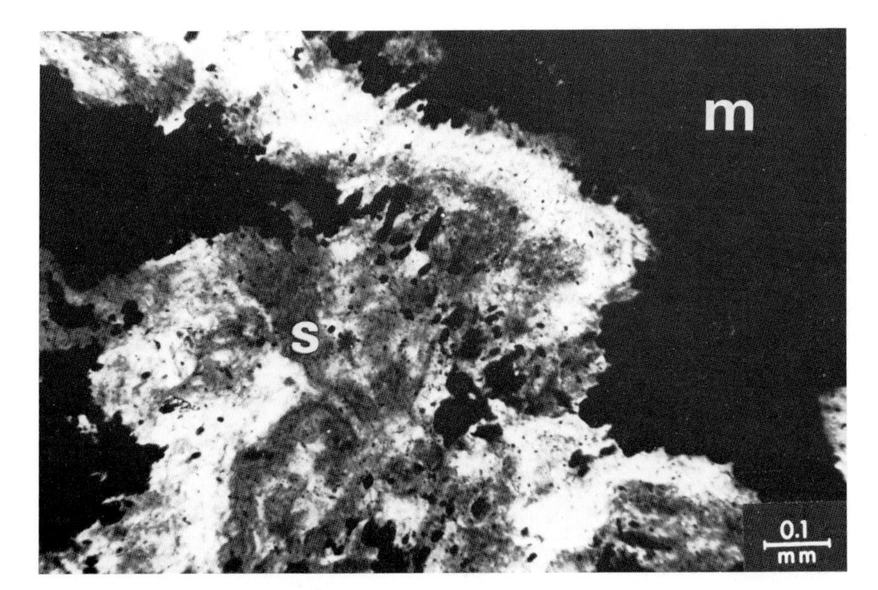

Fig. 136 Synkinematic mobilisation of metasomatic serpentine and of magnetite blastesis.
m = magnetite
s = serpentine
Vounies, Seriphos, Greece.
With crossed nicols.

Figs. 138 and 139 Show epidotised initial marbles fragmented (mega breccia) and surrounded by "hydrothermal" colloform bands of alternating radially arranged hedenbergite-lievrite.
m-b-e = mega breccia consisting of epidotised marbles.
c-sk = colloform skarns consisting of radiating hedenbergite-lievrite bands and surrounding mega-breccia fragments.
Kouduros, Seriphos, Greece.

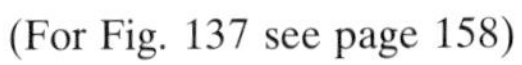
(For Fig. 137 see page 158)

Fig. 139

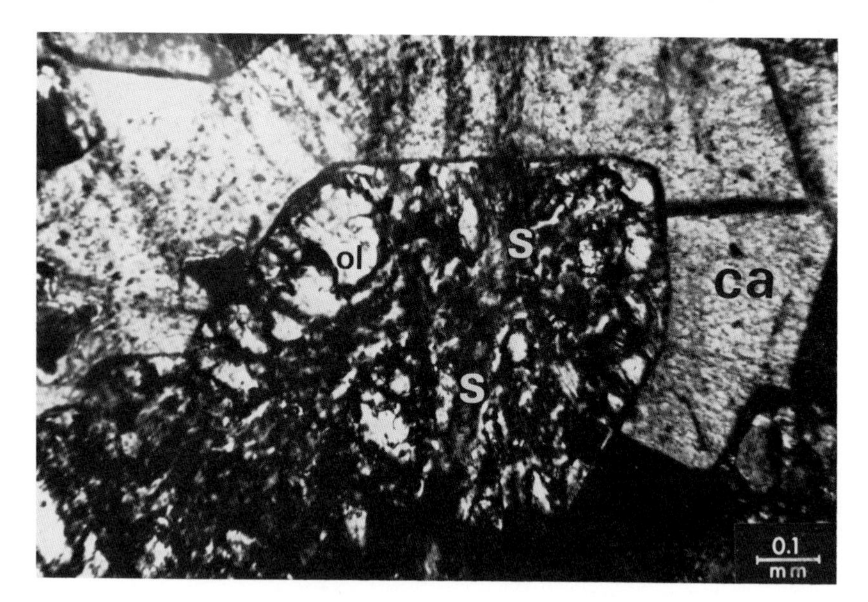

Fig. 137 Shows idioblastic forsterite in marble within the effects of the basic front of the Seriphos granite.
ca = calcite
ol = olivine idioblast
s serpentinisation of the olivine
Vounies, Seriphos, Greece.
With crossed nicols.

Fig. 140 Colloform structure consisting of radiating hedenbergite and lievrite crystals.
h = hedenbergite
l = lievrite
Handspecimen 1/2 natural size.
Kouduros, Seriphos, Greece.

Fig. 141 Geode consisting of quartz, agate, calcite, between spaces of the colloform structures consisting of hedenbergite and lievrite, (h-l).
ge = geode
Kouduros, Seriphos, Greece.

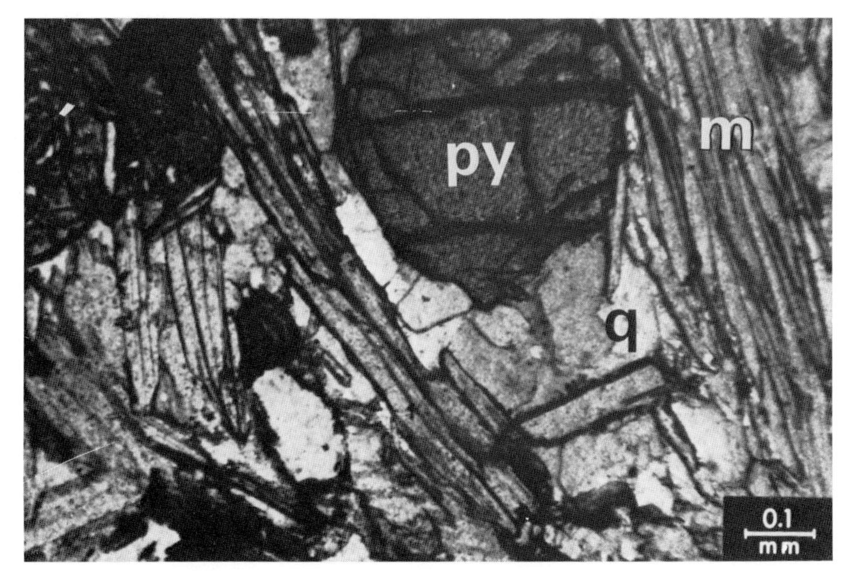

Fig. 142 Pyrope crystalloblast in quartz-mica texture.
py = pyrope crystalloblast
q = quartz
m = mica
Mica-quartzite (in pegmatoids)
Evros, N. Greece.
With crossed nicols.
(Sample courtesy Dr. A. Vgenopoulos)

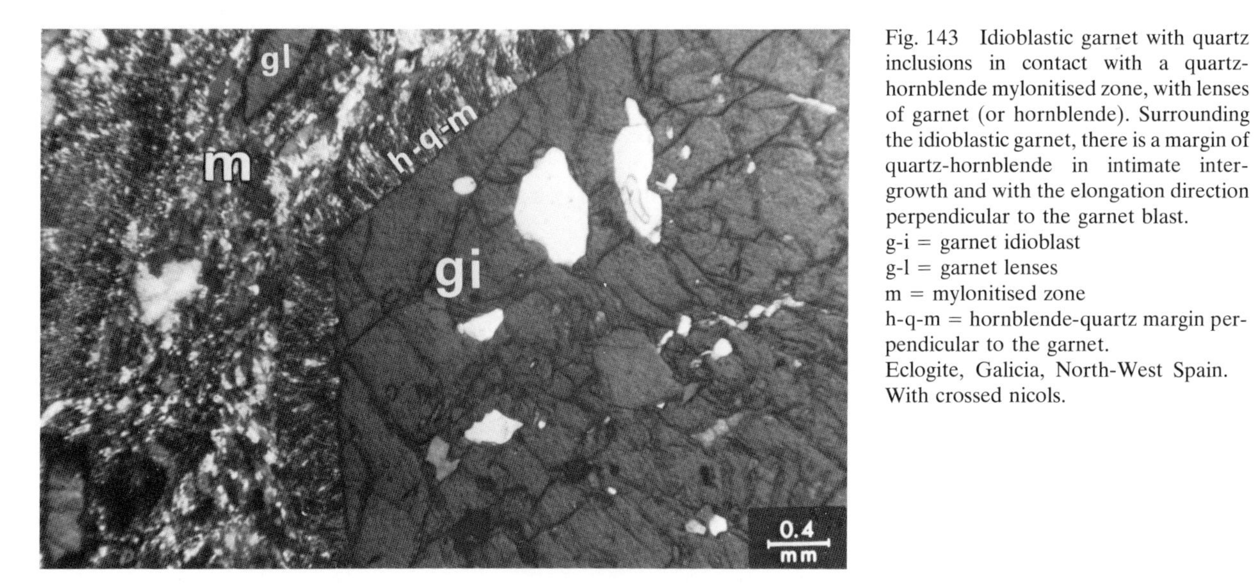

Fig. 143 Idioblastic garnet with quartz inclusions in contact with a quartz-hornblende mylonitised zone, with lenses of garnet (or hornblende). Surrounding the idioblastic garnet, there is a margin of quartz-hornblende in intimate intergrowth and with the elongation direction perpendicular to the garnet blast.
g-i = garnet idioblast
g-l = garnet lenses
m = mylonitised zone
h-q-m = hornblende-quartz margin perpendicular to the garnet.
Eclogite, Galicia, North-West Spain.
With crossed nicols.

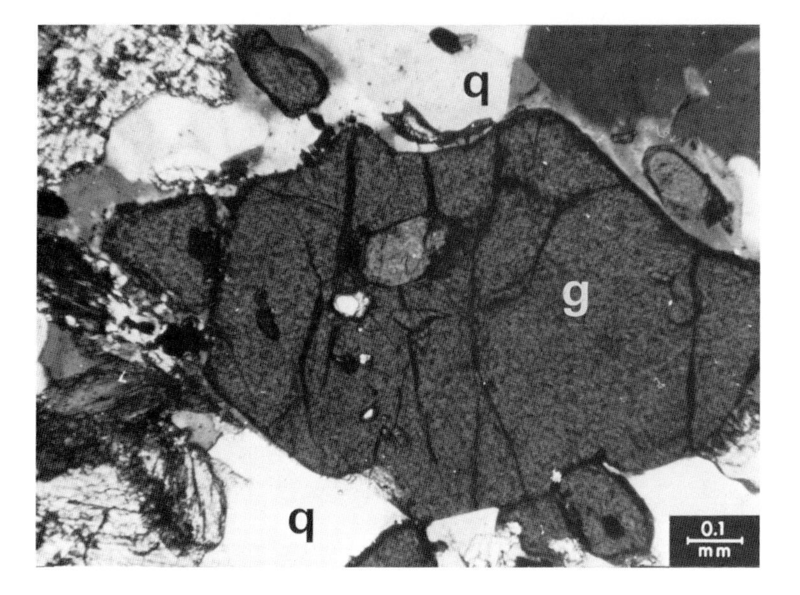

Fig. 144 Garnet crystalloblast with tectonically deformed quartz.
g = garnet (noticeable curved garnet outlines)
q = quartz
Eclogite. Galicia, North-West Spain.
With crossed nicols.

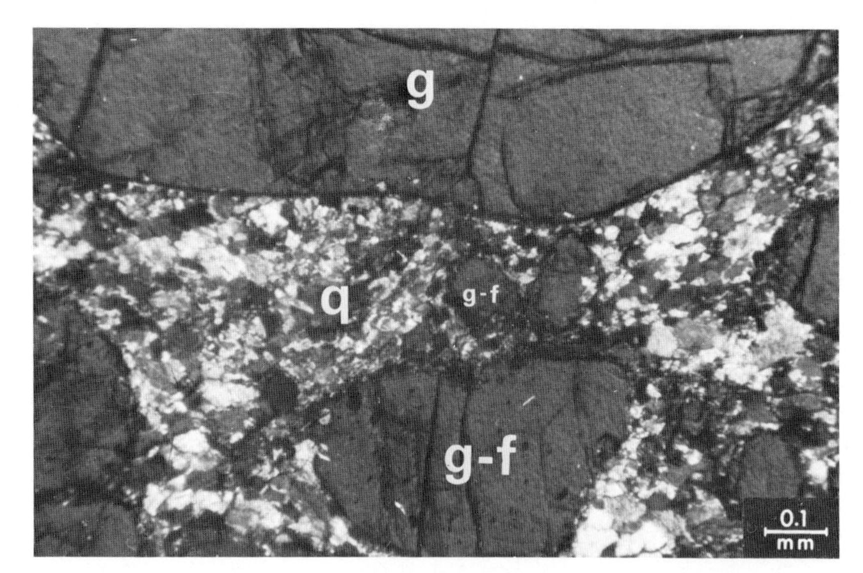

Fig. 145 Garnet crystalloblast with a crushed zone consisting of recrystallised quartz and having garnet fragments. Gentle curving outline of a garnet crystalloblast, most probably caused by deformation accompanied by the crushed-zone formation.
g = garnet
q = quartz recrystallised
g-f = garnet fragments
Eclogite. Galicia, North-West Spain.
With crossed nicols.

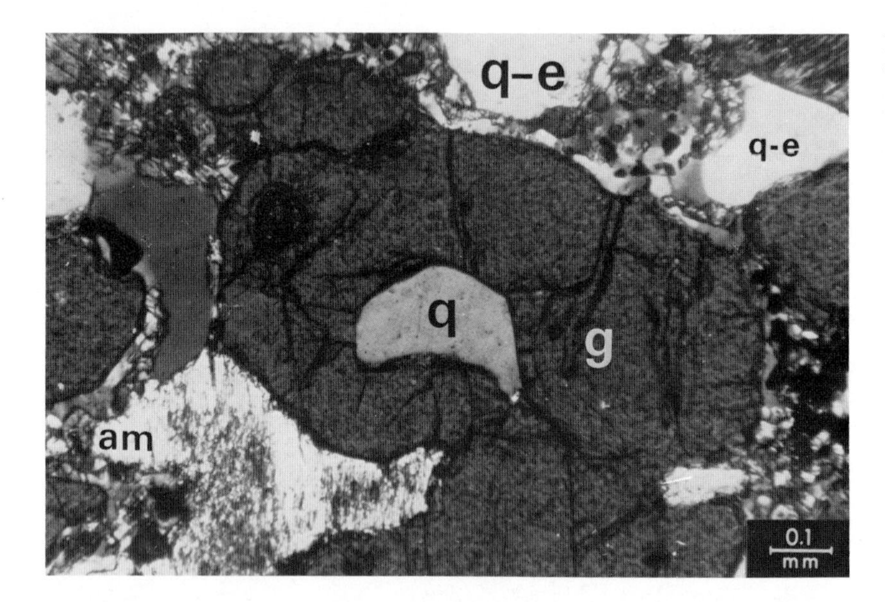

Fig. 146 Garnet crystalloblast with quartz included in the garnet. Also recrystallised quartz outside the crystalloblast with an amphibole.
g = garnet crystalloblast
q = quartz inclusion in the crystalloblast
q-e = quartz external to the garnet
am = amphibole
Eclogite. Galicia, North-West Spain.
With crossed nicols.

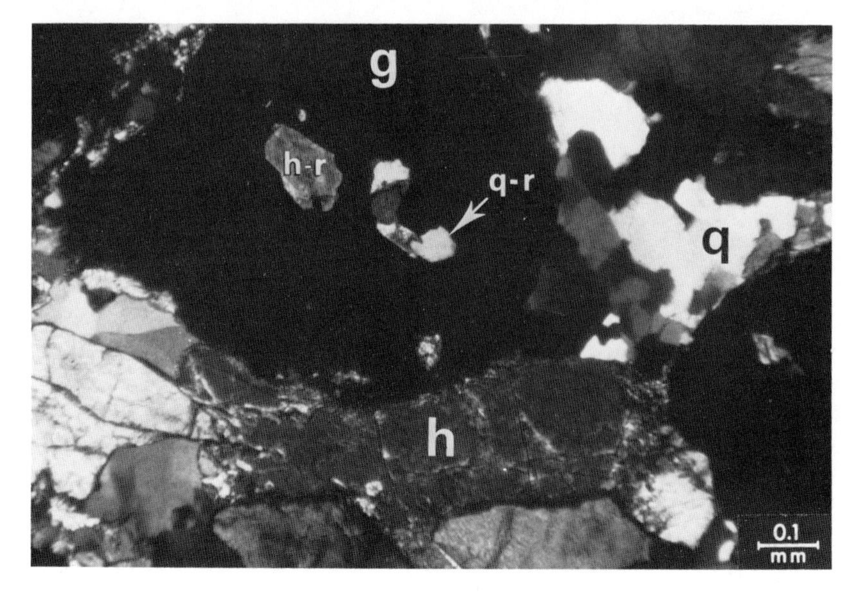

Fig. 147 Garnet crystalloblast enclosing quartz and hornblende. Hornblende and quartz also external to the garnet.
g = garnet crystalloblast
q = quartz
h = hornblende
q-r = quartz included in the garnet
h-r = hornblende included in the garnet crystalloblast
Eclogite. Galicia, North-West Spain.
With crossed nicols.

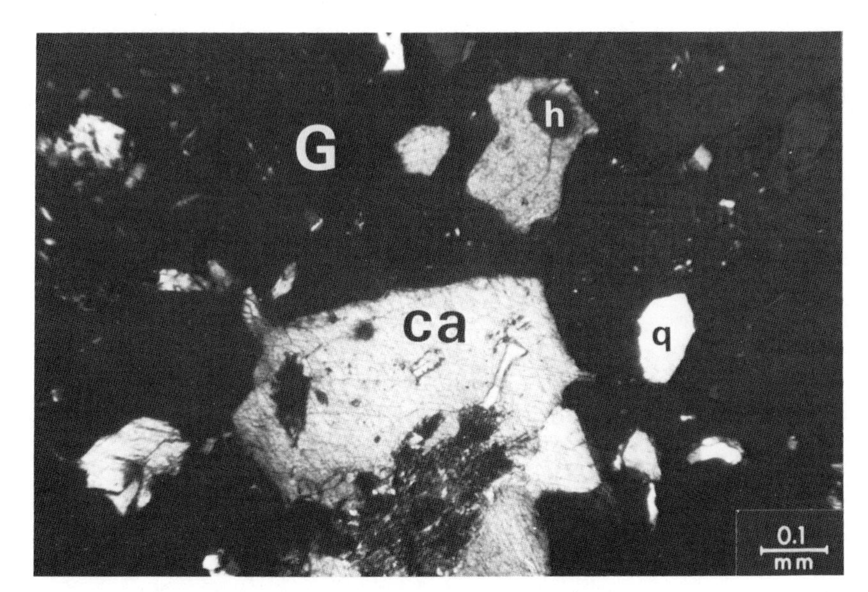

Fig. 148 Garnet crystalloblast enclosing calcite, hornblende and quartz.
G = garnet crystalloblast
h = hornblende
q = quartz
ca = calcite
Eclogite. Galicia, North-West Spain.
With crossed nicols.

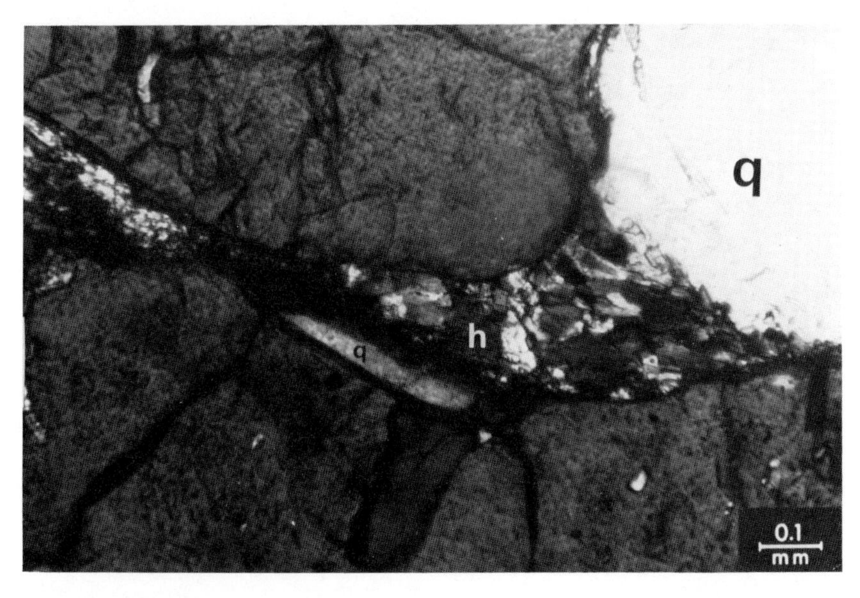

Fig. 149 A garnet crystalloblast with hornblende and quartz mobilised along garnet cracks.
h = hornblende
q = quartz
Eclogite. Galicia, North-West Spain.
With crossed nicols.

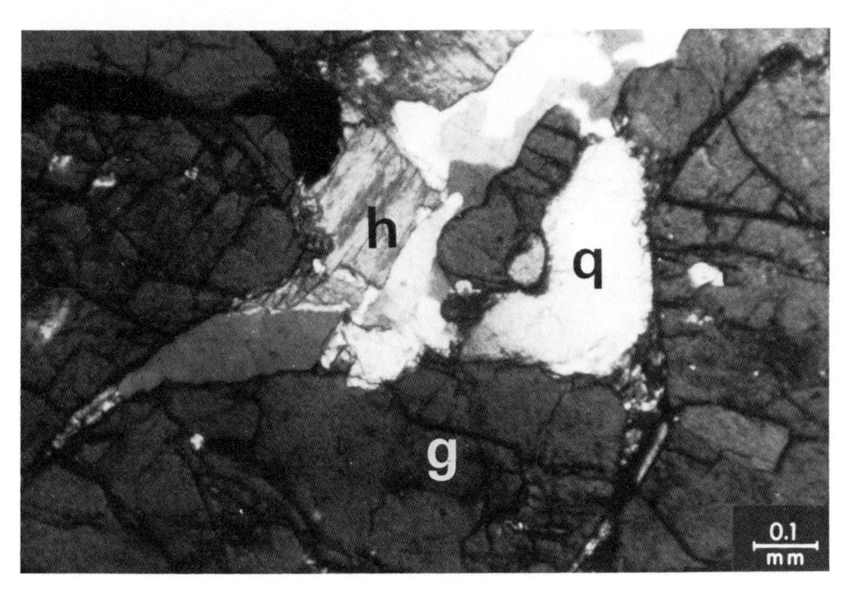

Fig. 150 A garnet crystalloblast with quartz and hornblende mobilised within the garnet.
g = garnet
q = quartz
h = hornblende
Eclogite. Galicia, North-West Spain.
With crossed nicols.

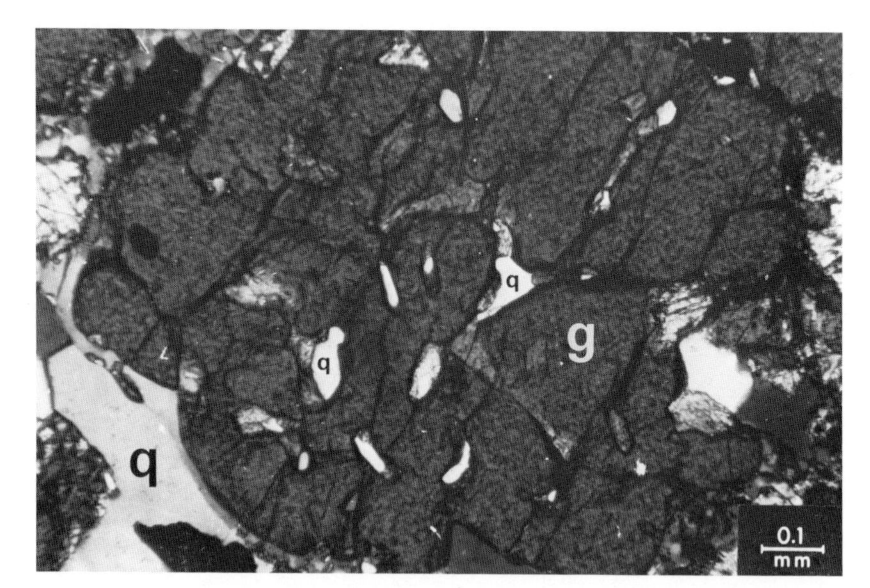

Fig. 151 Crystalloblastic garnet with quartz most probably mobilised. Also deformed quartz is present outside the garnet.
g = garnet
q = quartz
Eclogite. Galicia, North-West Spain.
With crossed nicols.

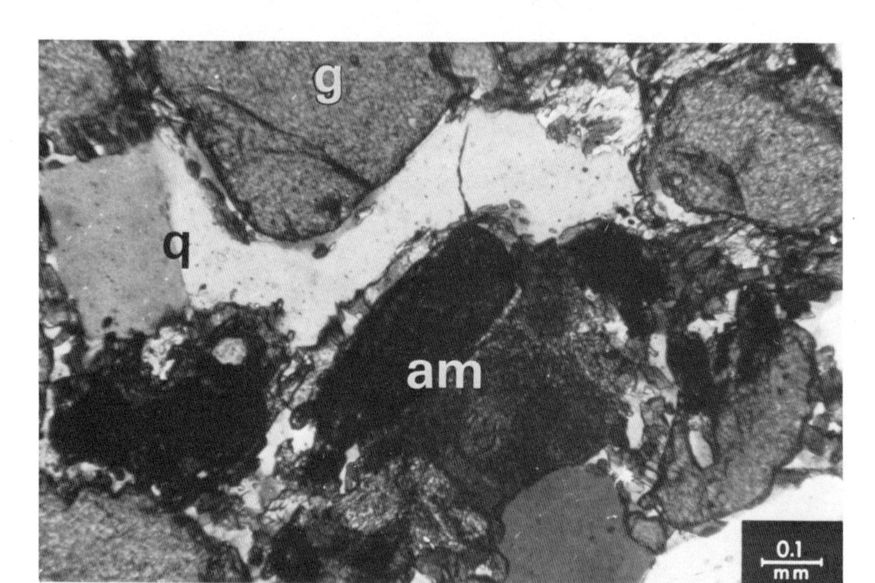

Fig. 152 Garnet crystalloblast with quartz and amphiboles.
g = garnet crystalloblast
q = quartz
am = amphible
Eclogite. Galicia, North-West Spain.
With crossed nicols.

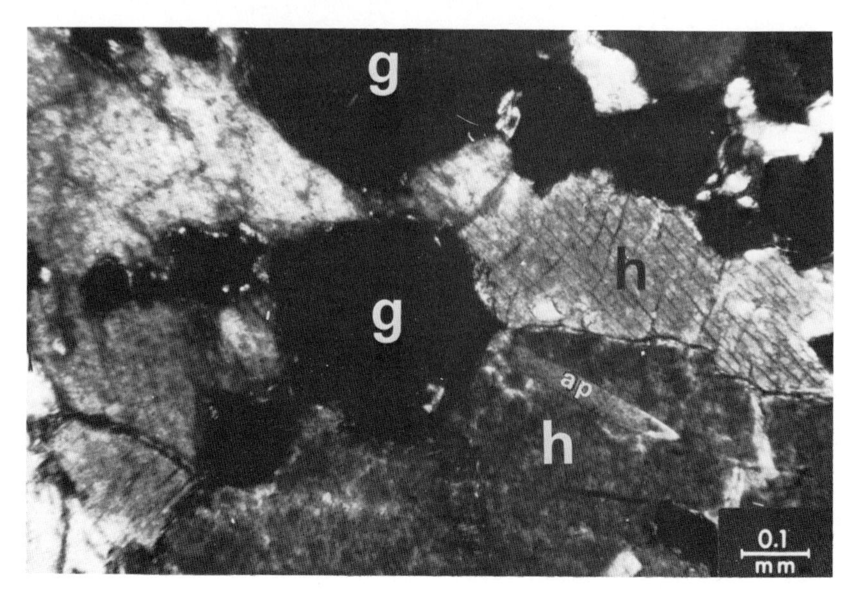

Fig. 153 Hornblende and garnet crystalloblasts.
g = garnet crystalloblast
h = hornblende
ap = apatite
Eclogite. Galicia, North-West Spain.
With crossed nicols.

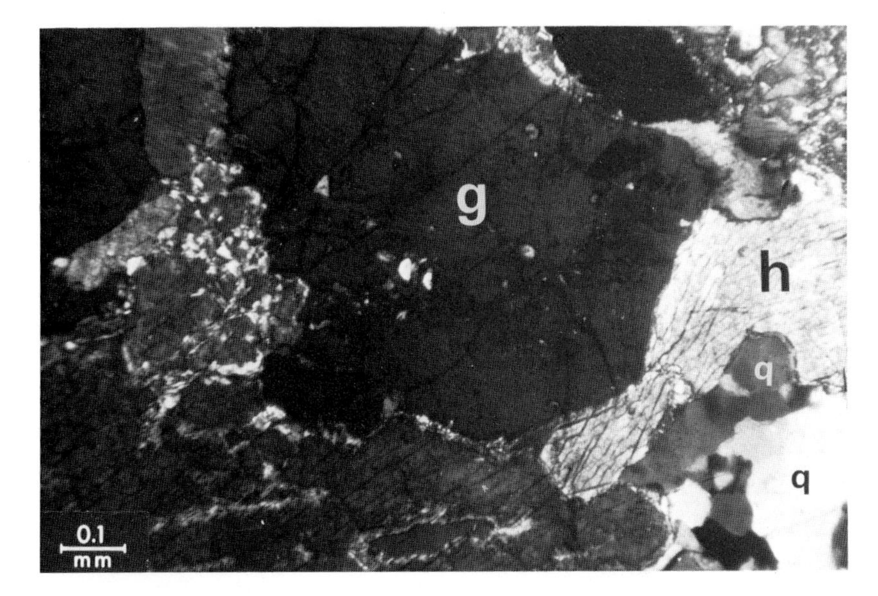

Fig. 154 Garnet crystalloblast surrounded with hornblende crystalloblasts and quartz.
g = garnet
h = hornblende
q = quartz
Eclogite. Galicia, North-West Spain.
With crossed nicols.

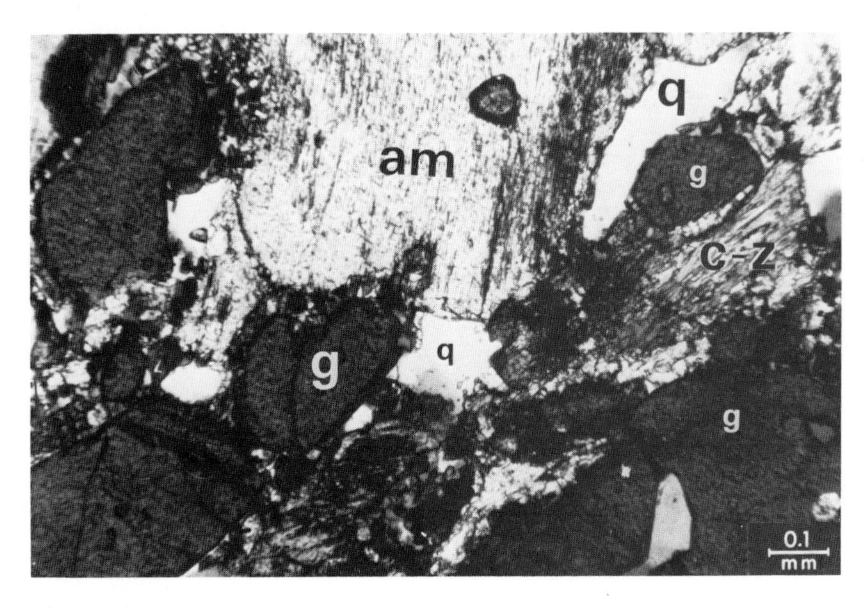

Fig. 155 Garnet crystalloblast, quartz and amphibole crystalloblast enclosing garnets.
g = garnet crystalloblast
q = quartz
am = amphibole
c-z = crushed zone consisting of quartz and mafic minerals.
Eclogite. Galicia, North-West Spain.
With crossed nicols.

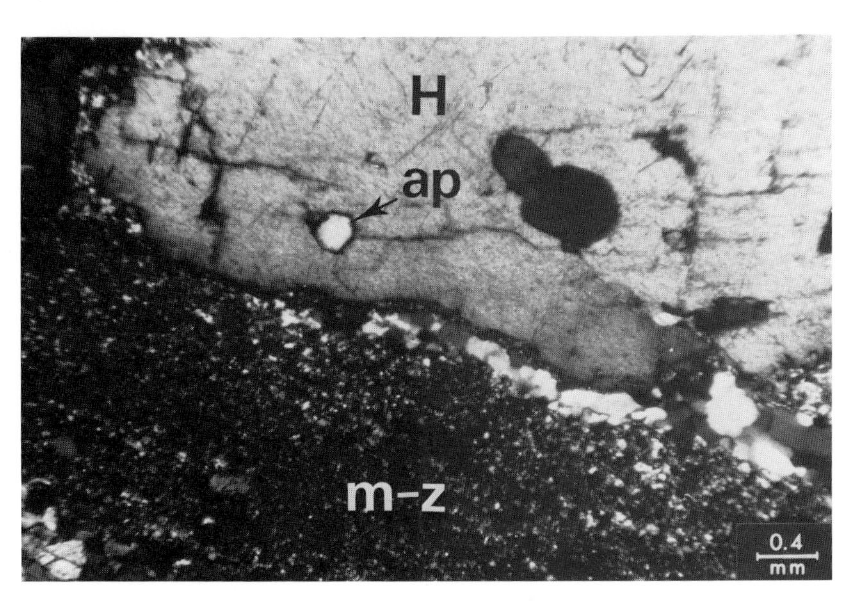

Fig. 156 A mylonitised zone, consisting mainly of fine quartz and amphibole, in sharp contact with a hornblende crystalloblast which includes an idiomorphic apatite.
m-z = mylonitised zone consisting of fine quartz and amphibole
H = crystalloblastic hornblende
ap = idiomorphic apatite
Ecologite. Galicia, North-West Spain.
With crossed nicols.

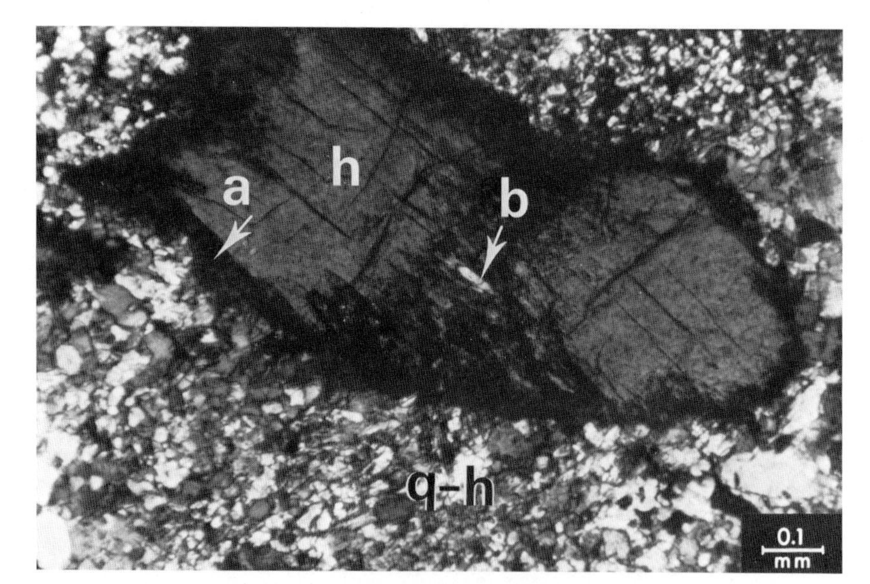

Fig. 157 A quartz-hornblende mylonitised zone, with a relic of hornblende, exhibiting a reaction margin (arrow "a").
q-h = quartz/hornblende mylonitised zone
h = hornblende in the mylonitised zone
Arrow "b" shows apatite present in the hornblende
Eclogite. Galicia, North-West Spain. With crossed nicols.

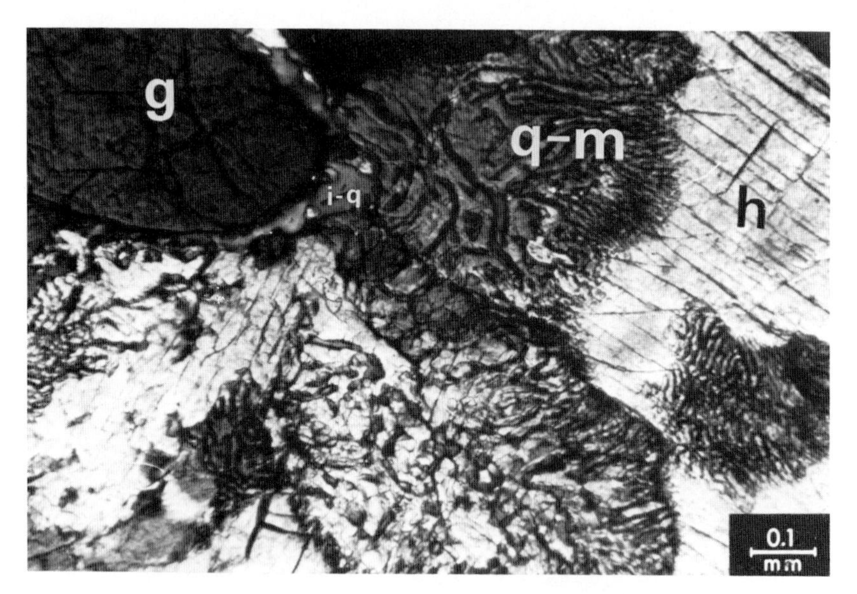

Fig. 159a and b Garnet with amphibole (hornblende). Quartz following the intergranular between the garnet and the amphibole extends into the amphibole taking advantage of the solution penetrability direction of the hornblende (e. g. cleavage etc.).
g = garnet
i-q = intergranular quartz
h = hornblende

(For Fig. 158 see page 165)

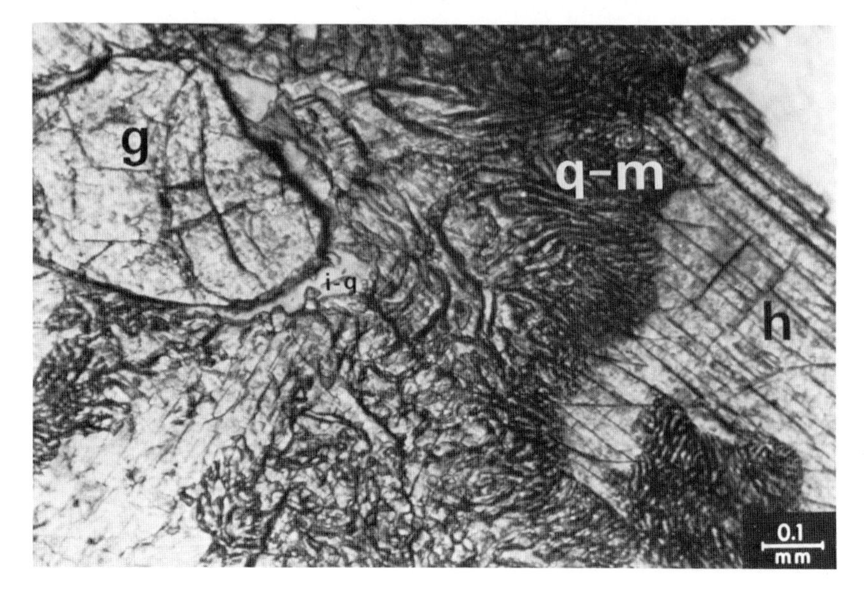

Fig. 159b

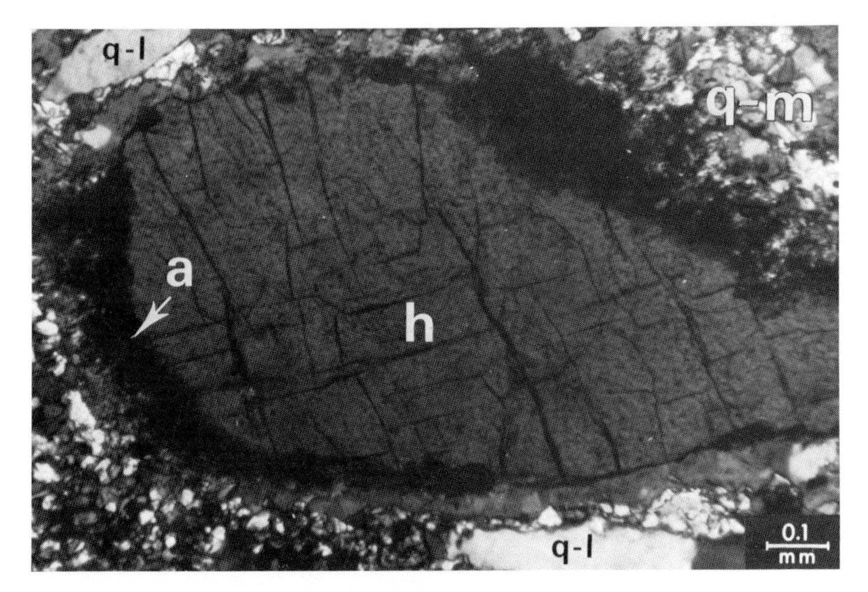

Fig. 158 A quartz-hornblende mylonitised zone with relic quartz lenses present and with a hornblende relic. Arrow "a" shows a margin surrounding the hornblende.
h = hornblende
q-l = quartz lenses
q-m = quartz-hornblende mylonitised
Eclogite. Galicia, North-West Spain.
With crossed nicols.

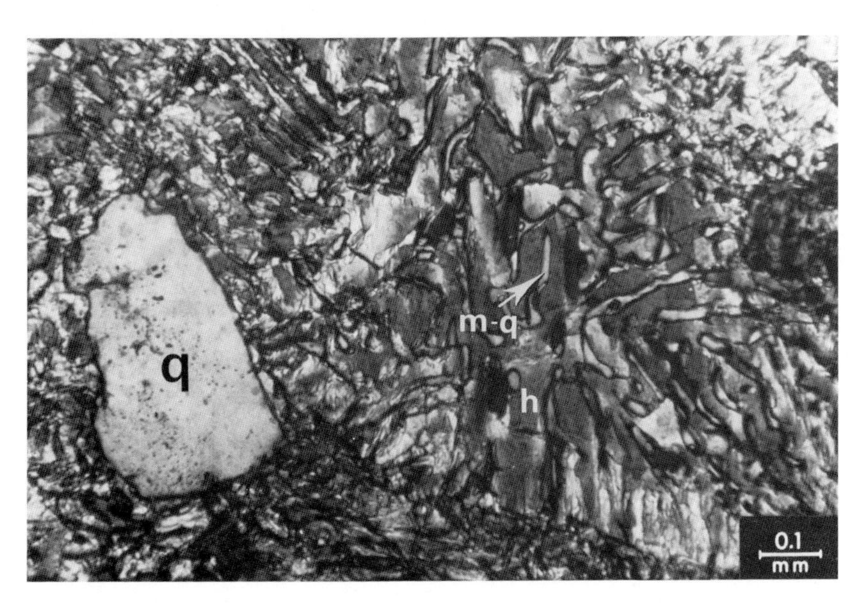

Fig. 160 Quartz included in a hornblende which in turn exhibits symplectic-myrmekitic intergrowth with quartz, which due to solution penetrability, infiltrated the hornblende.
q = quartz
m-q = myrmekitic quartz in symplectic intergrowth with hornblende
h = hornblende
Eclogite. Galicia, North-West Spain.
With crossed nicols.

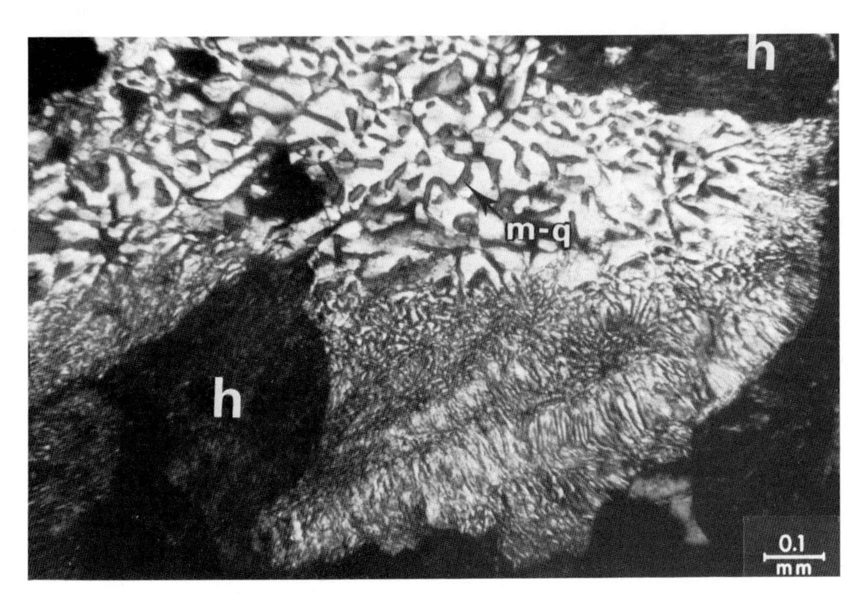

Fig. 161 Quartz/hornblende in symplectic intergrowth with the quartz, attaining infiltration worm-like bodies.
m-q = myrmekitic quartz
h = hornblende
Eclogite. Galicia, North-West Spain.
With crossed nicols.

Fig. 162 A textural pattern resulting from a symplectic quartz/hornblende intergrowth and tectonic mylonitisation; with coarse quartz grains as resistant forms.
q = coarse quartz
h = hornblende
t-s = tectonically affected symplectite
Eclogite. Galicia, North-West Spain.
With crossed nicols.

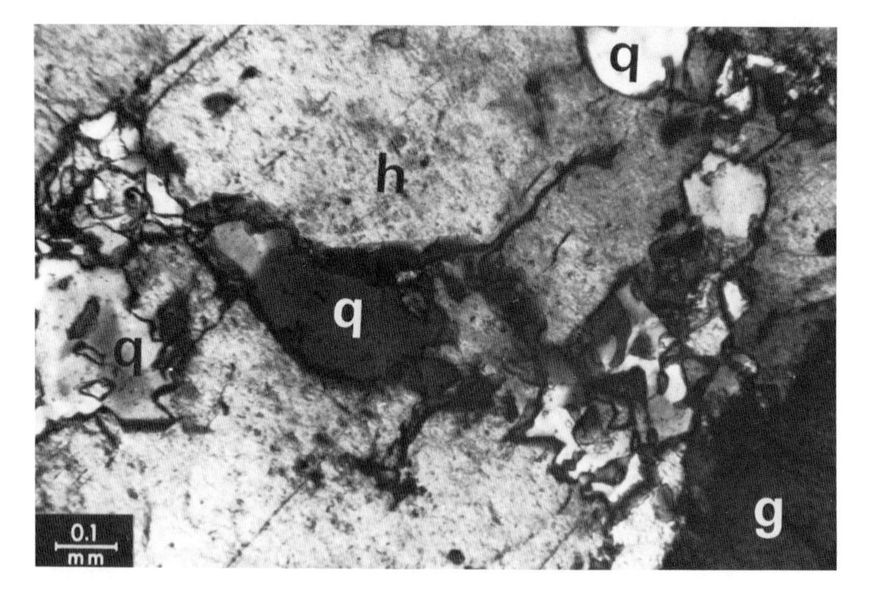

Fig. 163 Quartz replacing the hornblende.
q = quartz
h = hornblende
g = garnet
Eclogite. Galicia, North-West Spain.
With crossed nicols.

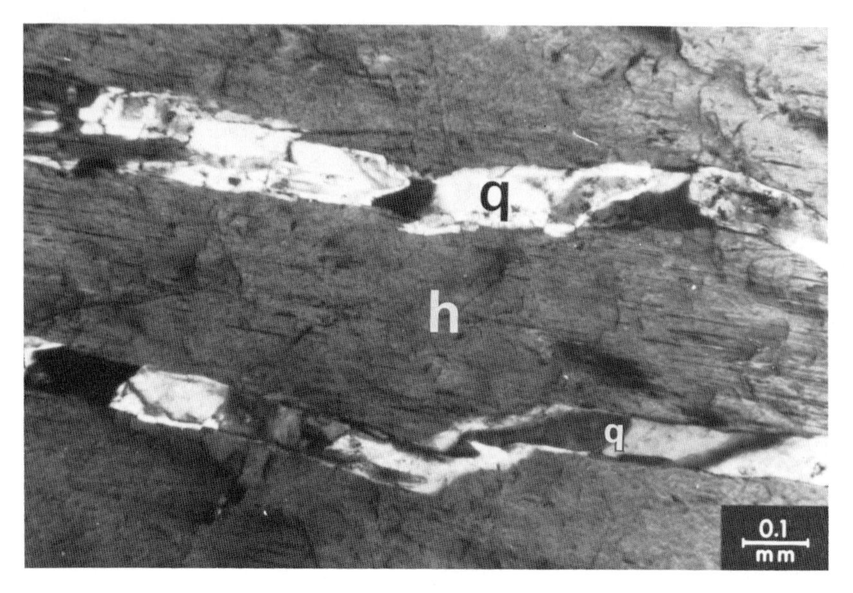

Fig. 164 Quartz veinlets cutting the hornblende and following its cleavage.
h = hornblende
q = quartz
Eclogite. Galicia, North-West Spain.
With crossed nicols.

Fig. 165 a and b Hornblende crystalloblast with included apatites.
h = hornblende crystalloblast
ap = apatite
Eclogite. Galicia, North-West Spain.
With crossed nicols.

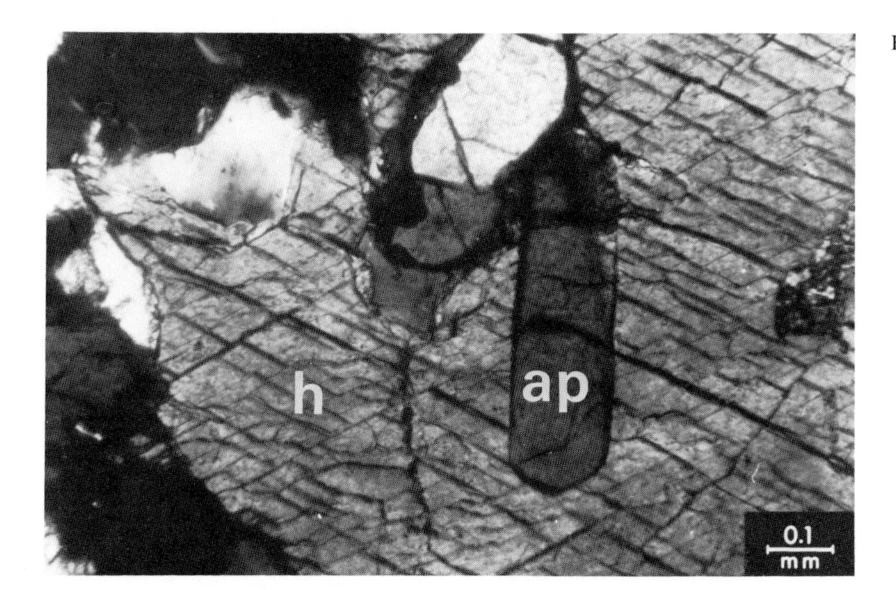

Fig. 165b

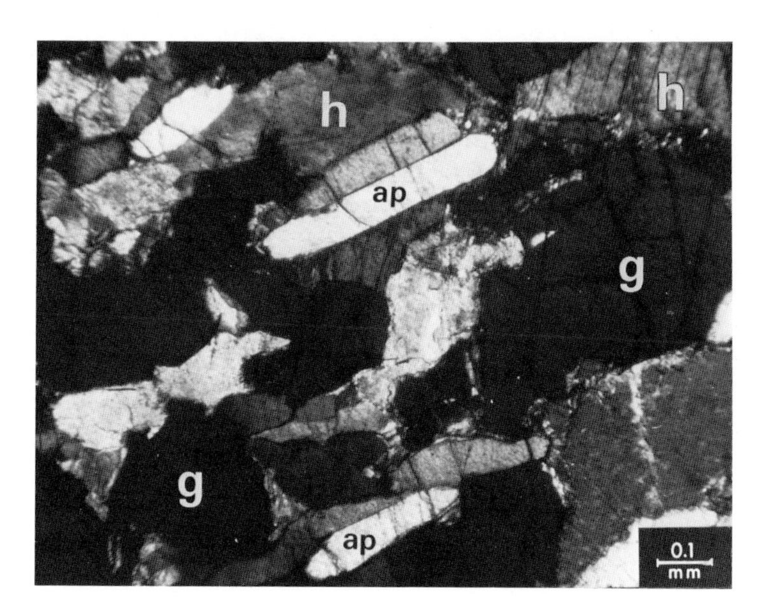

Fig. 166 Garnet crystalloblast with hornblende and blastic prismatic apatite.
g = garnet crystalloblast
ap = apatite
h = hornblende-blast
Eclogite. Galicia, North-West Spain.
With crossed nicols.

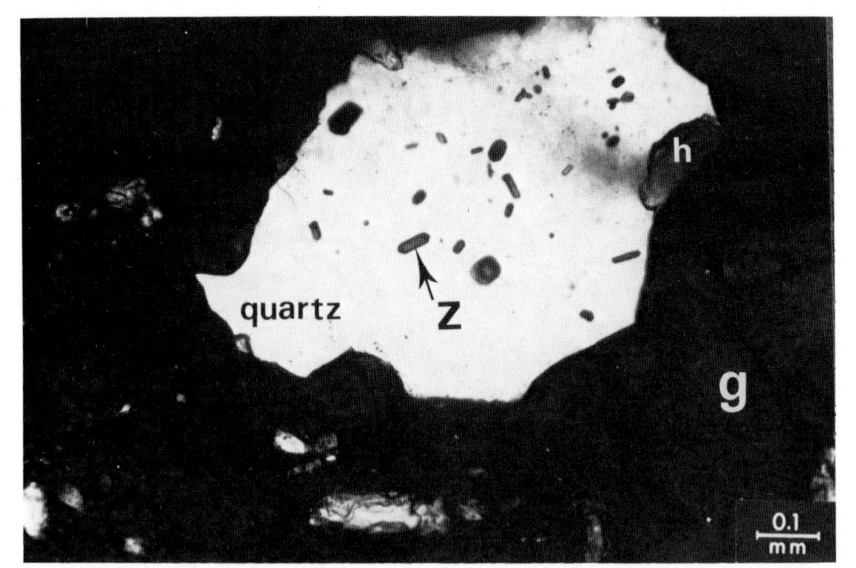

Fig. 167 Garnet crystalloblast enclosing quartz which in turn includes zoned and idiomorphic zircons.
g = garnet (isotropic-black)
h = hornblende
z = zircon (included in the quartz, which indicates undulating extinction).
Eclogite. Galicia, North-West Spain.
With crossed nicols.

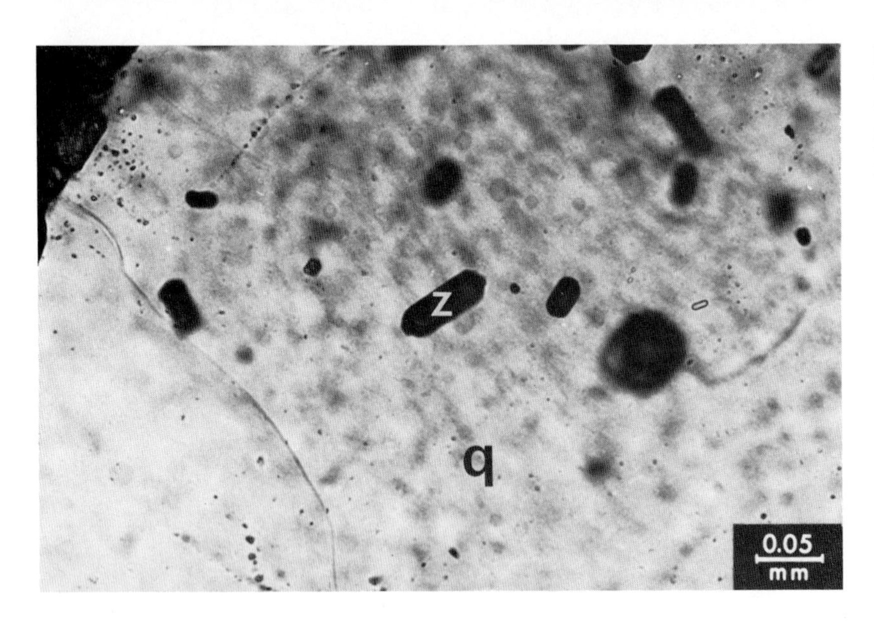

Fig. 168 Quartz included in the garnet with zoned idiomorphic zircons.
q = quartz
z = zircon
Eclogite. Galicia, North-West Spain.
With crossed nicols.

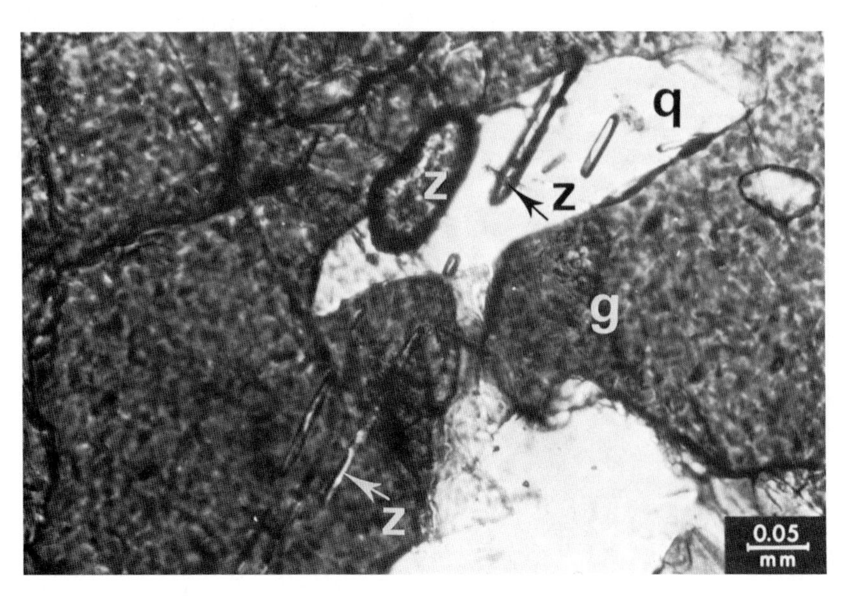

Fig. 169 Garnet with included quartz and with orientated zircons which, following the same orientation have grown as elongated needles into the garnet as well.
g = garnet (nicols not exactly crossed)
q = quartz
z = zircon
Eclogite. Galicia, North-West Spain.
With half-crossed nicols.

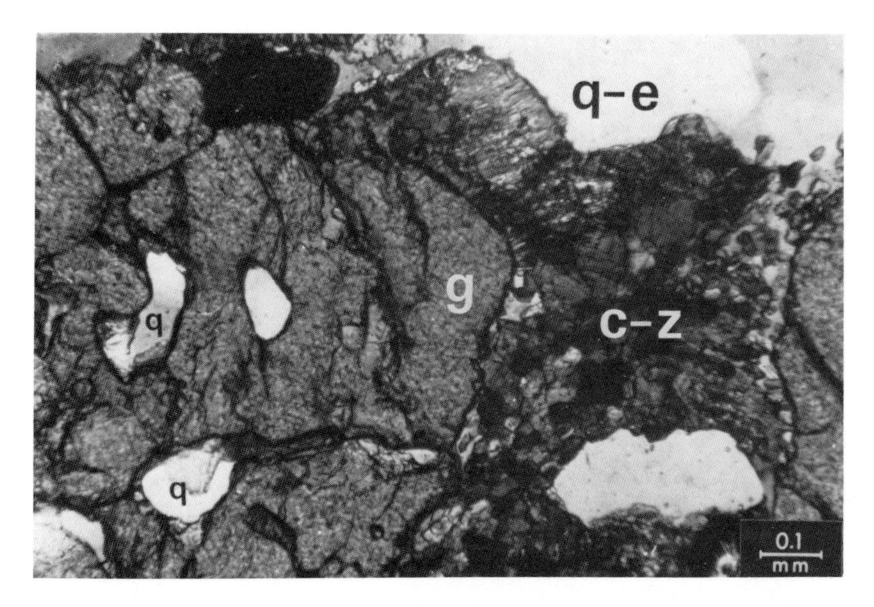

Fig. 170 Crystalloblastic garnet with quartz included (most probably mobilised) and a quartz/hornblende crushed zone between two garnets.
g = garnet
q = quartz (included in the garnet crystalloblast)
q-e = quartz in a crushed zone outside the garnet
c-z = crushed zone between two garnets
Eclogite. Glenelg, Inverness-shire, Scotland.
With crossed nicols.

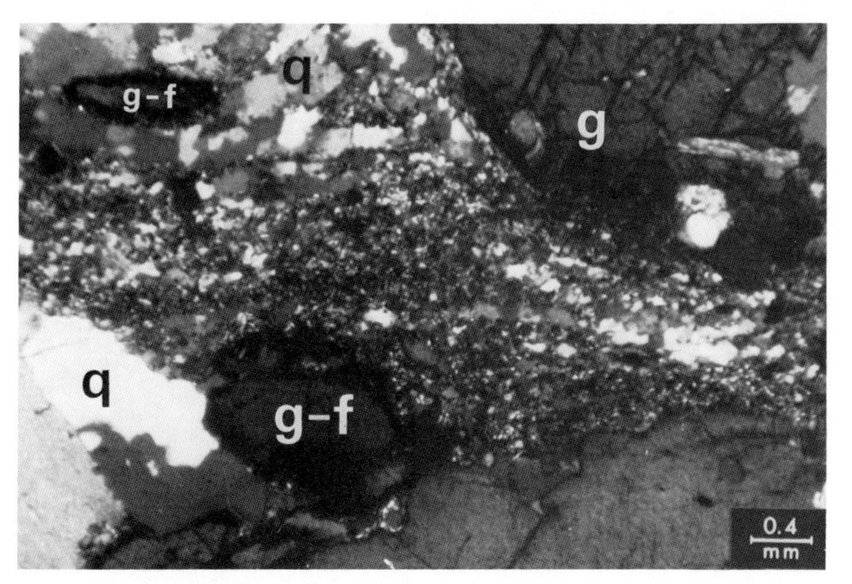

Fig. 171 A crushed zone consisting mainly of recrystallised quartz and garnet fragments between two garnet crystalloblasts.
q = quartz
g-f = garnet fragments
g = garnet
Eclogite. Galicia, North-West Spain.
With crossed nicols.

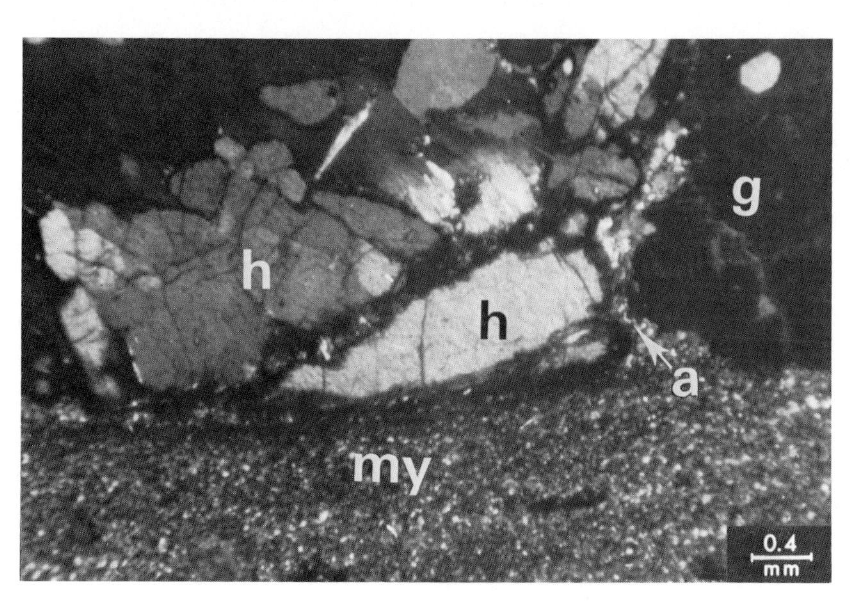

Fig. 172 A fine mylonitised zone consisting mainly of fine quartz with extension of it between the intergranular of adjacent hornblende and garnet.
g = garnet
h = hornblende
my = fine grained mylonitised zone
Arrow "a" shows extension of the mylonitic zone between the intergranular of hornblende and garnet.
Eclogite. Galicia, North-West Spain.
With crossed nicols.

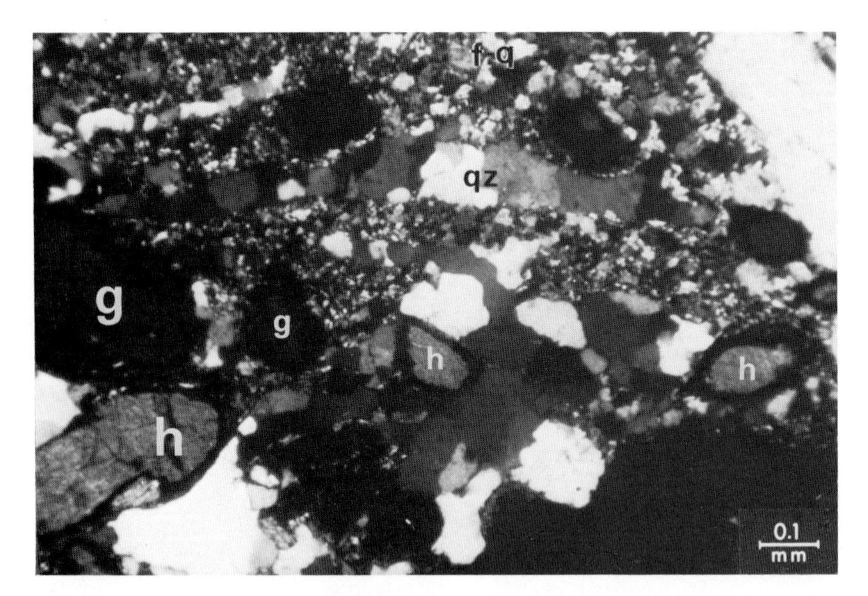

Fig. 173 Coarse grained quartzite, hornblende and garnet fragments as lenses within mylonitised fine grained zones.
qz = quartzitic lenses
h = hornblende lenses
g = garnet fragments
f-q = fine quartz of the mylonitic crushed zones.
Eclogite. Galicia, North-West Spain. With crossed nicols.

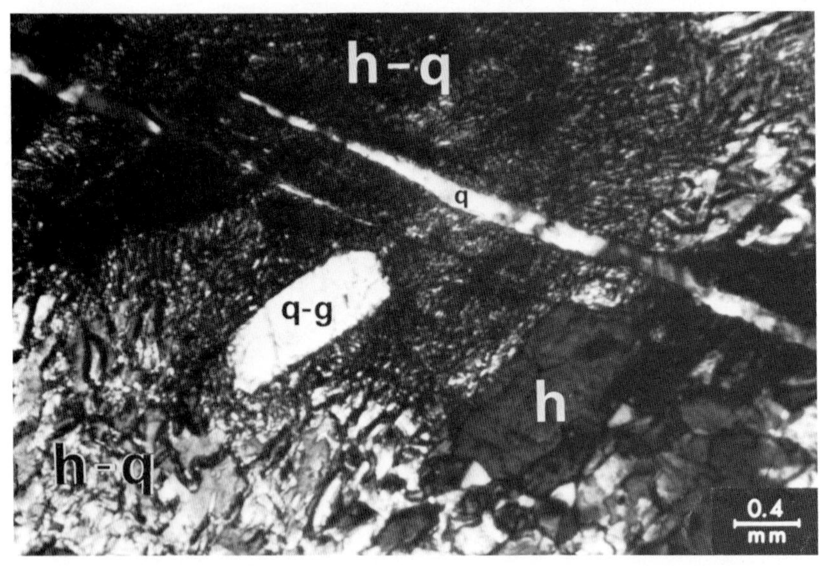

Fig. 175a,

(For Fig. 174 see page 171)

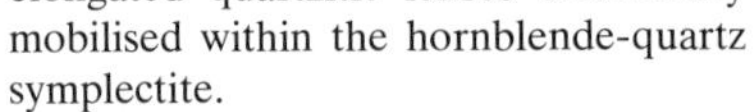

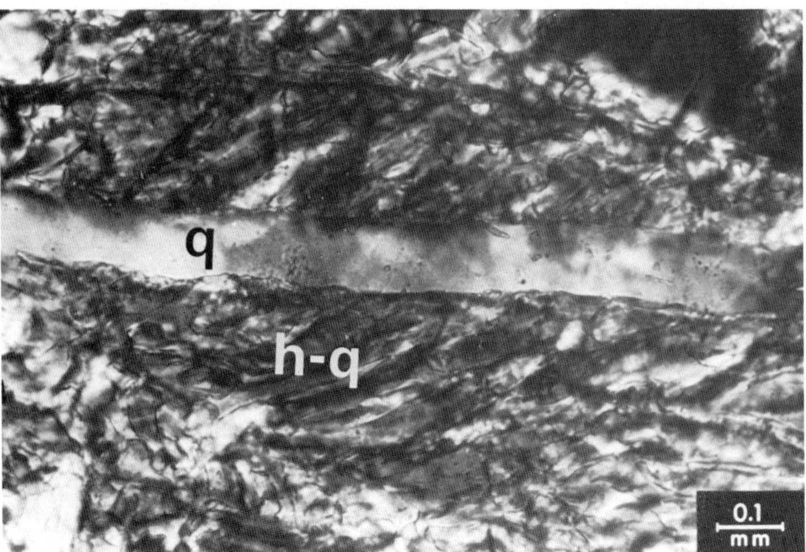

Fig. 175b Hornblende-quartz symplectic intergrowth (i. e. quartz in myrmekitic intergrowth with hornblende) and with elongated quartzitic lenses tectonically mobilised within the hornblende-quartz symplectite.
q = quartzitic lenses
h-q = hornblende quartz symplectite
h = hornblende
q-g = quartz grain
Eclogite. Galicia, North-West Spain. With crossed nicols.

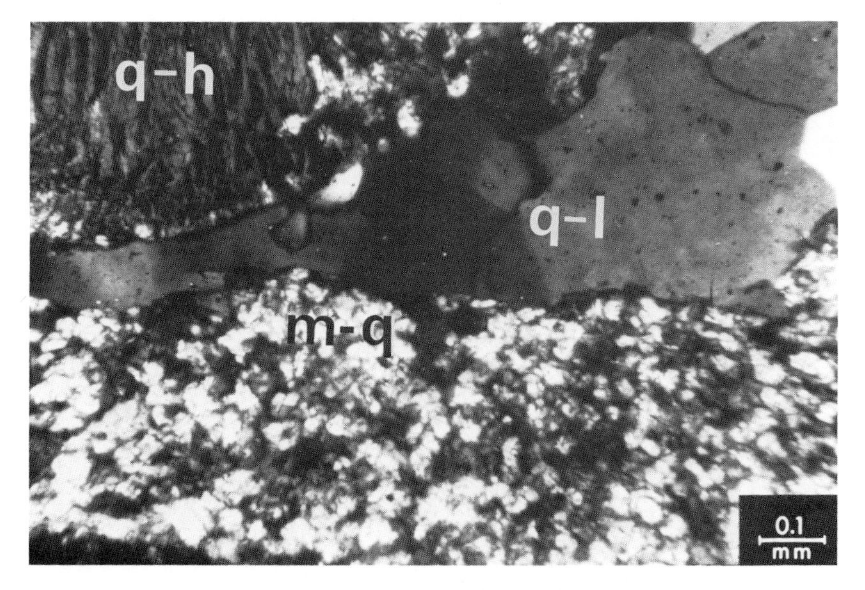

Fig. 174 Quartz-hornblende symplectic intergrowth (i. e. myrmekitic quartz in intergrowth with the hornblende) and a micro-mylonitised quartz prevailing zone with relics and lenses of coarse grained quartzite.
q-h = quartz-hornblende myrmekitic symplectite
m-q = mylonitised quartz prevailing zone
q-l = quartzitic lenses
Eclogite. Galicia, North-West Spain. With crossed nicols.

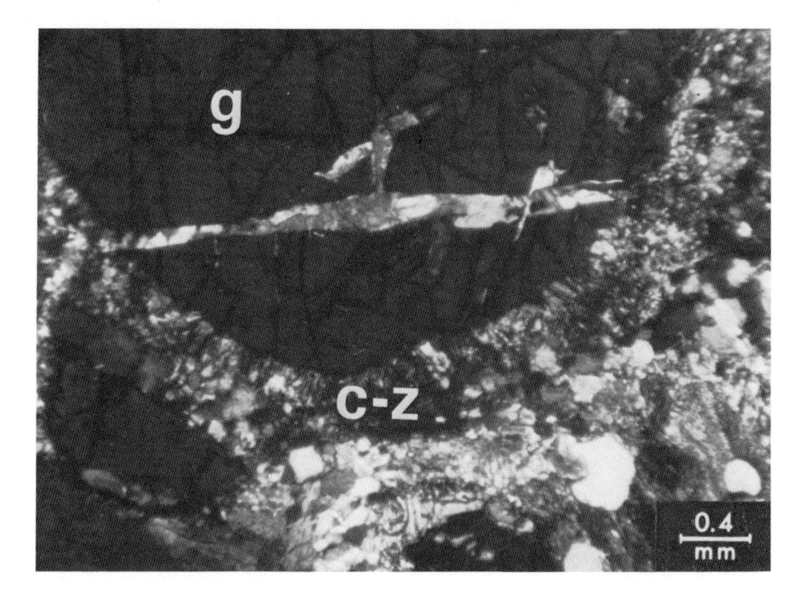

Fig. 176a and detail Fig. 176b Garnet crystalloblast with a crushed zone of quartz-hornblende with mobilisation and recrystallization of the crushed zone within cracks of the garnet.
g = garnet
c-z = crushed zone (mylonitised)
r-q = recrystallised quartz in the crack within the garnet
h = hornblende mobilised and occupying a part of the garnet crack.
Eclogite. Galicia, North-West Spain. With crossed nicols.

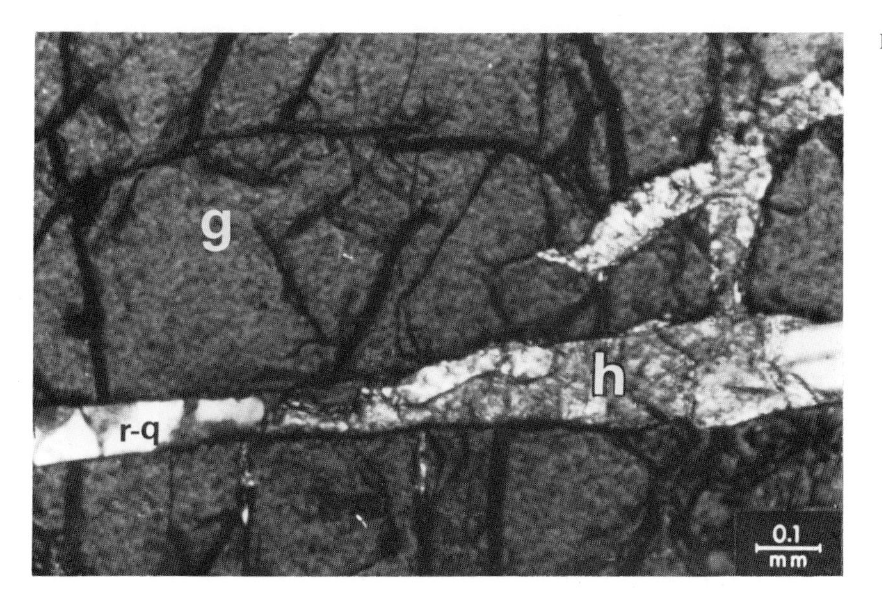

Fig. 176b

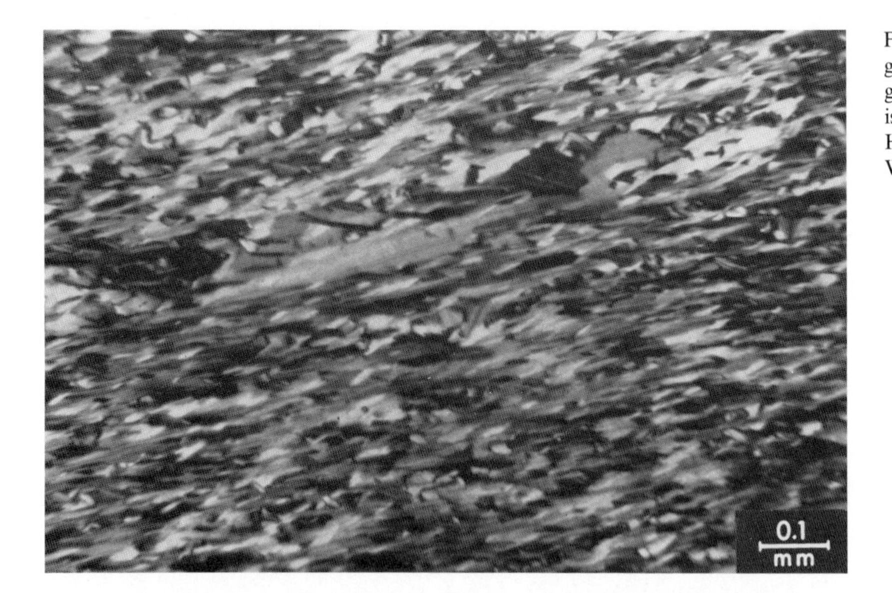

Fig. 177 Chlorite schist (leuchtenbergite) with recrystallised chlorite crystal grains. The direction of crystal elongation is parallel to the schistosity.
Hadabudussa, Gari-Boro, S. Ethiopia.
With crossed nicols.

Fig. 178 A chlorite (leuchtenbergite) "augen structure", in chlorite schist.
Hadabudussa, Gari-Boro, S. Ethiopia.
With crossed nicols.

Fig. 179 Chlorite schist transversed by a later chlorite veinlet. The veinlet consists of recrystallised leuchtenbergite (chlorite).
Hadabudussa, Gari-Boro, S. Ethiopia.
With crossed nicols.

Fig. 181 A chlorite "augen structure" along planes of chlorite recrystallization caused by deformation II. (The arrows mark the tectonic planes).
Hadabudussa, Gari-Boro, S. Ethiopia.
With crossed nicols.

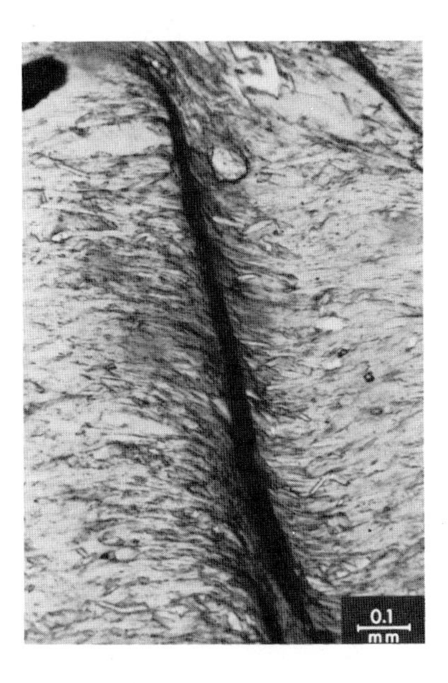

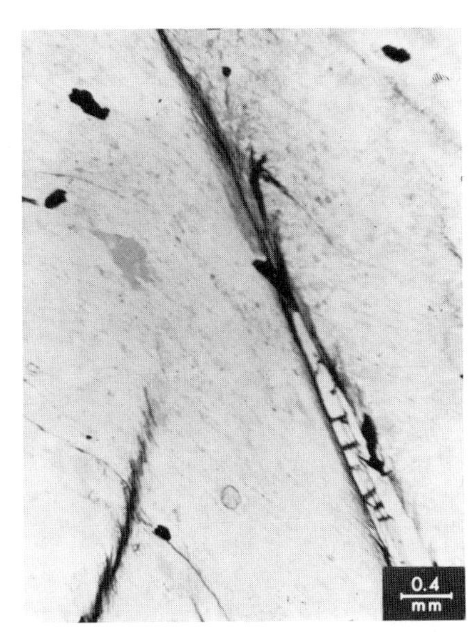

Fig. 180 Microcracks perpendicular to the schistosity of the chlorite schist. Iron concentration has taken place along the microfissures.
Hadabudussa, Gari-Boro, S. Ethiopia.
Without crossed nicols.

Fig. 182 Chlorite schist with microcracks (microfaults) partly occupied by nematoblastic anthophyllite and partly with mobilised iron. The ore minerals shown black are ilmenite with ex-solutions of haematite (see also Fig. 192).
Hadabudussa, Gari-Boro, S. Ethiopia.
Without crossed nicols.

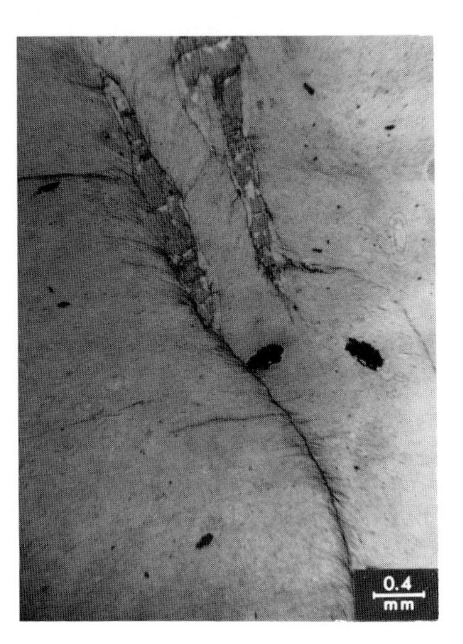

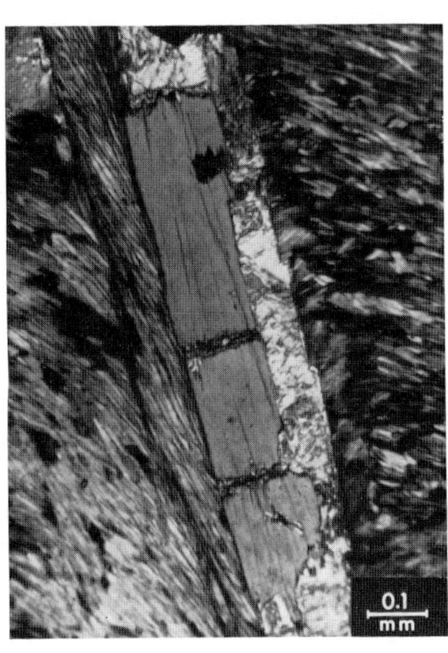

Fig. 183 Chlorite-schist (leuchtenbergite) with microcracks along which iron mobilisation has taken place. Anthophyllite nematoblasts following the tectonic planes are also shown. The ore-minerals are ilmenites with ex-solutions of hematite (see Fig. 192).
Hadabudussa, Gari-Boro, S. Ethiopia.
Without crossed nicols.

Fig. 184 Chlorite-schist cut across by nematoblastic anthophyllite. The direction of the chlorite schistosity is different on both sides of the nematoblast: thus indicating that the amphibole extends along a micro-tectonic displacement. The anthophyllite is partly changed into talc. A non-reaction contact between anthophyllite and chlorite is indicated.
Hadabudussa, Gari-Boro, S. Ethiopia.
With crossed nicols.

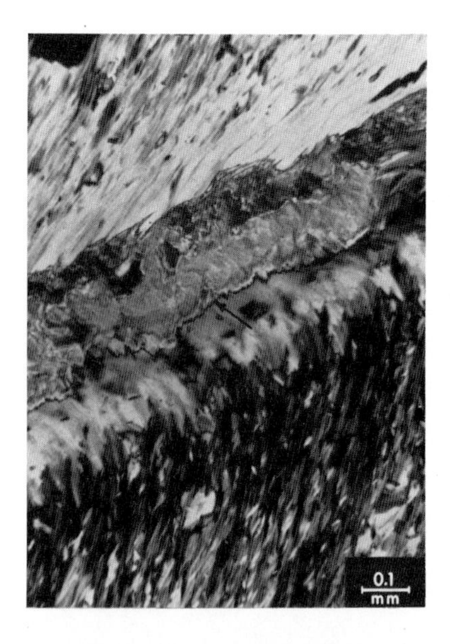

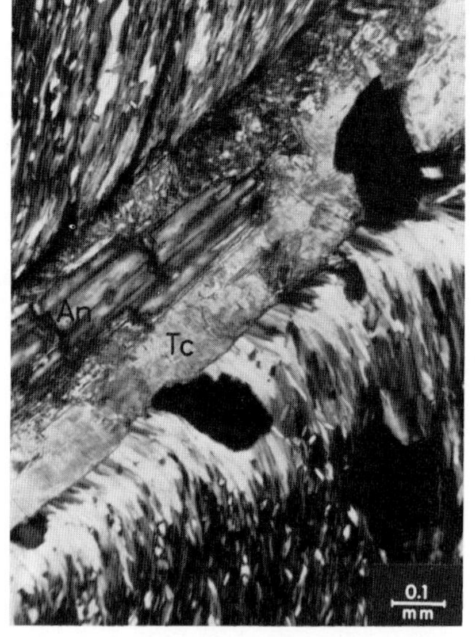

Fig. 185 Chlorite schist with a microfault initially occupied by anthophyllite: the anthophyllite is now entirely changed into talc. Metasomatic reactions have taken place between the talc and the recrystallised chlorite. (Arrow marks the reaction margin.)
Hadabudussa, Gari-Boro, S. Ethiopia.
With crossed nicols.

Fig. 186 Chlorite (leuchtenbergite) schist transversed along a tectonic line by blastic anthophyllite, (An). The amphibole is marginally altered to talc (Tc). The ore-minerals are ilmenite with haematite ex-solutions partly enclosed by the later blastic amphibole (now surrounded by amphibole's alteration, i. e. talc).
Hadabudussa, Gari-Boro, S. Ethiopia.
With crossed nicols.

Fig. 187 Holoblastic amphibolite. Anthophyllite blastic crystals showing a radiating crystallization of many centres of crystal growth.
Hadabudussa, Gari-Boro, S. Ethiopia.
With crossed nicols.

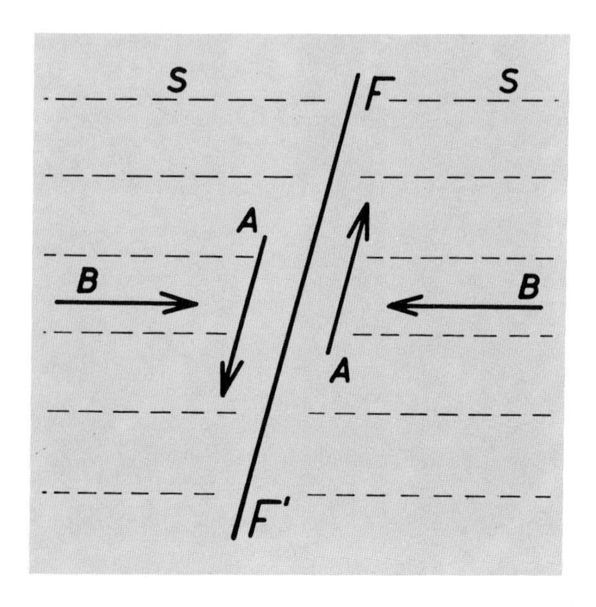

Fig. 188 (Diagram). Diagram showing the "hypothetical" conditions of the formation of anthophyllite. Arrows (A) mark the movement that has taken place along the microdisplacement. Arrows (B) mark the hypothetical movement of the solutions that built the anthophyllite, (S) marks the schistosity: (F-F') the microdisplacement.

Fig. 189a Blastic magnetite with chlorite inclusions and partly extending along the schistosity of the phyllite. The blastic magnetite also partly enclosed an anthophyllite crystal grain.
Hadabudussa, Gari-Boro, S. Ethiopia.
Without crossed nicols.

Fig. 189b Blastic magnetite partly enclosing chlorite schist components (following the direction of the schist exterior to the blastic crystal) and partly extending along the schistosity of the chlorite.
Hadabudussa, Gari-Boro, S. Ethiopia.
Without crossed nicols.

Fig. 189c Blastic magnetite with margins extending along the chlorite schistosity. Also chlorite schist grains are enclosed in the blastic magnetite.
Hadabudussa, Gari-Boro, S. Ethiopia.
With crossed nicols.

Fig. 189d Blastic magnetite in contact with recrystallised chlorite. It is difficult to know whether the leuchtenbergite recrystallization is influenced by the crystallization force exercised by the blastic magnetite.
Hadabudussa, Gari-Boro, S. Ethiopia.
With crossed nicols.

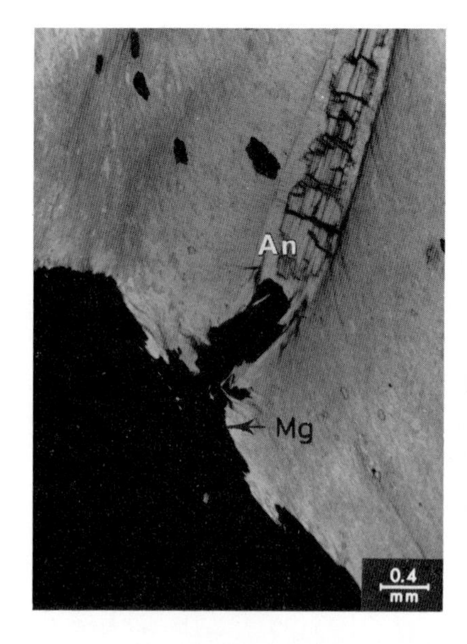

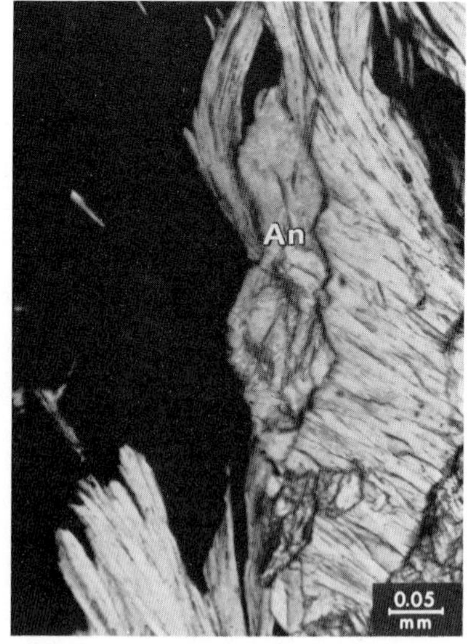

Fig. 190 Chlorite schist (leuchtenbergite) with small ore-minerals consisting of ilmenite with hematite (ex-solutions). A microfissure (microcrack) is occupied by anthophyllite (An) at margins changed into talc. A blastic magnetite (Mg) partly extends into the crack which is occupied by the amphibole.
Hadabudussa, Gari-Boro, S. Ethiopia.
Without crossed nicols.

Fig. 191 A detailed view of Fig. 189a. Blastic magnetite partly enclosed chlorite schist and being in contact with pre-existing anthophyllite (An). Magnetite (black).
Hadabudussa, Gari-Boro, S. Ethiopia.
Without crossed nicols.

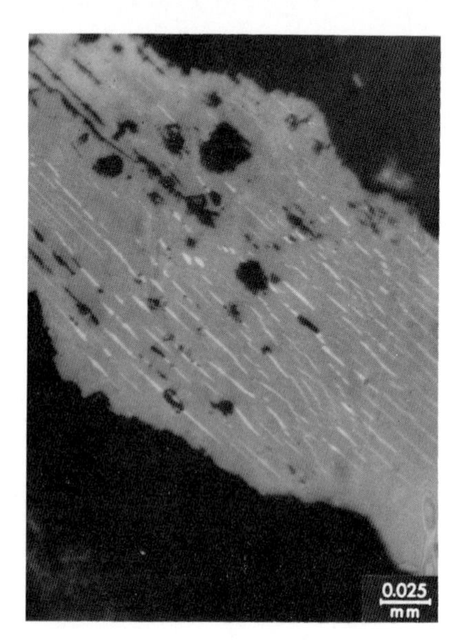

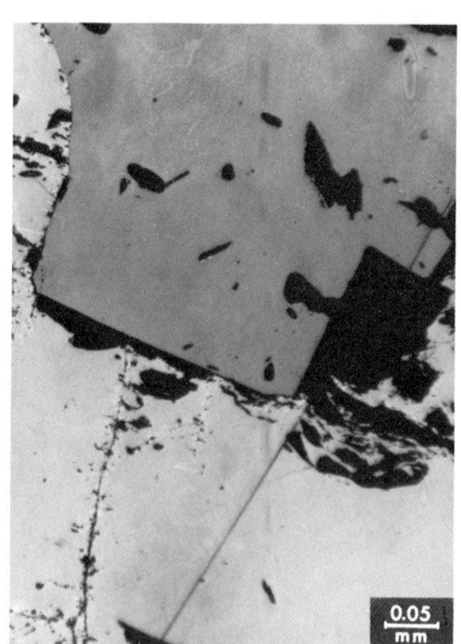

Fig. 192 Initial ilmenite with ex-solutions of haematite partly enclosed by anthophyllite (with internal reflection). Polished section, oil-immersion with one nicol.
Hadabudussa, Gari-Boro, S. Ethiopia.

Fig. 193 Blastic magnetite (with martite along cracks) partly enclosing ilmenite. (The ilmenite contains fine ex-solution lamellae of haematite not to be seen in the photomicrograph). Polished section, oil-immersion with one nicol.
Hadabudussa, Gari-Boro, S. Ethiopia.

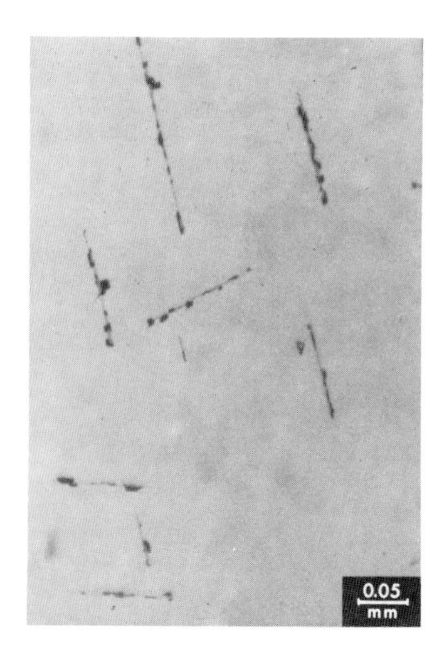

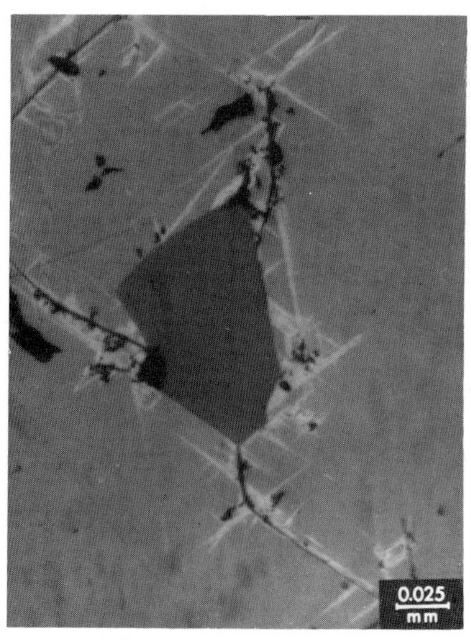

Fig. 194 Ilmenite (intergrowths) exsolution bodies orientated parallel to (111) face of the host magnetite. Spinel bodies are also present. Polished section, oil immersion with one nicol.
Hadabudussa, Gari-Boro, S. Ethiopia.
(Also see Augustithis (1964b)

Fig. 195 Blastic magnetite with ilmenite (dark grey) and with hematite martite (along cracks). Polished section, oil-immersion with one nicol.
Hadabudussa, Gari-Boro, S. Ethiopia.

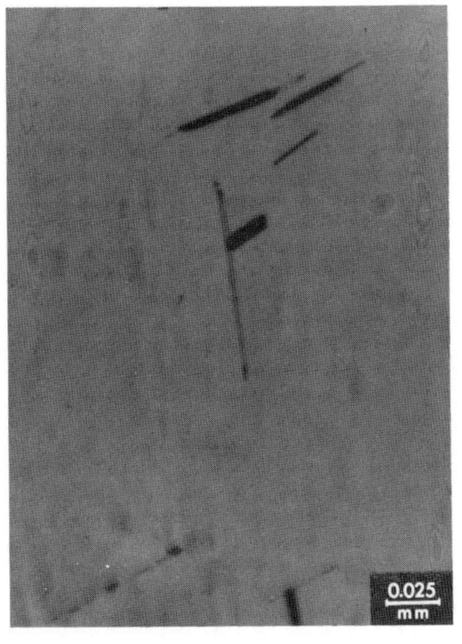

Fig. 196 Blastic magnetite with ilmenite lamellae. Along the margins of the magnetite with the ilmenite there are bodies of spinel (dark). Polished section oil-immersion with one nicol.
Hadabudussa, Gari-Boro, S. Ethiopia.

Fig. 197 Blastic magnetite (light grey) with ilmenite lamellae (middle grey) and lamellae of spinel (dark). Polished section, oil-immersion with crossed nicols.
Hadabudussa, Gari-Boro, S. Ethiopia.

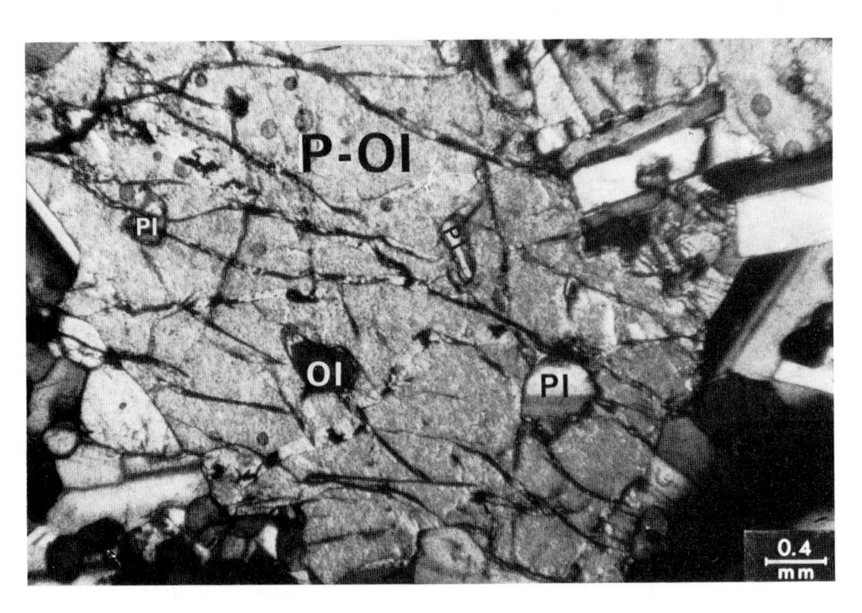

Fig. 198 Poikilitic olivine crystalloblast, enclosing and engulfing pre-olivine plagioclases. The olivine megablast also includes and corrodes pre-blastic olivine (components of the olivinefels) into which the olivineblast has grown.
P-Ol = poikiloblastic olivine
Ol = olivine included and corroded by the poikiloblastic olivine (megablast).
Pl = plagioclase
Gabbro picrite. 40 meters from northern margin at Stromstedet, Skaergaard, Greenland.
With crossed nicols.

Fig. 199 Plagioclase and olivines included in a pyroxene poikiloblast which is marginal to a plagioclase-olivine phase, components of which detached are included in the pyroxene mega-poikiloblast.
m-py =pyroxene megablast
pl-ph = plagioclase of the marginal plagioclase-olivine phase outside the pyroxene.
Pl = plagioclases corroded and included in the megablast.
Arrows "a" show extensions of the pyroxene into the intergranular of the external phase.
Skaergaard, Greenland.
With crossed nicols.

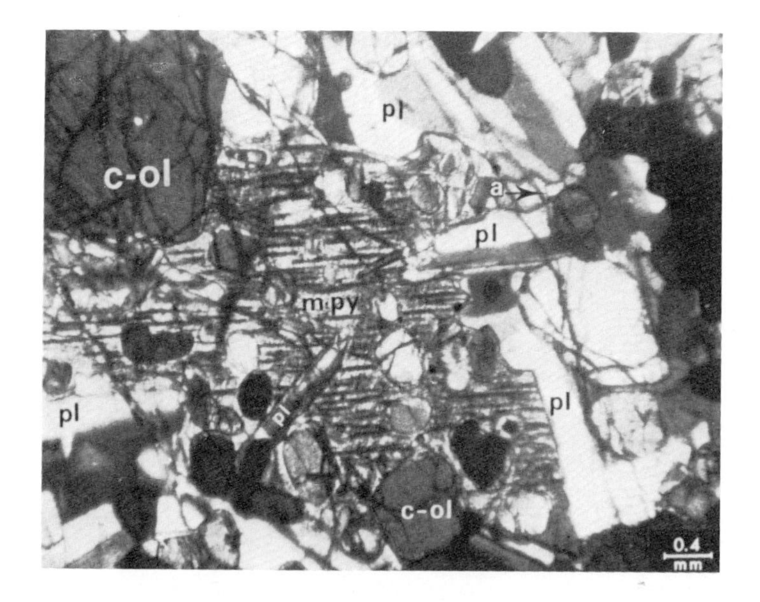

Fig. 200 Pyroxene poikiloblast with olivine and pyroxenes of a pre-blastic plagioclase-olivinefels phase and coarse olivine-blast engulfed, corroded and partly assimilated by the pyroxene megablast, prolongations of which also protrude (see arrow "a") into the adjacent plagioclase-olivine phase.
m-py = megablast pyroxene
c-ol = coarse olivine
pl = plagioclase
Skaergaard, Greenland.
With crossed nicols.

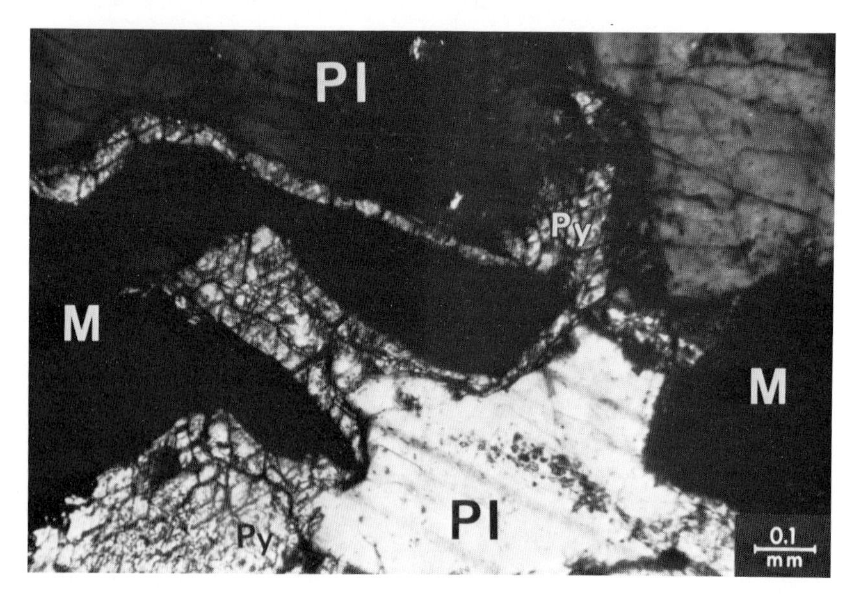

Fig. 201 Crystalloblastic pyroxene between corroded plagioclase and crystalloblastic magnetite (see also Fig. 567).
Pl = plagioclase
Py = pyroxene
M = magnetite crystalloblast
Average ferrogabbro (plagioclase-augite-olivine-magnetite "cumulate").
House area, Skaergaard, Greenland.
With crossed nicols.

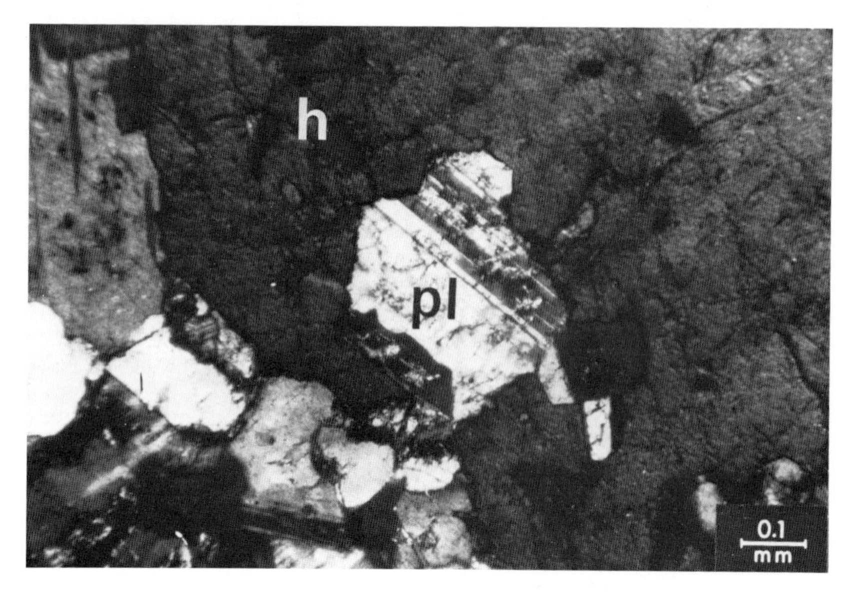

Fig. 202 Plagioclase corroded and included in hornblende crystalloblasts.
h = hornblende crystalloblast
pl = plagioclase
Mte Mattoni, Val Fredda, S. Adamello, Italy.
With crossed nicols.

Fig. 203 Plagioclases corroded and included in hornblende crystalloblasts.
h = hornblende crystalloblast
pl = plagioclase
Mte Mattoni, Val Fredda, S. Adamello, Italy.
With crossed nicols.

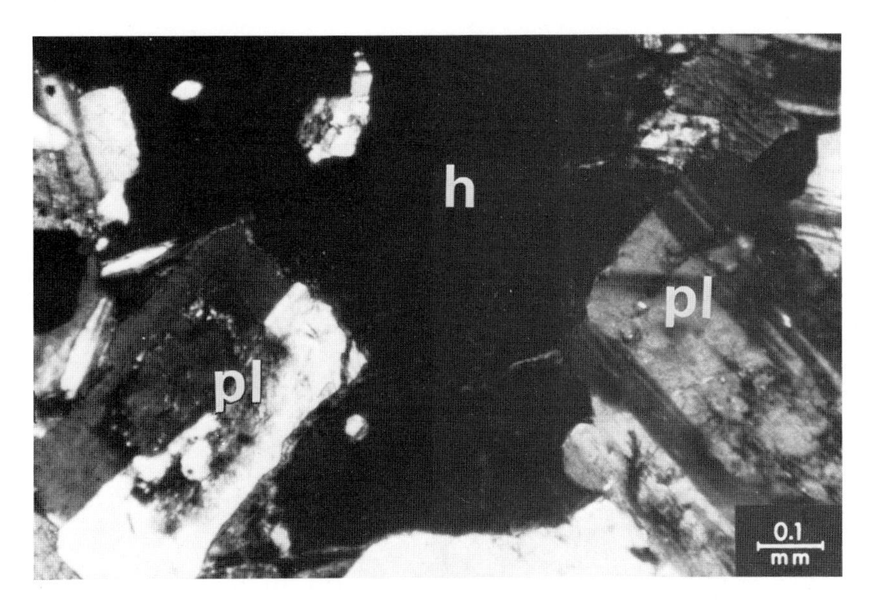

Fig. 204 Zoned plagioclase either enclosed or partly engulfed by hornblendeblasts.
h = hornblende
pl = plagioclase
Mte. Mattoni, Val Fredda, S. Adamello, Italy.
With crossed nicols.

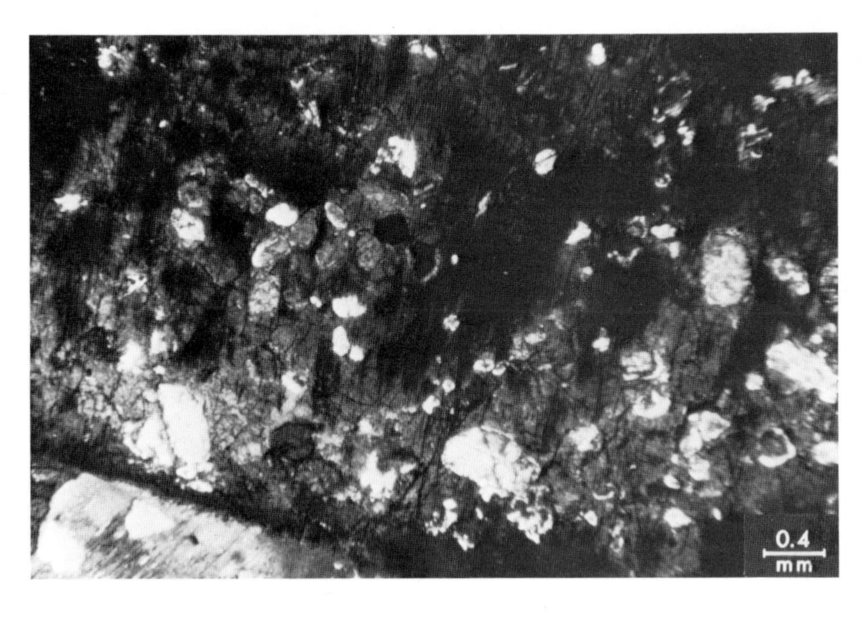

Fig. 205 Hornblende crystalloblast with plagioclase inclusions and local alterations of the hornblende.
Mte Mattoni, Val Fredda, S. Adamello, Italy.
With crossed nicols.

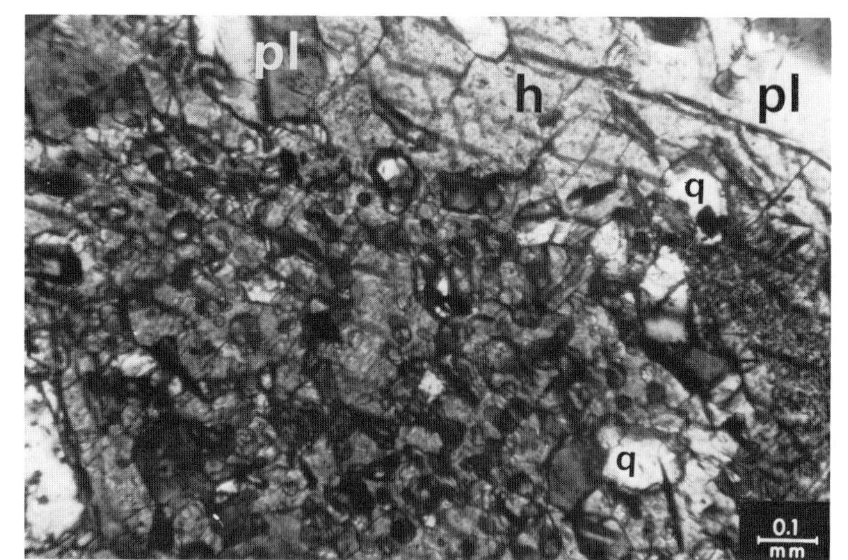

Fig. 206 Poikiloblastic hornblende with quartz and plagioclase inclusions.
h = hornblende poikiloblast
pl = plagioclase
q = quartz
Mte Mattoni, Val Fredda, S. Adamello, Italy.
With crossed nicols.

Fig. 207 Hornblende and plagioclase crystalloblasts.
h = hornblende crystalloblast
f = feldspar crystalloblasts
Mte Mattoni, Val Fredda, S. Adamello, Italy.
With crossed nicols.

Fig. 208 Hornblende crystalloblasts.
h = hornblende crystalloblasts
Mte Mattoni, Val Fredda, S. Adamello, Italy.
With crossed nicols.

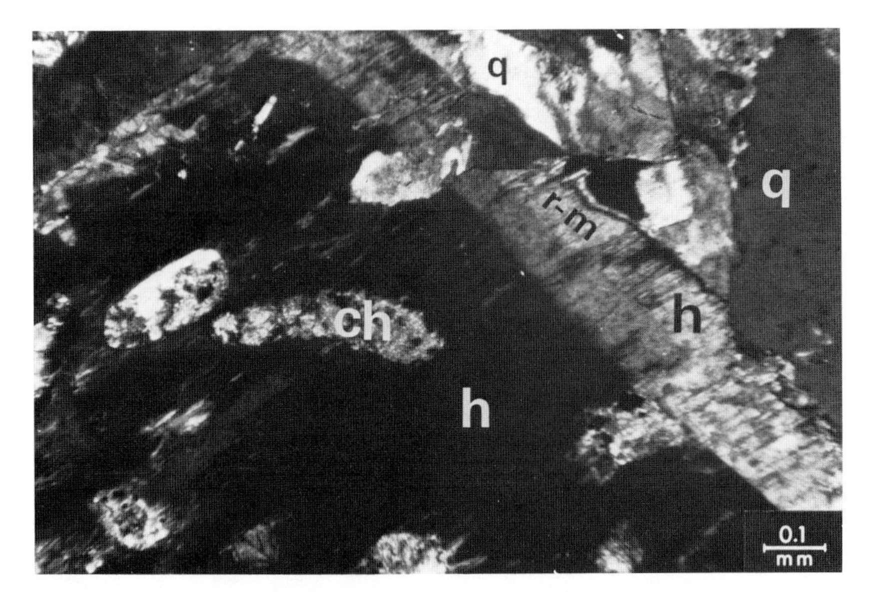

Fig. 209 Hornblende megablast with a reaction margin (r-m) and including initial feldspar now altered to chloritic aggregate.
h = hornblende
ch = chloritised initial plagioclase
q = quartz
Mte Mattoni, Val Fredda, S. Adamello, Italy.
With crossed nicols.

Figs. 210 and 211 Hornblende crystalloblasts enclosing altered plagioclases and locally itself altered.
h = hornblende crystalloblasts (Arrow "a" local alteration of hornblende crystalloblast)
Al-p = altered plagioclase included in the hornblende crystalloblast.
Mte Mattoni, Val Fredda, S. Adamello, Italy.
With crossed nicols.

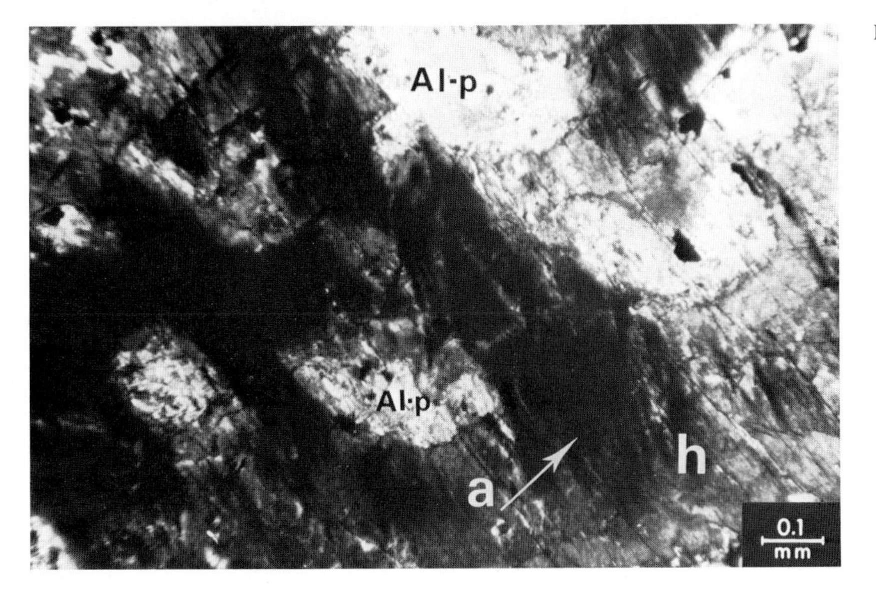

Fig. 211

Fig. 212 Hornblende crystalloblast enclosing altered plagioclase in contact with a later plagioclase crystalloblast enclosing older and altered plagioclases and hornblende crystal grains.
h = hornblende crystalloblast
pl = plagioclase crystalloblast
Pl-o = older and altered plagioclases included in the hornblende.
Mte Mattoni, Val Fredda, S. Adamello, Italy.
With crossed nicols.

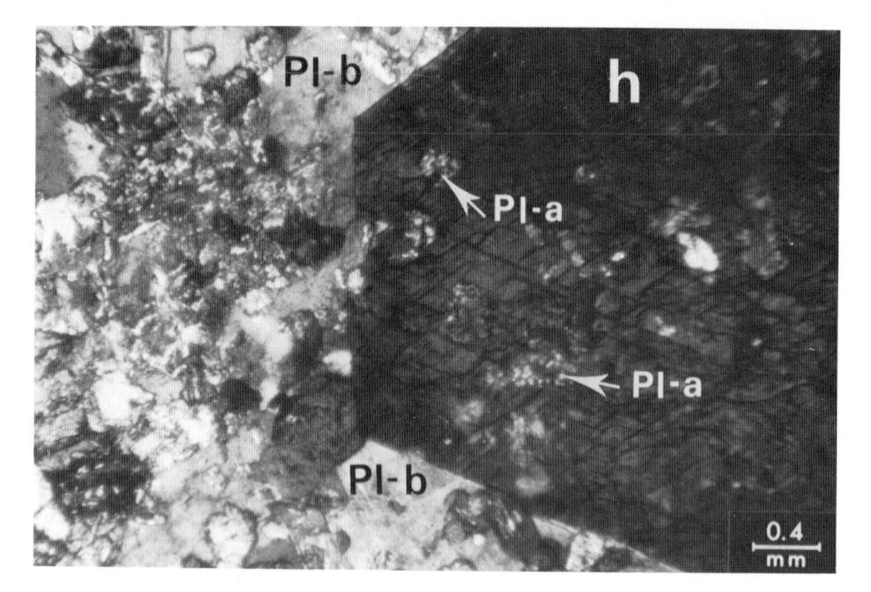

Fig. 213 Hornblende idioblast enclosing altered plagioclases, surrounded and enclosed by a later plagioclase crystalloblast which also encloses the altered plagioclase generation and altered hornblendes.
Pl-a = pre-hornblende plagioclases altered (sericitised, chloritised).
h = hornblende crystalloblast
Pl-b = plagioclase crystalloblast
Mte Mattoni, Val Fredda, S. Adamello, Italy.
With crossed nicols.

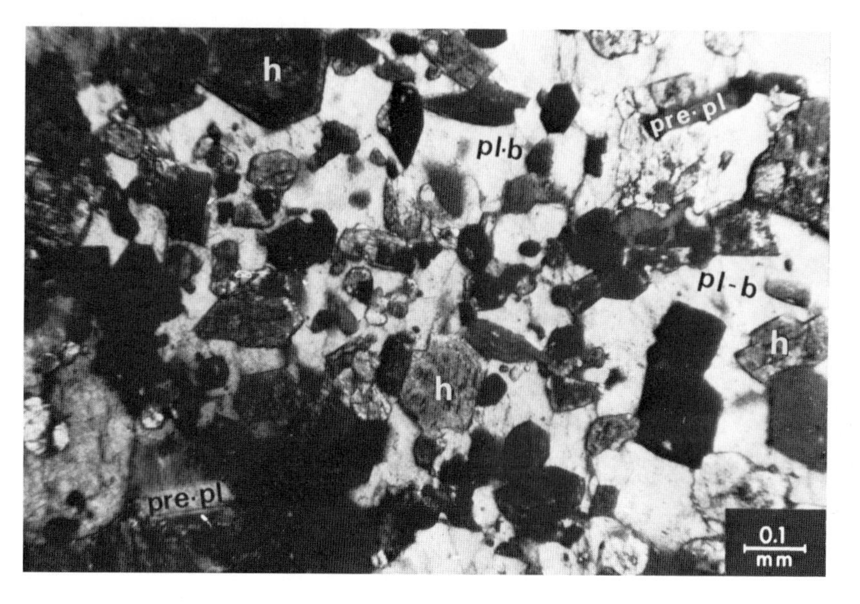

Fig. 214 Hornblendes and plagioclases included in a later plagioclase poikiloblast.
h = hornblende
pre-pl = pre-blastic plagioclases
pl-b = plagioclase crystalloblasts
Mte Mattoni, Val Fredda, S. Adamello, Italy.
With crossed nicols.

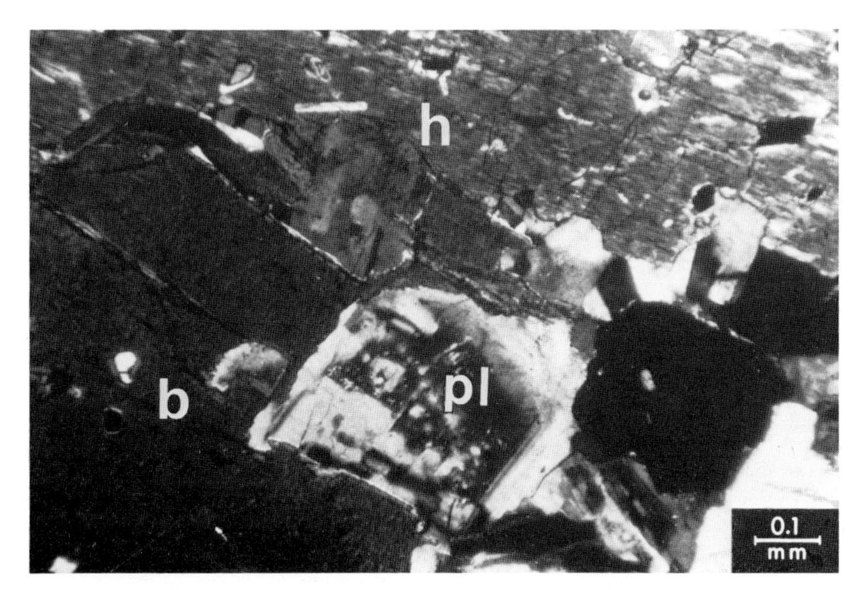

Fig. 215 Hornblende and biotite crystalloblast often enclosing zoned plagioclases.
h = hornblende
b = biotite
pl = plagioclases
Mte Mattoni, Val Fredda, S. Adamello, Italy.
With crossed nicols.

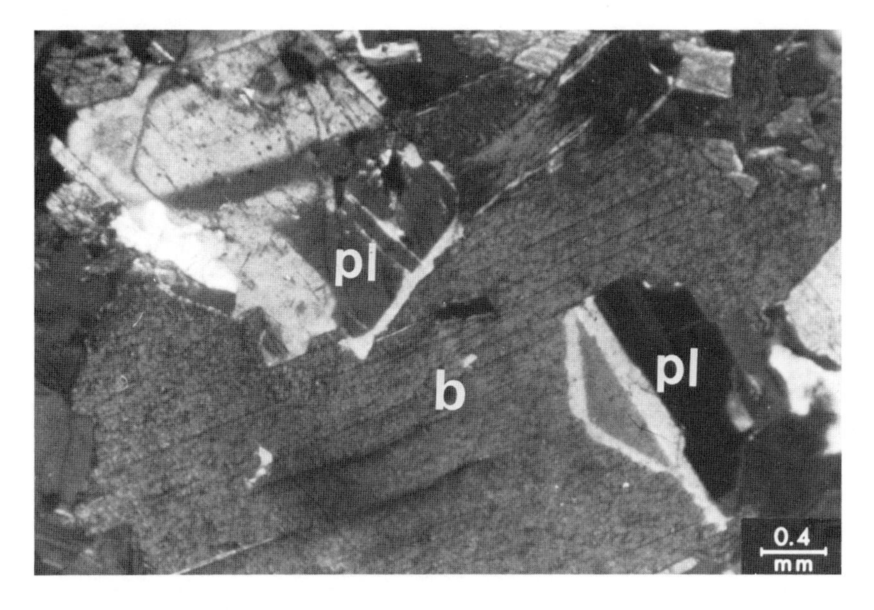

Fig. 216 Biotite crystalloblast (megablast) enclosing and corroding pre-blastic plagioclases.
b = biotite crystalloblast
pl = plagioclases corroded and enclosed by the biotite crystalloblast
Mte Mattoni, Val Fredda, S. Adamello, Italy.
With crossed nicols.

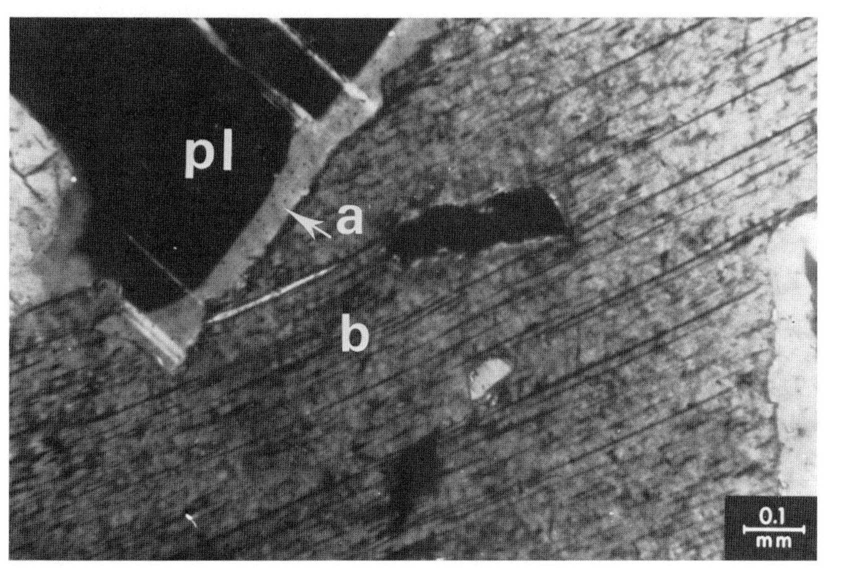

Fig. 217 Plagioclase enclosed and corroded by a later biotite crystalloblast. Arrow "a" shows a reaction margin of the enclosed and corroded plagioclase.
pl = plagioclase
b = biotite crystalloblast
Mte Mattoni, Val Fredda, S. Adamello, Italy.
With crossed nicols.

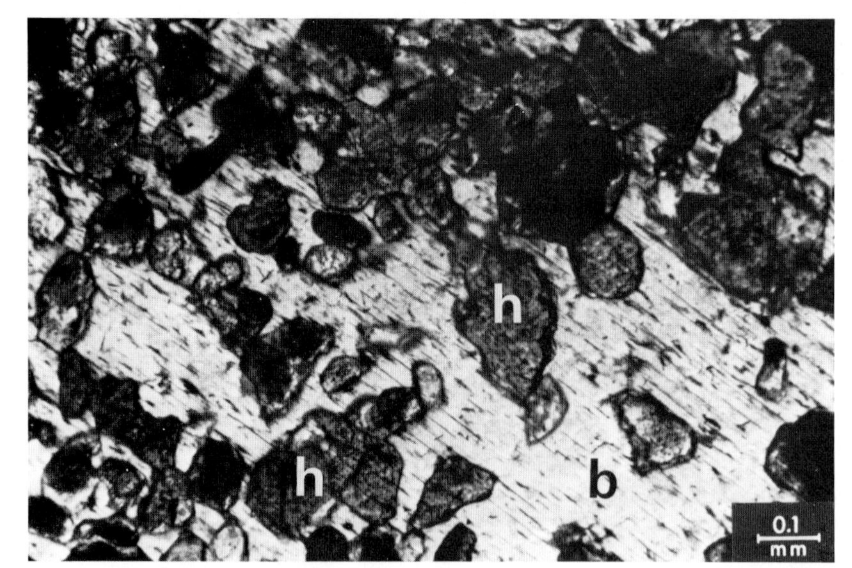

Fig. 218 Hornblendes included and corroded by biotite crystalloblasts.
h = hornblende
b = biotite crystalloblast (poikiloblast)
Mte Mattoni, Val Fredda, S. Adamello, Italy.
With crossed nicols.

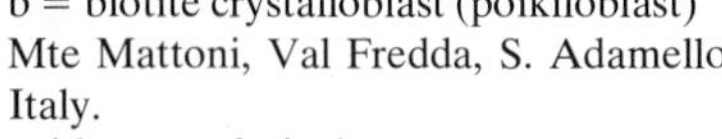

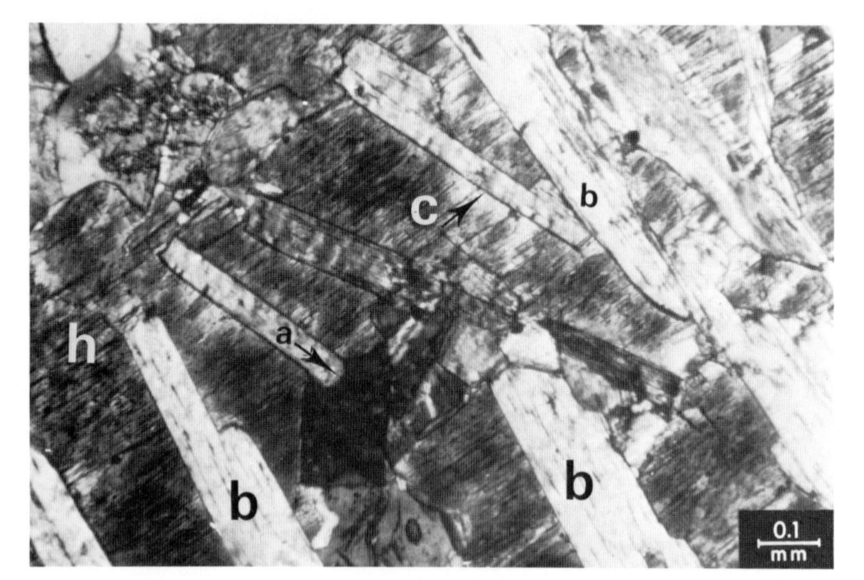

Fig. 219 Hornblende megablast with biotite laths perpendicular to cleavage direction of the hornblende crystalloblast. Also, arrow "a" shows biotite lath penetrating another biotite also blastically grown in the hornblende crystalloblast.
b = biotite lath orientated and almost perpendicular to the hornblende cleavage.
h = hornblende crystalloblast (arrow "c" shows the cleavage direction of the hornblende crystalloblast)
Mte Mattoni, Val Fredda, S. Adamello, Italy.
With crossed nicols.

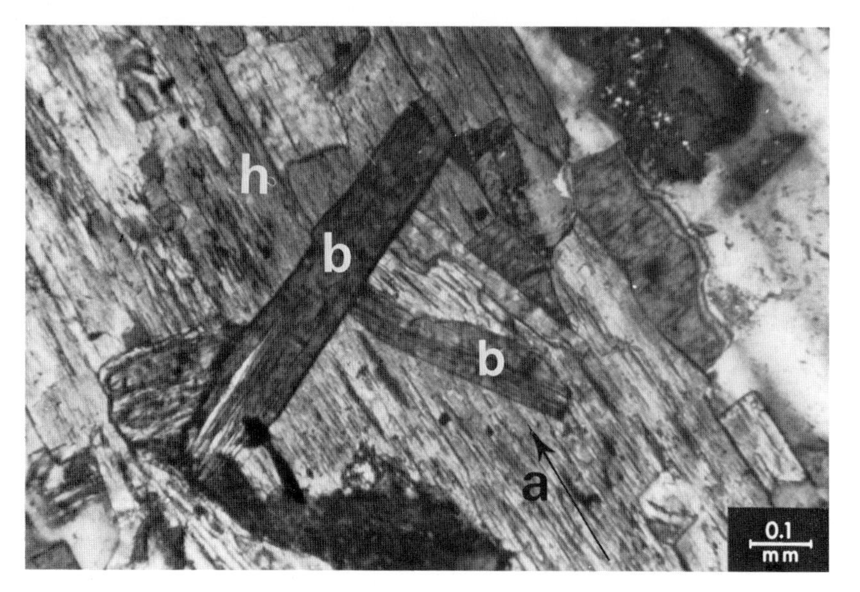

Fig. 220 Biotite crystalloblastic laths either parallel to cleavage of the hornblende or perpendicular to it.
h = hornblende crystalloblast
b = biotite laths
Arrows "a" show the cleavage of the hornblende
Mte Mattoni, Val Fredda, S. Adamello, Italy.
With crossed nicols.

Fig. 221 Hornblende megablast with later biotite crystalloblastic laths orientated within the hornblende crystalloblast. Arrow "a" shows a biotitic crystalloblastic lath penetrating another biotite crystalloblast in the hornblende.
Mte Mattoni, Val Fredda, S. Adamello, Italy.
Without crossed nicols.

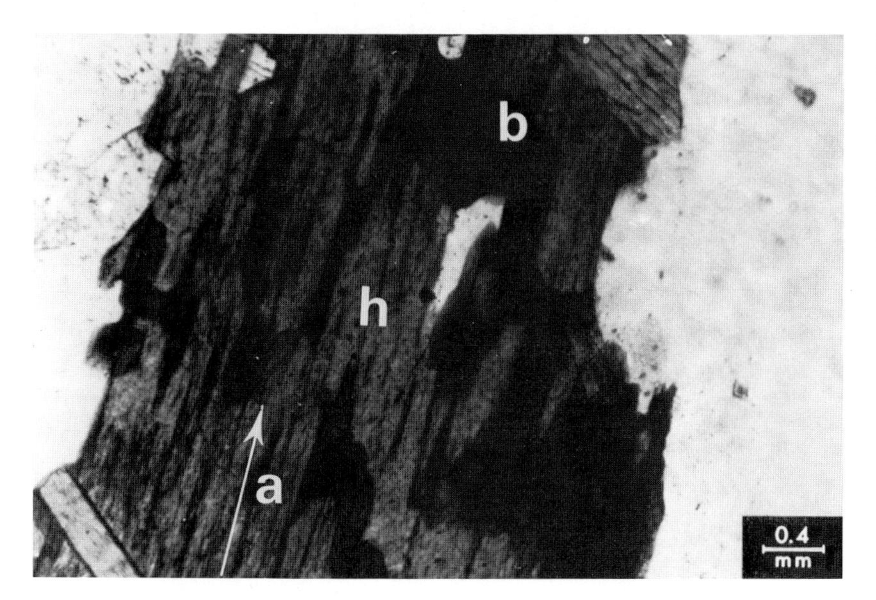

Fig. 222 A hornblende megablast replaced by orientated biotite which, as arrow "a" shows, follows the hornblende cleavage.
h = hornblende
b = biotite
Mte Mattoni, Val Fredda, S. Adamello, Italy.
Without crossed nicols.

Fig. 223 Garnet (almandine) crystalloblasts in the granite and transversing the boundary of the granite and an included basic xenolith.
G = granite
g-g = almandine crystalloblast in the granite
g-x = almandine crystalloblast transversing the boundary granite/xenolith.
x = basic xenolith included in the granite.
Rapakivi granite, Finland.
2/3 natural size.

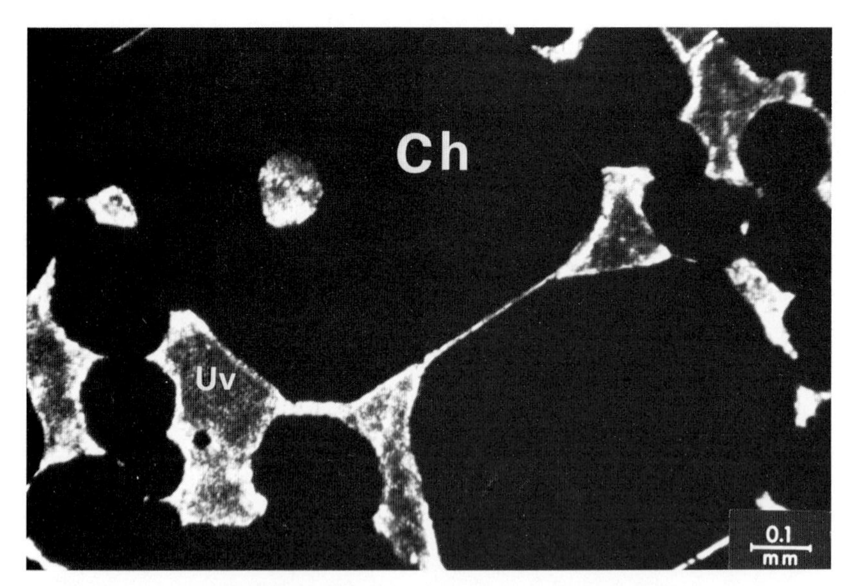

Fig. 224 Chromite grains partly rounded and surrounded by uvarovite crystalloblastesis (see also Fig. 673).
Ch = chromite
Uv = uvarovite occupying the spaces between the chromite.
Chromite rich band Bushveld, Transvaal, South Africa.
Without crossed nicols.

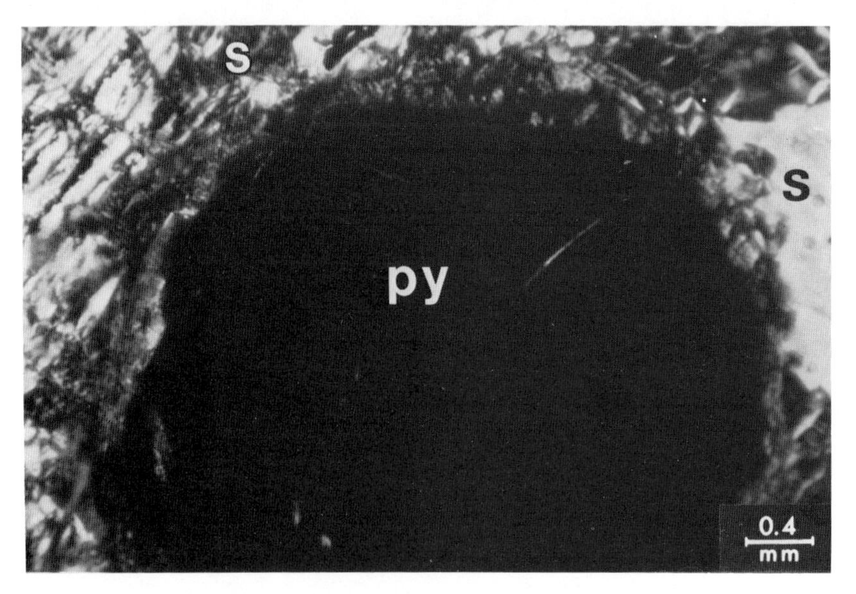

Fig. 225 Rounded pyrope "xenocryst" in serpentinised peridotite (also, see Fig. 707).
py = pyrope (isotropic)
s = serpentine
Serpentine with pyrope. Zoblitz, Saxony, Germany.
With crossed nicols.

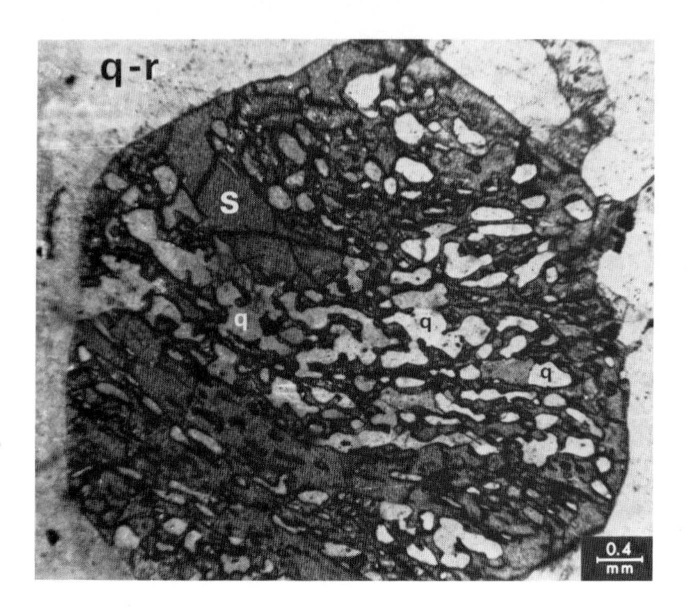

Fig. 226 Spessartite streptoblast in a quartzitic groundmass.
S = spessartite crystalloblast (poikiloblast)
q = quartz grains, included in distorted bands within the garnet crystalloblast.
q-r = recrystallised quartz (exhibiting undulating extinction) of the quartzitic groundmass.
Chios Island, Greece.
With half crossed nicols.

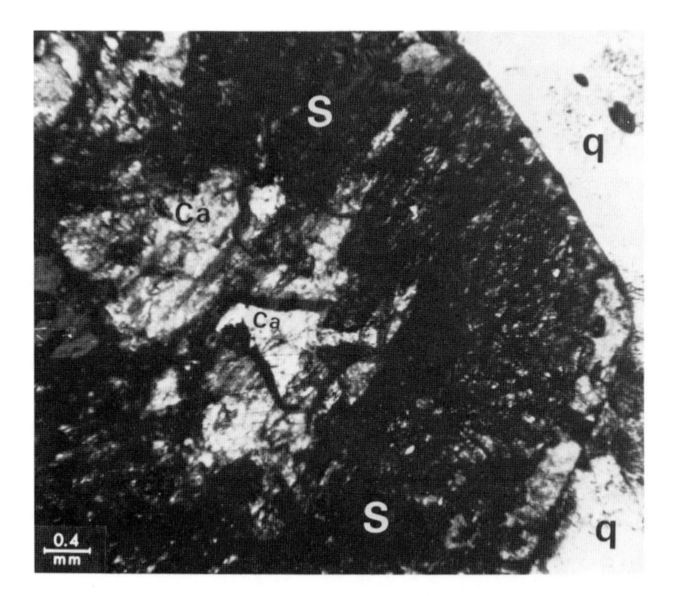

Fig. 227 Spesartite crystalloblast (streptoplast) with calcite inclusion in the process of "assimilation" by the garnet crystalloblast.
s = spessartite
ca = calcite (included relics in the spessartite)
q = quartz
Chios Island, Greece.
With crossed nicols.

Fig. 228 Ophitic plagioclase-pyroxene intergrowth texture. A pyroxene phenocryst with cleavage and cracks invaded by plagioclase laths which extend from the outside of the pyroxene into the host. Dolerite (coarse-grained basalt). Karroo series, S. Africa.
With crossed nicols.

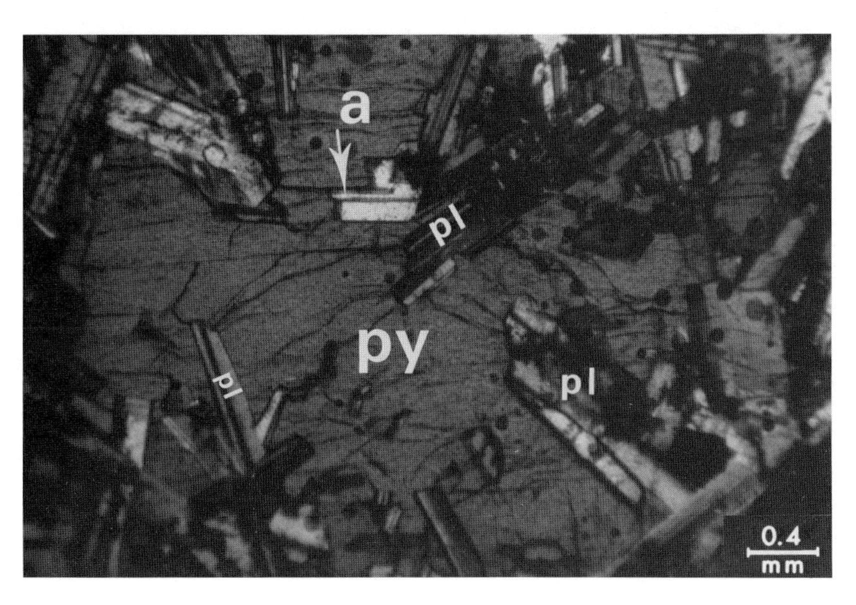

Fig. 229 Ophitic plagioclase-pyroxene intergrowth texture. Often the plagioclases extend from the margin inwards the pyroxene, transversing the cleavage and also as arrow "a" shows a plagioclase lath follows the cleavage.
py = pyroxene
pl = plagioclase
Microgabbro (dolerite). Palisade Sill, N. Jersey, U.S.A.
With crossed nicols.

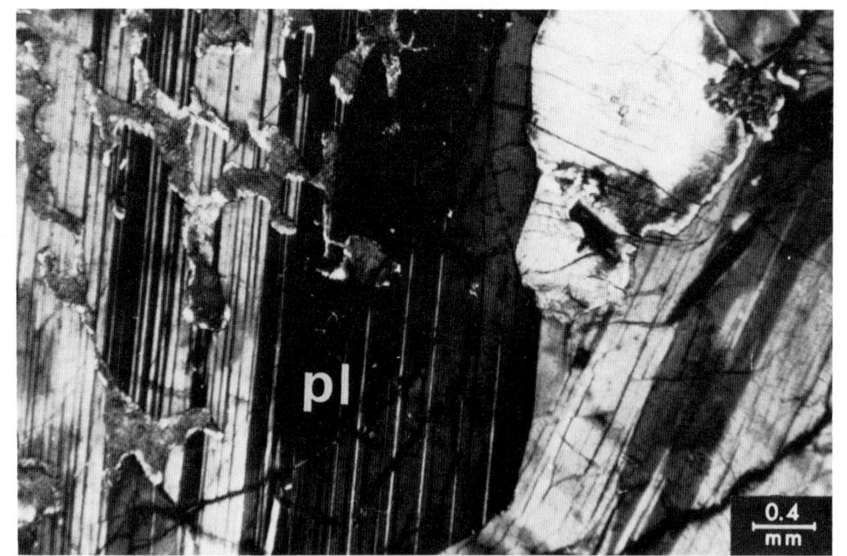

Fig. 230a and b Pyroxene in intergrowth with plagioclase. The pyroxene invades the feldspar and as Fig. 230(b) shows, it partly follows the plagioclase twinning and partly is in graphic-like intergrowth with the feldspar.
pl = plagioclase
py-a = pyroxene part following the twinning of the plagioclase.
py-b = pyroxene in graphic intergrowth with the plagioclase.
Olivine gabbro. Oberkainsbach, Odenwald, Germany.
With crossed nicols.

Fig. 230b

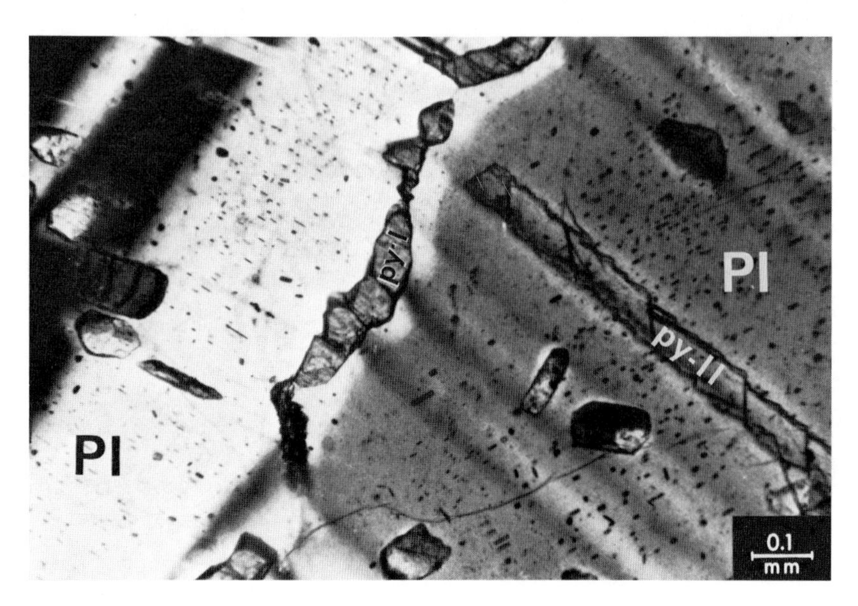

Fig. 231 Plagioclase invaded by pyroxene. The pyroxene may either follow the feldspars zoning or it may follow twin-planes of the plagioclase.
pl = plagioclase
py-I = pyroxene following the zoning of the feldspar
py-II = pyroxene parallel to the plagioclase's twinning.
Gabbroic rock (noritic in composition). Hitteroe, Norway.
With crossed nicols.

Fig. 232 Often the contacts pyroxene plagioclase are difficult to be interpreted. In this particular case they are dubious (it is therefore necessary to search and present other textural patterns where the relation of the one mineral to the other is elucidating, see Figs. 233, 234 and 235).
Magnetite-Gabbro. Estelien, Ringerike, Oslo, South Norway.
With crossed nicols.

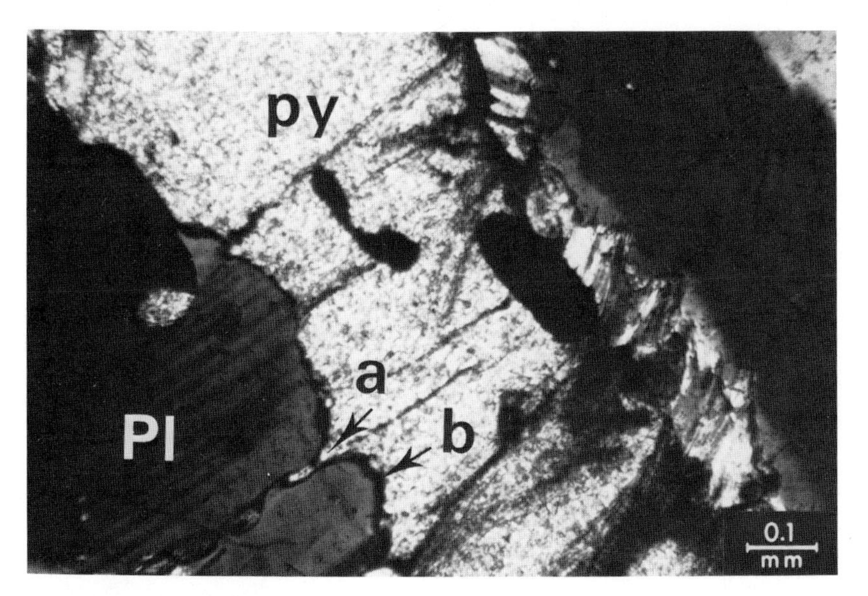

Fig. 233 Plagioclase corroded and invaded by pyroxene which clearly extends as a later growth into the feldspar.
Pl = plagioclase (arrow "a" shows corroded outline of the plagioclase)
py = pyroxene (arrow "b" shows extension of the pyroxene into the feldspar)
Magnetite-Gabbro. Estelien, Ringerike, Oslo, South Norway.
With crossed nicols.

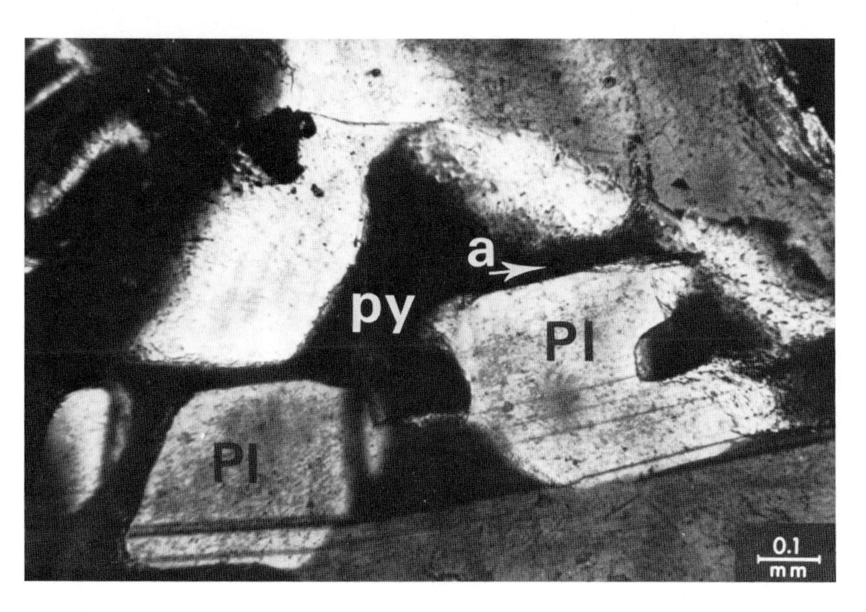

Fig. 234 Pyroxene invading and replacing the plagioclase, often clearly taking advantage of the feldspar twinning.
Pl = plagioclase
py = pyroxene (arrow "a" shows pyroxene following and replacing the feldspar along its twinning)
Magnetite-Gabbro. Estelien, Ringerike, Oslo, South Norway.
With crossed nicols.

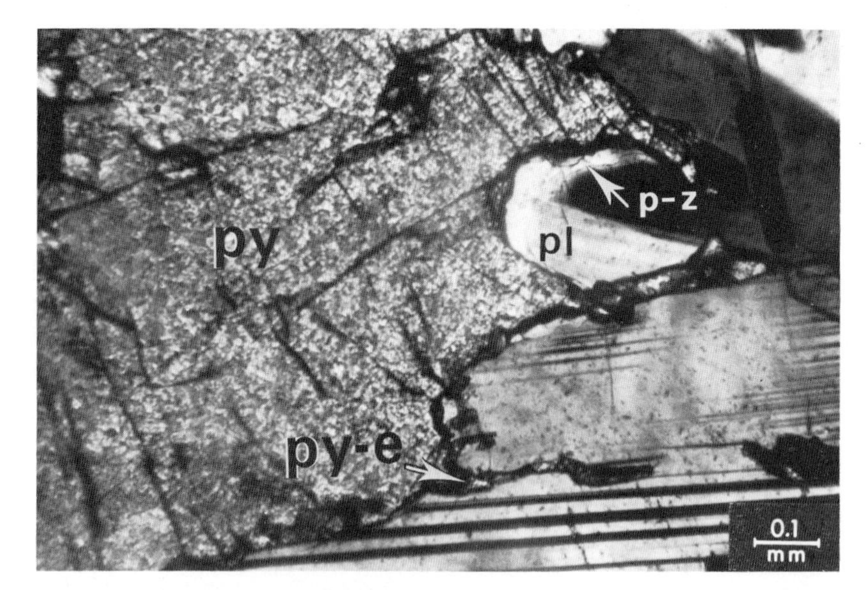

Fig. 235 Pyroxene invading the feldspars, either partly engulfing a plagioclase causing a reaction zoning in the feldspar or as intragranular extension following the inter-twin planes of the plagioclase.
py = pyroxene
pl = plagioclase
p-z = reaction zoning of the plagioclase engulfed by the pyroxene
py-e = pyroxene extension following the interlamellar twinning of the plagioclase
Magnetite-Gabbro. Estelien, Ringerike, Oslo, South Norway.
With crossed nicols.

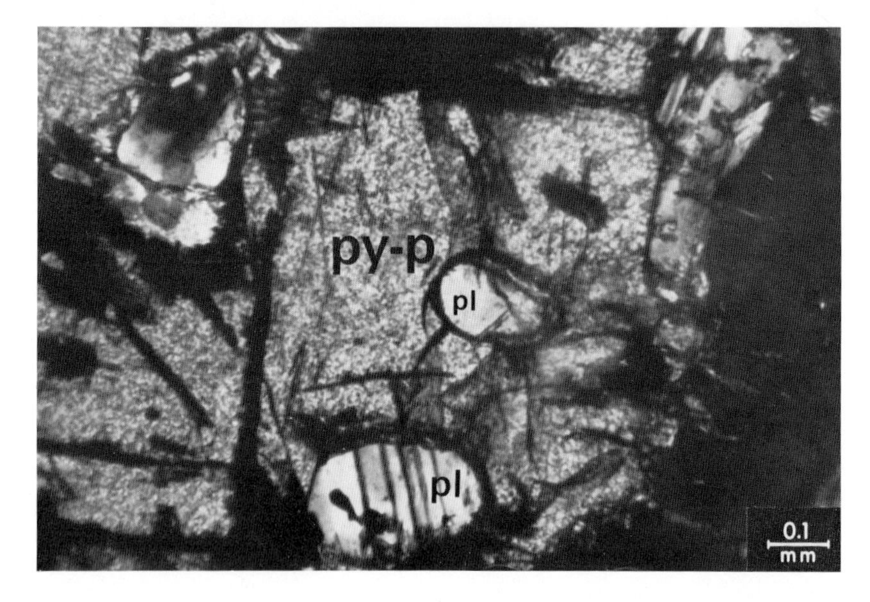

Fig. 236 Shows poikiloblastic pyroxene enclosing corroded and rounded plagioclases.
py-p = pyroxene poikiloblast
pl = plagioclase
Magnetite-Gabbro. Estelien, Ringerike, Oslo, South Norway.
With crossed nicols.

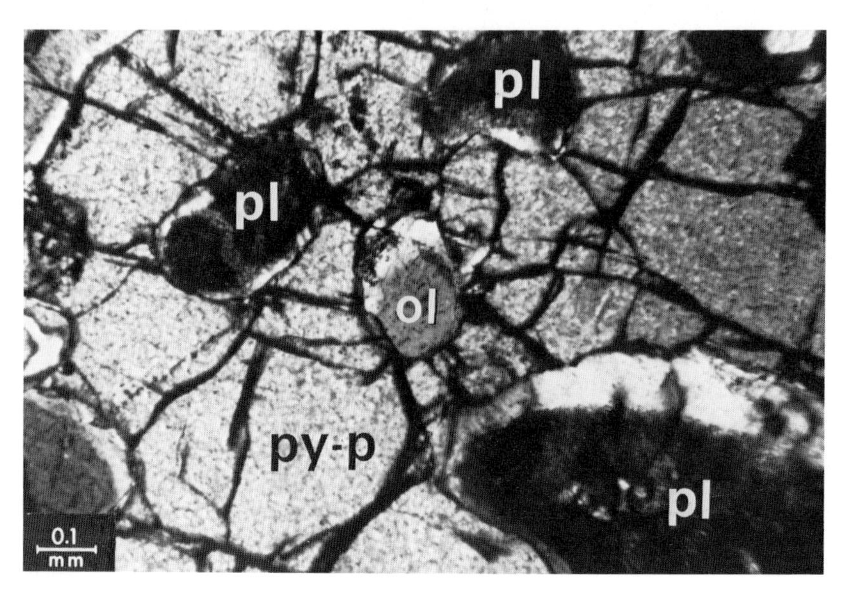

Fig. 237 Shows poikiloblastic pyroxene enclosing rounded and corroded plagioclases and olivines.
py-p = poikiloblastic pyroxene
pl = plagioclase
ol = olivine
Hypersthene-Norite. Deneykin Kamen, North Ural, USSR.
With crossed nicols.

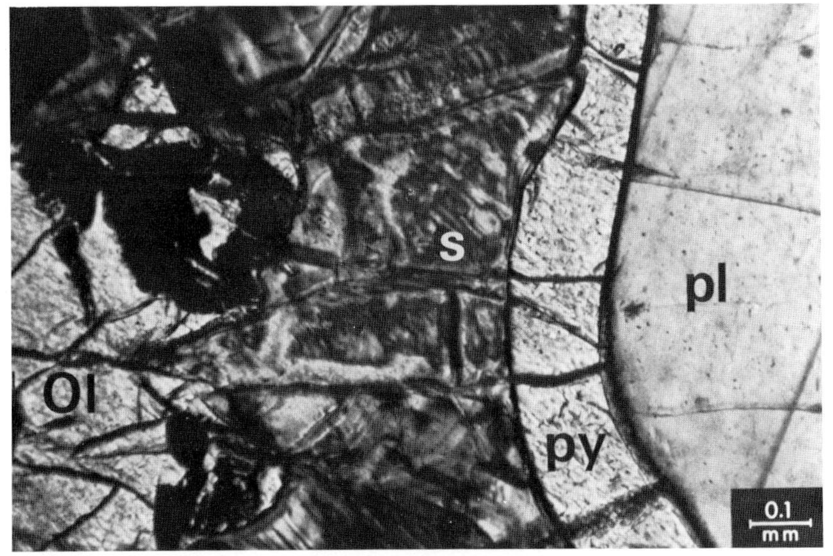

Fig. 238 Olivine serpentinised and with a marginal pyroxene following the boundary serpentinised olivine/feldspar.
ol = olivine
S = serpentinised olivine
pl = plagioclase
py = pyroxene (following the boundary serpentinised olivine/plagioclase).
Gabbro, Volpersdorf, Eulengebirge, Niederschlesien, a.p. Poland.
With crossed nicols.

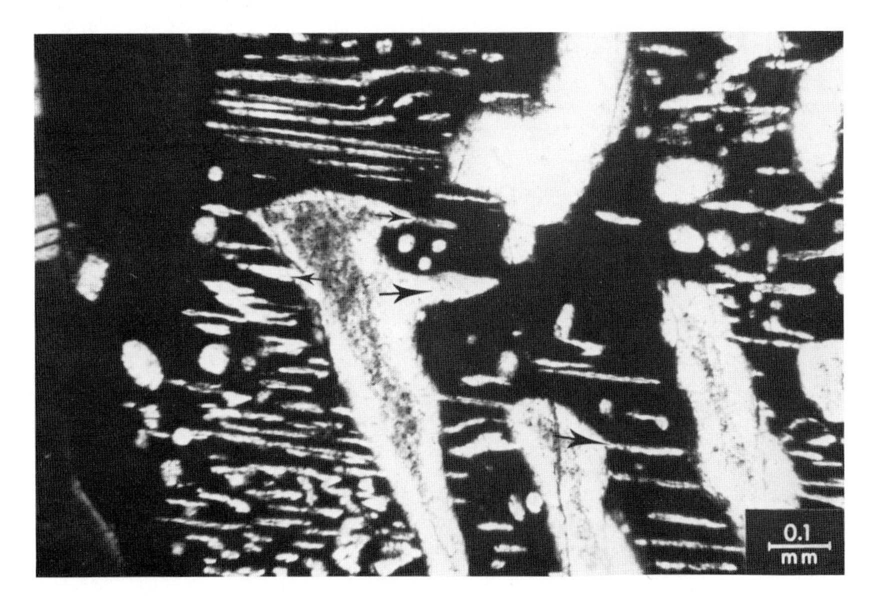

Fig. 239 Shows phase-a pyroxene (hypersthene) in the extinction position, invaded by diopsidic phase-b pyroxene.
These intergrowths of phase-a and phase-b are due to metasomatic replacements. Arrows show that phase-b clearly follows interleptonic spaces (cleavage planes) of phase-a.
Gabbro. Bushveld Massif, North of Pretoria, Transvaal, South Africa.
With crossed nicols.

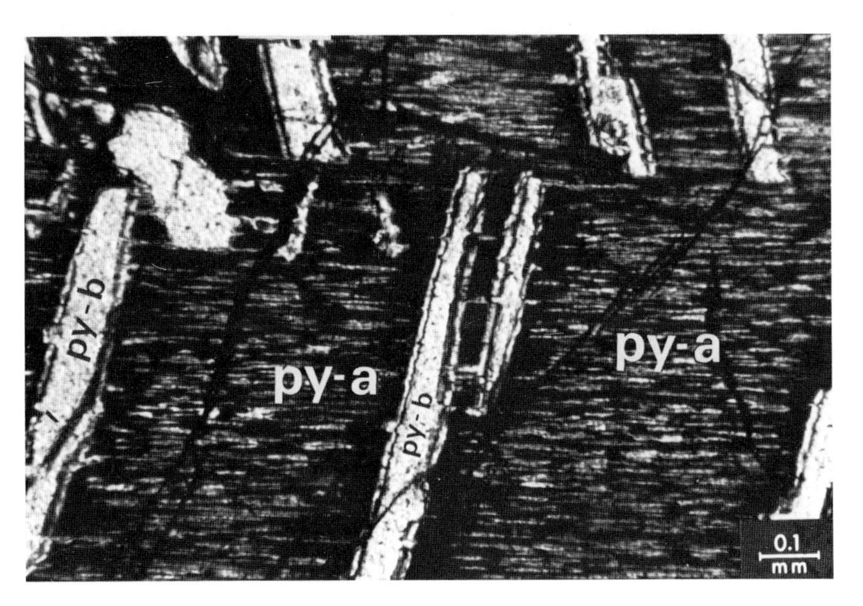

Fig. 240 Pyroxene phase-a (almost in the extinction position) is invaded by pyroxene phase-b following the interleptonic spaces (cleavage planes) of phase-a and as orientated lamellae transversing the cleavage of phase-a.
py-a = pyroxene phase-a with white lines representing phase-b.
py-b = phase-b as orientated lamellae in phase-a.
Gabbro. Bushveld Massif, North of Pretoria, Transvaal, South Africa.
With crossed nicols.

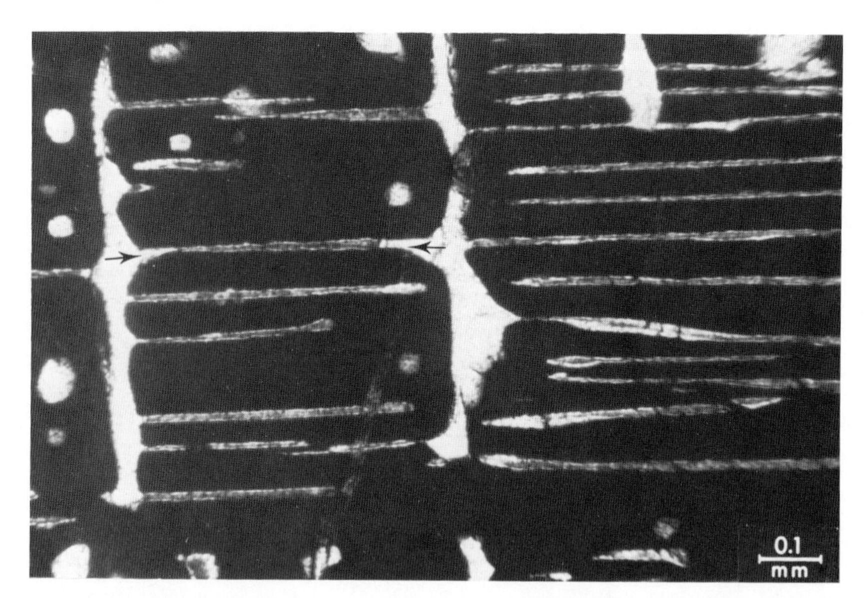

Fig. 241 Pyroxene phase-a (in the extinction position) with pyroxene phase-b, as orientated bodies sending extensions clearly invading phase-a along its cleavage planes.
Phase-a = in the extinction position (hypersthene)
Phase-b = as orientated bodies in phase-a ıdiopside)
Arrows show extensions of phase-b following as fillings the cleavage interleptonic spaces of phase-a.
Gabbro. Bushveld Massif, North of Pretoria, Transvaal, South Africa.
With crossed nicols.

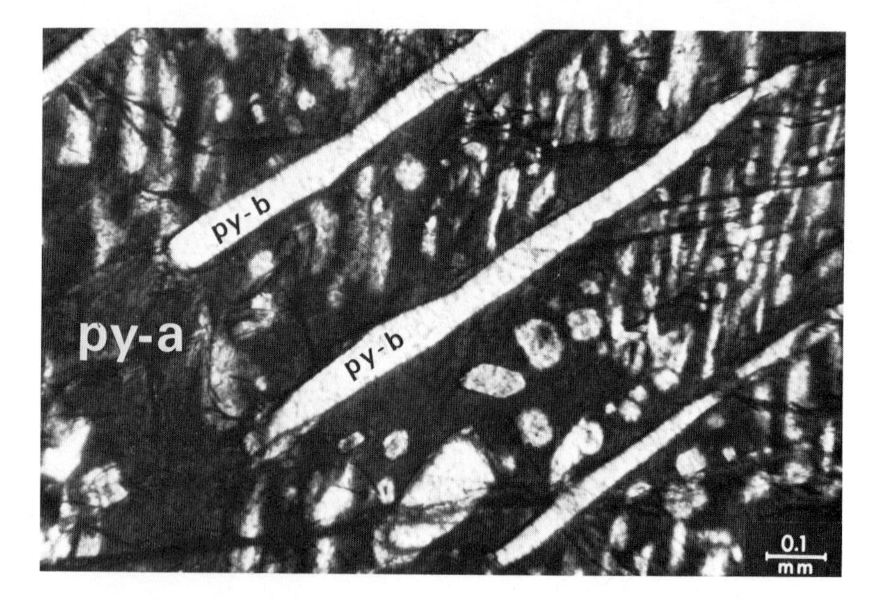

Fig. 242 Orientated intergrowths of pyroxene phase-b (diopside) in pyroxene phase-a (hypersthene)
py-a = pyroxene phase-a
py-b = pyroxene phase-b (diopside)
Gabbro. Bushveld Massif, North of Pretoria, Transvaal, South Africa.
With crossed nicols.

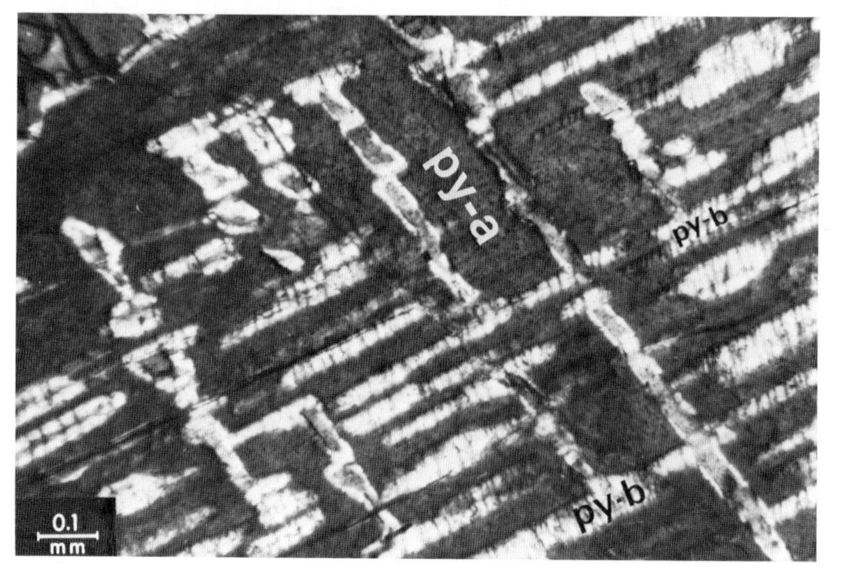

Fig. 243 Phase-a pyroxene (hypersthene, almost in the extinction position) with orientated bodies (phase-b diopsidic pyroxene) following two orientation directions of the host.
py-a = pyroxene phase-a (hypersthene, almost in the extinction position)
py-b = diopsidic pyroxene following two orientation directions within the host pyroxene.
Gabbro. Bushveld Massif, North of Pretoria, Transvaal, South Africa.
With crossed nicols.

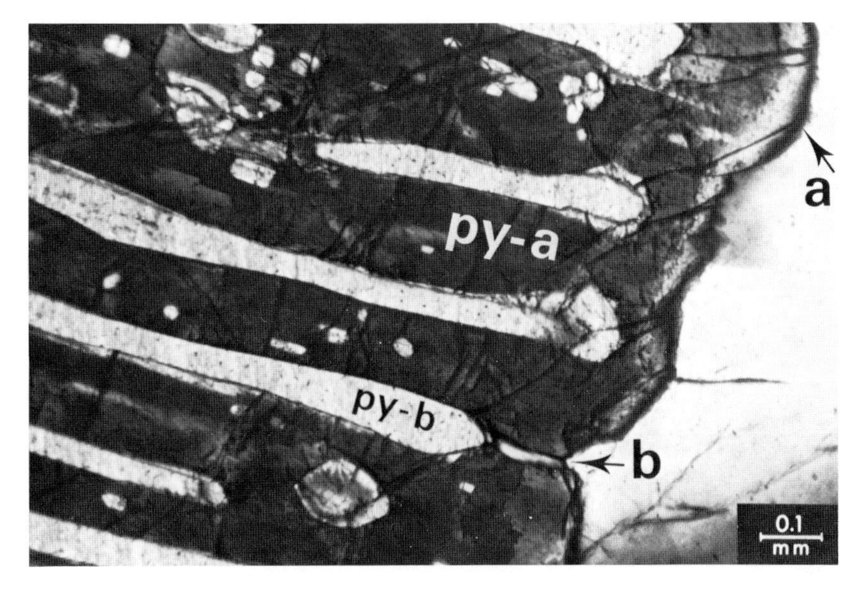

Fig. 244 Pyroxene phase-a, with orientated bodies of phase-b (diopside). Plagioclase is also present. Phase-a is corroded and invaded by the plagioclase.
py-a = phase-a (almost in the extinction position)
py-b = orientated diopsidic bodies in phase-a.
Arrow "a" shows corroded outline of pyroxene
Arrow "b" shows extension of plagioclase into pyroxene phase-a
Gabbro. Bushveld Massif, North of Pretoria, Transvaal, South Africa.
With crossed nicols.

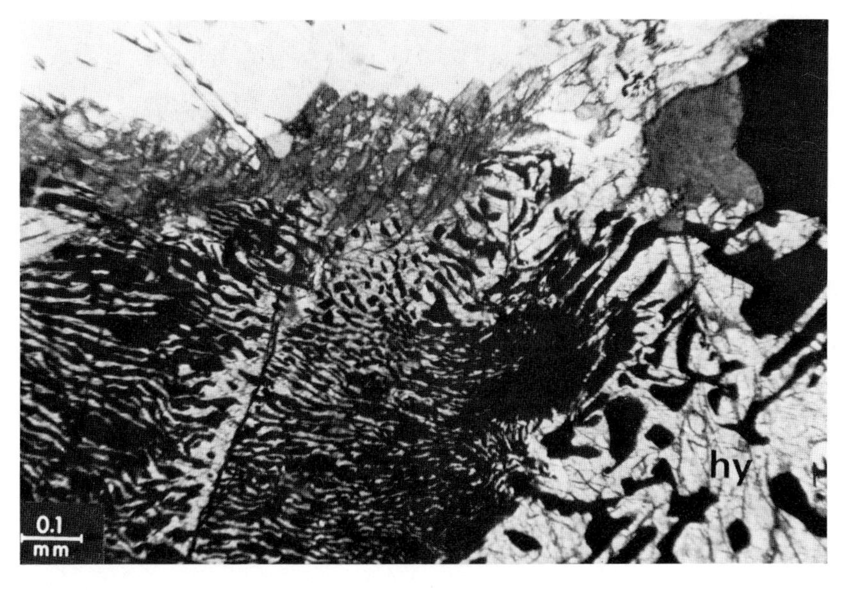

Fig. 246a Symplectic intergrowth structure of magnetite with hypersthene.
magnetite (black)
hy = hypersthene
With crossed nicols.
Norite (such intergrowths also occur in gabbroic rocks)
Hitteroe, Norway.
Without crossed nicols.

(For Fig. 245 see page 194)

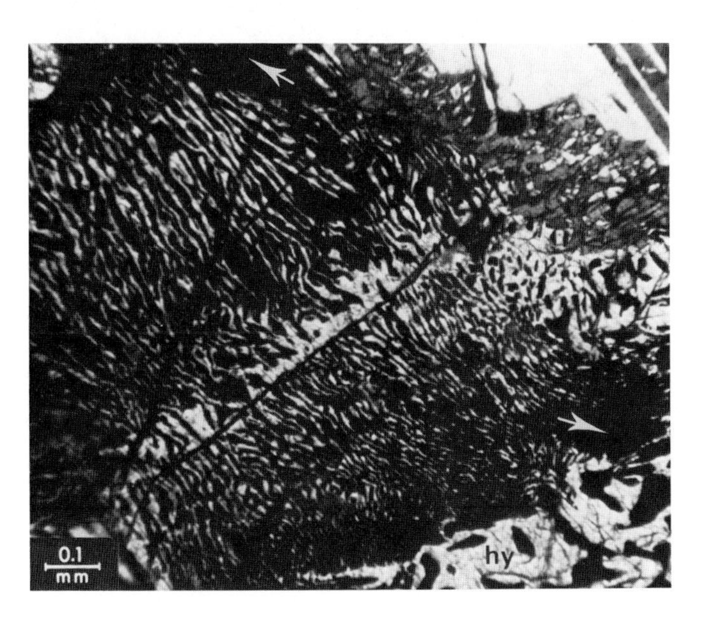

Fig. 246b Symplectic intergrowth structure of magnetite with hypersthene.
The magnetite transgresses from symplectic to more compact magnetite (see arrows).
hy = hypersthene
magnetite (black)
Norite. Hitteroe, Norway.
Without crossed nicols.

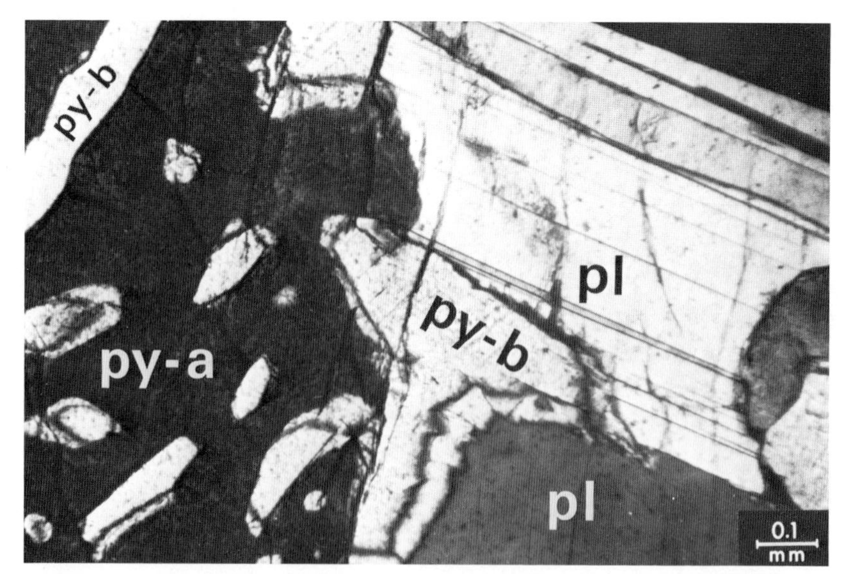

Fig. 245 Shows pyroxene phase-a with orientated intergrowth of phase-b. An individual of phase-b transverses the boundary between pyroxene phase-a and an adjacent plagioclase.
pl = plagioclase
py-a = pyroxene phase-a (almost in the extinction position)
py-b = pyroxene phase-b
Gabbro. Bushveld Massif, North of Pretoria, Transvaal, South Africa.
With crossed nicols.

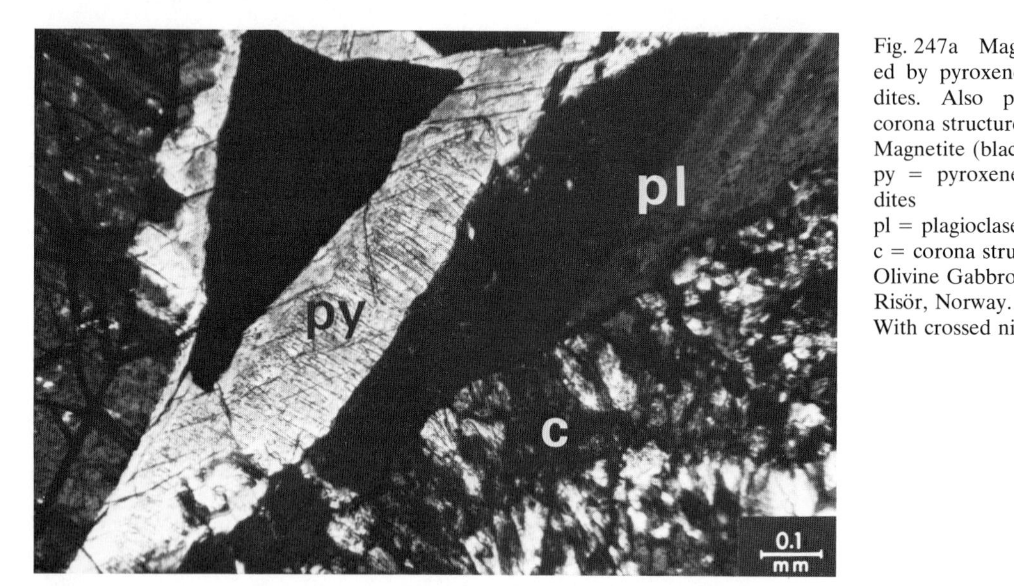

Fig. 247a Magnetite (black) surrounded by pyroxene with symplectic rhabdites. Also plagioclase present and corona structure indicated.
Magnetite (black)
py = pyroxene with symplectic rhabdites
pl = plagioclase
c = corona structure
Olivine Gabbro (hypersthene)
Risör, Norway.
With crossed nicols.

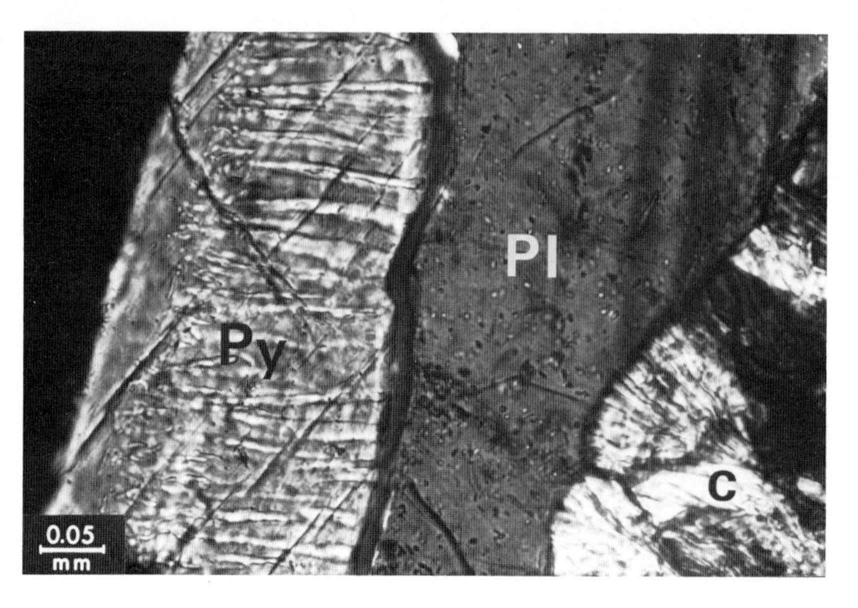

Fig. 247b detail of Fig. 247a Magnetite (black) and surrounding pyroxene (with symplectic rhabdite, comparable to myrmekitic intergrowths)
Py = pyroxene with symplectic rhabdites
Pl = plagioclase
c = corona structure
Olivine gabbro (hypersthene)
Risör, Norway.
With crossed nicols.

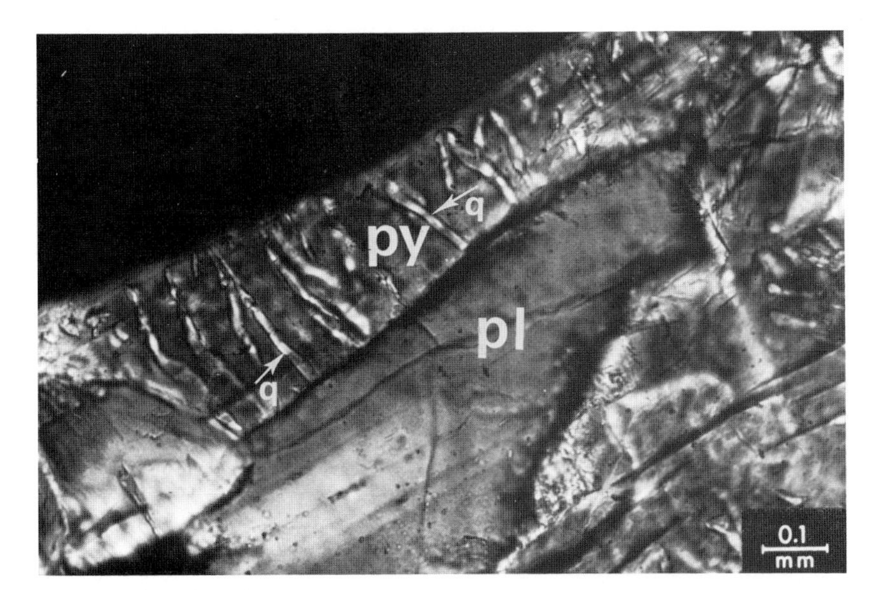

Fig. 248 Symplectic synantetic reaction structures. Magnetite (black) surrounded by pyroxene (with symplect myrmekitic structure) also plagioclase is indicated.
py = pyroxene
q = quartz
pl = plagioclase
Olivine gabbro (hypersthene).
Risör, Norway.
With crossed nicols.

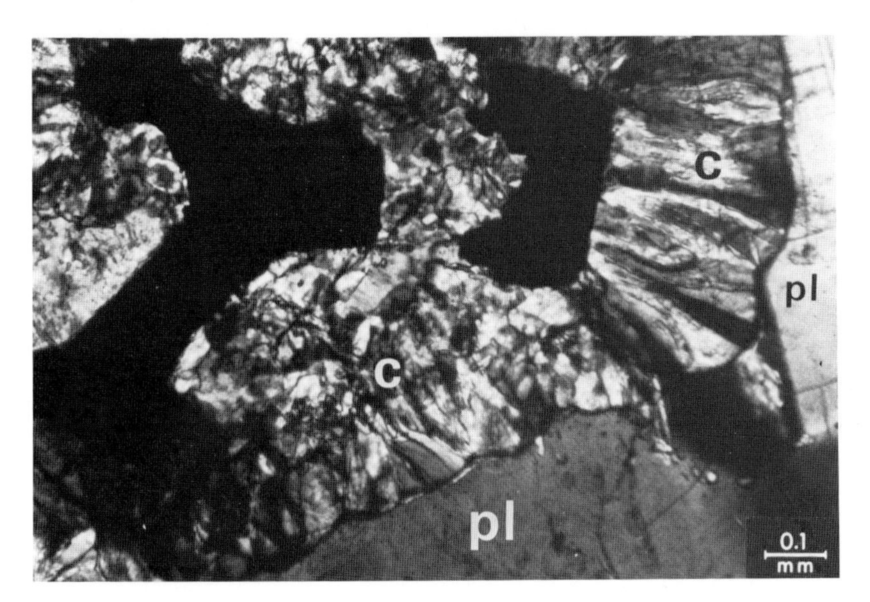

Fig. 249 Magnetite (black) surrounded by corona structure. Plagioclase is also present.
Magnetite (black)
c = corona structure
pl = plagioclase
Olivine gabbro (hypersthene)
Risör, Norway.
With crossed nicols.

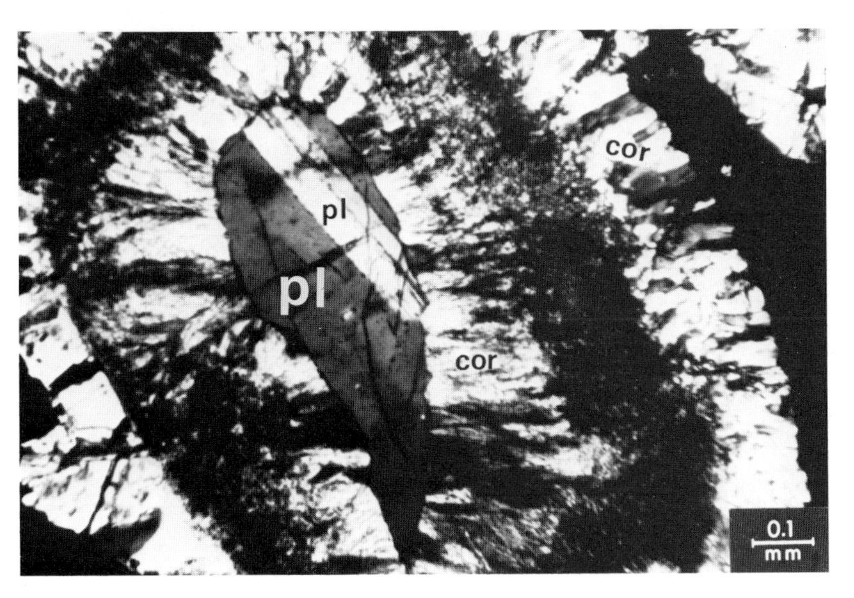

Fig. 250 Plagioclase surrounded by a corona structure of pyroxene.
pl = plagioclase
cor = corona structure consisting of pyroxene.
With crossed nicols.
Olivine Gabbro. Oberkainsbach, Odenwald, Germany.
With crossed nicols.

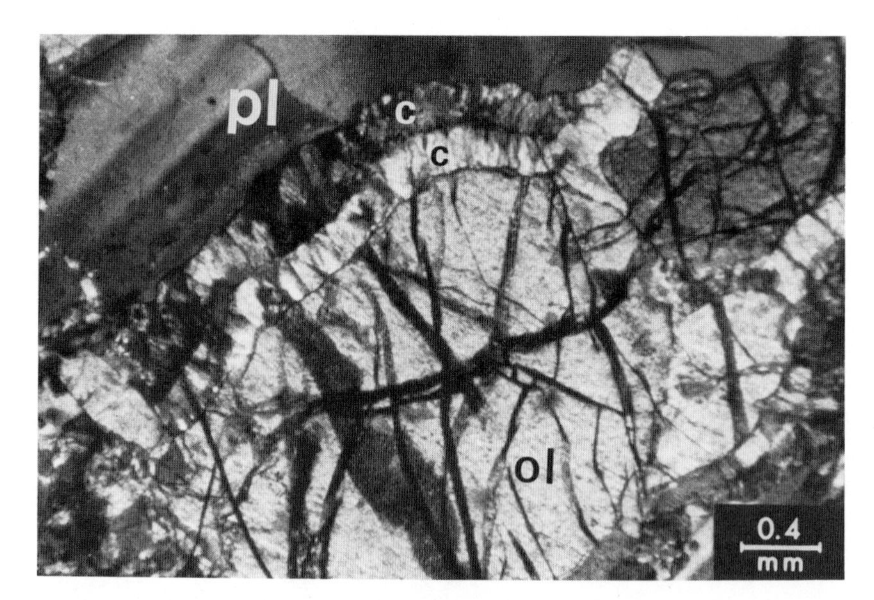

Fig. 251a Olivine surrounded by a double corona margin in contact with plagioclases.
ol = olivine
pl = plagioclase
c = corona structure (reaction corona structure consisting of pyroxene between olivine and plagioclase).
Olivine gabbro (hypersthene)
Risör, Norway.
With crossed nicols.

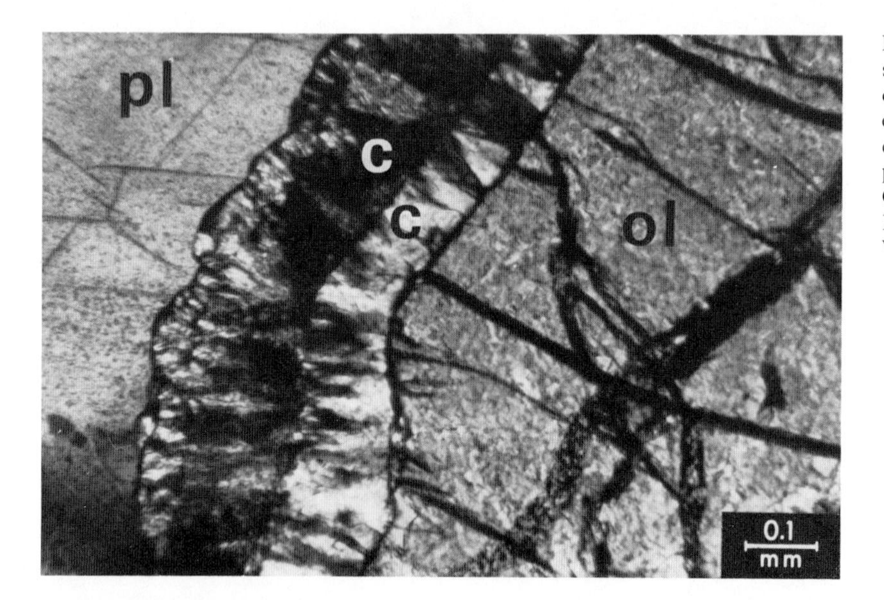

Fig. 251b Olivine with a double corona structure surrounded by feldspar.
ol = olivine
c = corona structure consisting of pyroxene (diopside)
pl = plagioclase
Olivine gabbro (hypersthene)
Risör, Norway.
With crossed nicols.

Fig. 252 Plagioclase crystalloblast consisting of two intimately intergrown phases. Ex-solution phases or infiltration and replacement phases as Fig. 253 might suggest.
Olivine gabbro (hypersthene)
Risör, Norway.
With crossed nicols.

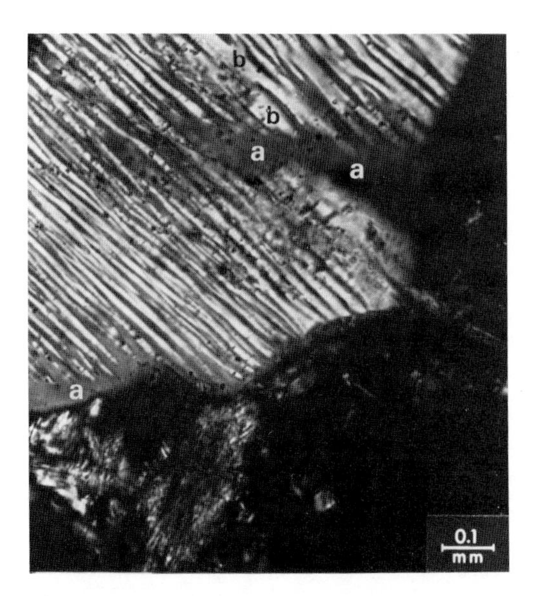

Fig. 253 A two phase plagioclase crystalloblast in which case phase-a is infiltrating and replacing phase-b in a comparable manner as perthitisation, in this case though we have two plagioclase phases.
a = plagioclase phase "a"
b = plagioclase phase "b"
Olivine gabbro (hypersthene)
Risör, Norway.
With crossed nicols.

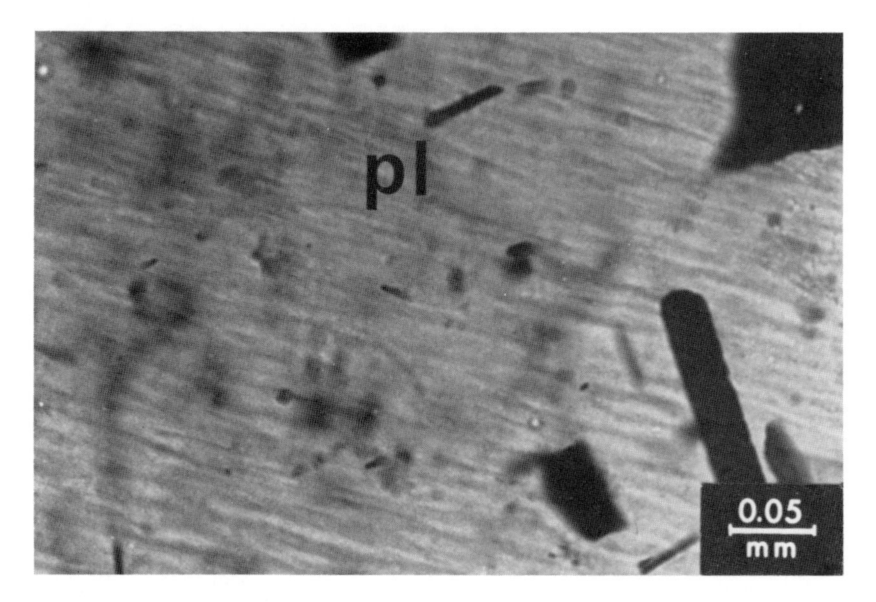

Fig. 254 A two phase plagioclase with orientated inclusions.
pl = plagioclase
Olivine gabbro (hypersthene)
Risör, Norway.
With crossed nicols.

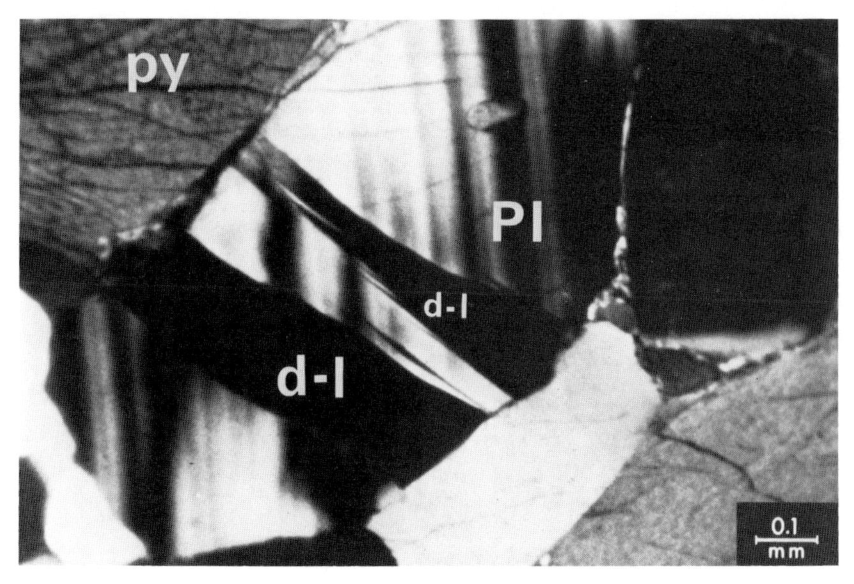

Fig. 255 Plagioclase with polysynthetic twinning and with deformation lamellae.
Pl = plagioclase
d-l = deformation lamellae
py = pyroxene
Hornblende-Gabbro. Aydat, Auvergne, France.
With crossed nicols.

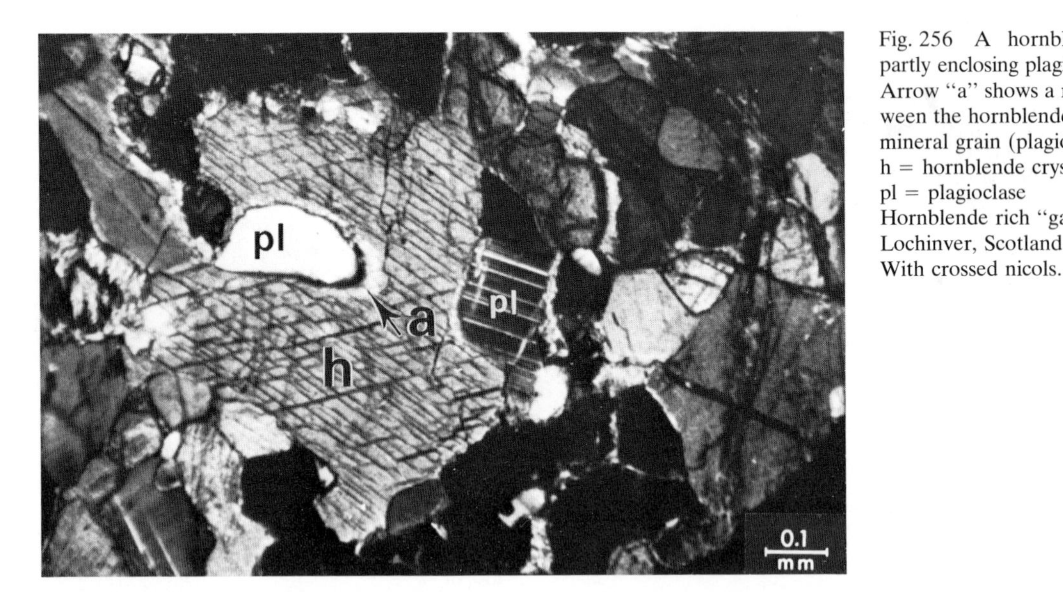

Fig. 256 A hornblende crystalloblast partly enclosing plagioclases and olivines. Arrow "a" shows a reaction margin between the hornblende and the pre-blastic mineral grain (plagioclase).
h = hornblende crystalloblast
pl = plagioclase
Hornblende rich "gabbroic" rock
Lochinver, Scotland.
With crossed nicols.

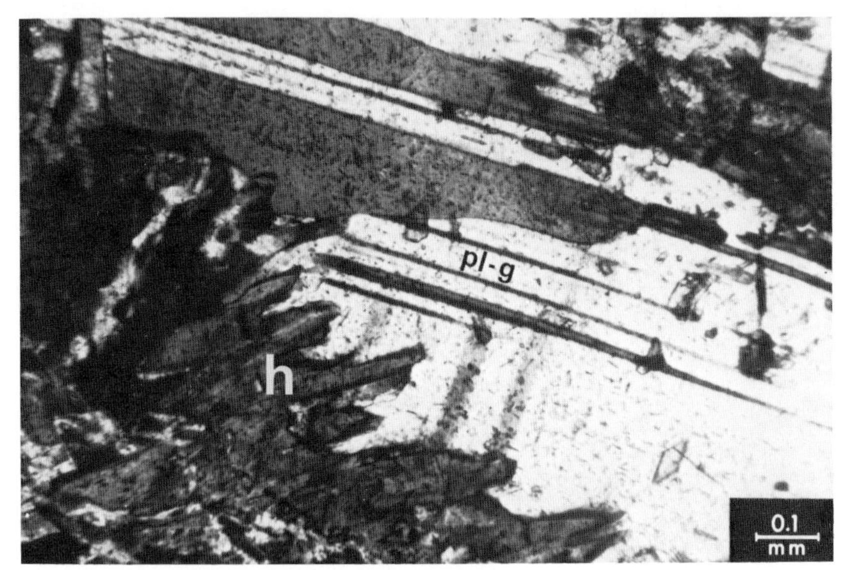

Fig. 257 Hornblende extending crystalloblastically into an adjacent plagioclase.
h = hornblende crystalloblasts
pl-g = Plagioclase host
Gabbro. Vinaric, Elbesteinitz, Middle Bohemia, CSSR.
With crossed nicols.

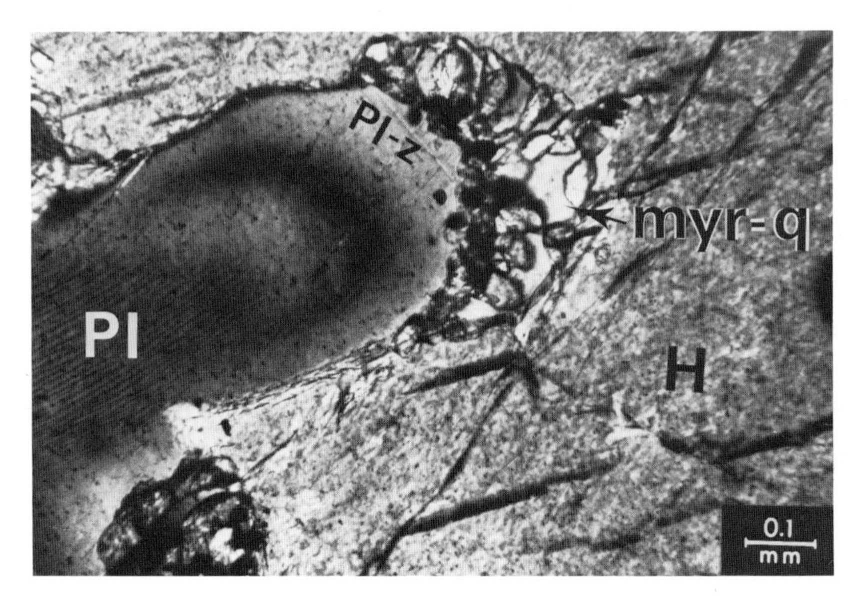

Fig. 258 Crystalloblastic hornblende enclosing, corroding and causing a reaction zoning on a pre-existing plagioclase. As a result of the reaction plagioclase/hornblende a synantetic reaction margin exists with myrmekitic quartz.
Pl = plagioclase
Pl-z = reaction zoning of plagioclase
H = hornblende crystalloblast
myr-q = myrmekitic quartz as a synantetic reaction product of plagioclase and later hornblende.
Hornblende-Gabbro. Anzola, Piemont, North Italy.
With crossed nicols.

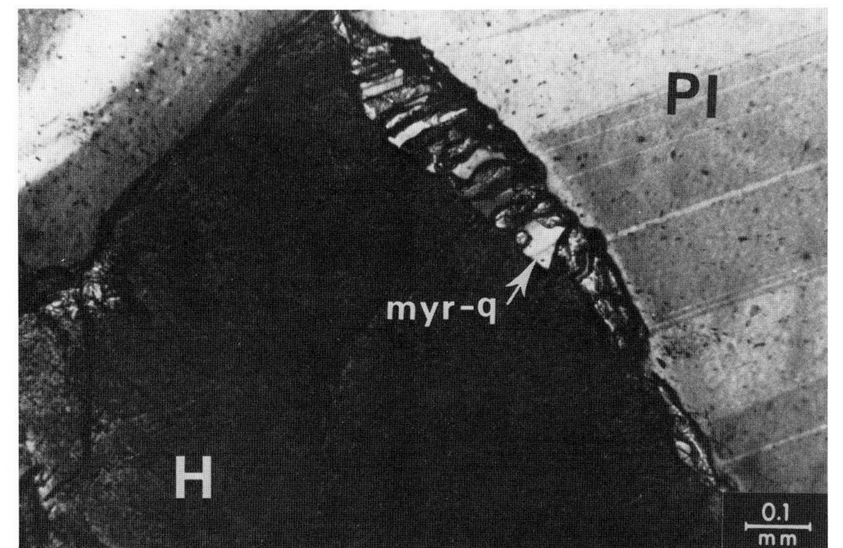

Fig. 259 Crystalloblastic hornblende in contact with plagioclase. As a result of the synantetic reaction of plagioclase and later crystalloblastic hornblende, a synantetic margin is produced with "myrmekitic" quartz.
Pl = plagioclase
H = hornblende crystalloblast
myr-q = myrmekitic quartz
Hornblende-Gabbro. Anzola, Piemont, North Italy.
With crossed nicols.

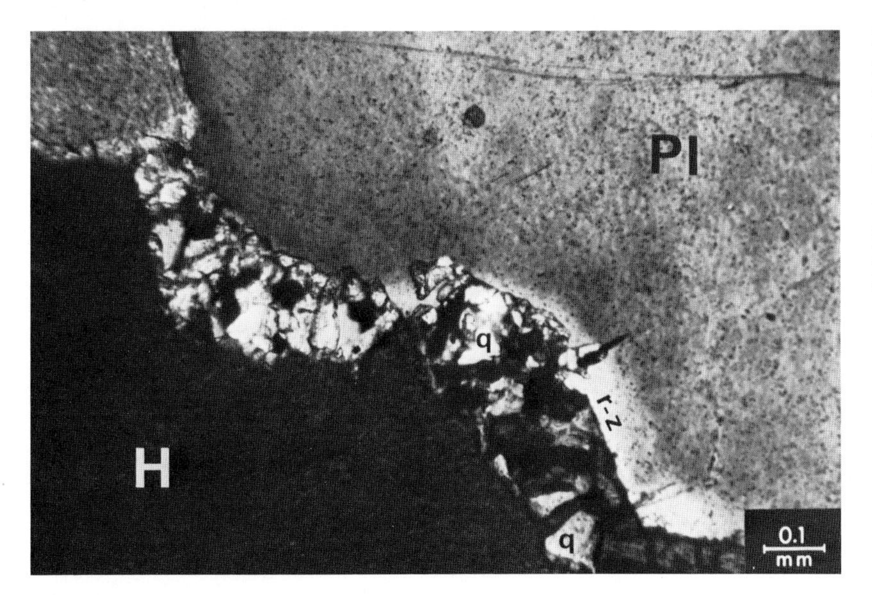

Fig. 260 Synantetic reaction margin of plagioclase and hornblende.
r-z = reaction zoning of the plagioclase
Pl = plagioclase
H = hornblende crystalloblast
q = myrmekitic quartz/hornblende intergrowth, as a synantetic reaction intergrowth of hornblende and plagioclase.
Hornblende-Gabbro. Anzola, Piemont, North Italy.
With crossed nicols.

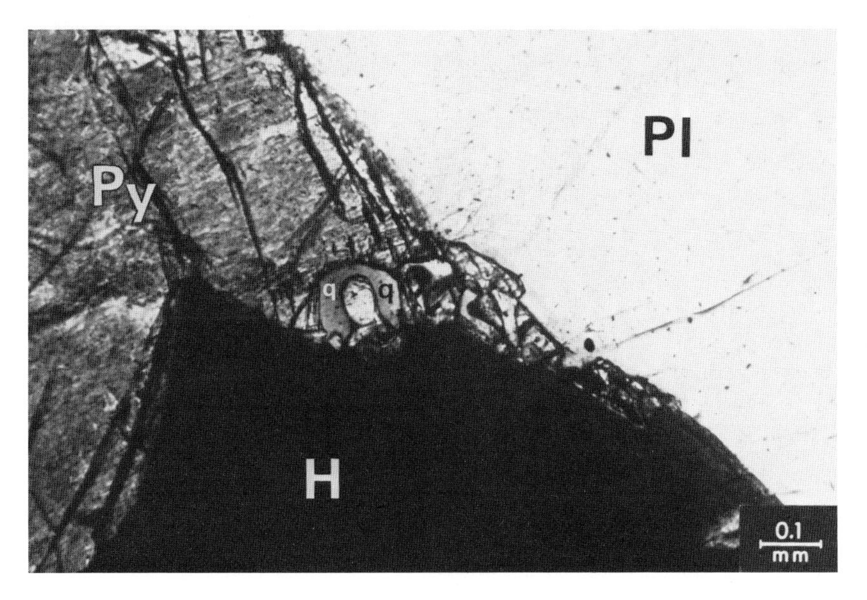

Fig. 261 Myrmekitic quartz as a synantetic reaction product of pyroxene and hornblende crystalloblast.
Py = pyroxene
H = hornblende crystalloblast
q = myrmekitic quartz
Pl = plagioclase
Hornblende-Gabbro. Anzola, Piemont, North Italy.
With crossed nicols.

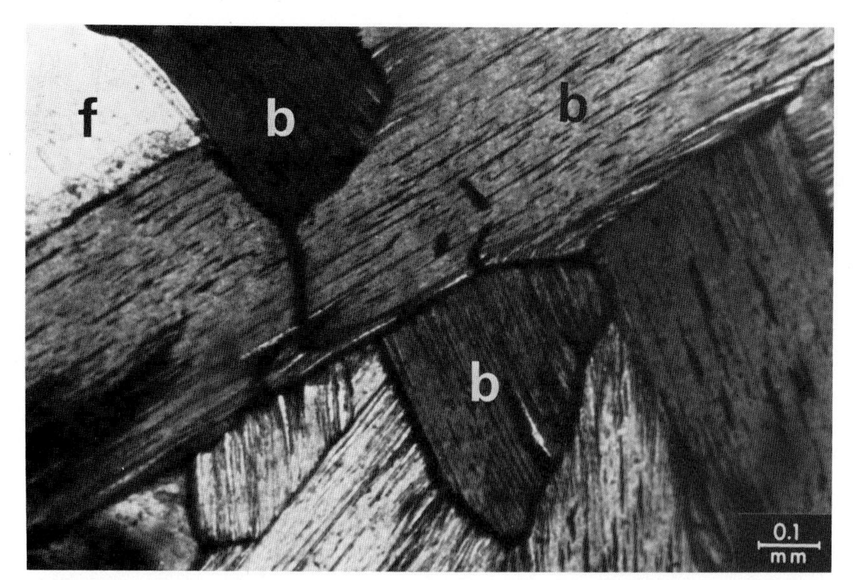

Fig. 262 Interpenetrating biotite crystalloblasts
b = biotite
f = feldspar
Gabbro. Gellivara, Lapland, Sweden.
With crossed nicols.

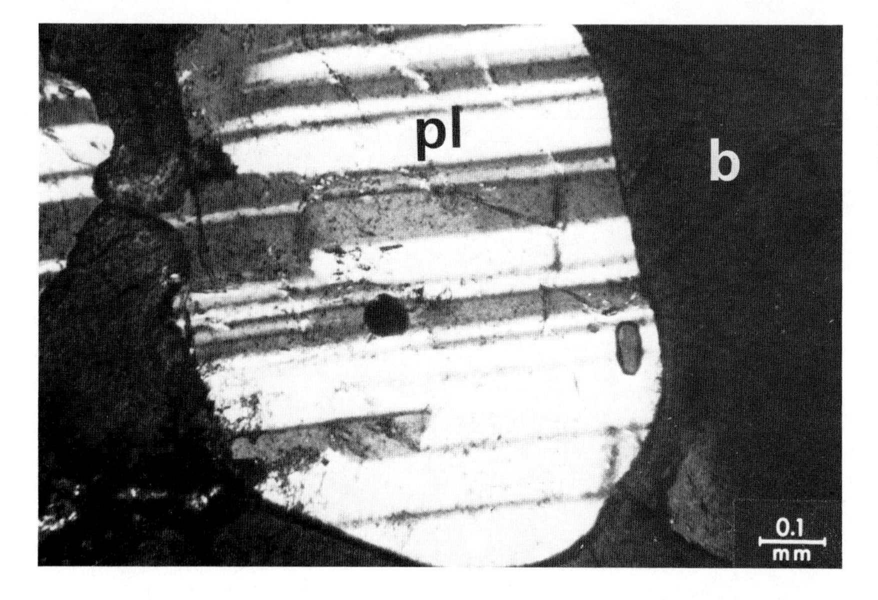

Fig. 263 Plagioclase rounded and enclosed by crystalloblastic biotite.
pl = plagioclase
b = biotite crystalloblast
Hornblende-Gabbro. Salem, Massachusetts, USA.

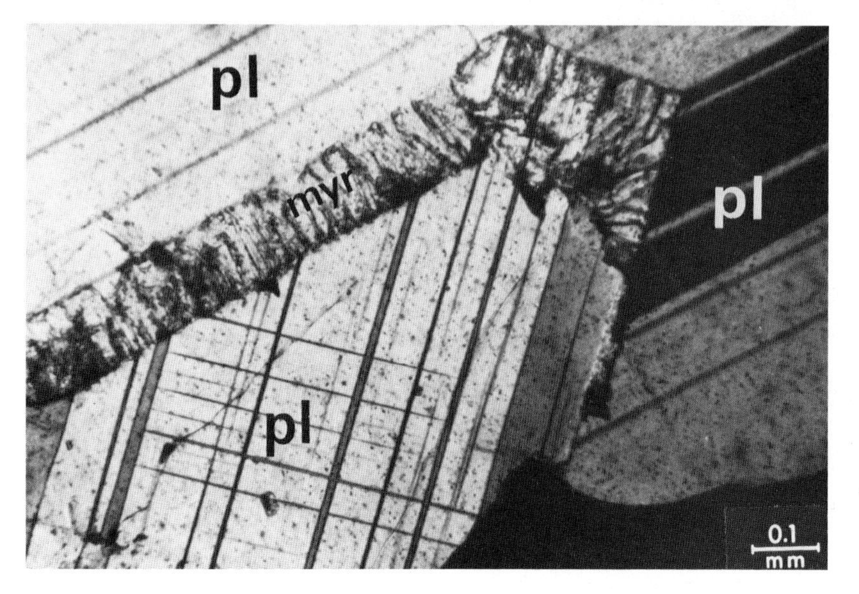

Fig. 264 Three differently orientated plagioclases. One of the plagioclases has a myrmekitic margin, which is due to the intracrystalline diffusion of quartz forming solutions.
pl = plagioclase
myr = myrmekitised plagioclase
Gabbro. Bushveld Massif, North of Pretoria, Transvaal, South Africa.
With crossed nicols.

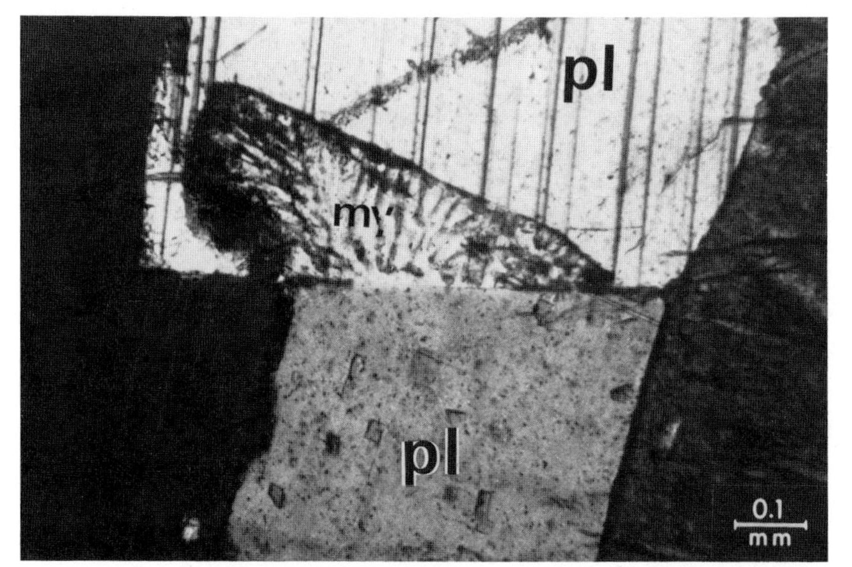

Fig. 265 A myrmekitised plagioclase entirely enclosed by another plagioclase.
pl = plagioclase
my = myrmekitised plagioclase
Gabbro. Bushveld Massif, North of Pretoria, Transvaal, South Africa.
With crossed nicols.

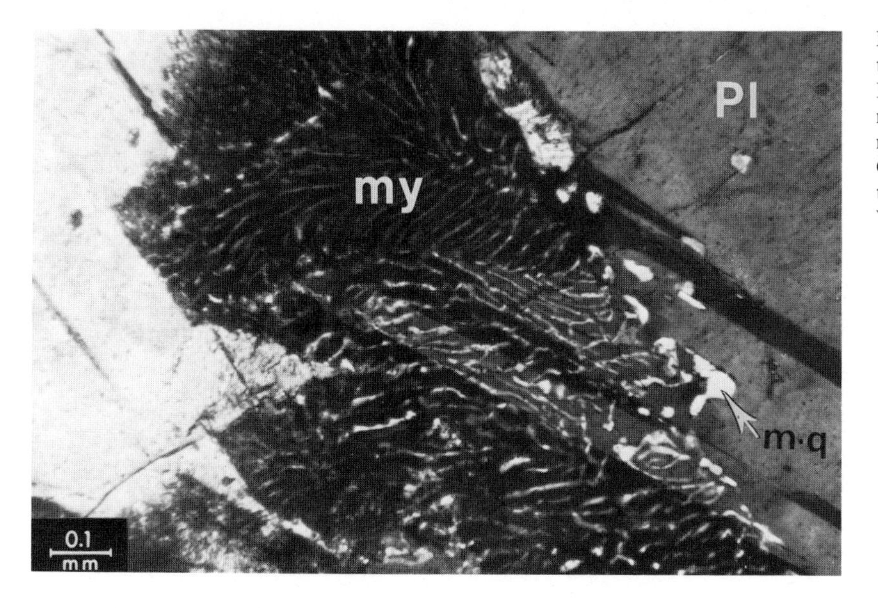

Fig. 266 Shows marginal myrmekitisation of the plagioclase.
Pl = plagioclase
my = myrmekitised plagioclase
m-q = myrmekitic quartz
Gabbro. Bushveld Massif, North of Pretoria, Transvaal, South Africa.
With crossed nicols.

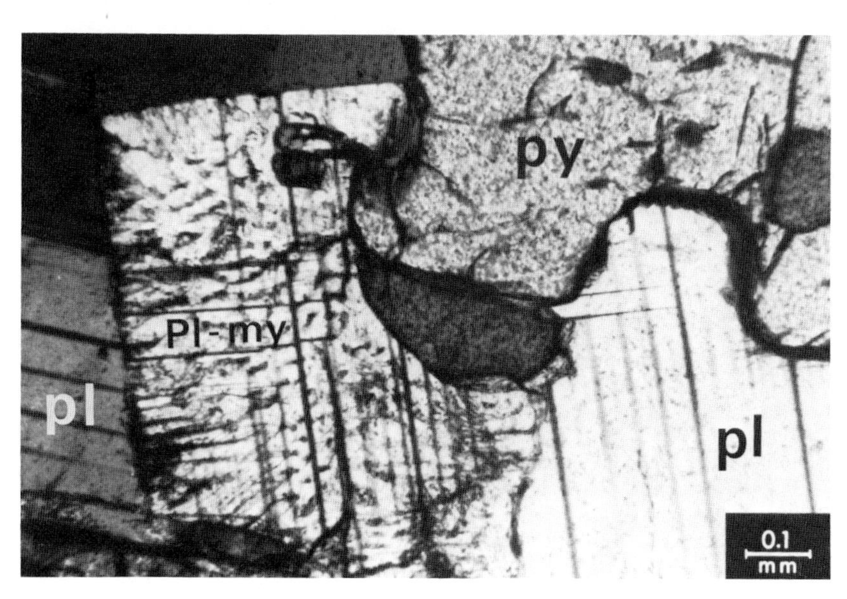

Fig. 267 Idiomorphic plagioclase myrmekitised by the infiltration of intracrystalline solution.
Pl-my = myrmekitised idiomorphic plagioclase
pl = plagioclase
py = pyroxene
Gabbro. Bushveld Massif, North of Pretoria, Transvaal, South Africa.
With crossed nicols.

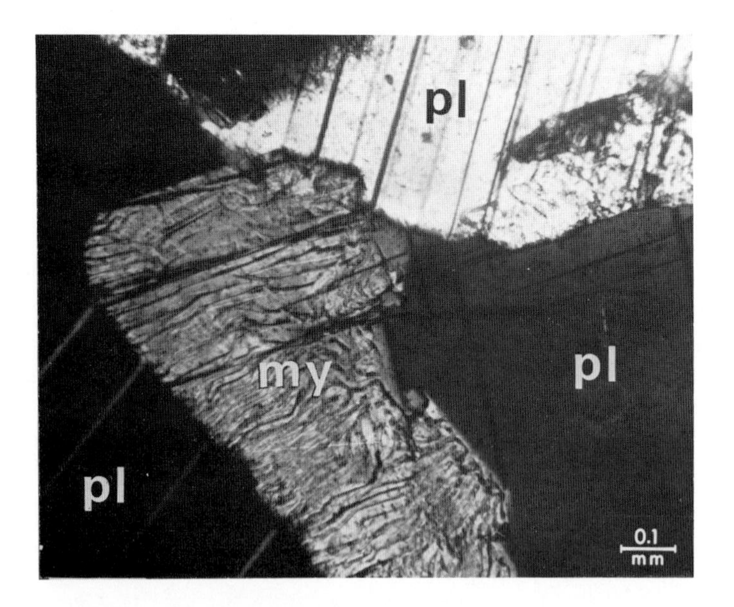

Fig. 268 Rounded and corroded plagioclase surrounded by other plagioclase myrmekitised by the infiltration of intracrystalline solutions which as intergranular diffusion have reached and myrmekitised the plagioclase.
The question that arises is why that particular plagioclase?
pl = plagioclase
my = myrmekitised plagioclase
Gabbro. Bushveld Massif, North of Pretoria, Transvaal, South Africa.
With crossed nicols.

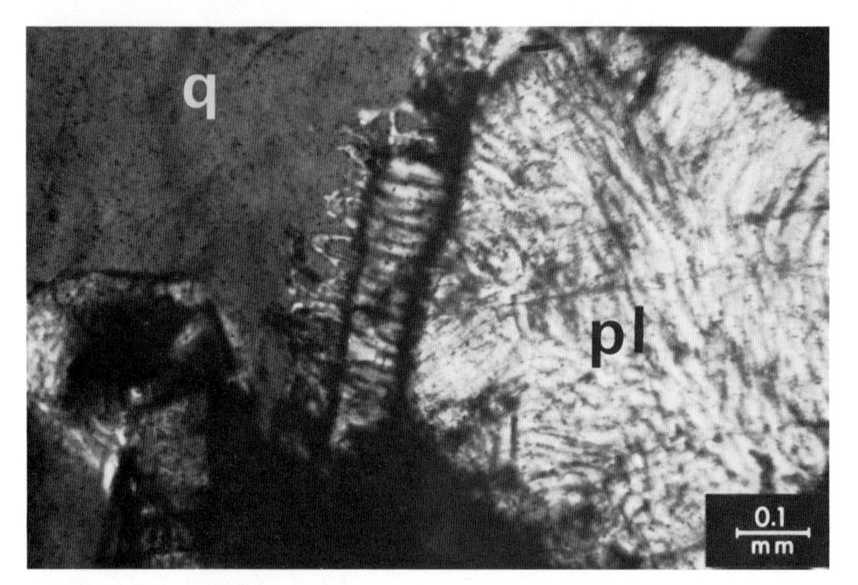

Fig. 269 Myrmekitised plagioclase (pl) with quartz (q) adjacent to it.
pl = plagioclase (myrmekitised)
q = quartz
With crossed nicols.
Gabbro. Bushveld Massif, North of Pretoria, Transvaal, South Africa.

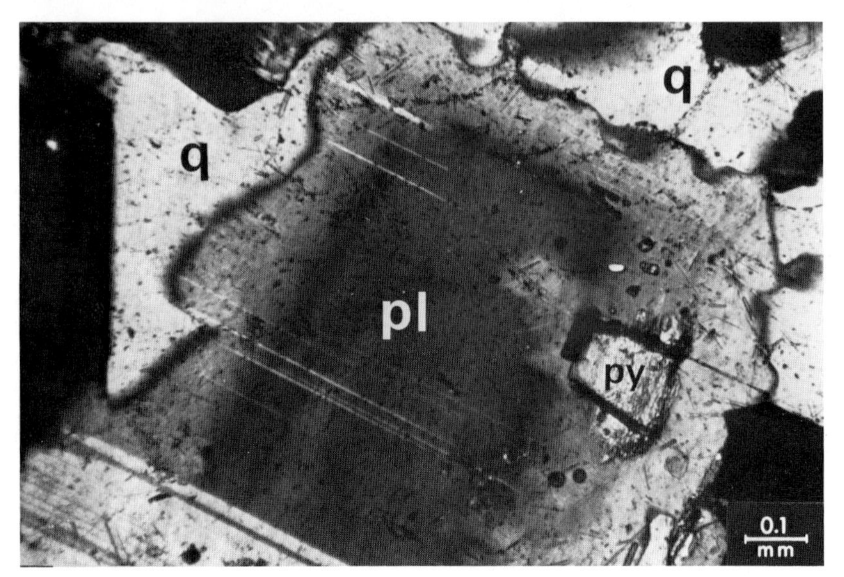

Fig. 270 Granophyric quartz invading zoned plagioclase enclosing pyroxene.
q = quartz
pl = plagioclase
py = pyroxene
Gabbro. Gross Bieberau, Odenwald, Germany.
With crossed nicols.

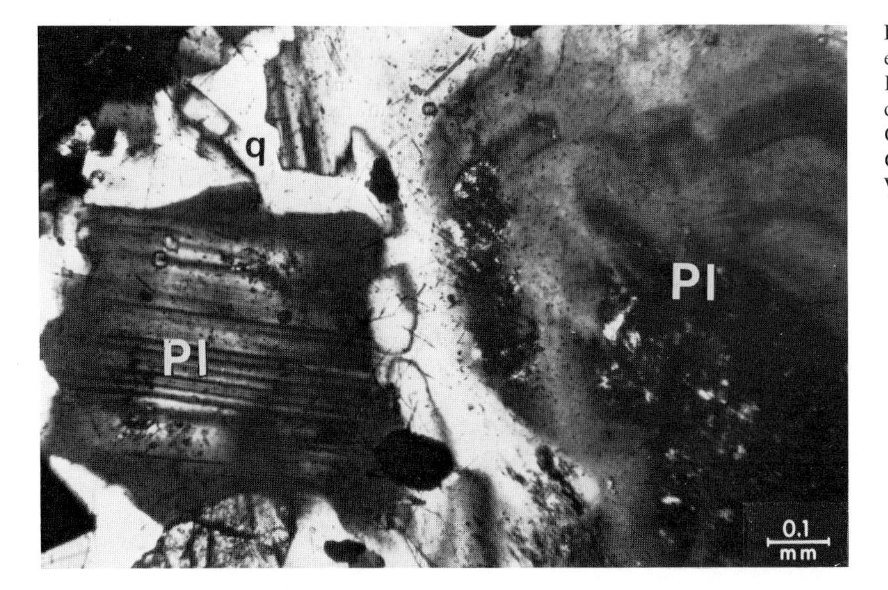

Fig. 271 Plagioclase marginally invaded by quartz.
Pl = plagioclase
q = quartz (granophyric)
Gabbro. Gross Bieberau. Odenwald, Germany.
With crossed nicols.

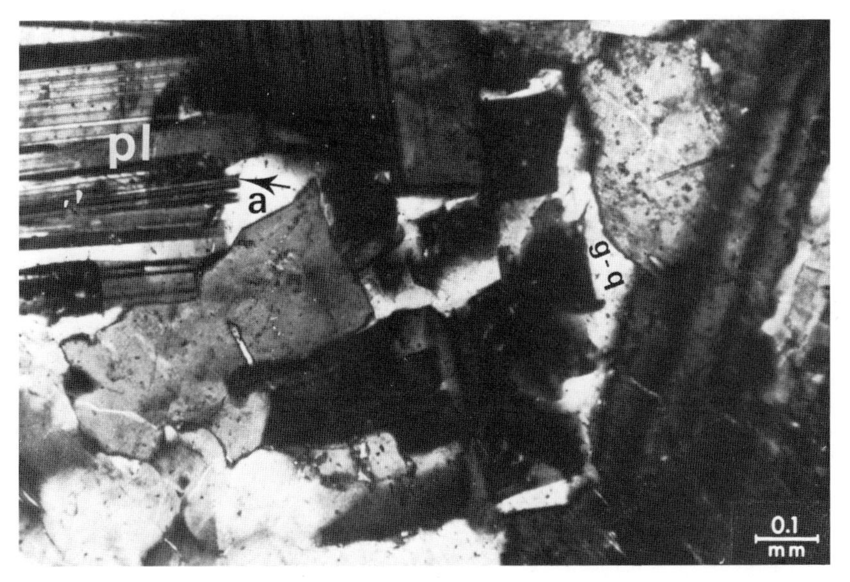

Fig. 272 Granophyric quartz following the intergranular between plagioclases and often corroding and replacing the feldspar. Arrow "a" shows remnants of protruding plagioclase twin lamellae surrounded and enclosed by the graphic quartz.
pl = plagioclase
g-q = granophyric quartz
Olivine-Gabbro. Kaltes Tal near Bad Harzburg, Harz, Germany.
With crossed nicols.

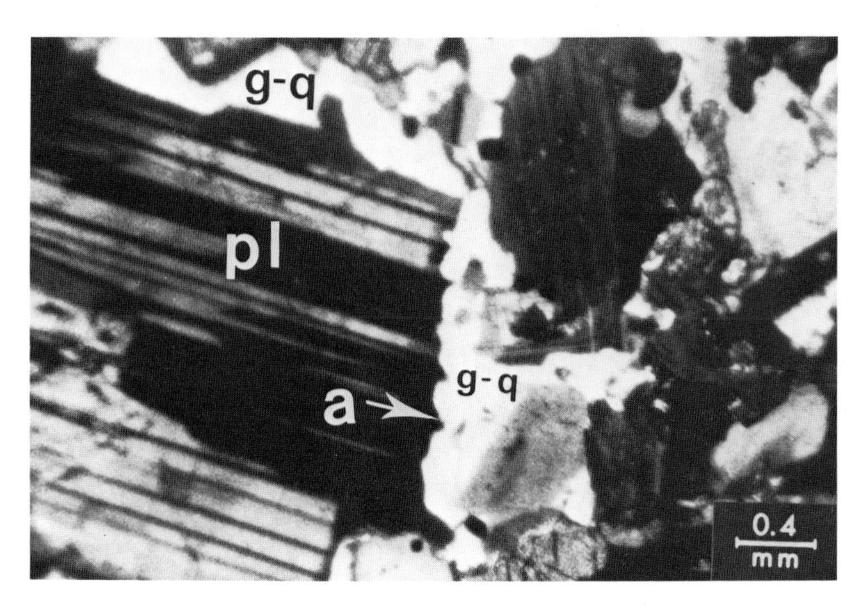

Fig. 273 Granophyric quartz following the intergranular spaces of the plagioclase and often corroding the feldspar.
g-q = granophyric quartz
pl = plagioclase
Arrow "a" shows the corroded outline of the plagioclase
Gabbro. Gross Bieberau, Odenwald, Germany.
With crossed nicols.

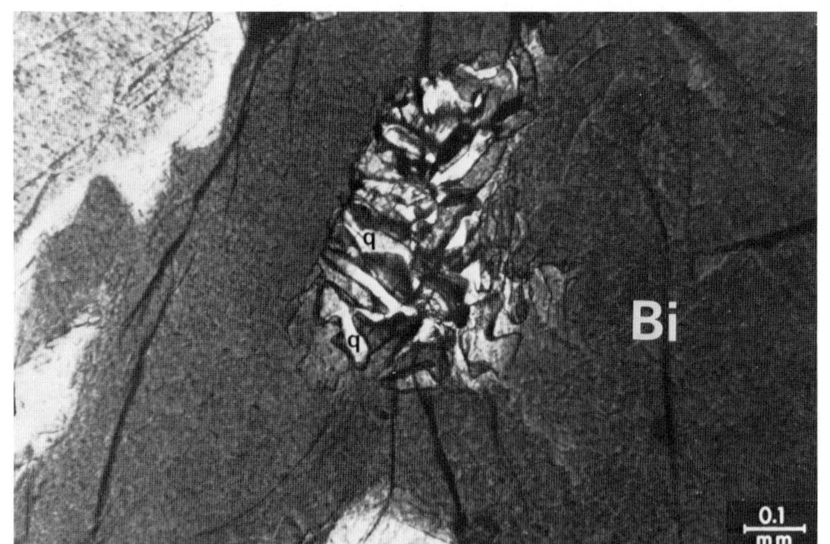

Fig. 274
Quartz granophyric intergrowths with biotite.
Bi = biotite
q = quartz
Hornblende-Gabbro. Aydat, Auvergne, France.
With crossed nicols.

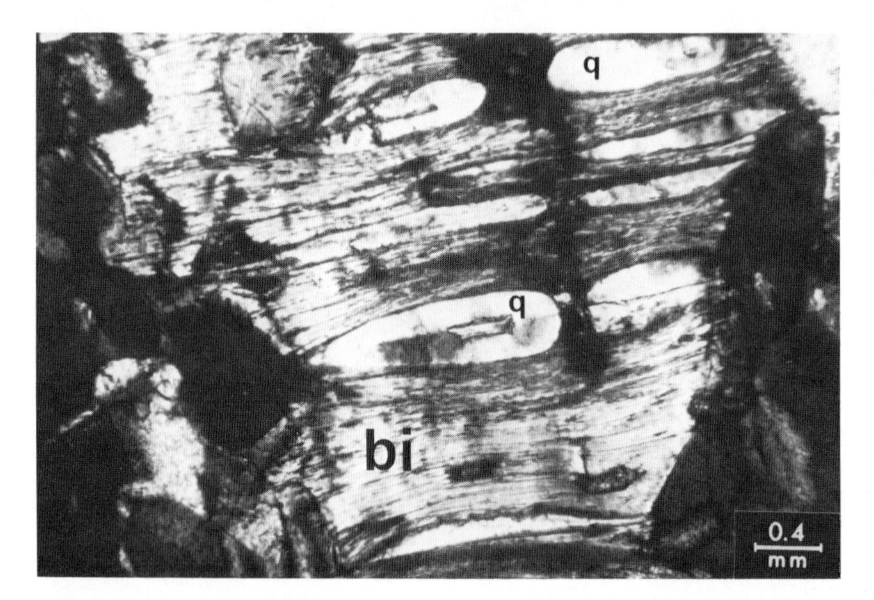

Fig. 275 Initial colloform quartz bodies following the cleavage of biotite-host into which the silica solution has diffused.
q = quartz
bi = biotite
Gabbro. Risör, South Norway.

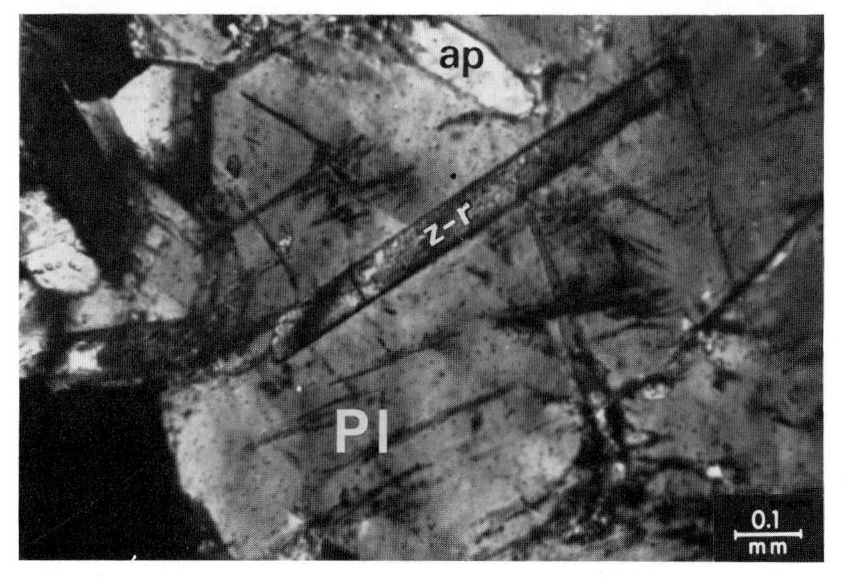

Fig. 276 Elongated zircon crystalloblasts and an apatite crystalloblast in plagioclases.
Pl = plagioclase
z-r = zircon crystalloblast
ap = apatite
Magnetite-Gabbro. Estelien, Ringerike, Oslo, South Norway.
With crossed nicols.

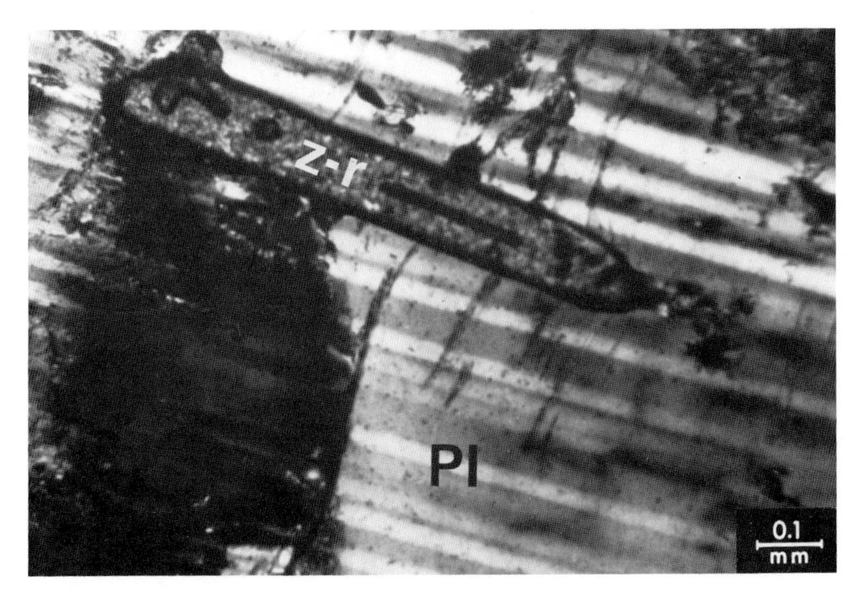

Fig. 277 Elongated zircon crystalloblast almost following the plagioclase polysynthetic twinning.
Pl = plagioclase
z-r = zircon crystalloblast
Magnetite-Gabbro. Estelien, Ringerike, Oslo, South Norway.
With crossed nicols.

Fig. 279a Gigantic oscillatory zoned augitic tecoblast including plagioclase laths of the groundmass.
au = augite tecoblast
pl = plagioclase lath parallel to the zoning of the augite.
G = essexitic groundmass.
Essexite, Crowford J., Lanarkshire, Scotland.
With crossed nicols.

(For Fig. 278 see page 206)

Fig. 279b Plagioclase phenocryst with fine rhythmical zoning in volcanic groundmass.
With crossed nicols.
Essexite. Bell Hill, Scotland.

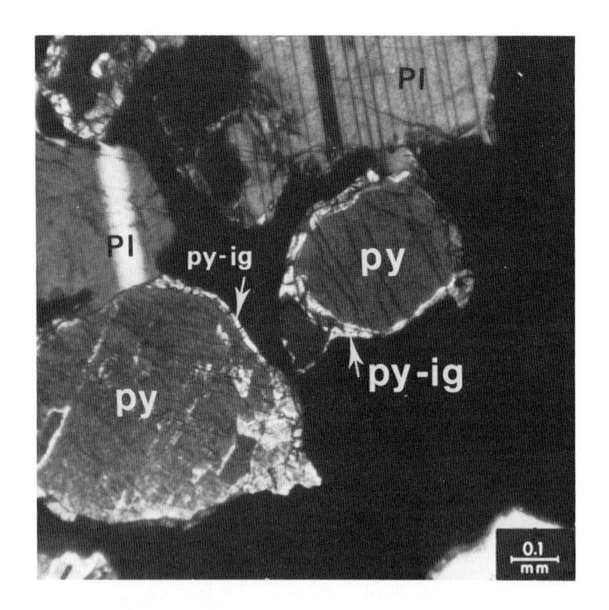

Fig. 278 Crystalloblastic magnetite enclosing pyroxenes and plagioclases. However, there is a phase-b pyroxene following the intergranular between plagioclase and magnetite.
Pl = plagioclase enclosed by the blastic magnetite
py = pyroxene enclosed by the blastic magnetite,
py-ig = pyroxene following the intergranular, between plagioclase and magnetite (diopside), magnetite (black)
Average ferrogabbro (plagioclase-augite-olivine-magnetite "cumulate"). House area, Skaergaard, Greenland. With crossed nicols. (Also see Fig. 577. Chapt. 16).

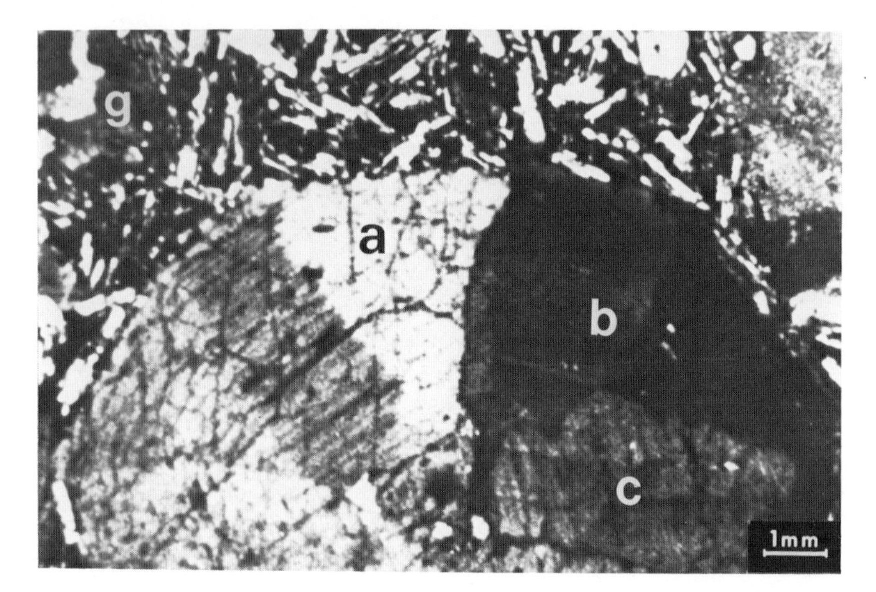

Fig. 280 Shows gigantic augitic phenocryst exhibiting "sectoral rotation", i. e. fracturing and rotation of parts of the augite phenocryst.
a, b, c = segment of the augite phenocryst, exhibiting different extinction due to sectoral rotation of the segments.
g = essexitic groundmass (mainly plagioclase laths).
Essexite, Crowford, J., Lanarkshire, Scotland.
With crossed nicols.

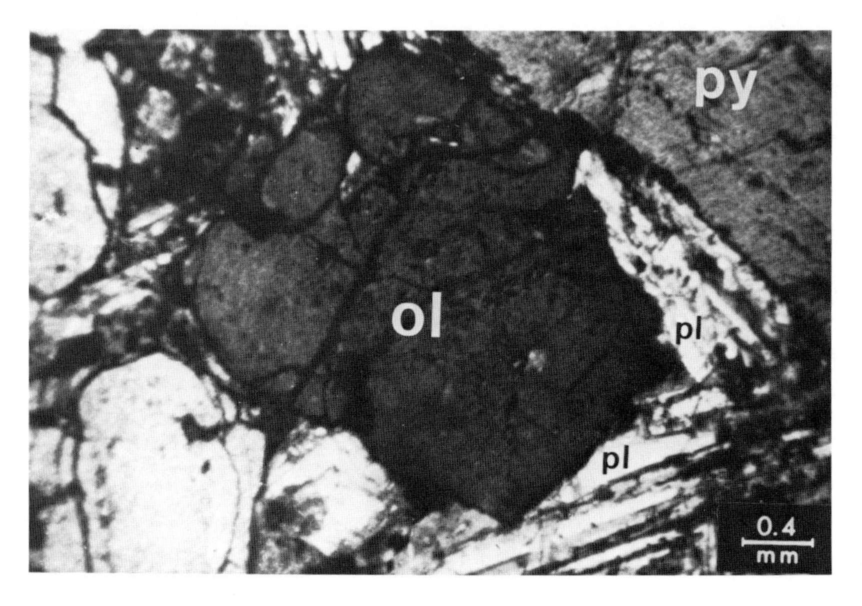

Fig. 281 Early olivine growth and tectoblastic oscillatory zoned augitic phenocrysts with interstitial sub-phenocrystalline plagioclase laths.
ol = olivine (almost in the extinction position)
py = oscillatory zoned augitic phenocrysts
pl = plagioclase laths.

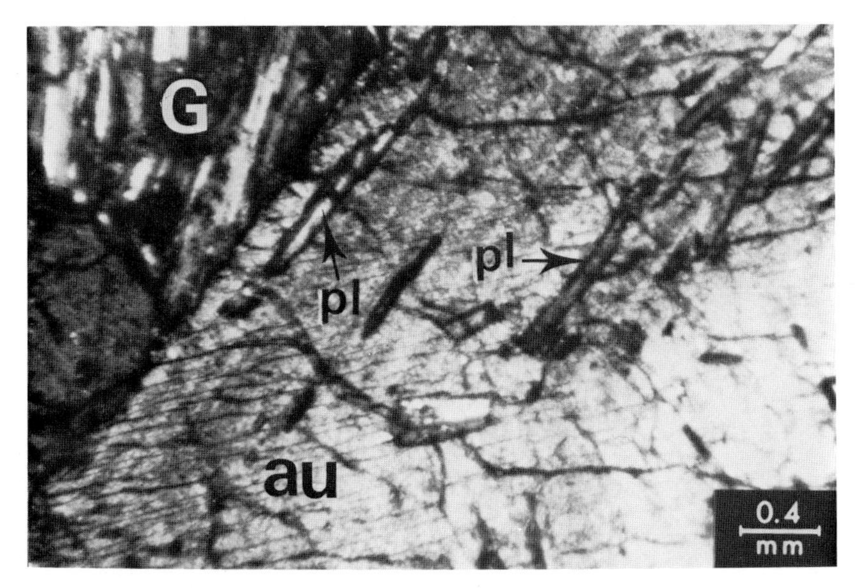

Figs. 282 and 283 Similar to Fig. 279a, showing details of the arrangement of the plagioclase laths parallel to the oscillatory zoning of the augites.
au = augite
pl = plagioclase
G = essexite groundmass
Essexite, Crowford J., Lanarkshire, Scotland.
With crossed nicols.

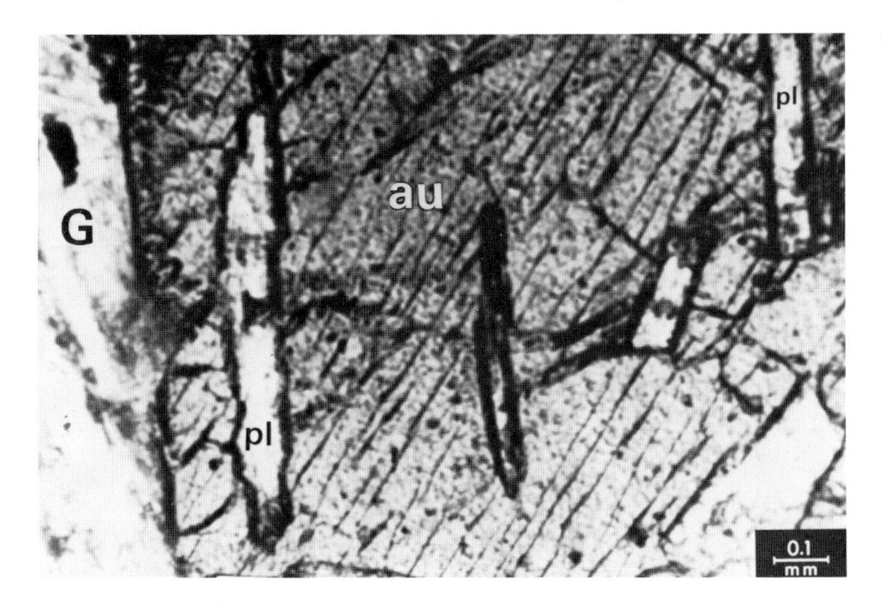

Fig. 283

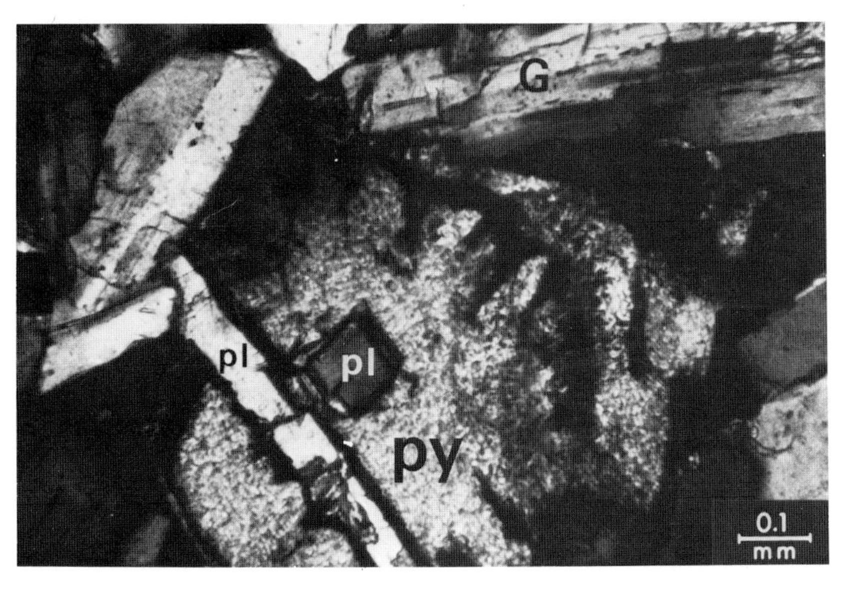

Fig. 284 Plagioclase in ophitic intergrowth with pyroxene.
py = pyroxene
pl = plagioclase
G = groundmass plagioclases
Essexite. Bell Hill, Scotland.
With crossed nicols.

Fig. 285 Plagioclase laths of the groundmass in ophitic intergrowth with rounded pyroxene.
Pl = plagioclase
py = pyroxene
Essexite. Bell Hill, Scotland.
With crossed nicols.

Fig. 286 Plagioclase in section apparently indicating that the pyroxene encloses the feldspar; actually the feldspar is in ophitic intergrowth with the pyroxene.
Pl = plagioclase
py = pyroxene
Essexite. Bell Hill, Scotland.
With crossed nicols.

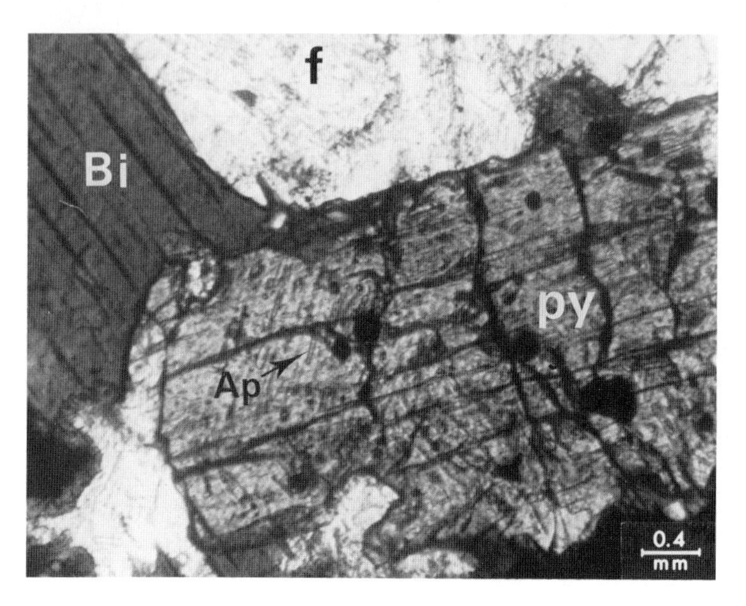

Fig. 287 Fine zoned pyroxene with interzonal apatite in contact with feldspar and biotite.
py = pyroxene (augite)
f = orthoclase
Bi = biotite
Ap = apatite interzonally in the pyroxene
Essexite (Monzodiorite). Rongstock, Bohemia, CSSR.
Without crossed nicols.

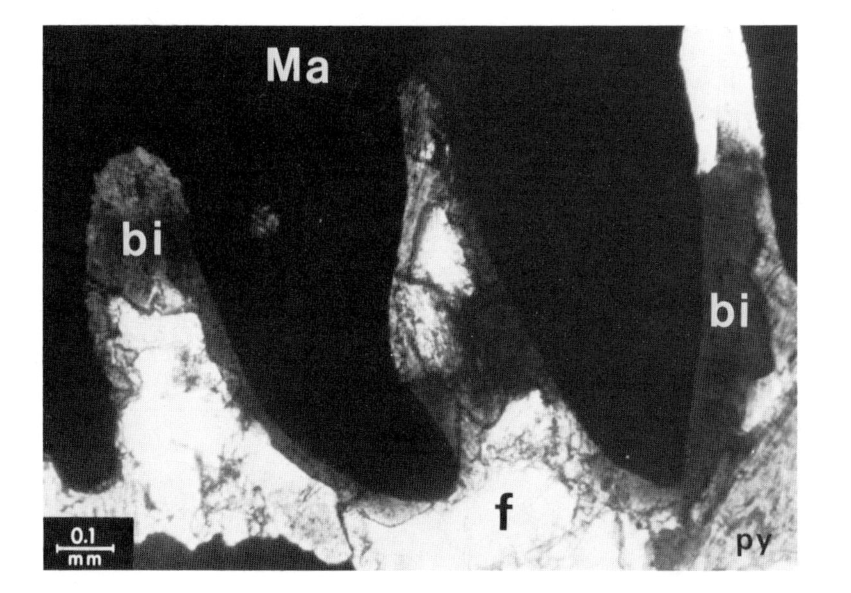

Fig. 288 Corroded magnetite with a margin of biotite.
Ma = magnetite
bi = biotite
f = feldspar
py = pyroxene
Essexite (Monzodiorite). Rongstock, Bohemia, CSSR.
Without crossed nicols.

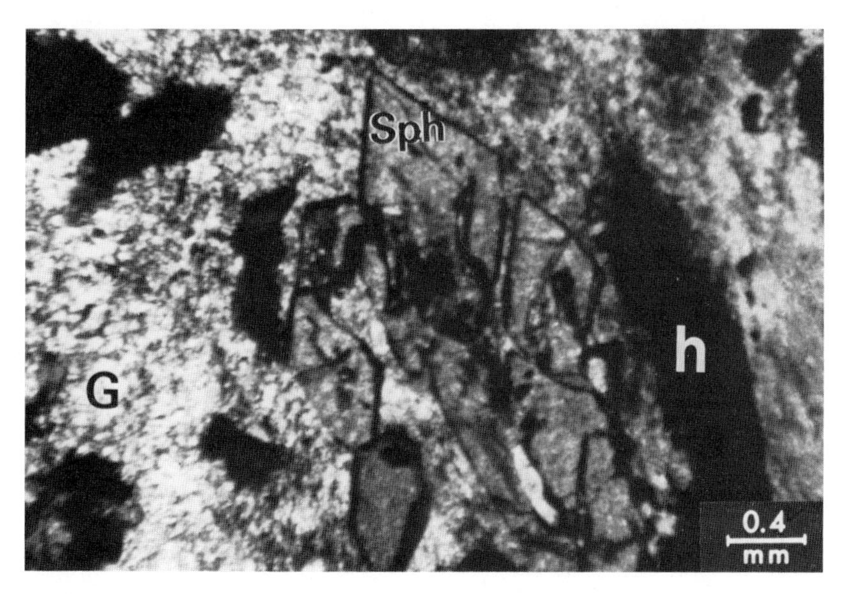

Fig. 289 Sub-idioblastic sphene with crystalloblastic hornblende in feldspar-predominating, essexitic groundmass.
Sph = sphene
h = hornblende crystalloblast
G = groundmass
Essexite. Alceria, Serra de Monchique, South Portugal.
With crossed nicols.

Fig. 290 Idioblastic sphene and crystalloblastic hornblende in essexitic groundmass.
Sph = sphene idioblast
h = hornblende
f = feldspar predominating groundmass
Essexite. Alceria, Serra de Monchique, South Portugal.
With crossed nicols.

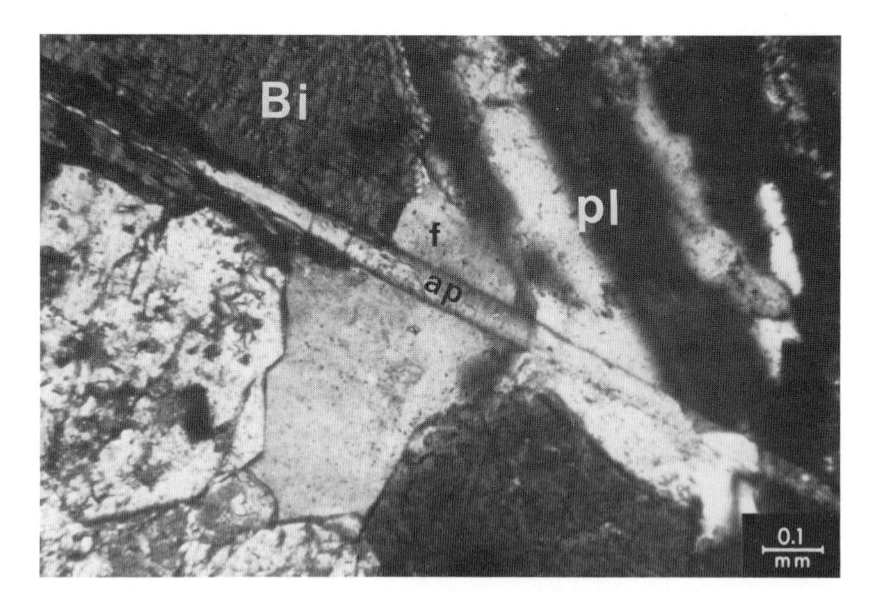

Fig. 291 Apatite nematoblast transversing plagioclase and biotite.
ap = apatite
f = orthoclase
Bi = biotite
pl = plagioclase
Essexite. Tofteholmen, South Norway.
With crossed nicols.

Fig. 292 Gigantic apatite crystalloblast.
ap = apatite
Amphibole-Essexite, fine grained. Alceria, Serra de Monchique, South Portugal.
With crossed nicols.

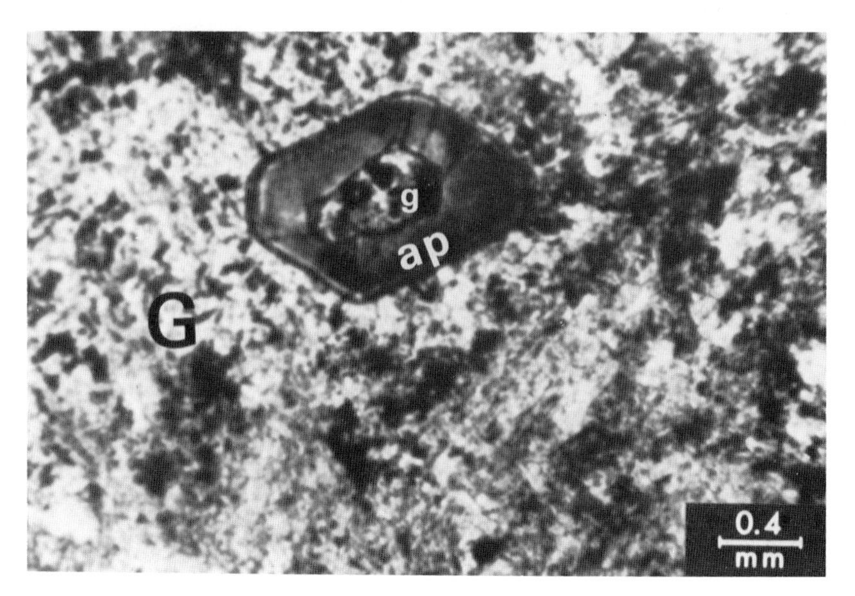

Fig. 293 Crystalloblastic apatite enclosing essexitic groundmass.
Ap = apatite
G = groundmass
g = groundmass enclosed in apatite
Amphibole-Essexite, fine grained. Alceria, Serra de Monchique, South Portugal.
With crossed nicols.

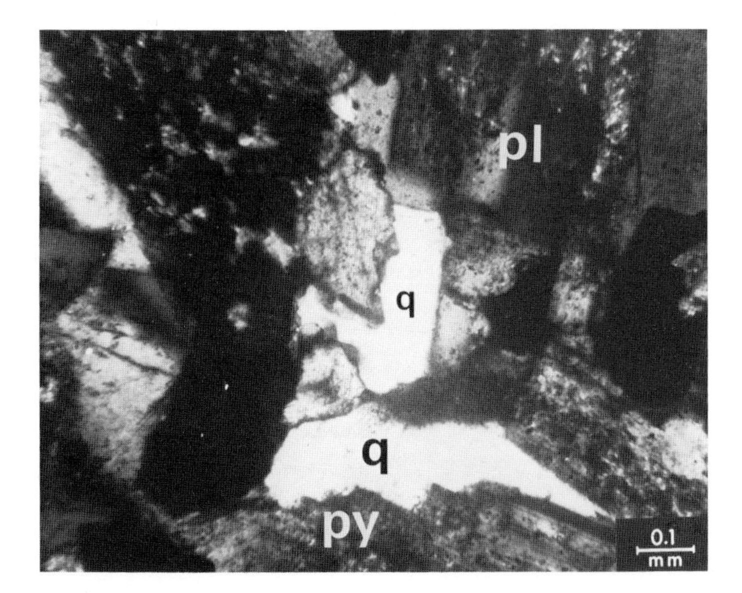

Fig. 294 Metasomatic quartz infiltrating and invading a pyroxene, taking advantage of the pyroxenes cleavage.
q = quartz
py = pyroxene
pl = plagioclase
Essexite. Tofteholmen, South Norway.
With crossed nicols.

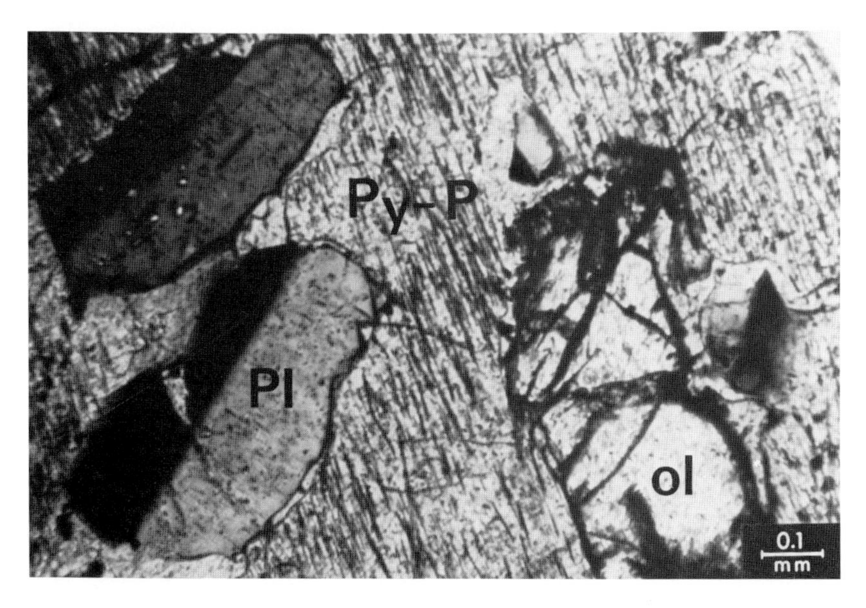

Fig. 295 Poikiloblastic pyroxene enclosing plagioclases and olivine.
Py-P = poikiloblastic pyroxene
Pl = plagioclase corroded and enclosed by poikiloblastic pyroxene
ol = olivine enclosed by the poikiloblastic pyroxene
Hypersthene-Norite. Deneykin Kamen, North Ural, USSR.
With crossed nicols.

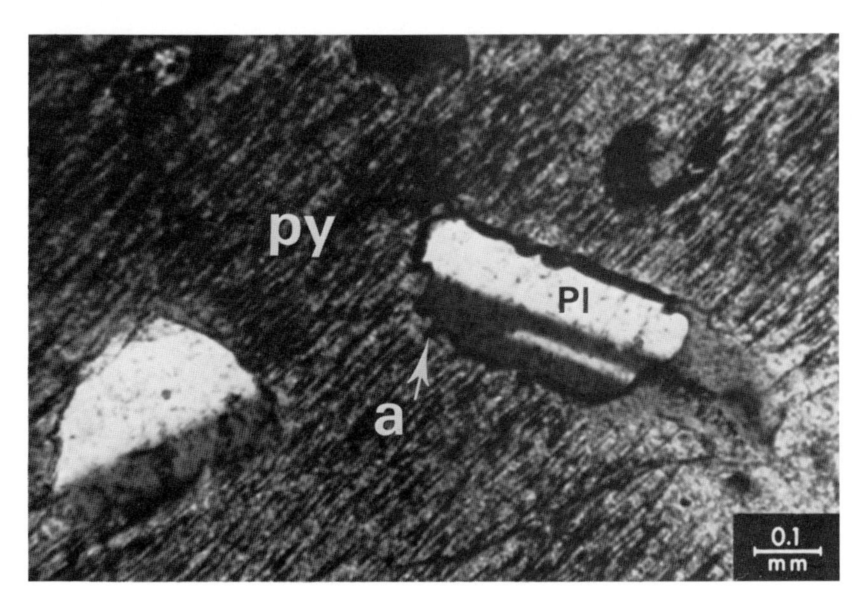

Fig. 296 Shows poikiloblastic pyroxene enclosing and corroding plagioclase.
py = pyroxene poikiloblast
pl = plagioclases, corroded and enclosed by pyroxene
Arrow "a" shows indentation of the plagioclase due to corrosion.
Hypersthene-Norite. Deneykin Kamen, North Ural, USSR.
With crossed nicols.

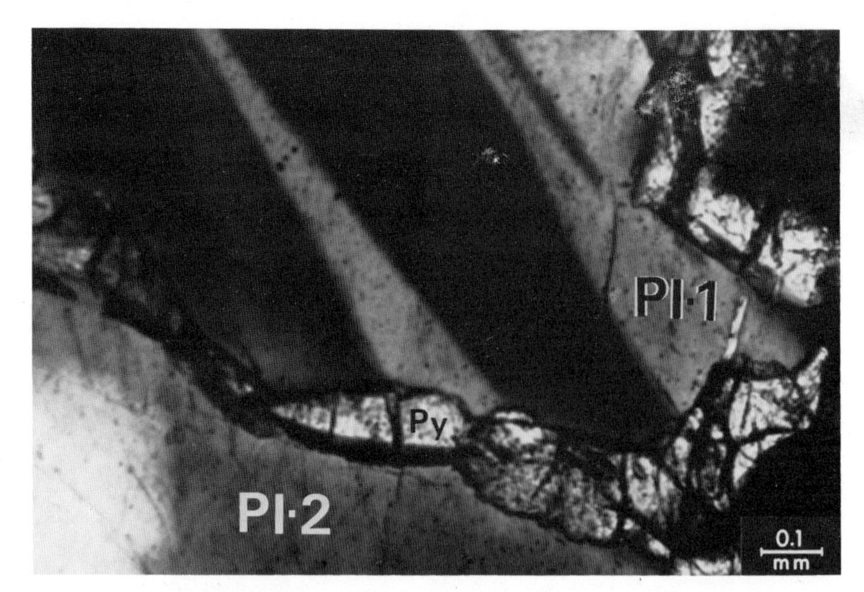

Fig. 297 Pyroxene intergranular between two plagioclases.
Pl-1 = plagioclase 1
Pl-2 = plagioclase 2
Py = pyroxene intergranular between Pl_1 and Pl_2
Biotite-Norite. Gravagliana, Mastalone Valley, Italy.
With crossed nicols.

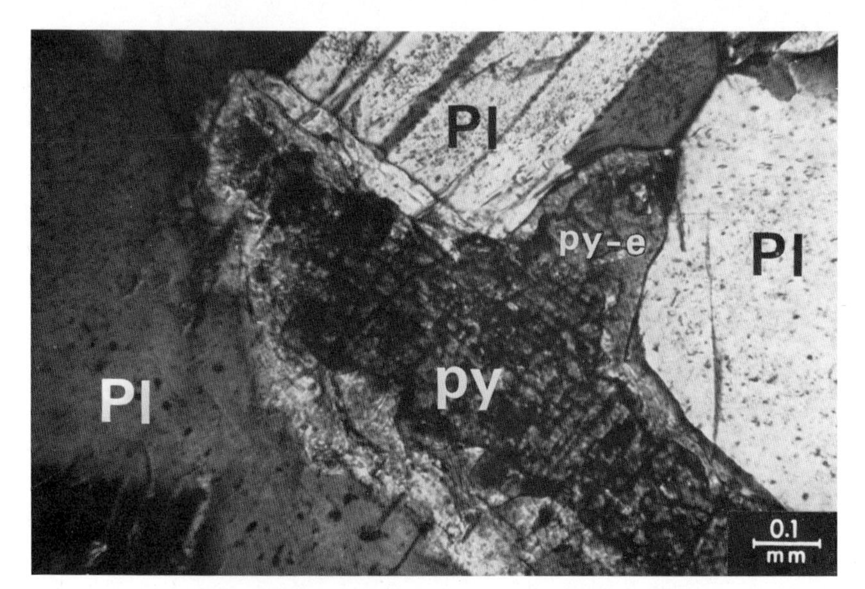

Fig. 298 Pyroxene with a "reaction" margin extending between two plagioclases.
py = pyroxene
py-e = extension of pyroxene between two plagioclases
Pl = plagioclase
Hypersthene-Norite. Deneykin Kamen, North Ural, USSR.
With crossed nicols.

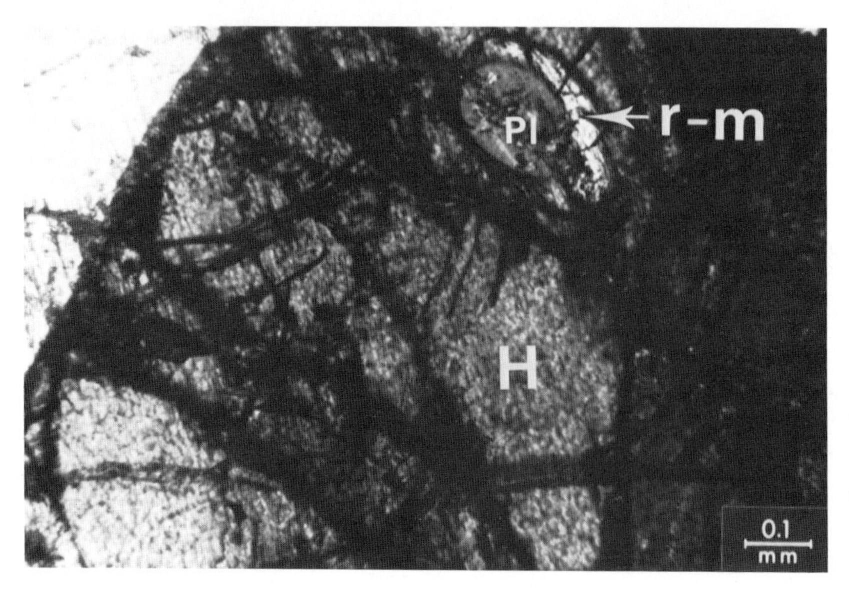

Fig. 299 Hornblende blastically grown enclosing plagioclase which shows a reaction margin between hornblende-plagioclase.
H = hornblende
Pl = plagioclase, rounded, enclosed by the hornblende
r-m = reaction margin between plagioclase and hornblende
H = hornblende
Norite. Samone, Val Sesia, North Italy.
With crossed nicols.

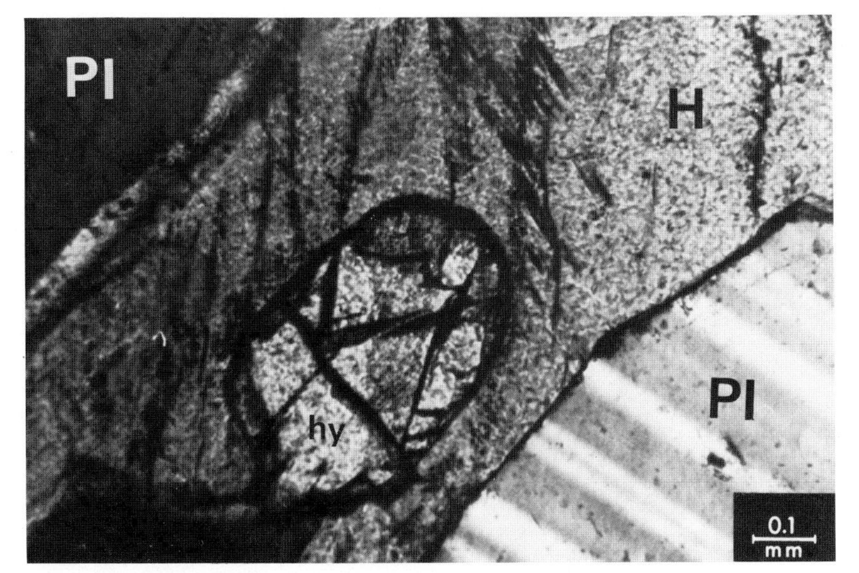

Fig. 300 Hypersthene enclosed by later blastic hornblende
H = hornblende
hy = hypersthene
Pl = plagioclase
Norite. Samone, Val Sesia, North Italy.
With crossed nicols.

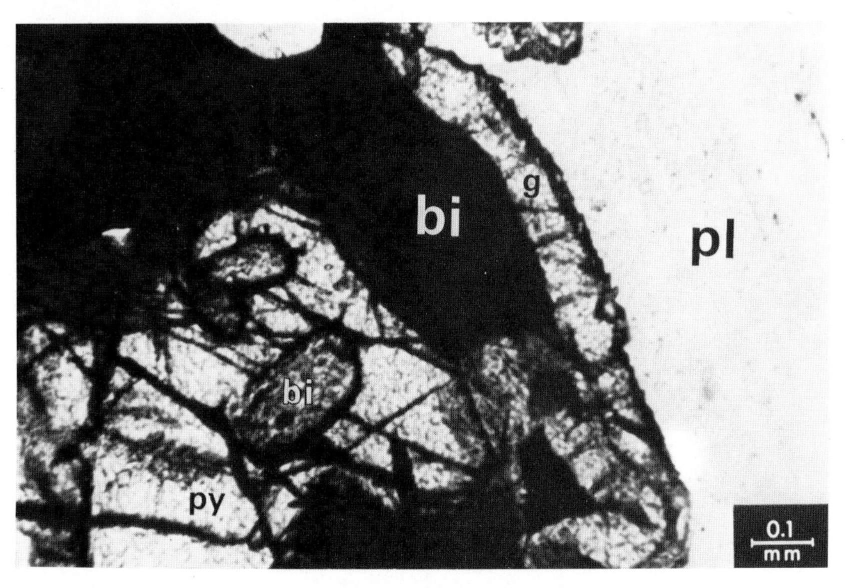

Figs. 301 and 302 Pyroxene with a margin of biotite which in turn is followed by a later margin of garnet crystalloblast.
py = pyroxene ("enclosing biotite")
bi = biotite margin surrounding the pyroxene
g = garnet crystalloblast, forming a layer surrounding the biotite
pl = plagioclase
Biotite-Norite. Gravagliana, Mastalone Valley, Italy.
Without and with crossed nicols.

Fig. 302

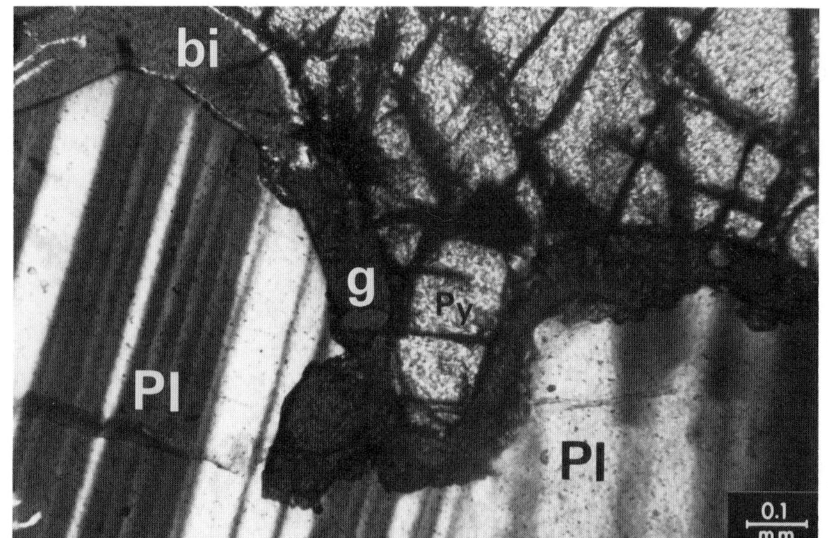

Fig. 303 Pyroxene with a margin of biotite and garnet crystalloblasts.
Py = pyroxene
bi = biotite
g = margin of garnet crystalloblast
Pl = plagioclase
Biotite-Norite. Gravagliana, Mastalone Valley, Italy.
With crossed nicols.

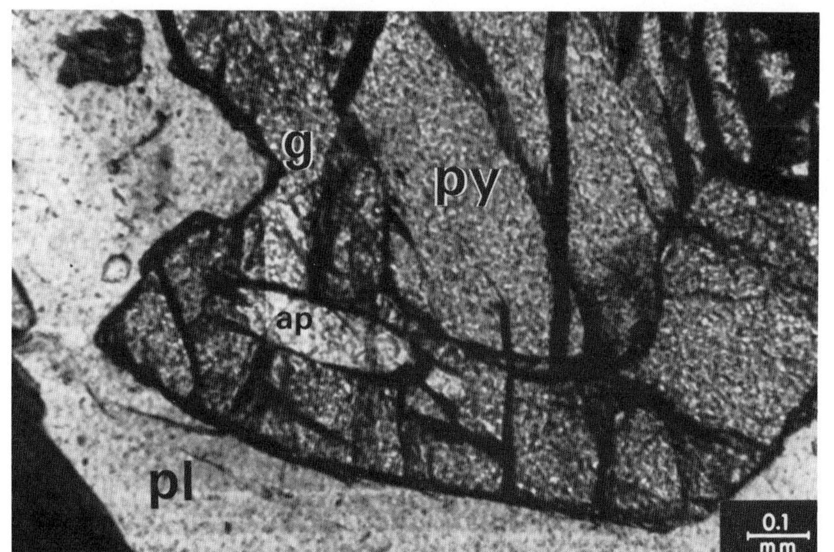

Figs. 304 and 305 Pyroxene with a margin of garnet in which apatite is also enclosed.
py = pyroxene
g = garnet (a crystalloblastic margin of the pyroxene and enclosing an apatite).
ap = apatite
pl = plagioclase
Biotite-Norite. Gravagliana, Mastalone Valley, Italy.
Without and with crossed nicols.

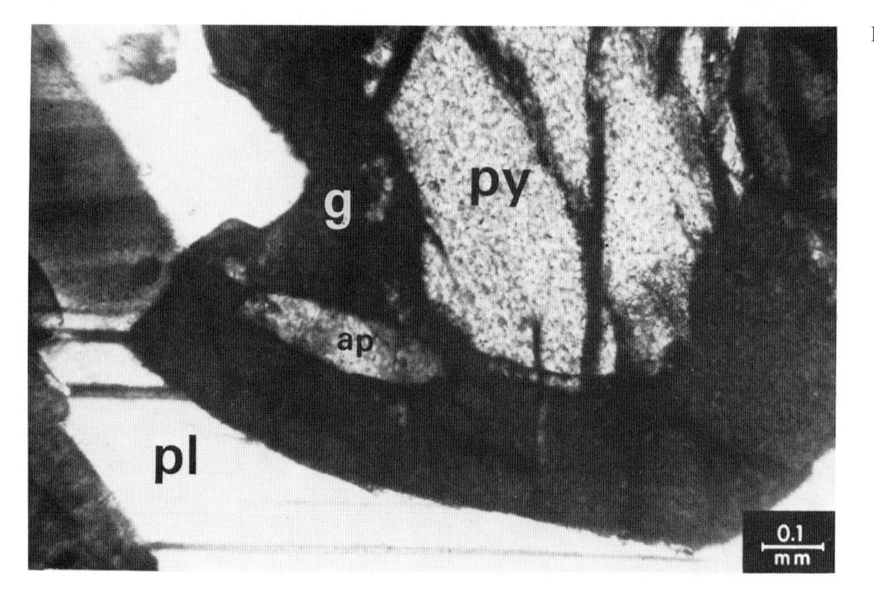

Fig. 305

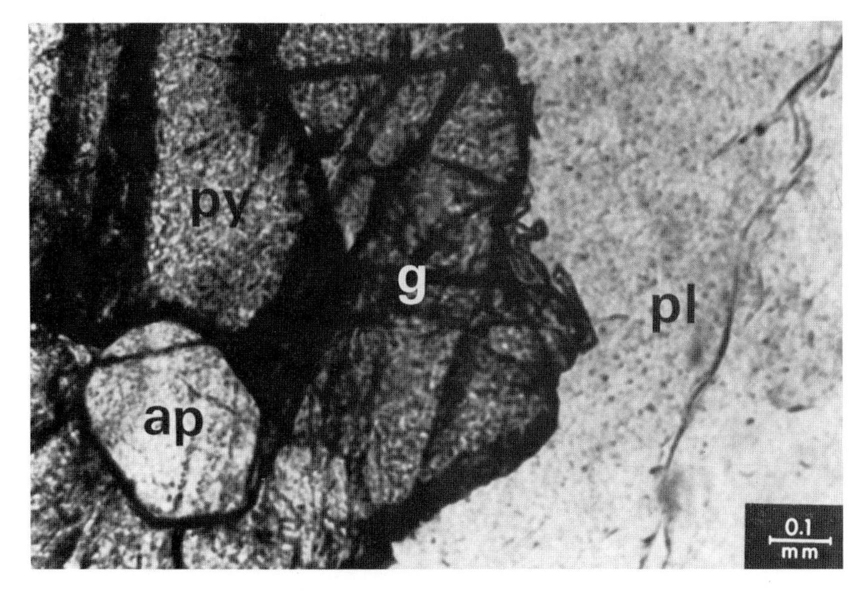

Figs. 306 and 307 Pyroxene with crystalloblastic margin of garnet. Between the pyroxene and the garnet crystalloblast there is an apatite.
py = pyroxene
g = garnet crystalloblast (margin around the pyroxene).
ap = apatite
pl = plagioclase
Norite. Mastalone Valley, Italy.
Without and with crossed nicols.

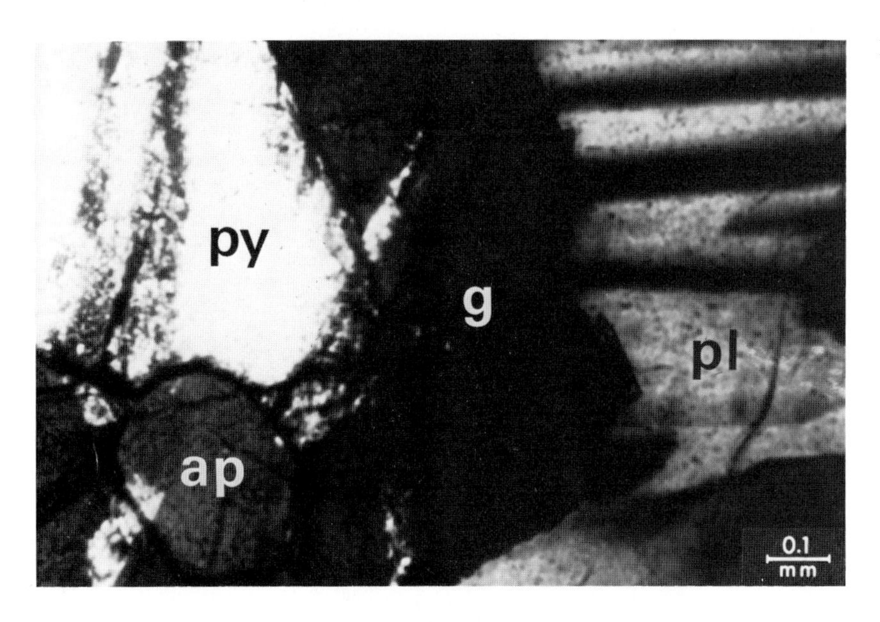

Fig. 307

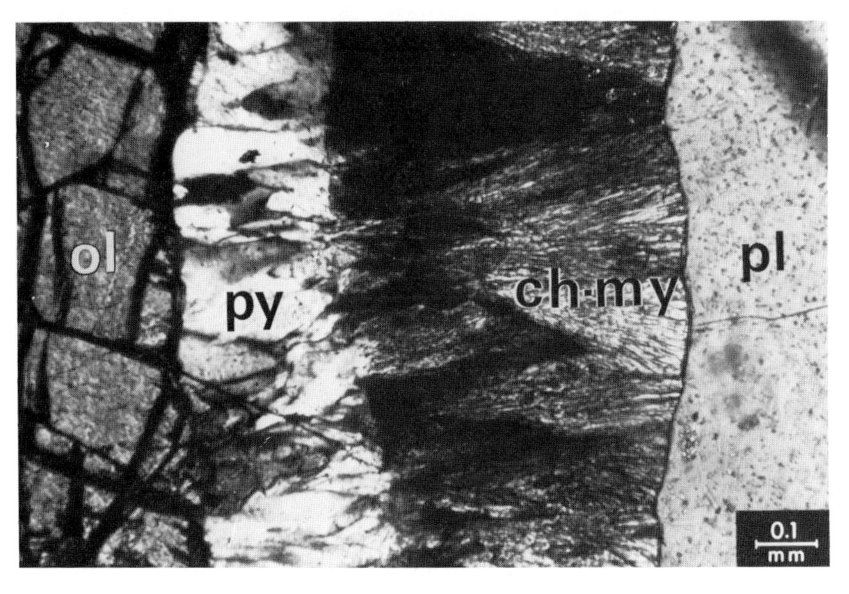

Fig. 308 A corona structure between olivine and plagioclase.
ol = olivine
py = pyroxene
ch-my = a myrmekitic chloritic intergrowth
pl = plagioclase
Hypersthene-Norite. Deneykin Kamen, North Ural, USSR.
With crossed nicols.

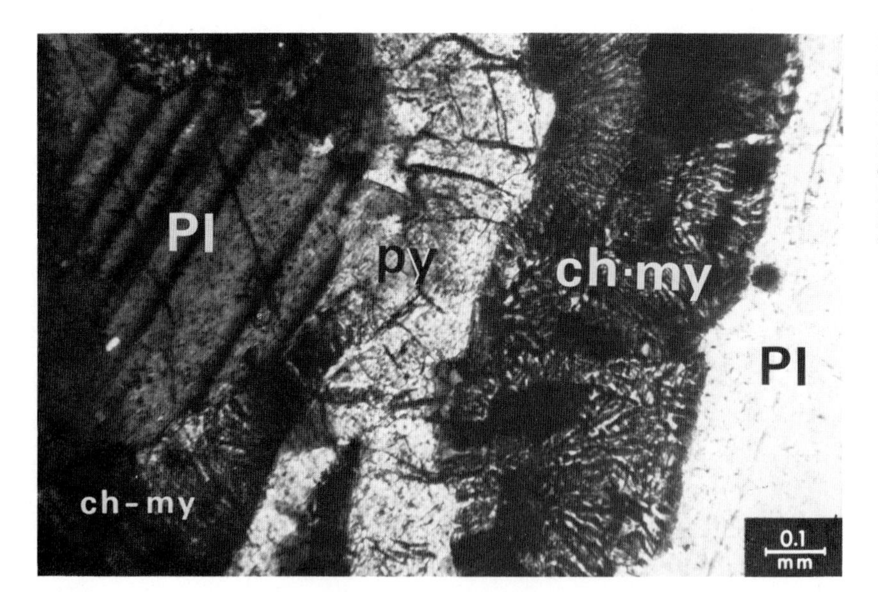

Fig. 309 Plagioclase with a pyroxene margin surrounded by a corona structure composed of chlorite-myrmekite.
Pl = plagioclase
py = pyroxene margin
ch-my = chlorite-myrmekite
Hypersthene-Norite. Deneykin Kamen, North Ural, USSR.
With crossed nicols.

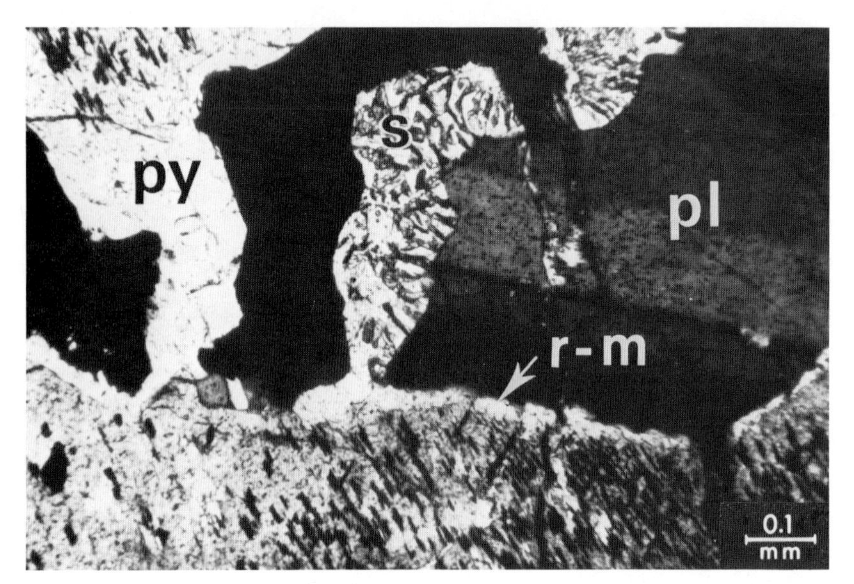

Fig. 310 Plagioclases corroded and with a reaction margin in contact with pyroxene or pyroxene/plagioclase symplectite.
pl = plagioclase
py = pyroxene
r-m = reaction margin
s = symplectite pyroxene/plagioclase
Hypersthene-Norite. Deneykin Kamen, North Ural, USSR.
With crossed nicols.

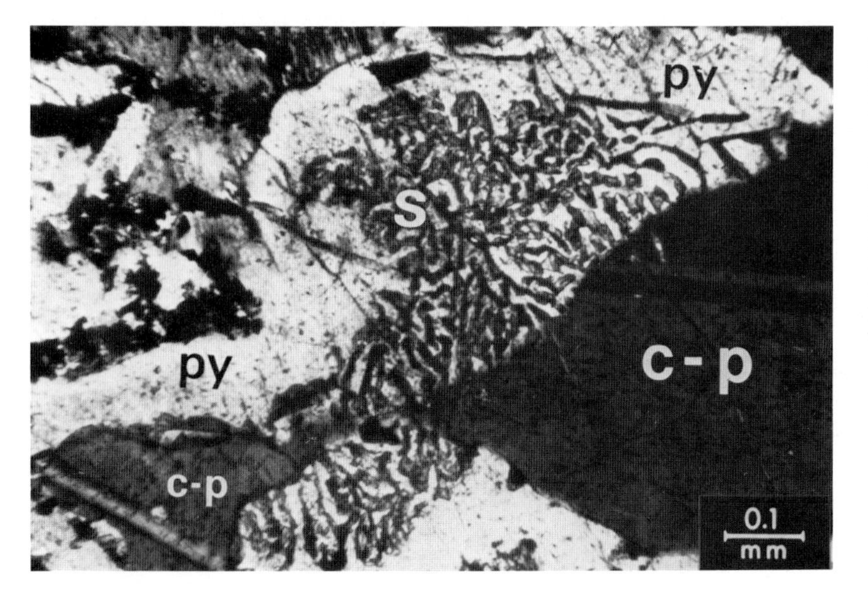

Fig. 311 Plagioclase corroded by pyroxene. The symplectite pyroxene/plagioclase results due to pyroxene "invading" and replacing the plagioclase. The corroded plagioclases (c-p) represent relics of this replacement process.
c-p = corroded plagioclase representing relics
py = pyroxene
s = pyroxene/plagioclase symplectite
Hypersthene-Norite. Deneykin Kamen, North Ural, USSR.
With crossed nicols.

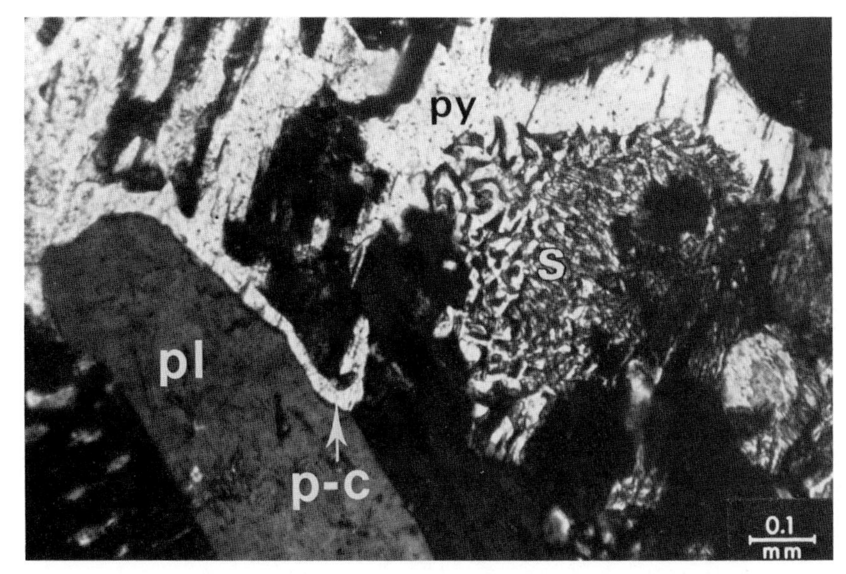

Fig. 312 Pyroxene/plagioclase symplectite with extensions of the pyroxene clearly "invading" the plagioclase.
py = pyroxene
pl = plagioclase
s = pyroxene/plagioclase symplectite.
p-c = extension of the pyroxene "invading" the plagioclase and causing a symplectic intergrowth.
Hypersthene-Norite. Deneykin Kamen, North Ural, USSR.
With crossed nicols.

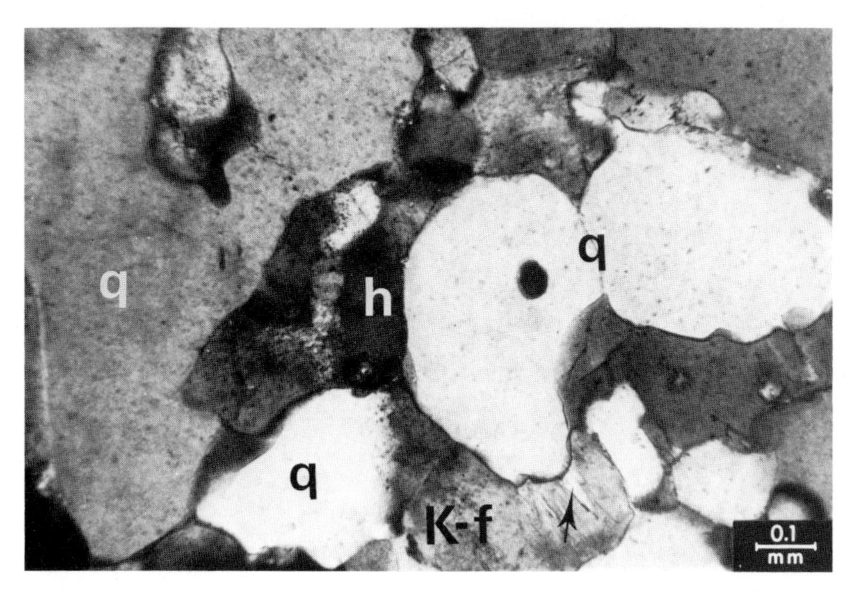

Fig. 313 Sedimentogenic-metamorphic granular quartz texture (the quartz exhibiting undulating extinction) with blastogenic hornblende intergranular. Also K-feldspar with perthite is shown.
q = sedimentogenic-metamorphic quartz
h = hornblende (intergranular)
K-f = K-feldspar
(arrow shows perthite)
Norite. Westport, New York, USA.
With crossed nicols.

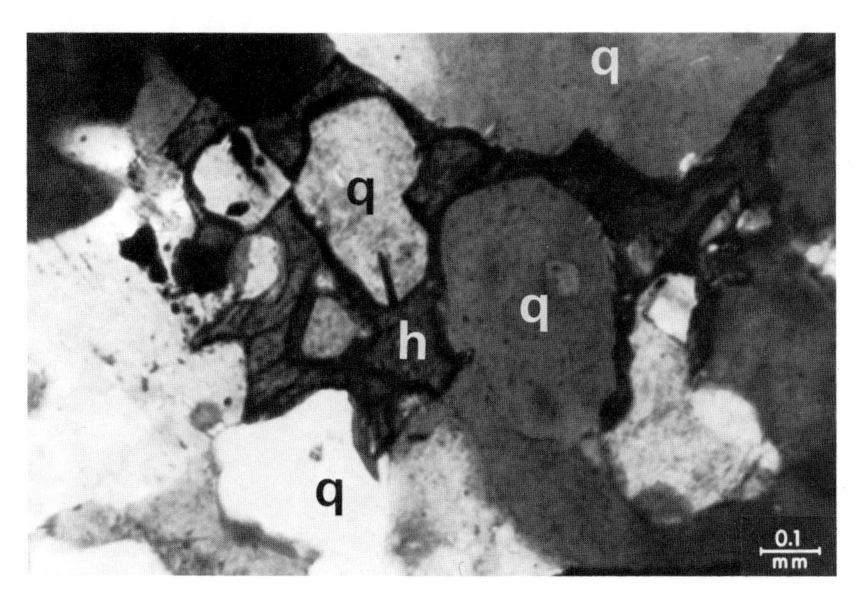

Fig. 314 Quartz grains showing undulating extinction and with brown hornblende as an intergranular-poikiloblast.
q = quartz
h = hornblende intergranular poikiloblast
Norite. Westport, New York, USA.
With crossed nicols.

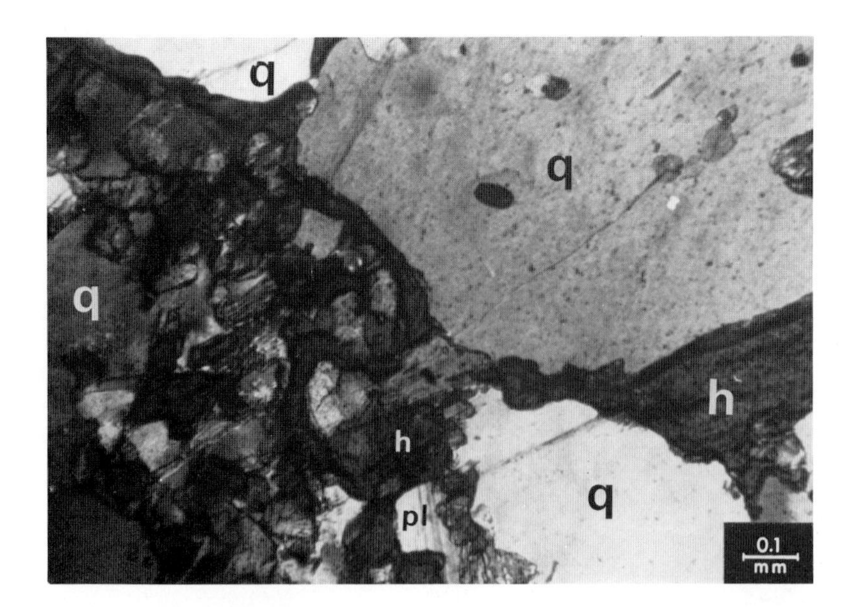

Fig. 315 Quartz grains showing undulating extinction (also plagioclase is present) with brown hornblende as an intergranular-poikiloblast.
q = quartz
h = hornblende intergranular poikiloblast
pl = plagioclase
Norite. Westport, New York, USA.
With crossed nicols.

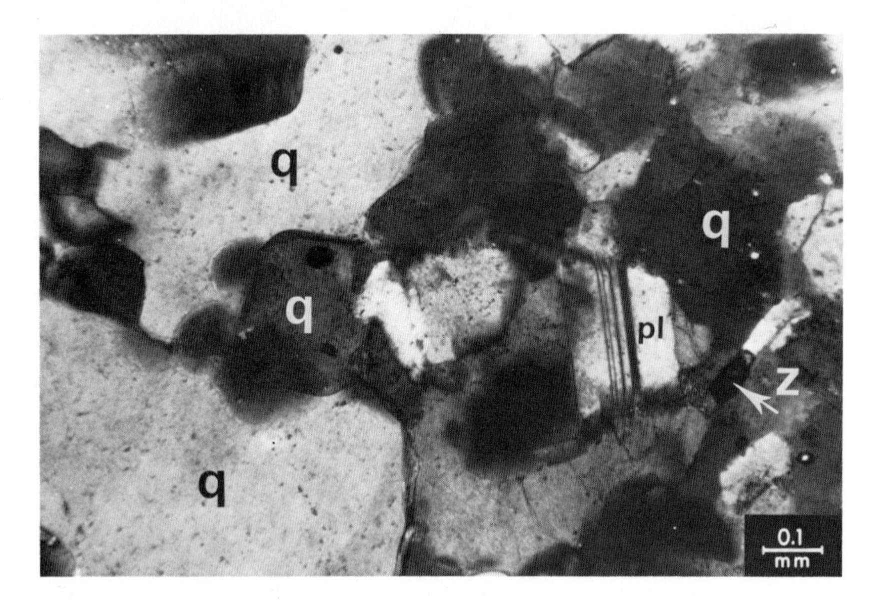

Fig. 316 Granular sedimentogenic-metamorph quartz texture with plagioclase and zircon.
q = quartz (indicating undulating extinction)
pl = plagioclase
z = zircon
Norite. Westport, New York, USA.
With crossed nicols.

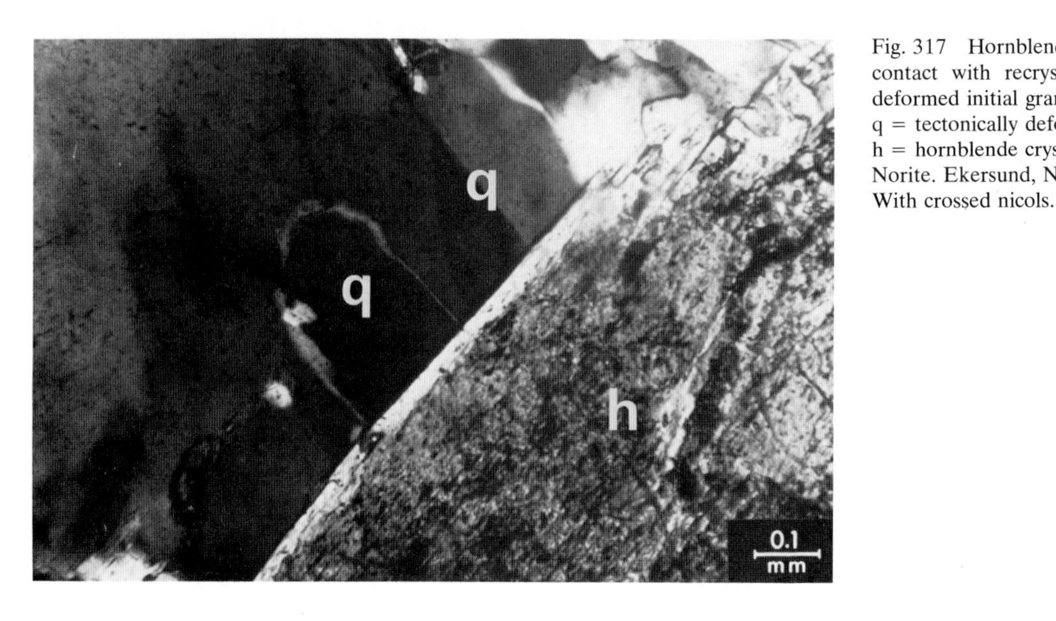

Fig. 317 Hornblende crystalloblast in contact with recrystallised tectonically deformed initial granular quartz.
q = tectonically deformed quartz
h = hornblende crystalloblast
Norite. Ekersund, Norway.
With crossed nicols.

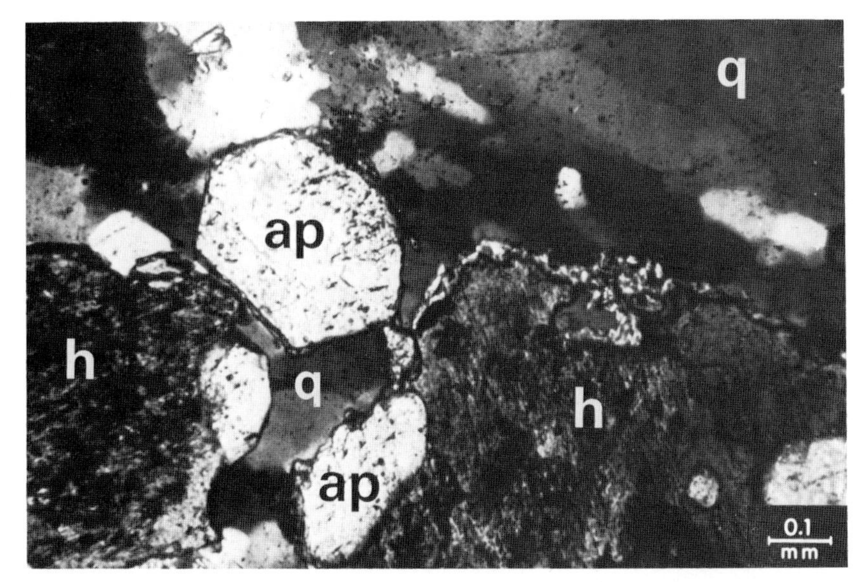

Fig. 318 Tectonically deformed quartz (undulating extinction) in contact with apatites and hornblende crystalloblasts.
q = tectonically deformed quartz
ap = apatite
h = hornblende crystalloblasts
Norite. Ekersund, Norway.
With crossed nicols.

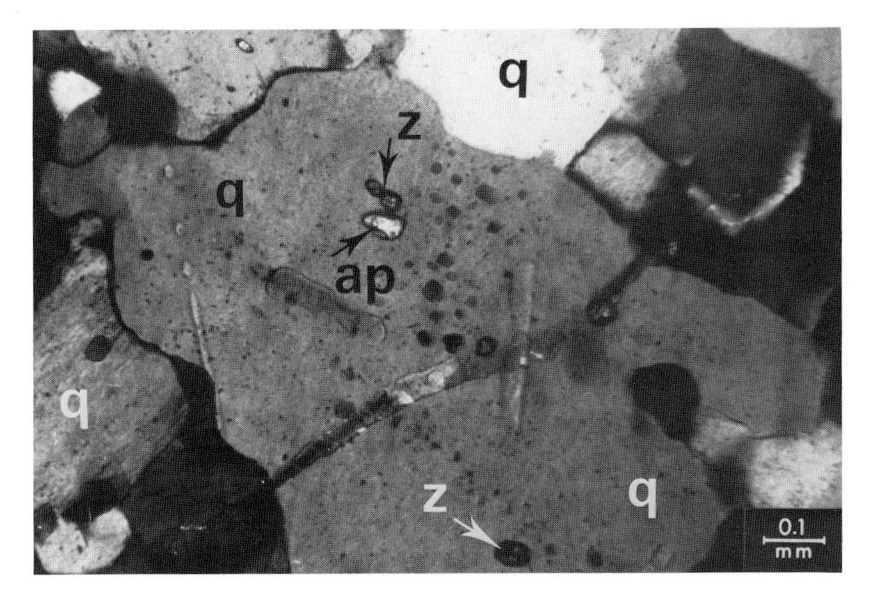

Fig. 319 Quartz recrystallised assimilating apatites and zircons.
q = quartz
ap = apatite
z = zircons
Norite. Westport, New York, USA.
With crossed nicols.

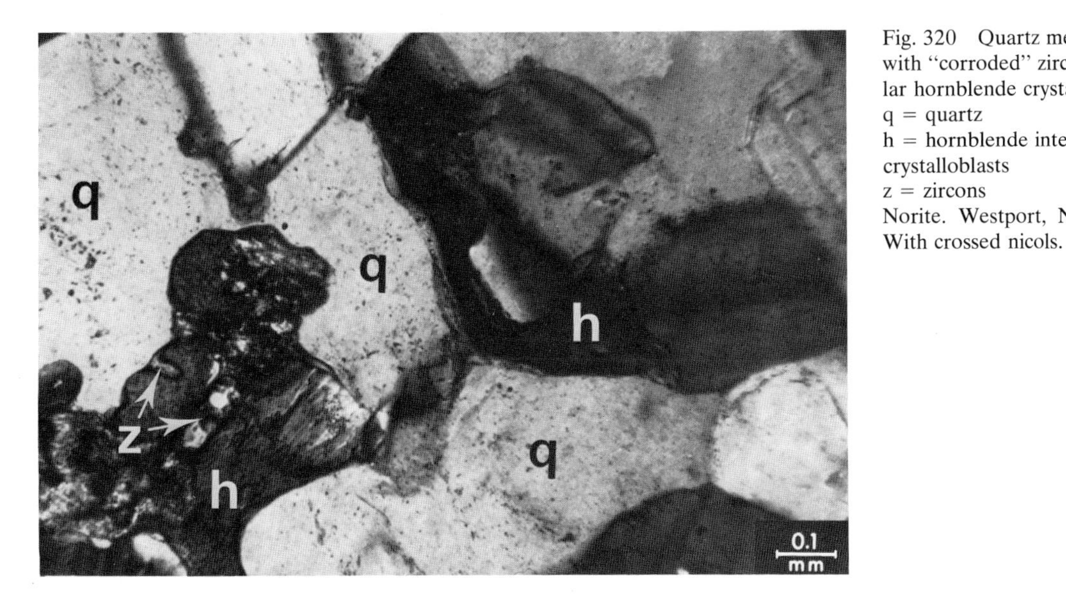

Fig. 320 Quartz metamorphic granular with "corroded" zircons and intergranular hornblende crystalloblasts.
q = quartz
h = hornblende intergranular crystalloblasts
z = zircons
Norite. Westport, New York, USA.
With crossed nicols.

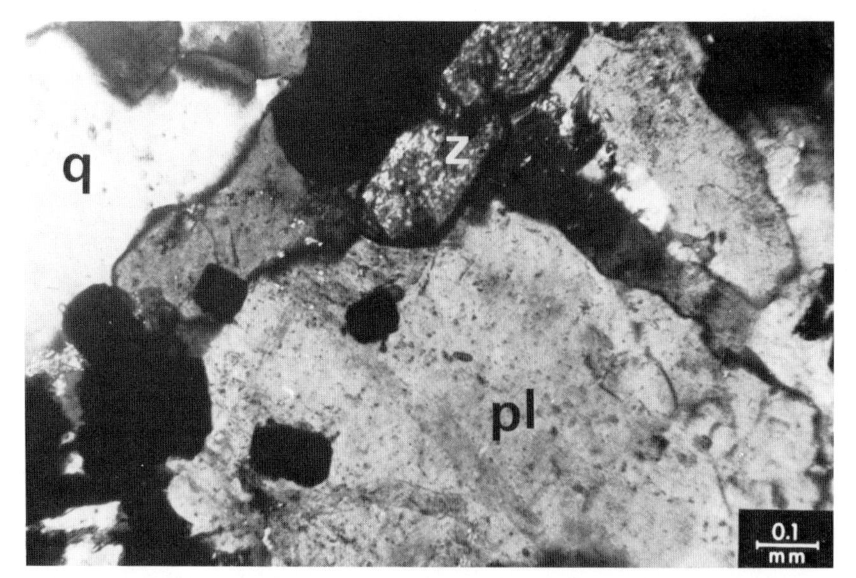

Fig. 321 Quartz-feldspar texture with zircons and ore-minerals
q = quartz
z = zircons
pl = plagioclase crystalloblasts
ore-minerals (black)
Norite. Westport, New York, USA.
With crossed nicols.

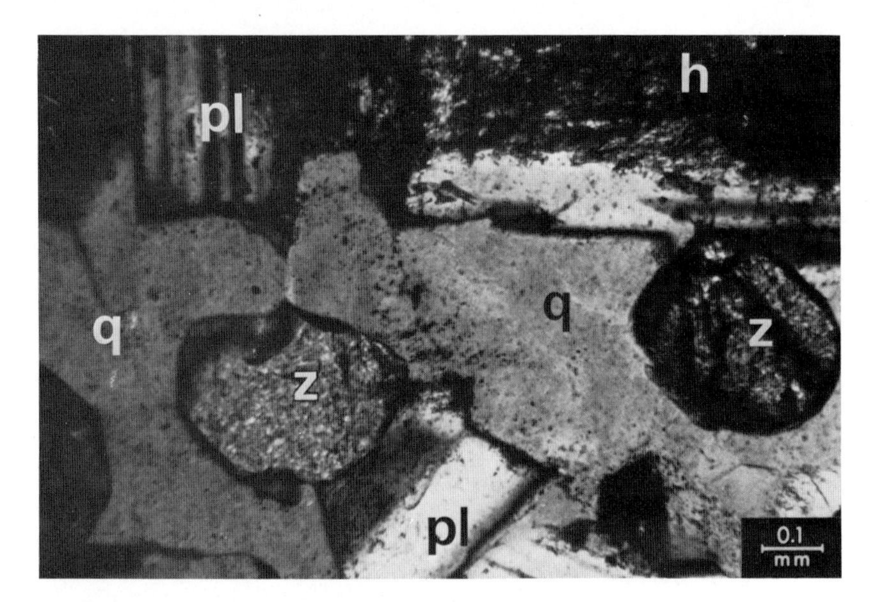

Fig. 322 Intergranular quartz between plagioclases and hornblende with corroded and rounded zircons.
q = quartz
pl = plagioclase
h = hornblende
z = zircon
Norite. Westport, New York, USA.
With crossed nicols.

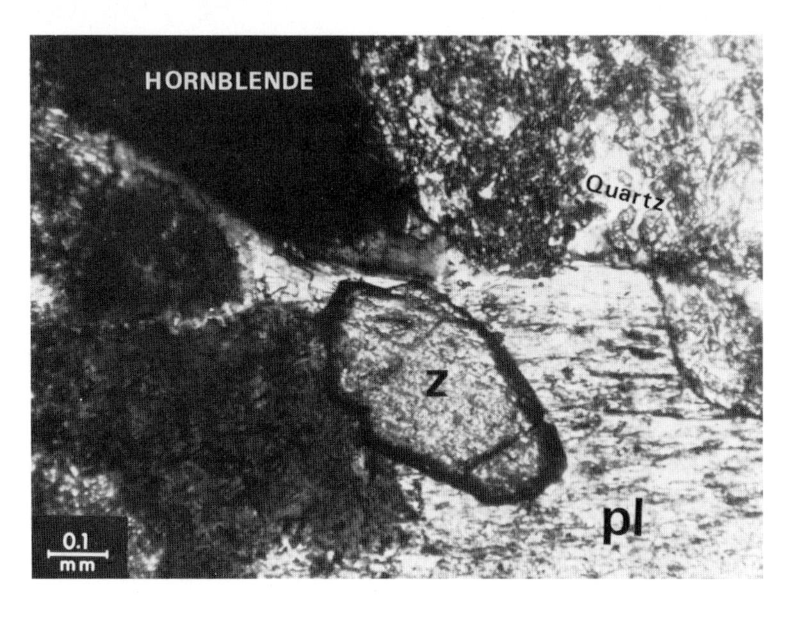

Fig. 323 Corroded zircon in feldspars and hornblende.
z = zircon (corroded)
pl = plagioclase
Norite. Ekersund, Norway.
With crossed nicols.

Fig. 324 Plagioclase crystalloblasts with an idiomorphic zircon orientated parallel to the polysynthetic twinning of the plagioclase.
Pl = plagioclase
z = zircon
Norite. Westport, New York, USA. With crossed nicols.

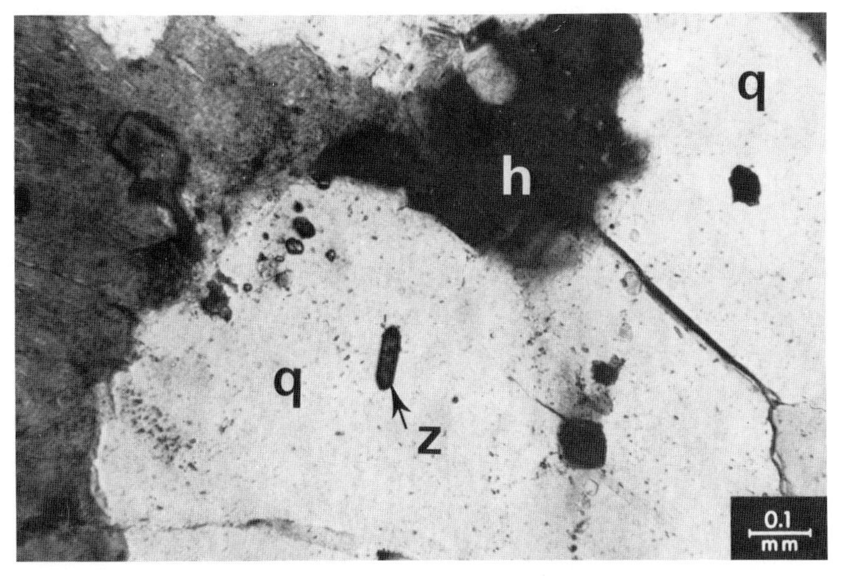

Fig. 325 Recrystallised quartz with idiomorphic zircon, and finer zircon grains within the quartz. Also intergranular crystalloblastic hornblende is shown.
q = quartz
z = zircon
h = hornblende
Norite. Westport, New York, USA. With crossed nicols.

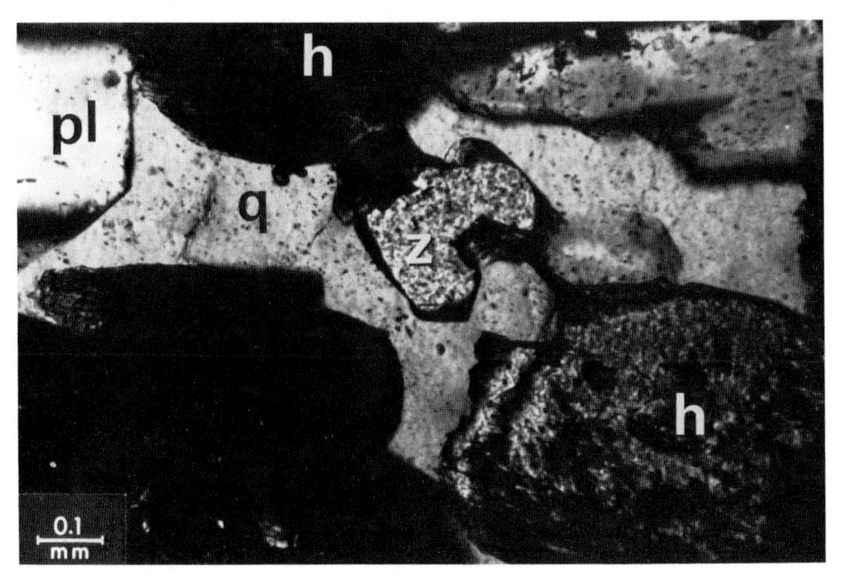

Fig. 326 Quartz with hornblende and idiomorphic zircon.
q = quartz
h = hornblende
z = zircon (idioblastic growth)
pl = plagioclase
Norite. Westport, New York, USA. With crossed nicols.

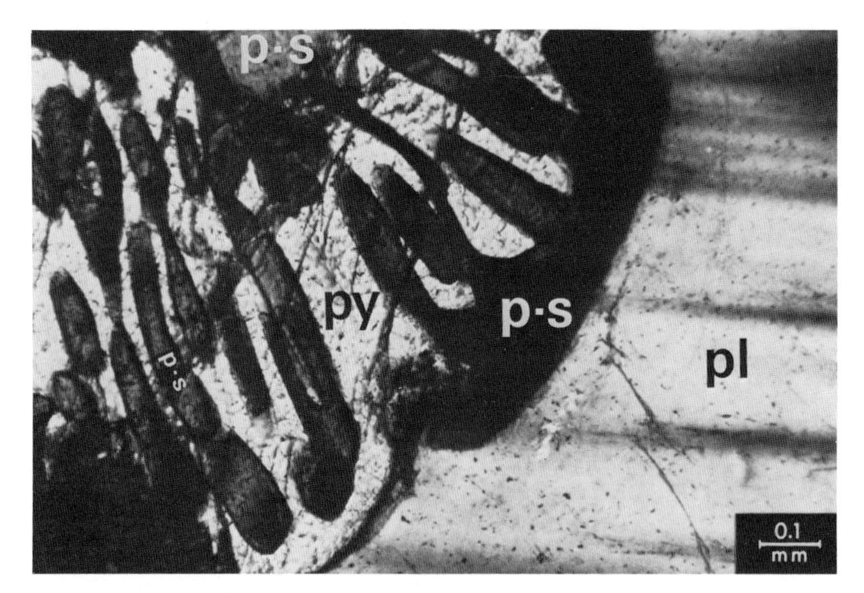

Fig. 327 Graphic-like intergrowth of pleonaste (spinel) with pyroxene. The pleonaste follows partly the margin between pyroxene and plagioclase.
py = pyroxene
pl = plagioclase
p-s = pleonaste (spinel)
A xenocryst or xenolith of mantle in paranorite. The spinel-pyroxene boundary is corroded as the plagioclase crystalloblast comes in contact with them.
Pleonaste-Norite. Veccia, Val Sesia, North Italy.
With crossed nicols.

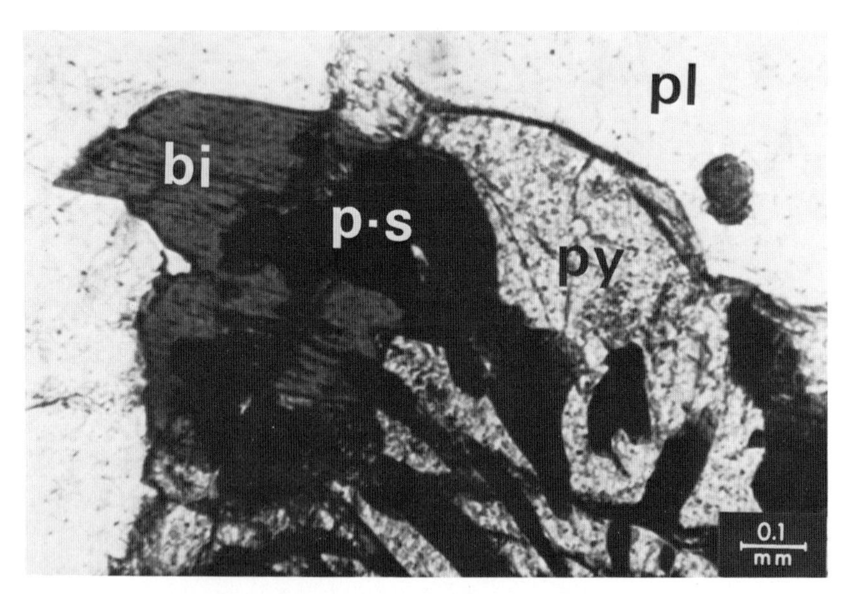

Fig. 328 Pleonaste (spinel) in intergrowth with pyroxene and with biotite.
py = pyroxene
p-s = pleonaste (spinel)
bi = biotite
pl = plagioclase
Pleonaste-Norite. Veccia, Val Sesia, North Italy.
Without crossed nicols.

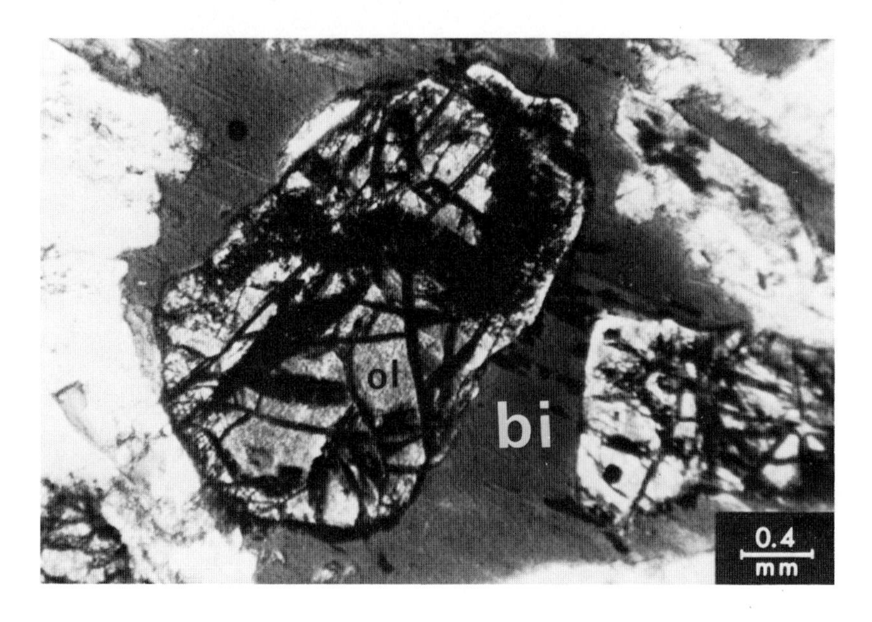

Fig. 331 Olivine surrounded by biotite crystalloblasts
ol = olivine
bi = biotite crystalloblasts
Troctolite. Radautal, Bad Harzburg, Harz, Germany.
Without crossed nicols.

(For Figs. 329 und 330 see page 223)

Figs. 329 and 330 Olivine crystalloblast (poikiloblast) enclosing corroded and rounded plagioclases.
ol = olivine poikiloblasts
pl = plagioclases, rounded and corroded
Troctolite. Volpersdorf, Niederschlesien, a.p. Poland.
With crossed nicols.

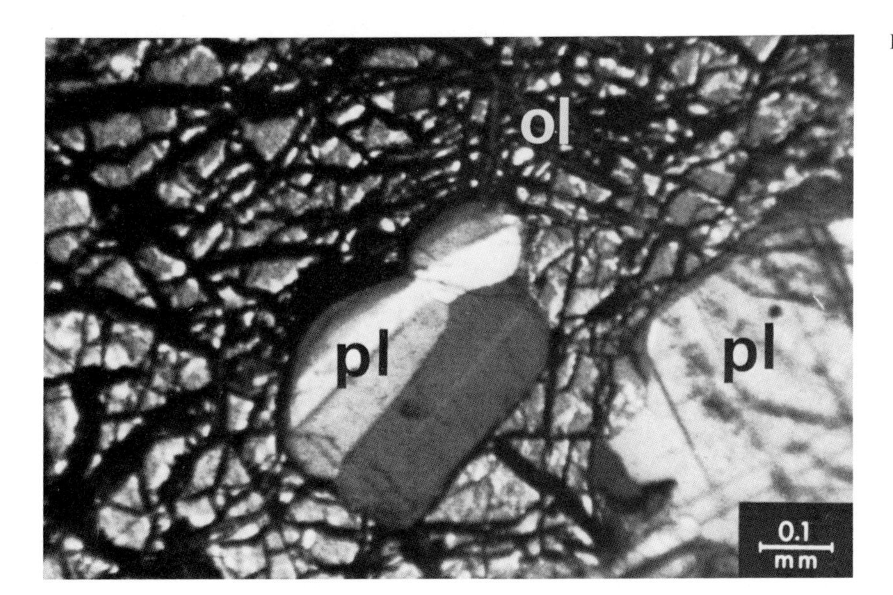

Fig. 330

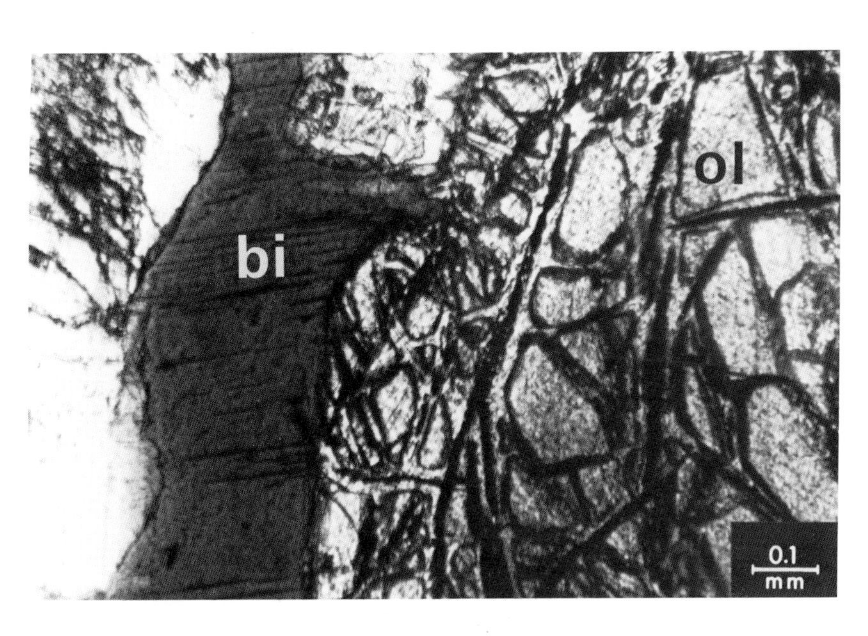

Fig. 332 Olivine with a crystalloblastic margin of biotite.
ol = olivine
bi = biotite crystalloblast
Troctolite. Radautal, Bad Harzburg, Harz, Germany.
Without crossed nicols.

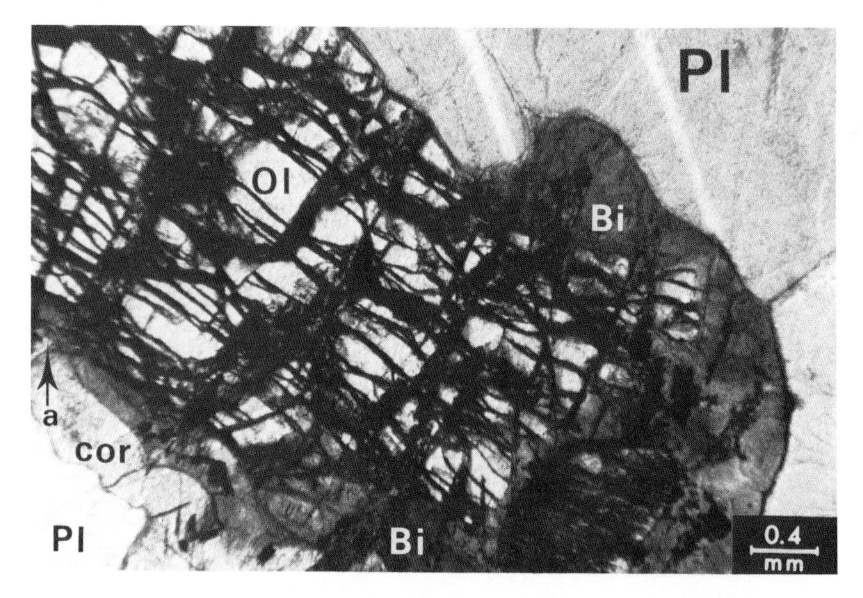

Fig. 333 Shows a general view of a troctolitic olivine and the adjacent plagioclases. A corona reaction structure partly develops between the olivine and the plagioclase and partly between the biotite and the plagioclase.
Ol = olivine with alterations following cracks
Pl = plagioclase
cor = corona structure
Bi = biotites
Arrow "a" shows the part of the corona structure between the olivine and the plagioclase.
Troctolite. Radautal, Bad Harzburg, Harz, Germany,
Without crossed nicols.

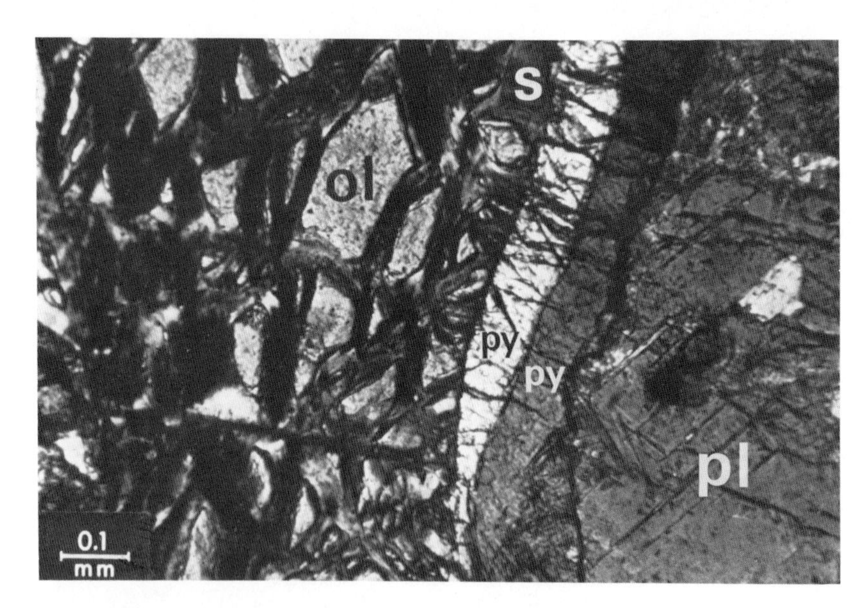

Fig. 334 Serpentinised olivine with a margin of pyroxene in contact with a plagioclase. The pyroxene margins to the olivine assume the form of corona structure.
ol = olivine (serpentinised)
s = serpentine
py = pyroxene margins
pl = plagioclase
Troctolite. Belhelvie, Aberdeenshire, Scotland.
With crossed nicols.

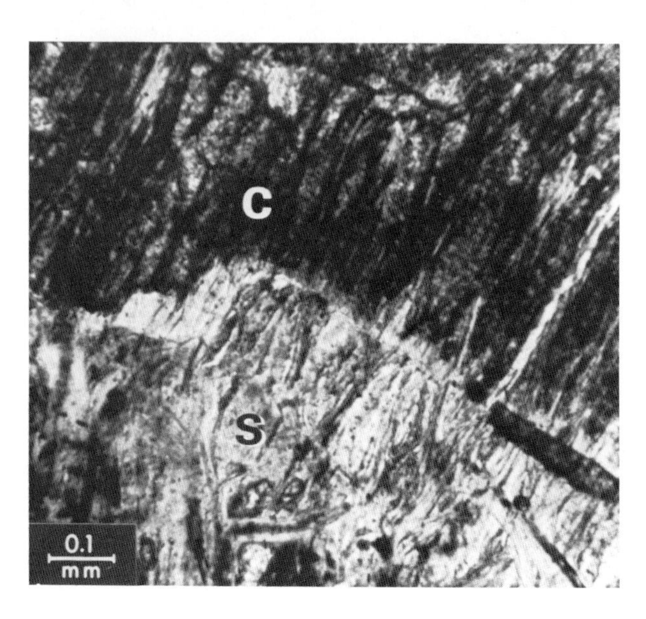

Fig. 335 Serpentinised olivine with pyroxene "corona".
s = serpentinised olivine
c = corona structure (consisting of pyroxene)
ch = chromite
pl = plagioclase
Troctolite. Belhelvie, Aberdeenshire, Scotland.
With crossed nicols.

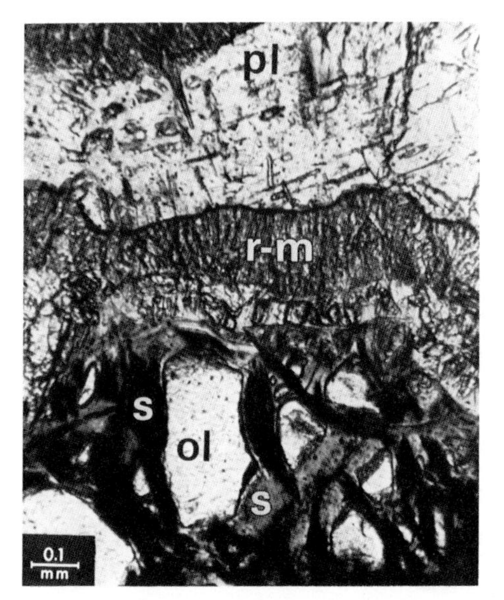

Fig. 336 Olivine serpentinised (antigorite) with a symplectic reaction margin (corona).
ol = olivine
s = serpentine (antigorite)
r-m = reaction margin (symplectic reaction margin of pyroxene)
pl = plagioclase
Troctolite. Belhelvie, Aberdeenshire, Scotland.
With crossed nicols.

Fig. 337 Twinned and zoned titano-augite corroded due to "magmatic" effects of the melts out of which the groundmass has been crystallised. As the textural pattern shows, the Shonkinite of Katzenbukkel, Odenwald, Germany, is a true magmatic crystallization.
Shonkinite, dense, Katzenbuckel, Odenwald, Germany.
With crossed nicols.

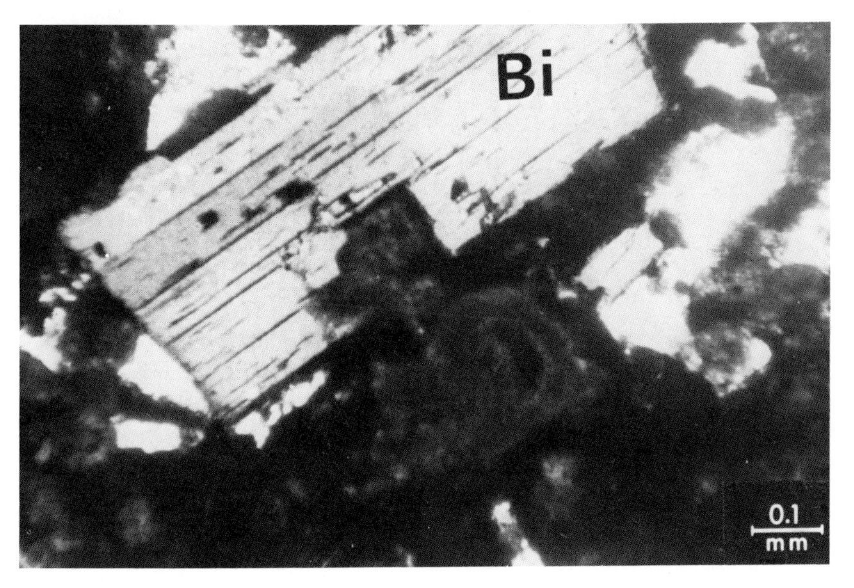

Fig. 338 Idioblastic biotite enclosing part of the groundmass.
Bi = biotite
Shonkinite, coarse grained. Katzenbuckel, Odenwald, Germany.
With crossed nicols.

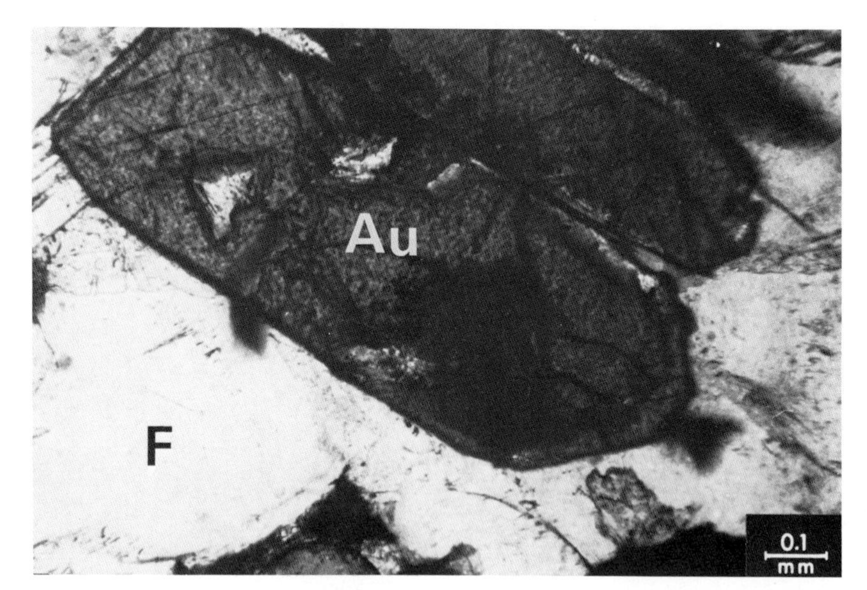

Fig. 339 Idioblastic augites including groundmass.
Au = augite
F = feldspar
Shonkinite. Podhorn, Moravia, CSSR.
Without crossed nicols.

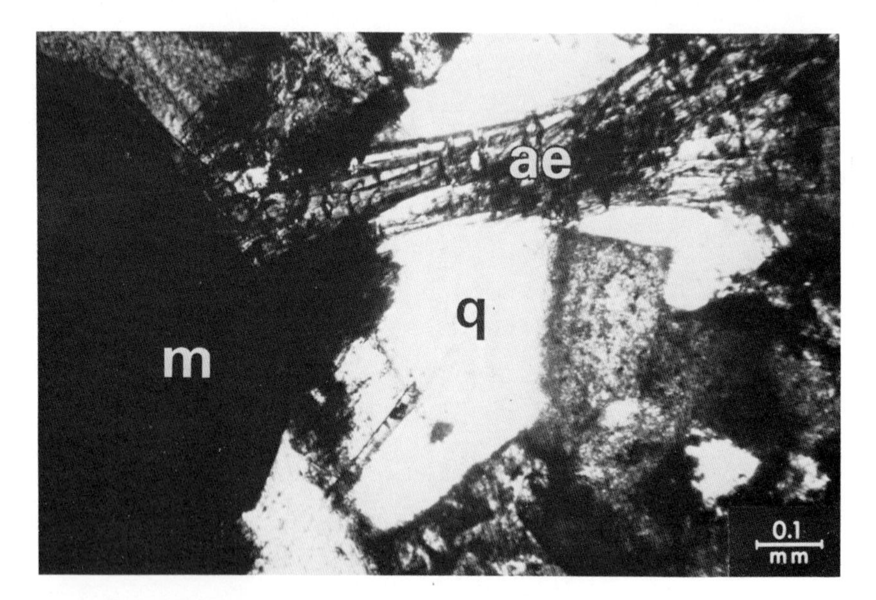

Fig. 340 Elongated prismatic aegerin crystalloblasts penetrating through feldspars and mafic minerals.
ae = aegerin prismatic crystalloblasts
m = mafic components
q = quartz
Natron-Shonkinite. Katzenbuckel, Odenwald, Germany.
With crossed nicols.

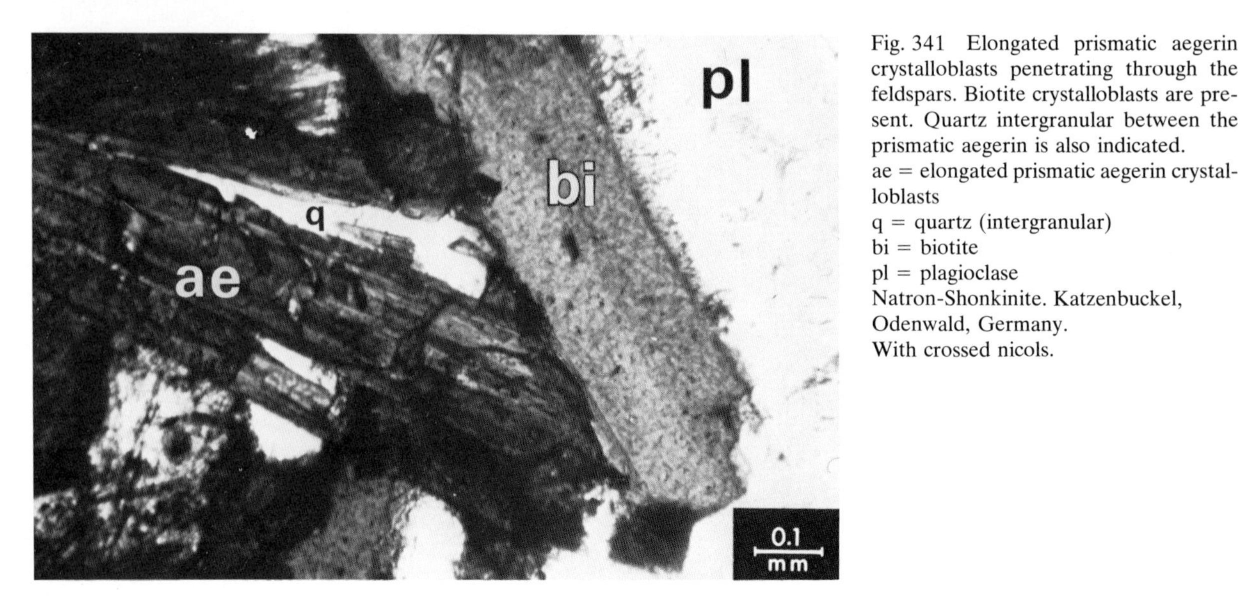

Fig. 341 Elongated prismatic aegerin crystalloblasts penetrating through the feldspars. Biotite crystalloblasts are present. Quartz intergranular between the prismatic aegerin is also indicated.
ae = elongated prismatic aegerin crystalloblasts
q = quartz (intergranular)
bi = biotite
pl = plagioclase
Natron-Shonkinite. Katzenbuckel, Odenwald, Germany.
With crossed nicols.

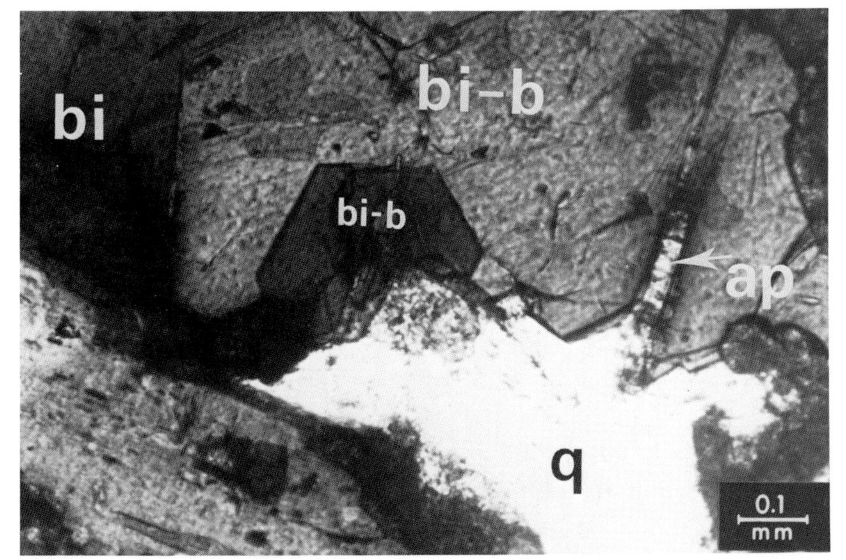

Fig. 342 Idioblastic biotite in larger blastic biotites. A prismatic apatite crystalloblast is also shown.
bi-b = biotite crystalloblast (idioblast)
bi = blastic biotite
q = quartz
ap = apatite prismatic blastic growths
Natron-Shonkinite. Katzenbuckel, Odenwald, Germany.
With crossed nicols.

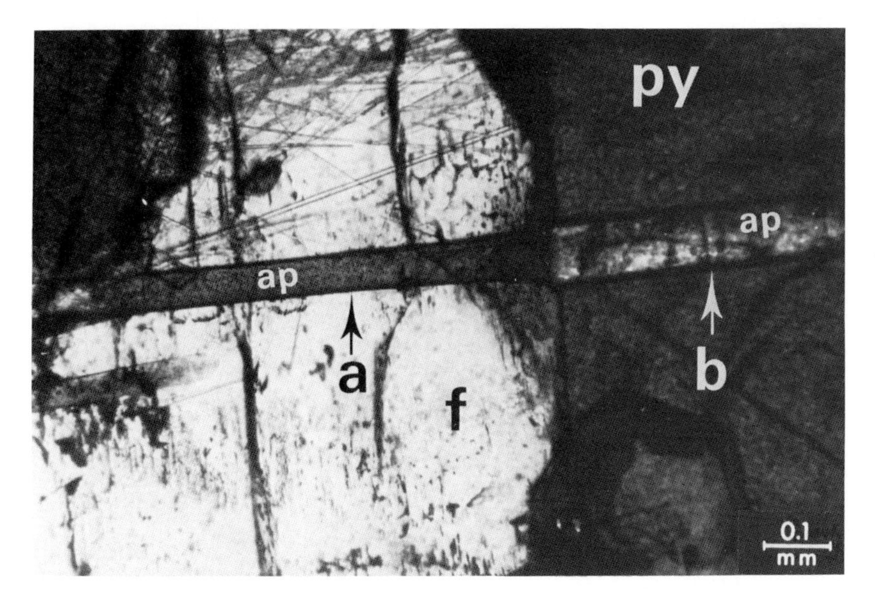

Fig. 343 Idioblastic apatite (elongated velonoblast), transversing pyroxenes and orthoclase. As arrows "a" and "b" show there is a change in the width of the apatite as it transverses the boundary pyroxene/feldspar.
py = pyroxene
f = feldspar orthoclase
ap = apatite (velonoblast)
Natron-Shonkinite. Katzenbuckel, Odenwald, Germany.
With crossed nicols.

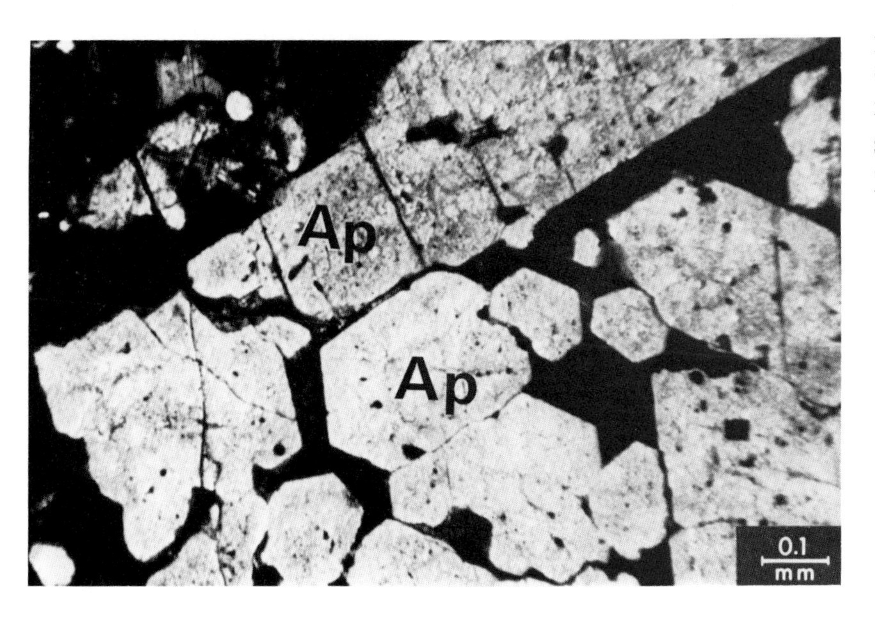

Fig. 344 Large apatite idioblasts in Shonkinite.
Ap = apatite idioblasts
With crossed nicols
Shonkinite-Porphyry. Katzenbuckel, Odenwald, Germany.
With crossed nicols.

Figs. 345, 346 and 347 Idioblastic apatite with magnetite occupying the inter-crystalline spaces. Iron-oxides granules are present in the apatite as shown in Fig. 346.
As Fig. 347 shows, transitions from the inter-crystalline to the intra-crystalline phase of the iron-oxide "granules" is indicated.
Ap = idioblastic apatites
i-o = iron-oxides
g = granules of iron-oxides in the apatites
Arrows show the transitions from inter-granular iron-oxides to intragranular iron-oxide granules.
Mica-Shonkinite. Katzenbuckel, Odenwald, Germany.
With crossed nicols.
(Comparable textural patterns of apatite with magnetite intergrowths are also observed in basalts, Augustithis (1978)).

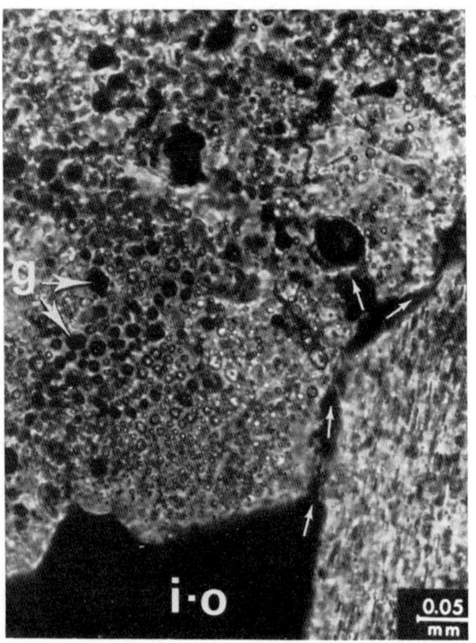

Figs. 346 and 347

Fig. 348 Apatite with regrowth margins in Shonkinite.
Ap = apatite
re-ap = regrowth margins of apatite
Shonkinite brecciated. Katzenbuckel, Odenwald, Germany.
With crossed nicols.

Fig. 349 Olivine phenocrysts in a groundmass clearly exhibiting the characteristics of a volcanic groundmass.
ol = olivine
pl = plagioclase laths
ore-minerals (black)
Theralite. Pald, Comitat Hont, Hungary.
Without crossed nicols.

Fig. 350 Oscillatory zoned augite phenocryst exhibiting hourglass structure; plagioclase phenocrysts, early olivine crystallizations in magmatogenic groundmass consisting mainly of plagioclase laths.
h-a = augite exhibiting hourglass structure
pl = plagioclase phenocrysts
ol = olivine
g = groundmass
Theralite. Pald Comitat, Hont, Hungary.
With crossed nicols.

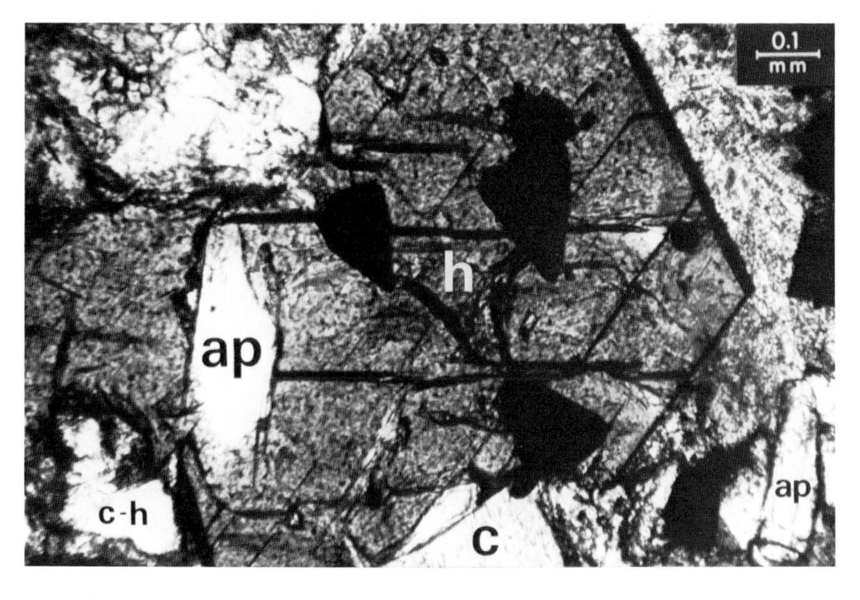

Fig. 351 Idioblastic hornblende crystalloblast in a groundmass of feldspar and calcite. The hornblende crystalloblast encloses ore-minerals, and calcite. Also a prismatic apatite is present.
h = hornblende crystalloblast
c-h = calcite in the hornblende
c = calcite outside the hornblende
ore-minerals (magnetite)
ap = apatite (prismatic)
Theralite. Söhle, Northern Moravia Highlands, CSSR.
With crossed nicols.

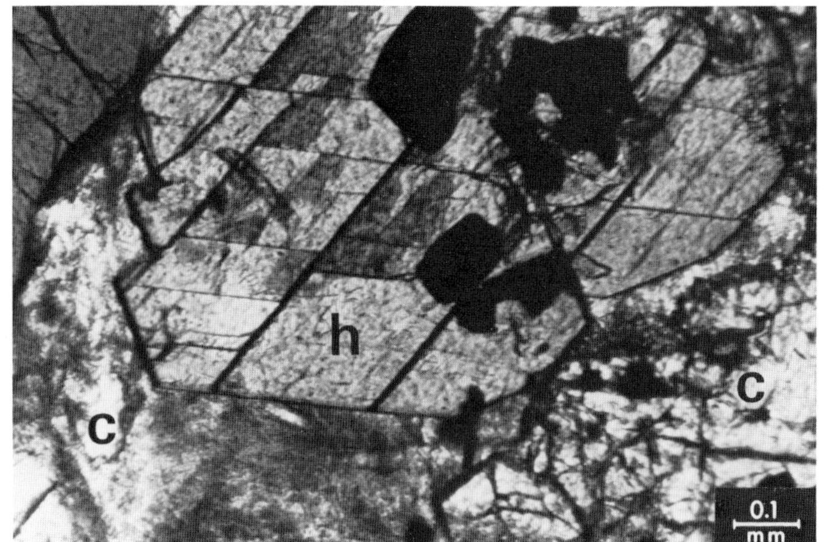

Fig. 352 An idioblastic hornblende exhibiting a perfect hornblende cleavage, and including ore-minerals (black).
h = hornblende
c = calcite of the groundmass
ore-minerals (black)
Theralite. Söhle, Northern Moravia Highlands, CSSR.
With crossed nicols.

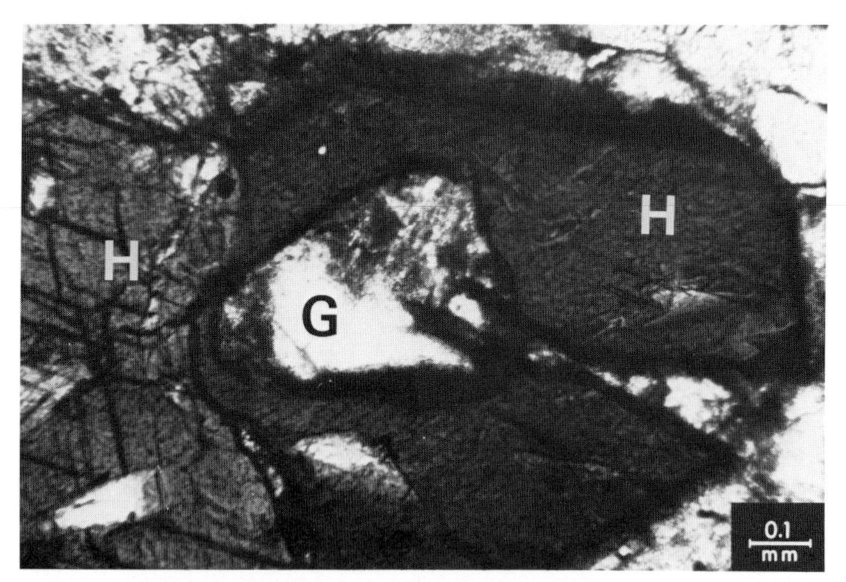

Fig. 353 Amphibole crystalloblast engulfing theralitic groundmass.
H = hornblende crystalloblast
G = theralitic groundmass partly enclosed by the amphibole.
Theralite. Söhle, Northern Moravia Highlands, CSSR.
With crossed nicols.

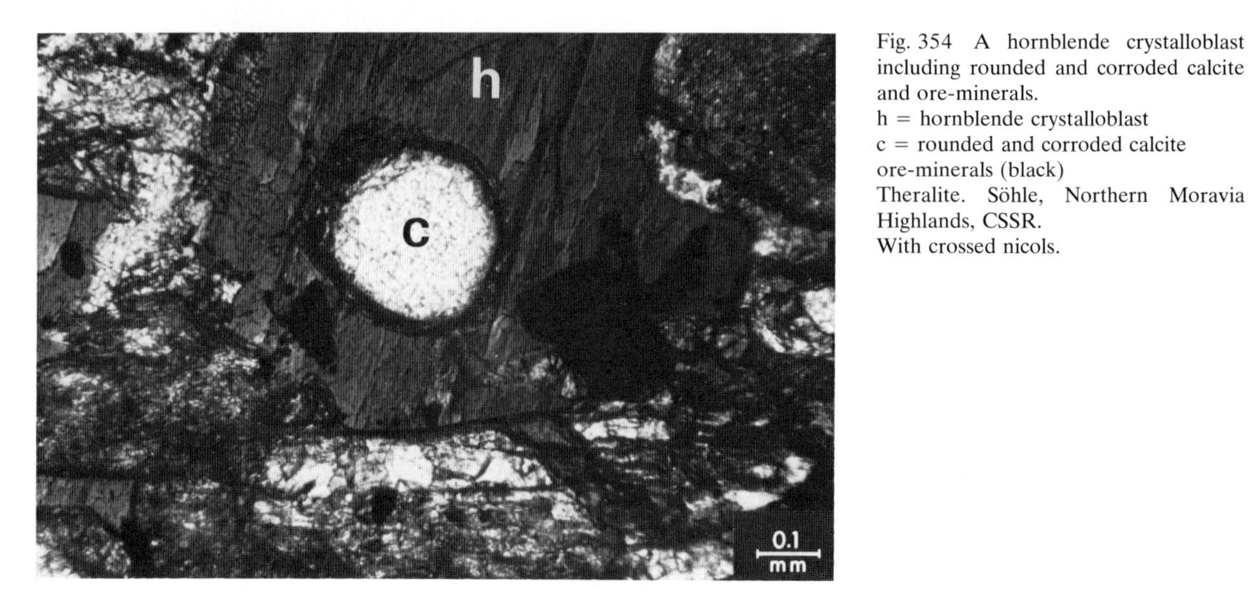

Fig. 354 A hornblende crystalloblast including rounded and corroded calcite and ore-minerals.
h = hornblende crystalloblast
c = rounded and corroded calcite
ore-minerals (black)
Theralite. Söhle, Northern Moravia Highlands, CSSR.
With crossed nicols.

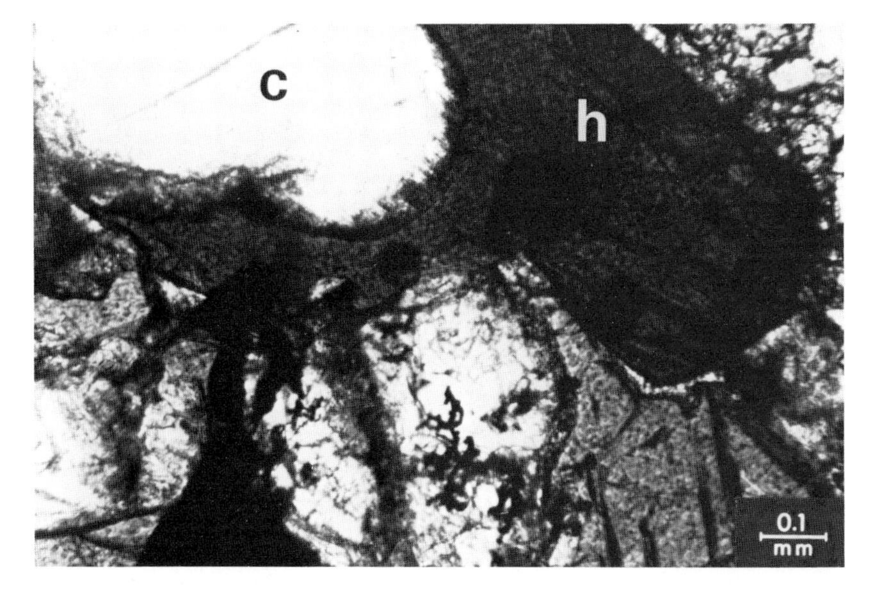

Fig. 355 Hornblende crystalloblast with rounded and corroded calcite. Also ore-minerals (black) are included in the hornblende.
c = calcite rounded, corroded and included in the hornblende.
h = hornblende crystalloblast
Theralite. Söhle, Northern Moravia Highlands, CSSR.
With crossed nicols.

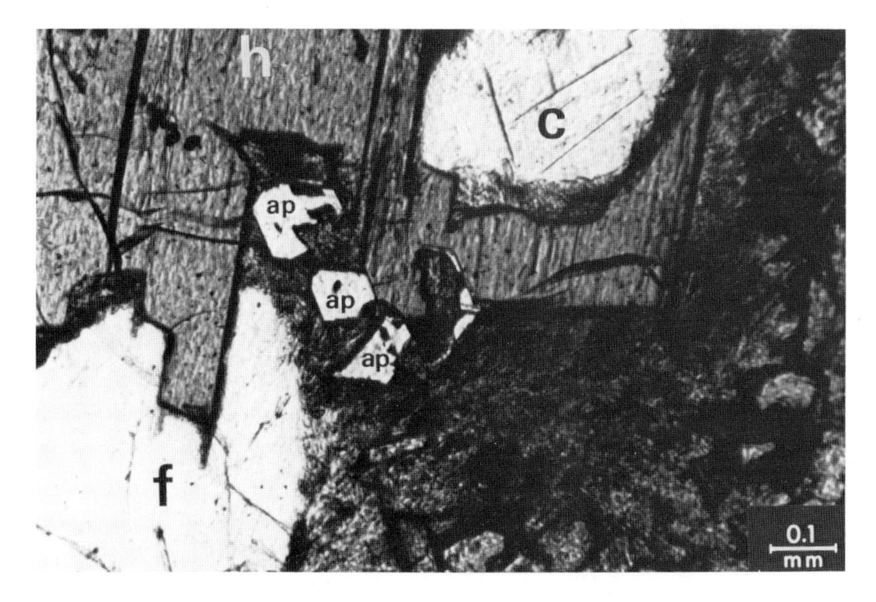

Fig. 356 A hornblende idioblast including rounded and corroded calcite. Also idioblastic apatites are present.
h = hornblende idioblast
c = calcite, rounded and included in the blastic hornblende.
ap = apatite idioblasts
f = feldspar
Theralite. Söhle, Northern Moravia Highlands, CSSR.
With crossed nicols.

Fig. 357 Pyroxenes included in crystalloblastic hornblende with later crystalloblastic prismatic apatite transversing the amphibole.
py = pyroxene
h = hornblende crystalloblasts
ap = apatite crystalloblast
Theralite. Söhle, Northern Moravia Highlands, CSSR.
With crossed nicols.

Fig. 358 Whereas in most of the studied cases, the hornblende crystalloblastesis is later in the sequence of crystalloblastic growth than the pyroxene (see Fig. 357), in the present case the pyroxene seems to extend into the hornblende. In reality though the pyroxene represents a relic part engulfed by the later amphibole growth, hornblende crystalloblast.
py = pyroxene part outside the hornblende
py-a = pyroxene part included in the hornblende
c = calcite of the groundmass
Theralite. Söhle, Northern Moravia Highlands, CSSR.
With crossed nicols.

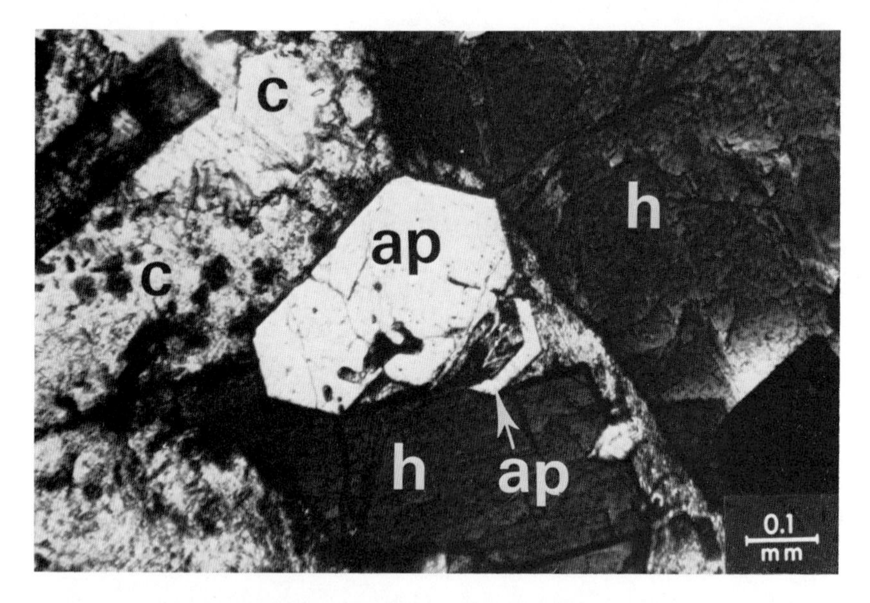

Fig. 359 Hornblende and apatite crystalloblasts.
h = hornblende crystalloblasts
ap = apatite crystalloblast
c = groundmass containing calcite
Theralite. Söhle, Northern Moravia Highlands, CSSR.
With crossed nicols.

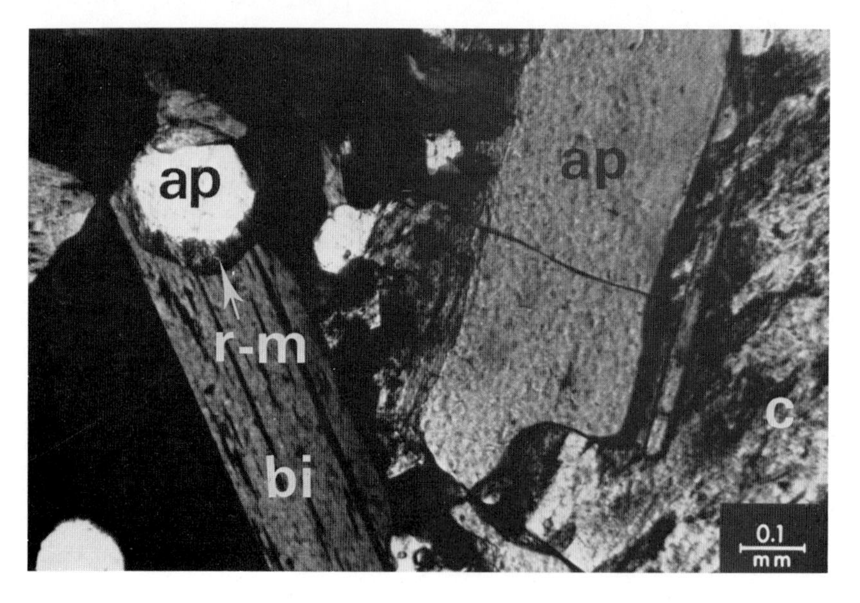

Fig. 360 Apatite rounded and included in the theralitic groundmass. Also idiomorphic apatite included in biotite with a reaction margin.
ap = apatite
bi = biotite
r-m = reaction margin between the idiomorphic apatite and the surrounding biotite.
c = calcite groundmass
Theralite. Picota, Serra de Monchique, South Portugal.
With crossed nicols.

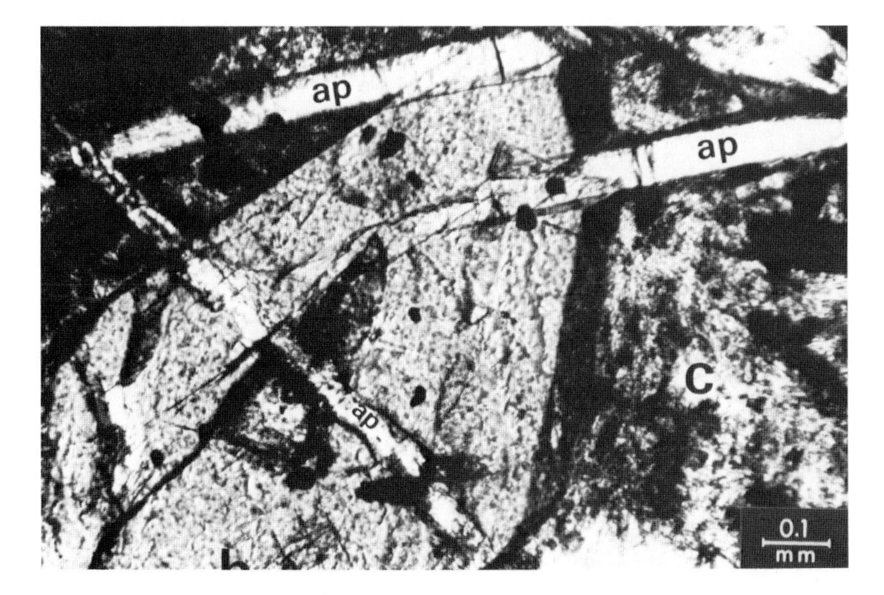

Fig. 361 Idioblastic hornblende in groundmass also containing calcite with later apatite crystalloblasts.
h = hornblende crystalloblast
ap = apatite crystalloblasts
c = calcite in the groundmass
Theralite. Söhle, Northern Moravia Highlands, CSSR.
With crossed nicols.

Fig. 362 Idioblastic hornblende crystalloblasts including calcite which is also a constituent of the groundmass. An idioblastic apatite is also indicated.
h = hornblende crystalloblast
c = calcite included in the amphibole
c-g = calcite in the groundmass
ap = apatite crystalloblast
Theralite. Söhle, Northern Moravia Highlands, CSSR.
With crossed nicols.

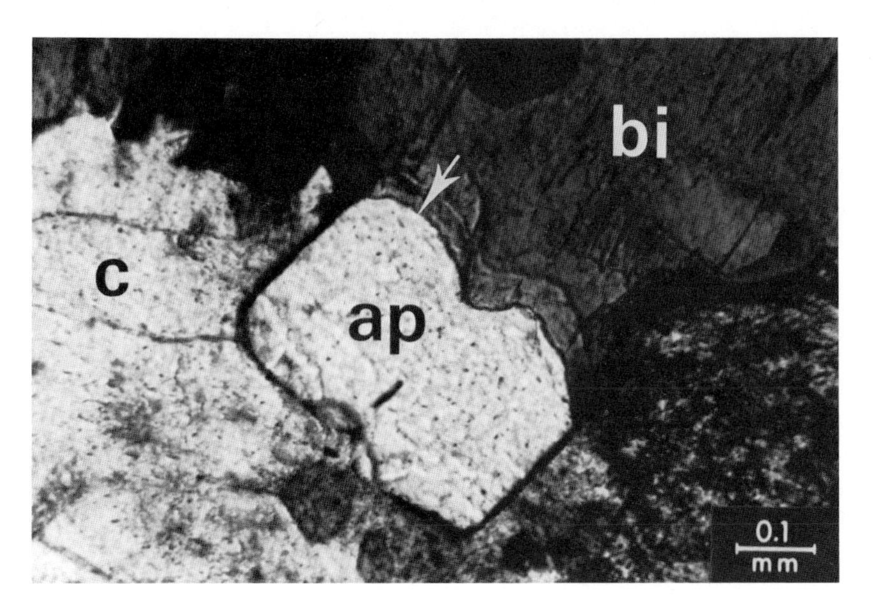

Fig. 363 Apatite crystalloblast corroded and surrounded by biotite.
bi = biotite (arrow shows corroded outline of apatite)
ap = apatite
c = calcite in the groundmass of the theralite.
Theralite. Picota, Serra de Monchique, South Portugal.
With crossed nicols.

Fig. 364 Well developed calcite crystals surrounded by the theralitic groundmass (also partly composed by calcite) and hornblende.
Ca = well developed calcite crystals
H = hornblende
G = groundmass
Theralite. Söhle, Northern Moravia Highlands, CSSR.
With crossed nicols.

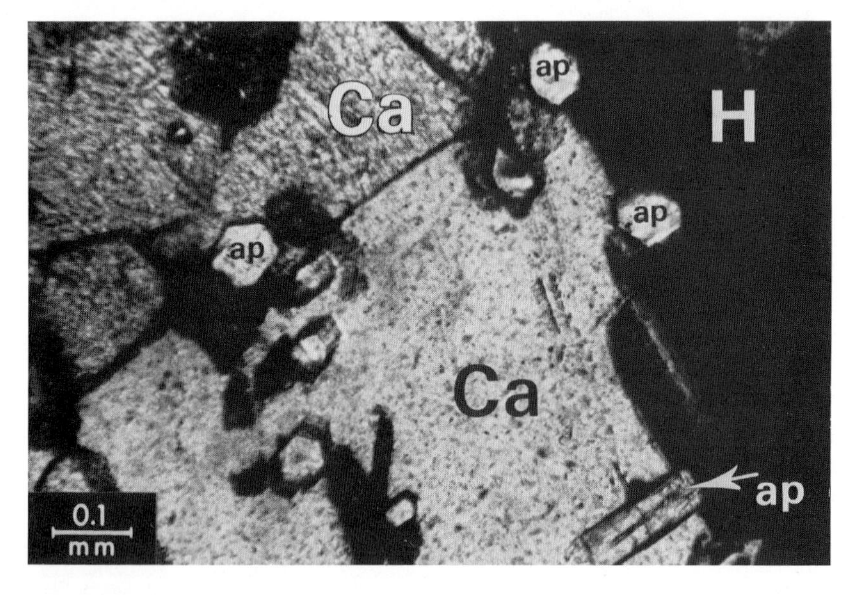

Fig. 365 Well developed calcite with idiomorphic apatite. Hornblende is also shown.
Ca = calcite
H = hornblende
ap = apatite
Theralite. Söhle, Northern Moravia Highlands, CSSR.
With crossed nicols.

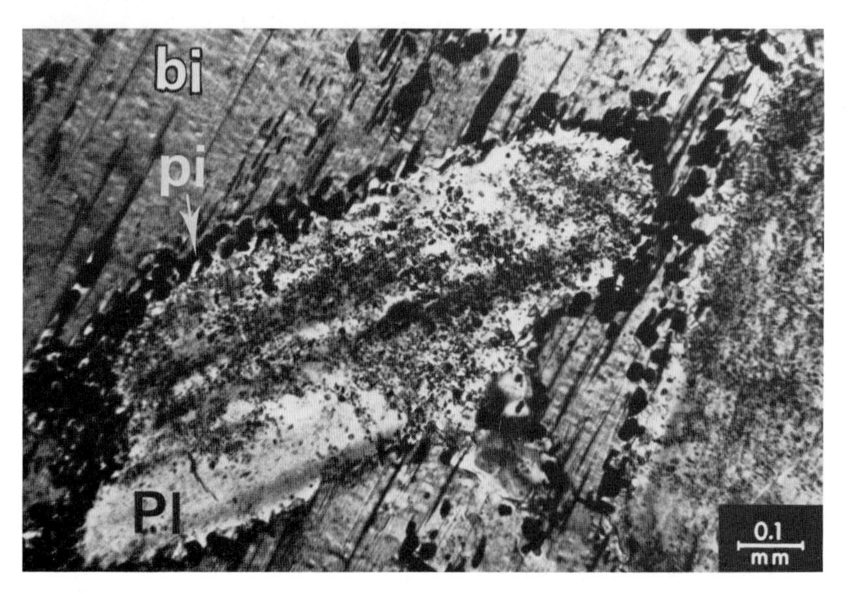

Fig. 366 Rounded and corroded plagioclase included in biotite.
Pl = plagioclase
bi = biotite
pi = ore pigments
Theralite. Alceria, Serra de Monchique, South Portugal.
With crossed nicols.

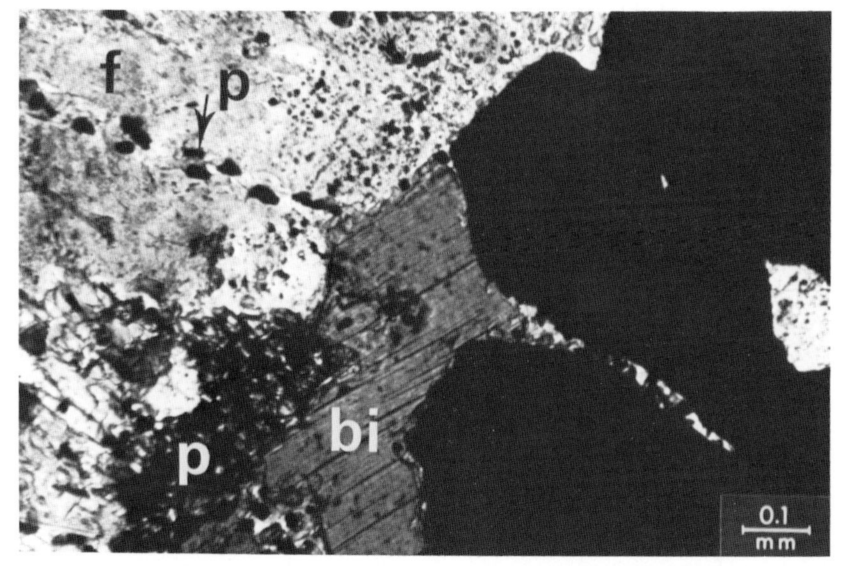

Fig. 367 Magnetite corroded and invaded by later biotite crystalloblasts. Feldspar is also indicated
magnetite (black)
bi = biotite crystalloblasts
f = feldspar
p = ore pigments
Theralite. Alceria, Serra de Monchique, South Portugal.
With crossed nicols.

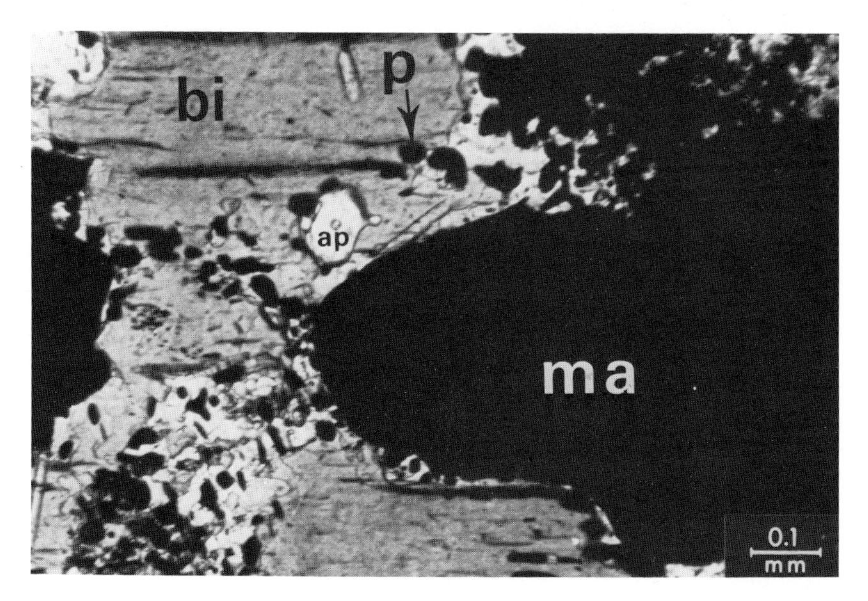

Fig. 368 Rounded magnetite (black) surrounded by blastic biotite also crystalloblastic apatite is present.
ma = magnetite
bi = biotite
ap = crystalloblastic apatite
p = ore-pigments
Theralite. Alceria, Serra de Monchique, South Portugal.
With crossed nicols.

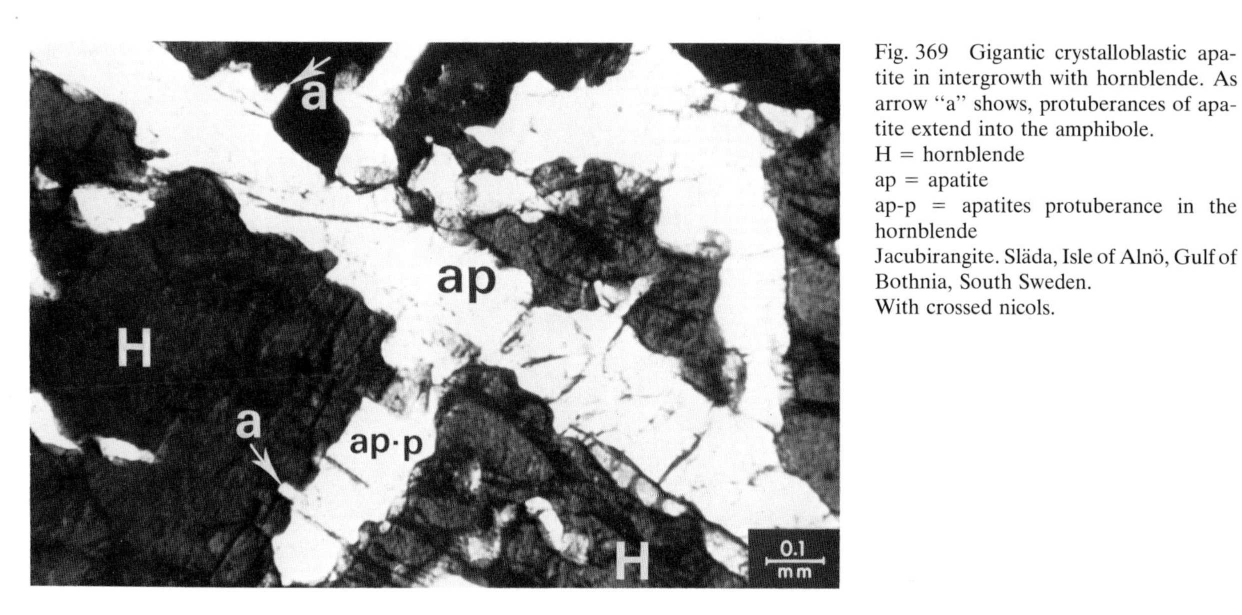

Fig. 369 Gigantic crystalloblastic apatite in intergrowth with hornblende. As arrow "a" shows, protuberances of apatite extend into the amphibole.
H = hornblende
ap = apatite
ap-p = apatites protuberance in the hornblende
Jacubirangite. Släda, Isle of Alnö, Gulf of Bothnia, South Sweden.
With crossed nicols.

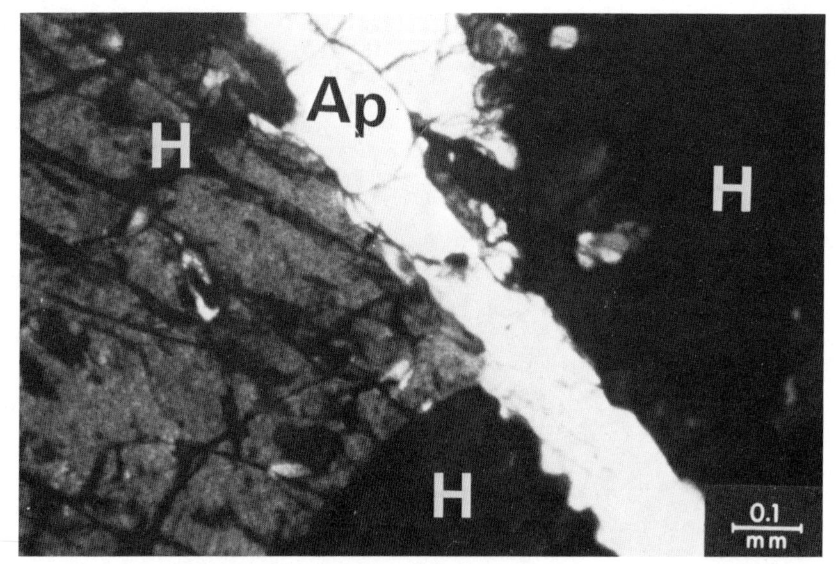

Fig. 370 Crystalloblastic apatite transversing two differently orientated hornblendes.
Ap = apatite
H = hornblende
Jacubirangite. Släda, Isle of Alnö, Gulf of Bothnia, South Sweden.
With crossed nicols.

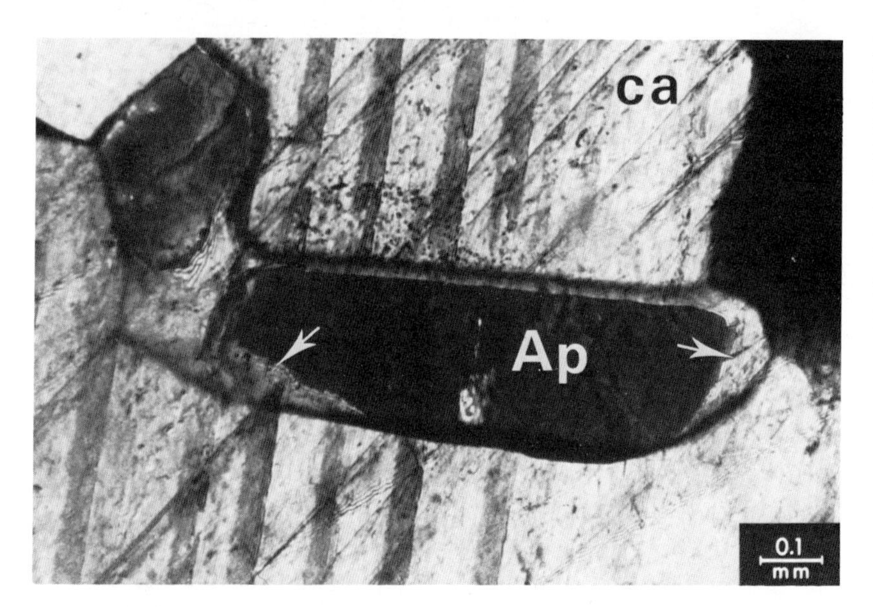

Fig. 371 Crystalloblastic apatite surrounded by crystalloblastic calcite. The apatite is corroded and replaced by the calcite.
ca = calcite with typical calcite twinning.
Ap = apatite
Arrows show corroded outline of apatite with replacement by calcite.
Jacubirangite. Släda, Isle of Alnö, Gulf of Bothnia, South Sweden.
With crossed nicols.

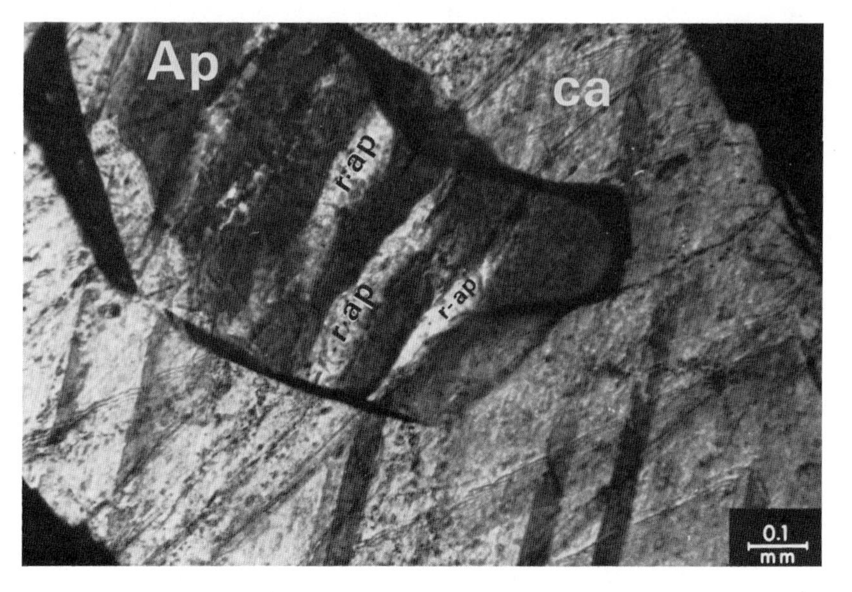

Fig. 372 Crystalloblastic apatite surrounded by later crystalloblastic calcite. The calcite replaces the apatite, producing an internal pattern of apatite replacement by calcite.
ca = calcite
Ap = apatite
r-ap = replaced apatite by calcite
Jacubirangite. Släda, Isle of Alnö, Gulf of Bothnia, South Sweden.
With crossed nicols.

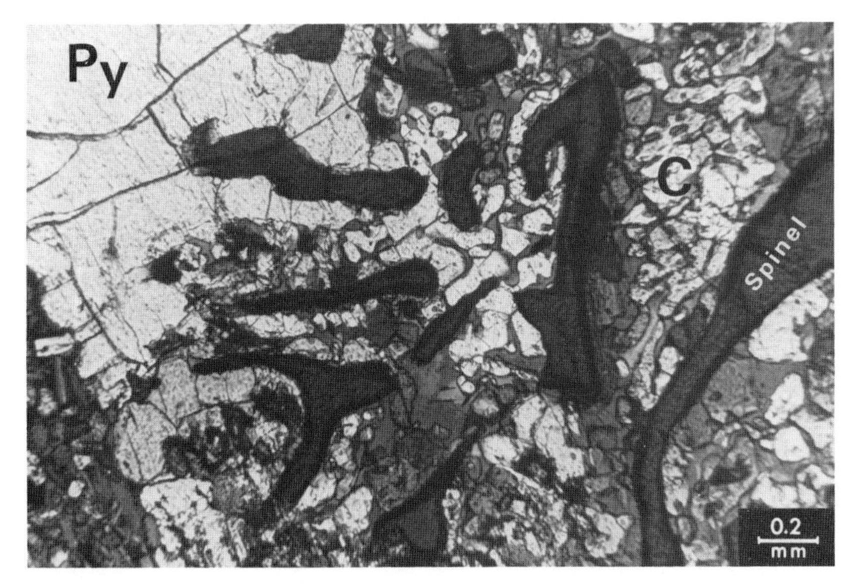

Fig. 373 Spinel in graphic intergrowth with bronzite and forsterite. The spinel represents crystal skeletons. Often a "crushed zone" is present between the spinel and the forsterite or pyroxene.
Py = pyroxene (bronzite)
C = "crushed zone"
Olivine bombs in basalt. Jato, Lekempti, W. Ethiopia.
With crossed nicols.

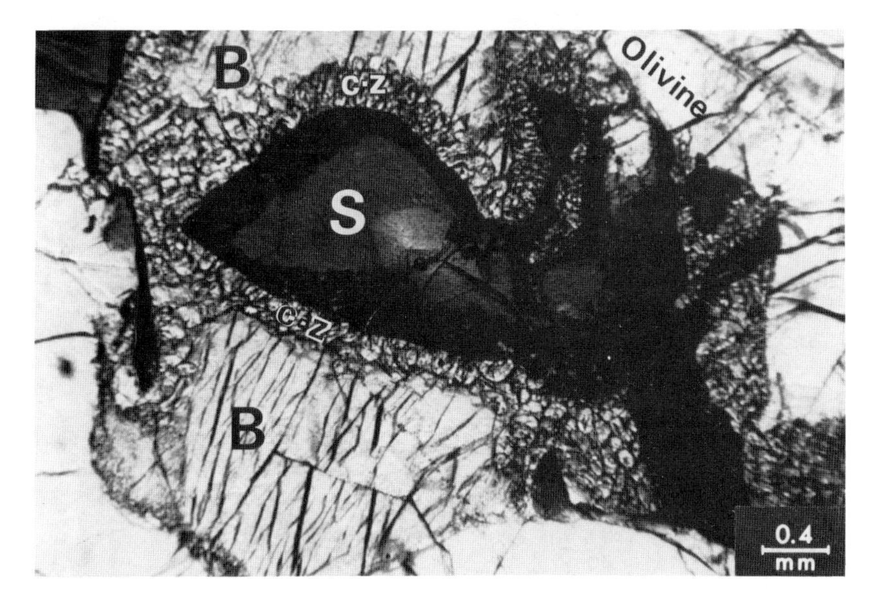

Fig. 374 Spinel with an oxidized zone surrounded by a crushed zone formed due to mantle deformation. Also adjacent pyroxene (bronzite) with a cleavage pattern due to mantle deformation.
S = spinel
c-z = crushed zone
B = bronzite with cleavage pattern due to mantle deformation.
Olivine bomb (mantle fragment) in basalt.
Jato, Lekempti, W. Ethiopia.
Without crossed nicols.

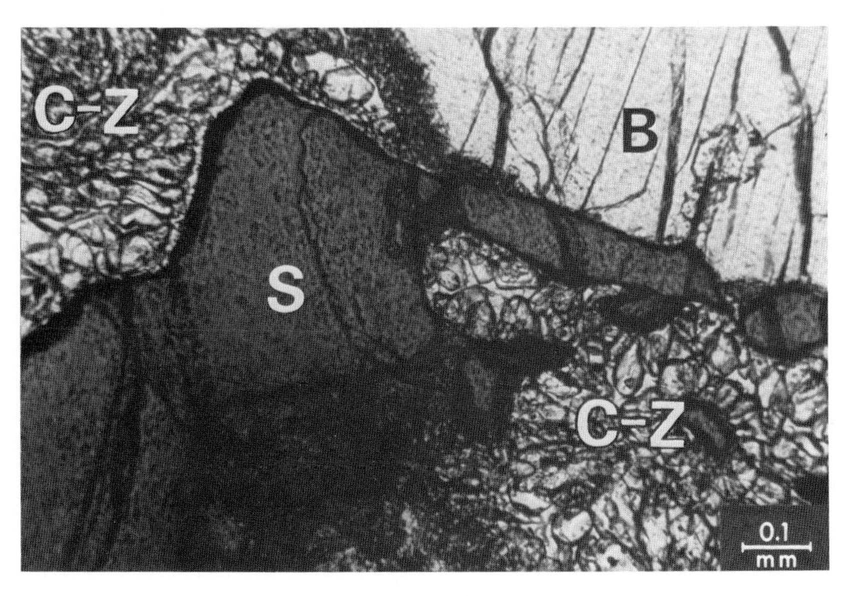

Fig. 375 Shows spinel in contact with bronzite and surrounded by a crushed zone.
B = bronzite
S = spinel
c-z = crushed zone
Olivine bombs in basalt. Jato, Lekempti, W. Ethiopia.
With crossed nicols.

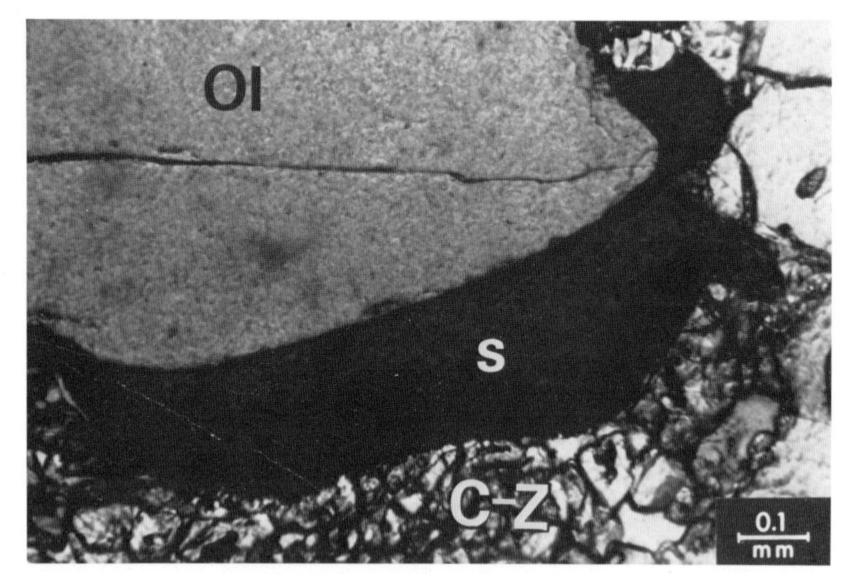

Fig. 376 Spinel (isotropic) in contact with adjacent olivine and with the crushed zone.
ol = olivine
s = spinel (chrome spinel)
c-z = crushed zone
Olivine bomb in basalt (mantle fragment in basalt)
Lekempti, W. Ethiopia.

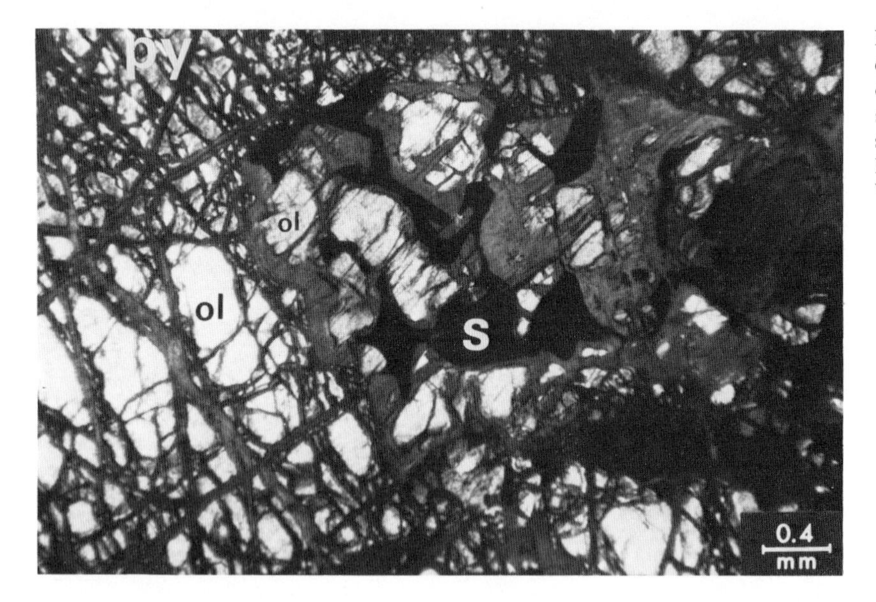

Fig. 378a, b Spinel (graphic in form) in contact with pyroxene and olivine.
ol = olivine
s =spinel
se = serpentine
Peridotite. Konitsa, N. Greece.
With crossed nicols.

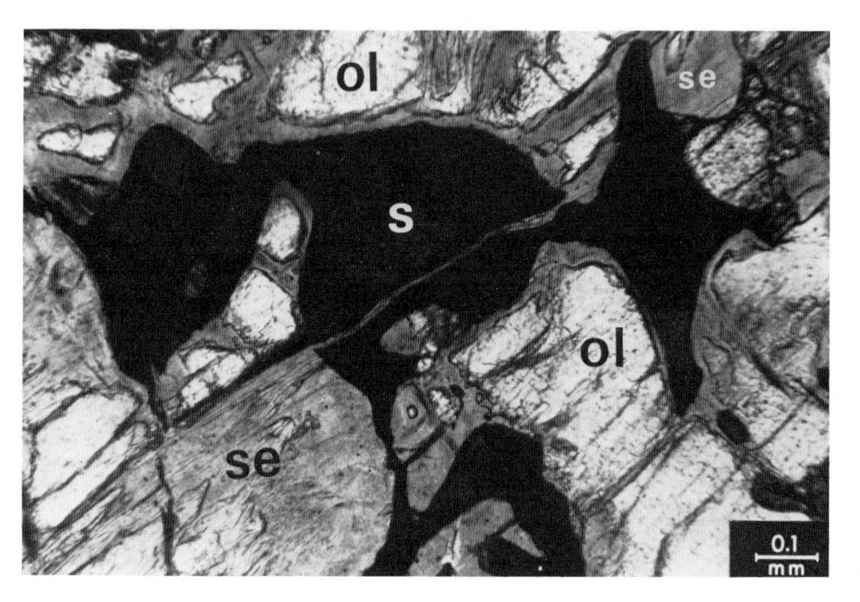

Fig. 387b

(For Fig. 387 see page 239)

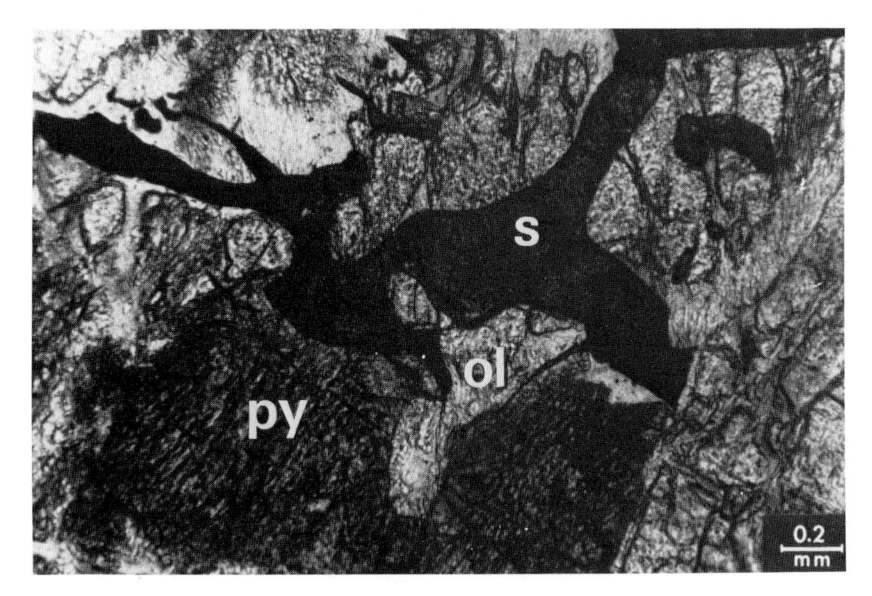

Fig. 377 Spinel, graphic in shape, in contact with pyroxene and olivine of the peridotite.
s = spinel
py = pyroxene
ol = olivine
Peridotite (mantle diapir in orogen) near Konitsa, N. Greece.
Peridotite. Konitsa, N. Greece.
With crossed nicols.

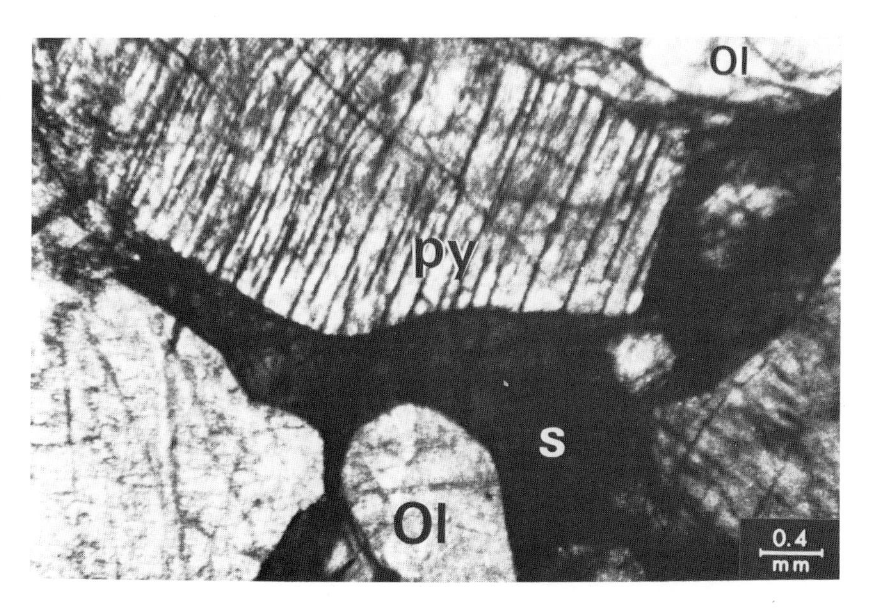

Fig. 379 Spinel graphic in form (actually a crystal skeleton) partly surrounding a twinned pyroxene and enclosing olivine.
s = spinel
py = pyroxene
Ol = olivine
Lherzolite. Castellamonte. Ivrea, Piemont, North Italy.
With crossed nicols.

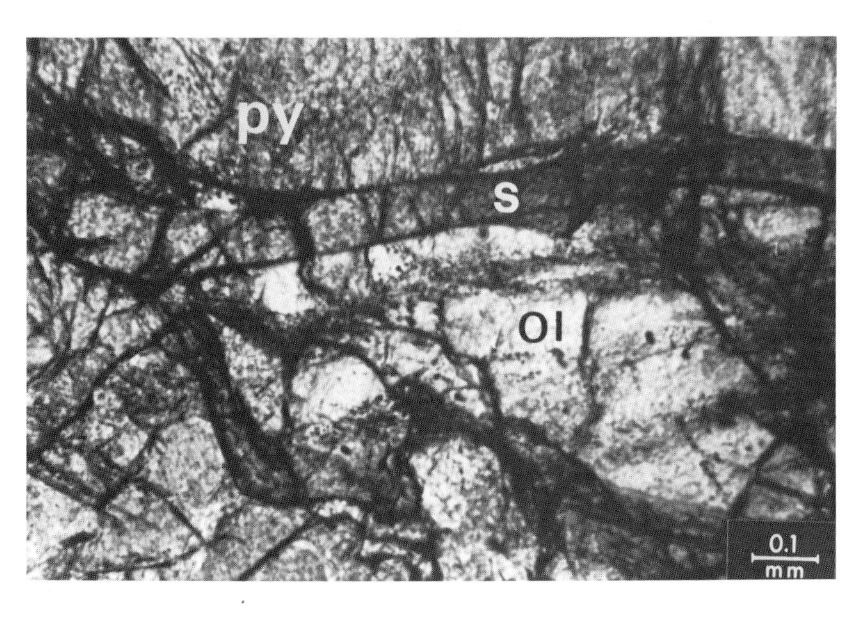

Fig. 380 Spinel (graphic in form) in intergrowth with olivine and pyroxene.
s = spinel
Ol = olivine
py = pyroxene
Lherzolite. Castellamonte. Ivrea, Piemont, North Italy.
With crossed nicols.

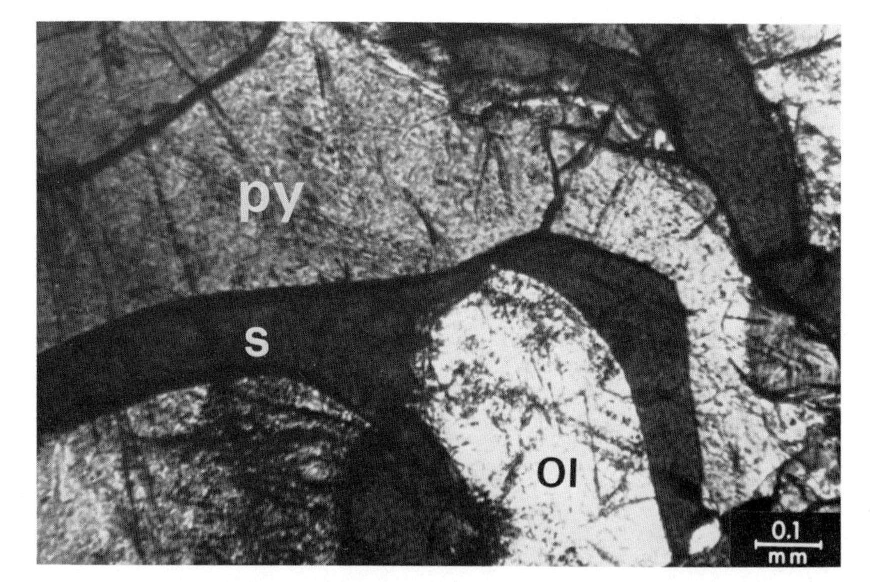

Fig. 381 Graphic spinel in intergrowth with olivine and pyroxene.
Ol = olivine
s = spinel
py = pyroxene
Lherzolite. Castellamonte. Ivrea, Piemont, North Italy.
With crossed nicols.

Figs. 383 and 384 Equigranular olivine in dunite.
Peridotite from Russia (USSR).
(Courtesy of the Gorni Institute, Leningrad).
With crossed nicols.

(For Fig. 382 see page 241)

Fig. 384

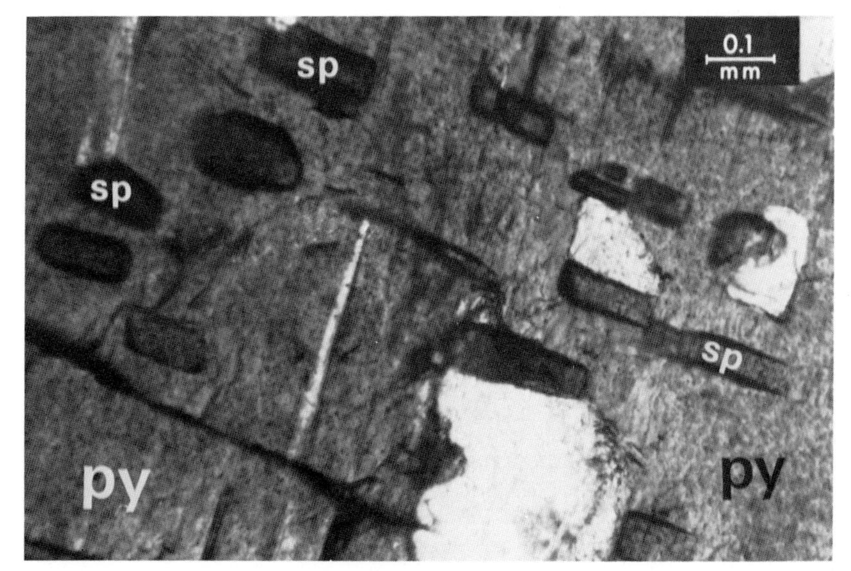

Fig. 382 Orientated spinels in pyroxene. Also calcite is present.
py = pyroxene
sp = spinels orientated in pyroxene
Dunite. Bettola. Ligurian Alps. Italy.
With crossed nicols.
(Similar intergrowths are observed in lherzolites.)

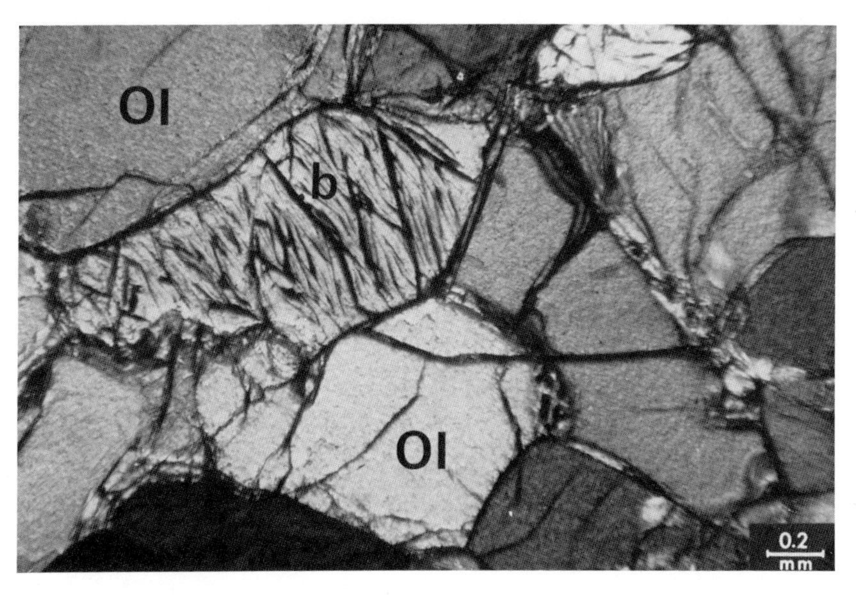

Fig. 385 Olivine and bronzite equigranular texture. The bronzite shows the development of a cleavage pattern, due to the overall pressure exercised by the overlaying crust on the mantle.
Ol = olivine
b = bronzite with cleavage pattern due to overall pressure.
Olivine bomb in basalt (mantle piece in basalt).
Lekempti, W. Ethiopia.
With crossed nicols.

Fig. 386 Mantle fragment in basalt, predominently consisting of equigranular olivine. Olivine bombs in basalt.
Eifel, Germany.
With crossed nicols.

Fig. 387 Almost radiating poly-twinning (deformation lamellae) in forsterite of the olivine bombs of the Canary Islands, developed due to sub-crustal tectonic influences and because of the great plastic deformation tendency of the olivine. Olivine bomb in basalt. Lanzarote, Canary Islands.
With crossed nicols.

Fig. 388 Olivine nodule (forsterite bomb in basalt), partly rounded and with a reaction margin with the basaltic groundmass. The forsterite shows an impressive and complex in orientation deformation-polysynthetic twinning. The forsterite nodules represent xenocrysts picked up by the basalt from the "dunite" which is partly covered by the olivine-nodules containing basalt.
T-ol = tectonically twinned (stressed) olivine nodule;
Black margin surrounding the olivine is a reaction of the olivine nodule and the basalt;
g = basaltic groundmass.
Olivine nodules containing basalt (mantle fragments in basalt).
Yubdo, Wollaga, W. Ethiopia.
With crossed nicols.

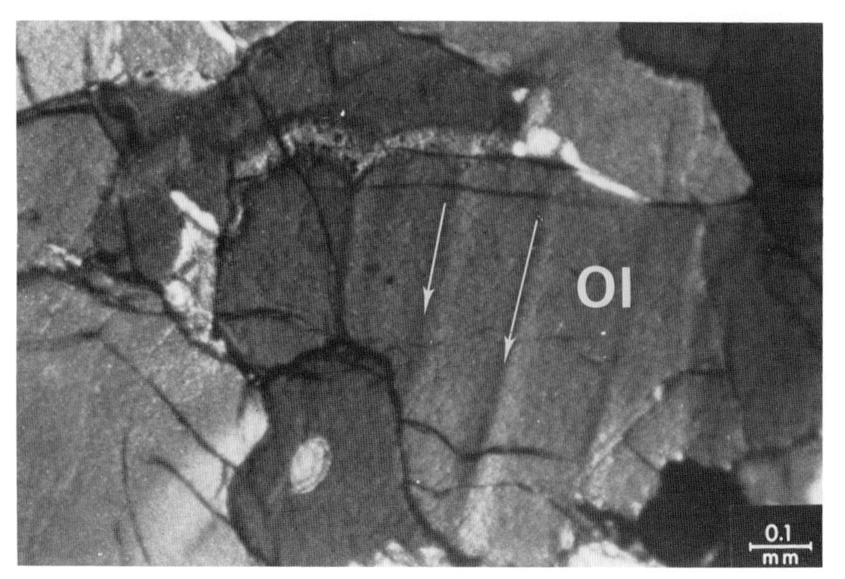

Fig. 389 Dunitic olivine with deformation lamellae (deformation twinning).
As is observed, deformation, twinning is exhibited by forsterite in olivine bombs in basalts which is attributed to mantle deformation. Commensurable is the tectonic deformation of olivine of dunitic "intrusions" which geologically do not exhibit post-intrusion deformation effects. However, in "intrusive" dunites that have not been affected by post-intrusive deformation, the tectonic deformation of the olivines may be synchronous or pre-intrusive.
Ol = olivine
Arrows show deformation lamellae.
Dunitic intrusion. Yubdo, Wollaga, Ethiopia.
With crossed nicols.

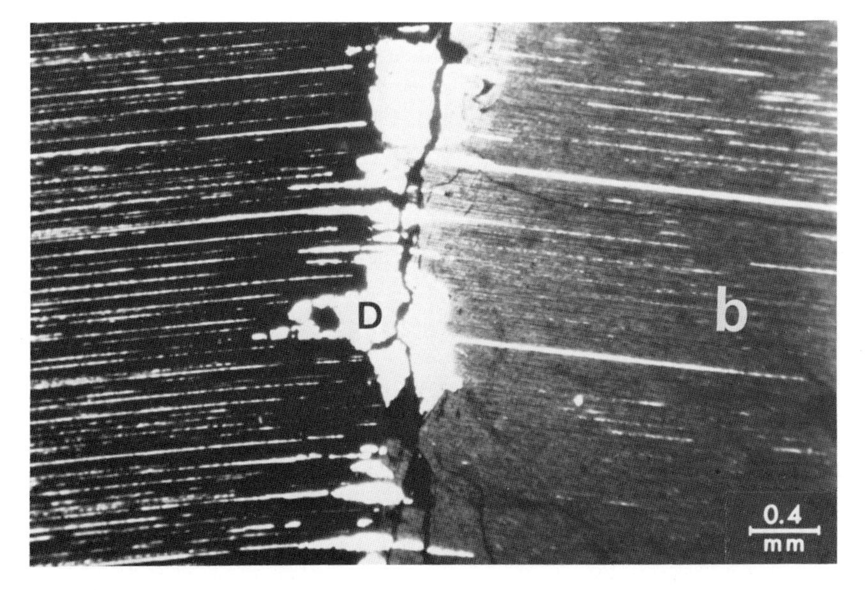

Fig. 390a and b Diopside intergranular between two bronzite crystal grains and with extension of the diopside following the bronzite's cleavage.
D = diopside
b = bronzite
Arrows show extension of the diopside following the cleavage of the bronzite.
Olivine bombs in basalt.
Lanzarote, Canary Islands.
With crossed nicols.
R. N. Thompson (Mineralogical Magazine Sept. 1978) ironically criticised the explanation (Augustithis 1978) that the diopside invades bronzite. Thompson insists that all such intergrowths (diopside/bronzite) are ex-solutions of diopside in bronzite.

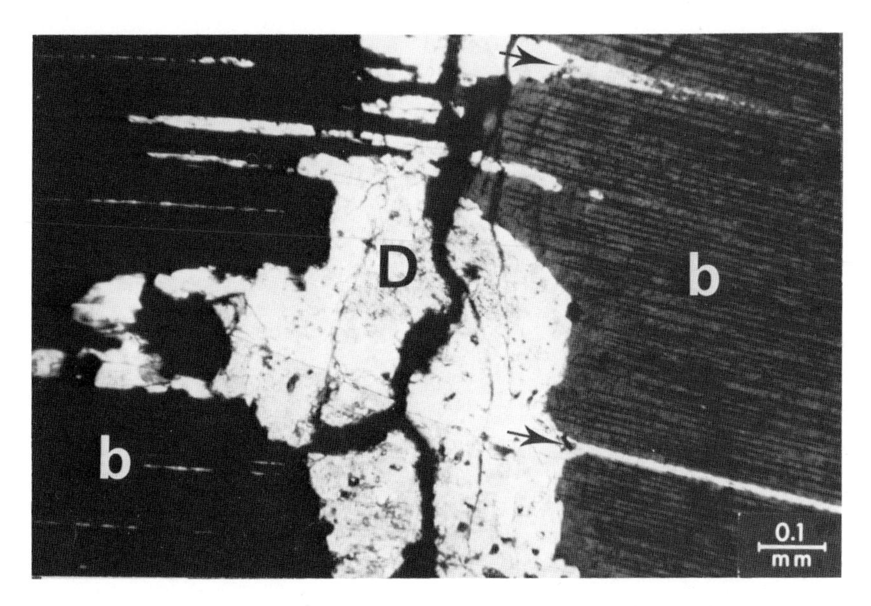

However, such an interpretation is contrary to the present observations in which case clearly the diopside occupies the intergranular between two differently orientated bronzites and sends prolongations invading the bronzites along the clearage.

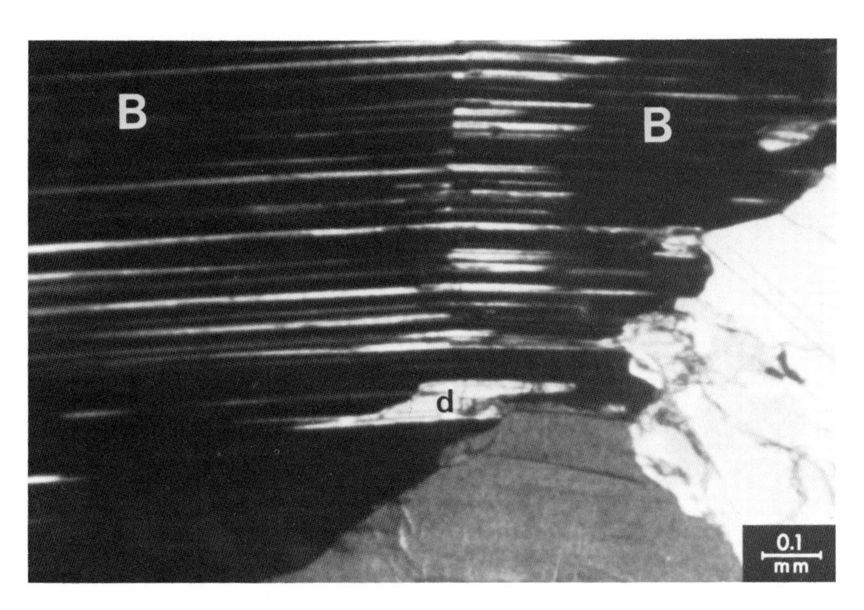

Fig. 391 Bronzite (in extinction position) with intergranular-interleptonic diopside, with extensions of it occupying the cleavage interleptonic spaces of the bronzite.
B = bronzite
d = diopside
Bronzitite. Jackson county, North Carolina, USA.
With crossed nicols.

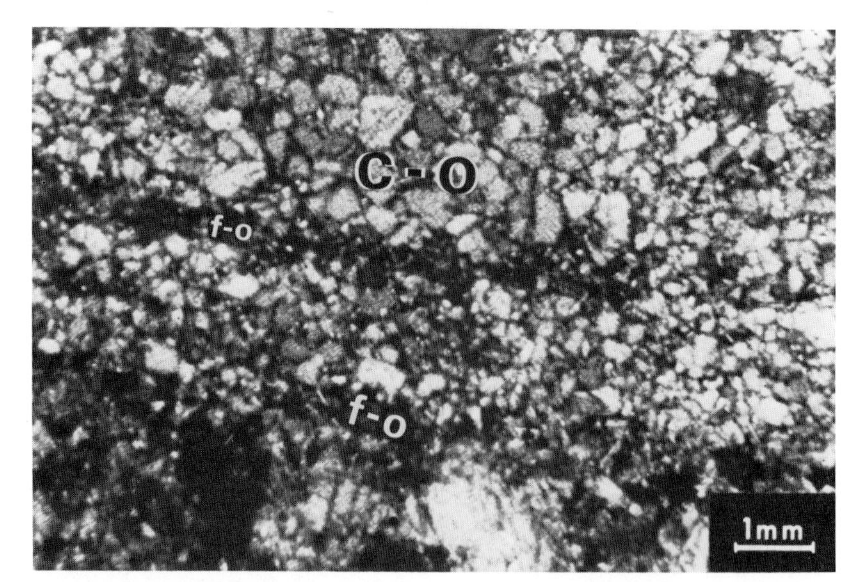

Fig. 392 Dunite exhibiting alternating coarse grained and fine grained olivine bands. As Fig. 393 shows, often tremolite crystalloblasts are associated with the fine grained olivine bands.
c-o = coarse grained olivine bands
f-o = fine grained olivine bands
Dunite. Corundum Hill, South California, USA.
With crossed nicols.

Fig. 394a, b Granular olivine, exhibiting undulating extinction and with tremolite crystalloblast occupying the olivine intergranular.
ol = olivine (granular olivine indicating undulating extinction due to deformation)
t = tremolite
Dunite. Rodberg, Norway.
With crossed nicols.

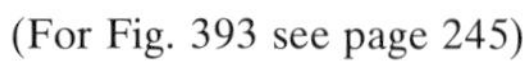
(For Fig. 393 see page 245)

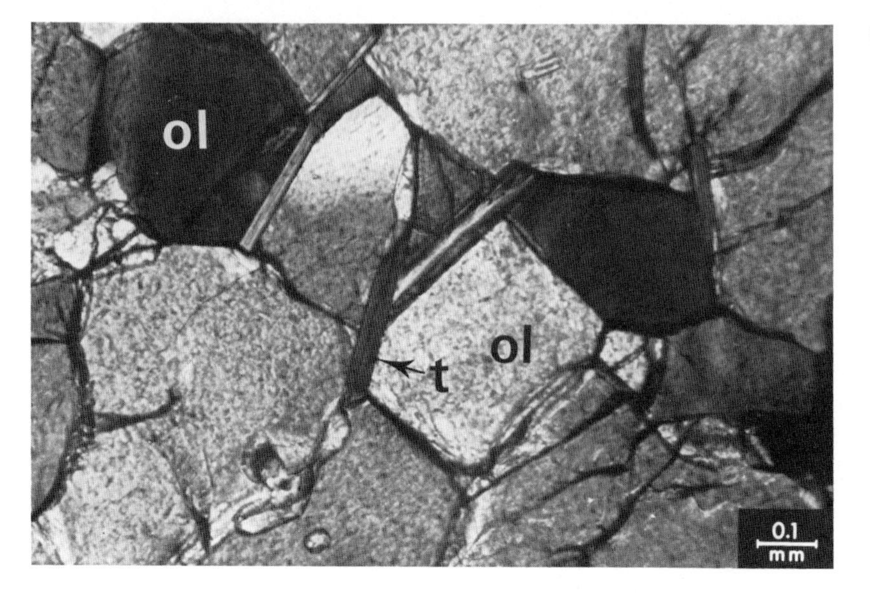

Fig. 394b

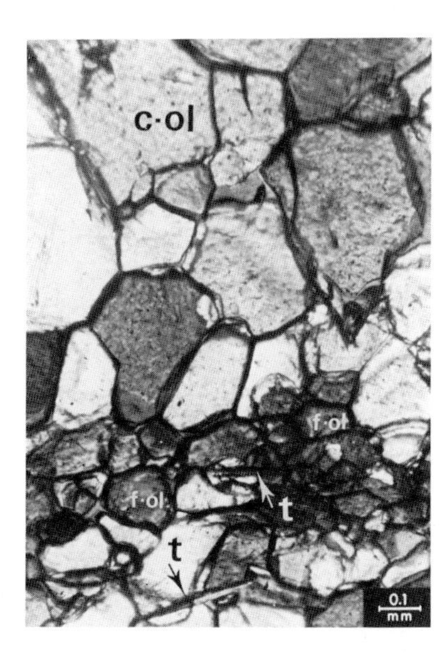

Fig. 393 Banded olivine, consisting of coarse grained olivine bands and bands of relatively finer grained olivines with tremolites often orientated parallel to the banding.
c-ol = coarse grained olivine
f-ol = fine grained olivine
t = tremolite orientated with their elongated direction parallel to the olivine banding.
Dunite. Corundum Hill, South Carolina, USA.
With crossed nicols.

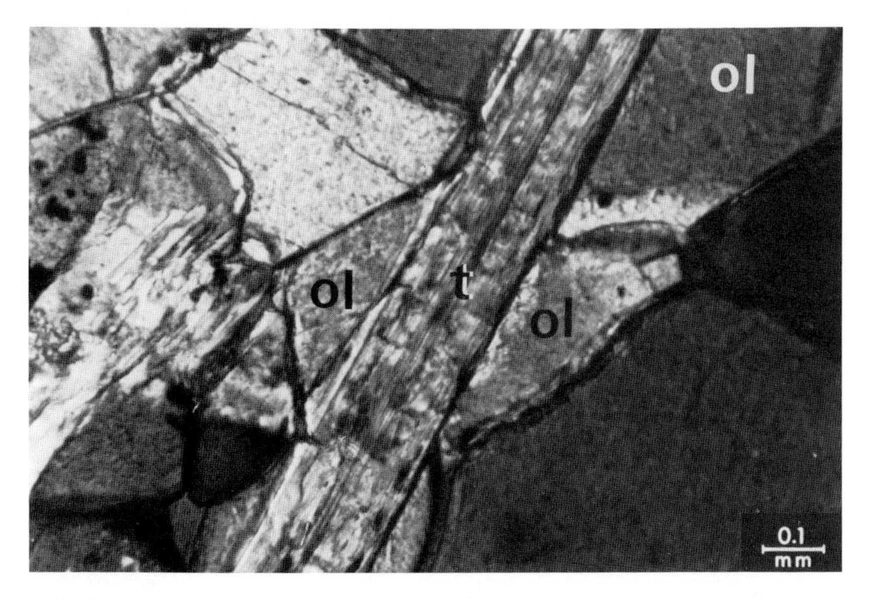

Fig. 395a Granular olivine with elongated tremolite crystalloblasts transversing the granular olivine texture.
ol = olivine
t = tremolite crystalloblasts
Dunite. Sommergraben near Kraubath, Styria, Austria.
With crossed nicols.

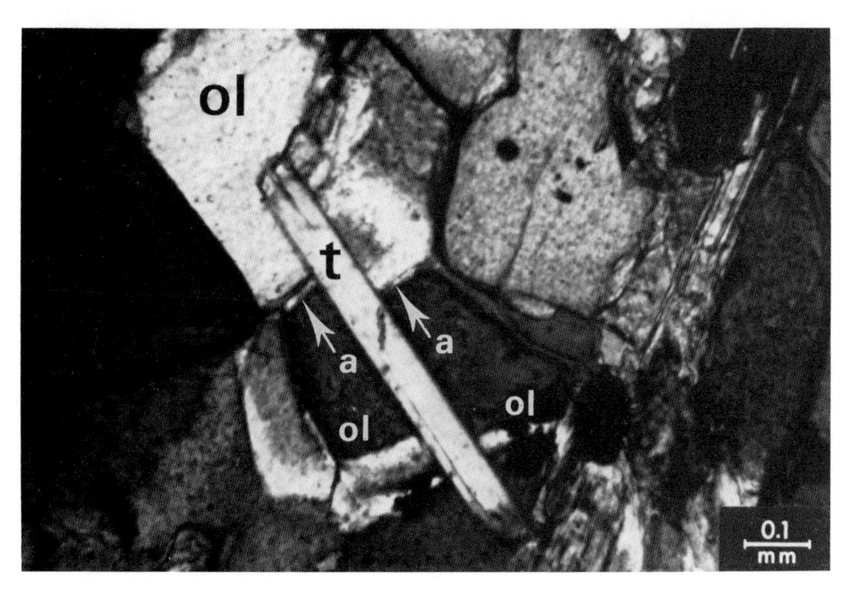

Fig. 395b Granular olivine with elongated tremolite crystalloblasts transversing the granular olivine texture.
Arrows "a" show that the olivine grain transversed by the tremolite nematoblast is displaced.
ol = olivine
t = tremolite crystalloblast
Dunite. Sommergraben near Kraubath, Styria, Austria.
With crossed nicols.

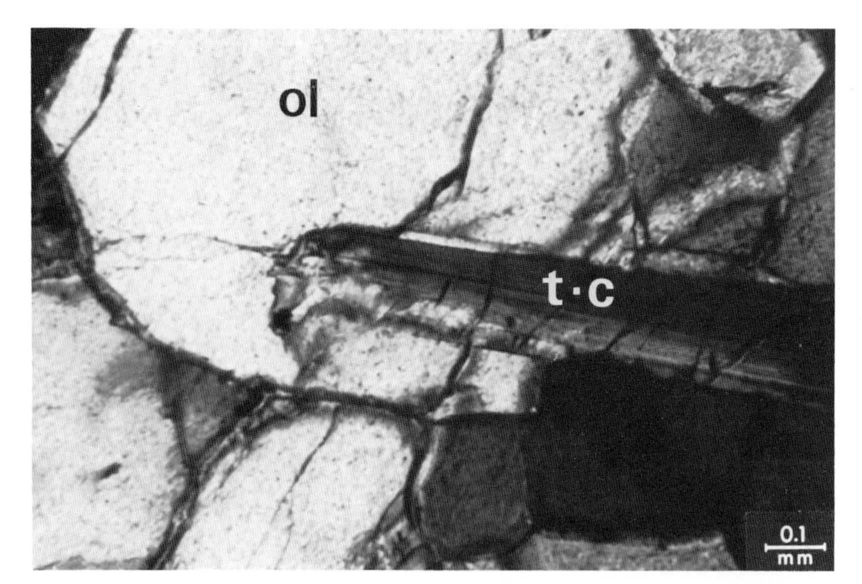

Fig. 396a Granular olivine with tremolite crystalloblast extending into the olivine.
ol = olivine (Granular)
t-c = tremolite crystalloblast
Dunite. Sommergraben near Kraubath, Styria, Austria.
With crossed nicols.

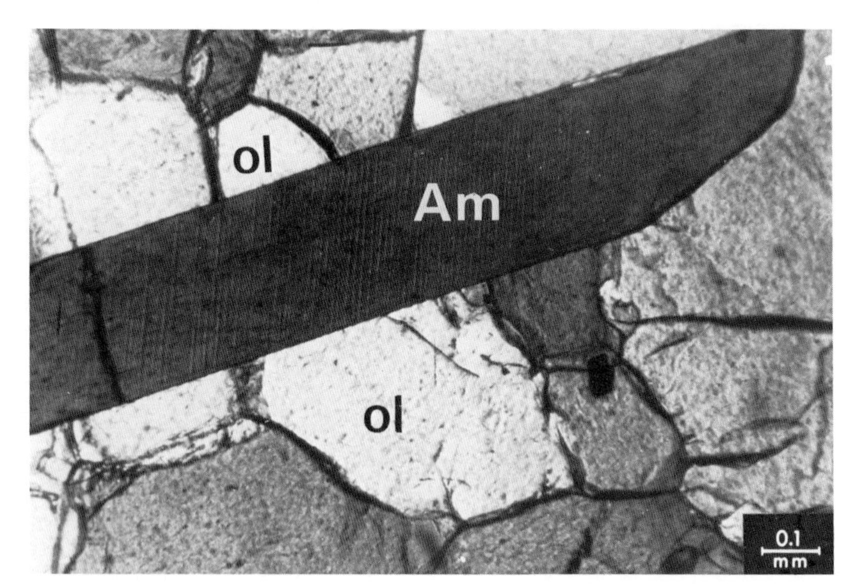

Fig. 396b Granular olivine with amphibole crystalloblast transversing the granular olivine texture.
ol = olivine
Am = amphibole crystalloblast with a lamination
Perpendicular to the crystal elongation.
Dunite. Rödberg, Norway.
With crossed nicols.

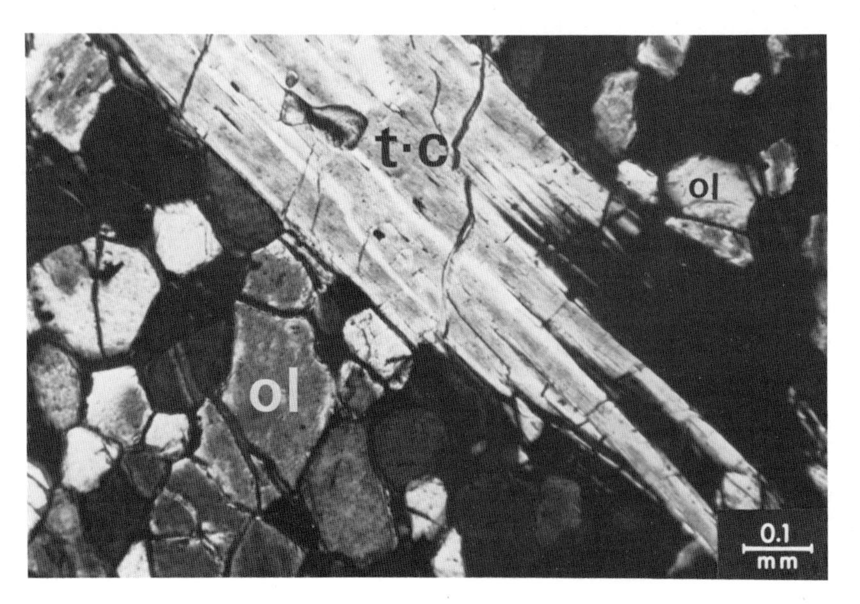

Fig. 397 Olivinefels texture transversed by tremolite crystalloblast.
ol = olivine of the olivinefels.
t-c = tremolite crystalloblast
Olivinefels. Prata, S. Chiavenna, Italy.
With crossed nicols.

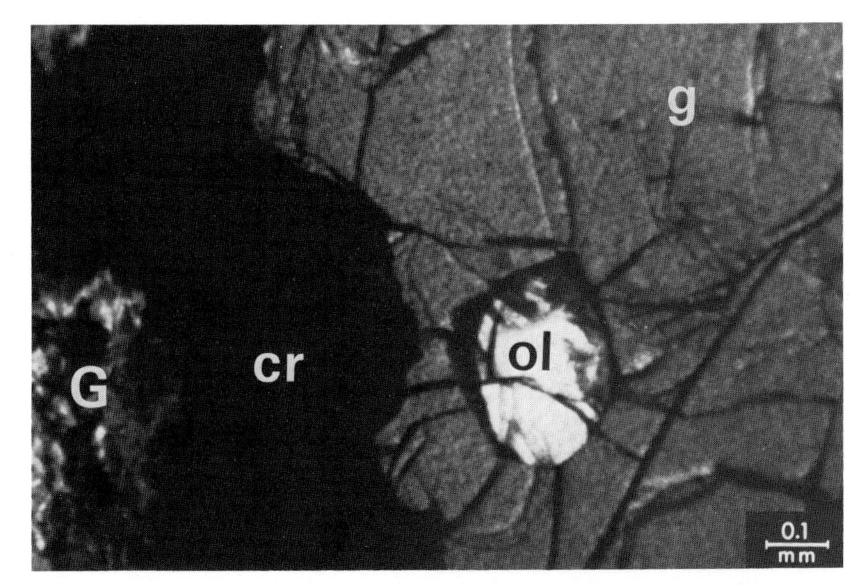

Fig. 398 Crystalloblastic garnet enclosing olivine in a garnet-peridotite. Also a corona reaction margin is formed between the garnet and the peridotite.
g = garnet
ol = olivine
cr = corona structure
G = "groundmass" of the peridotite
Garnet-Peridotite (Gordunite). Gorduno near Bellinzona, Tessin, Switzerland.
With crossed nicols.

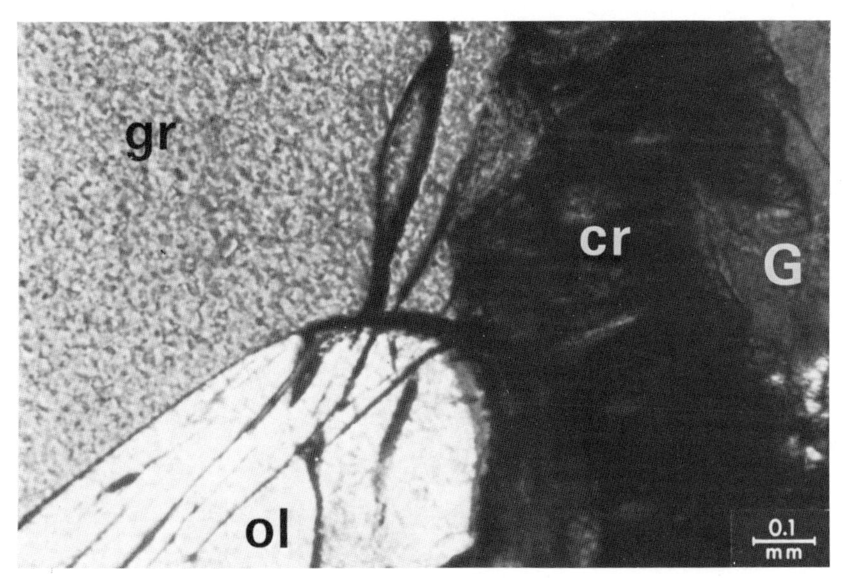

Fig. 399 Garnet and olivine both adjacent to each other with a common corona structure toward the peridotitic "groundmass".
gr = garnet
cr = corona structure
G = Peridotitic "groundmass"
ol = olivine
Garnet-Peridotite (Gordunite)
Gorduno near Bellinzona, Tessin, Switzerland.
Without crossed nicols.

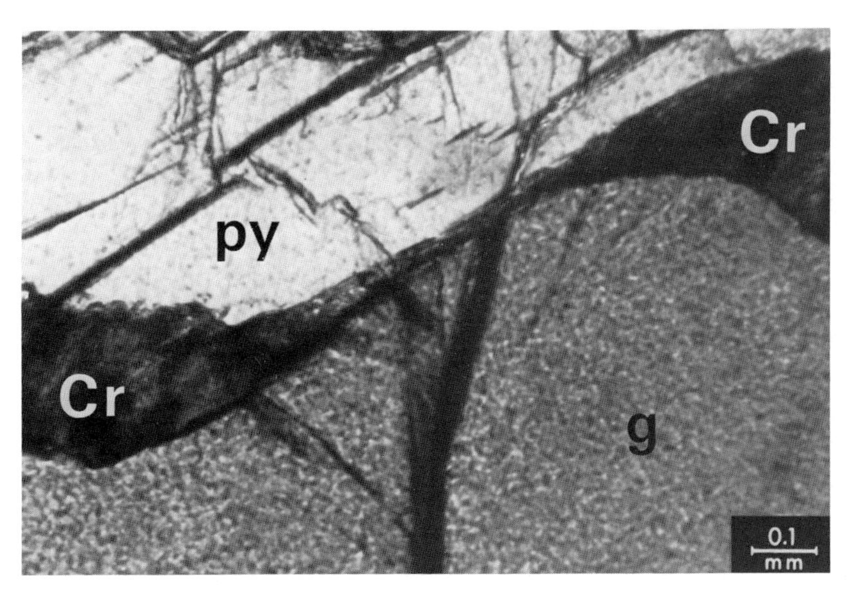

Fig. 400 Garnet crystalloblast with corona structure, which thins out between the garnet and the pyroxene of the garnet-peridotite.
g = garnet
Cr = corona structure
py = pyroxene
Garnet-Peridotite (Gordunite). Gorduno near Bellinzona, Tessin, Switzerland.
Without crossed nicols.

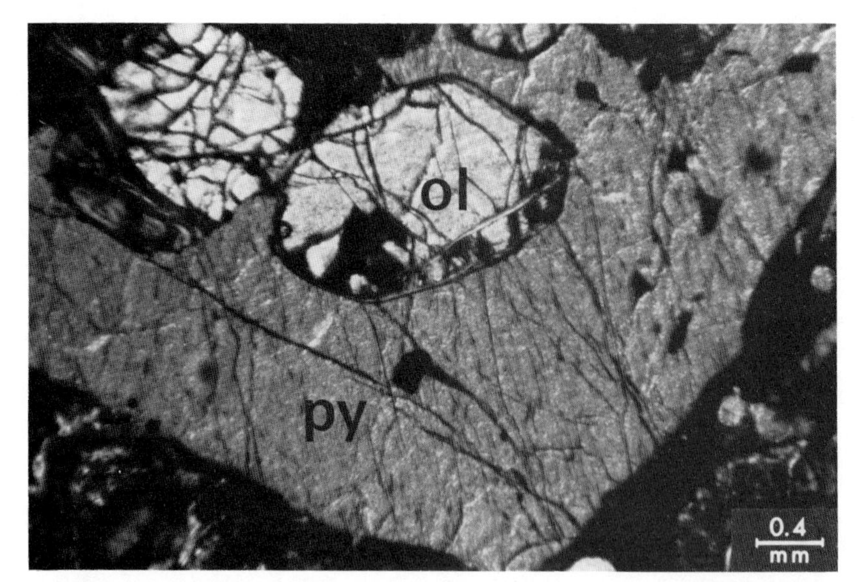

Fig. 401 Olivines enclosed by a later pyroxene poikiloblast.
ol = olivine
py = pyroxene poikiloblast
Dunitic peridotite from Russia (Courtesy of the Gorni Institute, Leningrad).
With crossed nicols.

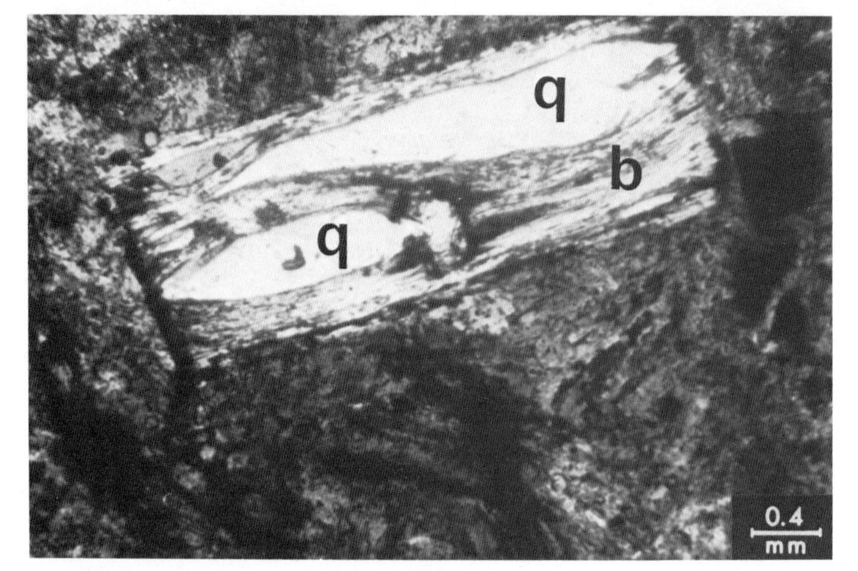

Fig. 402 Fragment of biotite with quartz lenses in the groundmass of a Minette.
b = biotite fragment
q = quartz lenses associated with the biotite
Minette. Steige, Vosges Mts. France.
Without crossed nicols.

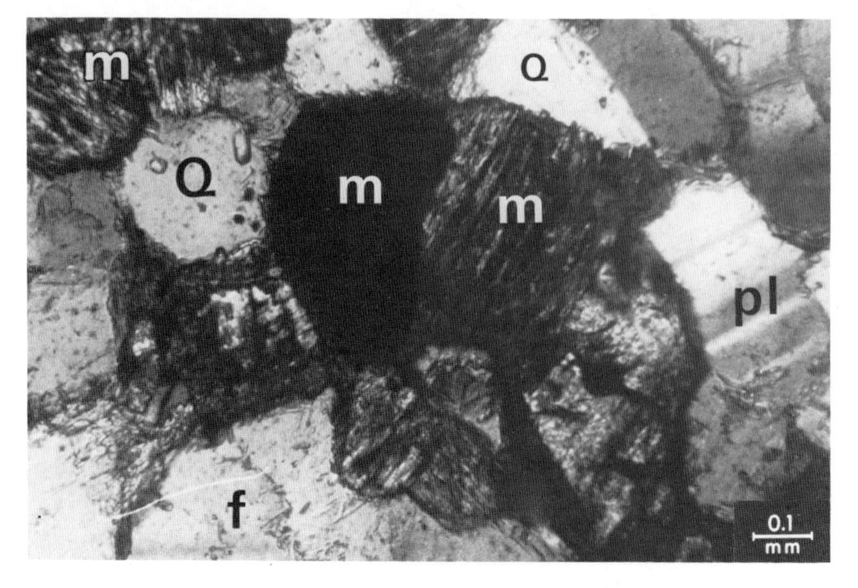

Fig. 403 Quartz, feldspar and mica granoblastic texture with recrystallised orthoclase granoblast.
Q = quartz
f = feldspar
m = micaceous grain
f-g = recrystallised orthoclase granoblast
pl = plagioclase
Minette. Steige, Vosges Mts., France.
With crossed nicols.

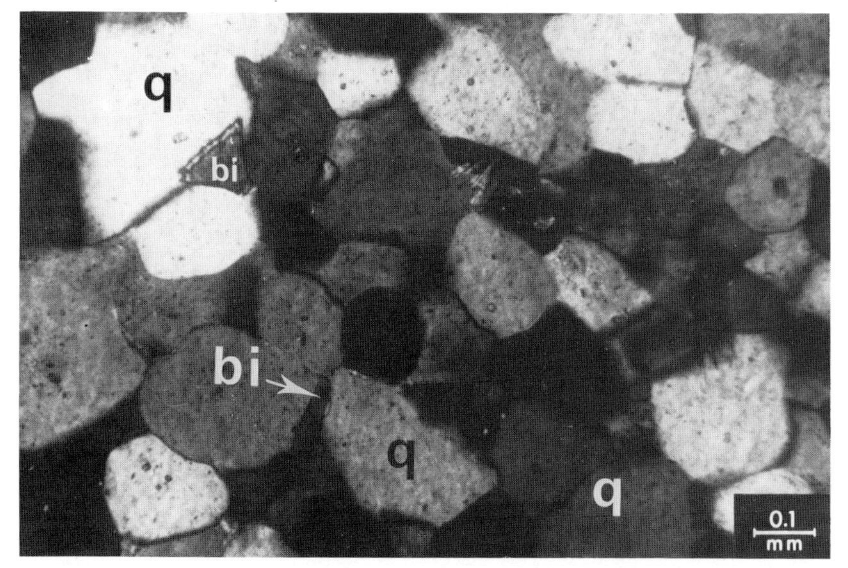

Fig. 404 Granoblastic quartz, metamorphogenic, with mica (biotite).
q = quartz
bi = biotite
Minette. Lindenfels, Odenwald, Germany.
With crossed nicols.

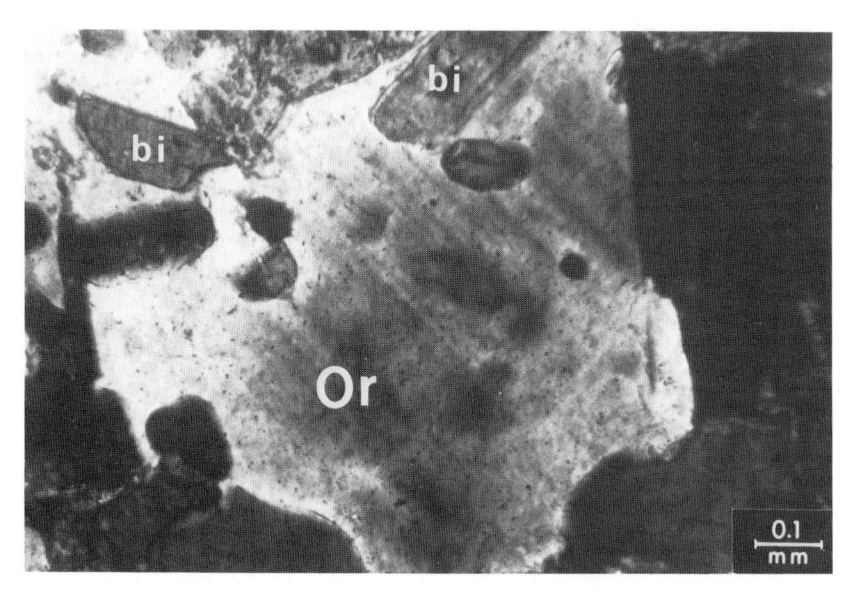

Fig. 405 Crystalloblastic orthoclase enclosing and corroding biotite laths.
Or = orthoclase crystalloblasts
bi = biotite laths enclosed and corroded by the feldsparblast.
Minette. Lindenfels, Odenwald, Germany.
With crossed nicols.

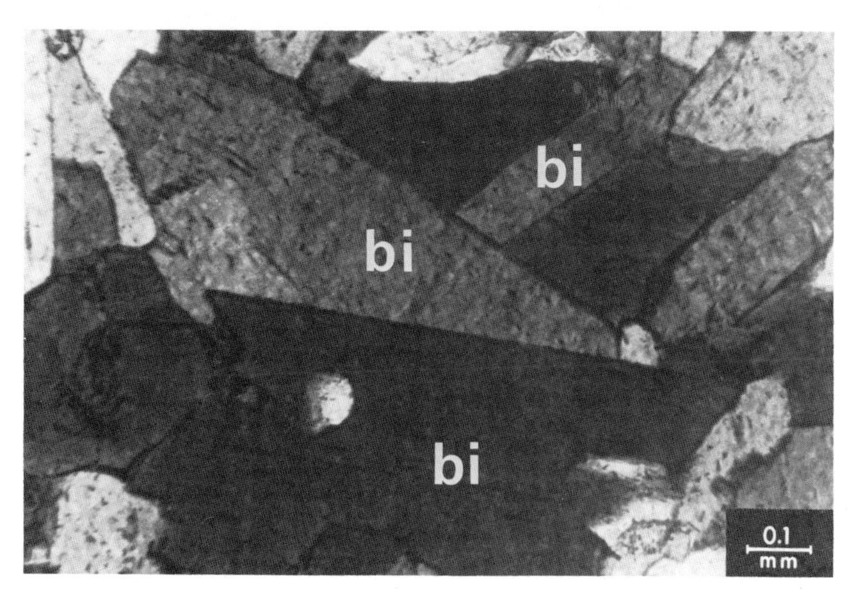

Fig. 406 Intersertal biotite crystalloblastic laths.
bi = biotite crystalloblasts
Minette. Lindenfels, Odenwald, Germany.
With crossed nicols.
(This textural pattern of intersertal biotites is a typical decussate metamorphic structure, see Harker, 1950).

Fig. 407 Idioblastic biotite and amphiboles in Minette.
Bi = idioblastic biotite
Am = amphiboles
G = groundmass
Minette. Orbishöhe, Odenwald, Germany.
Without crossed nicols.

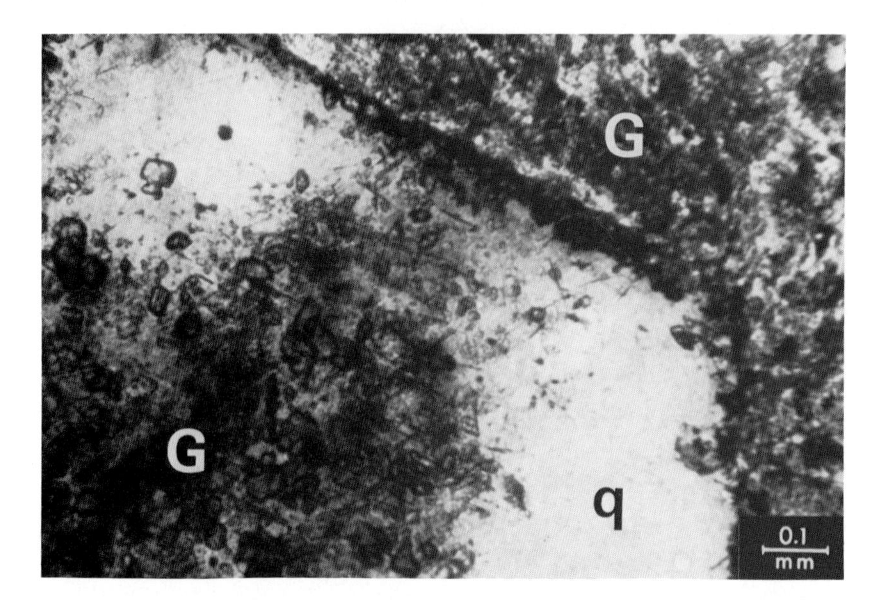

Fig. 408 Quartz idioblast enclosing groundmass components, in parts the quartz is autocathartically cleaned from inclusions.
q = quartz
G = groundmass
Minette. Steige, Vosges Mts., France.
Without crossed nicols.

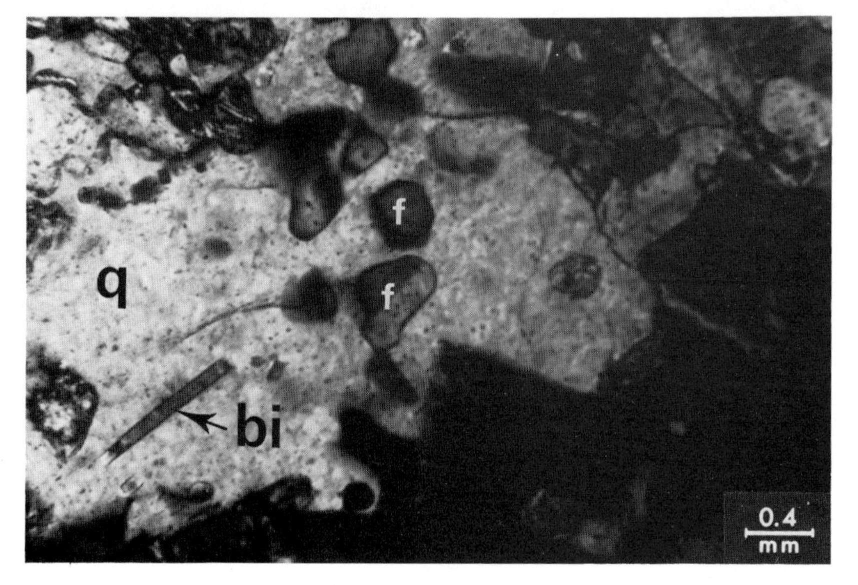

Fig. 409 Blastogenic quartz (poikiloblastic with feldspar and biotite included).
q = blastic quartz
f = feldspar corroded and included in the blastic quartz
bi = biotite
Minette. Lindenfels, Odenwald, Germany.
With crossed nicols.

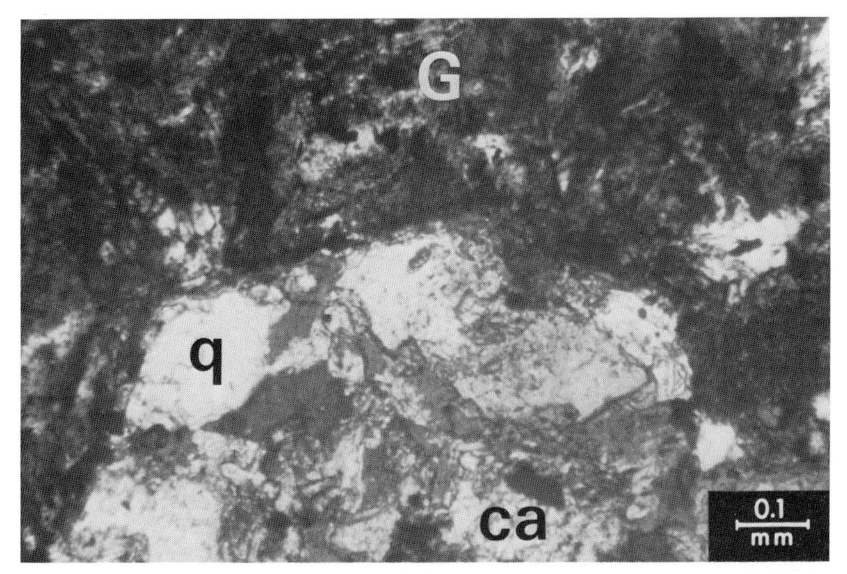

Fig. 410 Calcite-quartz aggregate in groundmass of Minette.
q = quartz
ca = calcite
G = groundmass
Minette. Steige, Vosges Mts., France.
With crossed nicols.

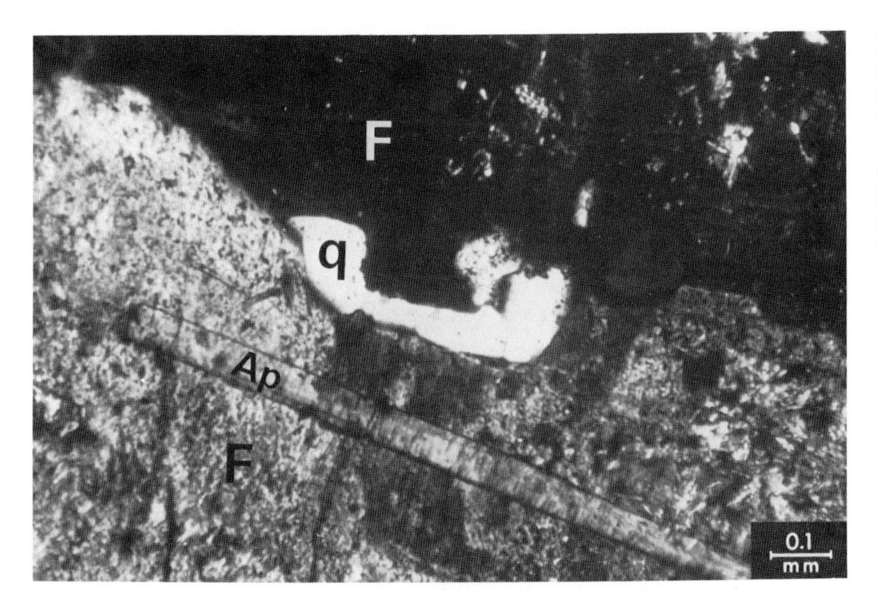

Fig. 411 Shows idioblastic apatite grown in feldspar and with quartz present.
F = feldspar
Ap = apatite
q = quartz
Minette. Sal Fell, Bassenthwaite, Cumberland, England.
With crossed nicols.

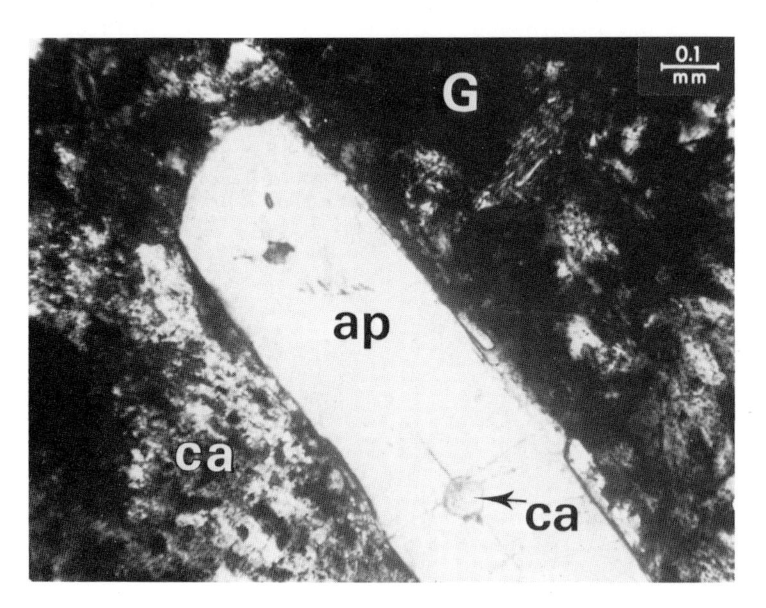

Fig. 412 Idioblastic gigantic apatite enclosing calcite.
ap = apatite
G = groundmass (orthoclase)
ca = calcite
Minette. Steige, Vosges Mts., France.
With crossed nicols.

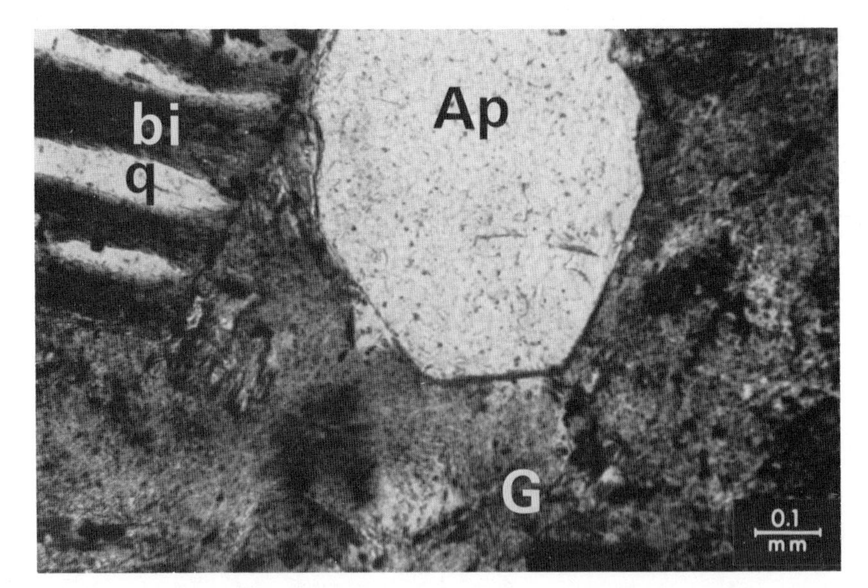

Fig. 413 Biotite fragment with quartz lenses in minette groundmass also crystalloblastic apatite sub-idioblastic in form.
bi = biotite fragment
q = quartz lenses in biotite
G = groundmass (orthoclase)
Ap = apatite sub-idioblastic
Minette. Steige, Vosges Mts., France.
With crossed nicols.

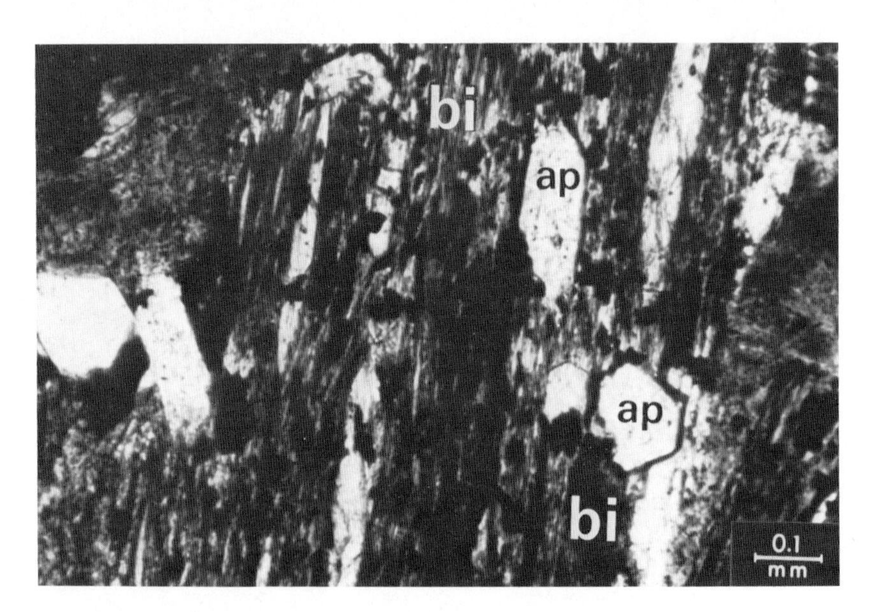

Fig. 414 Crystalloblastic apatite (idioblasts) in biotite. In cases, the apatite crystalloblasts follow the cleavage of the biotite.
bi = biotite
ap = apatite idioblasts
Minette. Welschbruch, Hohwald, Vosges Mts., France.
With crossed nicols.

Fig. 415 Biotite and aggregate of idioblastic apatites in the groundmass of Minette.
bi = biotite
ap = apatite aggregate
G = groundmass
Minette. Welschbruch, Hohwald, Vosges Mts., France.
With crossed nicols.

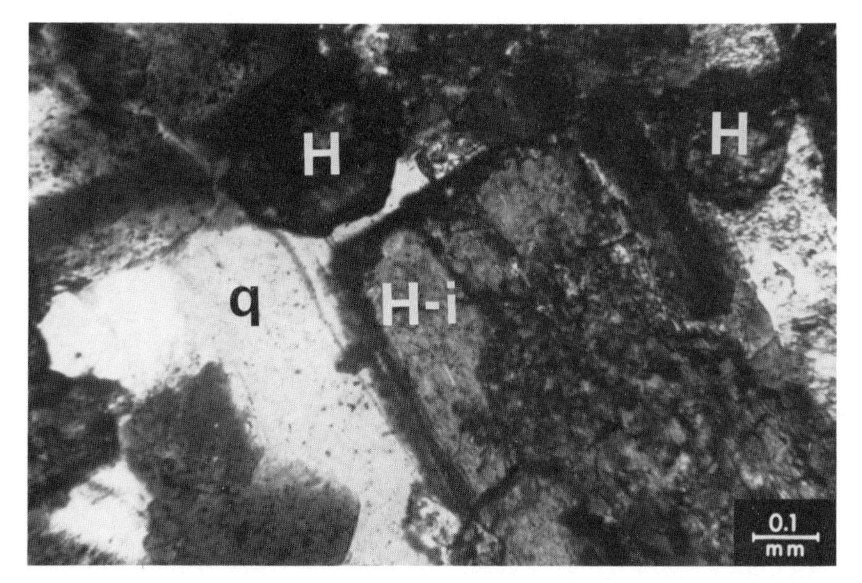

Fig. 416 Hornblende crystalloblasts with intergranular quartz.
H = hornblende
q = quartz
H-i = hornblende idioblast
Minette. Steige, Vosges Mts., France.
With crossed nicols.

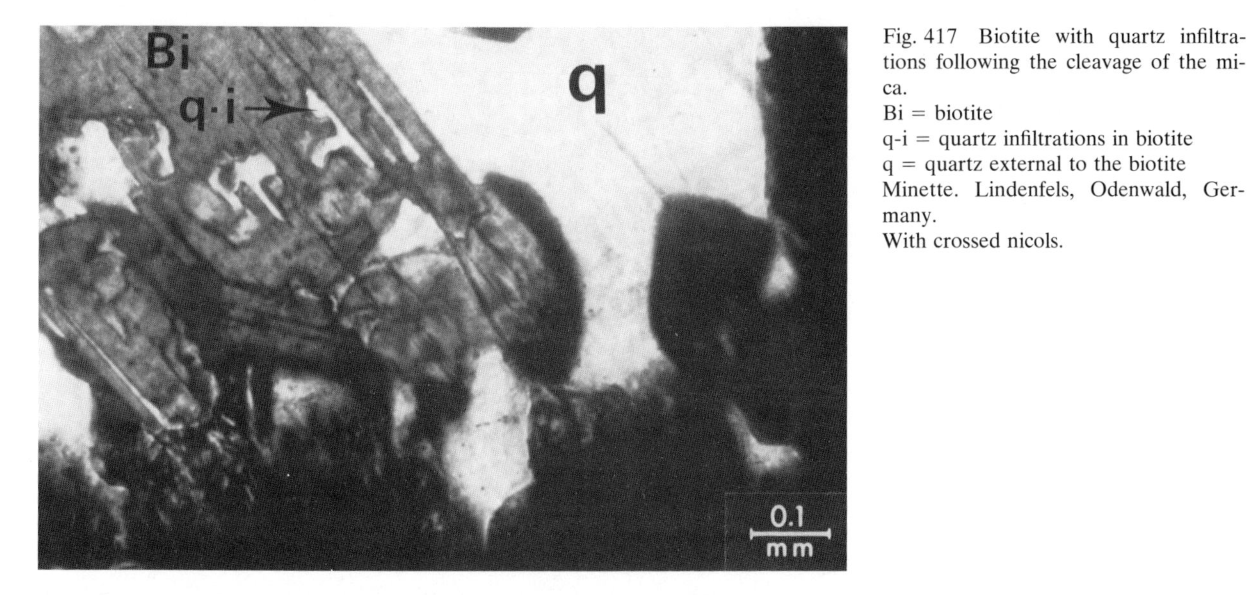

Fig. 417 Biotite with quartz infiltrations following the cleavage of the mica.
Bi = biotite
q-i = quartz infiltrations in biotite
q = quartz external to the biotite
Minette. Lindenfels, Odenwald, Germany.
With crossed nicols.

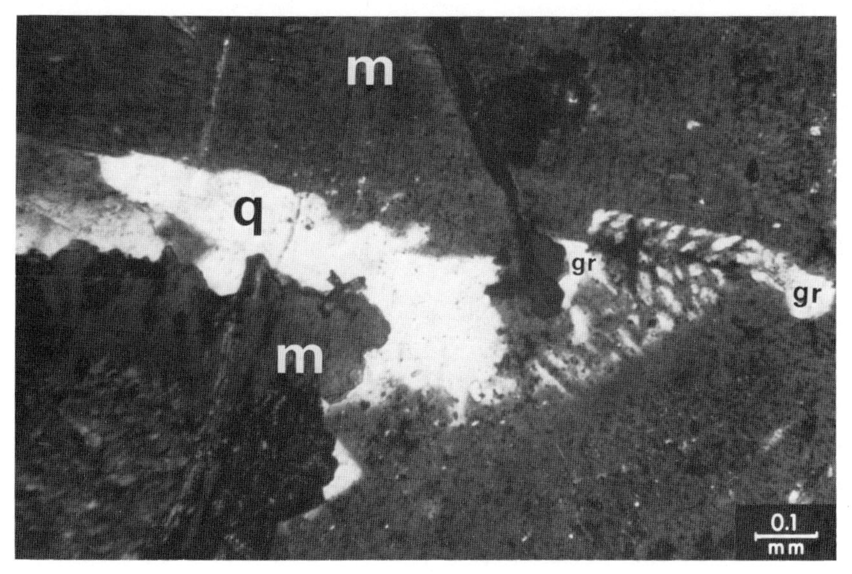

Fig. 418 Quartz marginal to mica extends into adjacent plagioclase and attains granophyric character.
m = mica
q = quartz
gr = granophyric quartz in orthoclase
Minette. Sal Fell, Bassenthwaite, Cumberland, England.
With crossed nicols.

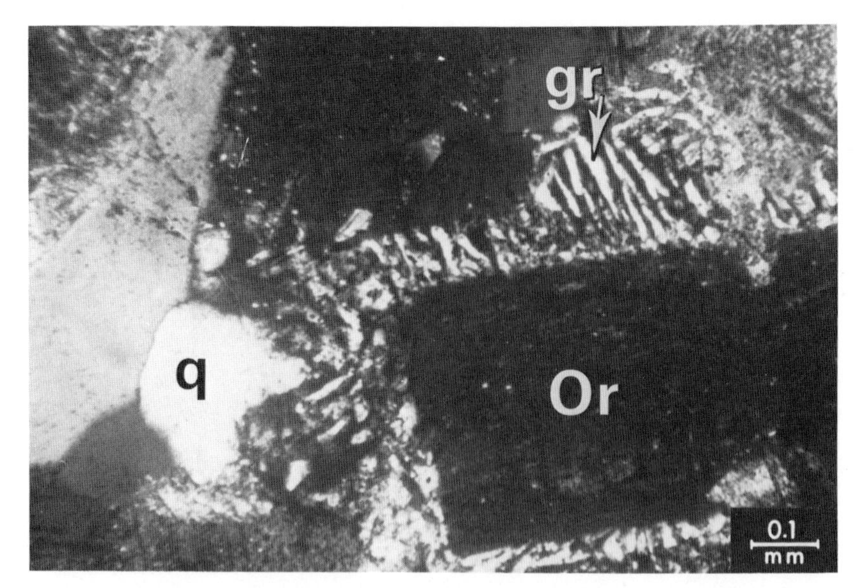

Fig. 419 Orthoclase with a margin of quartz in granophyric intergrowth with the outer orthoclase zone.
Or = orthoclase
gr = granophyric quartz in intergrowth with the outer orthoclase zone.
q = quartz
Minette. Sal Fell, Bassenthwaite, Cumberland, England.
With crossed nicols.

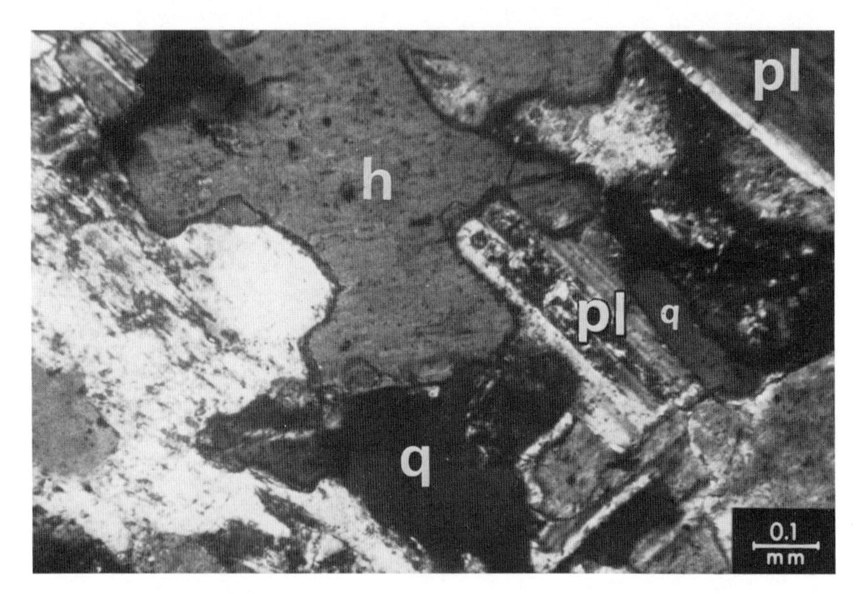

Fig. 420 Crystalloblastic hornblende partly enclosing and corroding plagioclases
h = hornblende
pl = plagioclase
q = quartz
Vogesite. Fraize, Meurthe river, Vosges Mts., France.
With crossed nicols.

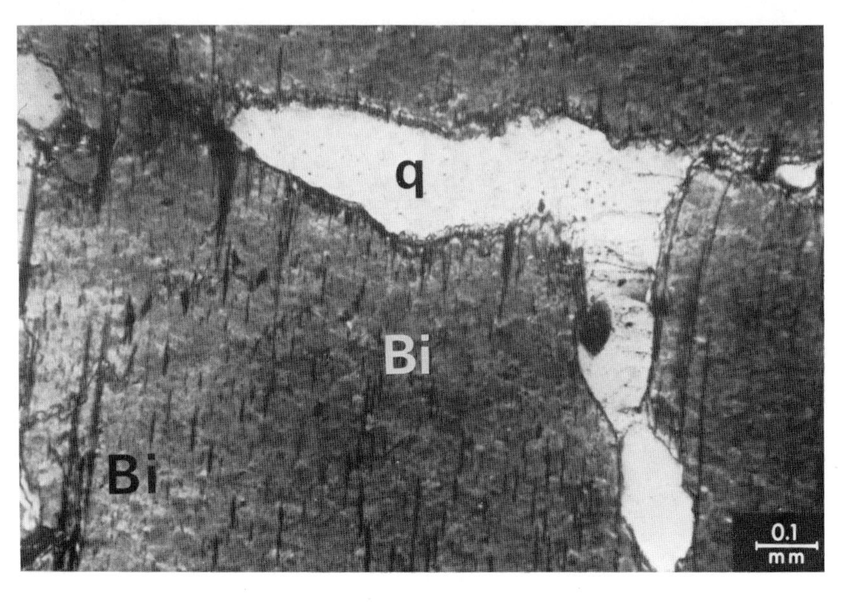

Fig. 421 Biotite showing undulating extinction (tectonic deformation) and invaded by later quartz, which follows partly the cleavage of the mica.
Bi = biotite
q = quartz
Vogesite. Fraize, Meurthe river, Vosges Mts., France.
With crossed nicols.

Fig. 422 Idioblastic hornblendes in vogesitic groundmass.
H = hornblende idioblasts
G = groundmass
Vogesite. Andlau valley, Vosges Mts., France.
Without crossed nicols.

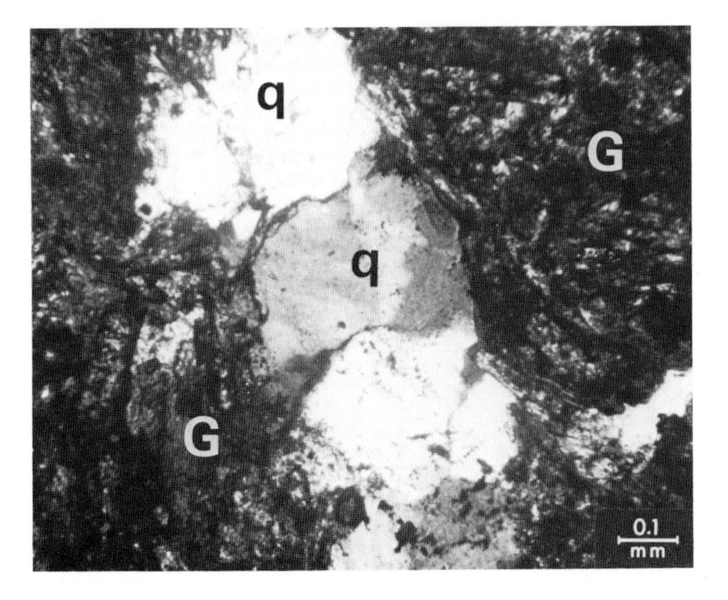

Fig. 423 Initially sedimentogenic quartz recrystallised and exhibiting undulating extinction.
q = quartz exhibiting undulating extinction (recrystallised)
G = groundmass
Vogesite. Andlau Valley, Vosges Mts., France.
With crossed nicols.

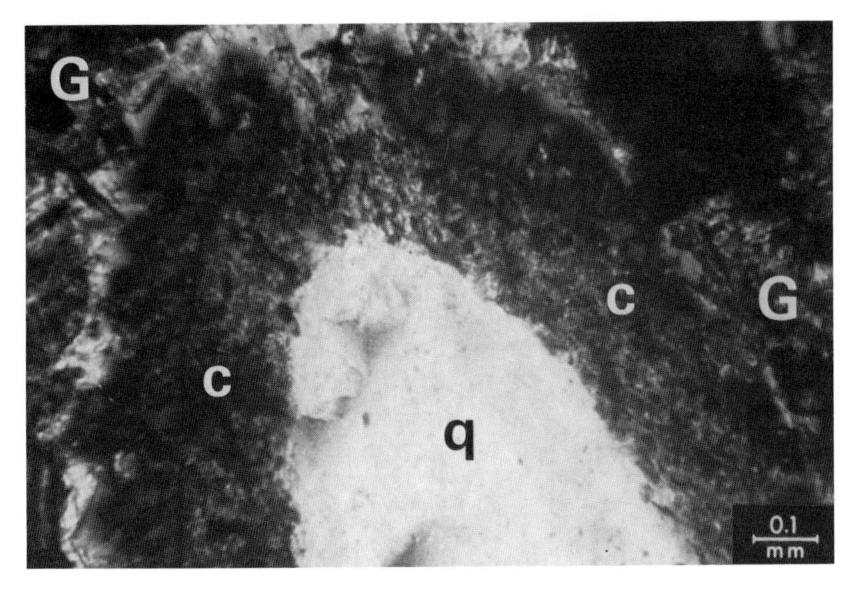

Fig. 424 Quartz xenocryst showing undulating extinction rounded and surrounded by a reaction margin (corona) in the kersantite's groundmass.
q = quartz xenocryst rounded and exhibiting undulating extinction.
c = corona reaction margin surrounding the quartz.
G = groundmass kersantite
Kersantite. Buchwald, Schlesien, a.p. Poland.
With crossed nicols.

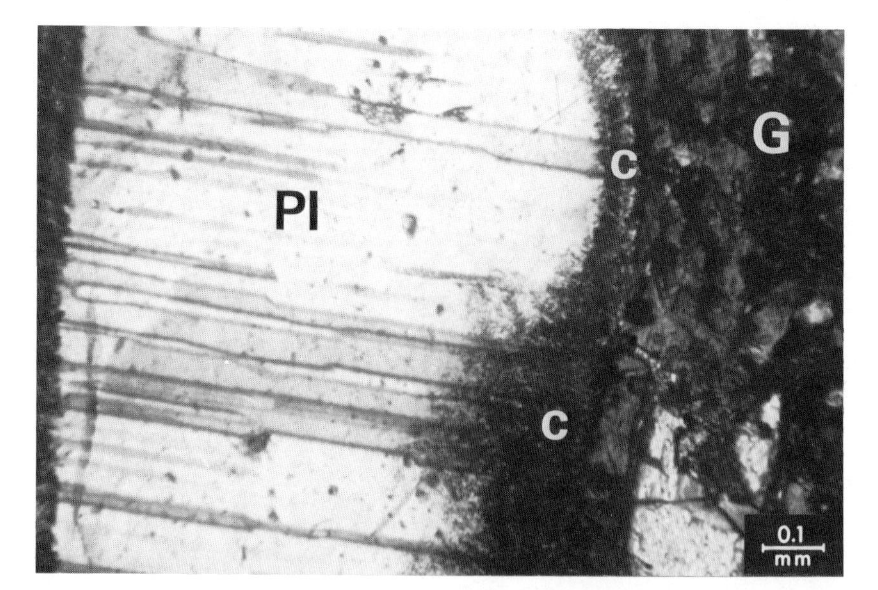

Fig. 425 Feldspar corroded with a reaction margin in the Kersantite groundmass.
Pl = plagioclase corroded
c = corona reaction margin
G = Kersantites groundmass
Kersantite. St. Diedler, Markirch, Vosges Mts., France.
With crossed nicols.

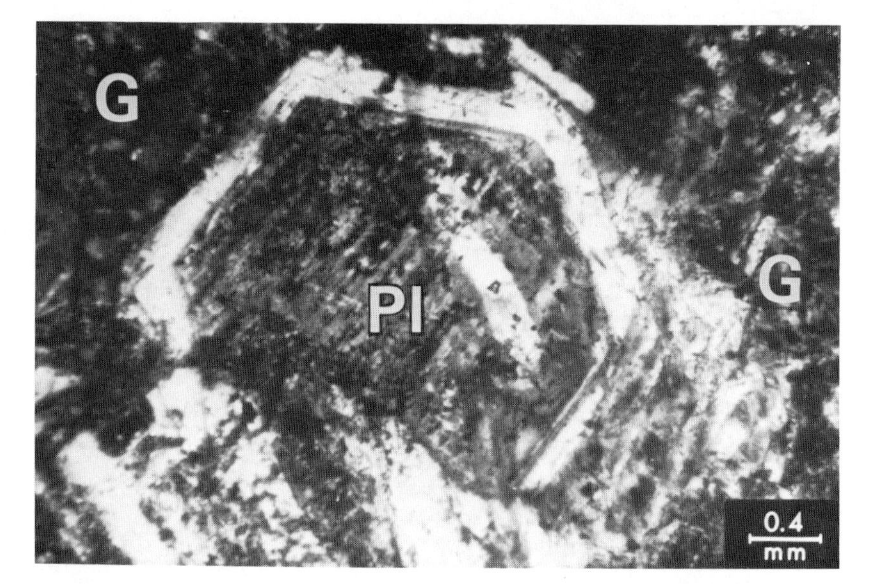

Fig. 426 Plagioclase tecoblast with margins free of inclusions.
Pl = plagioclase
G = groundmass
Kersantite. Treseburg, North Harz, Germany.
With crossed nicols.

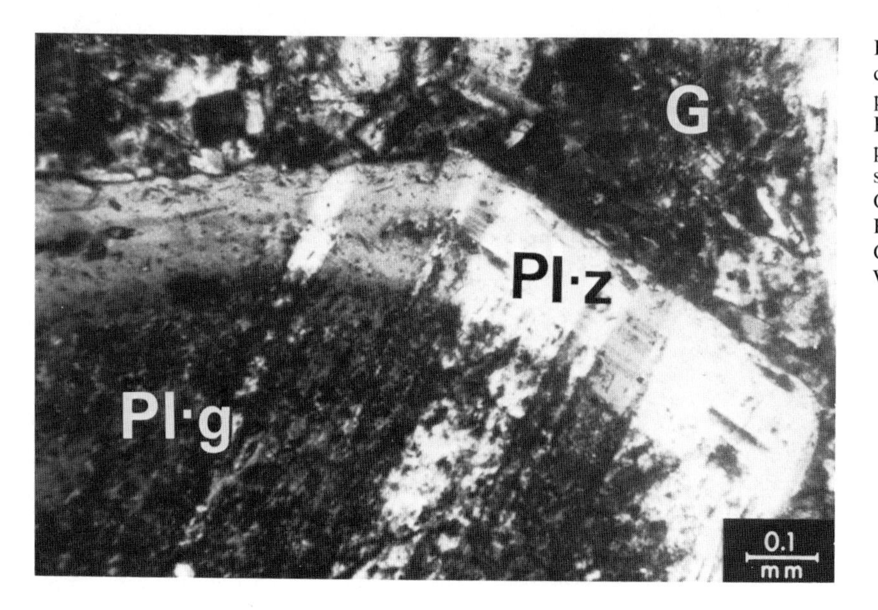

Fig. 427 Plagioclase tecoblast with the central part full of groundmass. The outer plagioclase margin is free of inclusions.
Pl-g = plagioclase full of groundmass
pl-z = plagioclase zone free of inclusions
G = groundmass
Kersantite. Treseburg, North Harz, Germany.
With crossed nicols.

Fig. 428 Prismatic hornblende as a predominant component of the Kersantites groundmass.
H = hornblende
F = feldspar
Kersantite. Buchwald, Schlesien, a.p. Poland.
With crossed nicols.

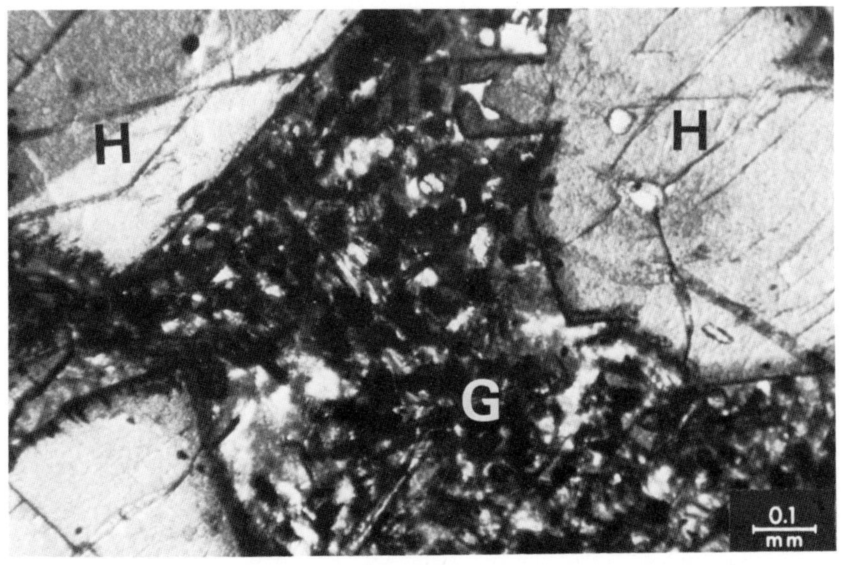

Fig. 429 Idioblastic hornblende with protuberances engulfing the camptonitic groundmass.
H = hornblende
G = camptonitic groundmass
Camptonite. Maena near Oslo, South Norway.
With crossed nicols.

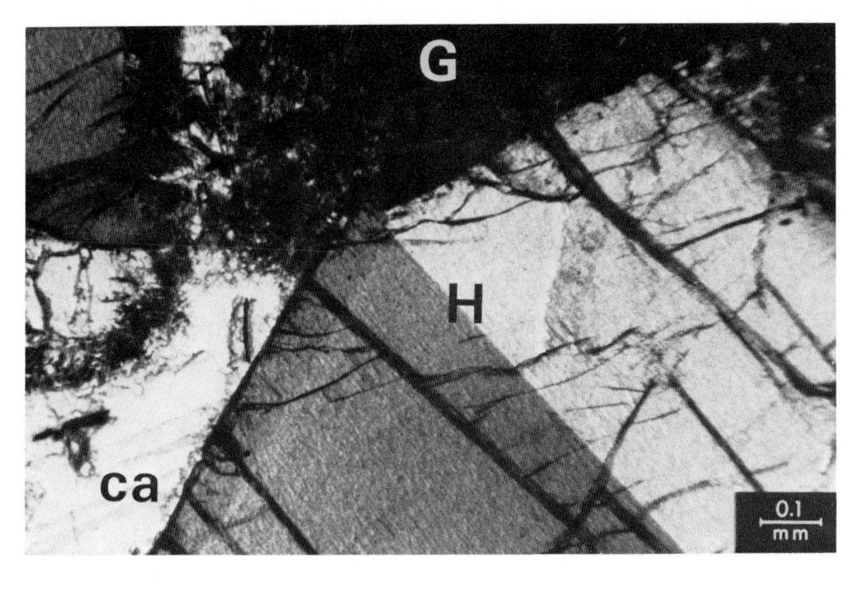

Fig. 430 Idioblastic twinned hornblende in camptonitic groundmass with calcite.
H = hornblende
G = camptonitic groundmass
ca = calcite in camptonitic groundmass
Camptonite. Maena near Oslo, South Norway.
With crossed nicols.

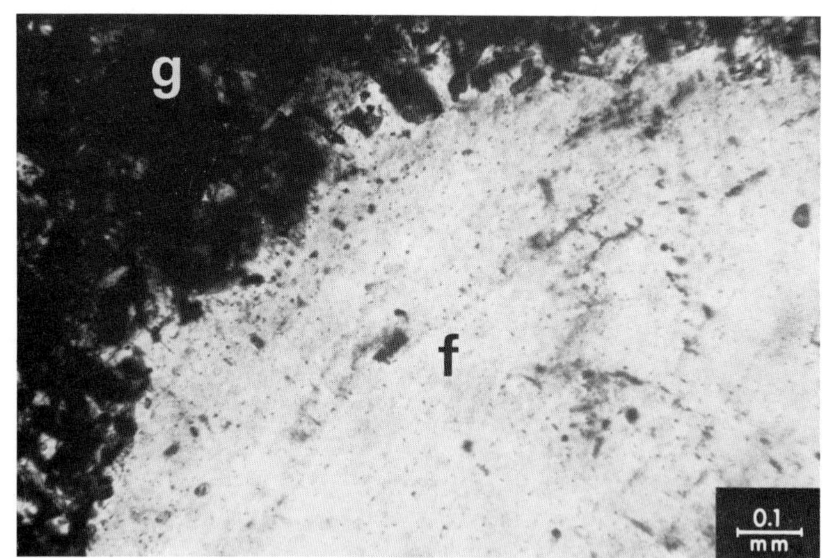

Fig. 431 Rounded and corroded feldspar surrounded by camptonitic groundmass
f = feldspar
g = groundmass
Biotite-Camptonite. Covanda, Serra de Monchique, South Portugal.
With crossed nicols.

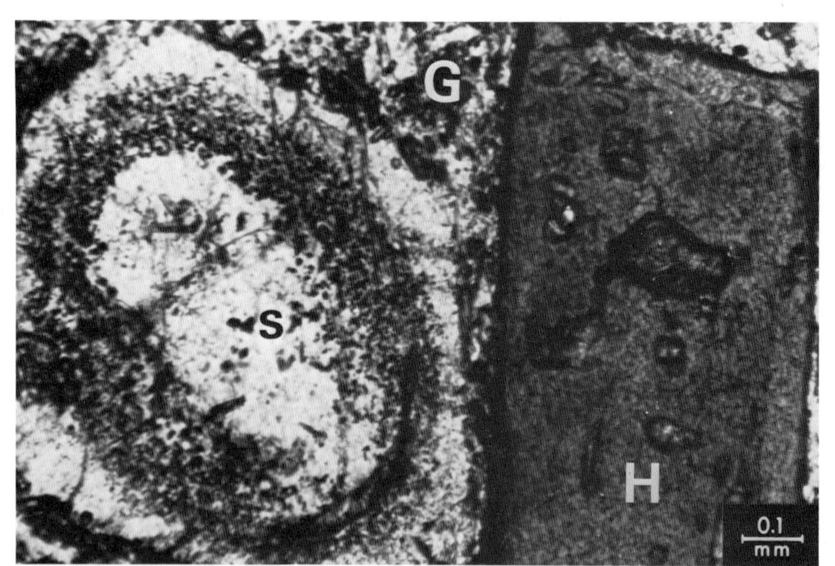

Fig. 432 Idioblastic hornblende including groundmass adjacent to idioblastic sodalite with a zone rich in relics of already assimilated groundmass.
H = idioblastic hornblende
S = sodalite
G = groundmass
Sodalite-Camptonite. Corte Grande, Serra de Monchique, South Portugal.
Without crossed nicols.

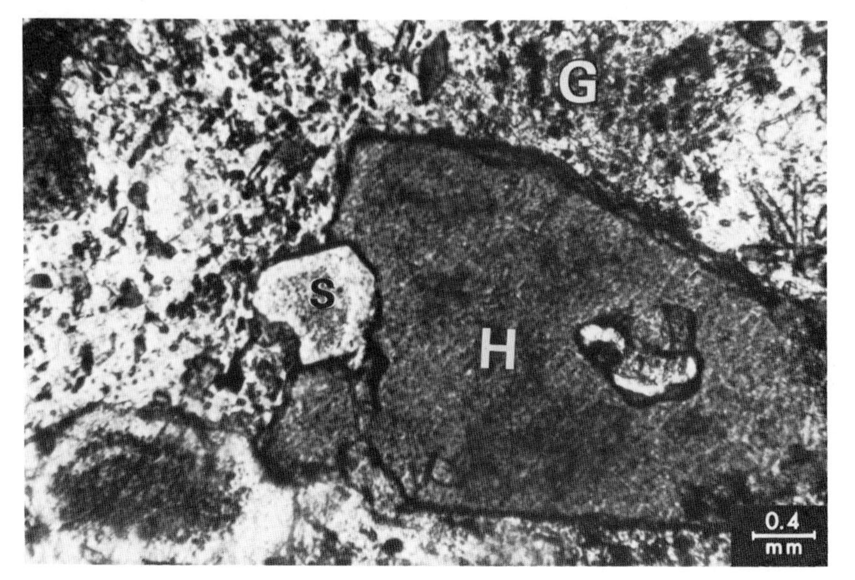

Fig. 433 Blastic hornblende including groundmass and partly engulfing idioblastic sodalite.
H = hornblendeblasts
S = sodalite
G = groundmass
Sodalite-Camptonite. Corte Grande, Serra de Monchique, South Portugal.
Without crossed nicols.

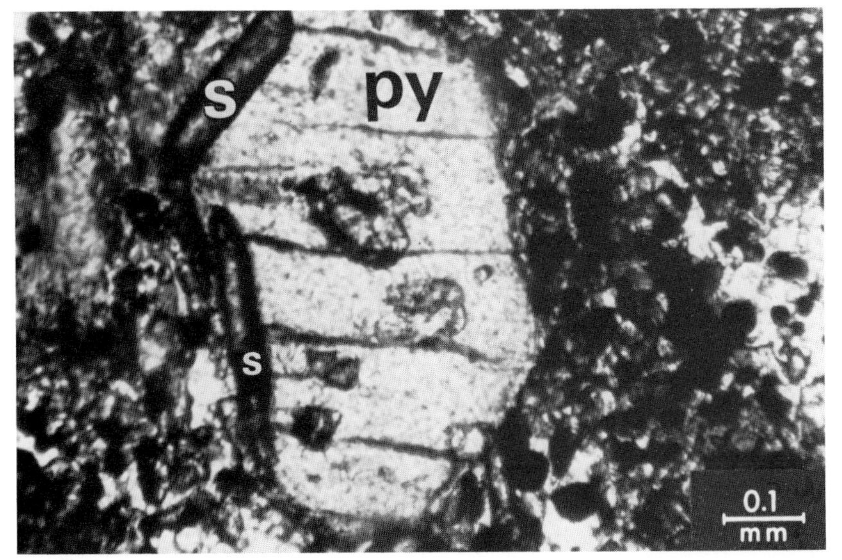

Fig. 434 Pyroxene enclosing sphenes with their elongated direction orientated parallel to the pyroxene crystal faces.
py = pyroxene
s = sphene orientated parallel to the crystal faces of the pyroxene.
Biotite-Camptonite. Covanda, Serra de Monchique, South Portugal.
With crossed nicols.

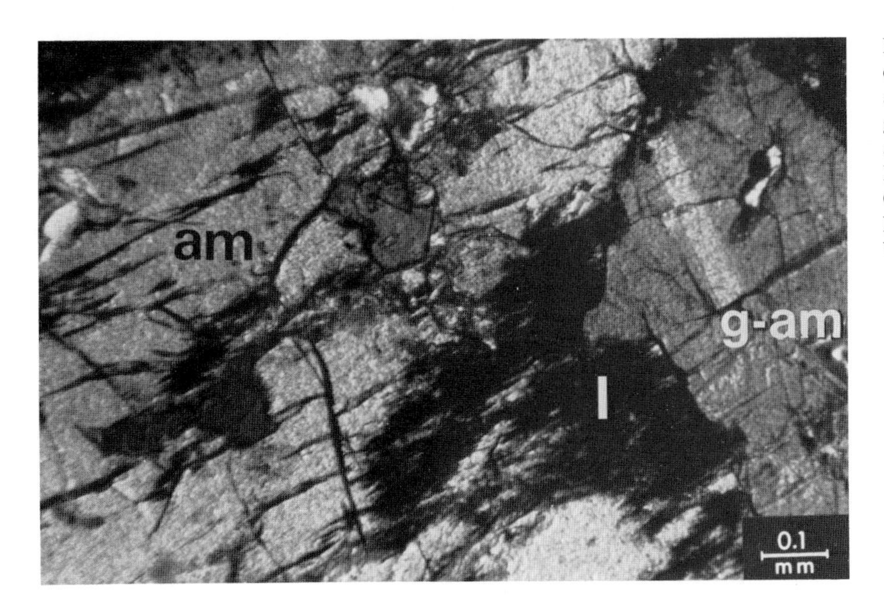

Fig. 435 Green hornblende with "iron-oxide" infiltrations together with normal amphibole.
g-am = green amphibole
am = normal amphibole (brown)
I = iron-oxide infiltrations
Camptonite, Maena, near Oslo, South Norway.
With crossed nicols.

Fig. 436 Idioblastic hornblende in camptonitic groundmass.
H = hornblende crystalloblast
f = feldspar of the groundmass
n = needles (apatite)
Pyroxene-Amphibole-Camptonite. Covanda, Serra de Monchique, South Portugal.
With crossed nicols.

Fig. 437 Idioblastic hornblende with sphene in feldspar "groundmass".
H = hornblende
S = Sphene
Pyroxene-Amphibole-Camptonite.
Covanda, Serra de Monchique, South Portugal.
With crossed nicols.

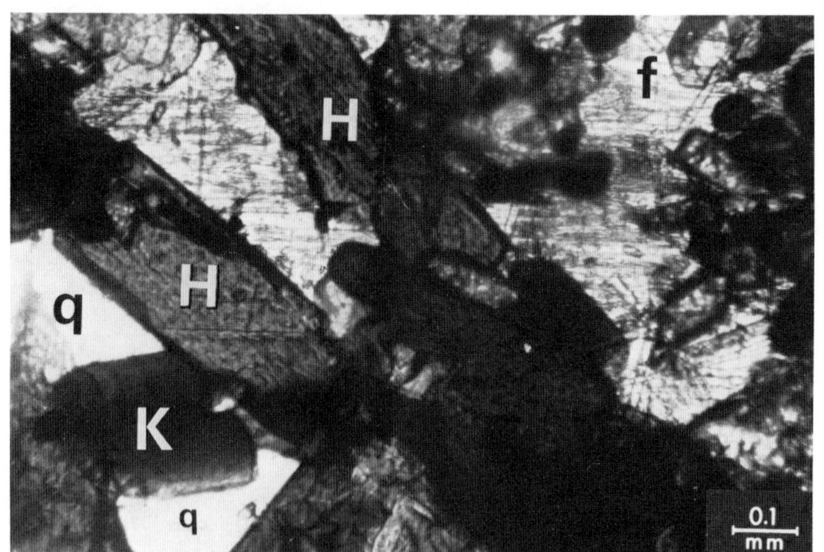

Fig. 438 Hornblende and Kyanite crystalloblasts in a quartz/feldspar background.
H = hornblende crystalloblast
K = kyanite crystalloblast
q = quartz
f = feldspar
Biotite-Amphibole-Camptonite.
Serra de Monchique, South Portugal.
With crossed nicols.

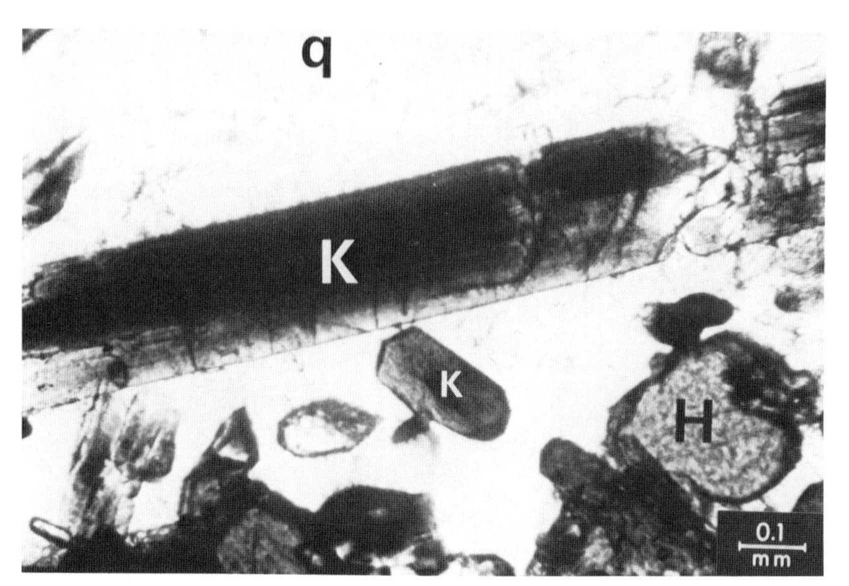

Fig. 439 Kyanite crystalloblast and hornblende crystalloblasts with quartz.
K = kyanite crystalloblast
H = hornblende crystalloblast
q = quartz
Amphibole-Camptonite.
Covanda, Serra de Monchique, South Portugal.
With crossed nicols.

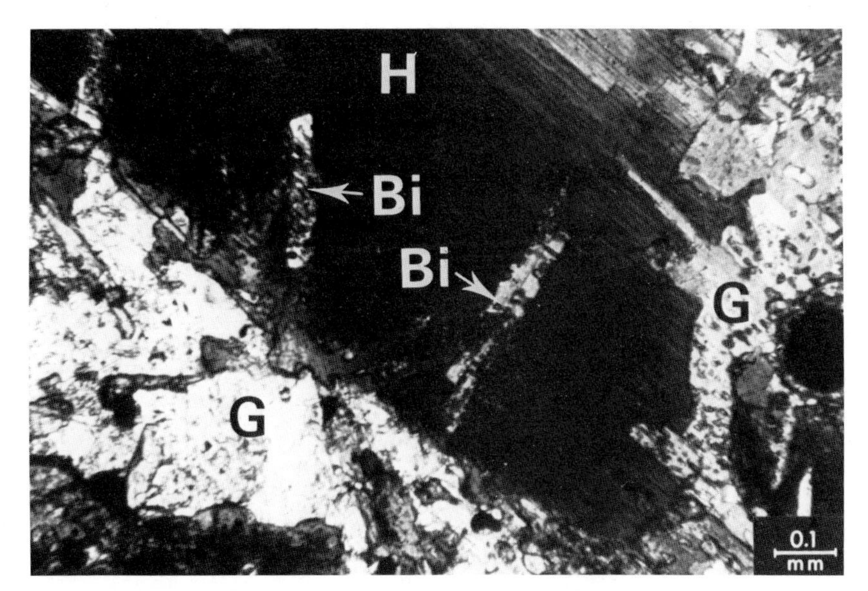

Fig. 440 Shows hornblende crystalloblast with mica crystalloblasts grown perpendicular to the amphibole.
H = hornblende
Bi = biotite
G = camptonitic groundmass
Biotite-Amphibole-Camptonite.
Covanda, Serra de Monchique, South Portugal.
With crossed nicols.

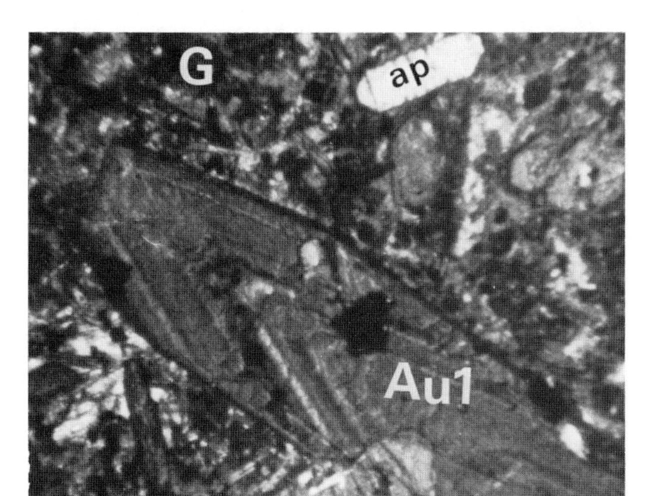

Fig. 441 Rhythmically zoned augite with another augite interpenetrating.
Au_1 = augite rhythmically zoned
Au_2 = augite penetrating into the zoned augite (Au_1)
G = groundmass (magmatogenic)
ap = apatite
Amphibole-Monchiquite. Picota, Serra de Monchique, South Portugal.
With crossed nicols.

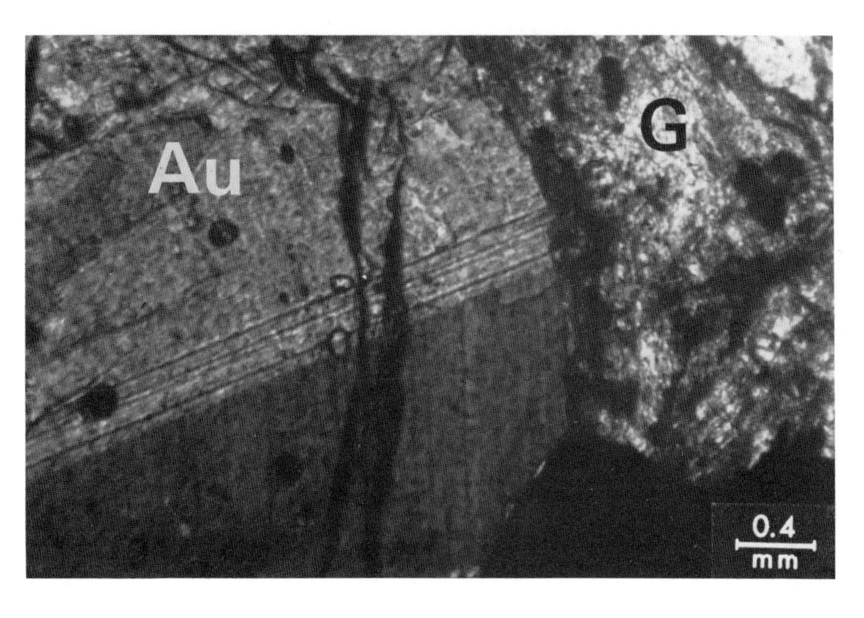

Fig. 442 Augite exhibiting fine rhythmical banding.
Au = augite
G = groundmass (magmatogenic)
Monchiquite. Jakuben, Bohemian Intermediate Mts., CSSR.
With crossed nicols.

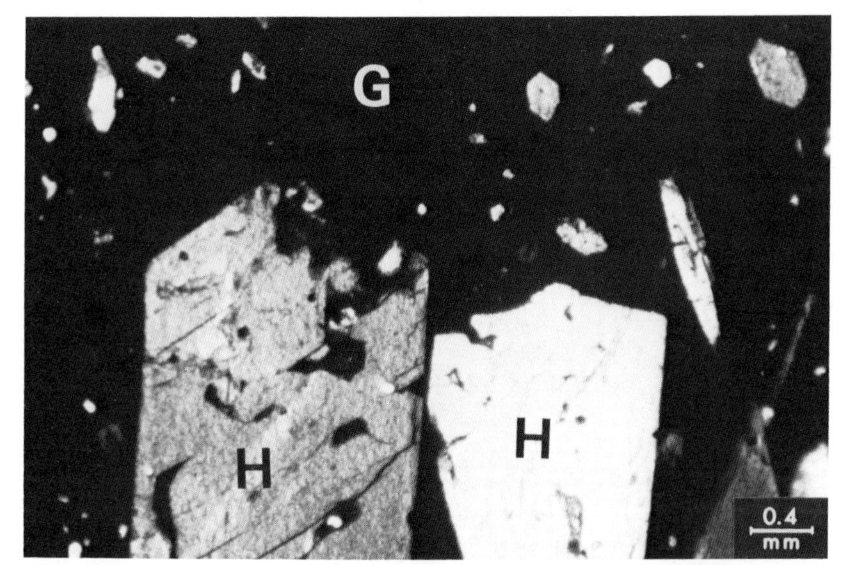

Fig. 443 Hornblende phenocrysts showing magmatic corrosion (in fine grained monchiquitic groundmass).
G = groundmass
H = hornblende
Monchiquite. Palmeira, Serra de Monchique, South Portugal.
With crossed nicols.

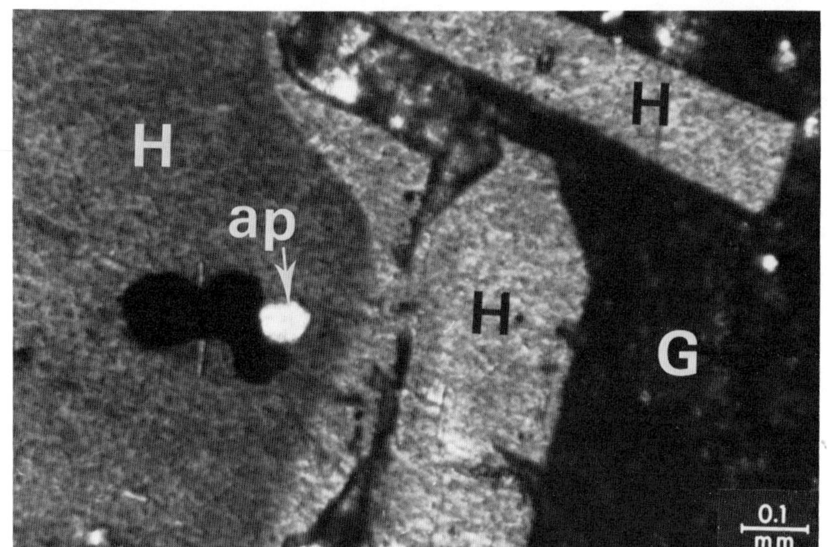

Fig. 444a Zoned hornblende tecoblast with a protuberance attaining idiomorphism.
H = hornblende
ap = apatite
G = groundmass
Black (ore-minerals)
Monchiquite. Palmeira, Serra de Monchique, South Portugal.
With crossed nicols.

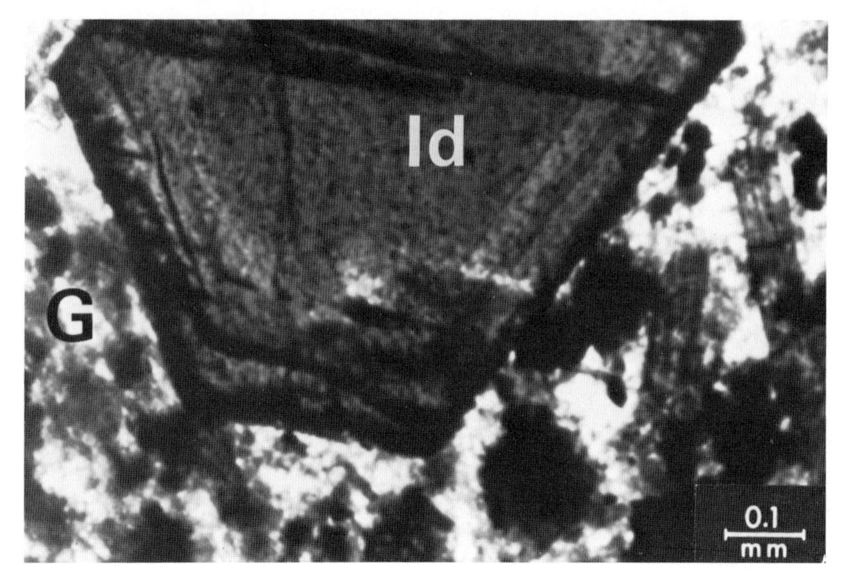

Fig. 444b Idioblastic zoned amphibole in monchiquitic groundmass.
G = monchiquitic groundmass
Id = idioblastic amphibole
Amphibole-Monchiquite. Ribeiro-di-Banho, Serra de Monchique, South Portugal.
With crossed nicols.

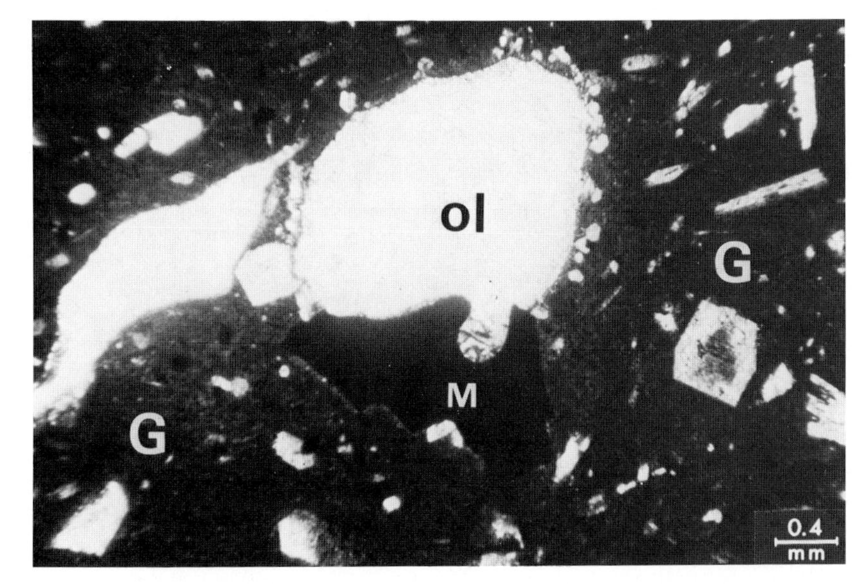

Fig. 445 Olivine exhibiting a corroded outline, in intergrowth with magnetite in monchiquitic groundmass.
ol = olivine
M = magnetite
G = Monchiquitic groundmass
Monchiquite. Palmeira, Serra de Monchique, South Portugal.
With crossed nicols.

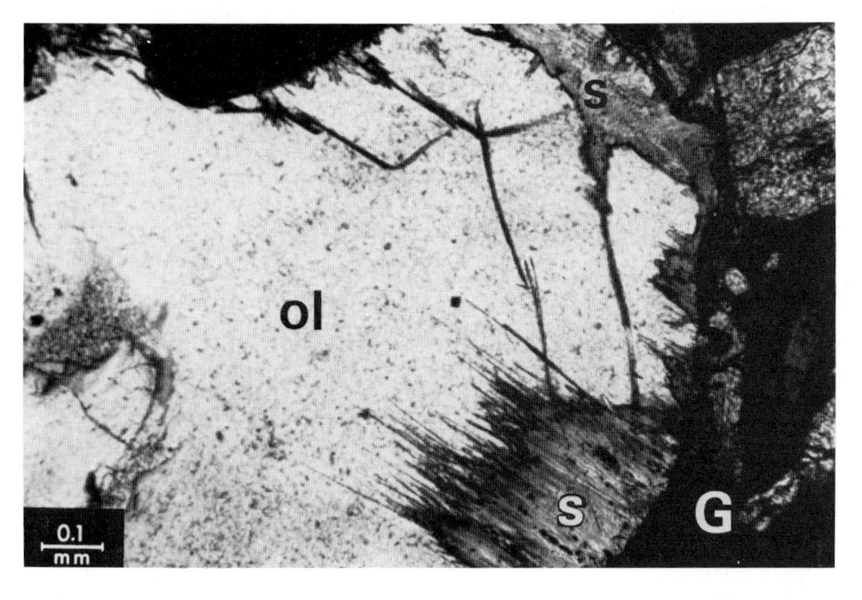

Fig. 446 Rounded olivine xenocrysts serpentinised, in Monchiquitic groundmass.
ol = olivine
s = serpentinised
G = Monchiquitic groundmass
Monchiquite. Palmeira, Serra de Monchique, South Portugal.
With crossed nicols.

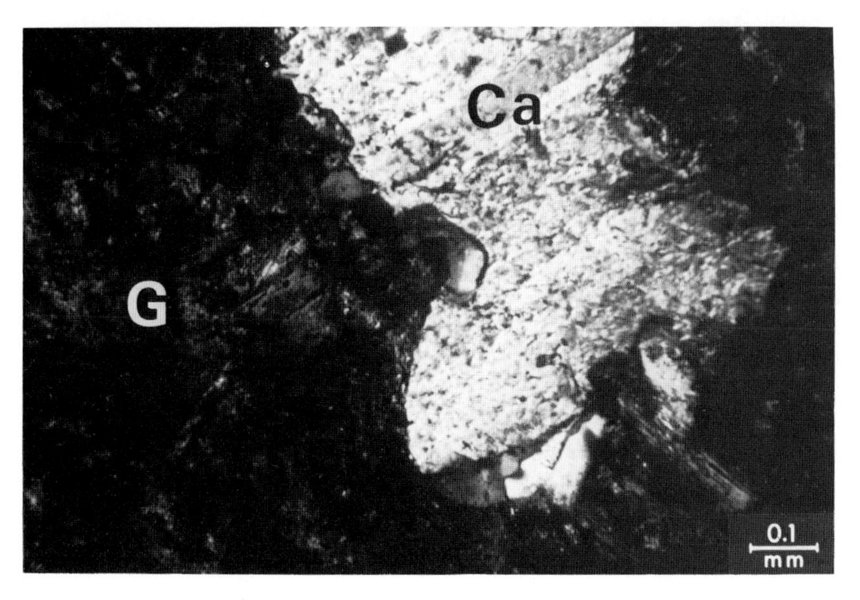

Fig. 447 Twinned calcite corroded and included in monchiquitic groundmass.
Ca = calcite (twinned)
G = Monchiquitic groundmass (analcime-isotropic)
Monchiquite (also observed in Camptonite).
Jakuben, Bohemian Intermediate Mts.
With crossed nicols.

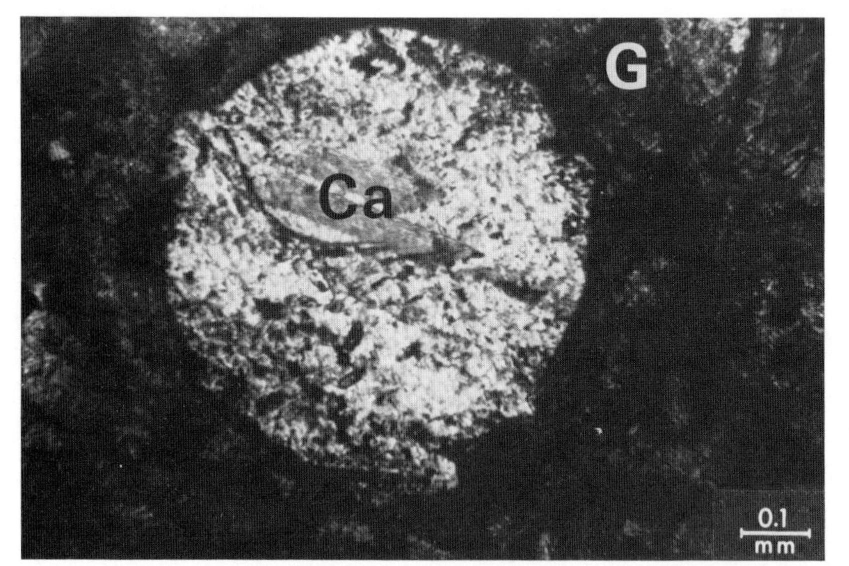

Fig. 448 Rounded calcitic structure with twinning, in monchiquitic groundmass.
Ca = calcite corroded (rounded) in monchiquitic groundmass.
G = monchiquitic groundmass
Monchiquite. Jakuben, Bohemian Intermediate Mts., CSSR.
With crossed nicols.

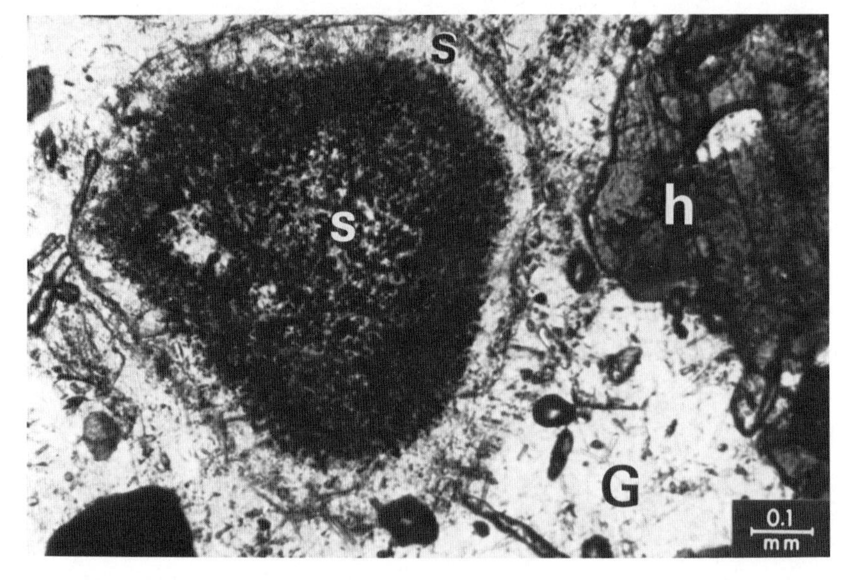

Fig. 449 Sodalite in monchiquitic groundmass.
s = sodalith (sodalite)
h = hornblende
G = groundmass
Biotite-Monchiquite. Picota, Serra de Monchique, South Portugal.
Without crossed nicols.

Fig. 450 Idioblastic hornblende often an abundant component of monchiquite.
H = hornblende
Amphibole-Pyroxene-Monchiquite. Picota, Serra de Monchique,
South Portugal.
Without crossed nicols.

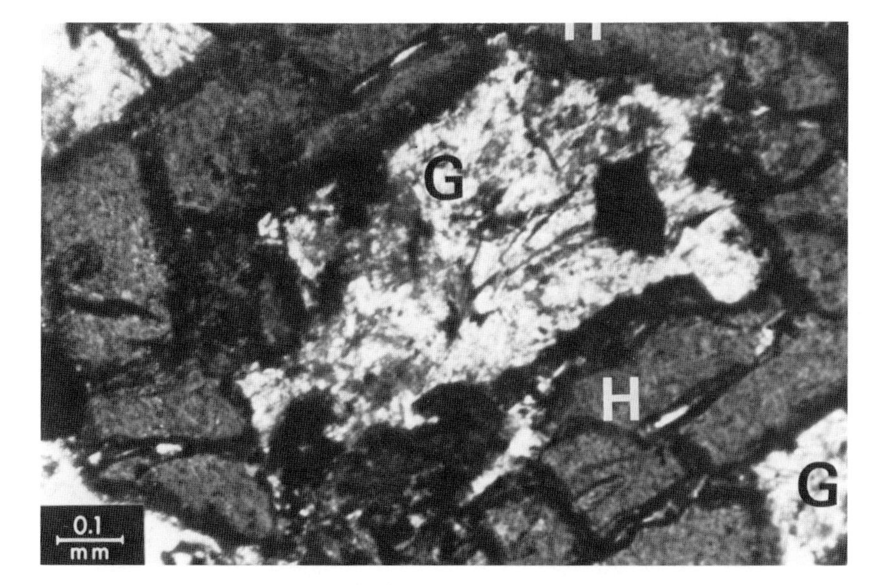

Fig. 451 Idioblastic hornblende enclosing components of the groundmass.
H = idioblastic hornblende
G = Monchiquitic groundmass
Amphibole-Monchiquite. Caldas, Serra de Monchique, South Portugal.
Without crossed nicols.

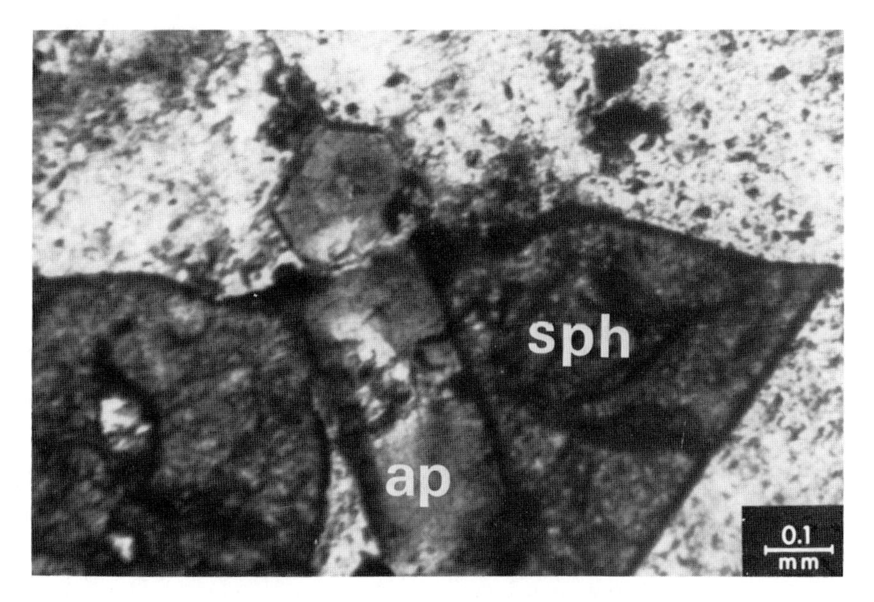

Fig. 452 Crystalloblastic apatite penetrating due to its crystalloblastic force, into a crystalloblastic sphene.
sph = sphene
ap = apatite
Amphibole-Monchiquite. Picota, Serra de Monchique, South Portugal.
With crossed nicols.

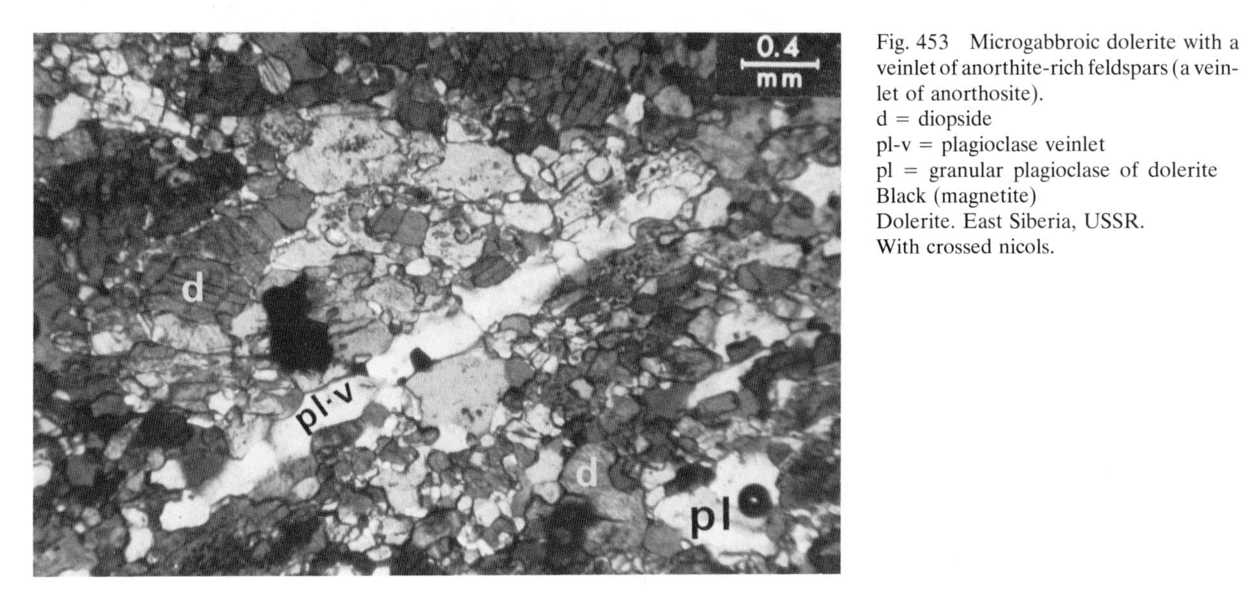

Fig. 453 Microgabbroic dolerite with a veinlet of anorthite-rich feldspars (a veinlet of anorthosite).
d = diopside
pl-v = plagioclase veinlet
pl = granular plagioclase of dolerite
Black (magnetite)
Dolerite. East Siberia, USSR.
With crossed nicols.

Fig. 454 Within the micro-gabbroic dolerite anorthite rich granular plagioclases "segregate" forming locally anorthosite with diopside intergranular to the feldspars.
Pl = granular plagioclase ("anorthite" actually An-rich plag.)
pl-p = polygonal plagioclase (anorthite)
d = intergranular diopside
Dolerite. East Siberia, USSR.
With crossed nicols.

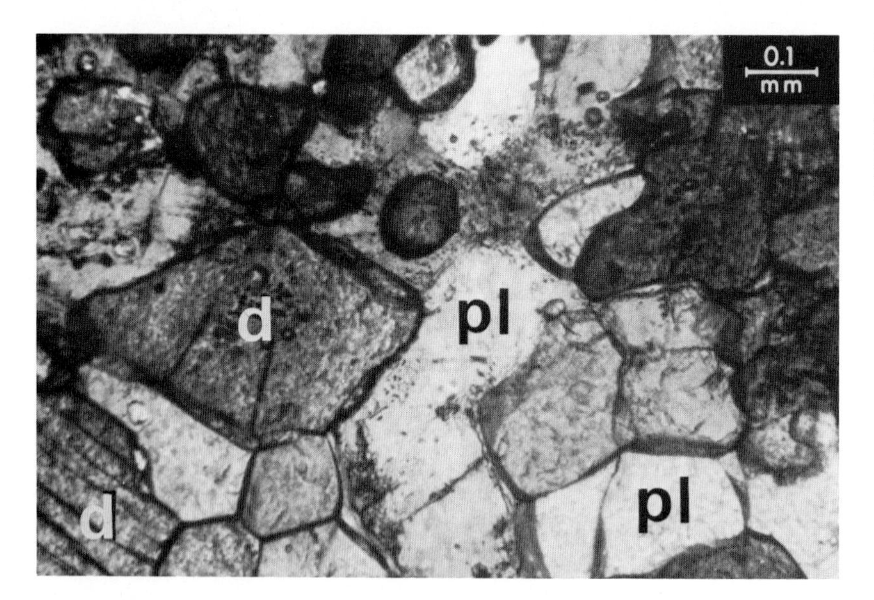

Fig. 455 Granular diopside (green diopside) and granular plagioclase.
d = diopside
pl = plagioclase
(Anorthosite-microgabbro)
Dolerite. East Siberia, USSR.
With crossed nicols.

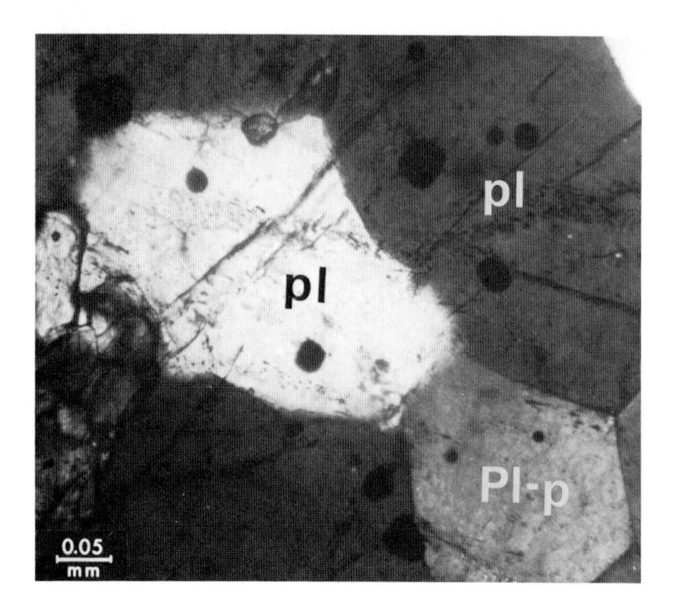

Fig. 456 Granular plagioclase. Also a polygonal anorthite rich plagioclase grain is exhibited.
pl = plagioclase
Pl-p = polygonal plagioclase
black (magnetite)
Dolerite. East Siberia, USSR.
With crossed nicols.

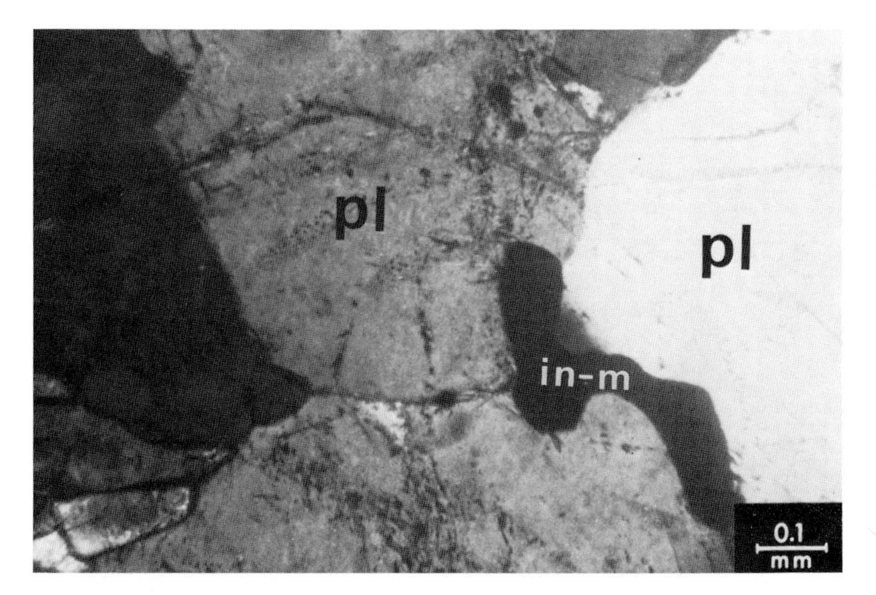

Fig. 457 Granular plagioclase with intergranular magnetite.
pl = plagioclase
in-m = intergranular magnetite
(anorthosite-microgabbro)
Dolerite. East Siberia, USSR.
With crossed nicols.

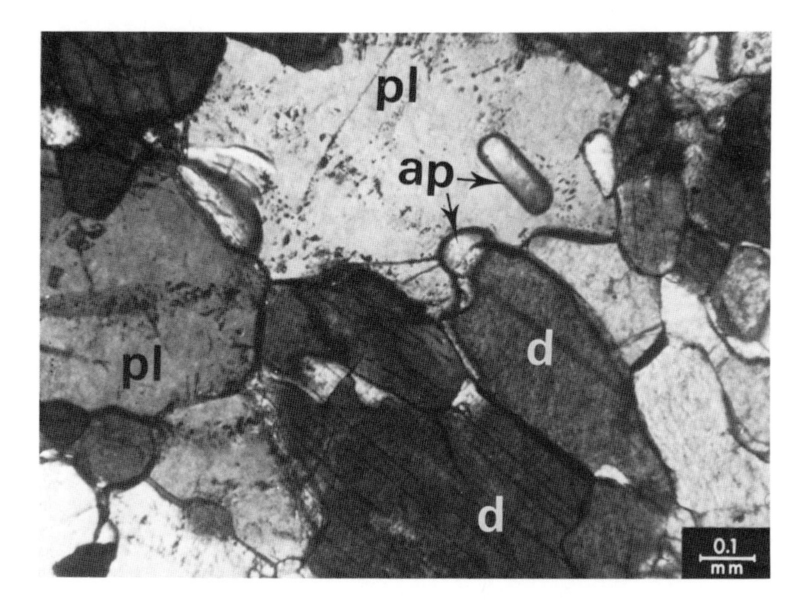

Fig. 458 Granular plagioclase and granular diopside. Also a rounded apatite is included in the plagioclase.
pl = plagioclase
d = diopside
ap = apatite
(Anorthosite-microgabbro)
Dolerite. East Siberia, USSR.
With crossed nicols.

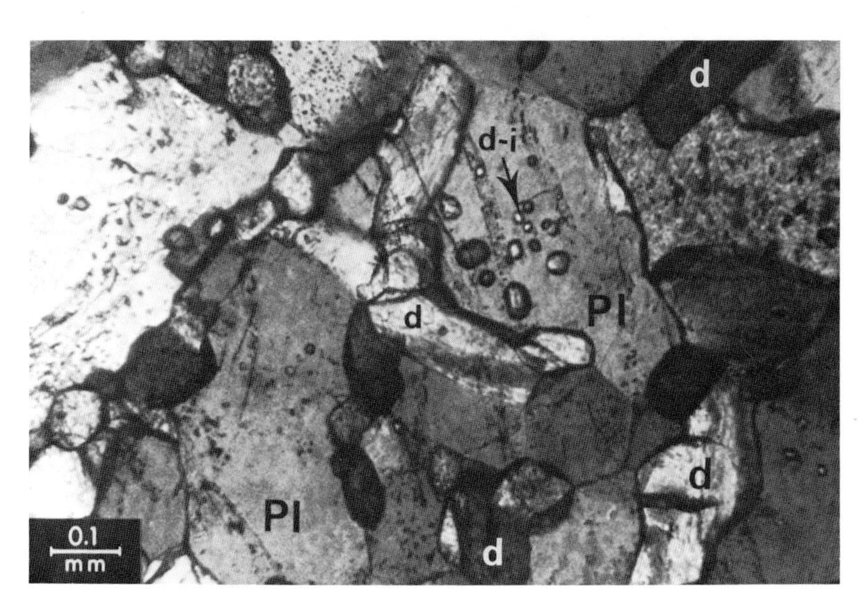

Fig. 459 A plagioclase diopside granular texture with rounded small diopsides included in a granular plagioclase.
pl = plagioclase (rich in An)
d = diopside
d-i = diopsides rounded and included in plagioclase
(Anorthosite-microgabbro)
Dolerite. East Siberia, USSR.
With crossed nicols.

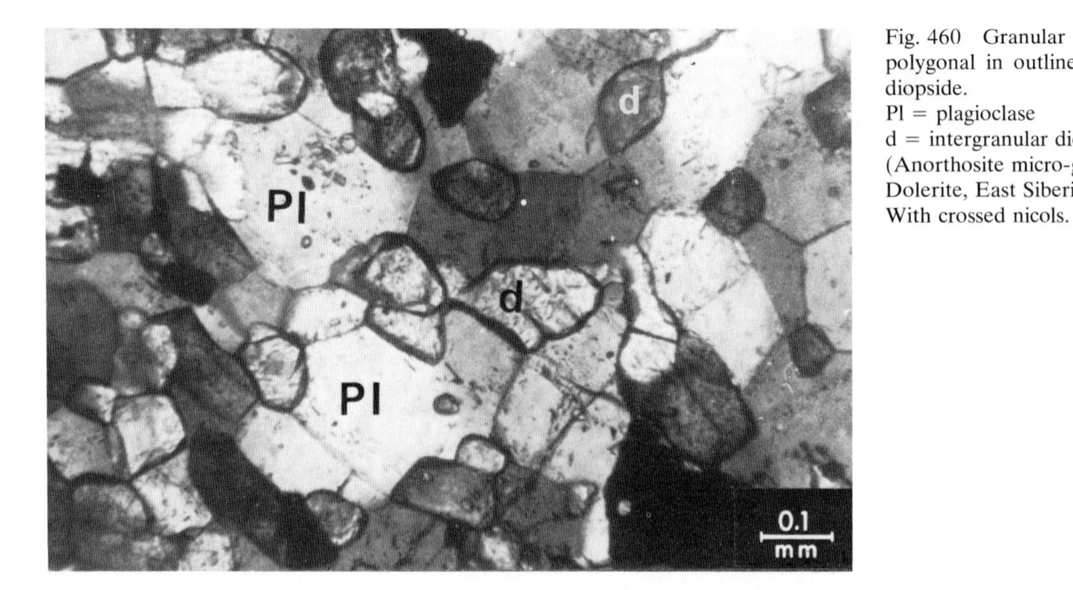

Fig. 460 Granular plagioclase. Often polygonal in outline with intergranular diopside.
Pl = plagioclase
d = intergranular diopside
(Anorthosite micro-gabbro)
Dolerite, East Siberia, USSR.
With crossed nicols.

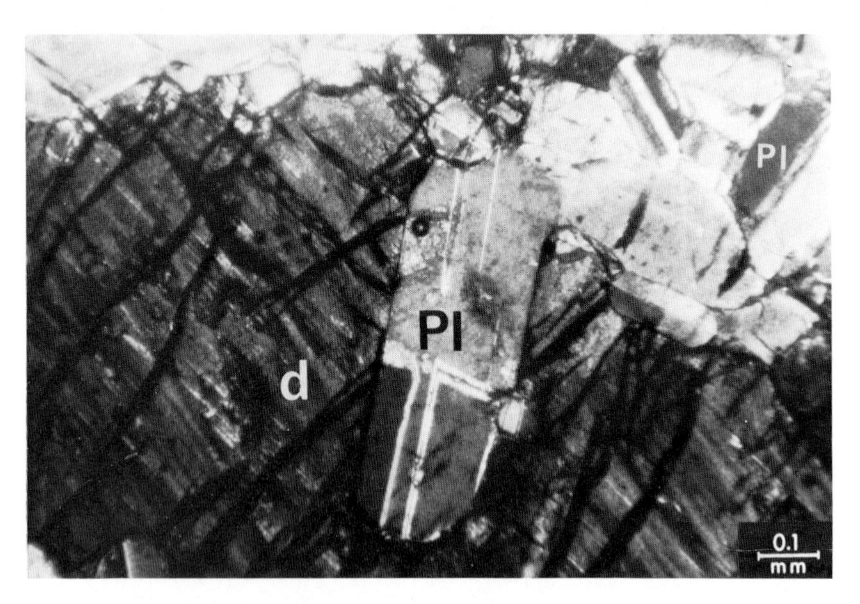

Fig. 462a, b Prismatic plagioclase crystalloblasts, enclosed and corroded by poikilitic diopside crystalloblast.
Pl = plagioclase
d = diopside crystalloblast
Anorthosite "layer"
Bushveld, Transvaal, South Africa.
With crossed nicols.

(For Fig. 461 see page 269)

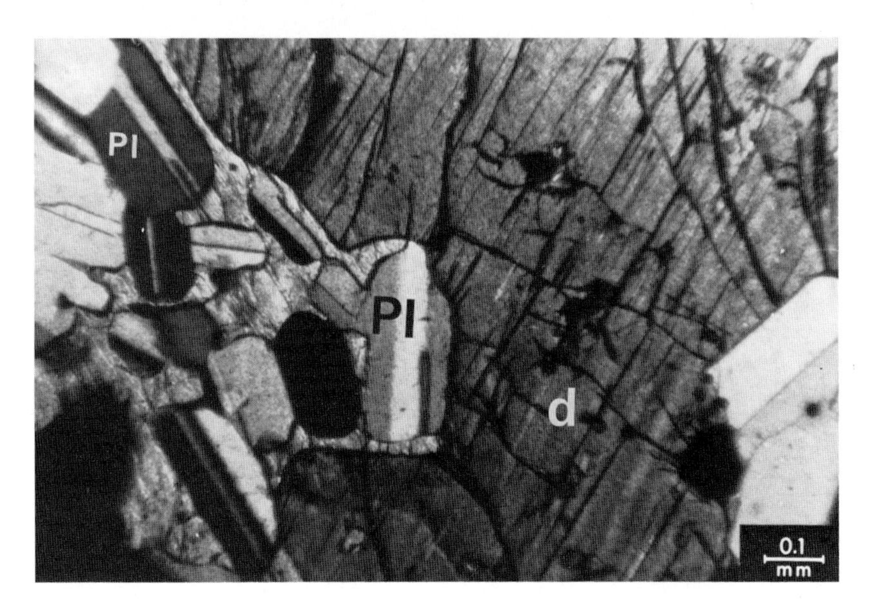

Fig. 462b

Fig. 461 Gabbroic anorthosite with clots of true anorthosite. Olivine crystals (high relief) outline polyhedra in the gabbroic areas. The rock is believed to be an igneous cumulate, but the fine-grained polygonal texture may have formed by recrystallization.
Lunar anorthosites (Apollo 11)
With crossed nicols.
By Wood, A. et al. (Science Vol. 167. No. 3918, 1970).

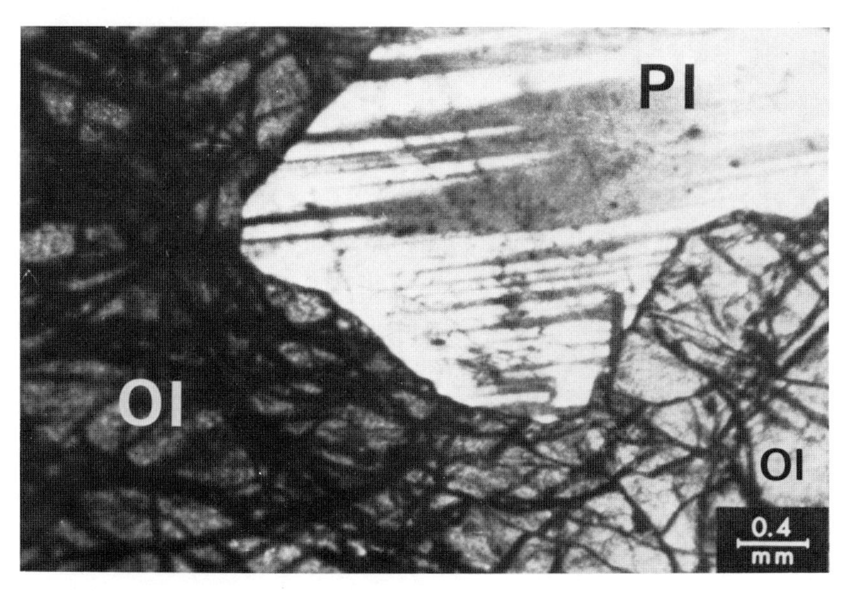

Fig. 463 Shows a "relic" olivine in contact with plagioclase crystalloblast (see also Fig. 464). The mafic mineral shows undulating extinction (as exhibited by the anomalous extinction of the olivine), and rounded, and corroded outline in contact with later plagioclaseblast.
Pl = plagioclase
ol = olivine (exhibiting anomalous extinction).
Anorthosite. Negra Massiva, Angola.
With crossed nicols.

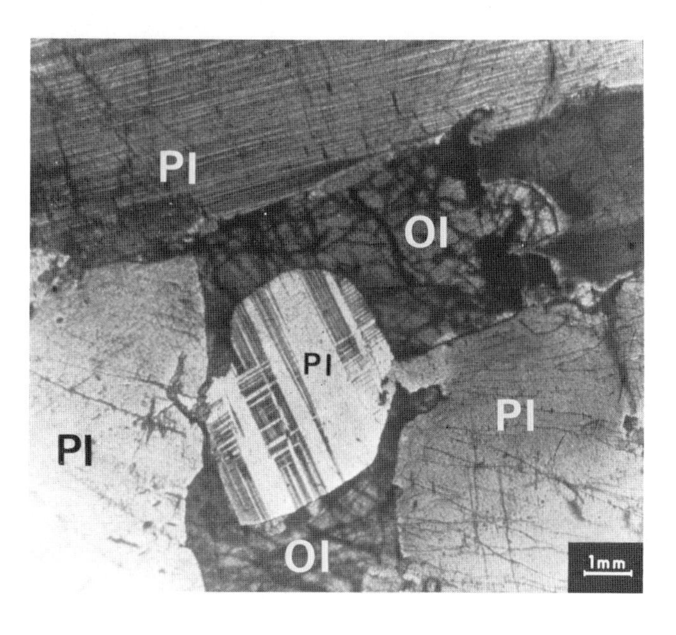

Fig. 464 Olivine relics restricted in the interspaces between plagioclase crystalloblasts.
Ol = olivine relics restricted between the plagioclase crystalloblasts.
Pl = plagioclase crystalloblasts
Anorthosite. Negra Massiva, Angola.
With crossed nicols.

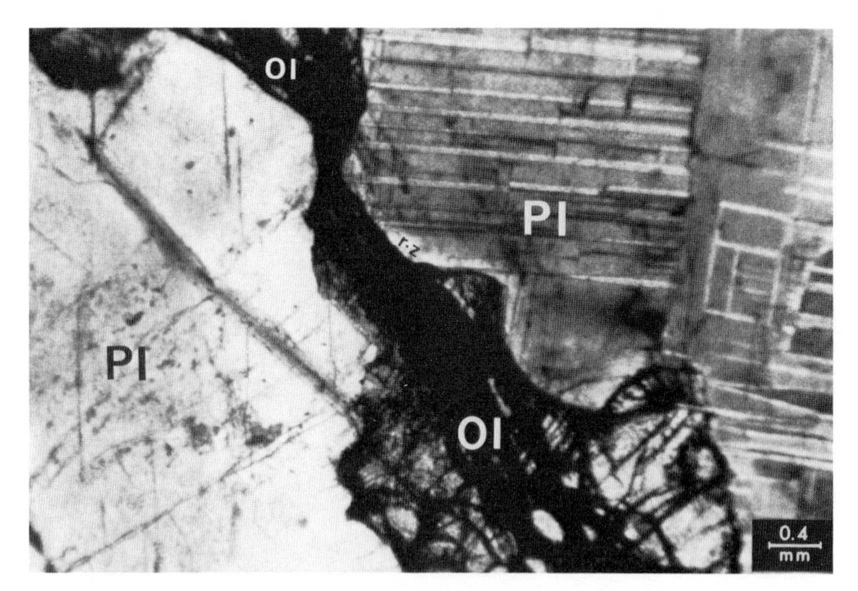

Fig. 465 Olivine (Fo 61 %) as an intergranular "relic" between plagioclase crystalloblasts. The olivine shows corroded outline and the plagioclase a reaction zone in contact with the olivine.
Pl = plagioclase
ol = olivine
r-z = reaction zone of the plagioclase as it comes in contact with the olivine.
Anorthosite. Negra Massiva, Angola. With crossed nicols.
(The olivine in the anorthosite of the Negra Massiva, Angola, is considered to represent (?) a mantle derivative, in contradistinction Griffin (1971) considers the Indre Sogn (Norway) anorthosite to be lower crust pushed upwards.)
The Fo content of the olivine has been determined by E. Mposkos.

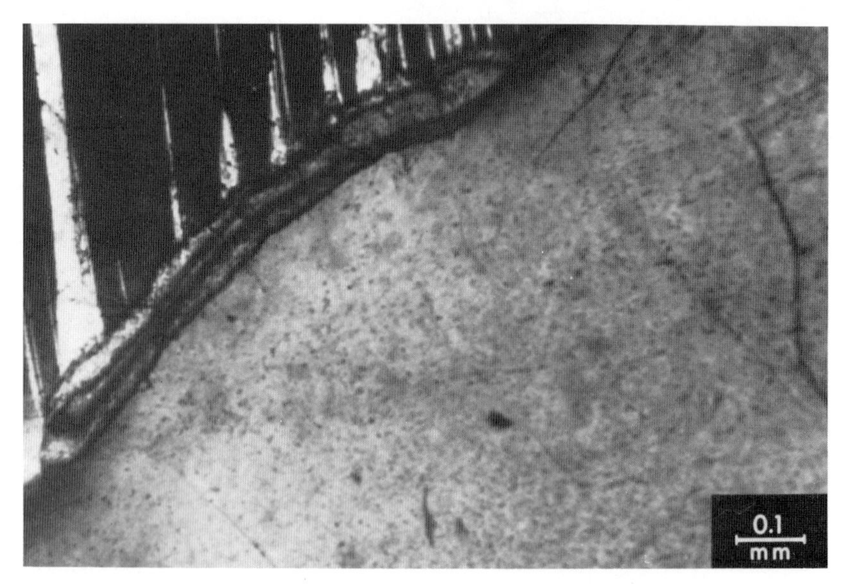

Figs. 466 and 467 Diopside intergranular between differently orientated plagioclases (An 70–80 %).
Pl = plagioclase
d = intergranular diopside
Anorthosite. Negra Massiva, Angola. With crossed nicols.

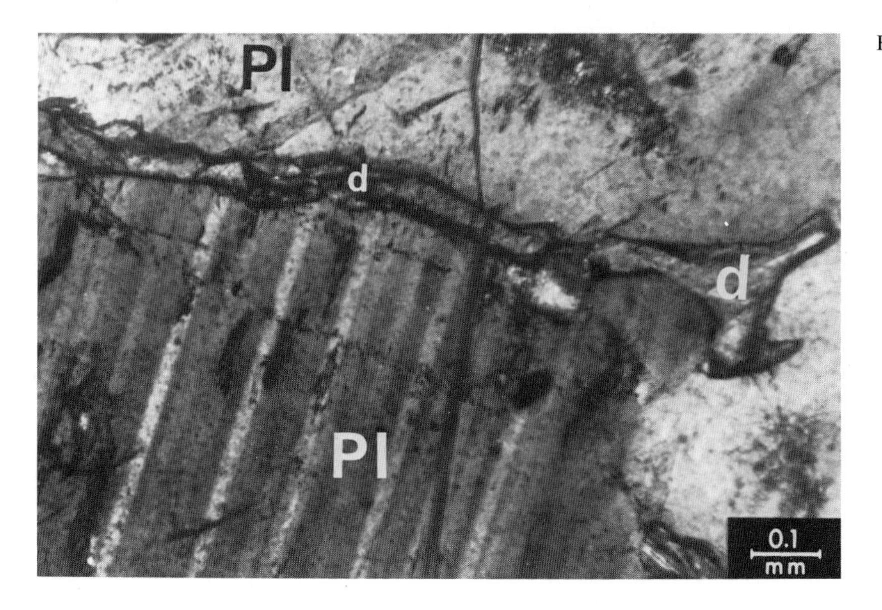

Fig. 467

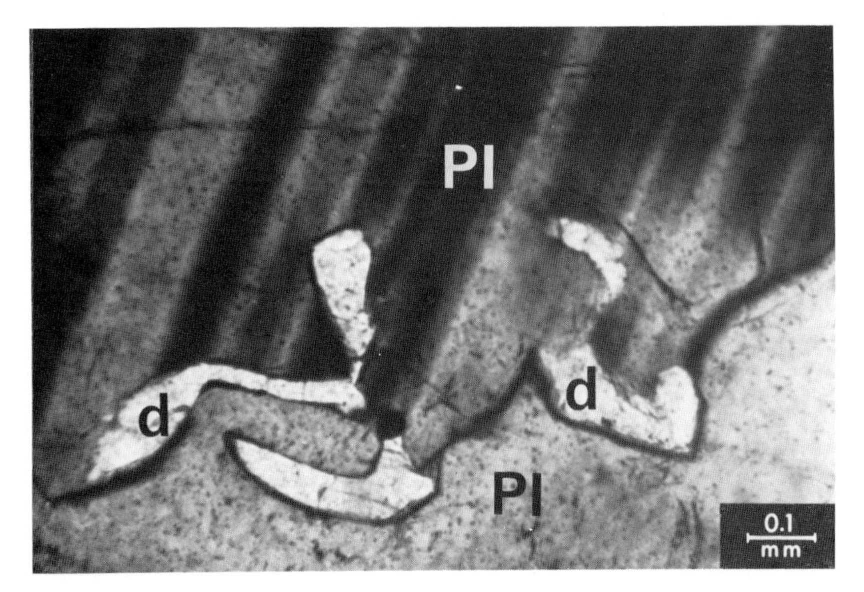

Fig. 468 Diopside intergranular between two differently orientated plagioclases (An content 70–80 %)
Pl = plagioclase
d = intergranular diopside
Anorthosite. Negra Massiva, Angola.
With crossed nicols.

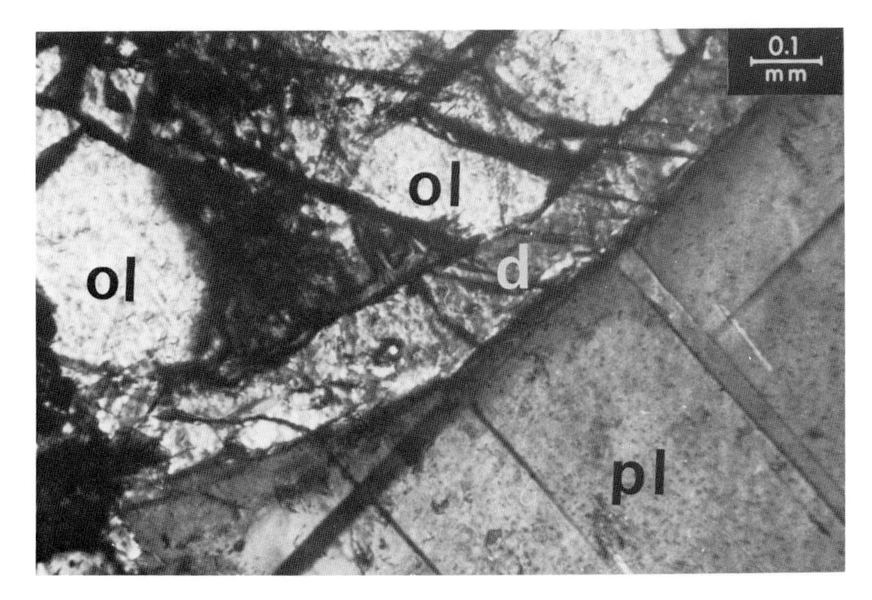

Fig. 469 A celyphytic diopside surrounding rounded olivine and being in contact with cross-lamellar polysynthetically twinned plagioclase.
ol = olivine
pl = plagioclase
d = diopside (celyphite) between olivine and plagioclase.
Anorthosite. Negra Massiva, Angola.
With crossed nicols.

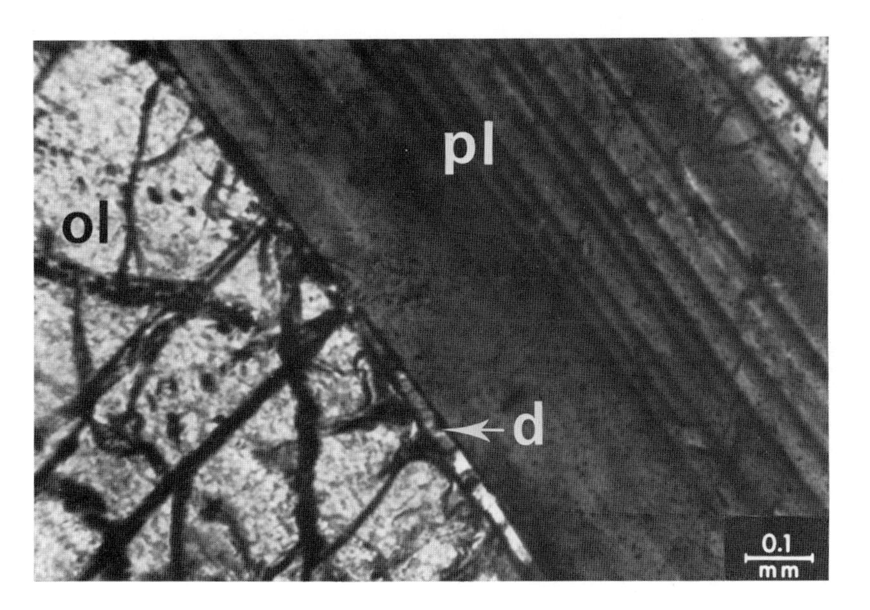

Fig. 470 A celyphitic diopside marginal to olivine and in contact with polysynthetically twinned plagioclase.
ol = olivine
d = diopside (celyphitic margin to olivine)
pl = plagioclase
Anorthosite, Negra Massiva, Angola.
With crossed nicols.

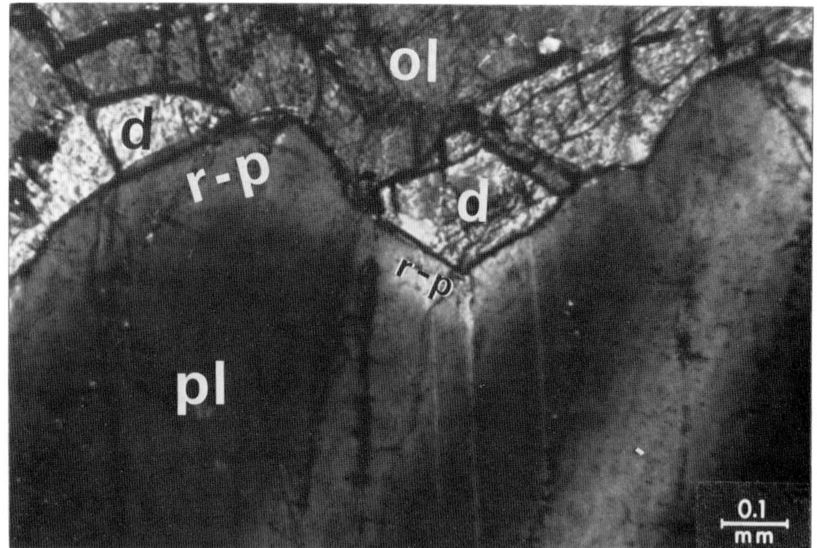

Fig. 471 Olivine in contact with plagioclase, often diopside is formed, in parts (later crystalloblastically) between the olivine and the plagioclase. The feldspar shows a reaction margin with the mafic minerals.
ol = olivine
d = diopside
pl = plagioclase
r-p = reaction margin of plagioclase.
Anorthosite, Negra Massiva, Angola.
With crossed nicols.

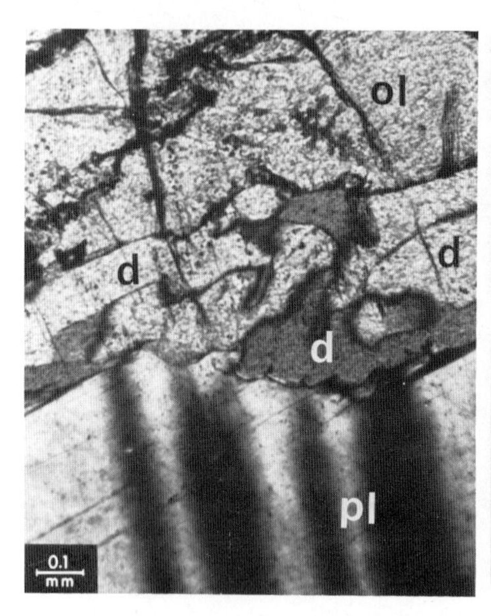

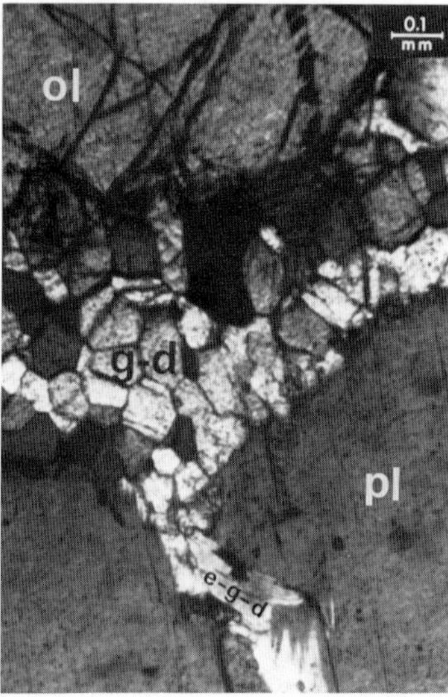

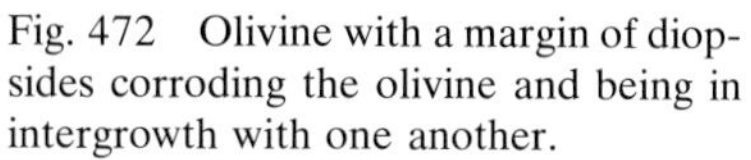

Fig. 472 Olivine with a margin of diopsides corroding the olivine and being in intergrowth with one another.
ol = olivine
d = diopsides (celyphitic intergrowth of two diopside phases).
pl = plagioclase
Anorthosite. Negra Massiva, Angola.
With crossed nicols.

Fig. 473 Olivine in contact with plagioclase and with a granular celyphitic margin of diopside.
The granular celyphitic diopside extends into the plagioclase.
ol = olivine
g-d = granular diopside
e-g-d = extension of the granular celyphitic diopside into the adjacent plagioclase (pl).
Anorthosite, Negra Massiva, Angola.
With crossed nicols.

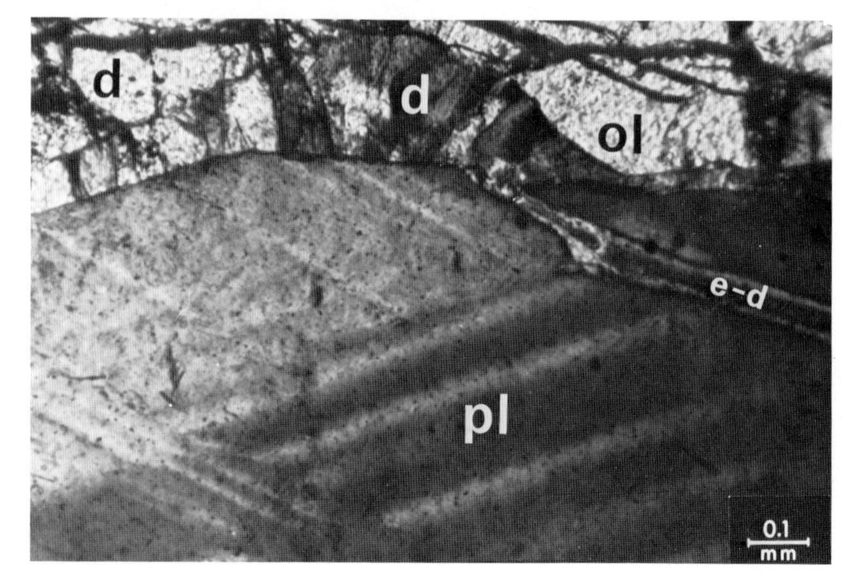

Fig. 474 Olivine with diopside celyphite consisting of crystal grains – differently orientated – and with a "vein-form" extension into an adjacent plagioclase.
ol = olivine
d = diopside (celyphite)
e-d = extension of diopside into the plagioclase
pl = plagioclase
Anorthosite, Negra Massiva, Angola.
With crossed nicols.

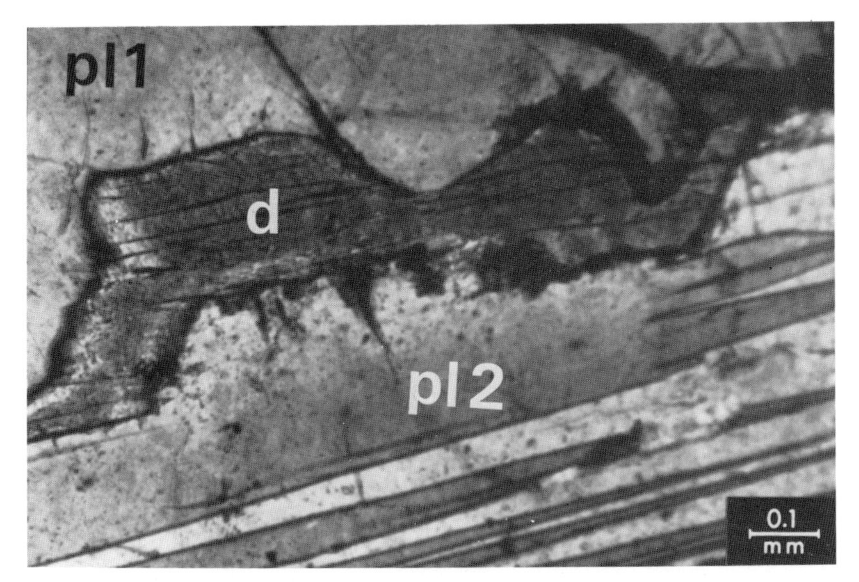

Fig. 475 Diopside crystalloblast following the intergranular (interleptonic) spaces between adjacent plagioclases.
d = diopside
pl_1 = plagioclase
pl_2 = plagioclase
Anorthosite, Negra Massiva, Angola.
With crossed nicols.

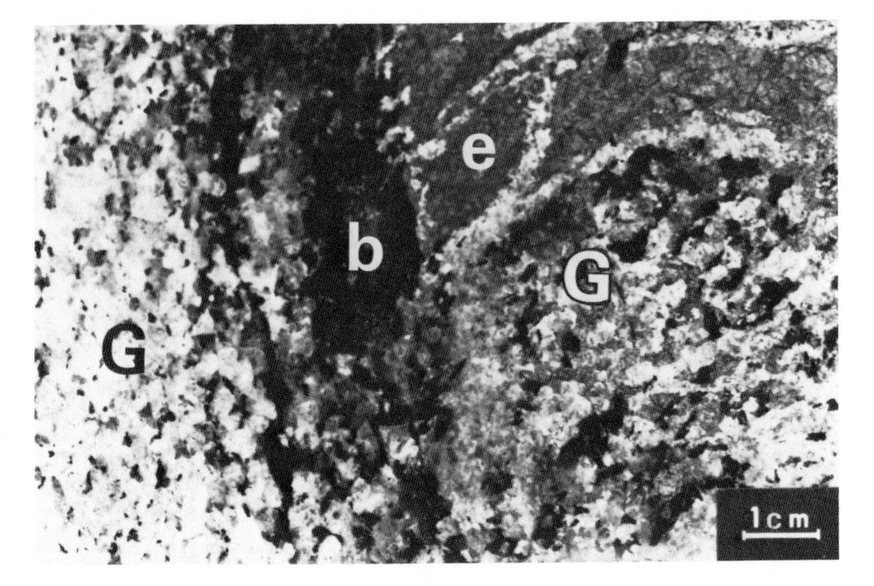

Fig. 476a Marble xenolith assimilated by the Harrar granite (Ethiopia).
G = granite
b = biotite enriched
e = epidote and with it remobilised calcite
Harrar, Ethiopia.
Hand specimen natural size.

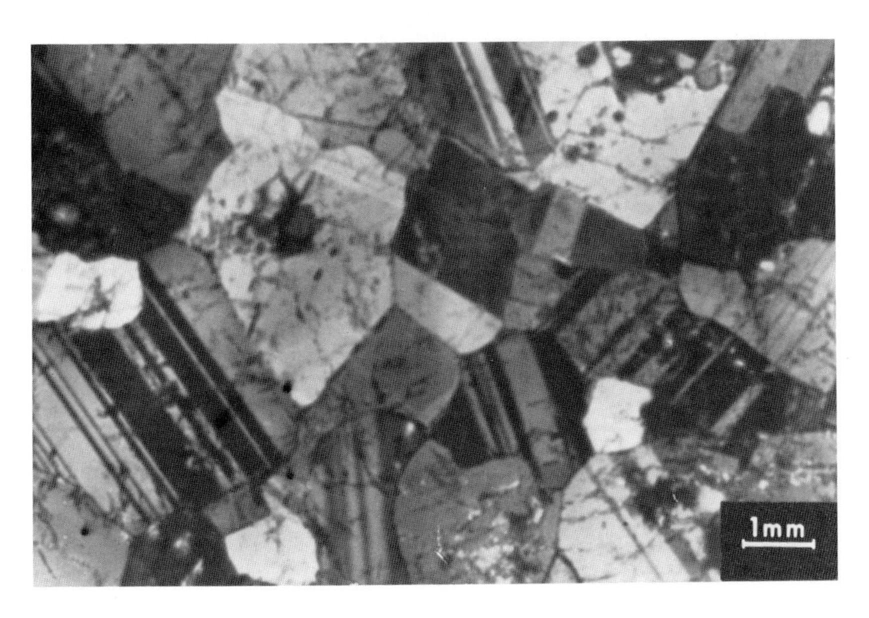

Fig. 476b As a result of marble-assimilation by the Harrar granite (see Fig. 476a), locally a plagioclase rich phase (An % 30–35) develops in the predominantly K-feldspar Harrar granite.
Harrar, Ethiopia.
With crossed nicols.

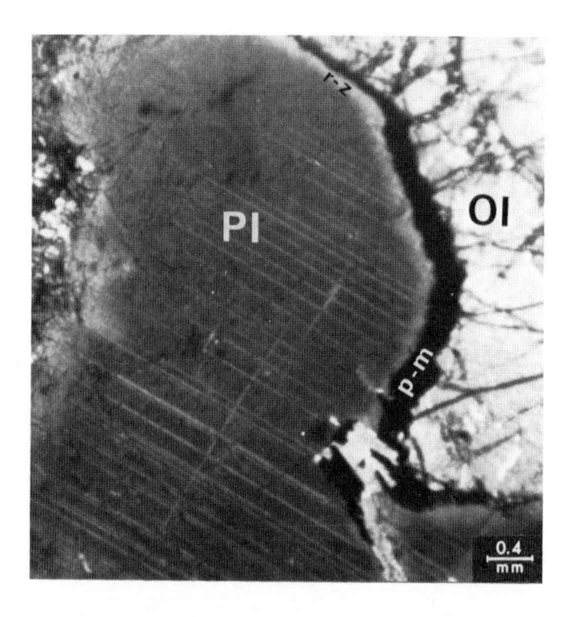

Fig. 477 Olivine "replaced" by later plagioclase crystalloblast, An 62%. At the front of the reaction, there is a margin of iron rich pigments (representing Fe originally present in the assimilated olivine). Also the plagioclase shows a reaction zone with different composition (An 82%), whereas the plagioclase itself shows An 62 % content.
Ol = olivine
Pl = plagioclase crystalloblast
r-z = reaction zone of plagioclase
p-m = iron-pigment margin
Anorthosite. Negra Massiva, Angola.
With crossed nicols.
(The increase of the anorthite content of the reaction zone (r-z) in contact with the mafic mineral (olivine, Fo 61%) is difficult to explain by the assimilation of the mafic phase involved, as was the case illustrated in Fig. 490 (a and b).)

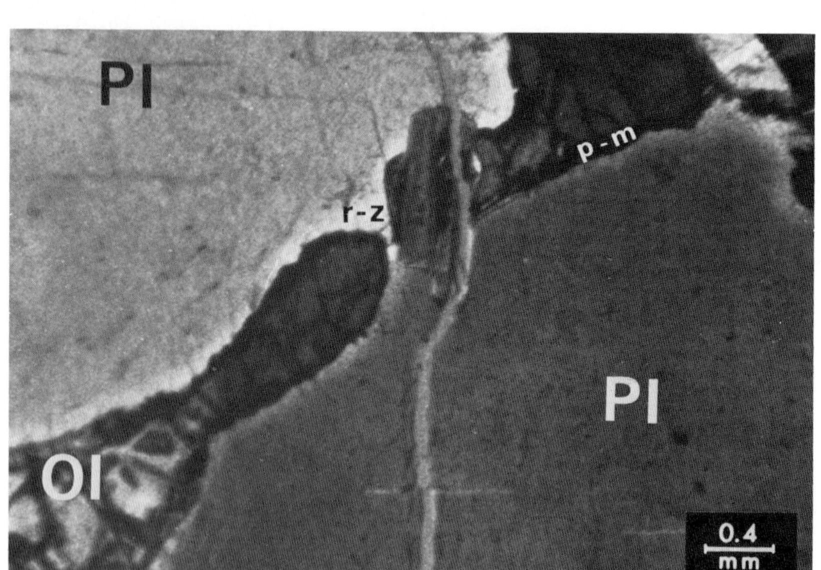

Fig. 478 Olivine relic between plagioclase crystalloblast. The plagioclase in contact with the olivine shows a reaction margin of different anorthite-content (see Fig. 477).
The olivine in contact with the plagioclase crystalloblast shows a margin rich in Fe-pigments (apparently the Fe that was originally present in the portion of the replaced olivine by the plagioclase crystalloblast and which could not be incorporated in the feldspar lattice).
Ol = olivine relic between plagioclase crystalloblasts
Pl = plagioclase crystalloblasts in contact with the olivine
p-m = pigment margin
r-z = reaction zone of the plagioclase.
Anorthosite. Negra Massiva, Angola.
With crossed nicols.

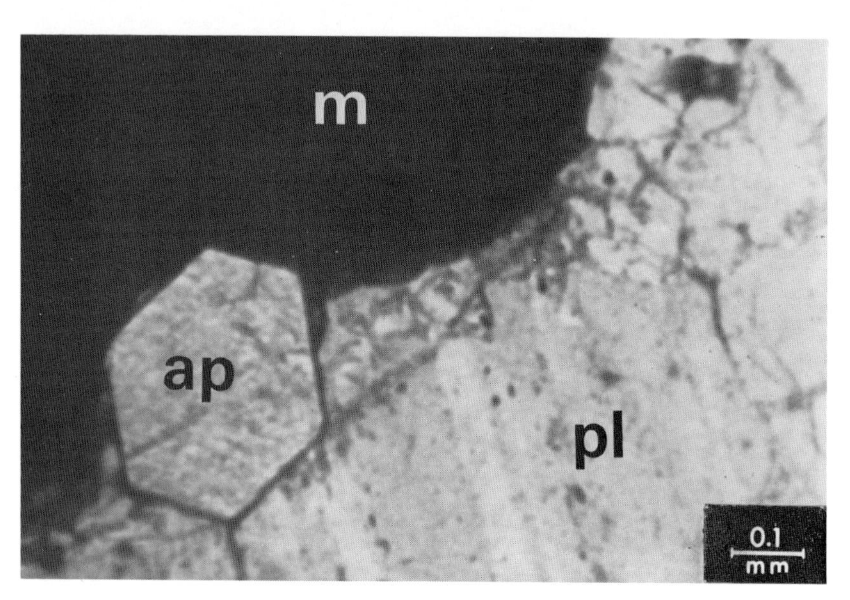

Fig. 479 Apatite idioblast transversing the boundary plagioclase-magnetite.
ap = apatite idioblast
pl = plagioclase
m = magnetite
Anorthosite, Negra Massiva, Angola.
With crossed nicols.

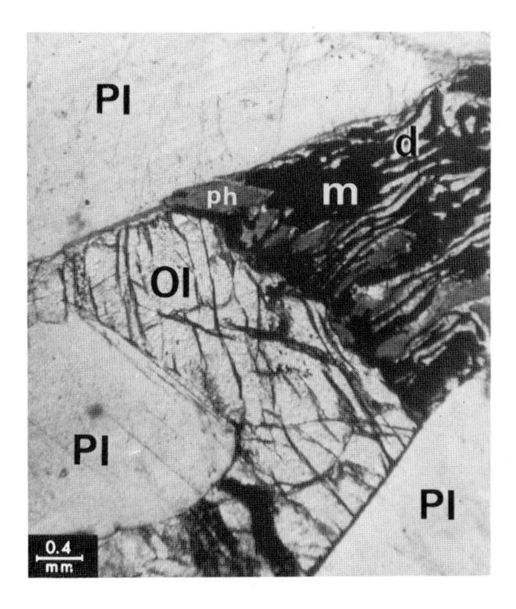

Fig. 480 Olivine relic restricted between plagioclase crystalloblasts. Also magnetite and diopside in symplectic intergrowth.
Phlogopite is also present.
Pl = plagioclase
Ol = olivine
m/d = magnetite in symplectic intergrowth with diopside.
ph = phlogopite
Anorthosite. Negra Massiva, Angola.
Without crossed nicols.

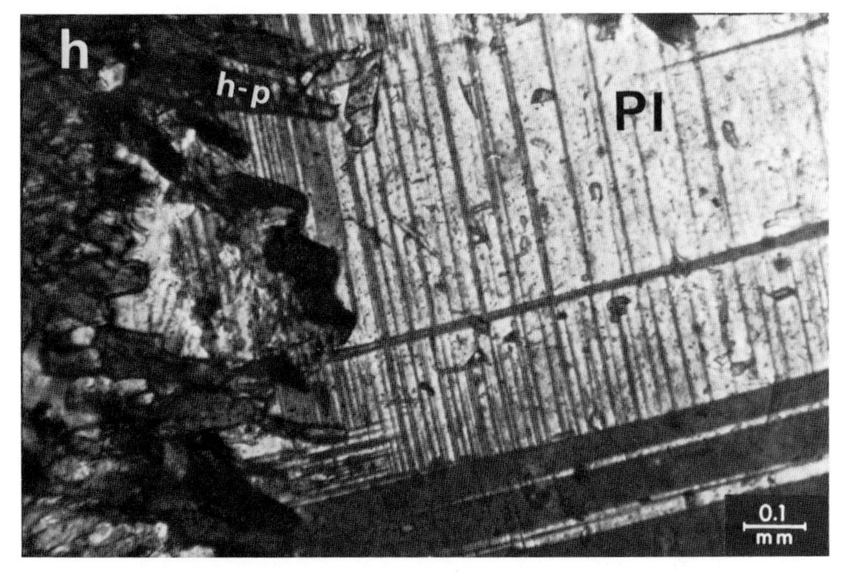

Fig. 481 Hornblende crystalloblasts "intercrystalline" between the plagioclase crystalloblasts extend into the feldspar and replace it.
Pl = plagioclase
h = hornblende
h-p = hornblende crystalloblast invading the plagioclase.
Anorthosite. Mount Royal, Canada.
With crossed nicols.

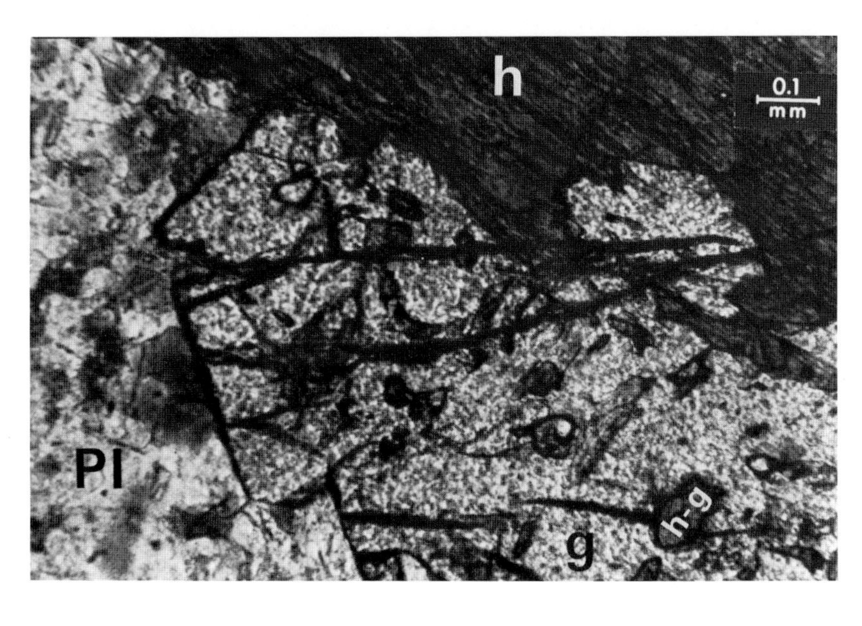

Fig. 482 Hornblende and garnet (grossular) crystalloblasts. The grossular encloses hornblende relics.
Pl = plagioclase
h = hornblende
g = grossular
h-g = hornblende enclosed in the garnet crystalloblast
Anorthosite. Mount Royal, Canada.
With crossed nicols.

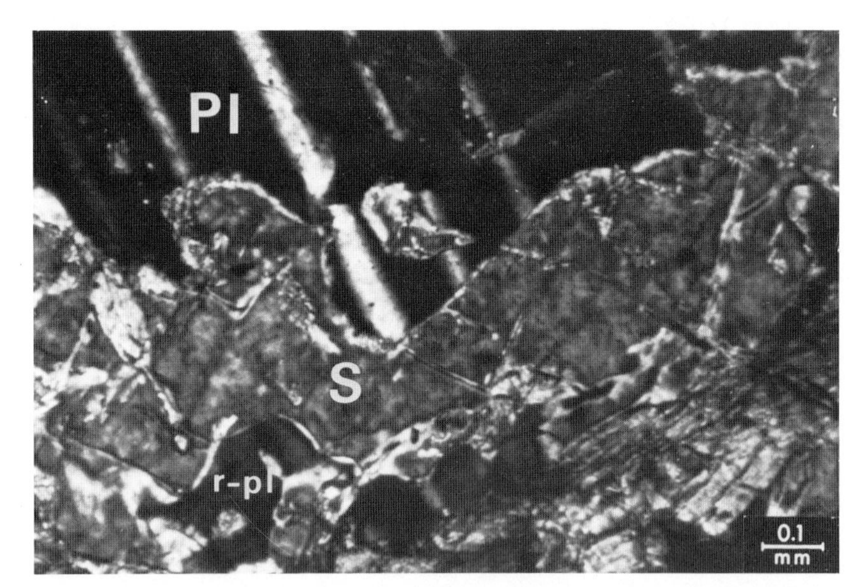

Fig. 483 Plagioclase replaced by scapolite. Relics of the plagioclase are included in the scapolite.
Pl = plagioclase
S = scapolite
r-pl = relics of plagioclase included in the scapolite.
Anorthosite. Mount Royal, Canada.
With crossed nicols.

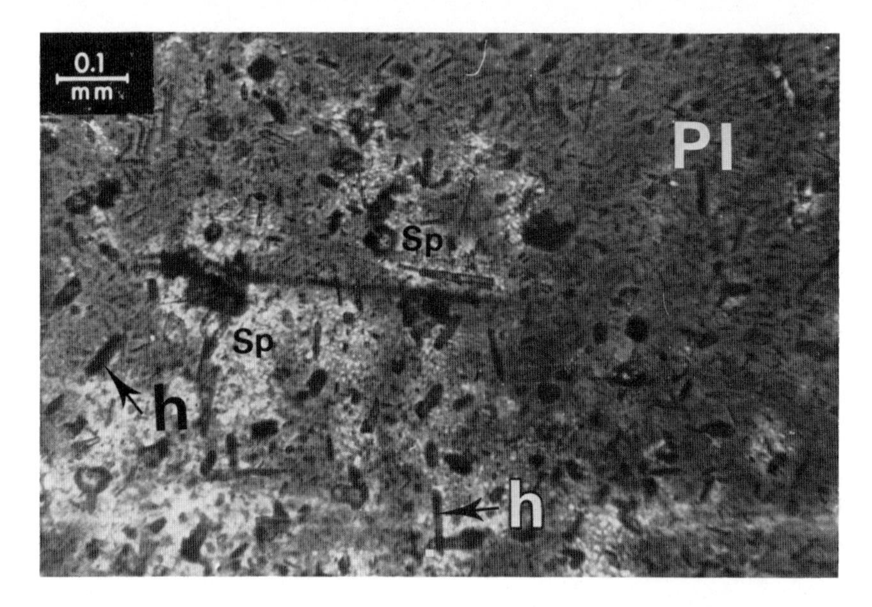

Fig. 484 Plagioclase indicating a spot-alteration, often accompanied by micaceous small crystals or hornblende formation.
Pl = plagioclase
Sp = Spot-alteration of the plagioclase.
h = small hornblendes (due to "metasomatism").
Anorthosite, Rogland, Norway.
With crossed nicols.

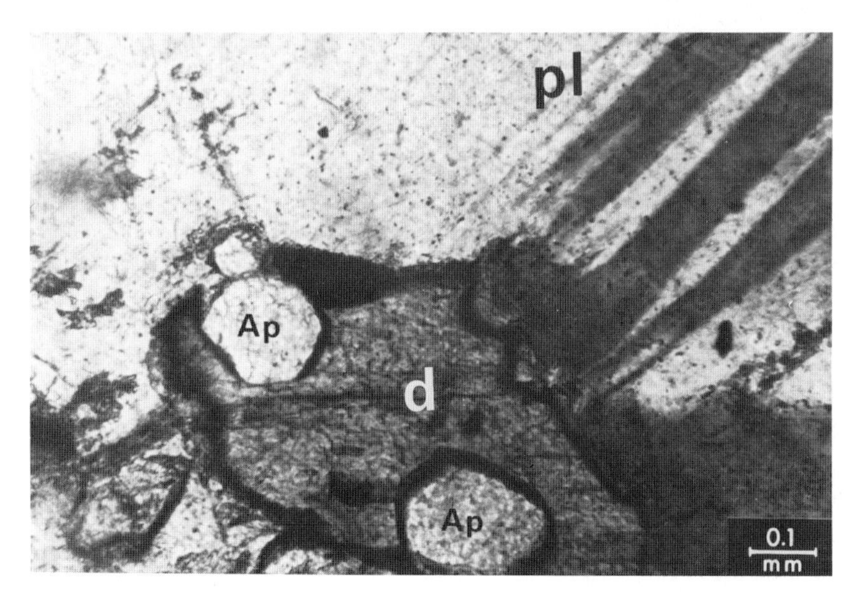

Fig. 485 Idiomorphic apatites partly enclosed by later formed diopside.
Ap = apatite
d = diopside
Pl = plagioclase
Anorthosite, Rogland, Norway.
With crossed nicols.

Fig. 486 Polysynthetic twinning of a plagioclase megablast transversed by a later more albitic plagioclase phase.
The plagioclase twinning is as "ghost" relic in the veinform plagioclase (see also Fig. 487).
Anorthosite, Rogland, Norway.
With crossed nicols.

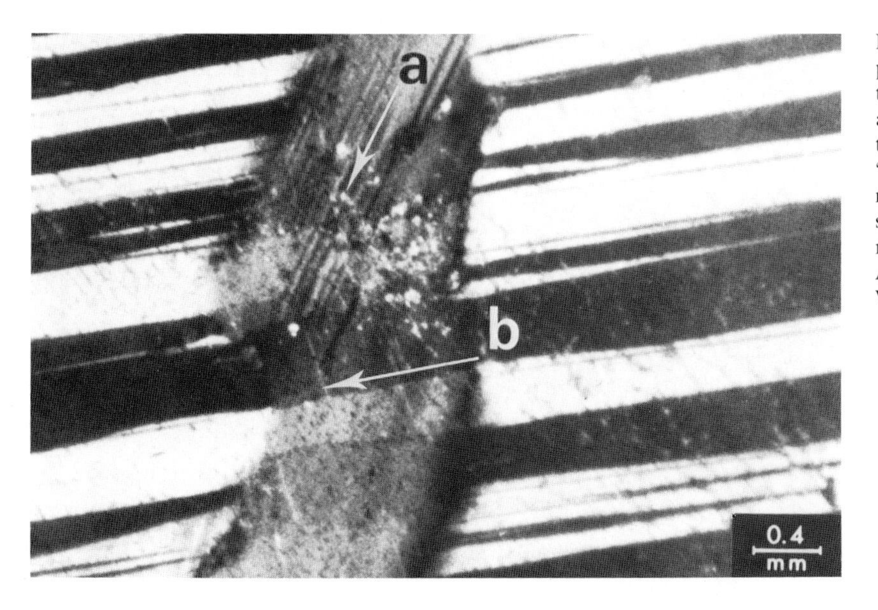

Fig. 487 (detail of Fig. 486) Veinform plagioclase transversing the polysynthetic twinning of a plagioclase megablast. In addition to the polysynthetic twinning of the veinform plagioclase (see arrow "a"), "ghost" relics of the polysynthetic twinning of the megablast plagioclase are present in the veinform plagioclase (see arrow "b").
Anorthosite. Rogland, Norway.
With crossed nicols.

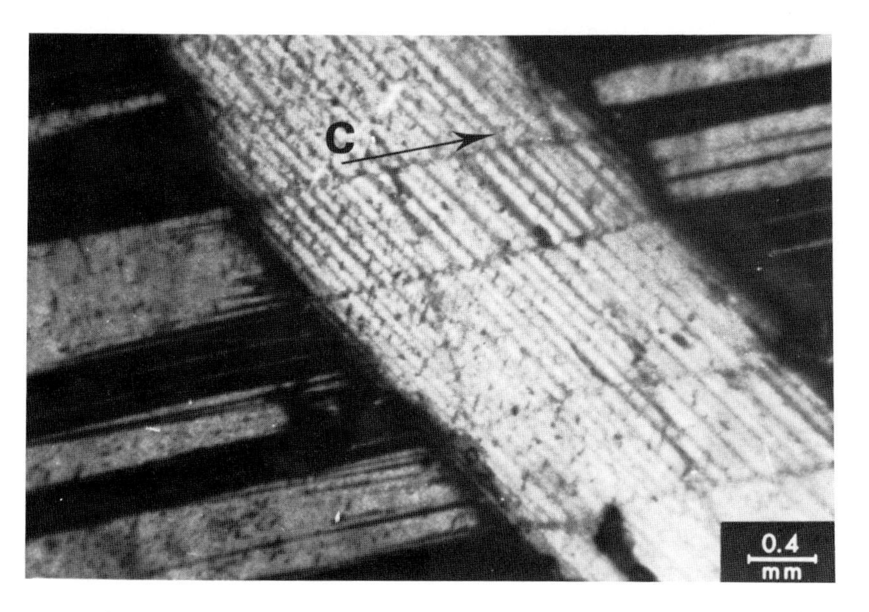

Fig. 488 The veinform plagioclase as it transverses the plagioclase megablast shows polysynthetic twinning exhibiting micro-displacement along directions which correspond to the "ghost" twinning of the megablast (see arrow "c"). Also see Figs. 486 and 487.
Anorthosite. Rogland, Norway.
With crossed nicols.

Fig. 489 Within the complex metasomatic processes of anorthitisation, antiperthitisation takes place in the anorthositic plagioclase megablasts.
Pl = plagioclase megablast
Ant = antiperthitisation (a late metasomatic phase in the plagioclase megablast).
Anorthosite. Rogland, Norway.
With crossed nicols.

Fig. 490 (a and b) Pyroxene phenocryst digested and assimilated along a crack by later tecoblastic plagioclase. The following zones are recognizable in the plagioclase: Zone I = 38–39 % An; II = 44 % An; III = 31–32 % An; IV = 34 % An. Zone II is as wide as the pyroxene. It also contains relics of the pyroxene not assimilated by the blastic feldspar. In addition pigment rests are present, representing elements which could not be incorporated in the lattice of the plagioclase.
Olivine basalt, Khidane Meheret (between Entoto and British Embassy), Addis-Abeba, Ethiopia.
With crossed nicols.

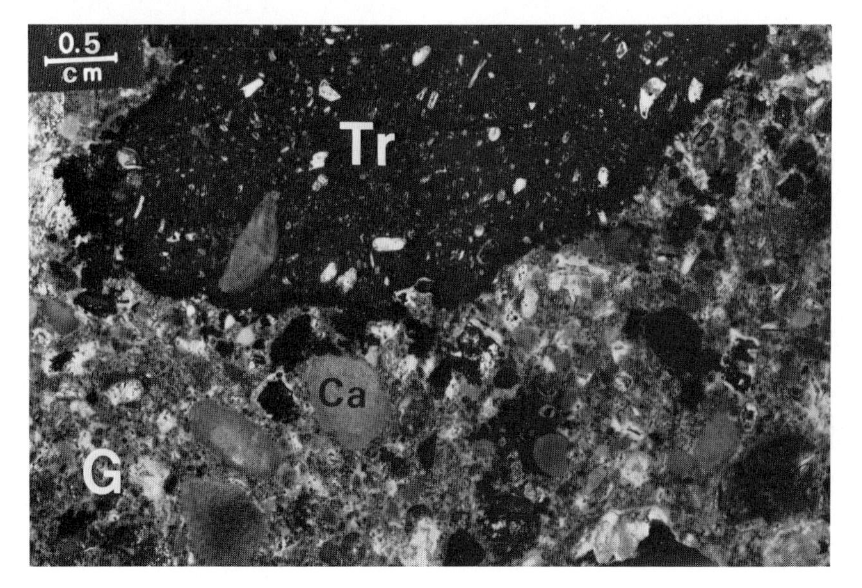

Fig. 491 Shows a kimberlitic tuffite (breccia) including a trachytic fragment and rounded marble pieces.
Tr = trachytic fragment (kimberlitic breccia)
Ca = marble piece (rounded fragment)
G = kimberlitic groundmass
Kimberlitic breccia. Yakutia, Siberia, USSR.

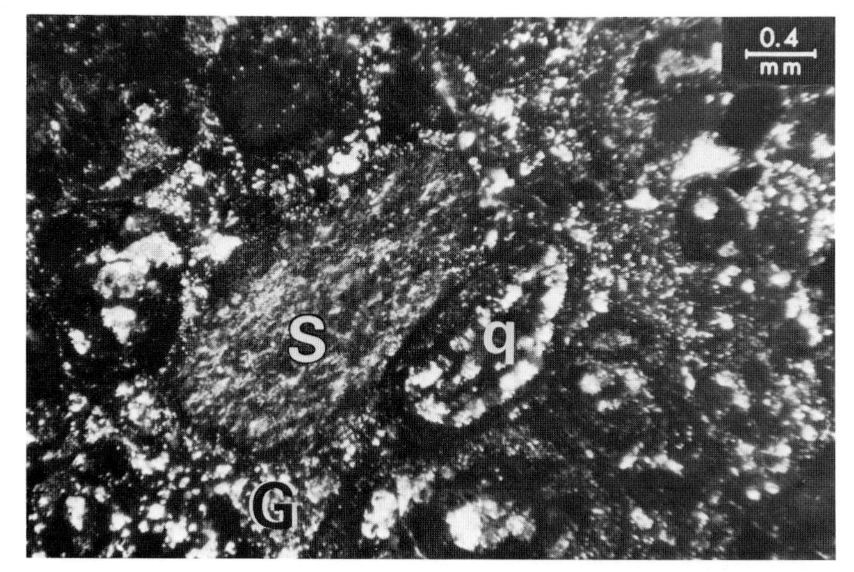

Fig. 492 Shows a general view of kimberlitic tuffite including quartzitic and serpentine "rounded" fragments.
S = serpentine
q = quartzitic fragment rounded
G = groundmass
Kimberlitic breccia.
Makkovik, Labrador, Canada.
With crossed nicols.

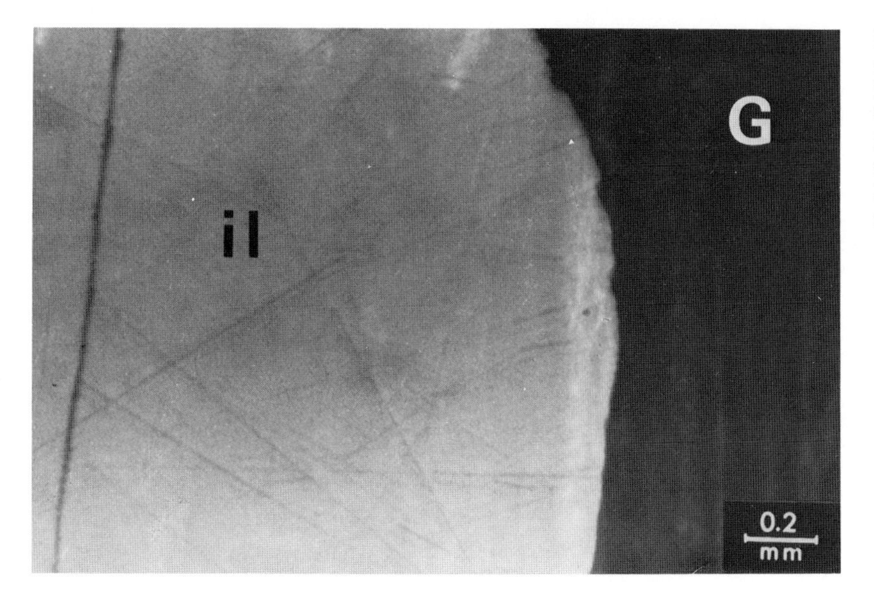

Fig. 493 Ilmenite with a "rounded" outline in kimberlitic groundmass.
il = ilmenite
G = kimberlitic groundmass
Kimberlite. M'sipashi plateau of Kundelungu, Zaire.
Polished section (with oil-immersion, one nicol).

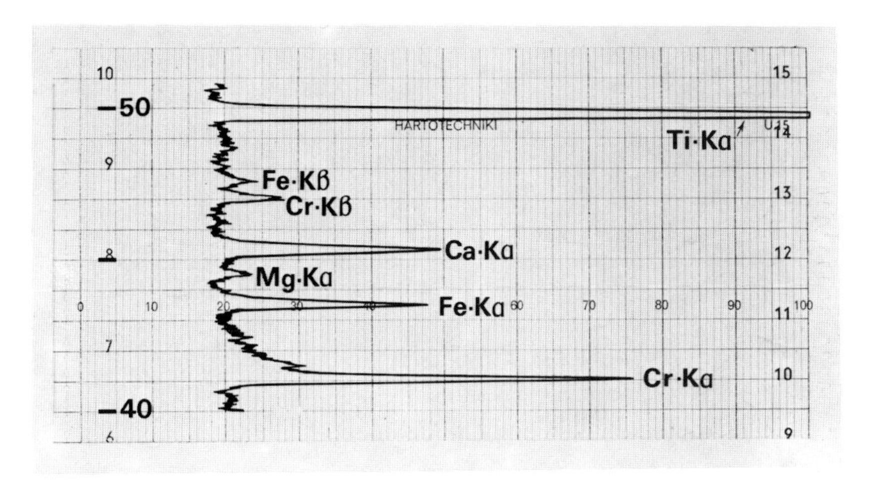

Fig. 494 X-ray fluorescence spectroanalysis of rounded ilmenite from the tuffitic kimberlite of M'sipashi plateau Kundelungu, Zaire.
Diagram by Dr. A. Vgenopoulos.

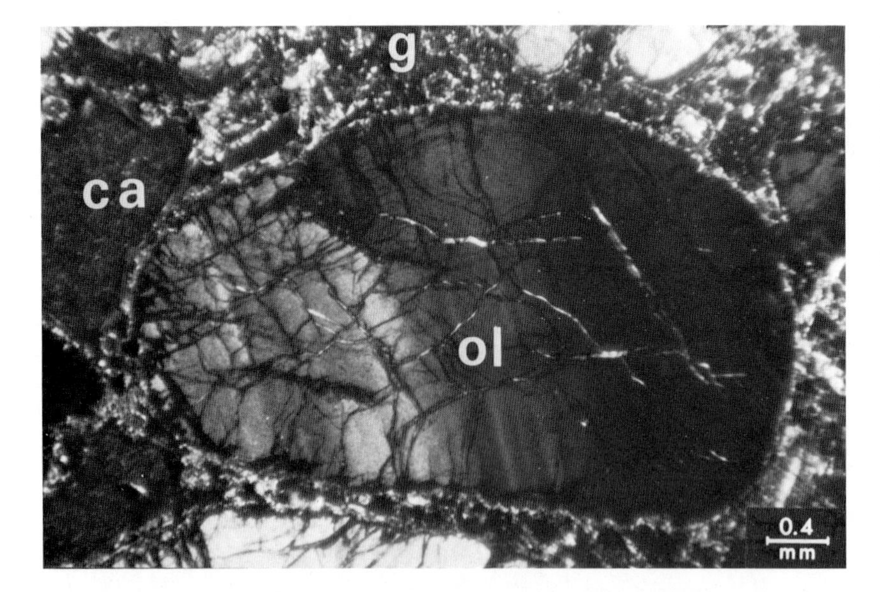

Fig. 495 Olivine xenocrysts rounded and indicating deformation lamellae in a tuffitic kimberlitic groundmass.
ol = rounded olivine xenocryst indicating deformation lamellae.
g = tuffitic kimberlitic groundmass
ca = calcite in the tuffitic groundmass
Tuffitic kimberlite.
Wesselton, South Africa.
With crossed nicols.

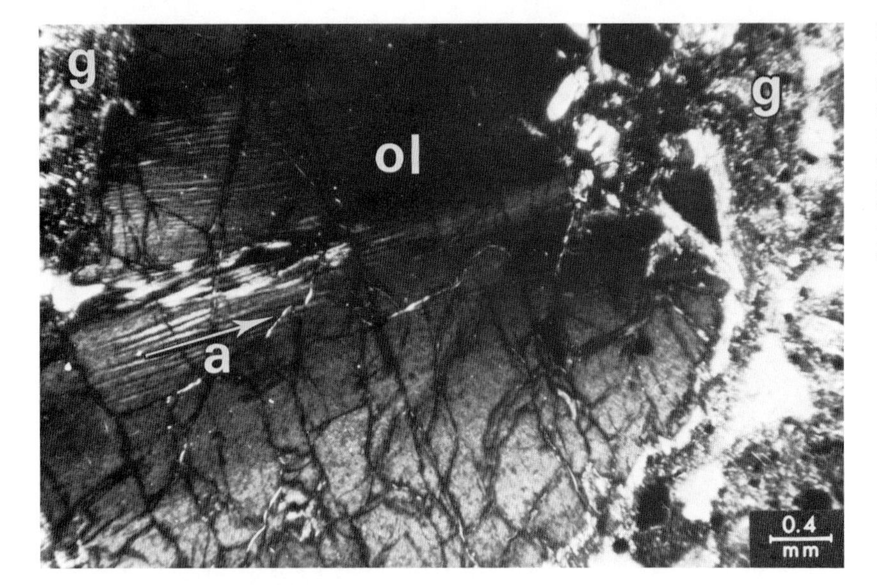

Fig. 496 An olivine xenocryst, actually a fragment, with a reaction margin in a tuffitic kimberlitic groundmass. The olivine xenocryst indicates deformation effects (arrow a).
ol = olivine xenocryst
g = tuffitic groundmass
Tuffitic kimberlite.
Kimberley, South Africa.
With crossed nicols.

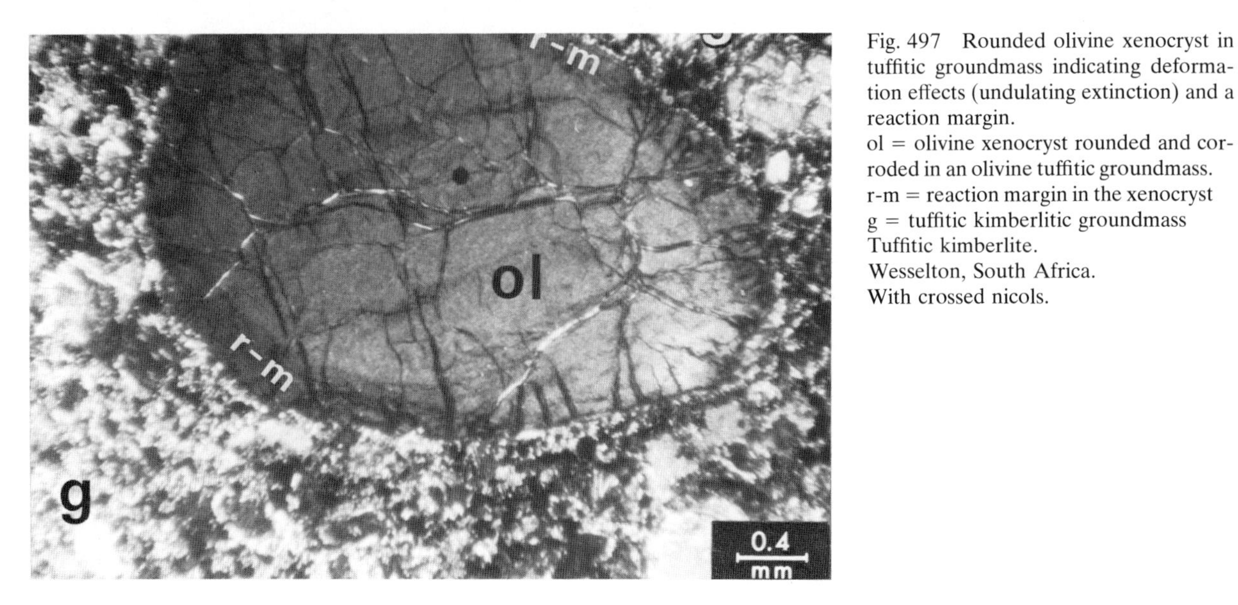

Fig. 497 Rounded olivine xenocryst in tuffitic groundmass indicating deformation effects (undulating extinction) and a reaction margin.
ol = olivine xenocryst rounded and corroded in an olivine tuffitic groundmass.
r-m = reaction margin in the xenocryst
g = tuffitic kimberlitic groundmass
Tuffitic kimberlite.
Wesselton, South Africa.
With crossed nicols.

Fig. 498 Olivine and spinel in graphic intergrowth with the olivine as rounded and corroded "fragment" in kimberlitic breccia. Deformation twinning is exhibited by the olivine fragment.
Ol = olivine
s = spinel in graphic intergrowth with the olivine.
G = kimberlitic groundmass
Tuffitic kimberlite.
Kimberley, South Africa.
With crossed nicols.

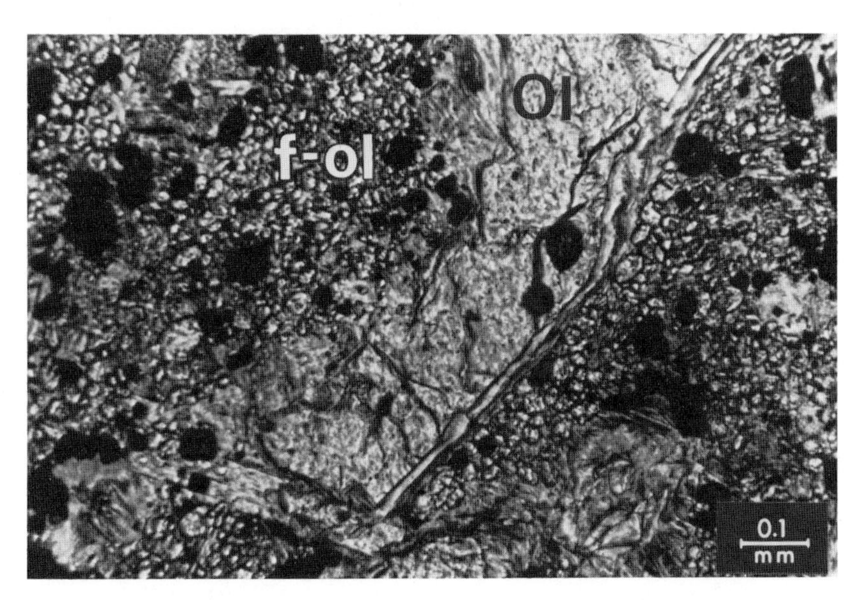

Fig. 499 Kimberlitic breccia exhibiting both large and fine olivine fragmentation.
Ol = olivine
f-ol = fine olivine fragments (comprising the greatest part of the kimberlitic groundmass).
ore-minerals (black).
Kimberlite breccia.
Kimberley, South Africa.
Without crossed nicols.

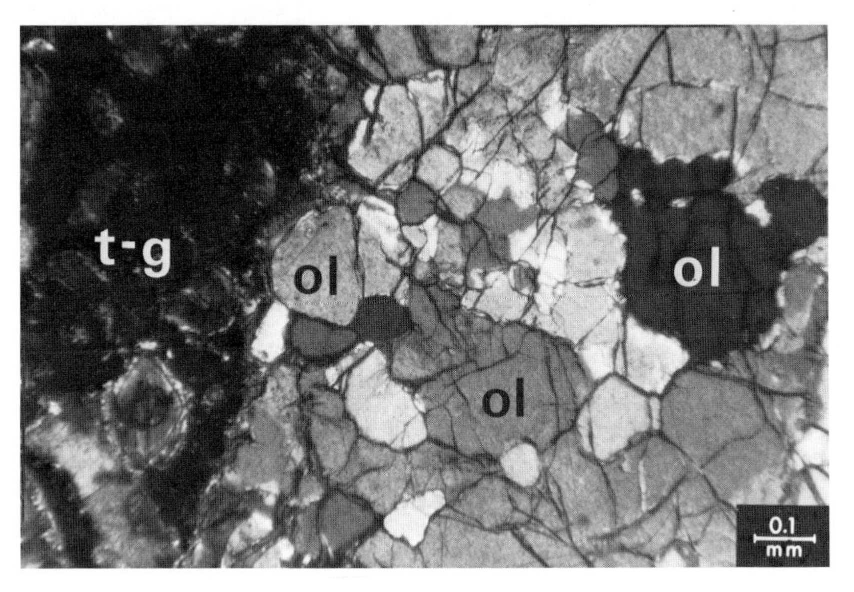

Fig. 500 A coarse grained olivine aggregate with olivine grains differently orientated, in tuffitic-kimberlitic groundmass.
ol = olivines
t-g = tuffitic kimberlitic groundmass
Tuffitic Kimberlite. Dufoitspan, South Africa.
With crossed nicols.

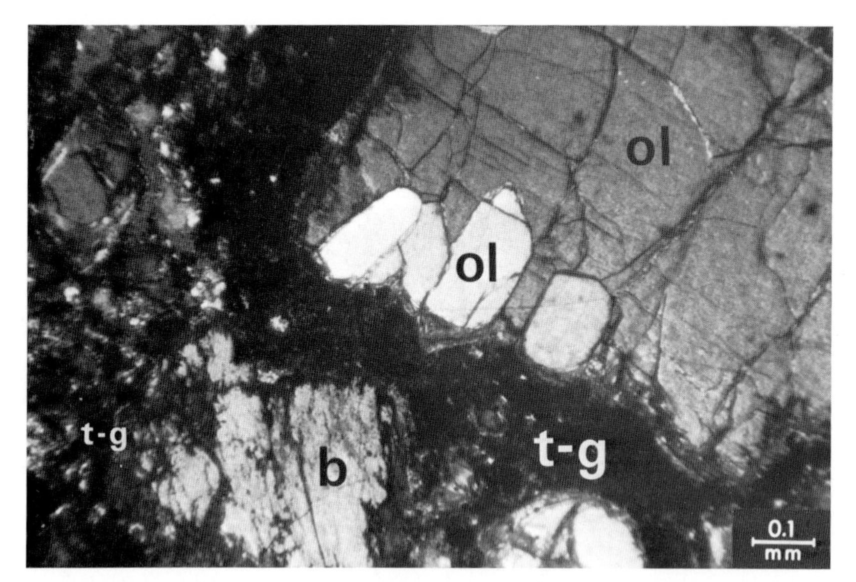

Fig. 501 Olivine phenocrysts with parts differently orientated in tuffitic-kimberlitic groundmass in which biotite is also present.
ol = olivine phenocryst
b = biotite
t-g = tuffitic kimberlitic groundmass
Tuffitic Kimberlite. Dufoitspan, South Africa.
With crossed nicols.

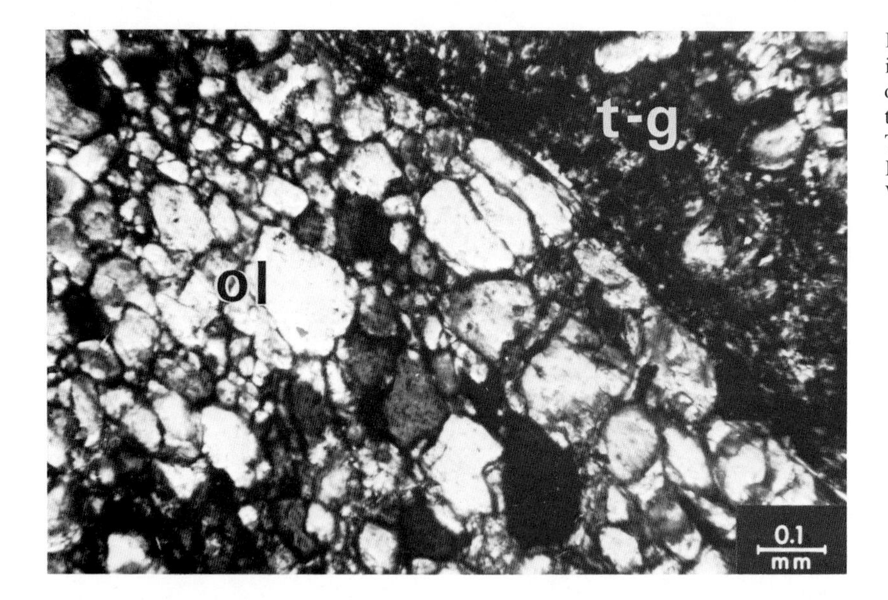

Fig. 502 A granular olivine aggregate in tuffitic kimberlitic groundmass.
ol = olivine aggregate
t-g = tuffitic kimberlitic groundmass
Tuffitic kimberlite.
Kimberley, South Africa.
With crossed nicols.

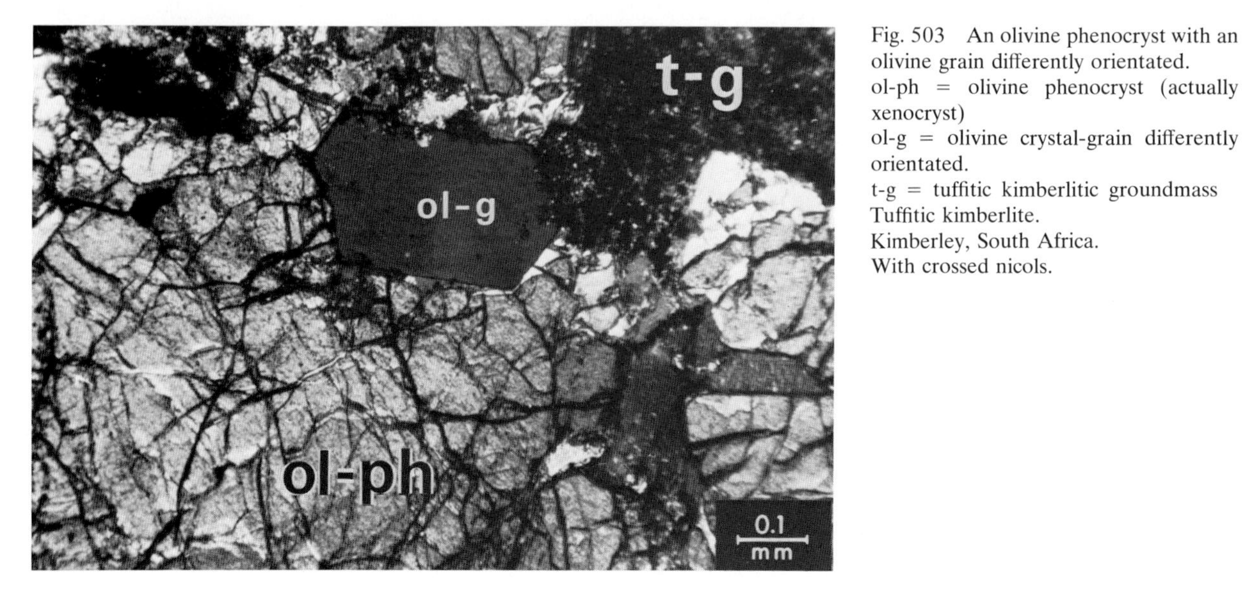

Fig. 503 An olivine phenocryst with an olivine grain differently orientated.
ol-ph = olivine phenocryst (actually xenocryst)
ol-g = olivine crystal-grain differently orientated.
t-g = tuffitic kimberlitic groundmass
Tuffitic kimberlite.
Kimberley, South Africa.
With crossed nicols.

Fig. 504 Rounded apatite and biotite fragments in tuffitic kimberlitic groundmass, containing calcite.
Ap = rounded apatite
bi = biotite fragment
G = kimberlitic groundmass
ap-I = idiomorphic apatite
Kimberlite.
Makkovik, Labrador, Canada.
With crossed nicols.

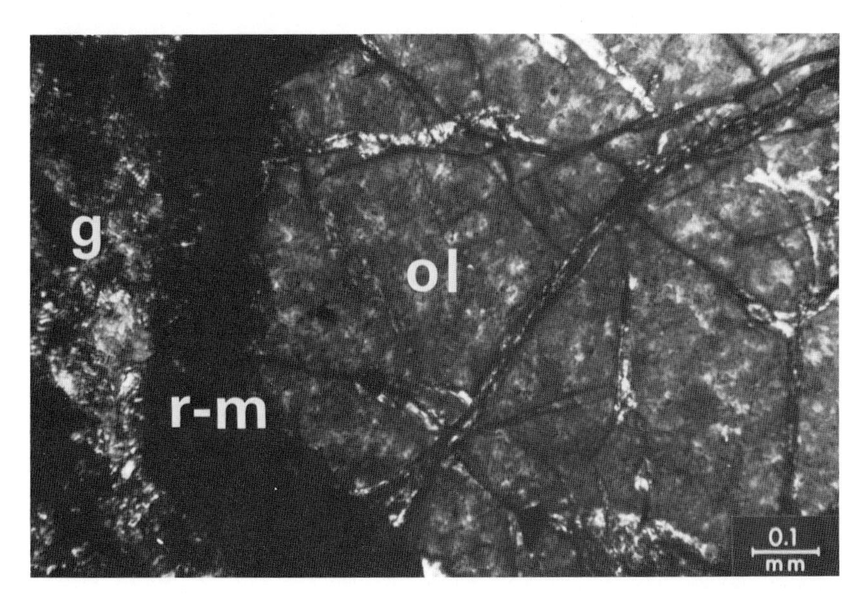

Fig. 505 An idiomorphic olivine with reaction margin (iron oxides) in a tuffitic kimberlitic groundmass.
ol = olivine
g = tuffitic groundmass
r-m = reaction margin
Kimberlite.
Makkovik, Labrador, Cannada.
With crossed nicols.

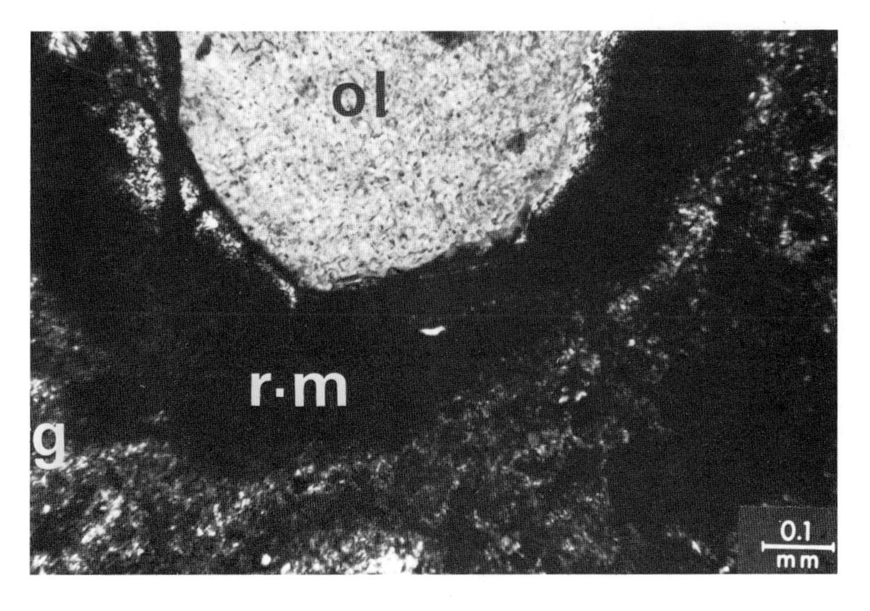

Fig. 506 A corroded olivine with a reaction margin (iron-oxides) in a tuffitic kimberlitic groundmass.
ol = olivine
g = tuffitic kimberlitic groundmass
r-m = reaction margin
Kimberlite.
Makkovik, Labrador, Canada.
With crossed nicols.

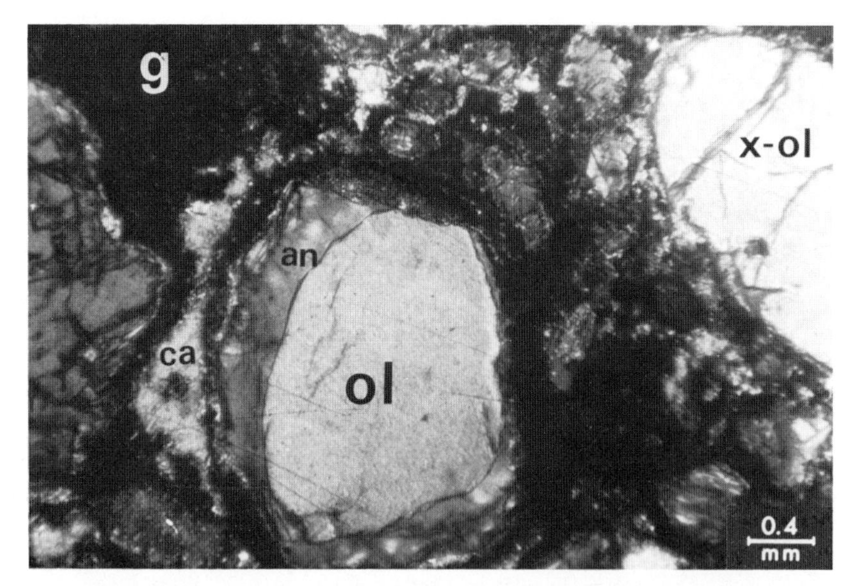

Fig. 507 Rounded olivine xenocryst (x-ol) and an olivine with an antigoritic margin in tuffitic kimberlitic groundmass with calcite.
ol = olivine
an = antigorite margin surrounding olivine
ca = calcite
g = tuffitic groundmass
Kimberlite. Kimberley. South Africa.
With crossed nicols.

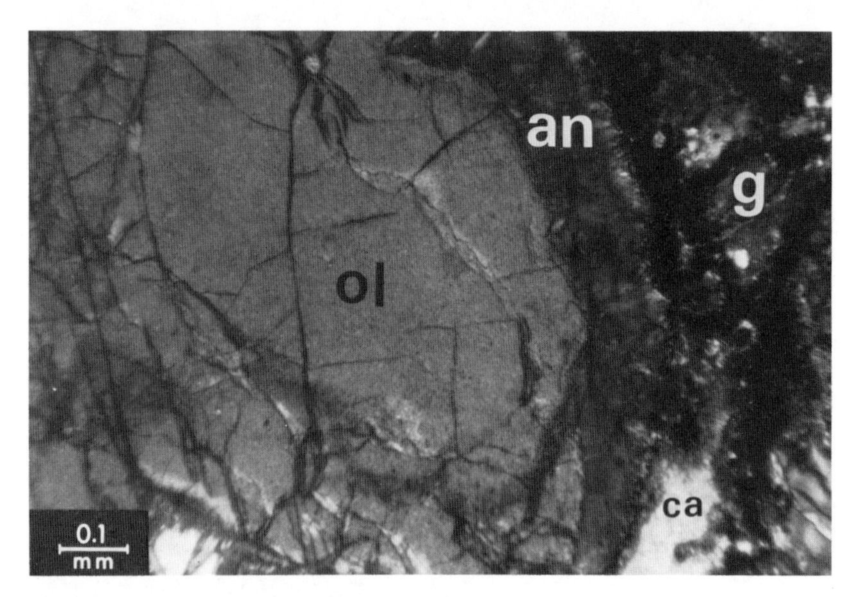

Fig. 508 Olivine with an antigorite margin in tuffitic kimberlite.
ol = olivine
an = antigorite
ca = calcite
g = tuffitic groundmass
Kimberlite. Kimberley, South Africa.
With crossed nicols.

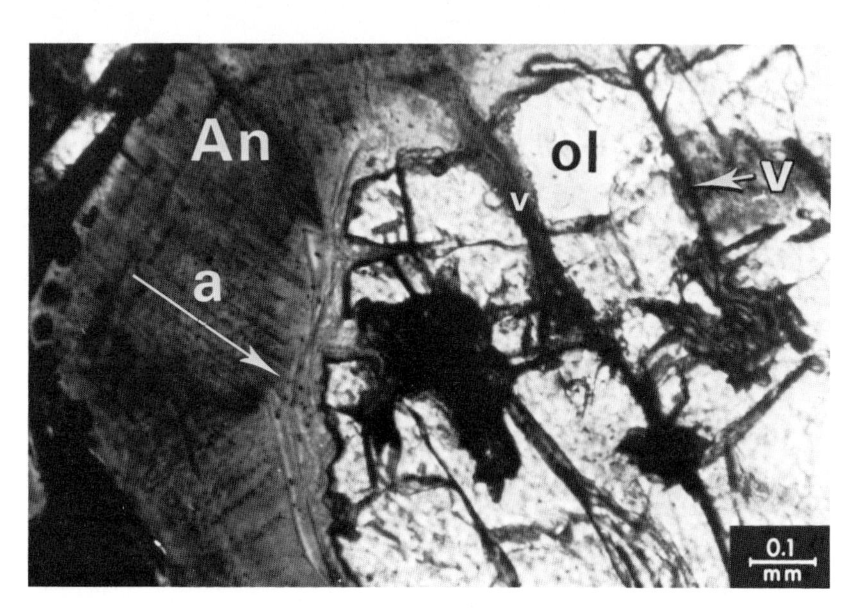

Fig. 509 Idiomorphic olivine with an antigorite margin. The antigorite fibres are perpendicular to the olivine margin (see arrow "a"). Also antigorite veinlets continue into the unaffected central olivine.
An = antigorite margin
ol = olivine idiomorph
v = veinlets of antigorite extending into the olivine.
Kimberlite.
Makkovik, Labrador, Canada.
With crossed nicols.

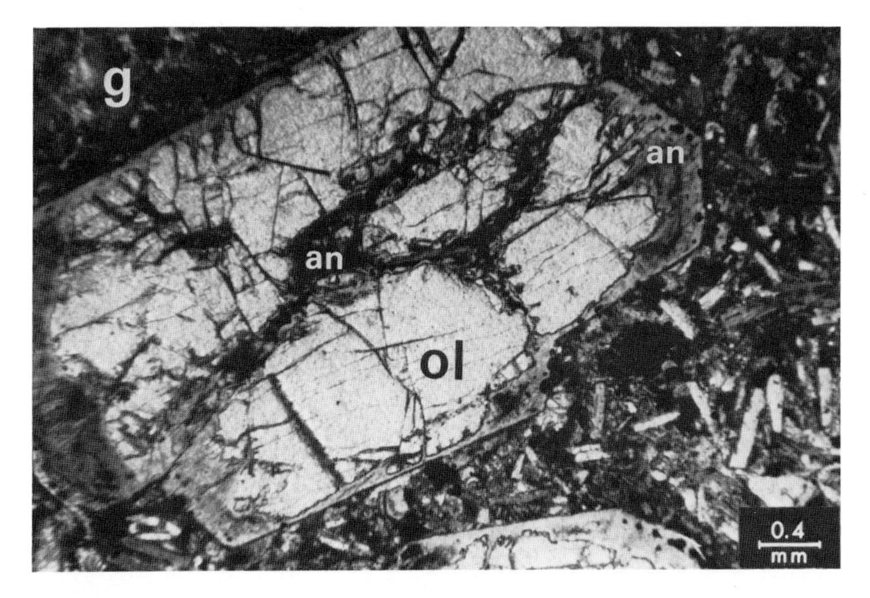

Fig. 510 Idiomorphic olivine in kimberlitic groundmass with an antigorite margin often extending inside the olivine.
ol = idiomorphic olivine
g = kimberlitic groundmass
an = antigorite, with extension in the olivine Kimberlite. Makkovik, Labrador, Canada.
With crossed nicols.

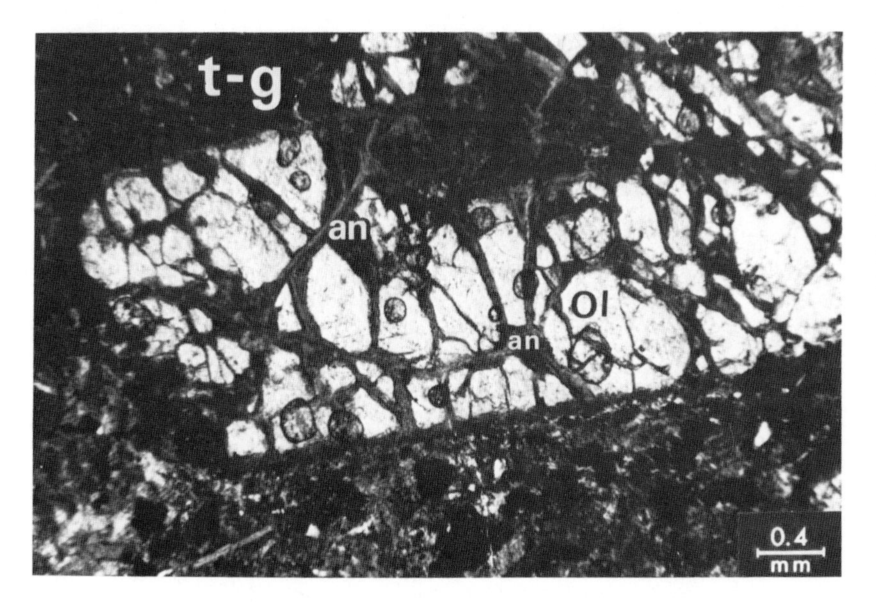

Fig. 511 Olivine phenocrysts in tuffitic-kimberlitic groundmass. The olivine phenocryst is transversed by antigorite veinlets often branching within the olivine.
ol = olivine phenocryst
an = antigorite veinlets
t-g = tuffitic kimberlitic groundmass
Kimberlite.
Makkovik, Labrador, Canada.
With crossed nicols.

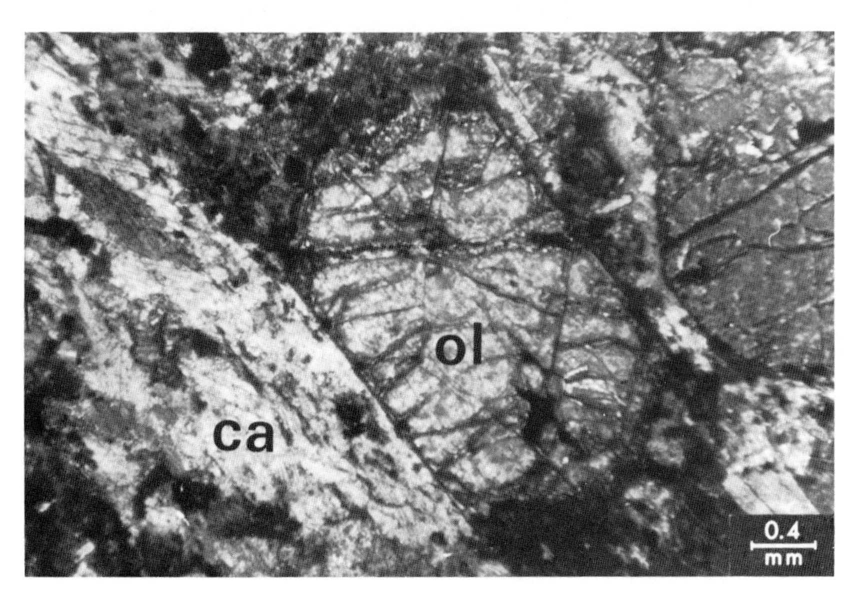

Fig. 512 Idiomorphic olivine in tuffitic-kimberlitic groundmass, in which calcite is predominant.
ol = idiomorphic olivine
ca = calcite, tuffitic-kimberlitic groundmass
Kimberlite.
Makkovik, Labrador, Canada.
With crossed nicols.

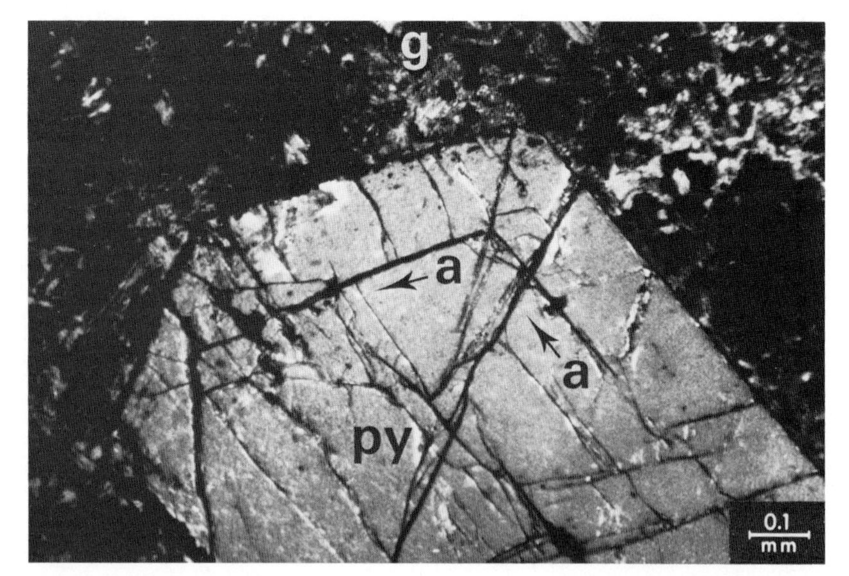

Fig. 513 Idiomorphic pyroxene in kimberlitic groundmass.
py = pyroxene
g = groundmass
Arrows "a" show zoning of the pyroxene
Kimberlite.
Makkovik, Labrador, Canada.
With crossed nicols.

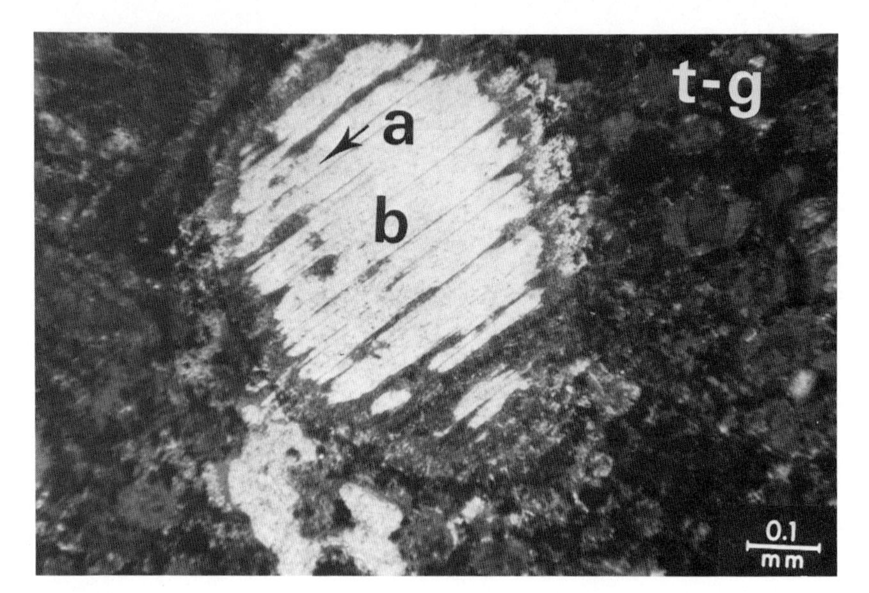

Fig. 514 A biotite xenocryst rounded, corroded and invaded along the cleavage by the tuffitic kimberlitic groundmass.
b = biotite
t-g = tuffitic kimberlite groundmass
Arrow "a" shows tuffitic groundmass invading the biotite xenocryst.
Tuffitic Kimberlite.
Wesselton, South Africa.
With crossed nicols.

Fig. 515 A rounded biotite aggregate with crystals differently orientated and associated with iron ore minerals in a tuffitic kimberlitic groundmass.
b = biotite aggregate
t-g = tuffitic kimberlitic groundmass
Kimberlite.
Makkovik, Labrador, Canada.
With crossed nicols.

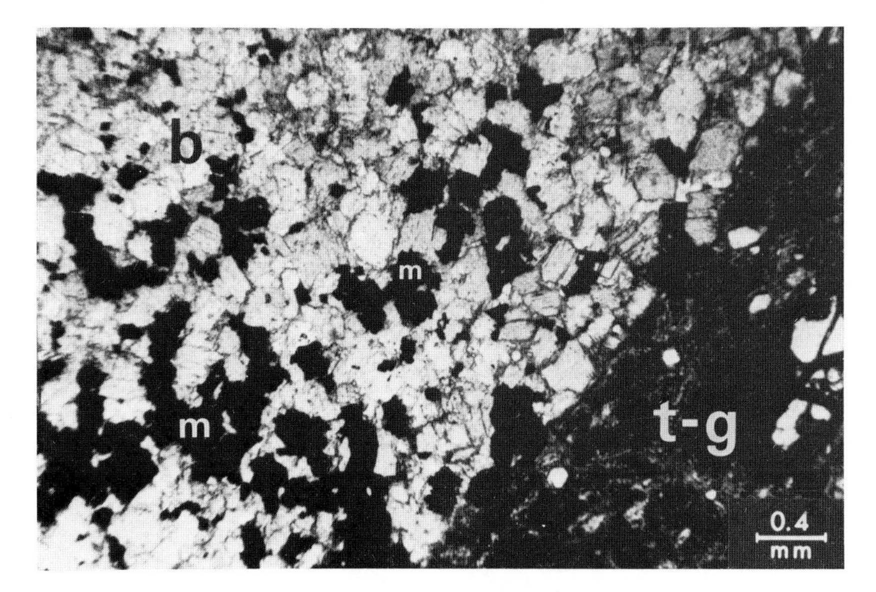

Fig. 516 A biotite aggregate consisting of differently orientated biotite interlocking grains and with later magnetite also in tuffitic-kimberlitic groundmass.
b = biotite
m = magnetite
t-g = tuffitic kimberlitic groundmass
Tuffitic kimberlite. Makkovik, Labrador, Canada.
With crossed nicols.

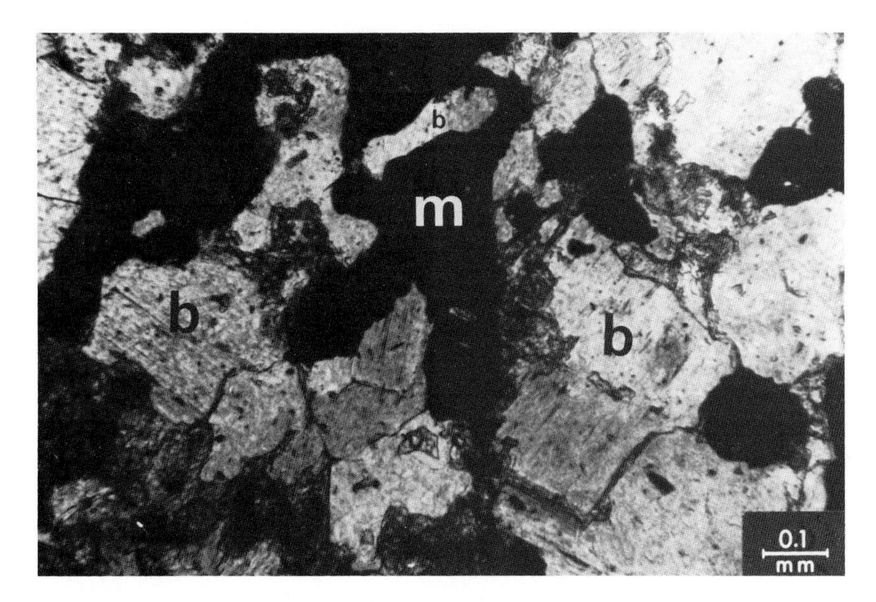

Fig. 517 Detail of comparable textural pattern as exhibited in Fig. 516. Blastogenic magnetite enclosing and engulfing biotites differently orientated.
b = biotite
m = magnetite
Tuffitic Kimberlite.
Wesselton, South Africa.
With crossed nicols.

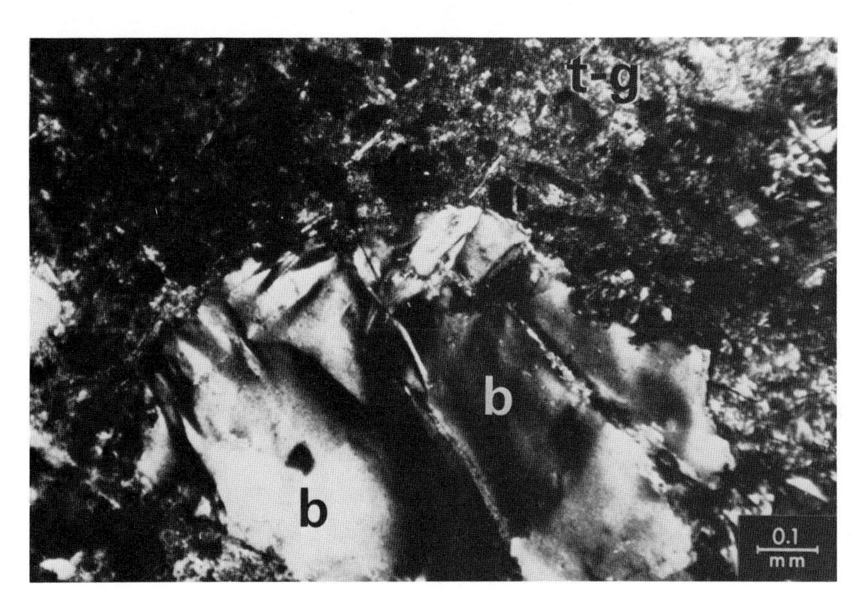

Fig. 518 Deformed biotite exhibiting undulating extinction in tuffitic kimberlitic groundmass.
b = deformed biotite, exhibiting undulating extinction.
t-g = tuffitic kimberlitic groundmass
Tuffitic Kimberlite.
Makkovik, Labrador, Canada.
With crossed nicols.

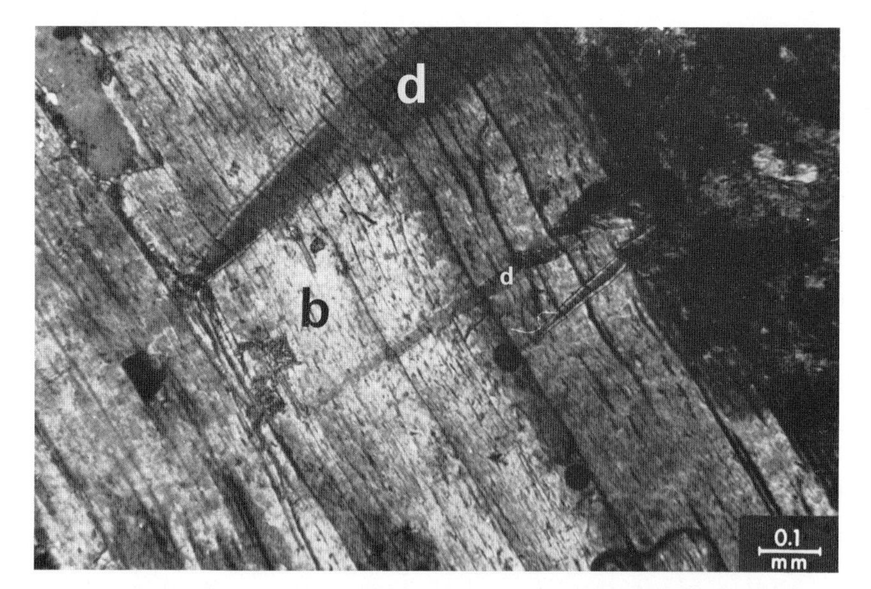

Fig. 519 Biotite in kimberlite showing deformation structure.
b = biotite
d = deformation effects on biotite
Tuffitic Kimberlite.
Makkovik, Labrador, Canada.
With crossed nicols.

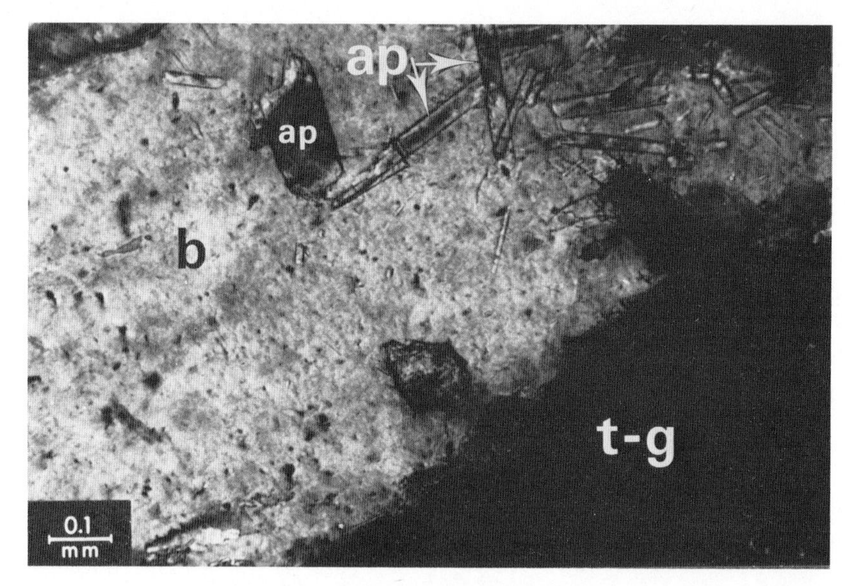

Fig. 520 Biotite with apatites in tuffitic-kimberlitic groundmass.
b = biotite
ap = apatites
t-g = tuffitic groundmass
Tuffitic Kimberlite.
Makkovik, Labrador, Canada.
With crossed nicols.

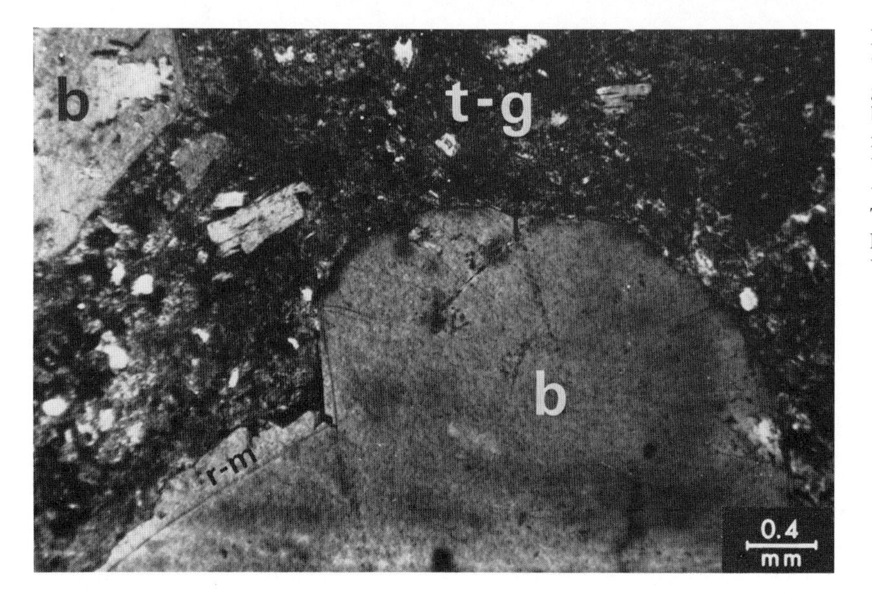

Fig. 521 Biotite corroded and with a "reaction margin" in tuffitic-kimberlitic groundmass.
b = biotite
r-m = reaction margin (actually initial biotite zonal growth)
t-g = tuffitic groundmass
Tuffitic Kimberlite.
Makkovik, Labrador, Canada.
With crossed nicols.

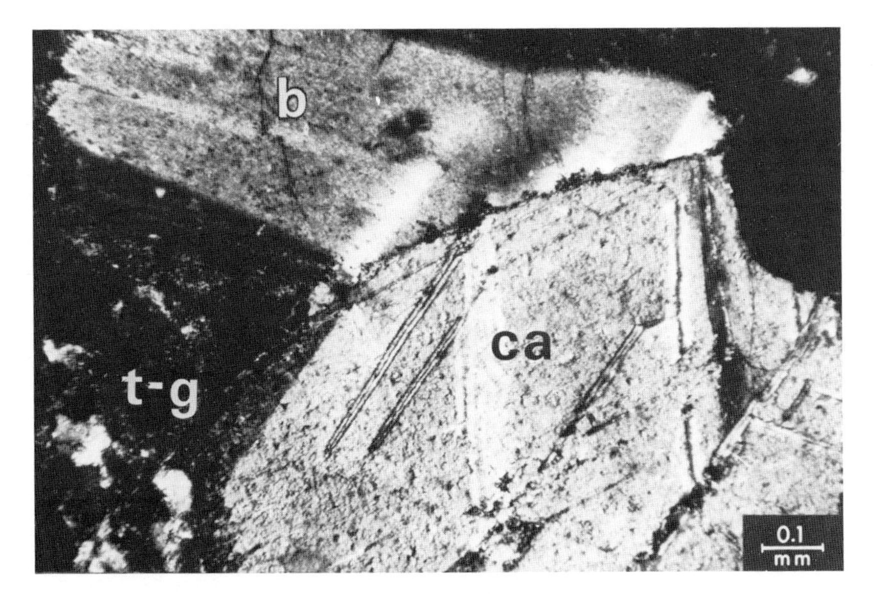

Fig. 522 Biotite fragment in contact with calcite in tuffitic-kimberlitic groundmass.
b = biotite
ca = calcite
t-g = tuffitic kimberlitic groundmass
Kimberlite. Kimberley, South Africa.
With crossed nicols.

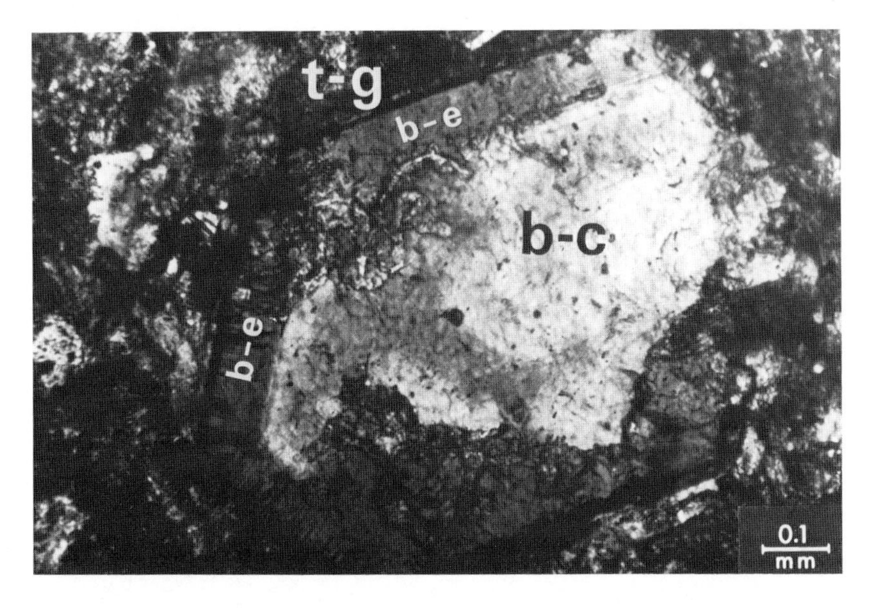

Fig. 523 Idiomorphic and zoned biotite with the biotite external zone attaining idiomorphic form.
b-c = biotite central zone
b-e = biotite external zone attaining idiomorphic form
t-g = tuffitic kimberlitic groundmass
Kimberlite.
Makkovik, Labrador, Canada.
With crossed nicols.

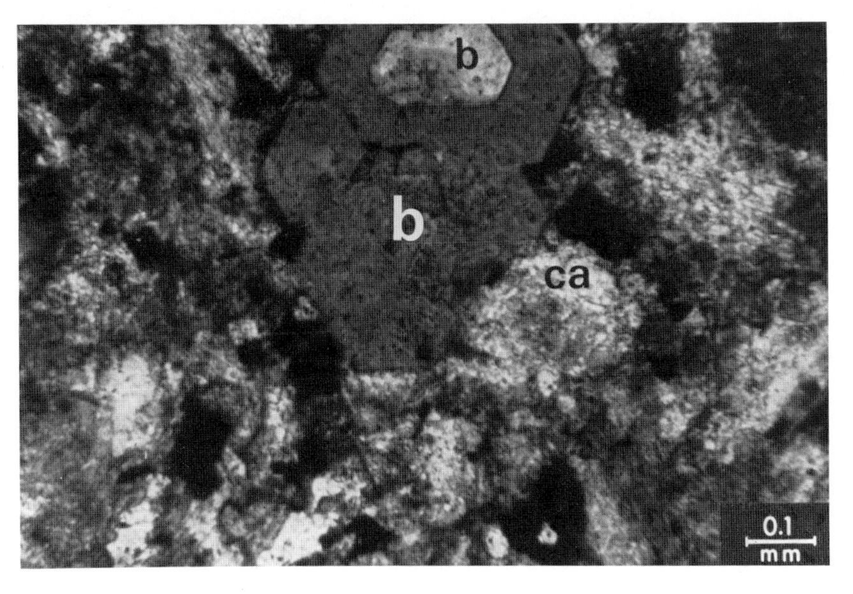

Fig. 524 A zoned biotite with the external zone attaining idiomorphic form.
b = biotite
ca = calcite in the tuffitic kimberlitic groundmass.
Kimberlite.
Makkovik, Labrador, Canada.
With crossed nicols.

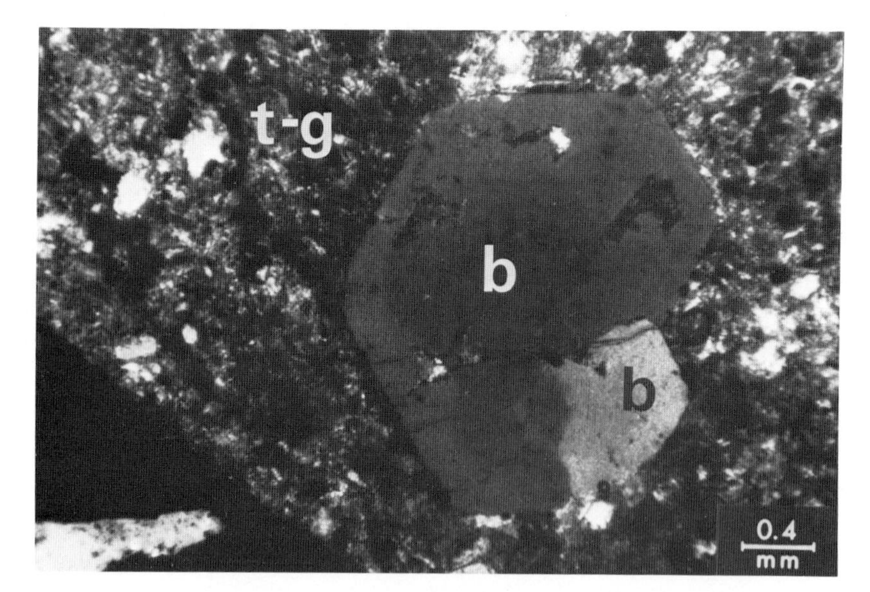

Fig. 525a Idiomorphic biotite phenocryst with a differently orientated part in tuffitic kimberlitic groundmass.
b = biotite phenocryst
t-g = tuffitic kimberlitic groundmass
Kimberlite.
Makkovik, Labrador, Canada.
With crossed nicols.

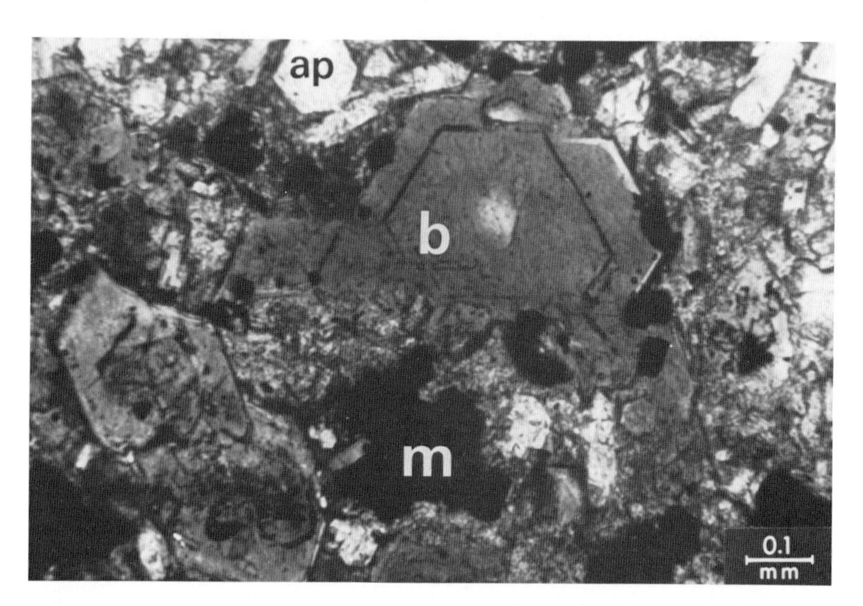

Fig. 525b Zoned idiomorphic biotite phenocryst in tuffitic kimberlitic groundmass with magnetite and apatite.
b = biotite
m = magnetite (black)
ap = apatite
Kimberlite.
Makkovik, Labrador, Canada.
With crossed nicols.

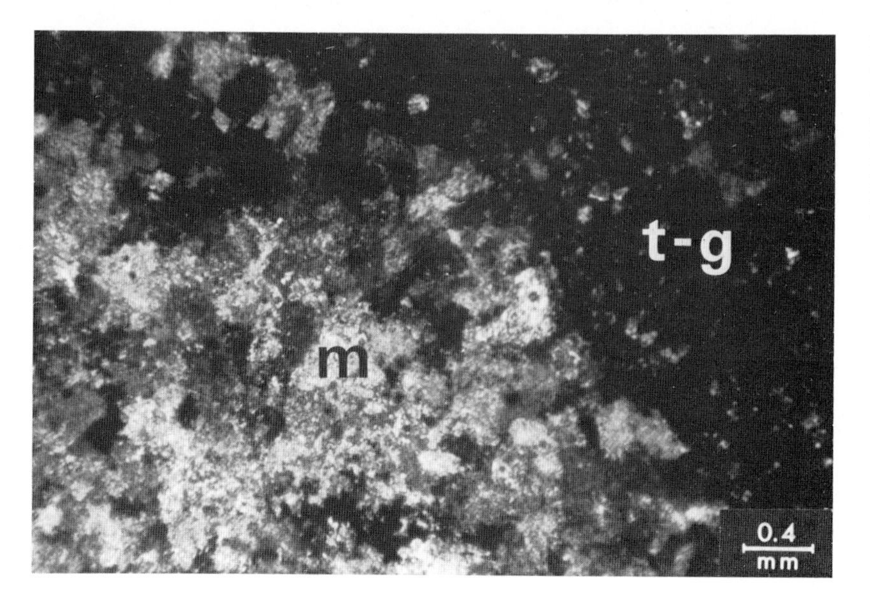

Fig. 526 A rounded marble fragment consisting of interlocking calcite, in tuffitic-kimberlitic groundmass.
m = marble
t-g = tuffitic kimberlitic groundmass
Kimberlite.
Makkovik, Labrador, Canada.
With crossed nicols.

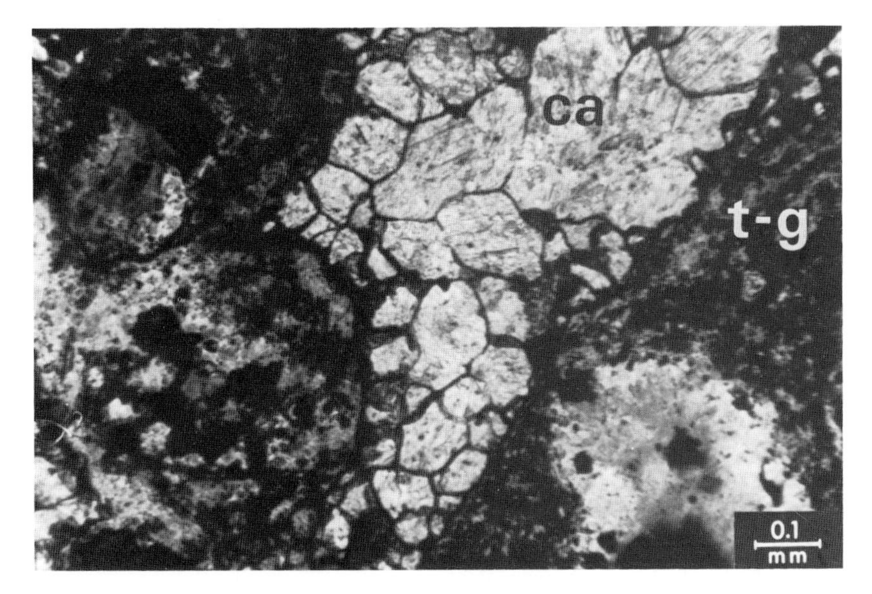

Fig. 527 A fragment composed of granular calcite in tuffitic kimberlitic groundmass.
ca = calcite
t-g = tuffitic kimberlitic groundmass
Kimberlite.
Makkovik, Labrador, Canada.
With crossed nicols.

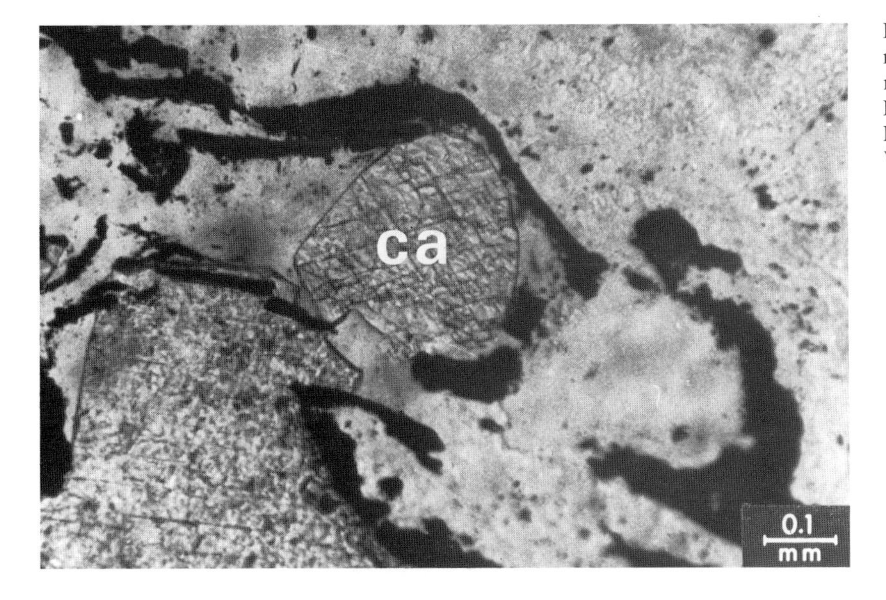

Fig. 528 Tuffitic-kimberlitic groundmass with iron-ore (black) and with a rounded calcite (ca).
Kimberlite.
Makkovik, Labrador, Canada.
Without crossed nicols.

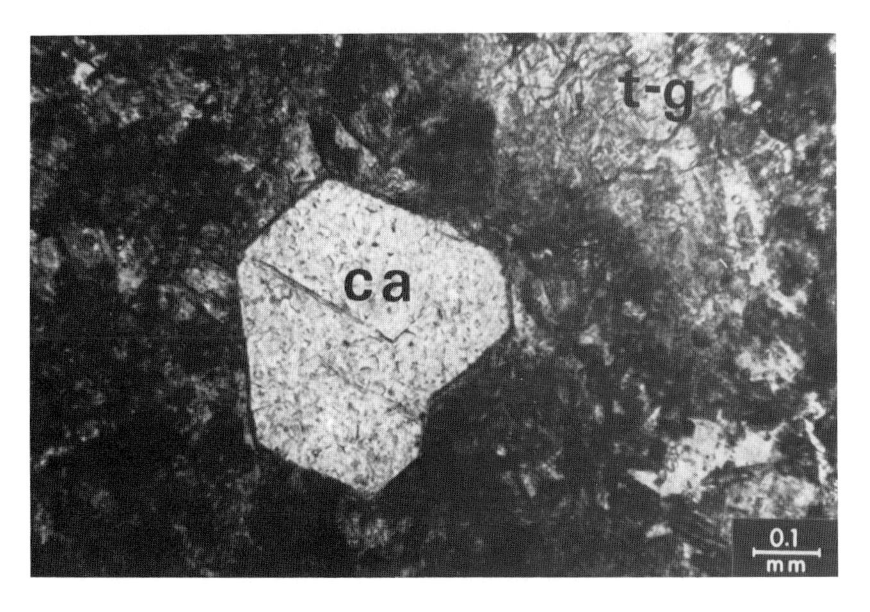

Fig. 529 An idiomorphic calcite phenocryst in tuffitic kimberlitic groundmass.
ca = calcite
t-g = tuffitic kimberlitic groundmass
Tuffitic Kimberlite.
Makkovik, Labrador, Canada.
With crossed nicols.

Fig. 530a, b A tuffitic kimberlitic groundmass with diffused calcite and with a local development of idiomorphic calcites and zeolites. As a micro-geode filling within the tuffitic kimberlitic groundmass.
ca = calcite
z = zeolite
t-g = tuffitic kimberlitic groundmass
Kimberlite.
Makkovik, Labrador, Canada.
With crossed nicols.

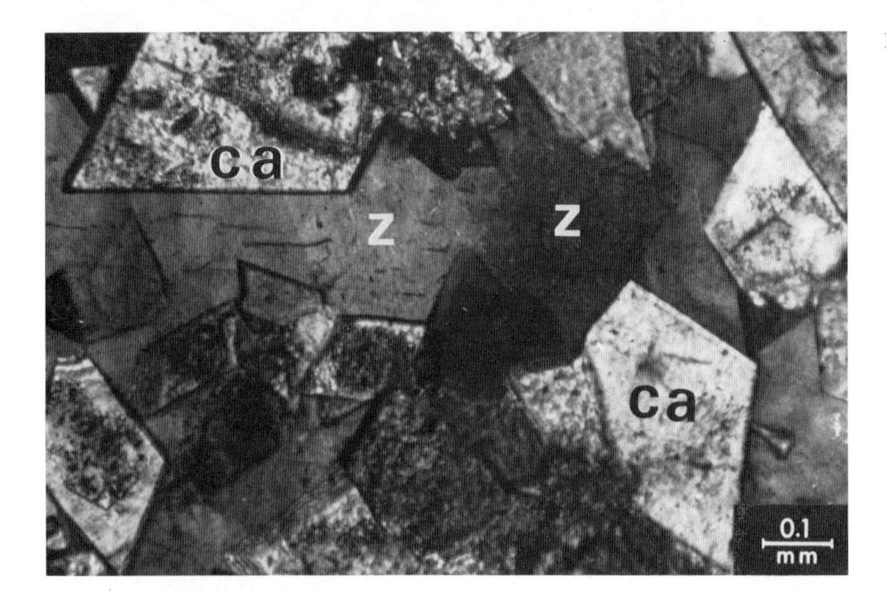

Fig. 530b

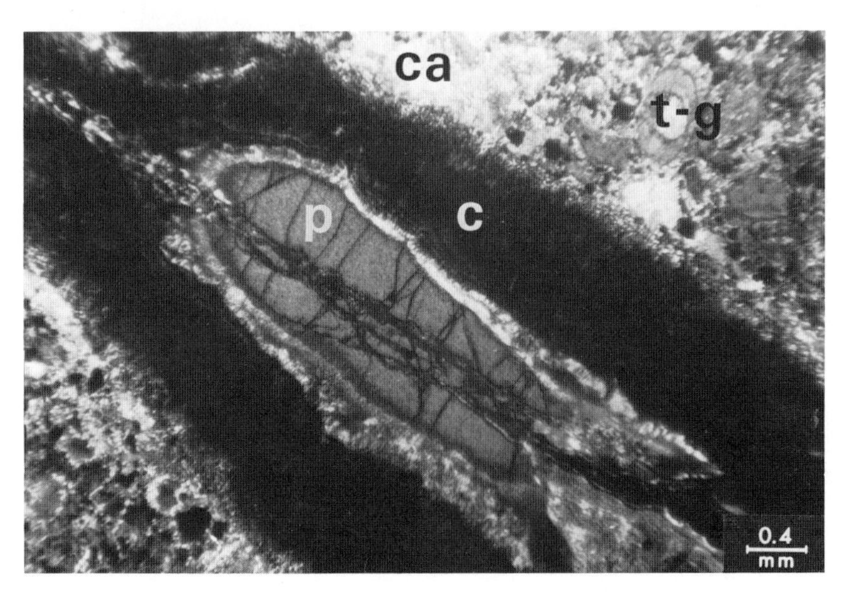

Fig. 531a Garnet in tuffitic kimberlitic groundmass. A celyphitic reaction margin is exhibited surrounding the garnet and is the reaction product between the pyrope and the tuffitic groundmass. Calcite is predominant in the groundmass.
p = pyrope
c = celyphitic reaction structure surrounding the garnet.
ca = calcite
t-g = tuffitic kimberlitic groundmass
Tuffitic Kimberlite.
Wesselton, South Africa.
With crossed nicols.

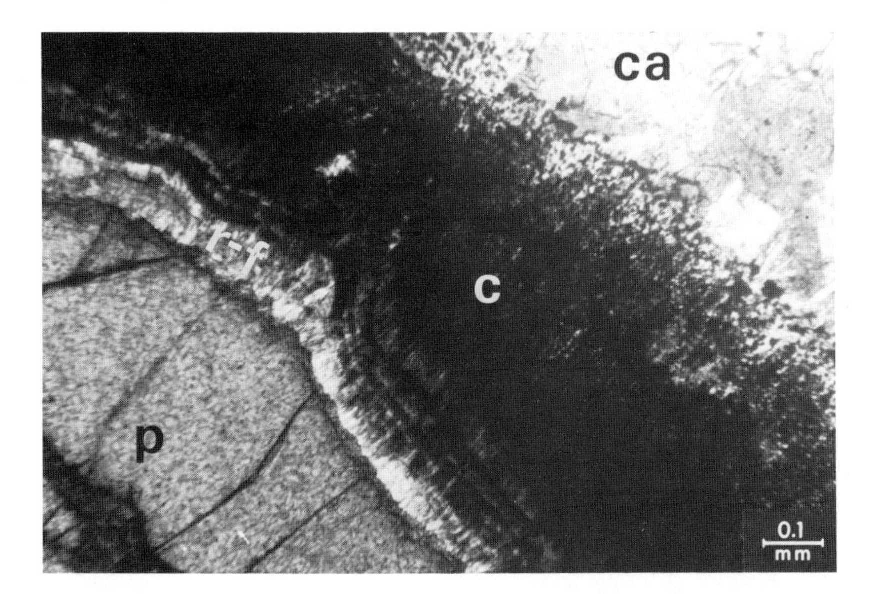

Fig. 531b Detail of Fig. 531a A detailed view of the celyphite structure. A reaction front is exhibited as a part of the celyphite.
p = pyrope
r-f = reaction front
c = celyphite structure
ca = calcite
Tuffitic Kimberlite.
Wesselton, South Africa.
With crossed nicols.

Fig. 532a Tremolite interpenetrating crystalloblasts within the tuffitic kimberlitic groundmass.
t = tremolite crystalloblast
t-g = tuffitic kimberlitic groundmass
ca = calcite of the groundmass
Tuffitic Kimberlite.
Wesselton, South Africa.
With crossed nicols.

Fig. 532b Radiating tremolitic crystalloblasts in tuffitic kimberlitic groundmass.
t = tremolite crystalloblast
Tuffitic Kimberlite.
Wesselton, South Africa.
With crossed nicols.

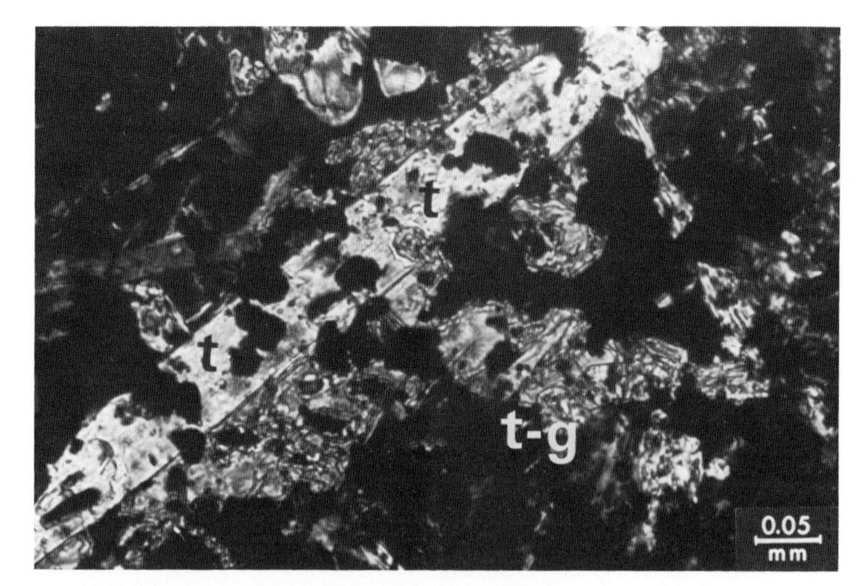

Fig. 533 A tremolite nematoblast enclosing magnetite and crystalloblastically penetrating through the tuffitic kimberlitic groundmass.
t = tremolite crystalloblast (nematoblast)
magnetite (black)
t-g = tuffitic kimberlitic groundmass
Tuffitic Kimberlite.
Makkovik, Labrador, Canada.
With crossed nicols.

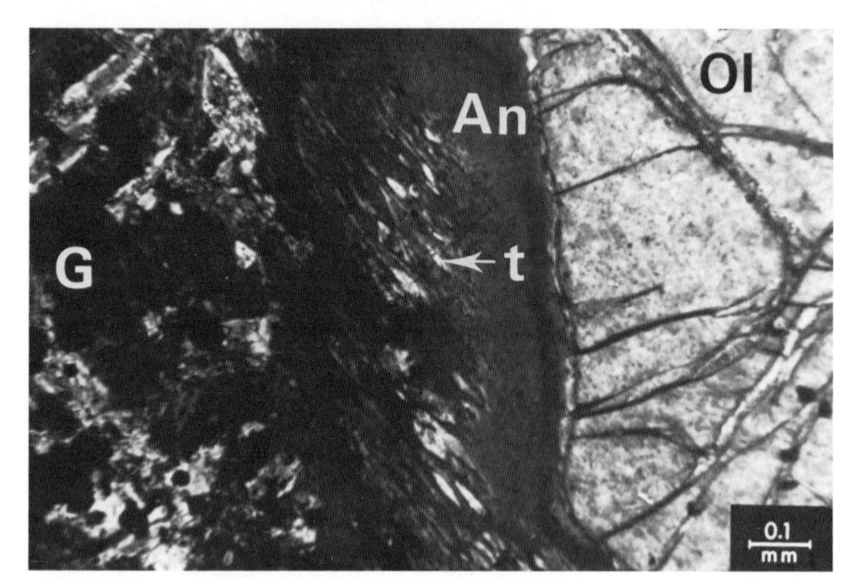

Fig. 534 Olivine with an alteration margin (antigoritic alteration margin of olivine) with tremolite crystalloblasts in the kimberlitic "groundmass" and also in the antigoritic margin of the olivine.
G = kimberlitic groundmass
Ol = Olivine
An = antigoritic alteration margin of olivine
t = tremolitic crystalloblasts.
Kimberlite. Kimberley, South Africa.
With crossed nicols.

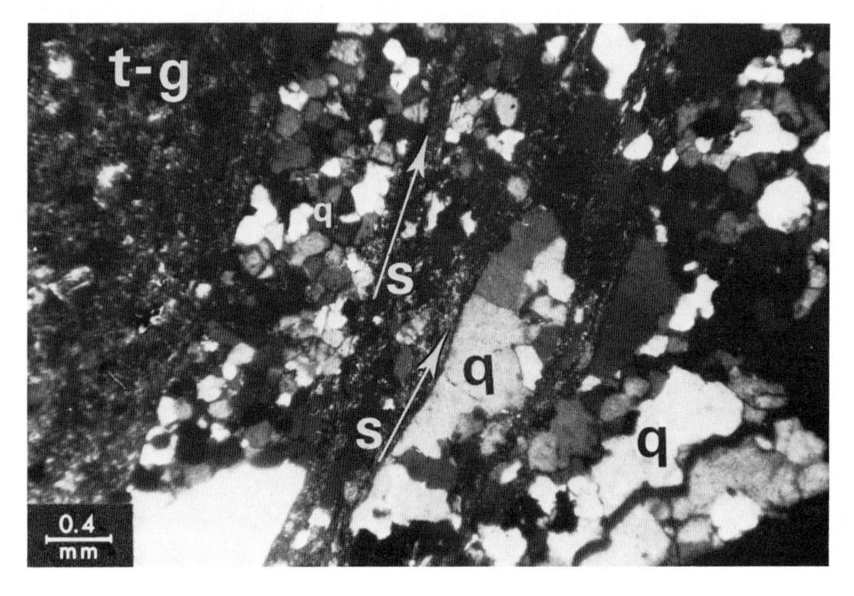

Fig. 535 Quartzite lenses delimited by shear planes in tuffitic kimberlitic groundmass.
q = quartzite
t-g = tuffitic kimberlitic groundmass
s = shear planes
Tuffitic Kimberlite.
Makkovik, Labrador, Canada.
With crossed nicols.

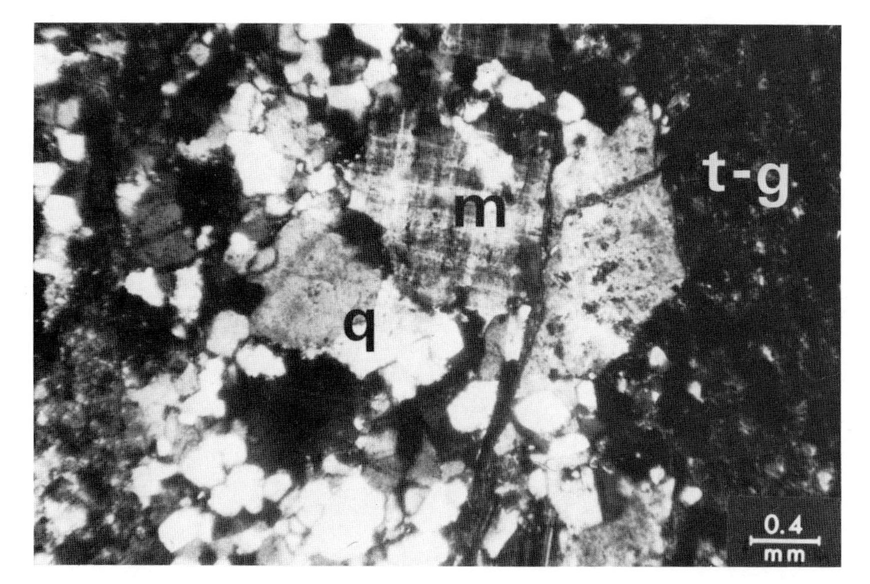

Fig. 536 Microcline-quartz granitic fragment in the form of a lens in tuffitic-kimberlitic groundmass.
m = microcline
q = quartz
t-g = tuffitic groundmass
Tuffitic Kimberlite.
Makkovik, Labrador, Canada.
With crossed nicols.

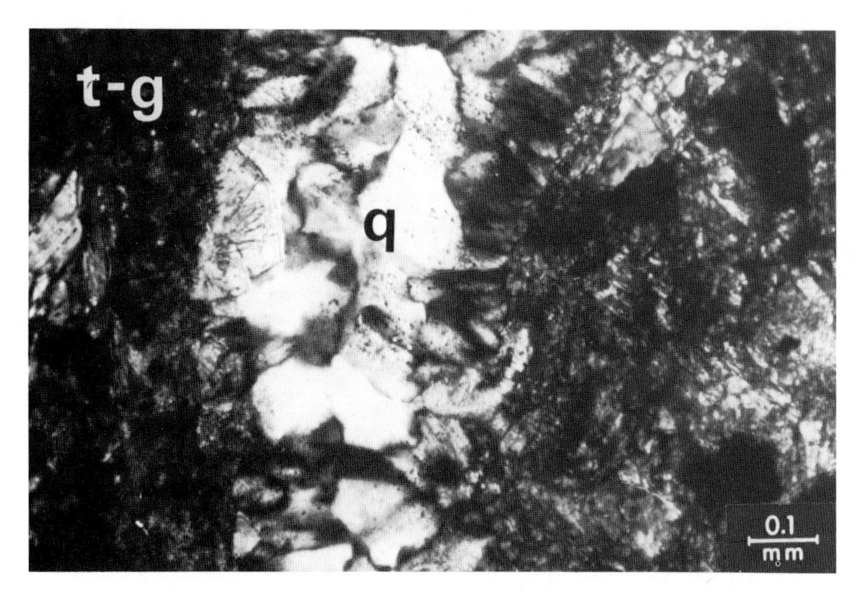

Fig. 537 A microgeode filled with granular quartz, probably a product of recrystallization from colloform silica.
q = quartz microgeode
t-g = tuffitic kimberlitic groundmass
Tuffitic Kimberlite.
Makkovik, Labrador, Canada.
With crossed nicols.

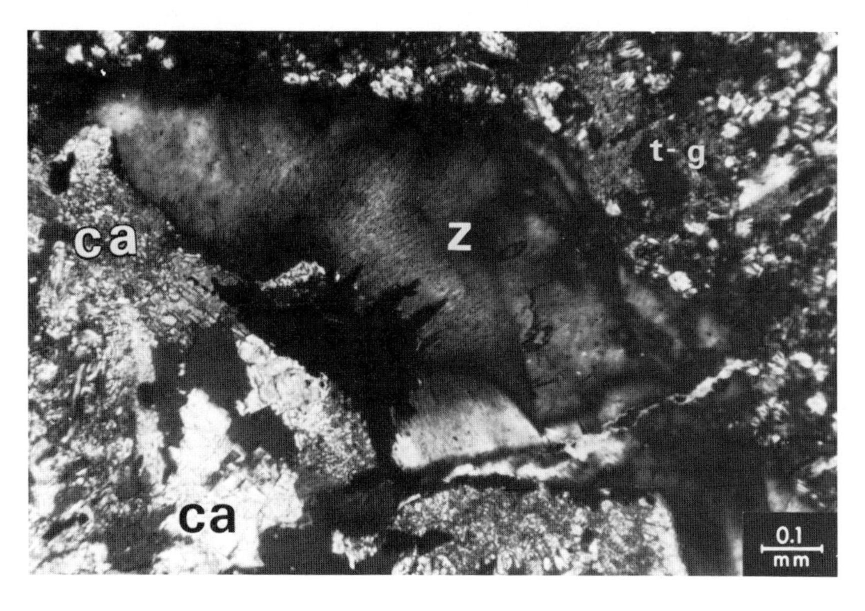

Fig. 538 A zeolite within the tuffitic kimberlitic groundmass.
z = zeolite
t-g = tuffitic kimberlitic groundmass
ca = calcite in the groundmass
Tuffitic Kimberlite. Kimberley, South Africa.
With crossed nicols.

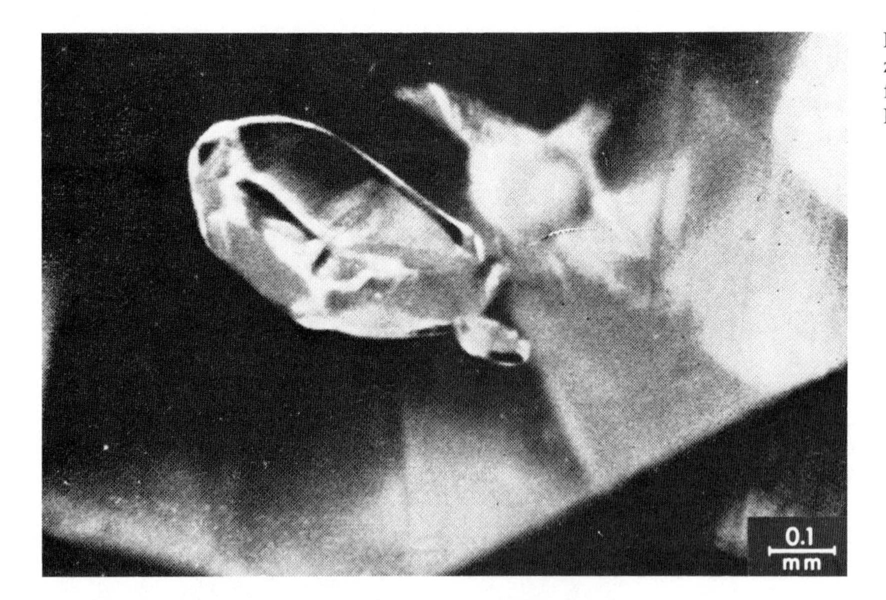

Fig. 539 Well developed idiomorphous zircon crystal embedded in a diamond from S. Africa.
By Gübelin, E. 1952.

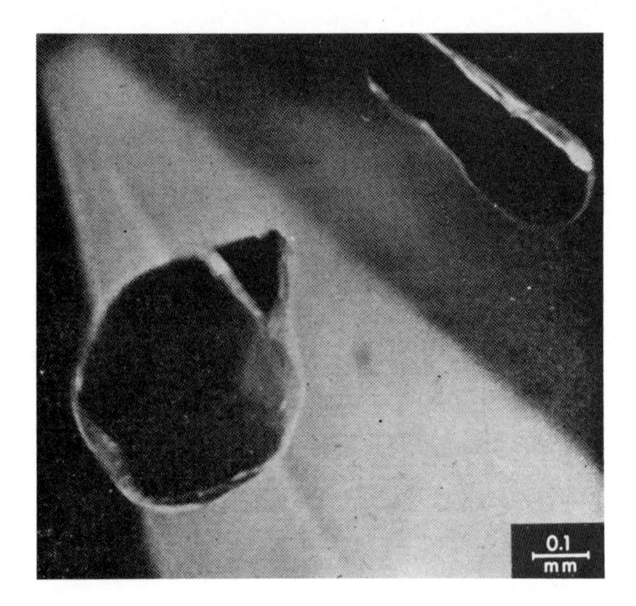

Fig. 540 Pebbles of garnet in diamond.
By Gübelin, 1948.

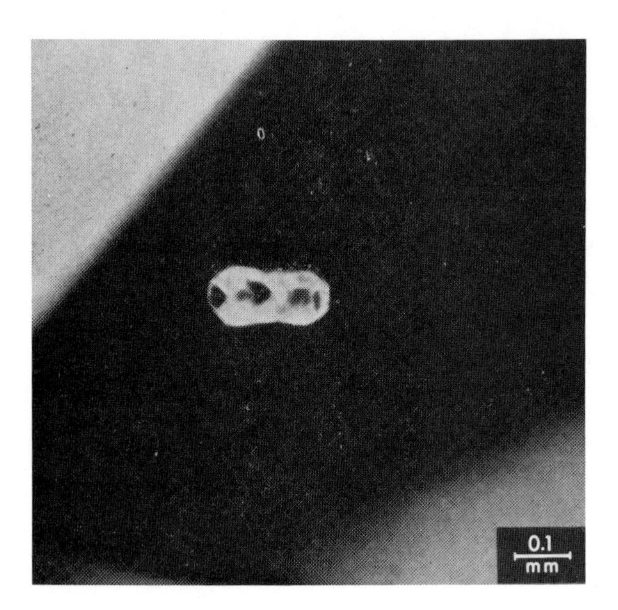

Fig. 541 Resorbed olivine in diamond.
By Gübelin, 1948.

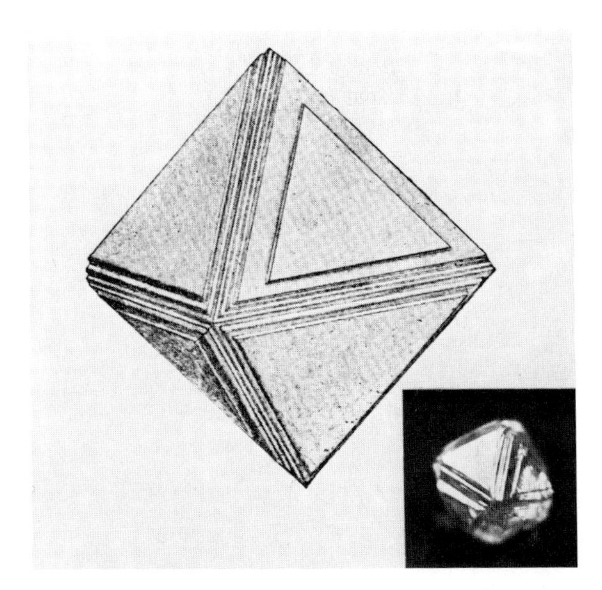

Fig. 542 A diamond crystalloblast? Drawing; from: Atlas der Kristallformen von Victor Goldschmidt (1916)
Lower right corner: Diamond octahedron from Urals, USSR.

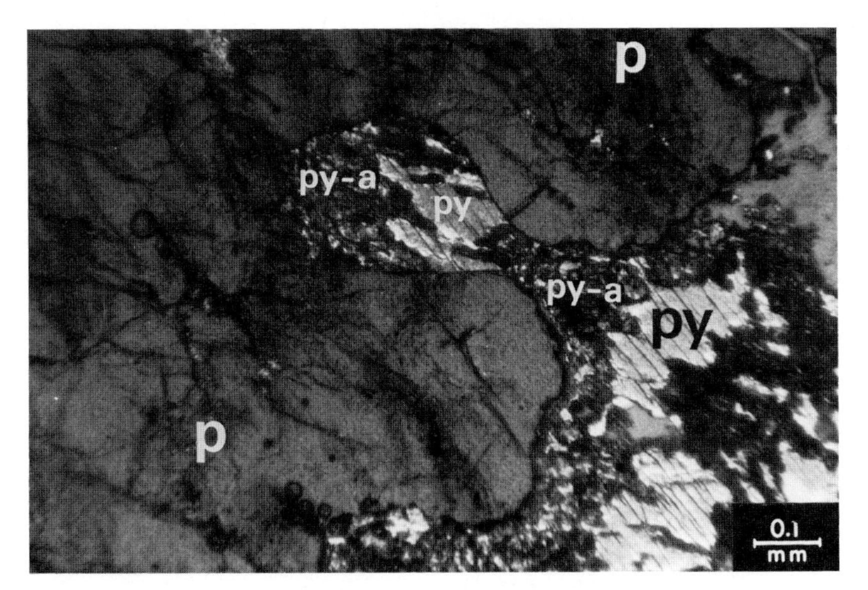

Fig. 543 Eclogitic xenolith in kimberlite. Pyrope garnet in intergrowth with pyroxene.
p = pyrope
py = pyroxene
py-a = pyroxene alteration
Eclogitic bomb in kimberlite. Roberts Vict. Mine, South Africa.
With crossed nicols.

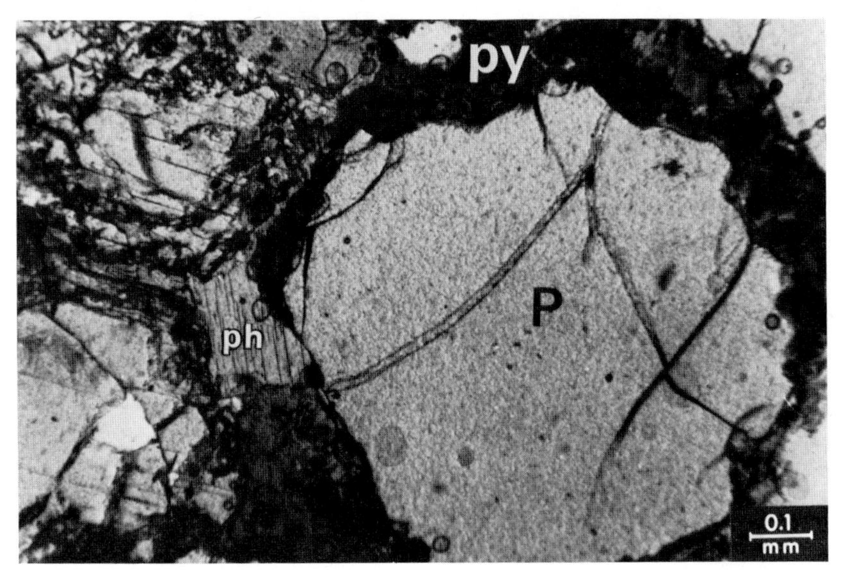

Fig. 544 Eclogite xenolith in kimberlite. Pyrope garnets in intergrowth with pyroxene (diopside) and intergranular phlogopite.
P = pyrope garnet
py = pyroxene
ph = phlogopite
Eclogitic bomb in kimberlite.
Roberts Vict. Mine, South Africa.
Without crossed nicols.

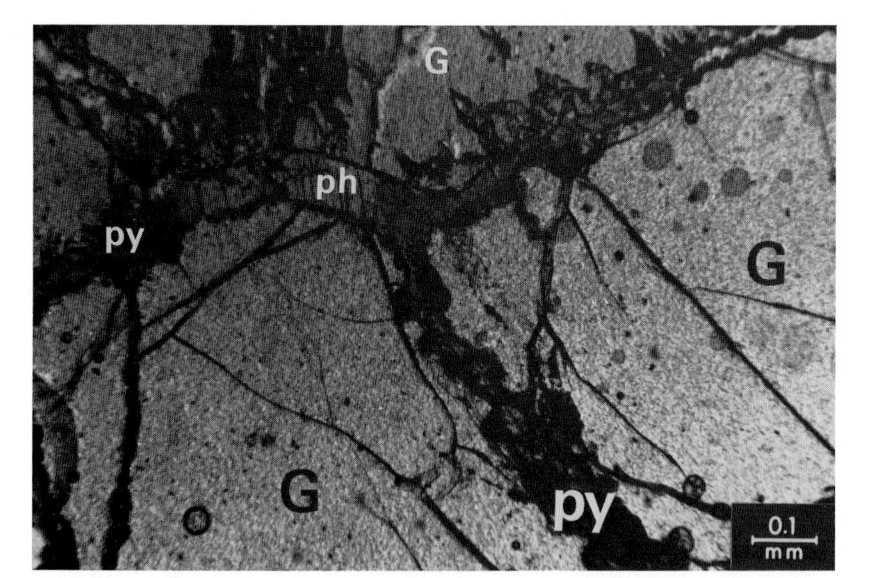

Fig. 545 Garnet (pyrope) with intergranular diopside and phlogopite.
G = garnet (pyrope)
py = pyroxene (diopside)
ph = phlogopite
Eclogitic bomb in kimberlite.
Roberts Vict. Mine, South Africa.
Without crossed nicols.

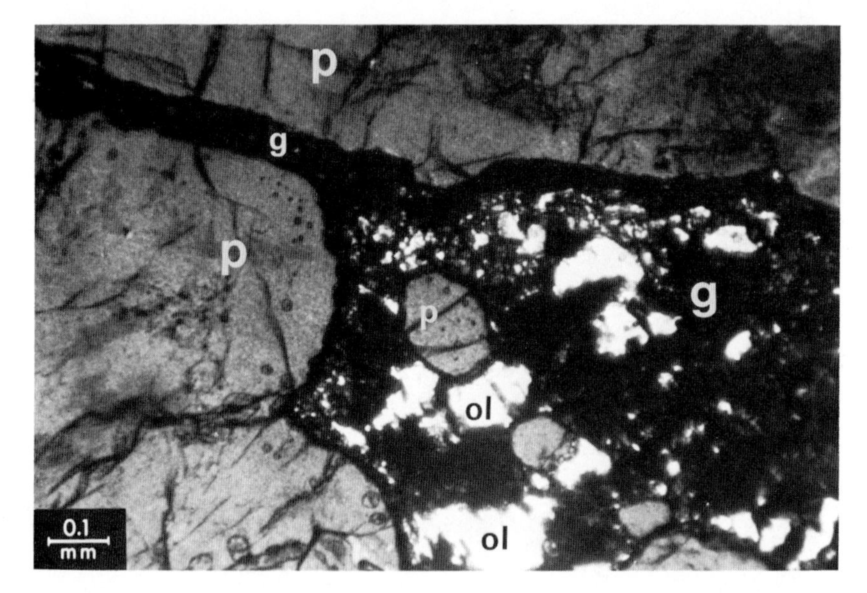

Fig. 546 Pyrope garnet with mobilised kimberlitic groundmass invading the garnet. Olivine is also shown.
p = pyrope garnet
g = kimberlitic groundmass
ol = olivine
Eclogitic bomb in kimberlite. Roberts Vict. Mine, South Africa.
Without crossed nicols.

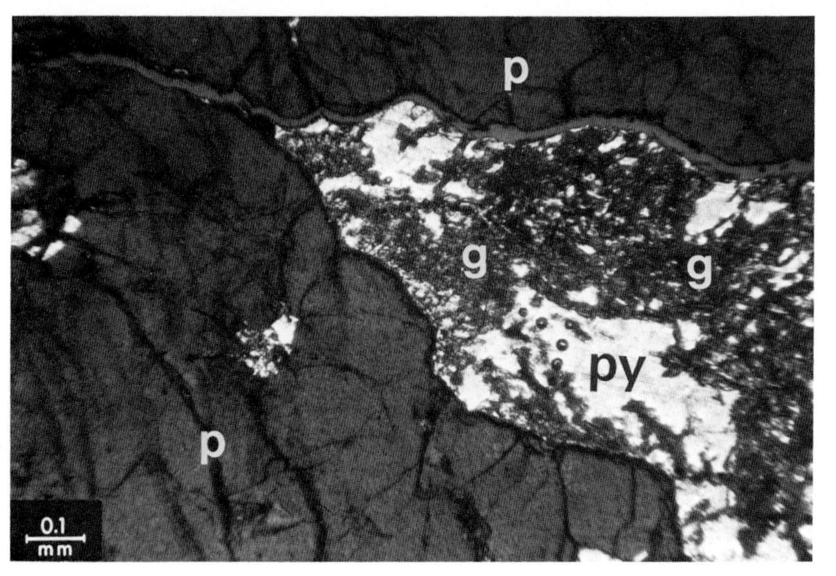

Fig. 547 Mobilised kimberlitic groundmass invading and replacing pyrope. Pyroxene is also present.
py = pyroxene
p = pyrope
g = kimberlitic groundmass
Eclogitic bomb in kimberlite. Roberts Vict. Mine, South Africa.
With crossed nicols.

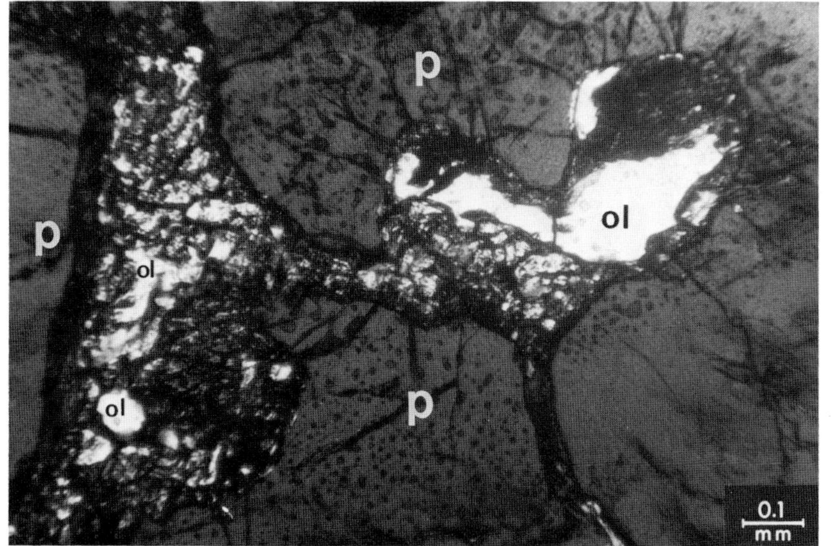

Fig. 548 Pyrope garnet invaded by kimberlitic groundmass which includes olivine.
p = pyrope
ol = olivine
Eclogitic bomb in kimberlite. Roberts Vict. Mine, South Africa.
Without crossed nicols.

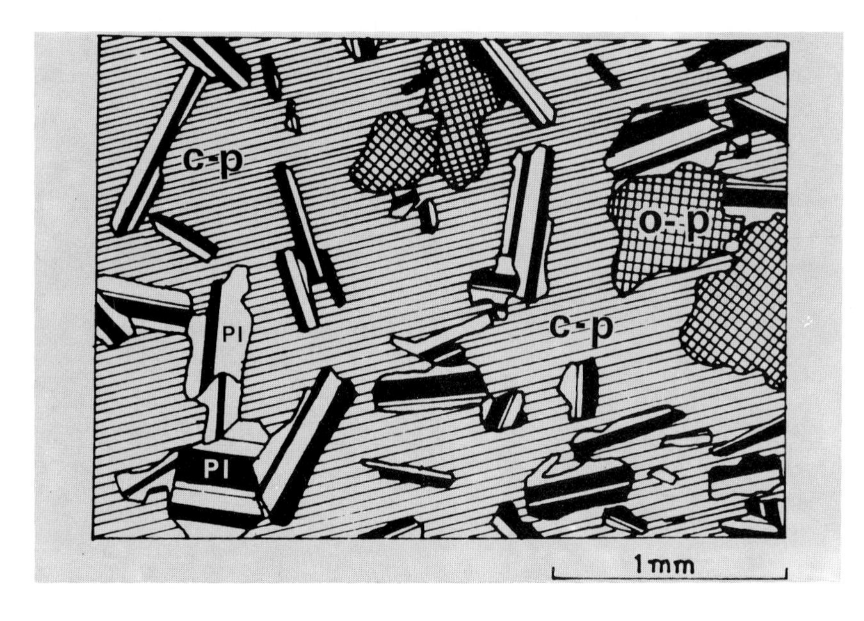

Fig. 549 Shows according to Scharlau, T. A. (1972) cumulate plagioclase and orthopyroxene enclosed by intracumulate (ophitic to sub-ophitic) clinopyroxene.
Pl = plagioclase
c-p = clinopyroxene
o-p = orthopyroxene
Bushveld Complex,
Transvaal, S. Africa.
Diagram by Scharlau, 1972.

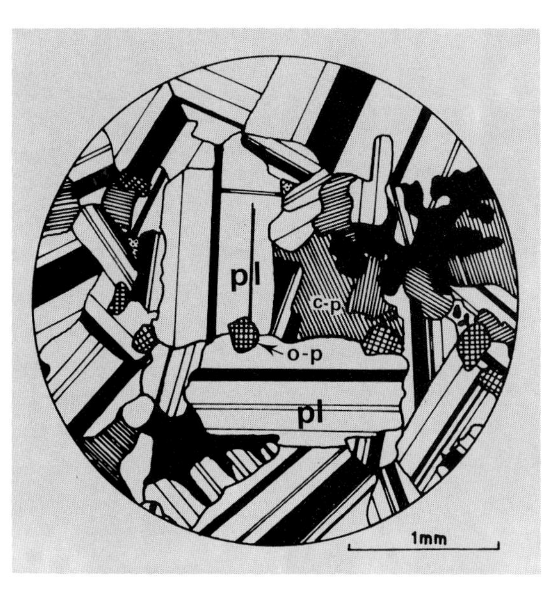

Fig. 550 Shows cumulate plagioclase, orthopyroxene and clinopyroxene enclosed by intracumulate magnetite (or ilmenite).
magnetite or ilmenite (black)
pl = plagioclase
c-p = clinopyroxene
o-p = orthopyroxene
Bushveld Complex,
Transvaal, S. Africa.
Diagram by Scharlau, 1972.

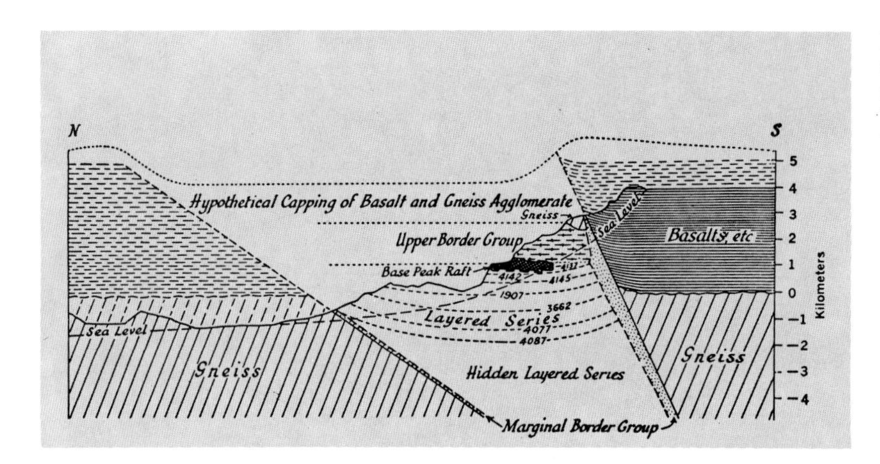

Fig. 551 North-south section of the original Skaergaard intrusion before the flexuring and subsequent erosion. (After Wager and Deer, Medd. om Grönland, Vol. 105, 1939).

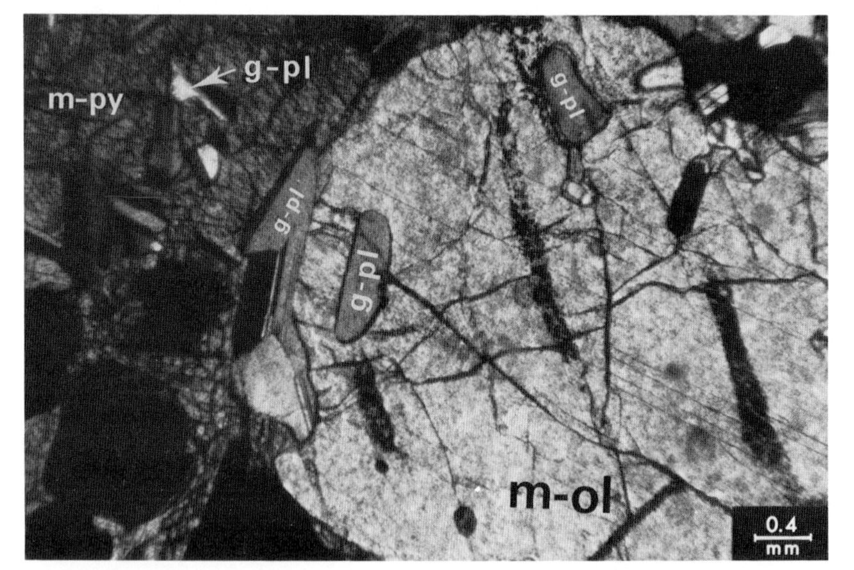

Fig. 552a Megablastic pyroxene and olivine including rounded and corroded groundmass plagioclases.
m-py = megablastic pyroxene (poikiloblastic in character).
m-ol = megablastic olivine.
g-pl = groundmass rounded and included in megablast.
Perpendicular feldspar rock (a peculiar eucrite with feldspars roughly perpendicular to margin of intrusion).
55 meters from western margin on Mellemo, Skaergaard, Greenland.
With crossed nicols.

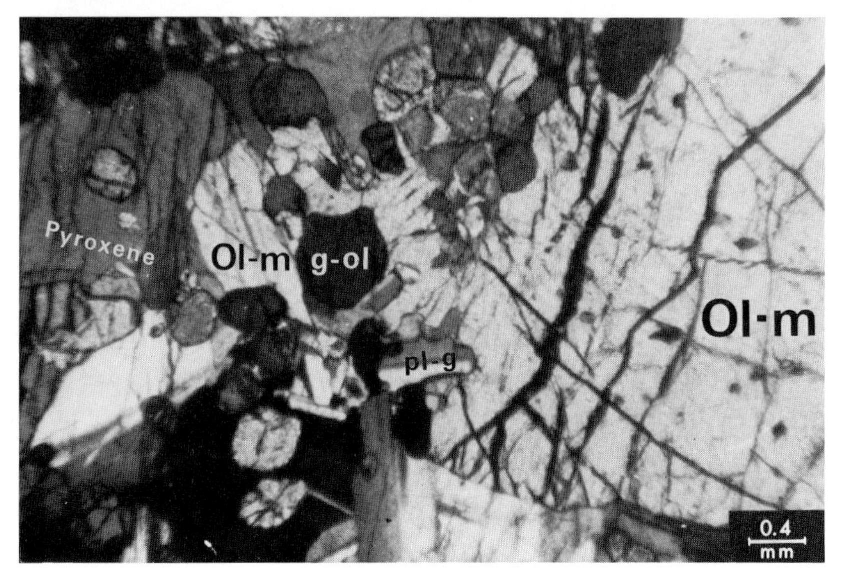

Fig. 552b Olivine megablast poikiloblastic in character with groundmass (a pre-phenocrystalline, i. e. olivine-plagioclase-fels) included, rounded and "assimilated" by the megablastic olivine poikiloblasts.
Ol-m = olivine megablast
g-ol = olivine groundmass
pl-g = plagioclase groundmass.
Gabbro picrite. 40 meters from northern margin at Stromstedet, Skaergaard, Greenland.
With crossed nicols.

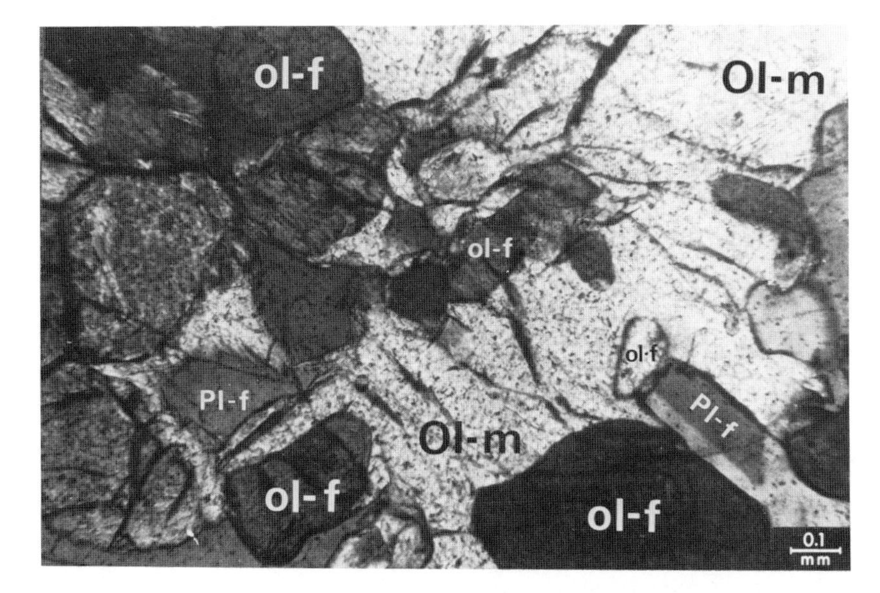

Fig. 553 Olivine megablast poikiloblastic in character including corroded, rounded and partly assimilated olivine and plagioclase belonging to a pre-blastic olivine-plagioclase-fels.
Ol-m = olivine megablast
ol-f = preblastic olivine grains belonging to olivinefels phase.
Pl-f = pre-blastic plagioclase grains belonging to the olivine plagioclase-fels.
Gabbro picrite. 40 meters from northern margin at Stromstedet, Skaergaard, Greenland.
With crossed nicols.

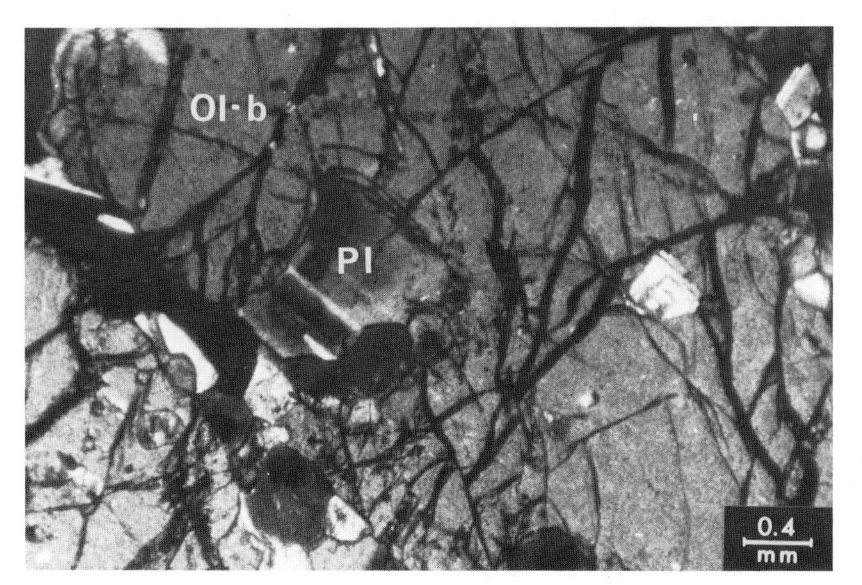

Fig. 554a and detail Fig. 554b Pre-blastic plagioclase and olivine grains included in a later olivine megablast.
A plagioclase grain showing corroded outline and reaction margins (see arrow "a") is indicated.
Ol-b = olivine blast
Pl = pre-blastic plagioclase exhibiting corroded outline.
Gabbro picrite. 40 meters from northern margin at Stromstedet, Skaergaard, Greenland.
With crossed nicols.

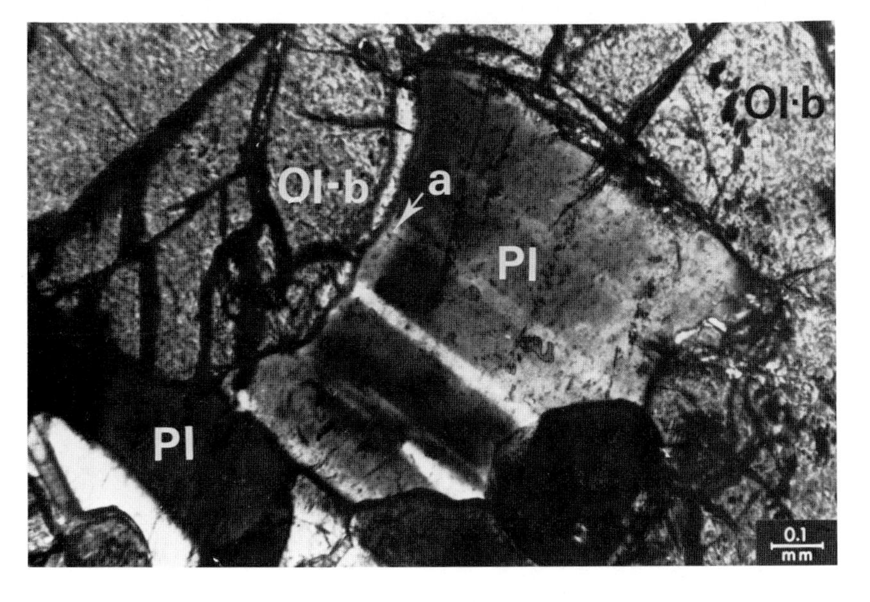

Fig. 554b

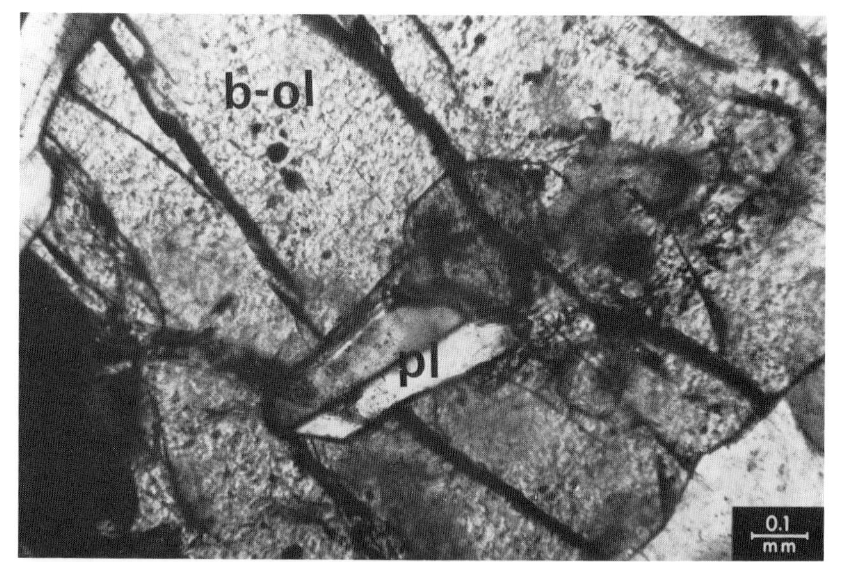

Fig. 555 Blastic olivine with olivine and plagioclase inclusions.
b-ol = blastic olivine
pl = plagioclase
Gabbro picrite. 40 meters from northern margin at Stromstedet, Skaergaard, Greenland.
With crossed nicols.

Fig. 556 Olivine megablast with plagioclase inclusions corroded and exhibiting distinct reaction margins effected on the feldspar by the later olivine growth.
Ol-m = olivine megablast
pl = plagioclase. Arrow "a" indicates reaction margin on plagioclase included by the olivineblast.
Gabbro picrite. 40 meters from northern margin at Stromstedet, Skaergaard, Greenland.
With crossed nicols.

Fig. 557 Olivine megablast with plagioclase either autocathartically pushed at the margins (corroded and rounded, see arrow "a") or included within the megablast and exhibiting rounded and corroded outlines, arrow "b".
ol-m = olivine megablast
pl = plagioclase
Average rock (plagioclase-augite-olivine "cumulate")
West side of Uttentals Plateau, Skaergaard, Greenland.
With crossed nicols.

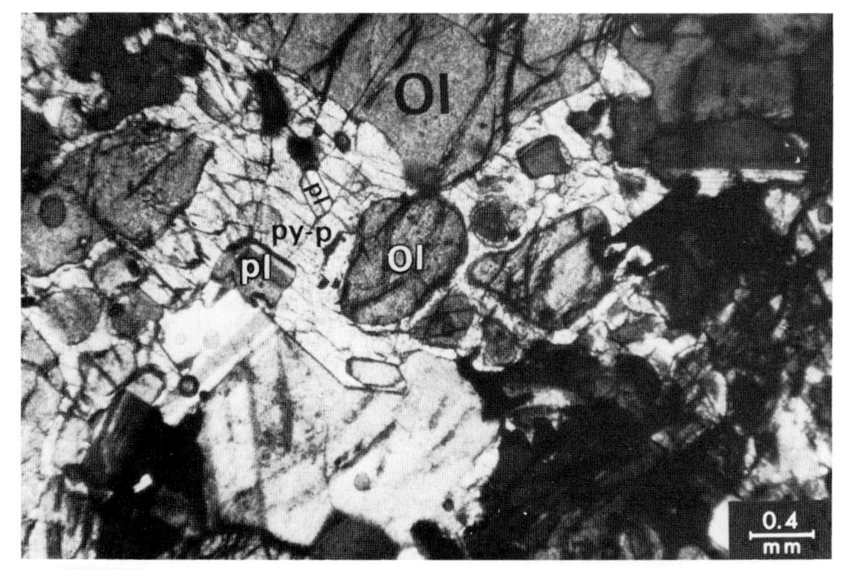

Fig. 558 Early olivineblasts partly engulfed and surrounded by a late pyroxene poikiloblast which included detached olivines and also plagioclase of the initial plagioclase-olivinefels phase into which have blastically grown both the olivine blast and the later pyroxene poikiloblasts.
Ol = olivineblast
py-p = pyroxene poikiloblast
pl = plagioclase included in the pyroxene, poikiloblast
Gabbro picrite. 40 meters from northern margin at Stromstedet, Skaergaard, Greenland.
With crossed nicols.

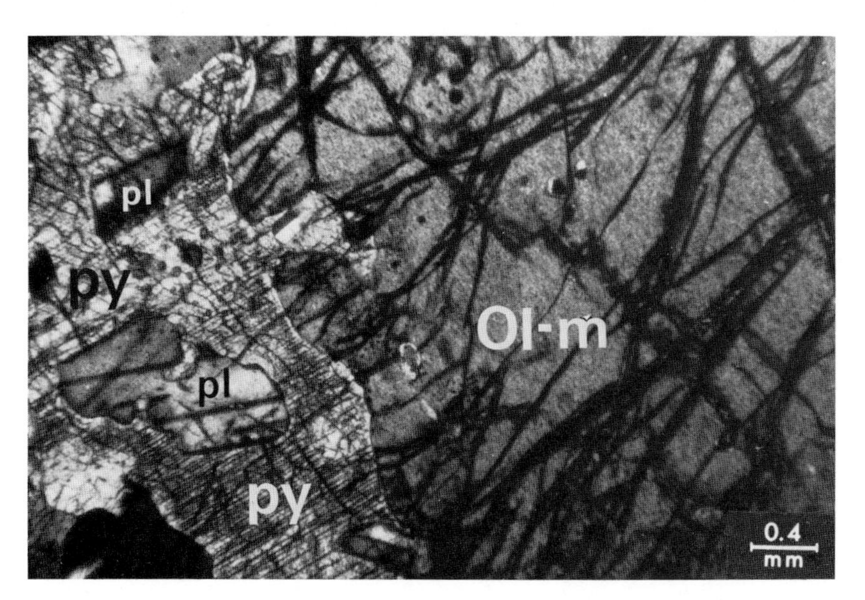

Fig. 559 Olivine megablast in contact with a later pyroxeneblast corroding both the olivine blast and including a plagioclase indicating clearly a corroded and assimilated outline.
Ol-m = olivine megablast
py = pyroxene poikiloblast
pl = plagioclase corroded and assimilated by the blastic pyroxene.
Average rock (plagioclase-augite-olivine "cumulate")
West side of Uttentals Plateau, Skaergaard, Greenland.
With crossed nicols.

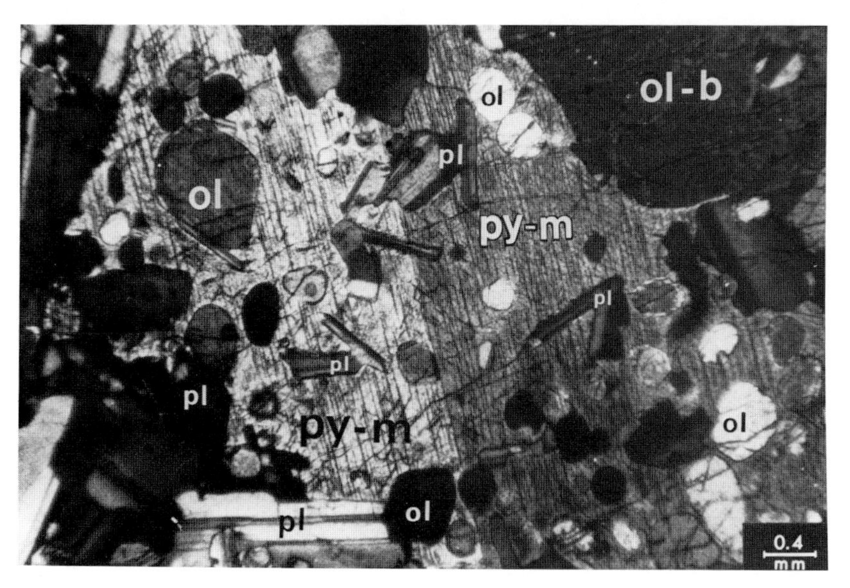

Fig. 560 Pyroxene mega-poikiloblast surrounding an olivine blast and including plagioclase and olivine of the preblastic plagioclase olivinefels phase.
py-m = pyroxene megapoikiloblast
ol-b = olivineblast
pl = plagioclases
ol = olivines
Average rock (plagioclase-augite-olivine "cumulate")
West side of Uttentals Plateau, Skaergaard, Greenland.
With crossed nicols.
(As a corollary to the crystalloblastic growth of pyroxenes is the synthesis of acmite by Prof. Takashi Fujii (per. com.) at temperatures below 100°C.)

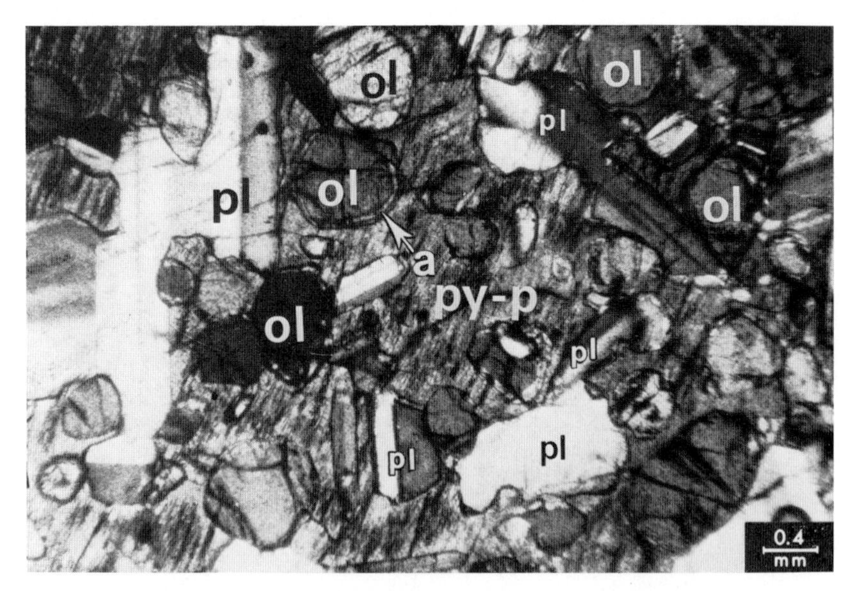

Fig. 561 Olivine and plagioclases of the olivine-plagioclasefels phase corroded, rounded and partly assimilated by the pyroxene poikiloblast.
In cases a reaction margin is exhibited between an olivine inclusion and the surrounding pyroxene poikiloblast (arrow "a").
py-p = pyroxene poikiloblast
ol = olivine
pl = plagioclase
Perpendicular feldspar rock (a peculiar eucrite with feldspars roughly perpendicular to margin of intrusion).
55 meters from western margin on Mellemo, Skaergaard, Greenland.
With crossed nicols.

Fig. 562 Plagioclases and olivines enclosed often corroded and rounded by a later pyroxene poikiloblast.
Perpendicular feldspar rock (a peculiar eucrite with feldspars roughly perpendicular to margin of intrusion).
55 meters from western margin of Mellemo, Skaergaard, Greenland.
With crossed nicols.

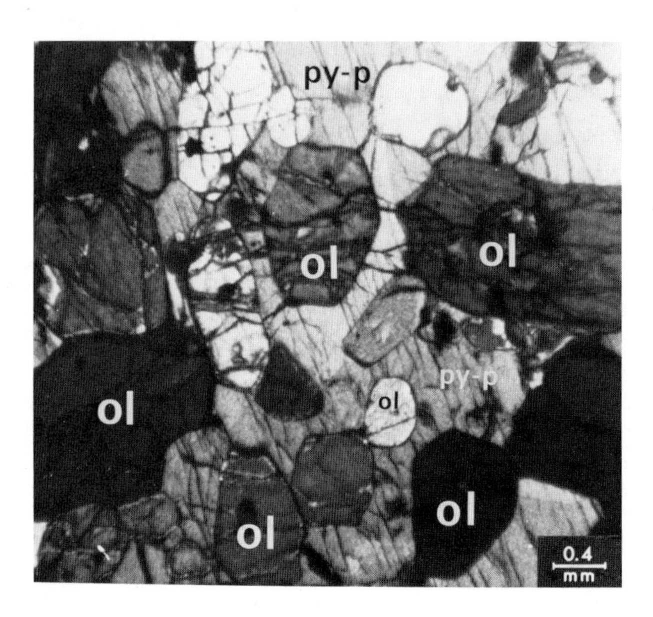

Fig. 563 Sub-idiomorphic and partly rounded olivines of an initial olivinefels phase included and partly rounded and corroded by a later pyroxene poikiloblast
Ol = olivine
py-p = poikiloblastic pyroxene
Kaersutite-hornblendite. Kaersut, Greenland.
With crossed nicols.

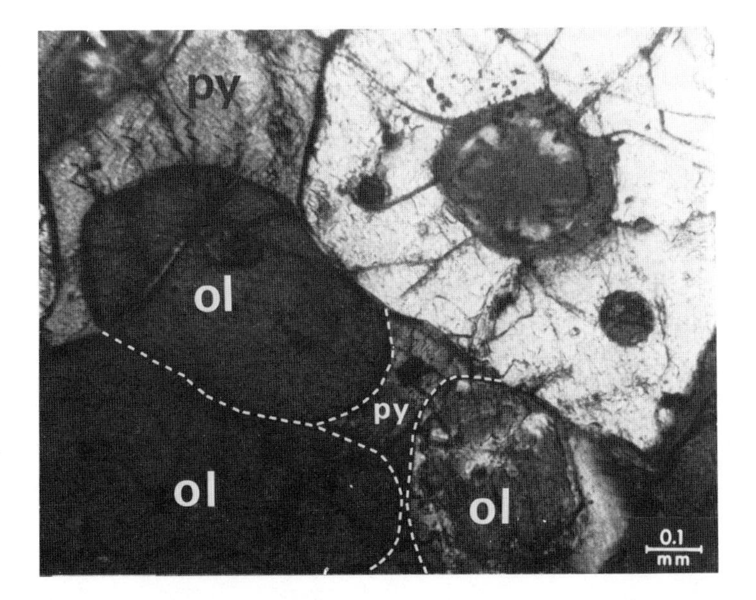

Fig. 564 Detail of Fig. 563 Olivines included in poikiloblast.
Ol = olivines
py = pyroxene poikiloblast
Kaersutite-Hornblendite.
Kaersut, Greenland.
With crossed nicols.

Fig. 565 Magnetite with pyroxene and a late generation of diopside occupying the interleptonic spaces between the magnetite and the pyroxene.
Magnetite (black)
py = pyroxene (augite)
n-p = new generation of pyroxene between magnetite and pyroxene.
Melanocratic average rock (plagioclase-pyroxene-ore "cumulate").
Foot of Boubouca Ridge at sea level.
Skaergaard, Greenland.
With crossed nicols.

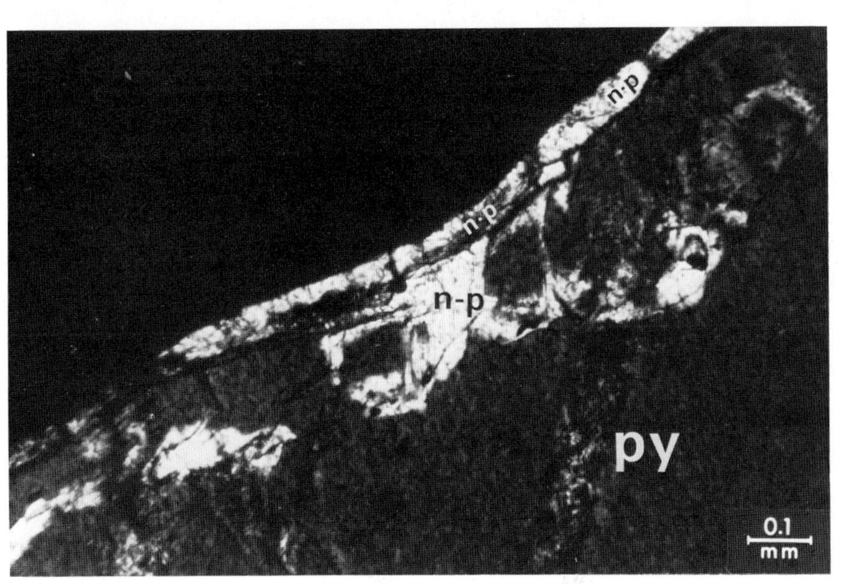

Fig. 566 Magnetite (black) with a new generation of diopside occupying the interleptonic spaces between magnetite and pyroxene.
py = pyroxene (augite)
n-p = diopside (new generation of pyroxene)
Average ferrogabbro (plagioclase-augite-olivine-magnetite "cumulate").
Skaergaard, Greenland.
With crossed nicols.

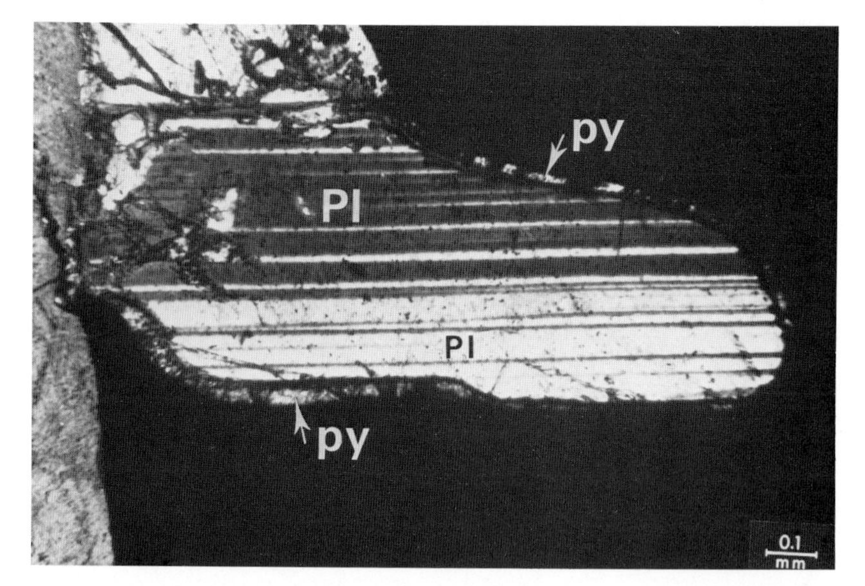

Fig. 567 Plagioclase corroded and rounded by a late phase magnetite (black). A late phase pyroxene crystallization occupies the interleptonic spaces between the magnetite and the included plagioclase.
Pl = plagioclase
py = pyroxene (diopside)
Average ferrogabbro (plagioclase-augite-olivine-magnetite "cumulate").
House area, Skaergaard, Greenland.
With crossed nicols.

Fig. 568 The intergranular between the plagioclase and the adjacent pyroxene is occupied by a new crystallization of pyroxene which also invades the magnetite.
Pl = plagioclase
magnetite (black)
py = pyroxene occupying the space between magnetite and plagioclase.
Arrows "a" show pyroxene invading the magnetite. Average ferrogabbro (plagioclase-augite-olivine-magnetite "cumulate").
House area, Skaergaard, Greenland.
With crossed nicols.

Fig. 569 Pseudo-ophitic plagioclase/pyroxene symplectic intergrowth. Plagioclase laths corroded and partly enclosed by pyroxene megablast.
Pl = plagioclase
m-py = megablast-pyroxene
Average rock (plagioclase-augite-olivine "cumulate")
West side of Uttentals Plateau, Skaergaard, Greenland.
With crossed nicols.

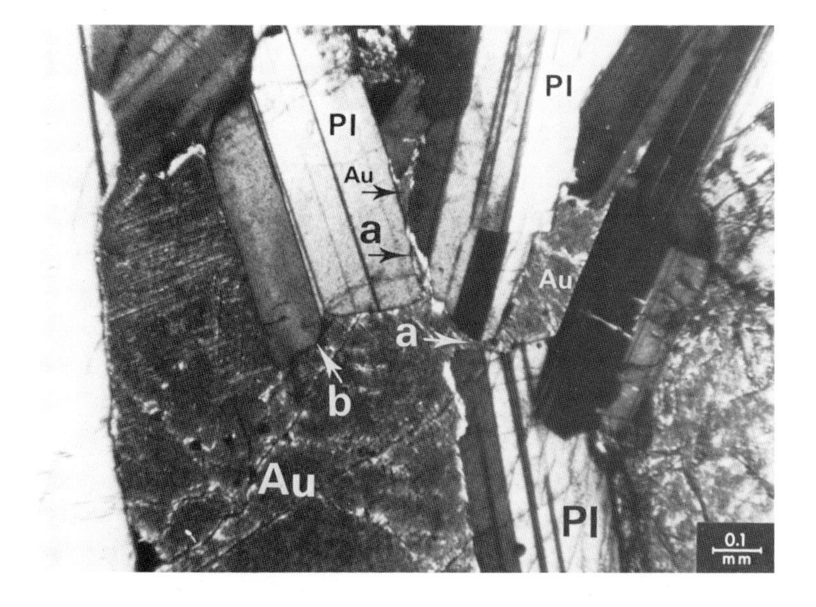

Fig. 570 Gabbroic pseudo-ophitic intergrowth. The poikiloblastic augitic megablast extends and occupies the spaces between the plagioclase laths (clearly corroding the pre-pyroxene plagioclases) and producing a pseudo-ophitic pyroxene/plagioclase symplectic intergrowth.
Pl = plagioclase
Au = augitic pyroxene
Arrows "a" show extensions of the pyroxene occupying the spaces between the plagioclase laths. Arrow "b" shows rounded and corroded plagioclase laths.
Average rock (plagioclase-augite-olivine "cumulate")
West side of Uttentals Plateau, Skaergaard, Greenland.
With crossed nicols.

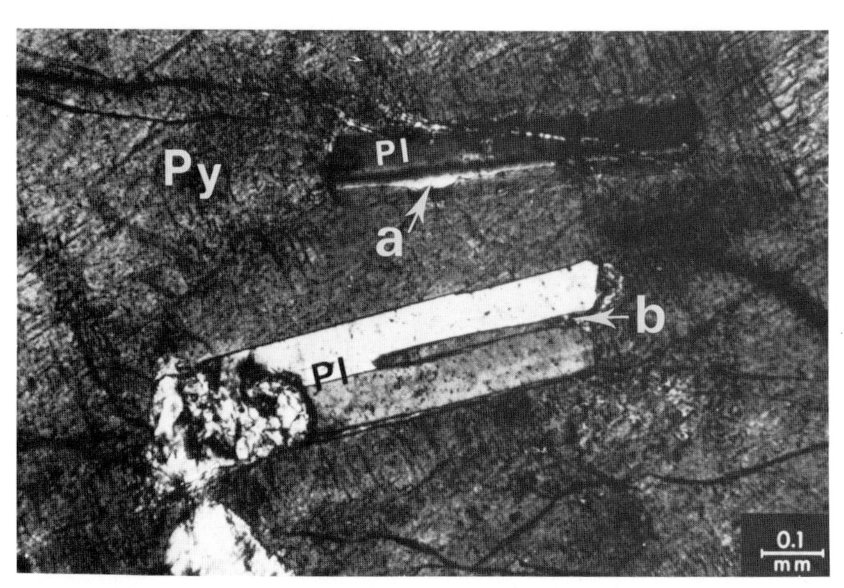

Fig. 571 Plagioclase laths enclosed by a later pyroxene. Arrow "a" shows corroded outline of the plagioclase lath enclosed in the pyroxene. Similarly arrow "b" shows plagioclase lath enclosed by the pyroxene which extends into the twinning plane of the plagioclase.
Py = pyroxene
Pl = plagioclase
Average rock (plagioclase-augite-olivine "cumulate")
West side of Uttentals Plateau, Skaergaard, Greenland.
With crossed nicols.

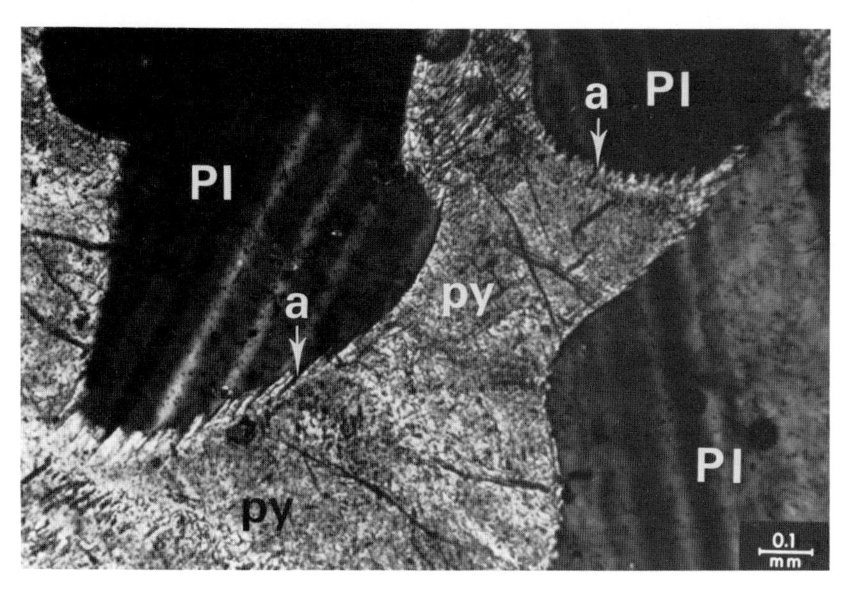

Fig. 572 Pseudo-ophitic plagioclase/pyroxene intergrowth textures. Plagioclases included in the pyroxene and showing corroded outlines. Arrows "a" show indentations of the corroded plagioclase, dependant on the different resistance to corrosion of the plagioclase lamellae.
Pl = plagioclase
py = pyroxene
Average rock (plagioclase-augite-olivine "cumulate")
West side of Uttentals Plateau, Skaergaard, Greenland.
With crossed nicols.

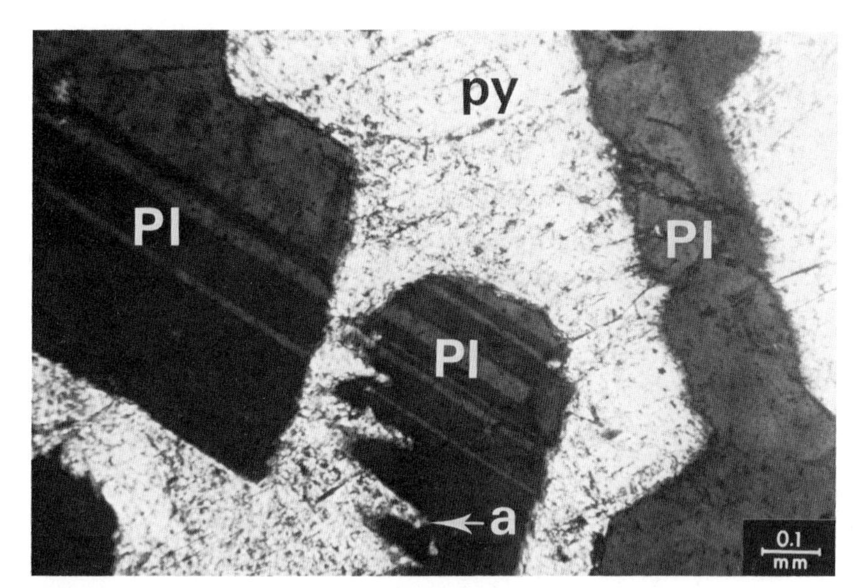

Fig. 573 Pseudo-ophitic plagioclase/pyroxene symplectic intergrowth textures.
Pl = plagioclase laths
py-m = pyroxene megablast enclosing and corroding the included plagioclase laths.
Arrow "a" shows the corroded outline of an included plagioclase invading it parallel to its polysynthetic twinning.
py = pyroxene
Pl = plagioclase.
Olivine gabbro. Southern margin, Skaergaard, Greenland.
With crossed nicols.

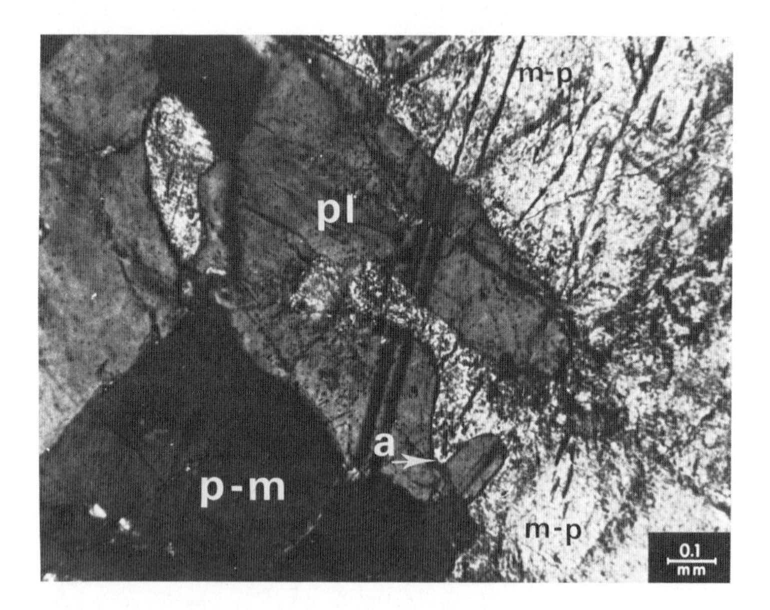

Fig. 574 Pyroxene megablast including corroding and replacing plagioclases.
m-p = pyroxene megablast
pl = plagioclase corroded by pyroxene megablast.
Arrow "a" shows the corroded outline of the plagioclase and replacement of plagioclase by the pyroxene.
Olivine gabbro. Southern margin, Skaergaard, Greenland.
With crossed nicols.

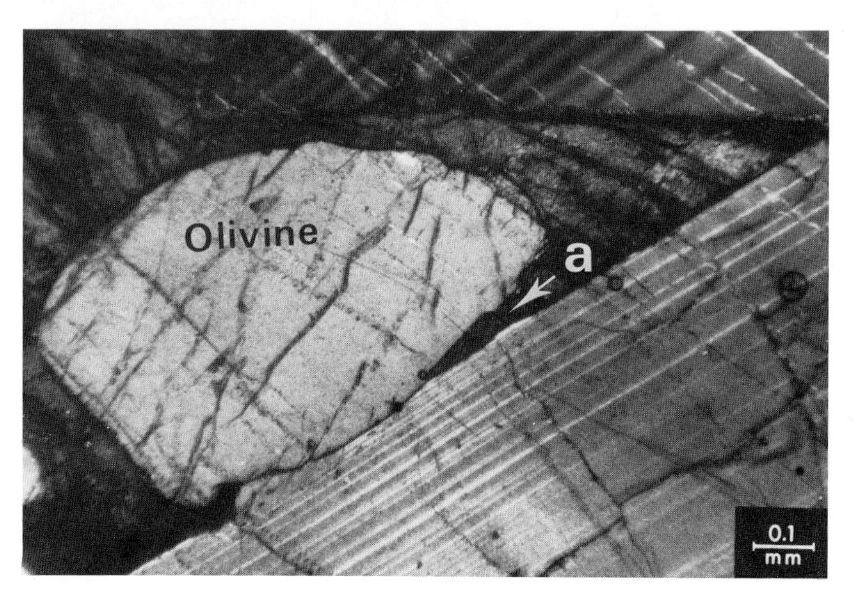

Fig. 575 Pyroxene megablast corroding plagioclase and including an olivine. Arrow "a" shows the intergranular extensions of the pyroxene megablast. Banded melanocratic and leucocratic rock. Skaergaard, Greenland.
With crossed nicols.

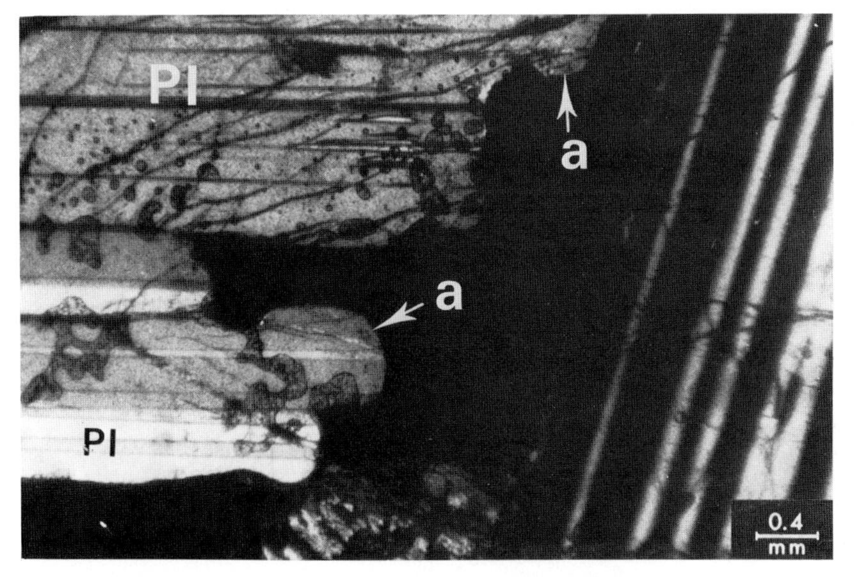

Fig. 576 Late phase magnetite crystallization (blastic magnetite phase) including rounded and corroded plagioclase laths.
Magnetite (black)
Pl = plagioclase
Arrows "a" show rounded, corroded plagioclase outlines.
Coarse average rock (plagioclase olivine-augite "cumulate").
West side of Uttentals Plateau, Skaergaard, Greenland.
With crossed nicols.

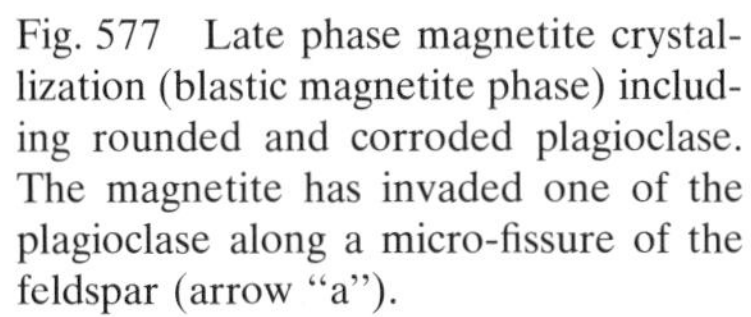

Fig. 577 Late phase magnetite crystallization (blastic magnetite phase) including rounded and corroded plagioclase. The magnetite has invaded one of the plagioclase along a micro-fissure of the feldspar (arrow "a").
"Purple band" ferrogabbro (iron wollastonite-plagioclase-olivine-magnetite-apatite "cumulate"), Skaergaard, Greenland.
With crossed nicols.
(The presence of wollastonite indicates a metamorphic-metasomatic genesis.)

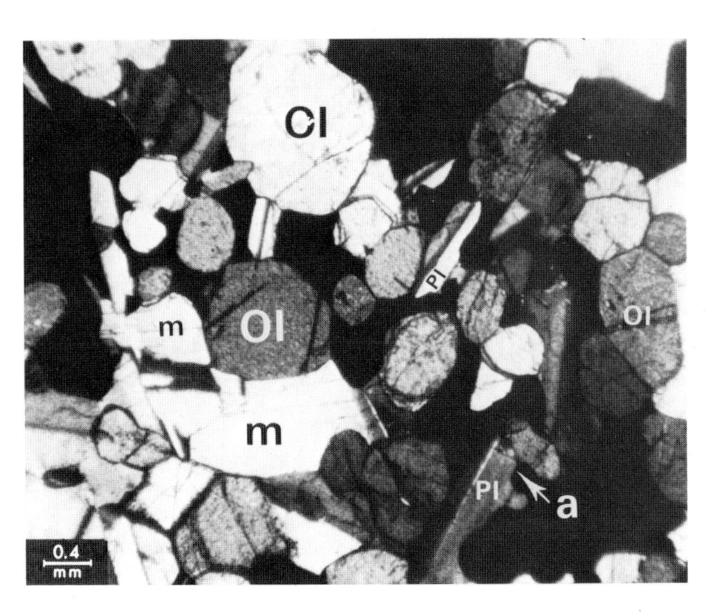

Fig. 578 Plagioclase olivinefels invaded by a late blastogenic magnetite phase which occupies the interleptonic spaces of the intergranular and corrodes olivine and pyroxene crystal grains.
Ol = olivine
Pl = plagioclase
m = mica
magnetite (black)
Arrow "a" shows corroded outline of the plagioclase. Perpendicular feldspar rock (a peculiar eucrite with feldspars roughly perpendicular to margin of intrusion).
55 meters from western margin on Mellemo, Skaergaard, Greenland.
With crossed nicols.

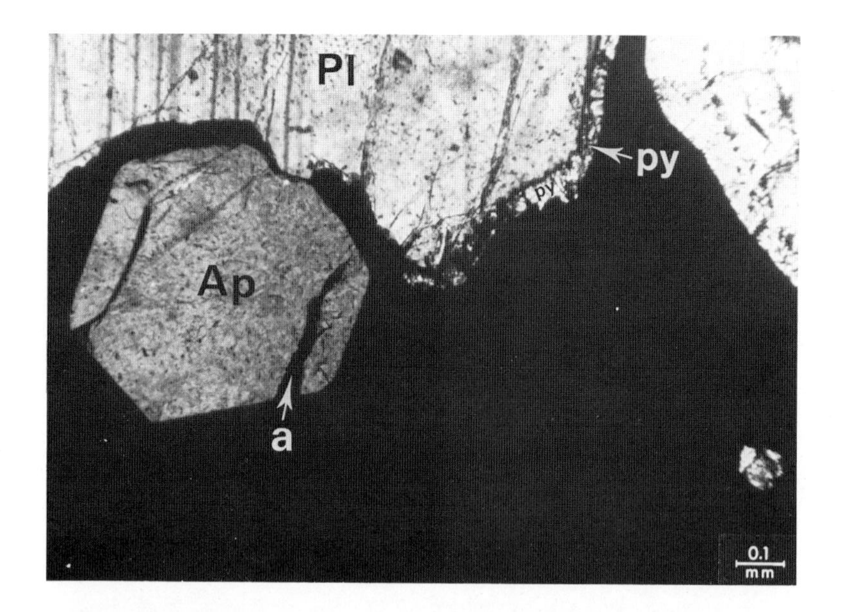

Fig. 579 Blastic magnetite including apatite and partly invading it along a crack. Between the magnetite and an adjacent plagioclase pyroxene "infiltrations" have taken place. Magnetite (black).
Pl = plagioclase
Ap = apatite (arrow "a" shows apatite crack invaded by magnetite).
py = pyroxene
Average ferrogabbro (plagioclase-augite-olivine-magnetite "cumulate").
House area, Skaergaard, Greenland.
With crossed nicols.

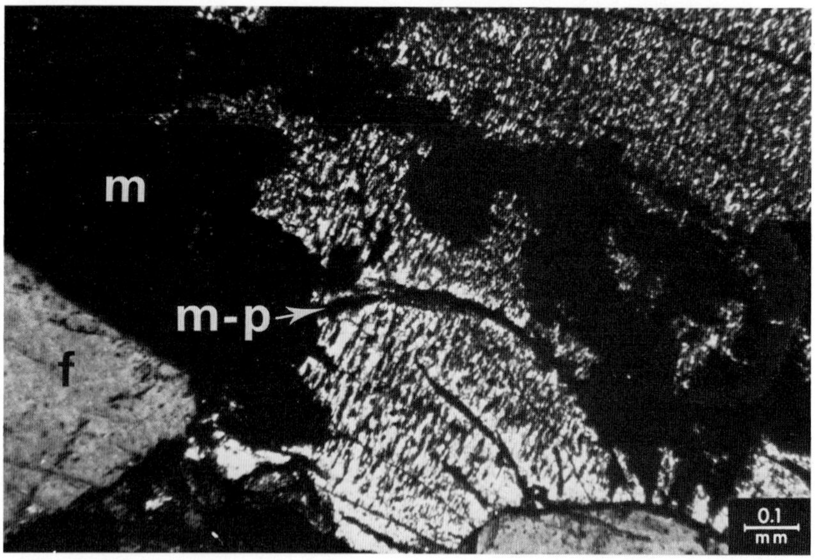

Fig. 580a Magnetite with infiltration extending and occupying the interleptonic spaces of the pyroxene cleavage.
m = magnetite replacements of the pyroxene
m-p = magnetite interleptonically within the pyroxene host.
f = feldspar
"Purple band" ferrogabbro (iron wollastonite-plagioclase-olivine-magnetite-apatite "cumulate").
Skaergaard, Greenland.
With crossed nicols.

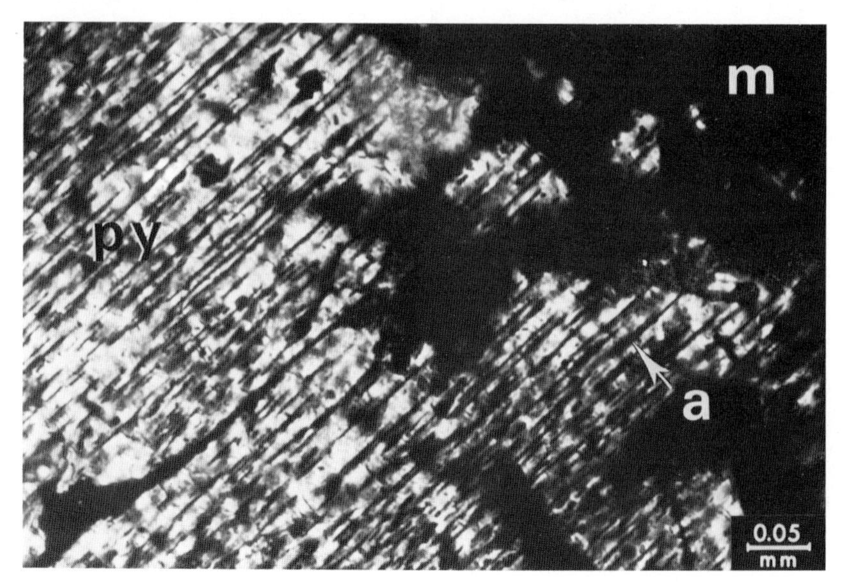

Fig. 580b Pyroxene invaded by magnetite with extension of the ore-mineral following the interleptonic spaces of the pyroxene cleavage.
m = magnetite
py = pyroxene
Arrows "a" show the magnetite following the interleptonic cleavage spaces of the pyroxene.
"Purple band" ferrogabbro (iron wollastonite-plagioclase-olivine-magnetite-apatite "cumulate").
Skaergaard, Greenland.
With crossed nicols.

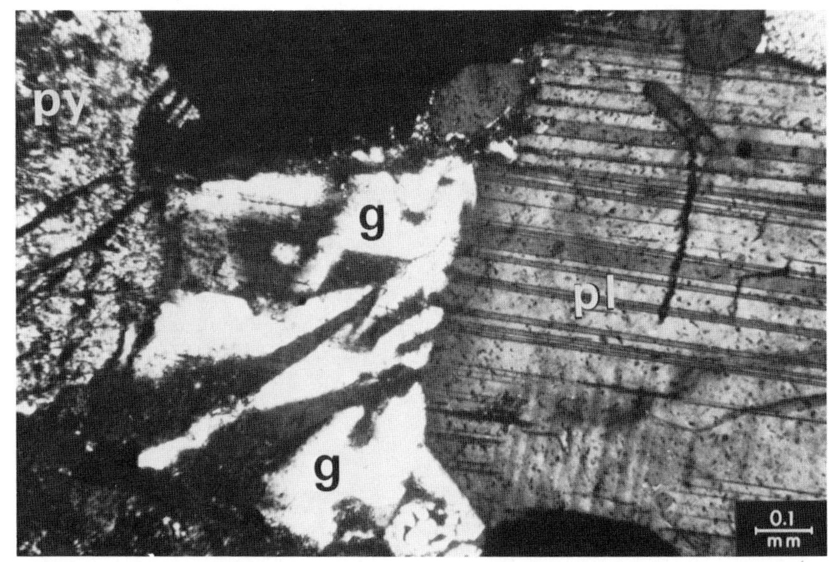

Fig. 581 Granophyric quartz (micrographic) in symplectic intergrowth with plagioclase.
g = granophyric quartz
pl = plagioclase
py = pyroxene
Fayalite ferrogabbro. "Upper purple Band".
Basistoppen, Skaergaard, Greenland.
With crossed nicols.

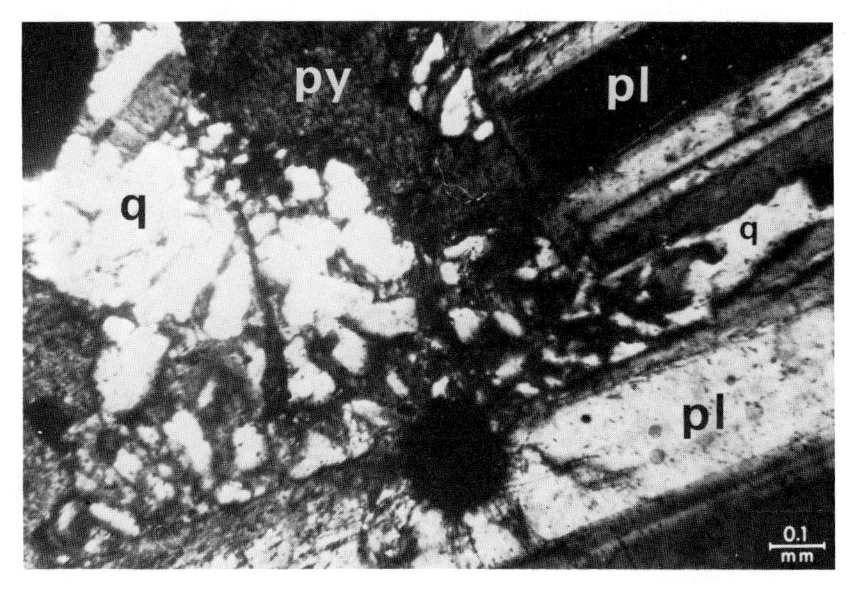

Fig. 582 Granophyric quartz (q) in symplectic intergrowth with pyroxene and with extensions of the micrographic quartz following the twinning of an adjacent plagioclase.
py = pyroxene
pl = plagioclase
Fayalite ferrogabbro "Upper purple Band".
Basistoppen, Skaergaard, Greenland.
With crossed nicols.

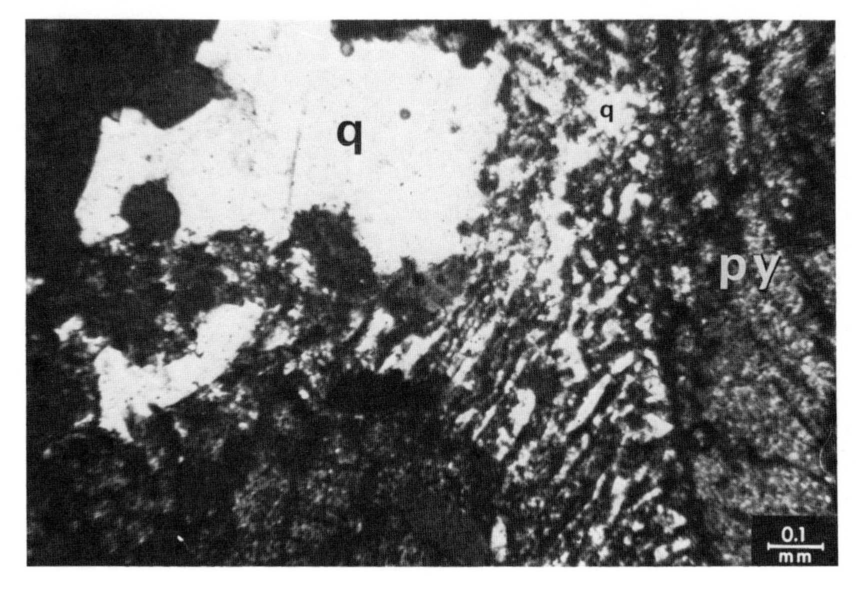

Fig. 583 Quartz (q) with transitions into quartz infiltrations (attaining a granophyric resemblance) into an adjacent pyroxene (py).
Fayalite ferrogabbro "Upper purple Band".
Basistoppen, Skaergaard, Greenland.
With crossed nicols.

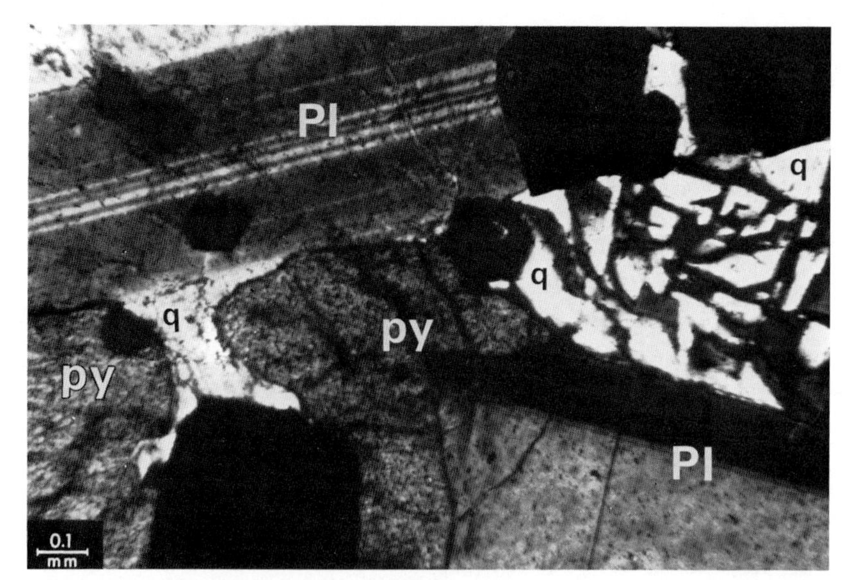

Fig. 584 Micrographic quartz (q) as interstitial intergrowth occupying spaces between large plagioclases (Pl). The granophyric quartz extends as infiltration into the interleptonic spaces between plagioclases and pyroxenes and invading an adjacent magnetite along cracks.
Pl = plagioclases
py = pyroxene
Magnetite (black)
Fayalite ferrogabbro "Upper purple Band".
Basistoppen, Skaergaard, Greenland.
With crossed nicols.

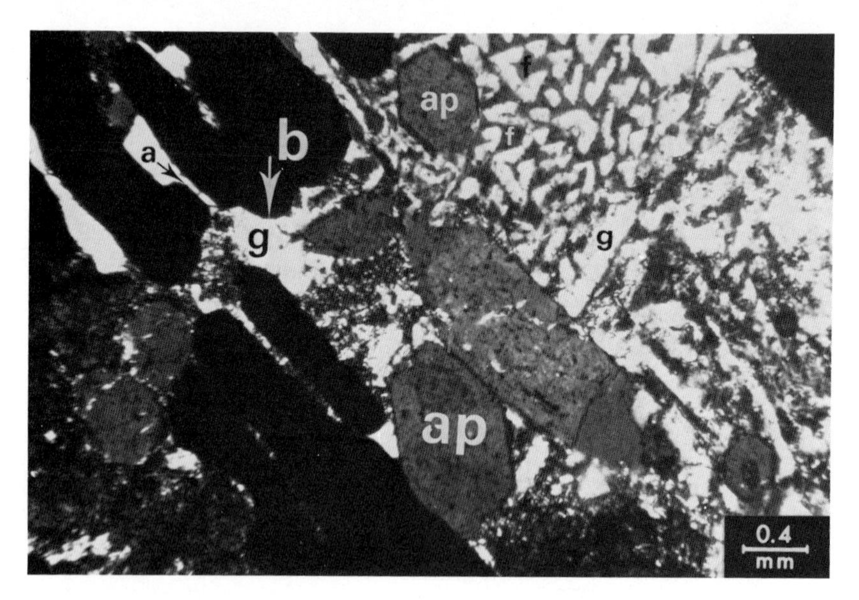

Fig. 585 Micro-graphic quartz/plagioclase intergrowth with granophyric quartz invading into adjacent magnetites which are exhibiting rounded outlines due to corrosion by the granophyric symplectite.
Magnetite (black)
g = granophyric quartz
f = plagioclase
ap = apatite
Arrows "a" show granophyric quartz invading magnetite.
Arrow "b" shows corroded outline of the magnetite.
Fayalite ferrogabbro, "Upper purple Band".
Basistoppen, Skaergaard, Greenland.
With crossed nicols.

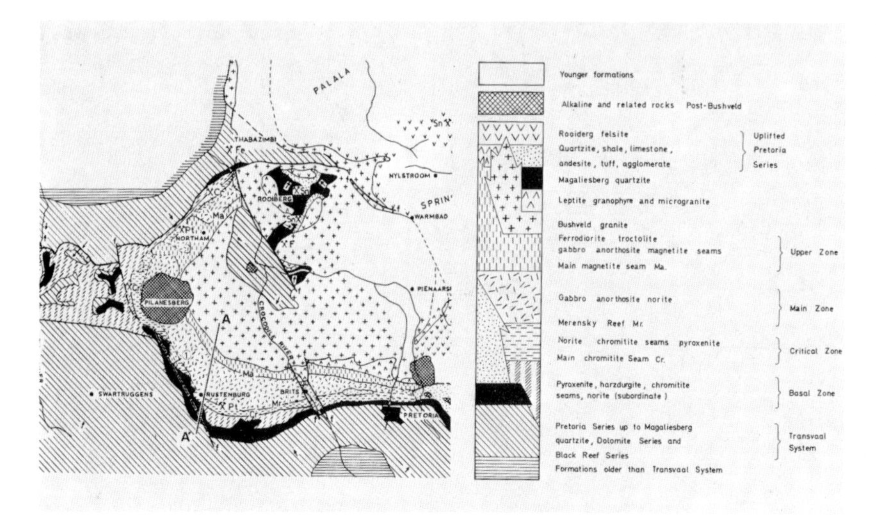

Fig. 586 Geological sketch map showing part of the ultrabasic Bushveld Complex based on the geological map of Willemse (1969), Scale 1:4 700 000.
A-A´ shows traverse along which samples have been studied.

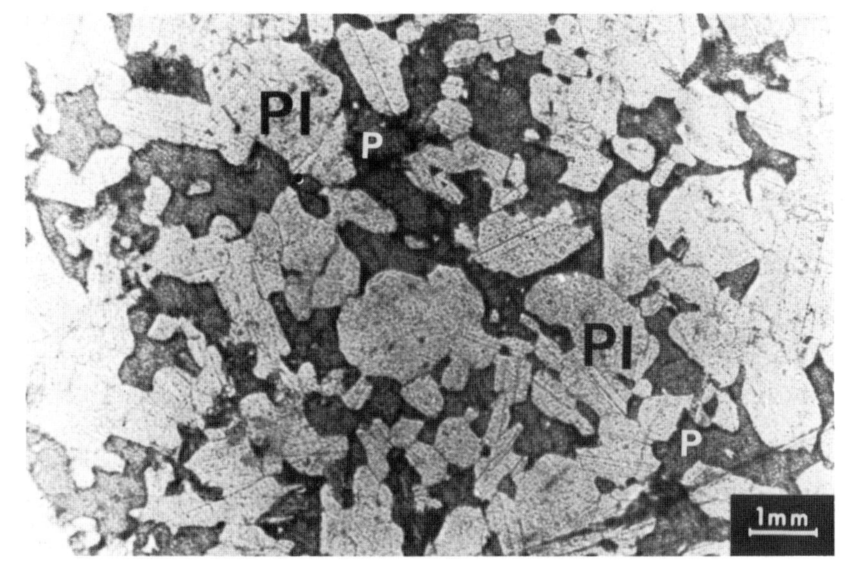

Fig. 587 Plagioclases (often rounded and corroded) included in a poikiloblastic pyroxene megablast.
Pl = plagioclase
P = poikiloblastic pyroxene
Norite-leuconorite-anorthosite.
Near Bapong, Bushveld, Transvaal, South Africa.
Without crossed nicols.

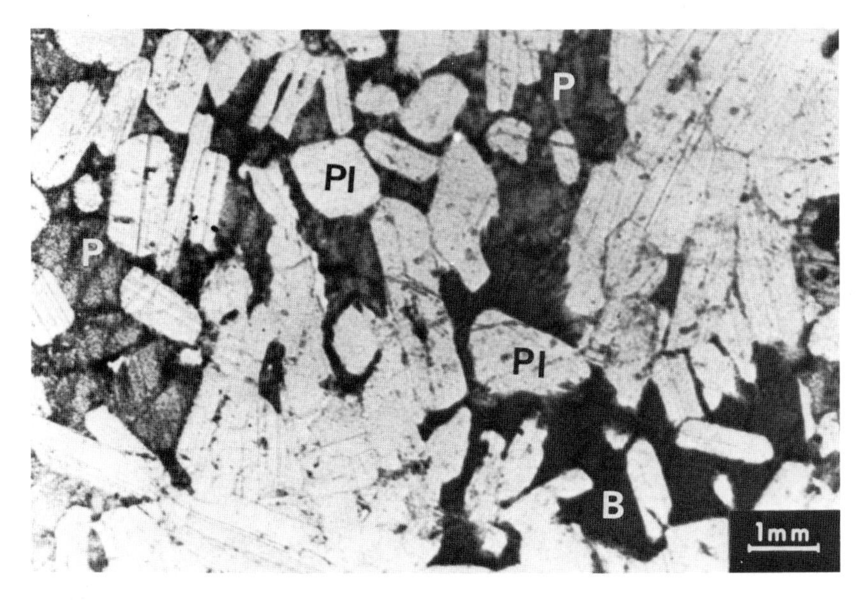

Fig. 588 Plagioclases included in a poikiloblastic pyroxene megablast which "transgresses" into biotite-poikiloblast.
Pl = plagioclases
P = pyroxene megablast (poikiloblast)
B = biotite poikiloblast
Norite-Gabbro. Between Hex River and Nooitgedacht, Bushveld, Transvaal, South Africa.
Without crossed nicols.

Fig. 589 Plagioclases included in a pyroxene poikiloblast (megablast).
Pl = plagioclase
P = pyroxene
Norite Gabbro. Between Hex River and Nooitgedacht, Bushveld, Transvaal, South Africa.
With half crossed nicols.

Fig. 590 Pyroxene including plagioclases and sending protuberances into the plagioclase texture. The textural pattern simulates ophitic intergrowths.
Pl = plagioclase
P = pyroxene
Norite-Gabbro. Between Hex River and Nooitgedacht, Bushveld, Transvaal, South Africa.
With crossed nicols.

Fig. 591 Pyroxene and amphibole granoblasts included in a plagioclase poikiloblast (megablast).
Py = pyroxene
A = amphibole
Pl = plagioclase poikiloblast
Norite-Gabbro. Waterkloof, Bushveld, Transvaal, South Africa.
With crossed nicols.

Fig. 592 Amphiboles included in a plagioclase megacrystalloblast.
A = amphibole
Pl = plagioclase poikiloblast
Norite-gabbro. Boschpoort, Bushveld, Transvaal, South Africa.
With crossed nicols.

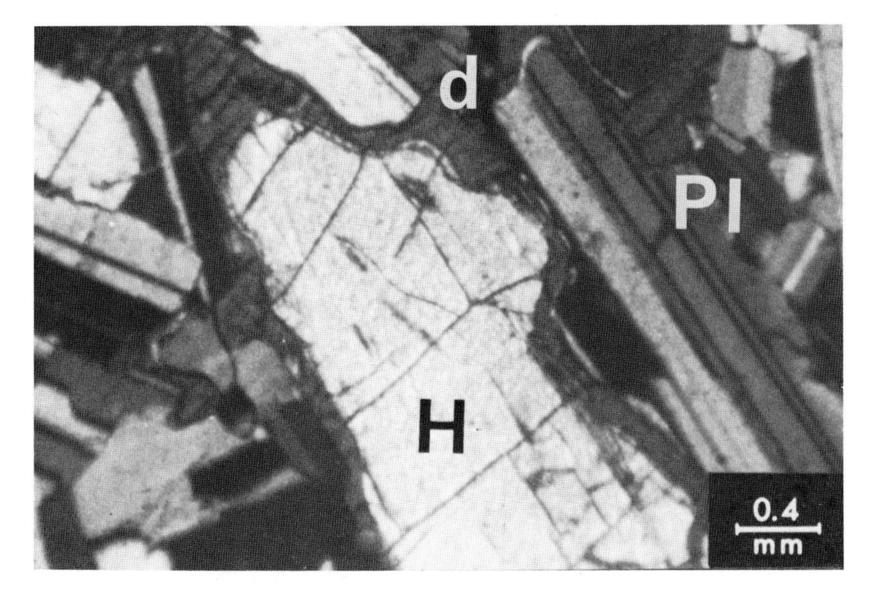

Fig. 593 Orthopyroxene (hypersthene) with a diopsidic margin which extends also into the intergranular of the plagioclases.
H = hypersthene
d = diopsidic margin of the hypersthene extending also into the intergranular of the plagioclases.
H = hypersthene
d = diopsidic margin
Pl = plagioclase
Anorthosite-leuconorite. Near Bapong, Bushveld, Transvaal, South Africa.
With crossed nicols.

Fig. 594 Orientated, elongated plagioclases in cases tectonically bent, and pyroxene delimited by the plagioclases and sending protuberances into the plagioclase intergranular.
Pl = plagioclase
Py = pyroxene (arrow "a" shows extension of the pyroxene into the plagioclase intergranular).
"Ferrogabbro". Vorentoe (South East of Bethane), Bushveld, Transvaal, South Africa.
With crossed nicols.

Fig. 595 Olivine, pyroxene and amphibole granoblastic texture.
Norite-gabbro. Between Waterval and Waterkloof, Bushveld, Transvaal, South Africa.
With crossed nicols.

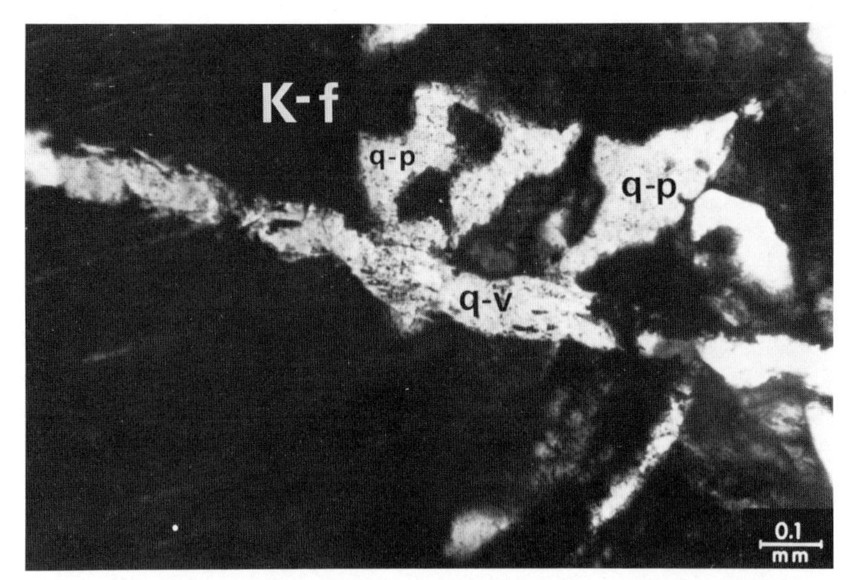

Fig. 596 and Fig. 597 K-feldspar transversed by a quartz-veinlet which sends protuberances (extensions) into the adjacent orthoclase and which attain granophyric form.
K-f = K-feldspar
q-v = quartz veinlet
q-p = protuberances of the quartz-veinlet extending into the orthoclase and attaining granophyric (micrographic) forms.
Granophyre. Langberg, Bushveld, Transvaal, South Africa.
With crossed nicols.

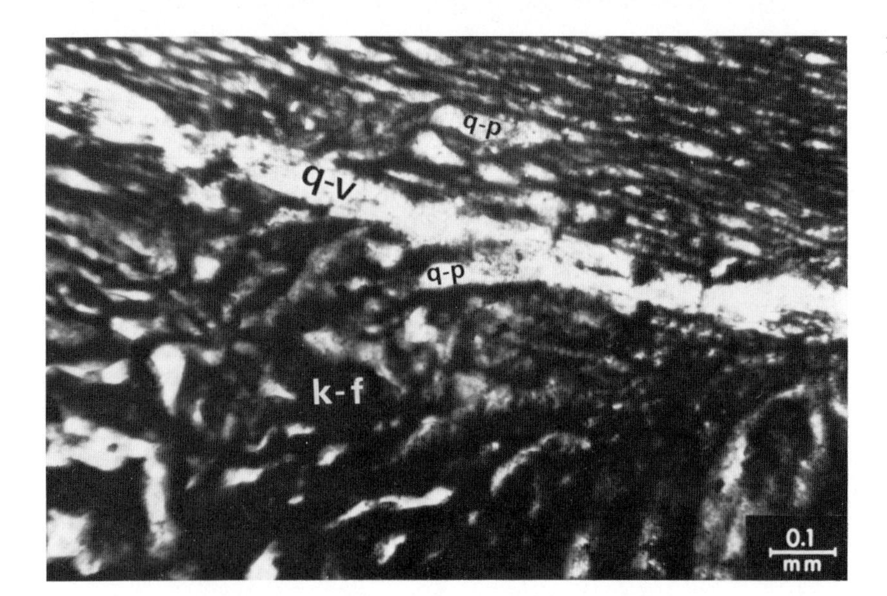

Fig. 597

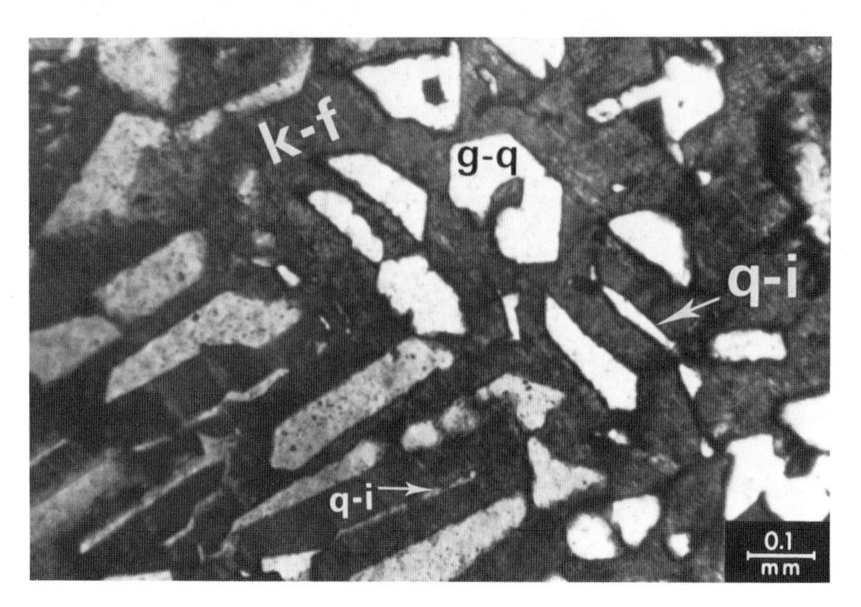

Fig. 598 Quartz occupying interleptonic spaces in the orthoclase, is similarly orientated to well developed graphic-quartz forms.
q-i = quartz occupying the interleptonic spaces of K-feldspar.
g-q = well developed graphic quartz.
K-f = orthoclase
Granophyre. Langberg, Bushveld, Transvaal, South Africa.
With crossed nicols.

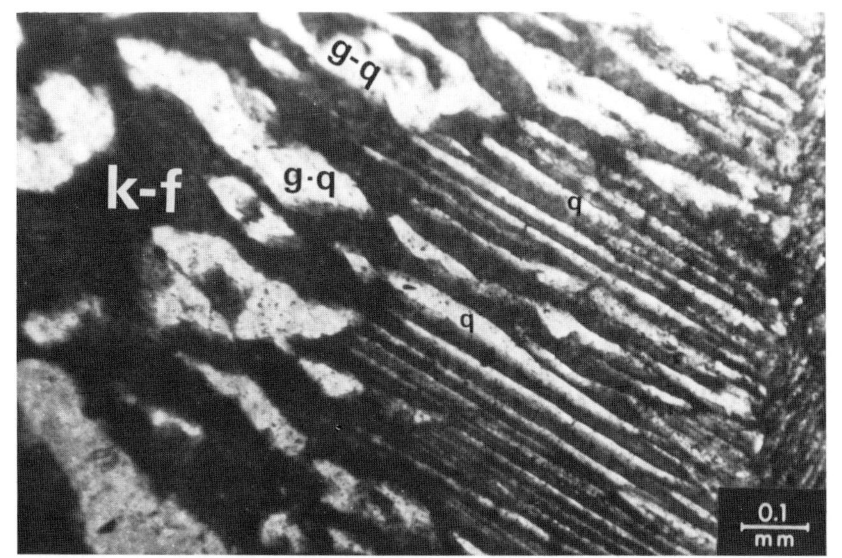

Fig. 599 Granophyric quartz transgressing into quartz occupying the interleptonic spaces of the K-feldspar.
g-q = granophyric quartz
q = quartz occupying the interleptonic spaces
K-f = orthoclase
Granophyre. Langberg, Bushveld, Transvaal, South Africa.
With crossed nicols.

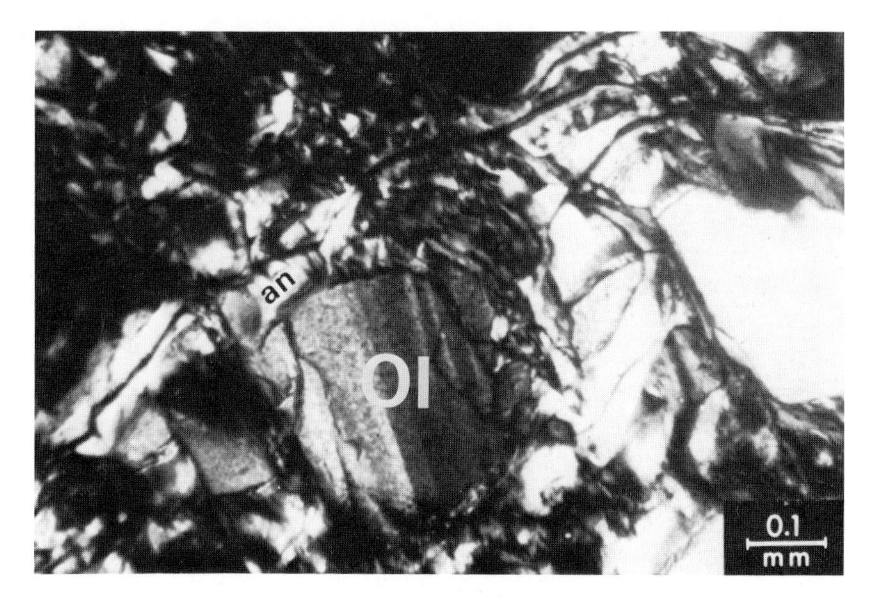

Fig. 601 Olivine relic in serpentinised mass, with the olivine exhibiting deformation lamellae.
ol = olivine
an = antigorite (due to the serpentinisation of the dunite)
Dunite with chromite bands.
Xerolivado, Vourinos, Greece.
With crossed nicols.

(For Figs. 600a, b, c see page 318)

Fig. 602 Olivine relics in a serpentinised mass. The olivine grains exhibit deformation lamellae. The deformation lamellae of adjacent olivines have a common orientation (probably indicating initial large olivines serpentinised).
ol = olivine exhibiting deformation lamellae
an = antigorite
Dunite with chromite bands.
Xerolivado, Vourinos, Greece.
With crossed nicols.

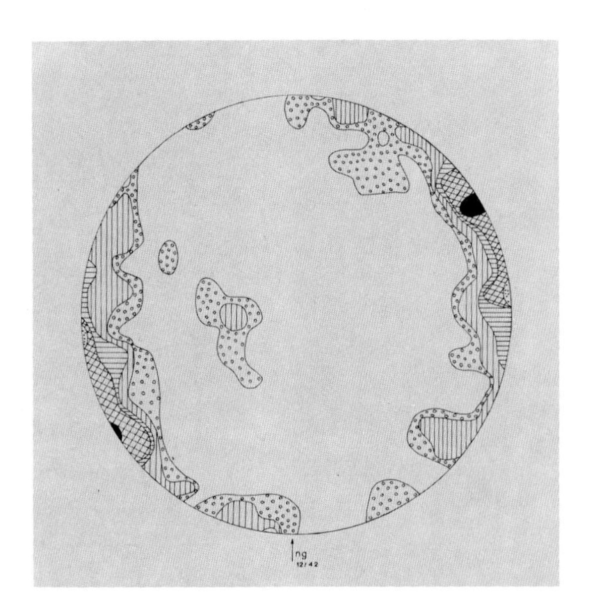

Fig. 600a, b, c Petrofabric analysis showing an example of the distribution of olivine grains in the chromite deposits of Vourinos.
By Spathi, K. 1966.

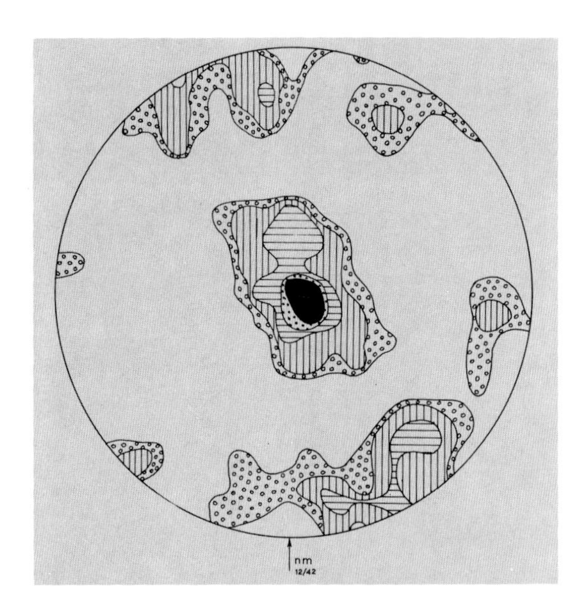

Fig. 600b

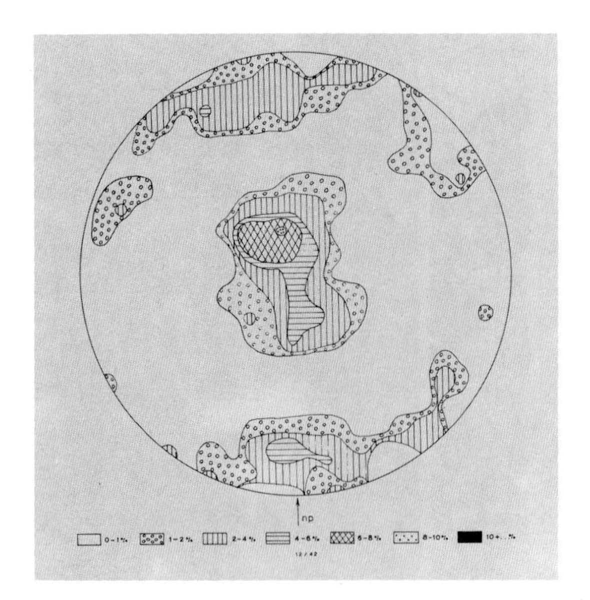

Fig. 600c

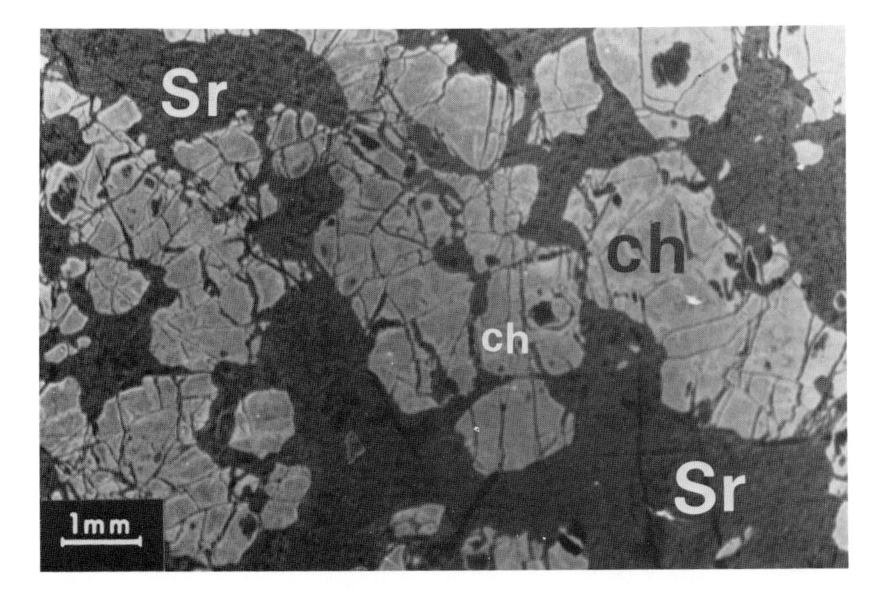

Fig. 603 Rather rounded "granular" cataclastic chromite rich band in banded dunite (consisting of alternating chromite/olivine bands). The olivine is to a great extent serpentinised.
Sr = serpentine
ch = chromite "rounded" with strong cataclastic effects.
Xerolivado, Vourinos, Greece.
Polished section, with one nicol.

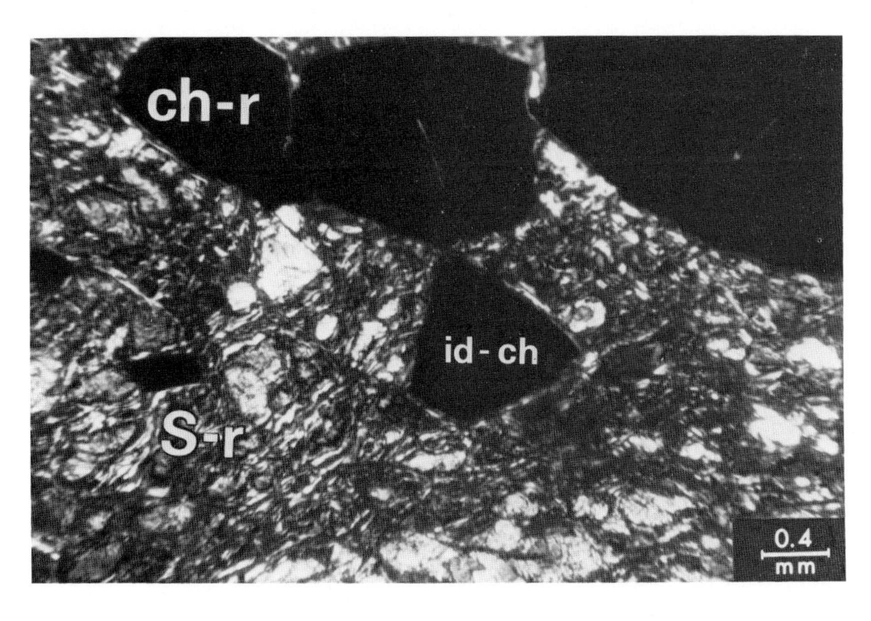

Fig. 604 Rounded granular and idiomorphic chromite in serpentine. chromite-bands in dunite (serpentinised).
Sr = serpentine
ch-r = rounded chromite
id-ch = idiomorphic chromite
Banded chromite. Xerolivado, Vourinos, Greece.
With crossed nicols.

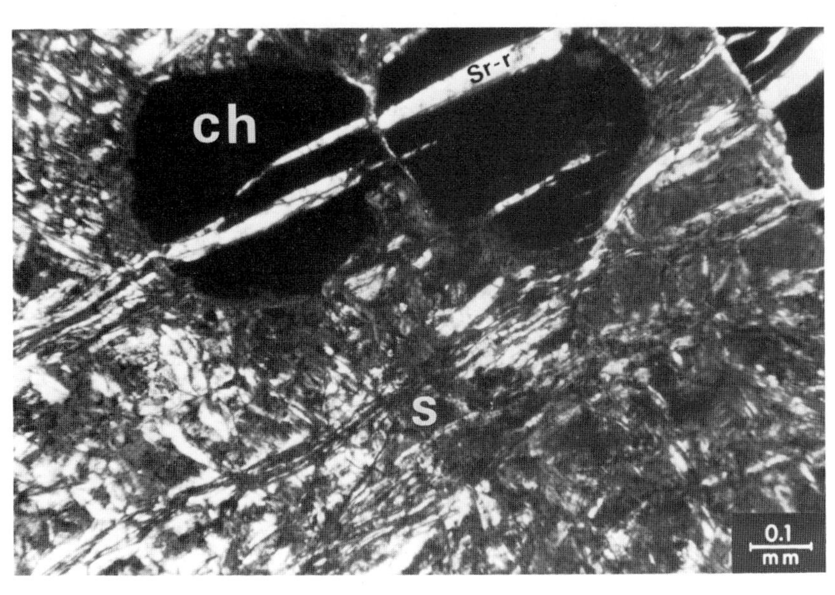

Fig. 605 Rounded chromite cataclastically affected with "re-binding" serpentine holding together the fractured chromite grains.
ch = chromite cataclastically affected.
Sr-r = re-binding serpentine holding together the fractured chromite grains.
s = serpentinised dunite.
Chromite-ore. Xerolivado, Vourinos, North Greece. With crossed nicols.

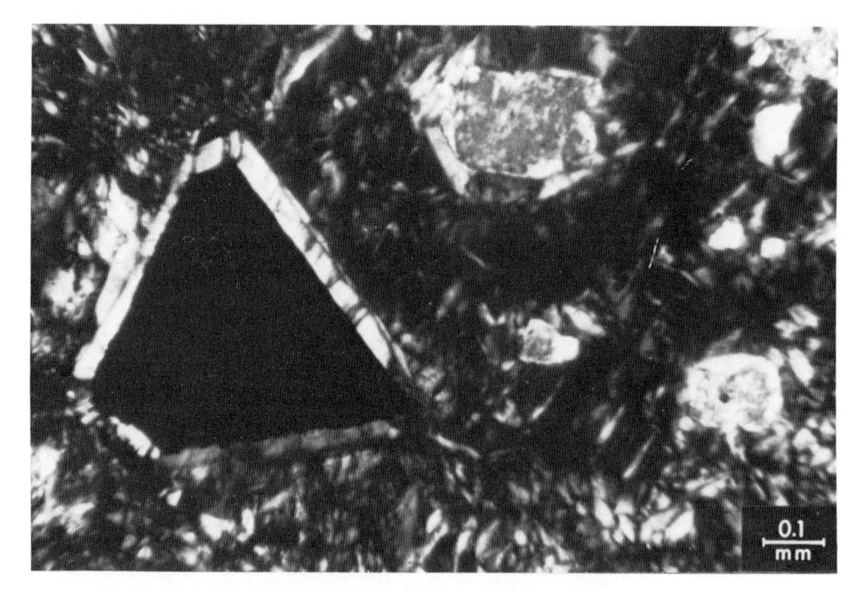

Fig. 606 Idiomorphic chromite with a margin of recrystallised serpentine in serpentine (due to the antigoritisation of initial dunite).
Dunite with bands of chromite.
Banded chromite. Xerolivado, Vourinos, Greece.
With crossed nicols.

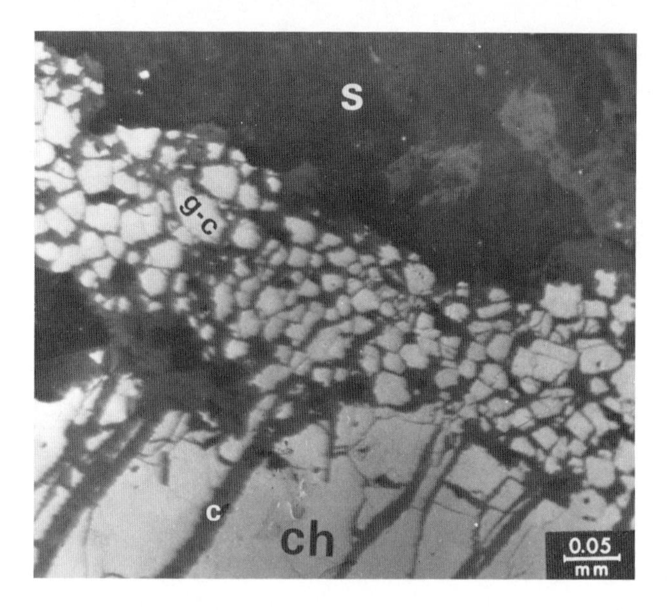

Fig. 607 Compact cataclastically affected chromite transgressing marginally into a zone of fractured granular-chromite with intergranular brown-serpentine (recrystallised serpentine).
ch = chromite cataclastically affected
c = cataclastic fractures of compact chromite
s = serpentine
g-c = granular chromite zone due to chromite fracturing.
Chromite-ore. Rodiani, North Greece.
Polished section (oil immersion, with one nicol).

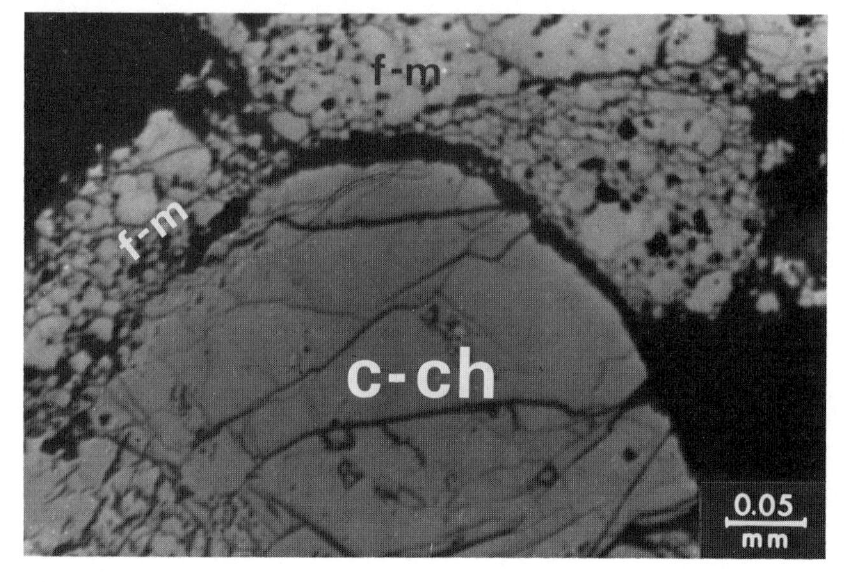

Fig. 608 Compact chromite with a fractured marginal chromite zone, consisting of granular chromite.
c-ch = compact chromite
f-m = fractured marginal chromite zone
Banded chromite. Xerolivado, Vourinos, Greece.
With crossed nicols.

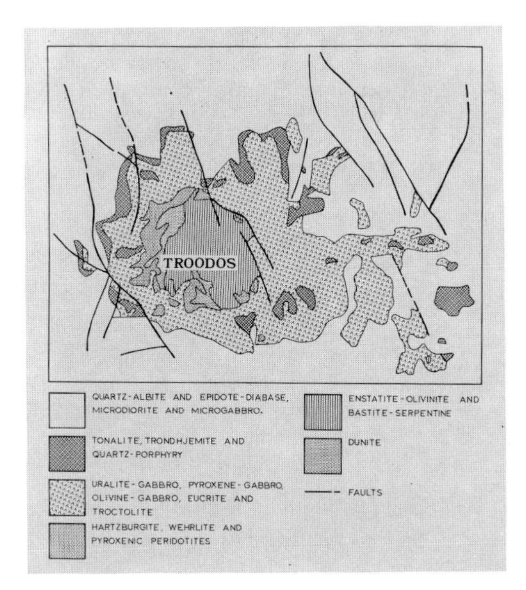

Fig. 609 A geological sketch map of the Troodos ophiolitic complex, based on the Geological map of Cyprus published by the Geol. Survey of Cyprus.
Scale 1:500000

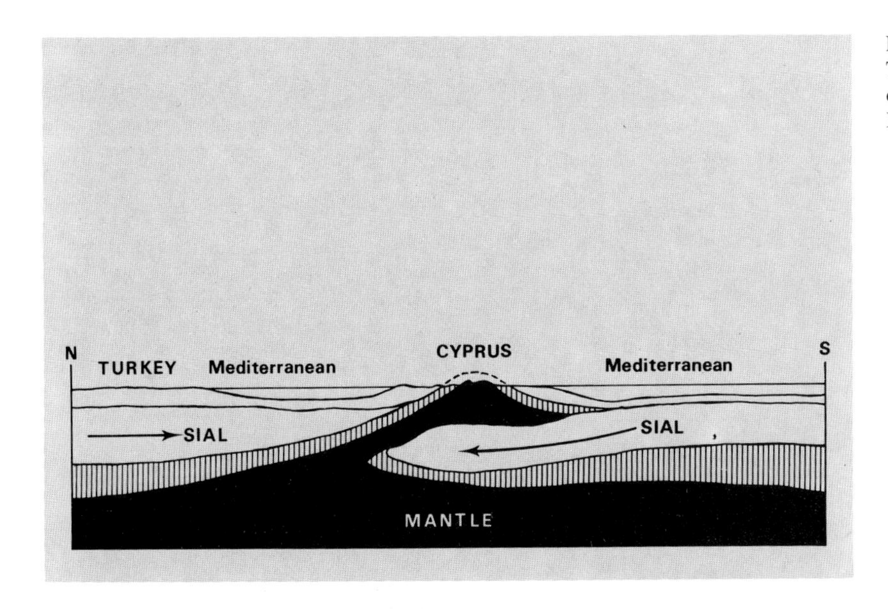

Fig. 610 Hypothetical diagram showing Troodos, Cyprus, to be due to mantle diapirism.
By Gass and Masson-Smith (1965)

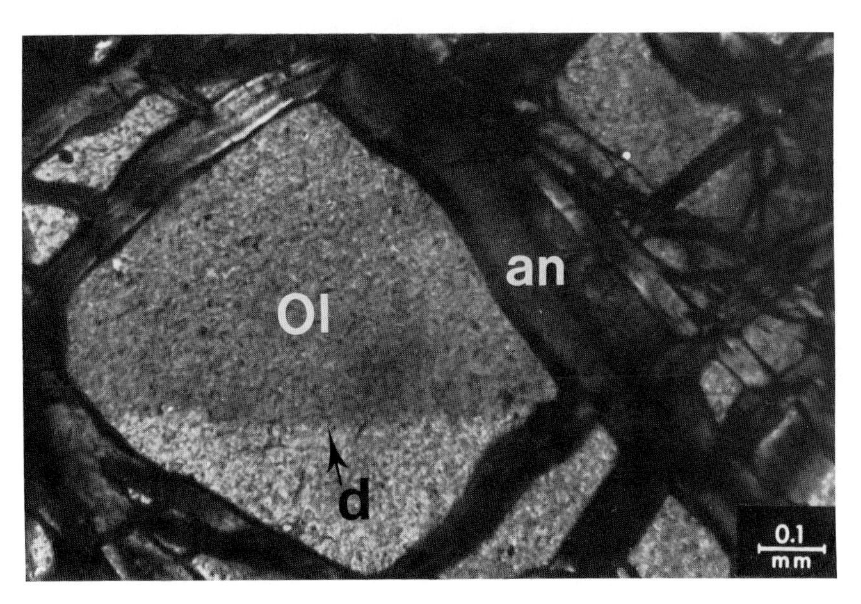

Fig. 611 Deformed forsteritic olivine antigorised. Often the deformation effects appear as "undulating" extinction or deformation twinning. The deformation effect of dunitic olivines could be synchronous with the banding of the dunite as was the case of Vourinos (see Figs. 601, 602).
Ol = olivine
d = deformation twinning of the forsterite
an = antigorite
Troodos, Cyprus.
With crossed nicols.

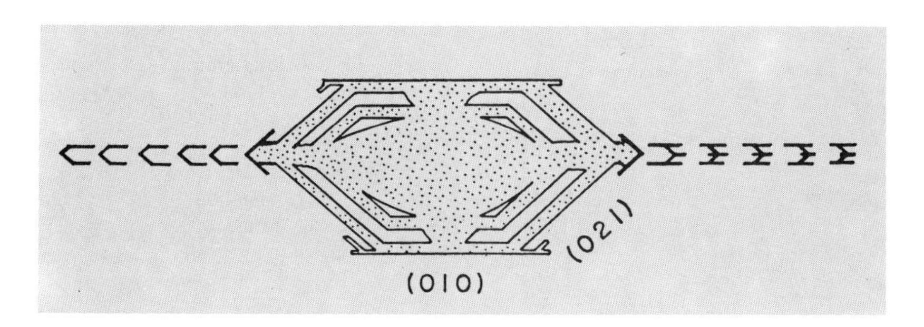

Fig. 612 Composite idealized sketch of a skeletal olivine crystal with the lantern and chain habit, (100) section; right-hand-side chain has H-shaped units, left-hand-side chain has U-shaped units.
Diagram by Fleet (1975).

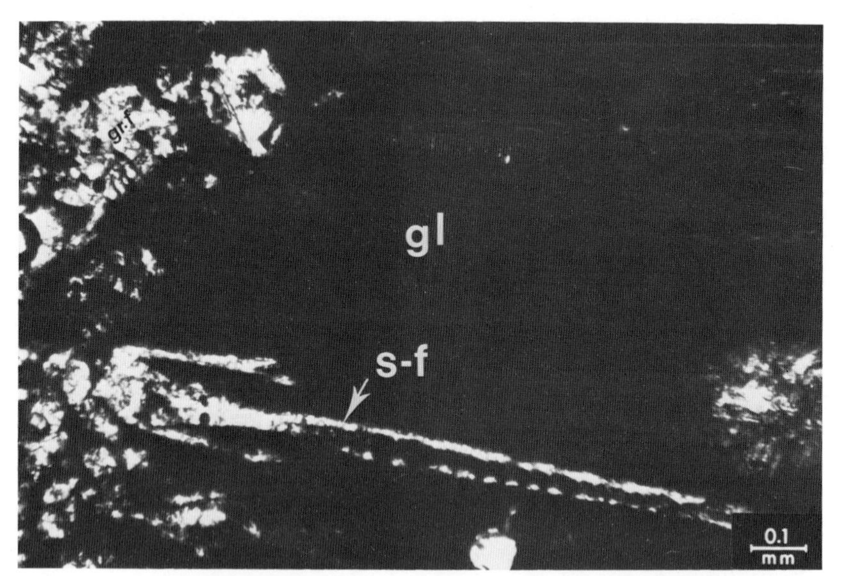

Fig. 613a, b Experimentally produced Spinifex Forsterite (Fo 95–97%)
Granular forsterite (Fo 95–97%) and pyroxene crystallised in the central part of the crucible, sends extensions into the chilled glassy margins.
gl = glassy margin (chilled margin)
gr-f = granular forsterite
s-f = spinifex forsterite
Quenched melt: (superheated (1450°C) crushed olivine bomb of the Gerona olivine basalt-Spain, and subsequently quenched in water).

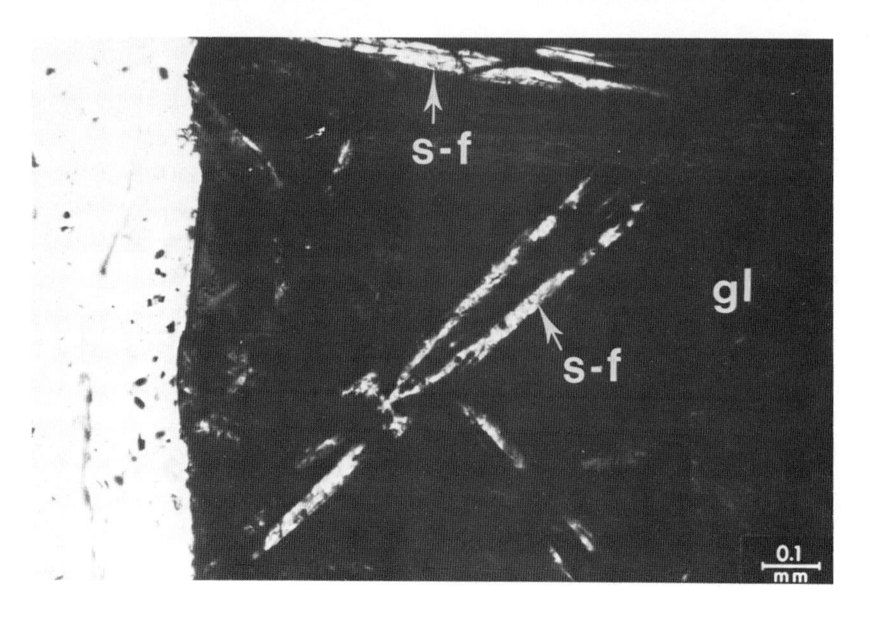

Fig. 613b

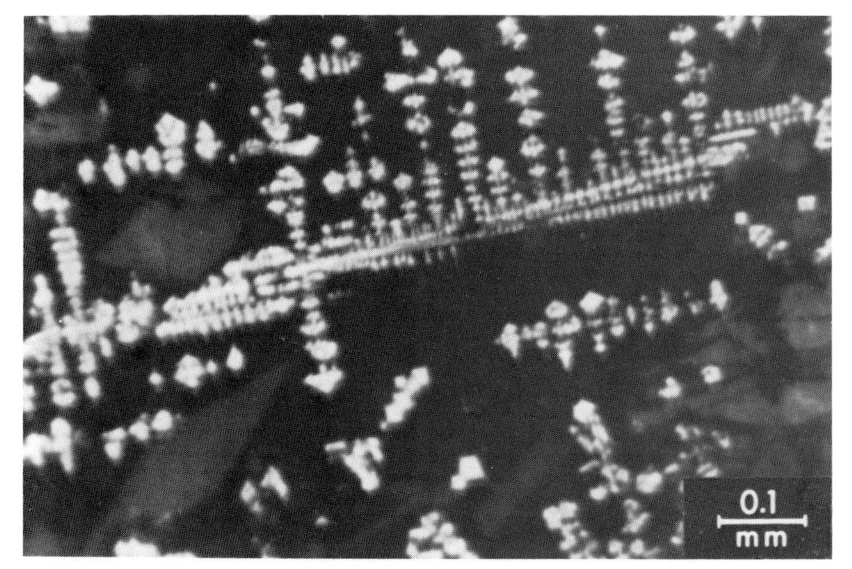

Fig. 614 Spinifex like hercynite or magnetite skeletal growths in slags.
Slag. Larymna, Greece.
Polished section (oil immersion with one nicol)
Sample courtesy Dr. E. Mposkos.
(Actually impregnations of slags into the bricks of the furnace.)

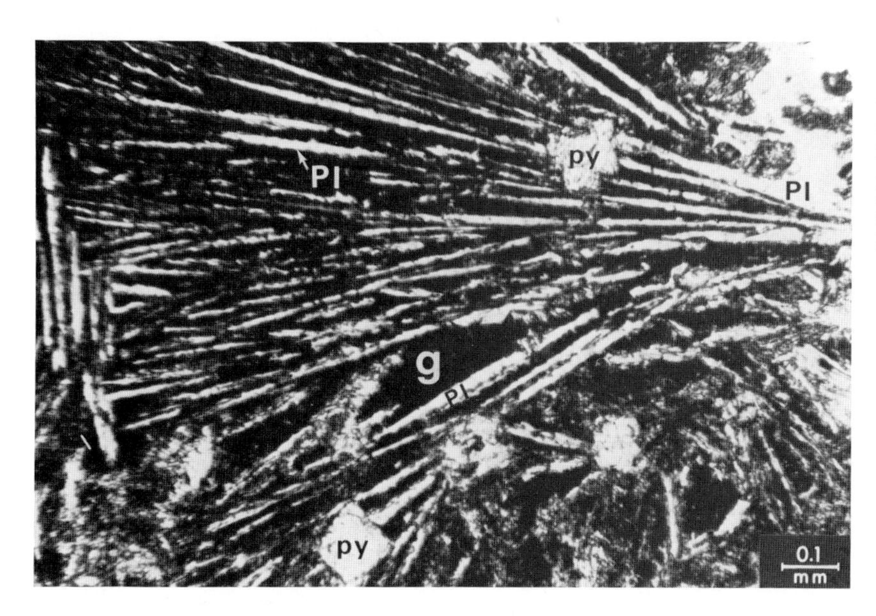

Fig. 615a Typical Spinifex texture. Elongated radiating spinifex plagioclases in glassy basaltic groundmass (due to rapid cooling).
Pl = plagioclase
g = glassy basaltic groundmass
py = pyroxenes
Basalt. Mid-Atlantic ridge.
With crossed nicols.

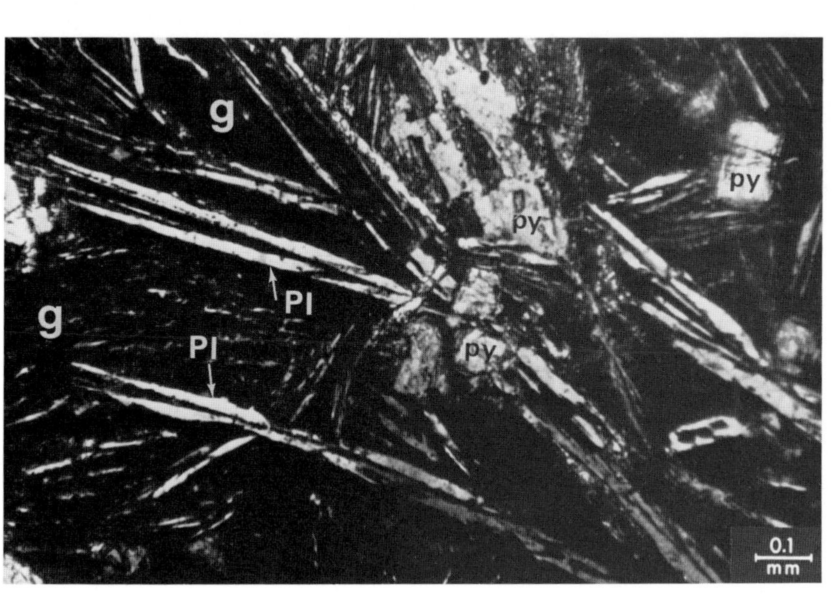

Fig. 615b Spinifex plagioclase radiating out of pyroxene nuclei.
Pl = plagioclase
g = glassy groundmass
py = pyroxene
Basalt. Mid-Atlantic ridge.
With crossed nicols.

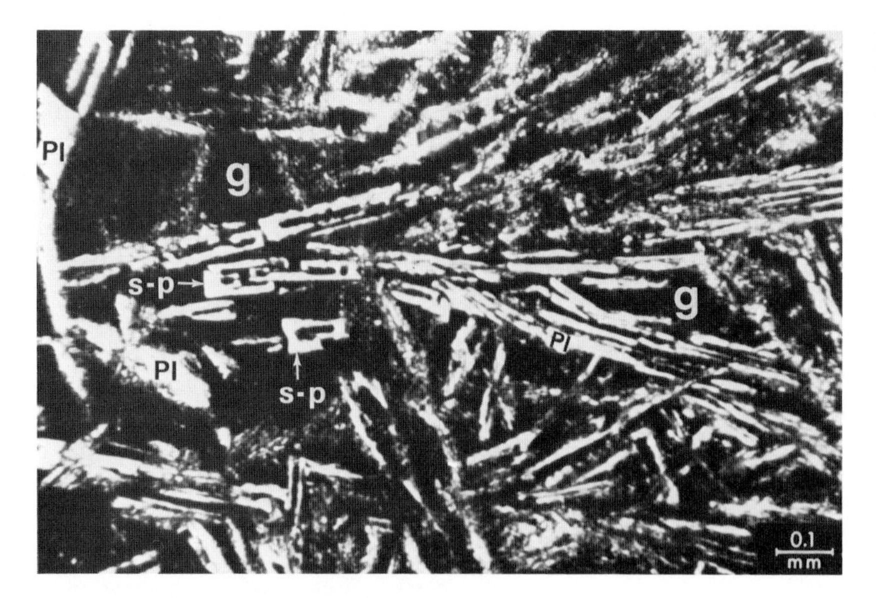

Fig. 615c Skeletal spinifex plagioclase crystal growths in glassy basaltic groundmass. Often the plagioclase show interpenetration.
Pl = plagioclase
g = glassy basaltic groundmass
s-p = skeletal plagioclase growths
Basalt. Mid-Atlantic ridge.
With crossed nicols.

Fig. 616 Photomicrograph of the flow top of a spinifex-textured peridotitic komatiite flow. Skeletal grains of olivine (serpentinised) set in a fine grained devitrified glass groundmass.
Arrow "a" shows olivine crystal form comparable to the skeletal olivine crystal shown in Fig. 612.
Photomicrograph by Arndt et al. (1977).
Lavas. Munro Township, Northeast Ontario.
With crossed nicols.

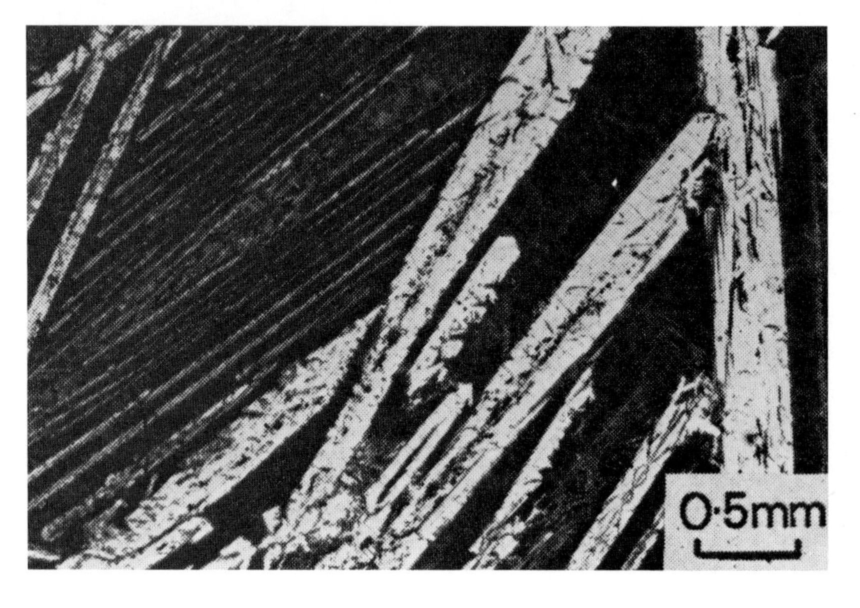

Fig. 617a Photomicrograph of the upper part of the spinifex-textured zone of a peridotitic komatiite flow.
Skeletal plates of olivine (serpentinised) are randomly orientated in a matrix (groundmass) of clinopyroxene and devitrified glass.
Photomicrograph by Arndt et al. (1977).
Lavas Munro Township, Northeast Ontario.
With crossed nicols.

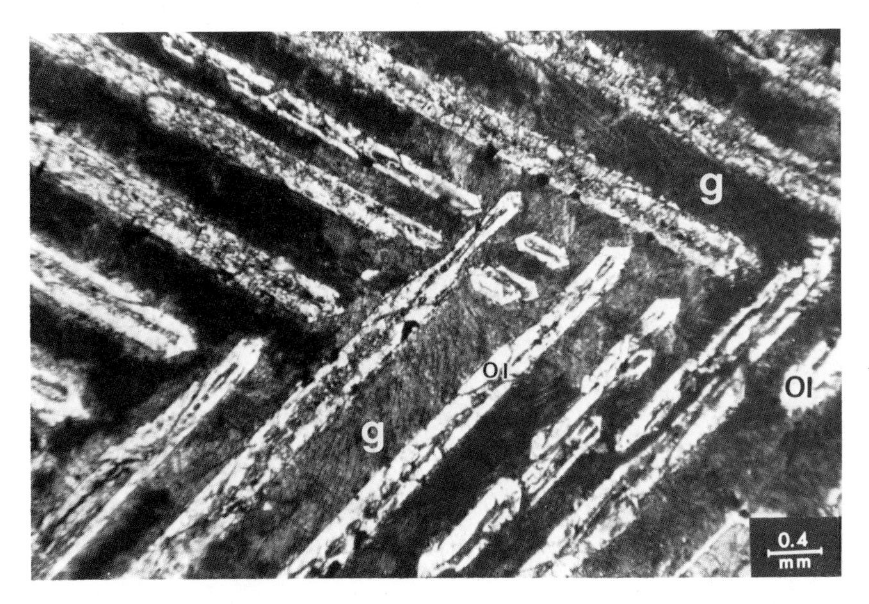

Fig 617b Two sets of parallel orientated spinifex olivines intersecting at an angle.
Ol = olivine (skeletal crystals of spinifex olivine)
g = glassy groundmass.
Spinifex Olivine. Archean Greenstone Abitibi belt, Quebec, Canada Sample courtesy Dr. K.D. Collerson.

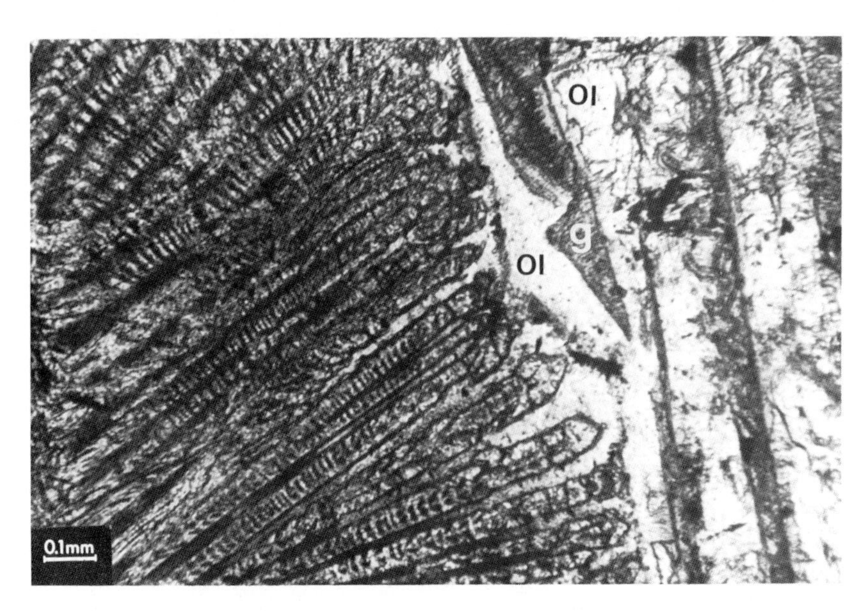

Fig. 617c Symplectic "granophyric" spinifex olivine in glassy groundmass.
Ol = olivine (Spinifex)
g = groundmass.
Spinifex Olivine. Archean Greenstone Abitibi belt, Quebec, Canada
Sample courtesy Dr. K.D. Collerson.

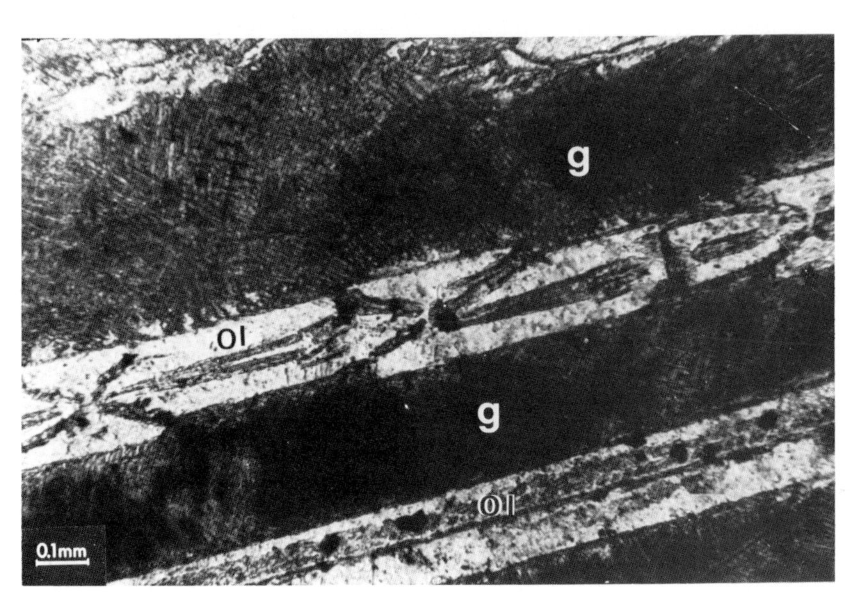

Fig. 617d Blade and chain skeletal crystals of spinifex olivine in glassy groundmass.
Ol = olivine
g = glassy groundmass
Spinifex Olivine. Archean Greenstone Abitibi belt, Quebec, Canada
Sample courtesy Dr. K.D. Collerson.

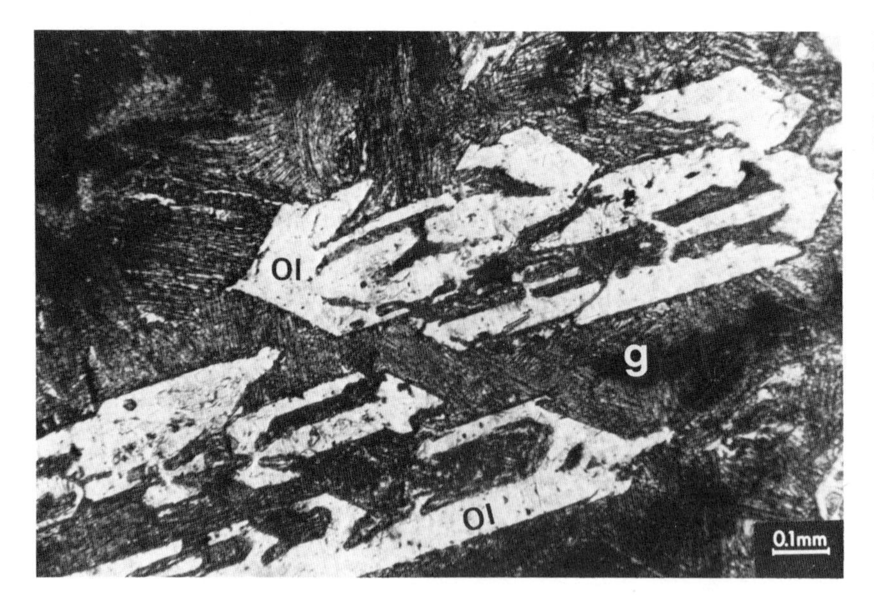

Fig. 617e Skeletal Crystals of Spinifex olivine in glassy groundmass.
Ol = olivine
g = glassy groundmass.
Spinifex Olivine. Archean Greenstone Abitibi belt, Quebec, Canada
Sample courtesy Dr. K. D. Collerson.

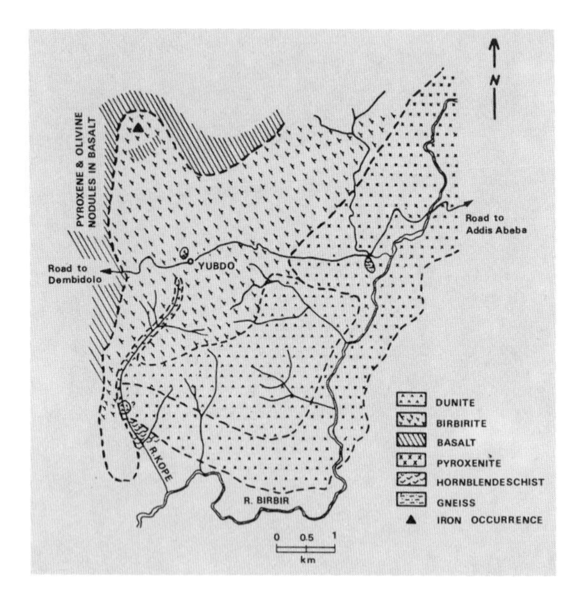

Fig. 618 Geological sketch-map of the ultrabasic complex of Yubdo, prepared by Ato Tasteye Gabre Hana. The Topography is based on aerial photographs.

Fig. 619 Forsterite (Fo 98%) with iron-oxide bodies orientated within the host.
ol = olivine
i-o = iron oxide bodies
an = antigorised olivine
Yubdo, Wollaga, W. Ethiopia.
With crossed nicols.

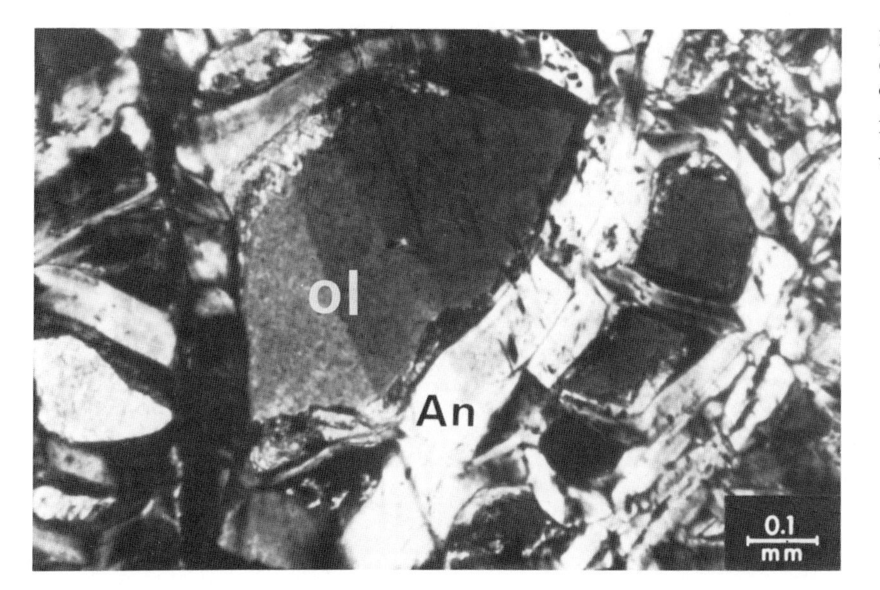

Fig. 620 Olivine tectonically affected exhibiting broad deformation lamellae.
ol = olivine
An = antigorite
Yubdo Wollaga, W. Ethiopia.
With crossed nicols.

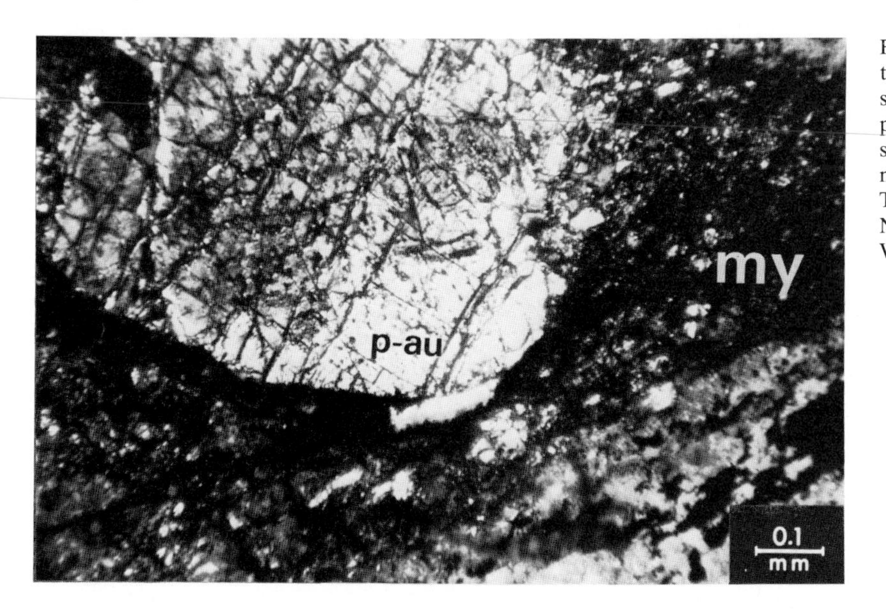

Fig. 621 Tectonically affected peridotite with a rounded pyroxene "augen" structure in a mylonitised zone.
p-au = pyroxene (rounded) "augen" structure
my = mylonitised zone
Tectonically affected peridotite. Edessa, Northern Greece.
With crossed nicols.

Fig. 622 Tectonically affected peridotite with a shear plane (a mylonitisation zone), resulting in the roundening of a relic pyroxene.
p-r = relic pyroxene with rounded outline due to tectonic effects.
my = mylonitised zone
Tectonically affected peridotite.
Edessa, Northern Greece.
With crossed nicols.

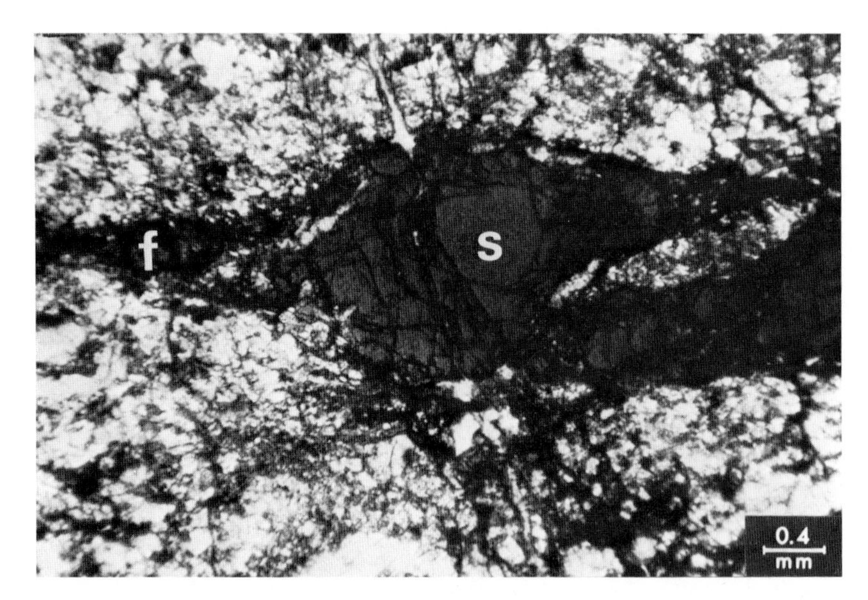

Fig. 623 While most of the spinels are unaffected and remain as resistant structure under mantle deformation conditions, spinels in mantle diapirs in the earth's crust when affected by directed strain are tectonically deformed.
s = spinel tectonically deformed and mylonitised
f = fragmentation of the spinel when the plastic deformation is surpassed.
Tectonically affected peridotite.
Edessa, Northern Greece.
With crossed nicols.

Fig. 624 Pyroxenes and spinels of mantle diapirs tectonically deformed in shear zones.
S = spinels tectonically deformed.
py = pyroxene tectonically deformed.
Tectonically affected peridotite.
Edessa, Northern Greece.
With crossed nicols.

Fig. 625 Deformed olivine with fine deformation twinning in a mass of mylonitised olivine.
ol-d = deformed olivine with fine deformation lamellae.
myl = mylonitised olivine
Dunite. Amita Bay, Milford Sound, New Zealand.
With crossed nicols.

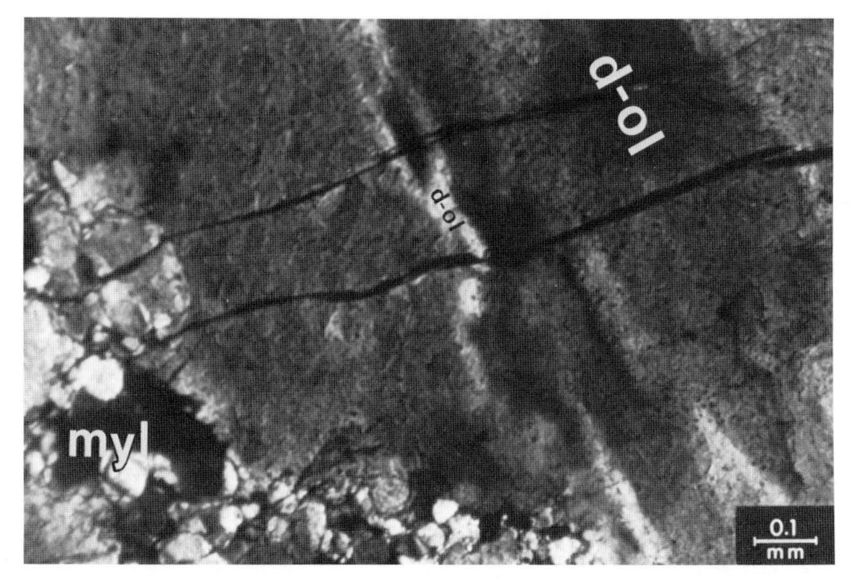

Fig. 626 Tectonically deformed olivine with crack pattern of olivine deformation which may represent an initial phase of olivine rupture and mylonitization.
d-ol = deformation "zones" within a tectonically deformed olivine.
myl = mylonitised texture
Dunite. Amita Bay, Milford Sound, New Zealand.
With crossed nicols.

Fig. 627 Deformed olivine exhibiting undulating extinction in a mass of mylonitised olivine.
d-ol = deformed olivine exhibiting undulating extinction.
myl = mylonitised mass
Dunite. Amita Bay, Milford Sound, New Zealand.
With crossed nicols.

Fig. 628 Tectonically deformed dunite, mylonitised and recrystallised indicating undulating extinction.
my-o = mylonitised and recrystallised olivine texture.
Dunite. Amita Bay, Milford Sound, New Zealand.
With crossed nicols.

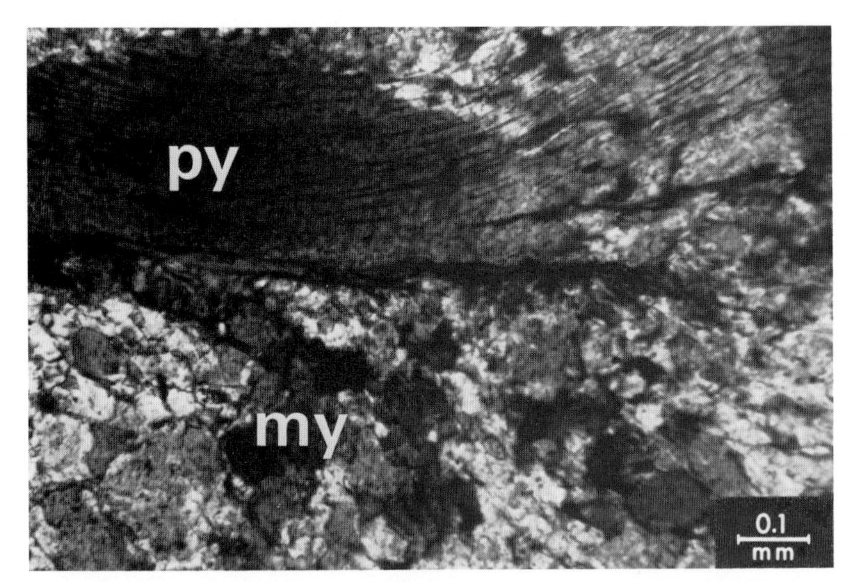

Fig. 629 Tectonically deformed dunite (mylonitised and recrystallised olivine texture) also deformed pyroxene is shown.
py = pyroxene
my = mylonitised and recrystallised olivine texture.
Dunite. Amita Bay, Milford Sound, New Zealand.
With crossed nicols.

Fig. 630 Polysynthetically twinned pyroxene indicating a bending of the twin lamellae due to tectonic deformation.
Tectonically affected peridotite.
Edessa, Northern Greece.
With crossed nicols.

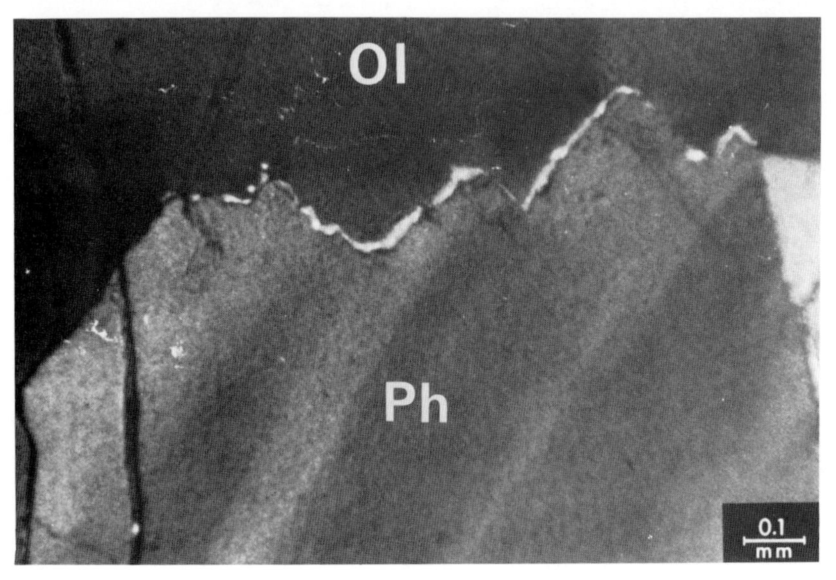

Fig. 631 Shows olivine exhibiting tectonic deformation (undulating extinction) and deformed phlogopite with an undulating tectonic deformation
Ol = olivine exhibiting undulating extinction.
Ph = phlogopite exhibiting deformation effect.
Dunite, coarse grained. Dreiser Weiher, Eifel Mts., Germany.
With crossed nicols.

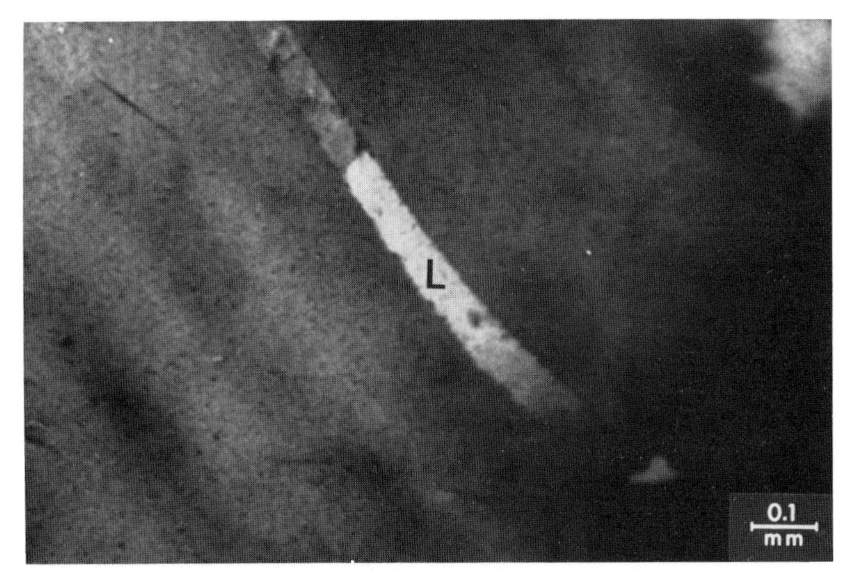

Fig. 632 Shows deformed phlogopite with a deformation "lamella" (L) and exhibiting undulating extinction.
Dunite, coarse grained. Dreiser Weiher, Eifel Mts., Germany.
With crossed nicols.

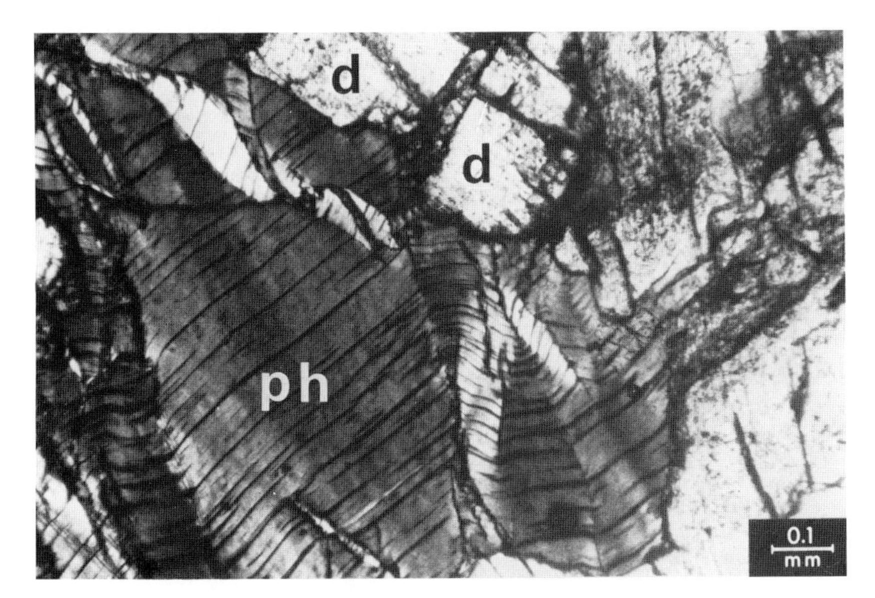

Fig. 633 Tectonically deformed phlogopite in peridotite, composed mainly of diopside and forsterite.
d = diopside
ph = phlogopite
Micaceous Peridotite (Courtesy Gorni Institute, Leningrad), Russia.
With crossed nicols.

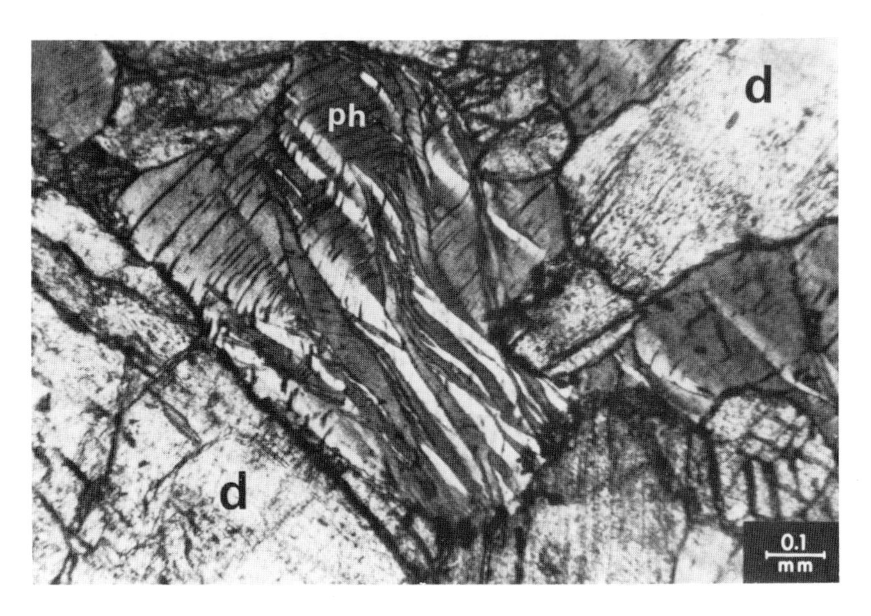

Fig. 634 Tectonically deformed phlogopite (a pattern of deformation lamellae is shown).
d = diopside
ph = phlogopite (with a pattern of deformation lamellae)
Micaceous peridotite (Courtesy Gorni Institute, Leningrad). Russia.
With crossed nicols.

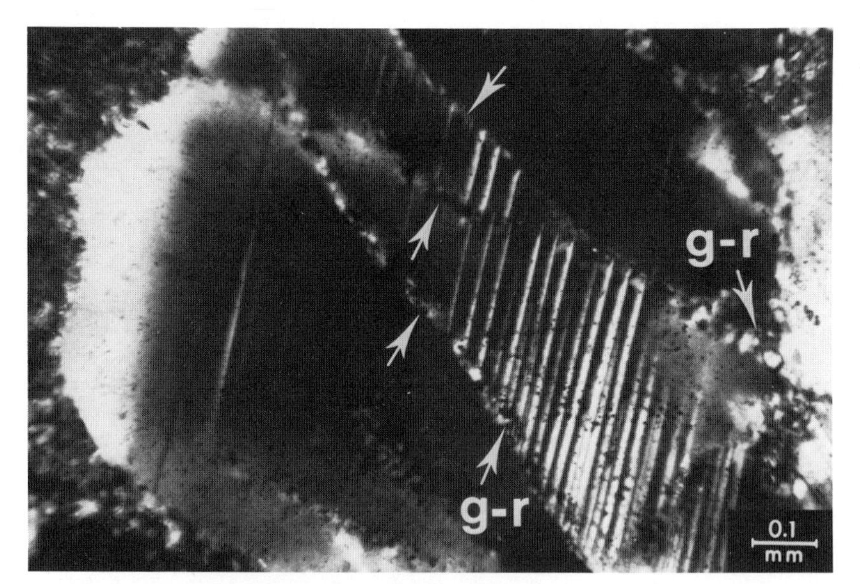

Fig. 635 Tectonically fractured plagioclase with a granular recrystallization following the ruptures.
Arrows indicate ruptures.
g-r = granular recrystallization following the rupture zone.
Gabbro. Rosswein, Saxony, Germany.
With crossed nicols.

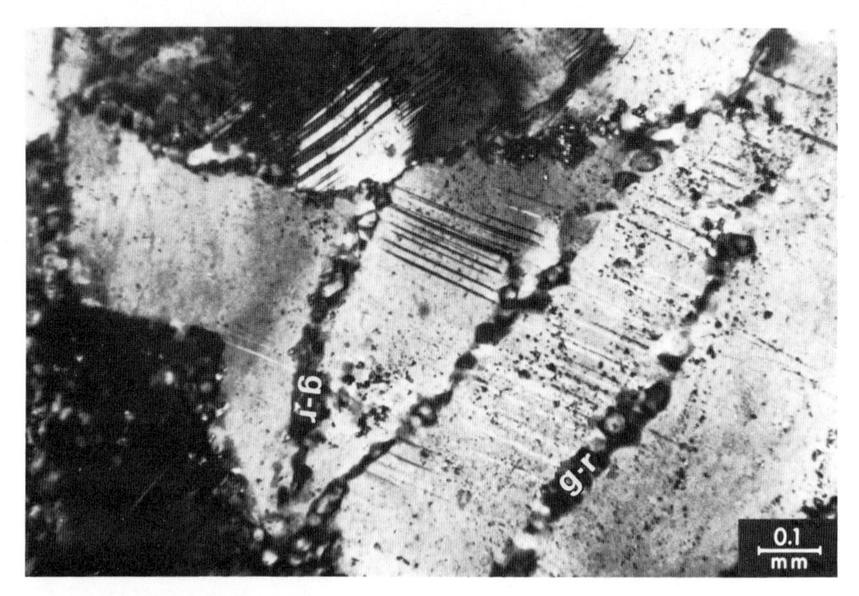

Fig. 636 Tectonically deformed plagioclase with granular recrystallization following the rupture zones.
g-r = granular recrystallization of the plagioclase.
Gabbro. Rosswein, Saxony, Germany.
With crossed nicols.

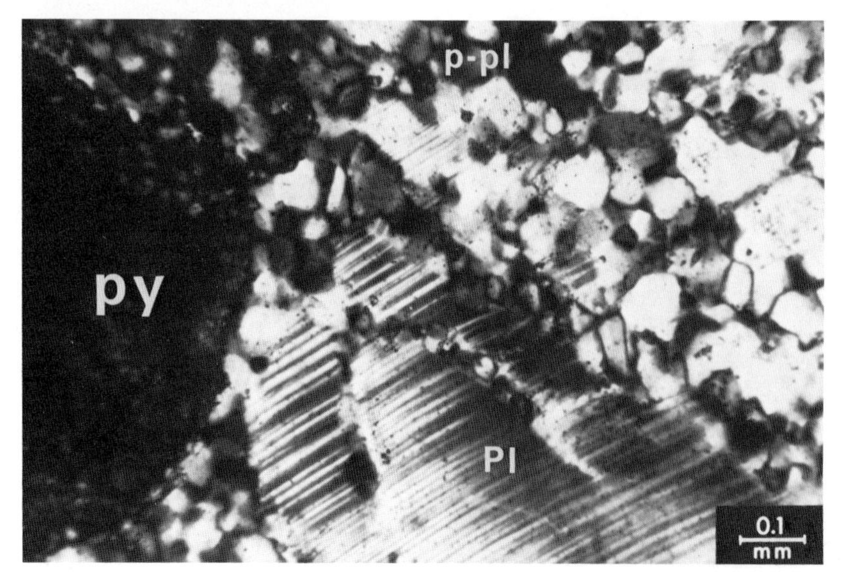

Fig. 637 Ruptured plagioclase indicating block displacement in a plagioclase mylonitised zone, consisting of paracrystalline granular plagioclase.
Also pyroxene is shown tectonically affected.
Pl = plagioclase showing block displacement.
p-pl = paracrystalline granular plagioclase (tectonically recrystallised plagioclase of a mylonitic zone).
py = pyroxene tectonically affected.
Gabbro. Dean Quarry, Falmouth, Cornwall, England.
With crossed nicols.

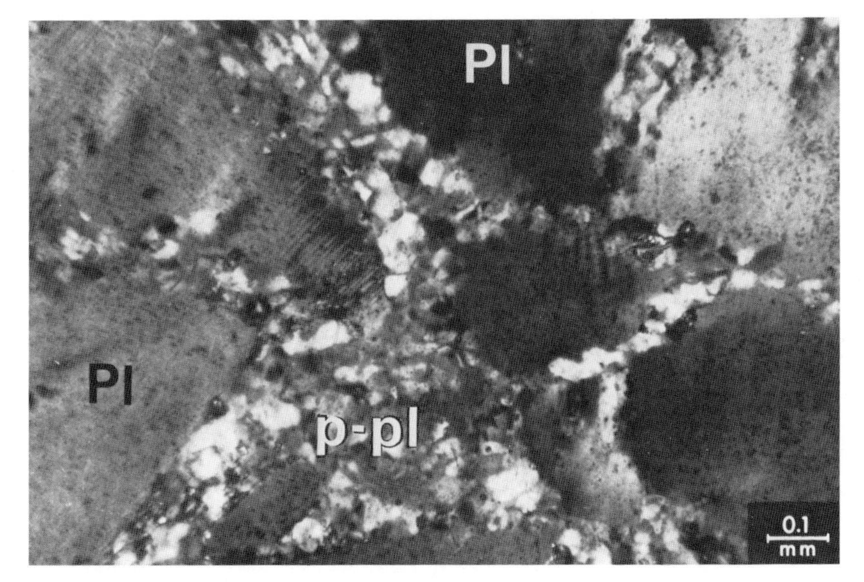

Fig. 638 Plagioclase showing undulating extinction and paracrystalline granular plagioclase following "healed" rupture zones.
Pl = undulating plagioclase
p-pl = paracrystalline granular plagioclase.
Gabbro. Bohringen, Saxony, Germany.
With crossed nicols.

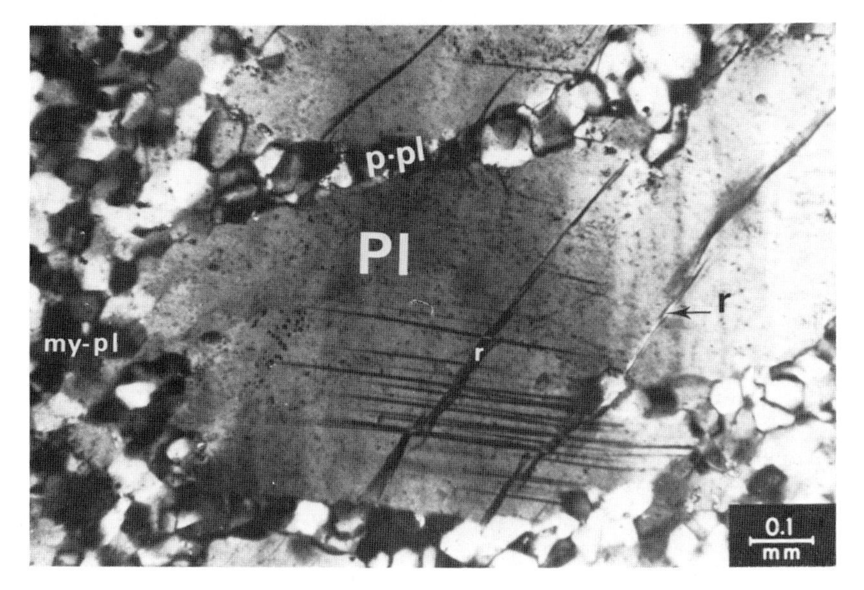

Fig. 639 Tectonically deformed plagioclase (indicating undulating extinction) with paracrystalline plagioclase invading it probably along zones of weakness due to deformation.
It should be indicated that a rupture system is free from recrystallization.
Pl = deformed plagioclase (undulating extinction)
p-pl = paracrystalline plagioclase, with feldspar recrystallisation
r = ruptures without recrystallization
my-pl = mylonitised plagioclase
Gabbro. Bohringen, Saxony, Germany.
With crossed nicols.

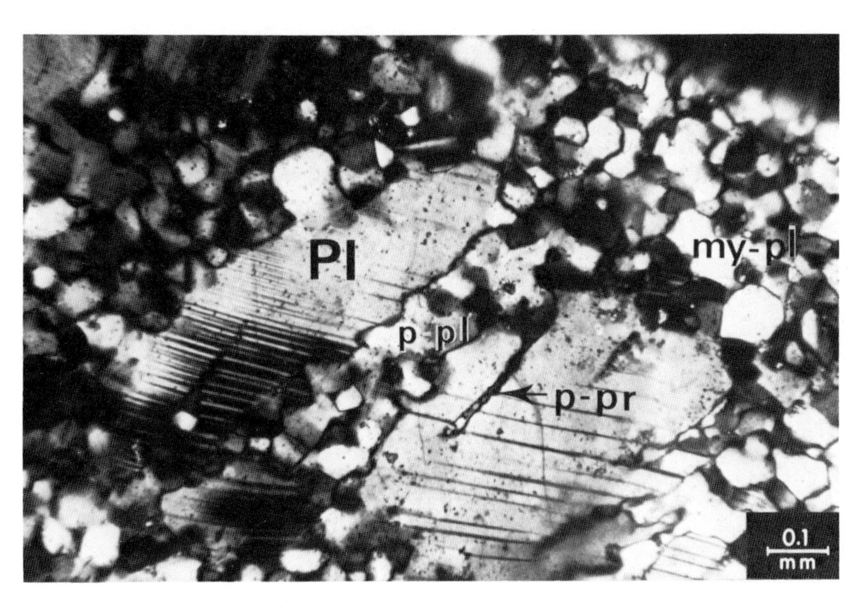

Fig. 640 Deformed plagioclase in a plagioclase mylonitised zone. Paracrystalline granular plagioclase "invades" plagioclase along weakness zones, in cases, sending protuberances.
my-pl = mylonitised plagioclase
p-pl = paracrystalline plagioclase recrystallized and "mobilised" along rupture zones.
p-pr = paracrystalline plagioclase with extensions.
Pl = plagioclase.
Gabbro. Dean Quarry, Falmouth, Cornwall, England.
With crossed nicols.

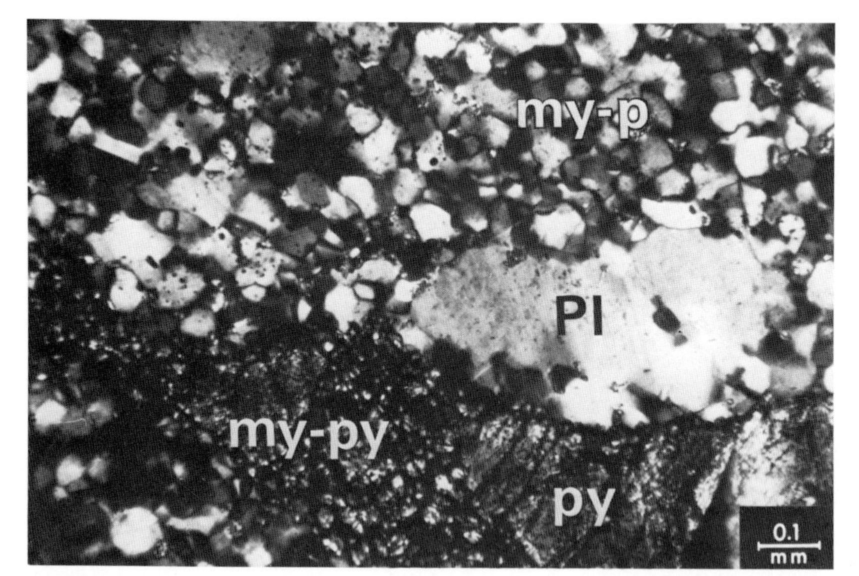

Fig. 641 Plagioclase and pyroxene both mylonitised with deformed pieces left.
Pl = plagioclase
my-p = mylonitised plagioclase
py = pyroxene
my-py = mylonitised pyroxene
Gabbro. Bohringen, Saxony, Germany.
With crossed nicols.

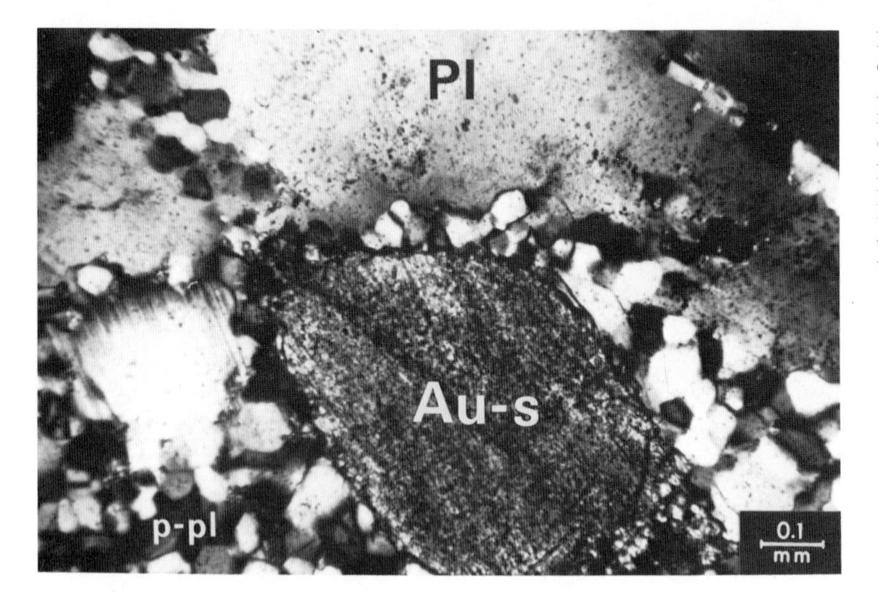

Fig. 642 "Augen-structure" of pyroxene in mylonitised plagioclase.
Au-s = augen structure of pyroxene in mylonitised and recrystallised plagioclase.
p-pl = paracrystalline plagioclase.
pl = plagioclase
Gabbro. Dean Quarry, Falmouth, Cornwall, England.
With crossed nicols.

Fig. 643 Deformed and mylonitised plagioclase and pyroxene.
Pl = deformed plagioclase
py = deformed twinned pyroxene with a mylonitised "marginal zone".
py-my = mylonitised and mobilised pyroxene (paracrystalline granular).
my-pl = mylonitised paracrystalline plagioclase recrystallization.
Gabbro. Bohringen, Saxony, Germany.
With crossed nicols.

Fig. 644 Fractured pyroxene with paracrystalline granular pyroxene following the rupture zones.
Also mylonitised recrystallised plagioclase (paracrystalline).
py = pyroxene fractured.
py-g = granular paracrystalline pyroxene following rupture zones of the pyroxene.
myl-r = mylonitised and recrystallised (granular) plagioclase.
Gabbro. Bohringen, Saxony, Germany. With crossed nicols.

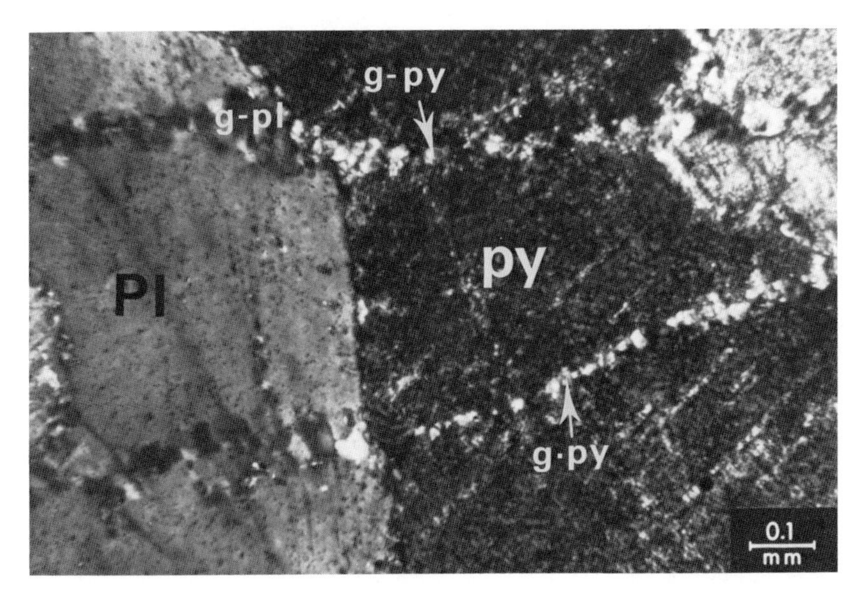

Fig. 645 Plagioclase and pyroxene tectonically ruptured and with paracrystalline granular plagioclase and pyroxene occupying the respective ruptures.
Pl = plagioclase
py = pyroxene
g-pl = paracrystalline plagioclase occupying rupture zones of the plagioclase.
g-py = paracrystalline (granular pyroxene) occupying rupture zones of the pyroxene.
Gabbro. Bohringen, Saxony, Germany. With crossed nicols.

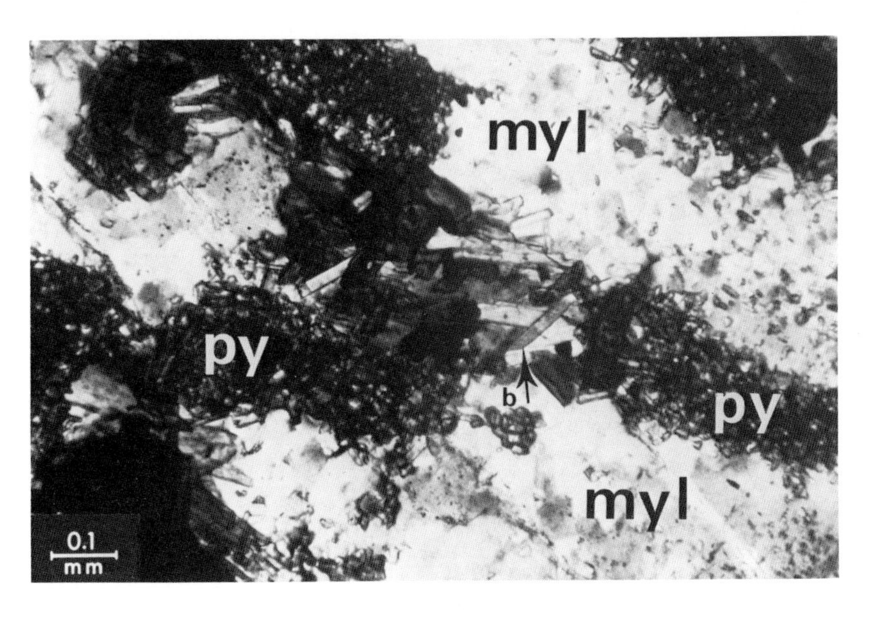

Fig. 646 Mylonitised and mobilised pyroxene in a mylonitic zone of recrystallised plagioclase. Blastic biotite laths are also formed.
py = pyroxene mylonitised and mobilized.
myl = mylonitised plagioclase-granular recrystallised.
b = blastic biotite associated with the mylonitised pyroxene zone.
Gabbro. Bohringen, Saxony, Germany. With crossed nicols.

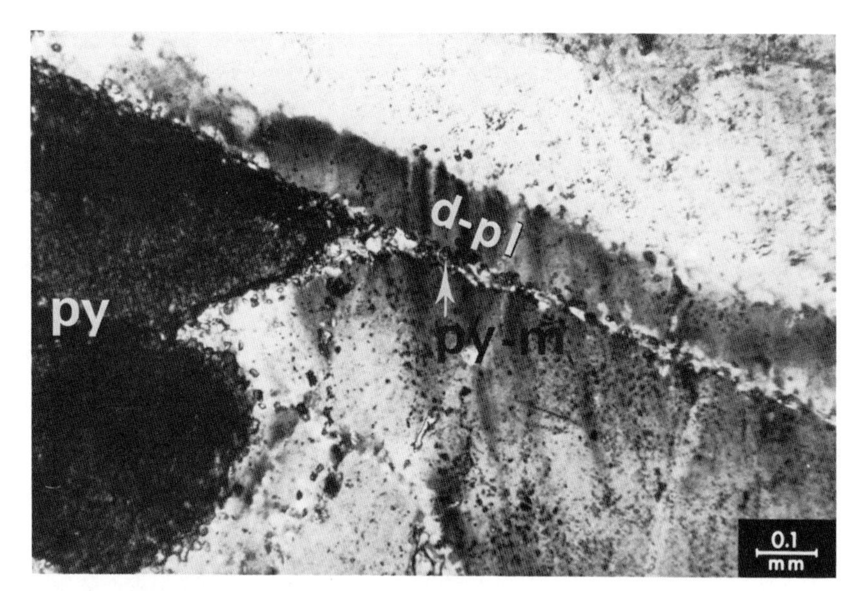

Fig. 647 Deformed pyroxene (with rounded outlines due to tectonic deformation) with mobilised tectonic extension in deformed plagioclase.
py = pyroxene
py-m = tectonically mobilised pyroxene (indicated as granules) in tectonic zones of the plagioclase.
d-pl = tectonically deformed plagioclase
Gabbro. Rosswein, Saxony, Germany.
With crossed nicols.

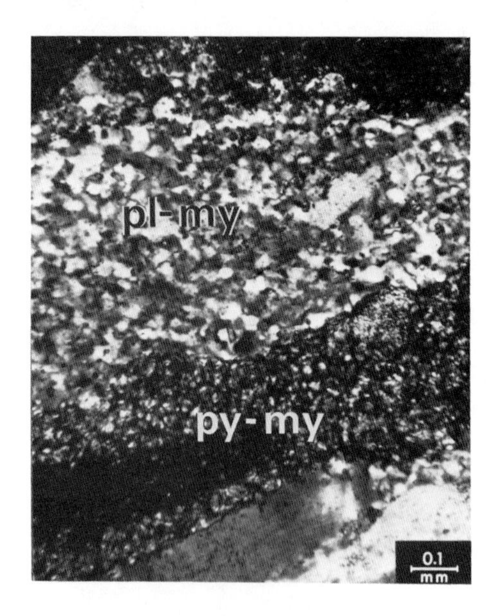

Fig. 648 Mylonitised pyroxene and plagioclase.
pl-my = mylonitised and recrystallised plagioclase.
py = mylonitised granular pyroxene.
Gabbro. Bohringen, Saxony, Germany.
With crossed nicols.

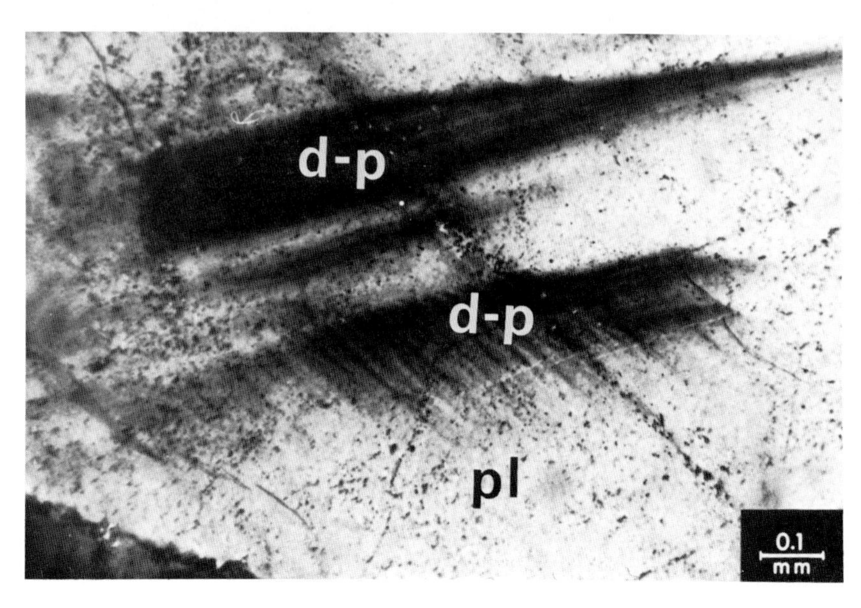

Fig. 649 Tectonically deformed plagioclase with undulating extinction "shadows" indicating fine lamellar structure (deformation twinning) due to tectonic causes.
pl = plagioclase
d-p = deformation patterns (indicating undulation and fine deformation twinning).
Gabbro. Rosswein, Saxony, Germany.
With crossed nicols.

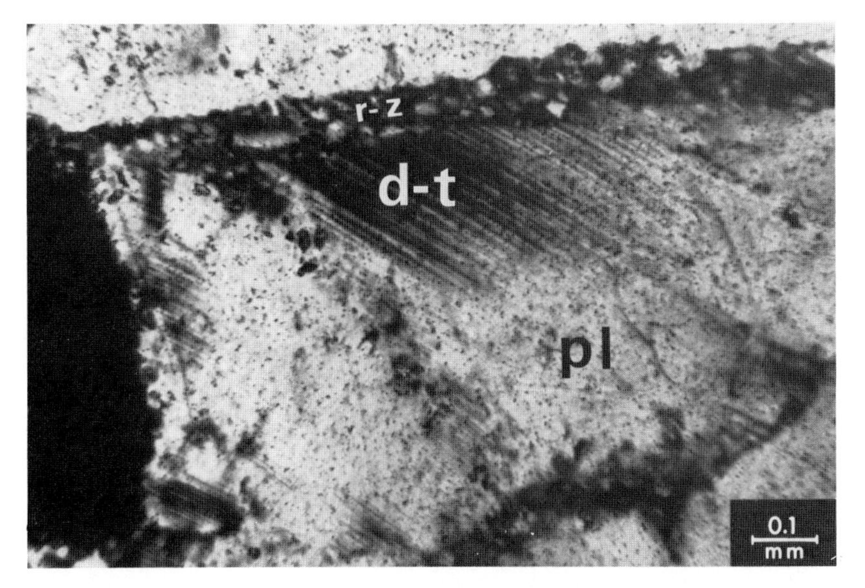

Fig. 650 A rupture zone with granular paracrystalline plagioclase in plagioclase showing fine deformation twinning and undulating extinction.
pl = plagioclase
r-z = rupture zone (with granular paracrystalline plagioclase).
d-t = deformation twinning in plagioclase.
Gabbro. Rosswein, Saxony, Germany.
With crossed nicols.

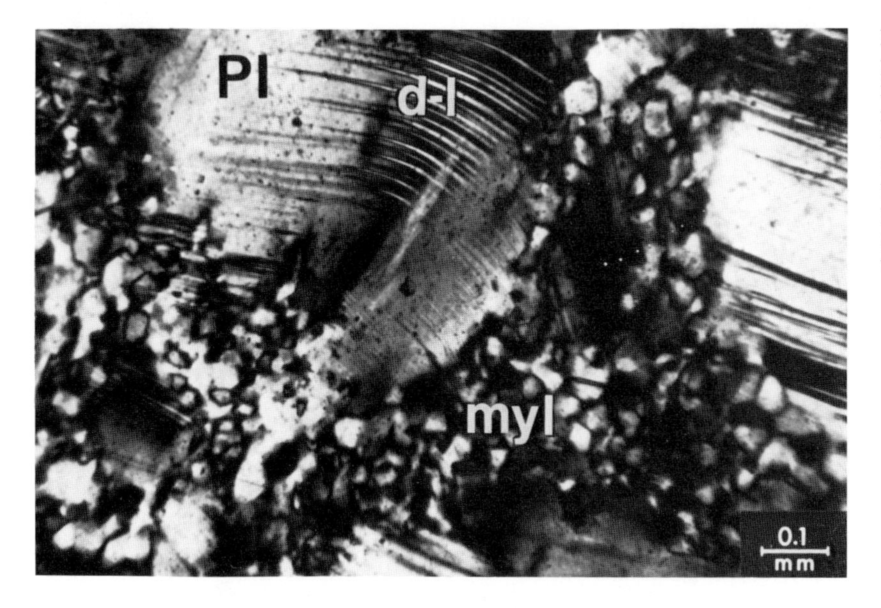

Fig. 651 Mylonitised and deformed plagioclase, indicating bending of the deformation twin lamellae.
myl = mylonitised plagioclase (paracrystalline)
d-l = deformation lamellae showing a bending
Pl = deformed plagioclase
Gabbro. Bohringen, Saxony, Germany.
With crossed nicols.

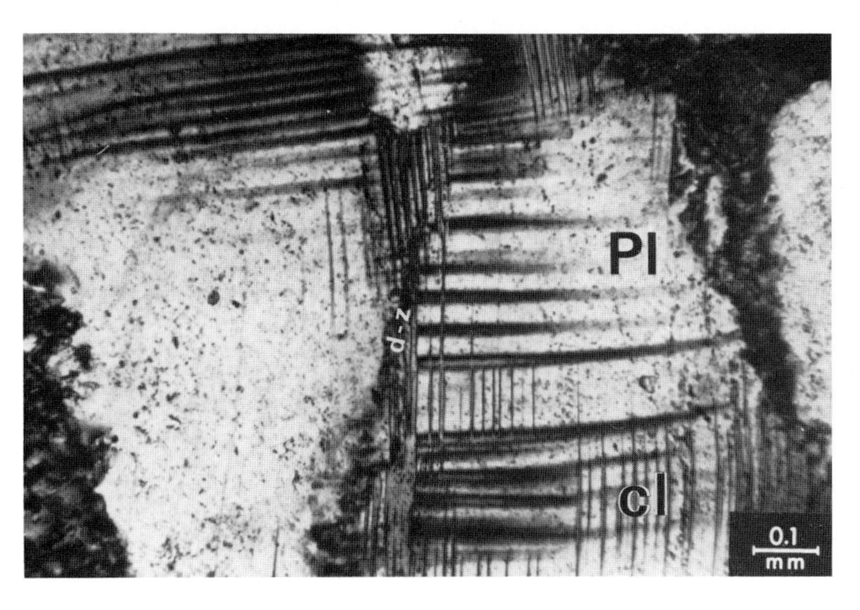

Fig. 654 Deformed plagioclase with a deformed zone out of which initiates a system of cross-lamellar tectonic twinning.
Pl = deformed plagioclase
d-z = deformation zone in the plagioclase
c-l = cross-lamellar deformation twinning, associated with the deformation zone.
Gabbro. Bohringen, Saxony, Germany.
With crossed nicols.

(For Figs. 652 and 653 see page 338)

Figs. 652 and 653 Tectonically deformed plagioclase showing a lamination parallel to mylonitisation zones.
Pl = plagioclase
t-l = tectonic lamination parallel to zones of mylonitisation
My = mylonitisation zones
p-myl = pyroxene mylonitised
Gabbro. Bohringen, Saxony, Germany. With crossed nicols.

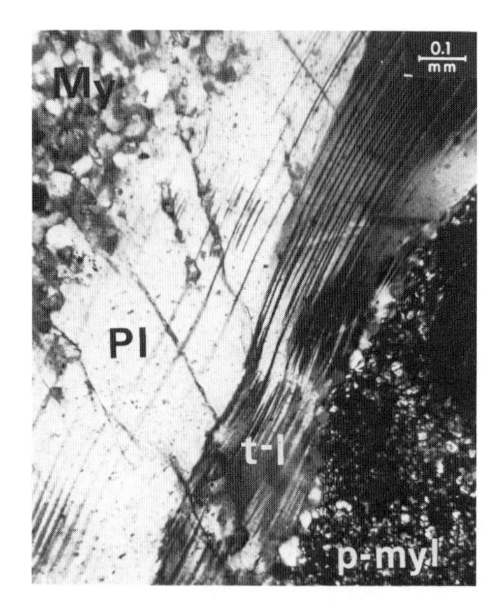

Fig. 653

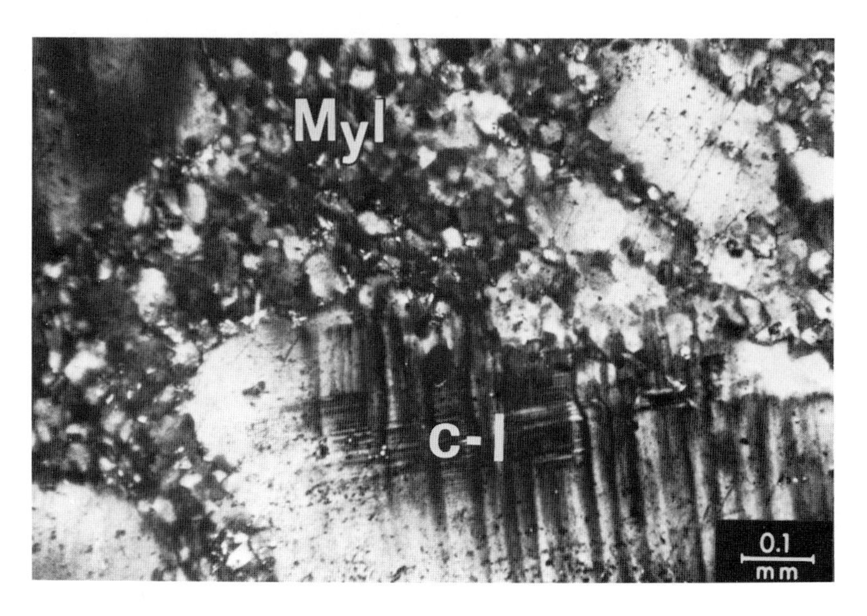

Fig. 655 Mylonitised plagioclase (granular paracrystalline) with a deformed relic (indicating cross-lamellar twinning).
Myl = mylonitised paracrystalline plagioclase
c-l = cross-lamellar twinning.
Gabbro. Dean Quarry, Falmouth, Cornwall, England.
With crossed nicols.

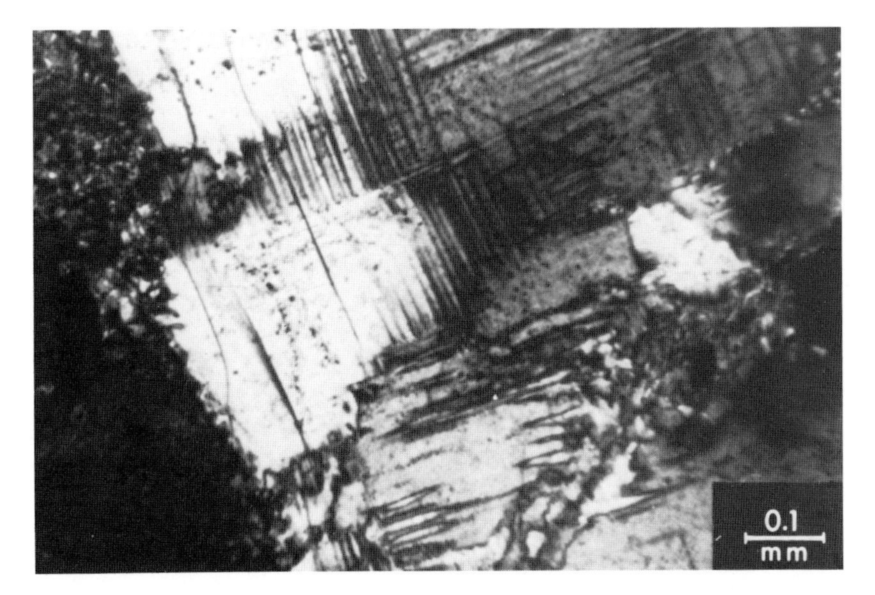

Fig. 656a, b Deformed plagioclase, in mylonitised zones, indicating cross-lamellar deformation twinning.
Gabbro. Bohringen, Saxony, Germany. With crossed nicols.

Fig. 656b

Fig. 657 Tectonically deformed plagioclase, showing two directions of deformation twinning.
Tectonically deformed plagioclase with deformation twinning in a paracrystalline granular-mylonitised plagioclase.
myl = mylonitised plagioclase
d-t = deformation twinning
Gabbro. Bohringen, Saxony, Germany. With crossed nicols.

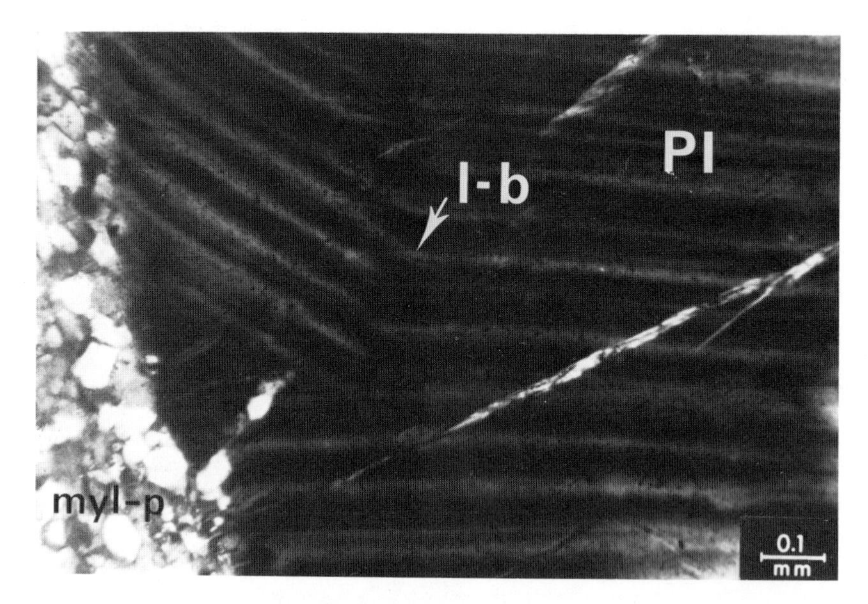

Fig. 658 Deformed plagioclase exhibiting a deformation "polysynthetic twinning", showing a bending of the deformation lamellae in a mylonitised granular plagioclase mass.
Pl = deformed plagioclase
l-b = deformation lamellae showing a bending
myl-p = mylonitised plagioclase mass.
Gabbro. Bohringen, Saxony, Germany.
With crossed nicols.

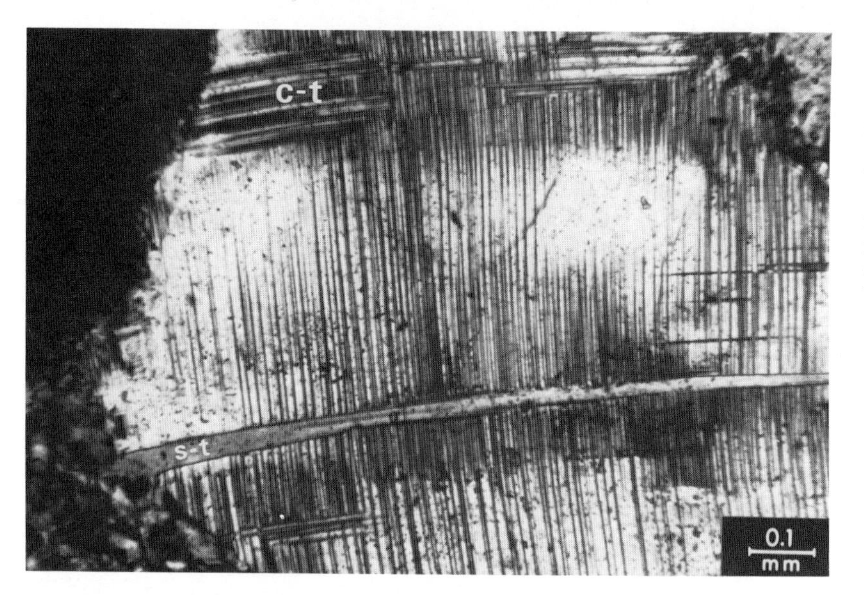

Fig. 659 Deformed plagioclase with deformation fine cross-lamellar twinning and with a secondarily developed twin lamella intersecting the deformation lamellae.
c-t = cross lamellar twinning
s-l = secondarily developed twin lamella intersecting the fine cross lamellar twinning.
Gabbro. Rosswein, Saxony, Germany.
With crossed nicols.

Fig. 660a Granular chromite with orientated ex-solutions of rutile (the rutile lamellae follow the octahedral face of the chromite).
ch = chromite
r = rutile lamellae
Polished section (oil-immersion, with one nicol).
Rodiani, North Greece.

Fig. 660b Granular chromite with fine rutile grains and with rutile and spinel "lamellae" orientated parallel to the octahedral face of the chromite.
ch = chromite
r-g = rutile grains
Polished section (oil-immersion, with one nicol).
Rodiani, North Greece.

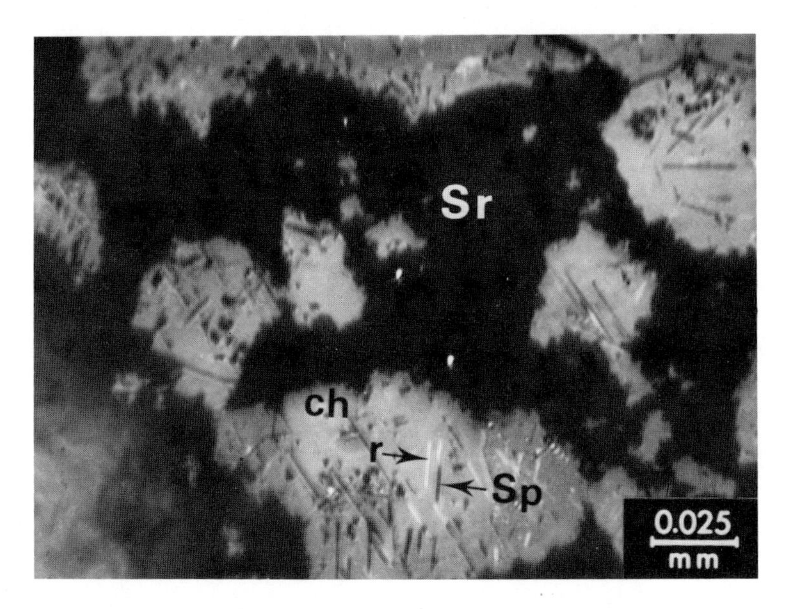

Fig. 661 Chromite with rutile and spinel lamellae orientated parallel to the octahedral face of the chromite.
ch = chromite
Sr = serpentine
Sp = spinel
r = rutile lamellae
Polished section (oil-immersion, with one nicol).
Rodiani, North Greece.

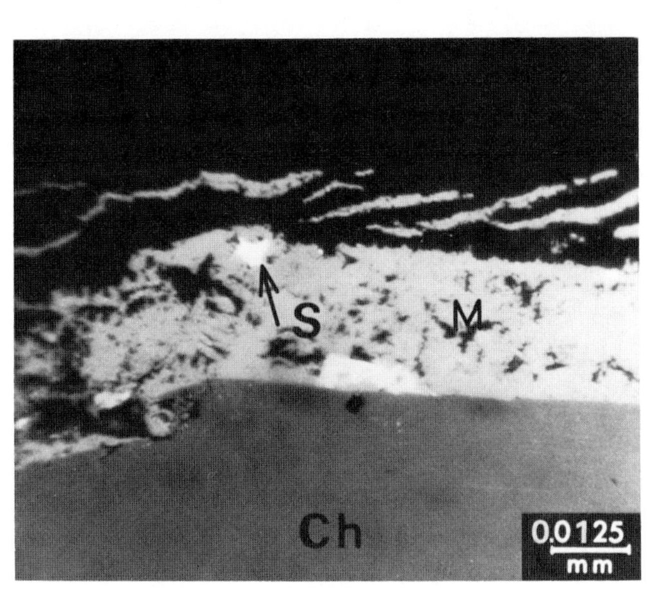

Fig. 662 Chromite (often idiomorphic, see Fig. 717a) with a magnetite margin free of ilmenite ex-solutions.
Ch = chromite
M = magnetite
S = platinoid mineral (often Ni, Co sulphides occur in this form).
Dunitic pipe. Yubdo, W. Ethiopia.
Polished section (with oil-immersion, one nicol).

Fig. 663 Zig-zag bending of the banded (layered) ultrabasics of the Ivrea zone Italy.
Arrows indicate the zig-zag folding.
Courtesy: Dr. Fred. Seligmann.

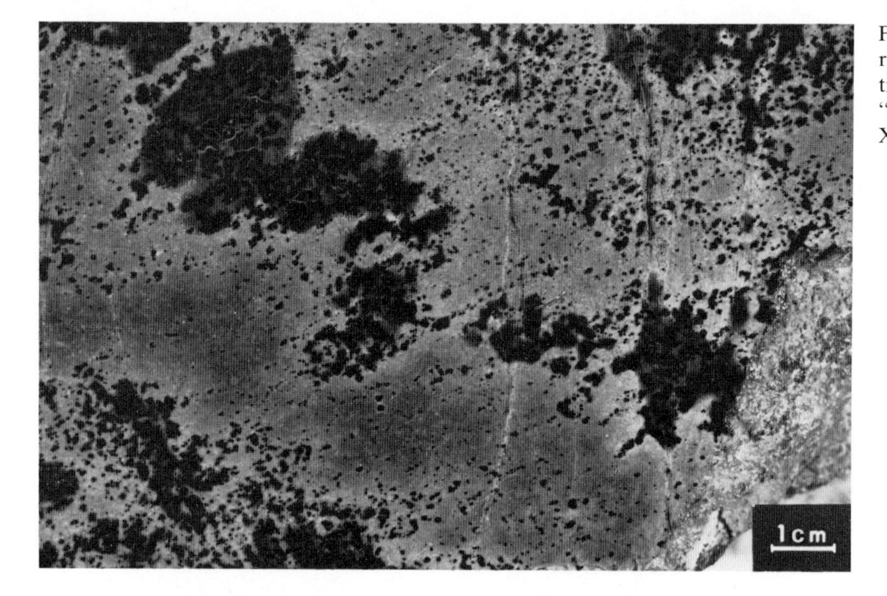

Fig. 664 "Ptygmatically" folded and ruptured chromite "band" in serpentinised dunite.
"Schlieren"-chromite.
Xerolivado, N. Greece.

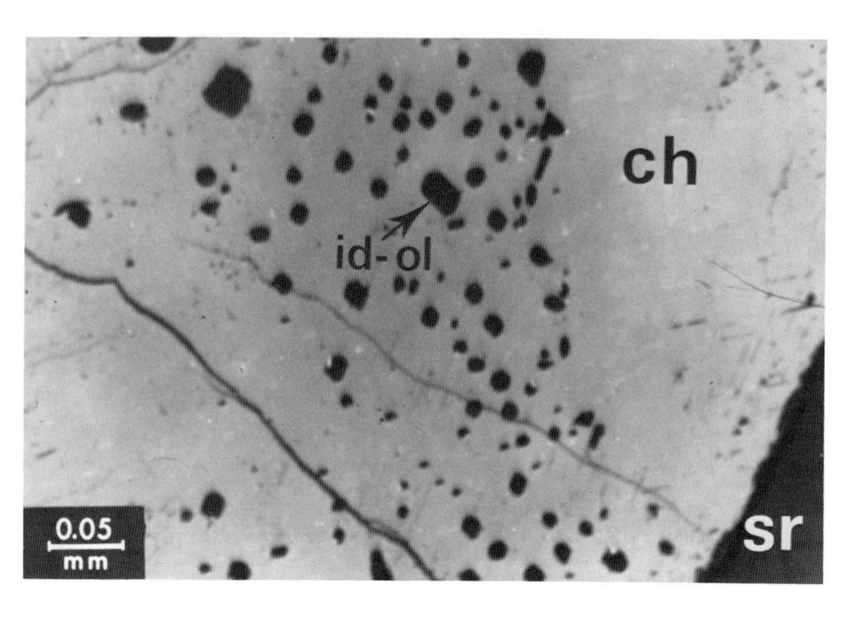

Fig. 665 Early crystallised fine grained olivine included in chromite.
id-ol = idiomorphic olivine
ch = chromite
sr = serpentine
Polished section (oil-immersion, with one nicol).
Rodiani, North Greece.

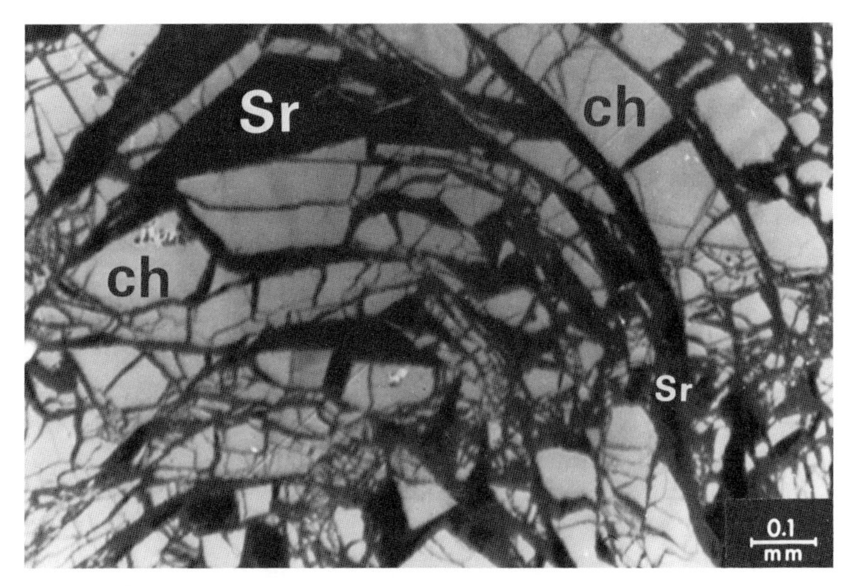

Fig. 666 Curved cataclasis in chromite
ch = chromite
Sr = serpentine
Polished section (oil-immersion, with one nicol).
Rodiani, North Greece.

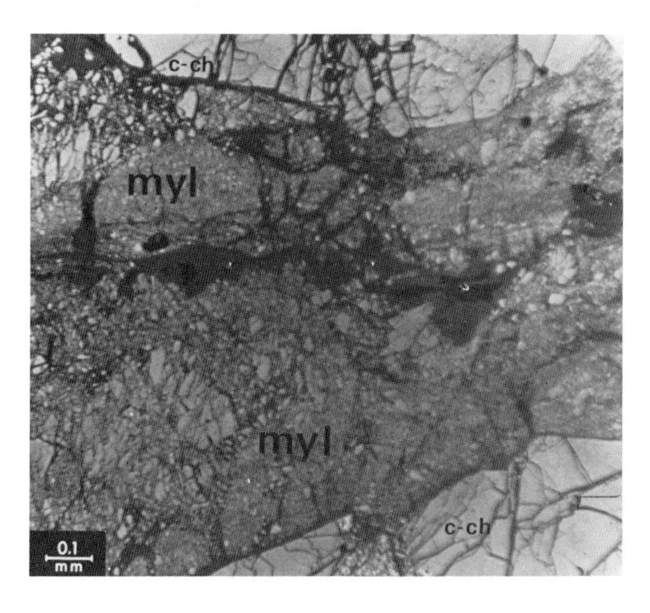

Fig. 667 Compact chromite, transversed by a micro-mylonitic zone.
myl = mylonitised chromite consisting of fine fragmented chromite.
c-ch = compact chromite
Polished section (oil-immersion, with one nicol).
Rodiani, North Greece.

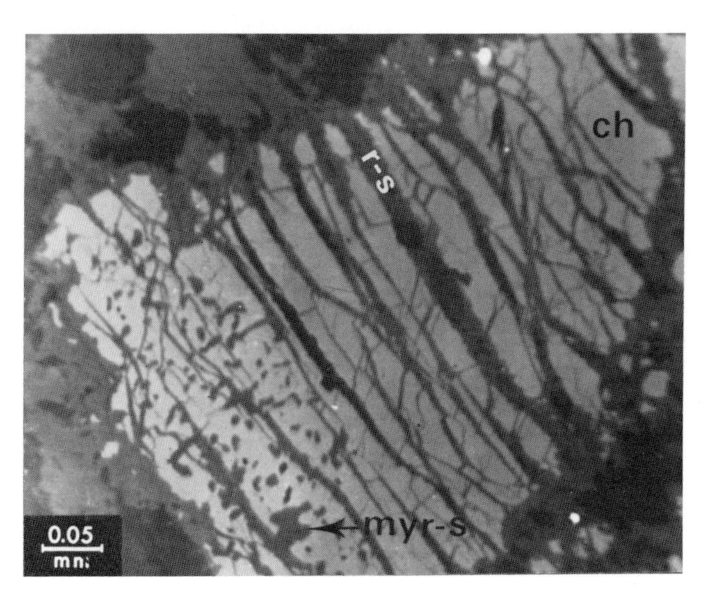

Fig. 668 Cataclastically affected chromite with "re-binding" serpentine holding together the chromite fragments.
ch = chromite
r-s = rebinding serpentine
myr-s = intergrowth of myrmekitic serpentine as extension of "re-binding serpentine".
Chromite ore.
Rodiani, N. Greece.
Polished section (oil-immersion, with one nicol).

Fig. 669 Chromite "potato", actually a tectonically fragmented chromite piece with polished faces due to tectonic effects.
Domokos, Greece.
Hand specimen approximately natural size.

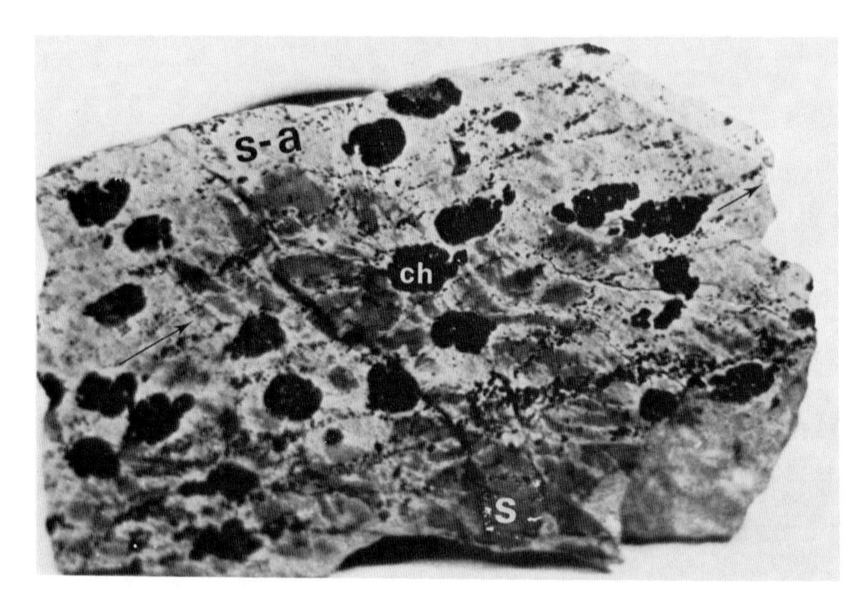

Fig. 670a Boudinage of chromite bodies in serpentine and altered-serpentine.
ch = chromite
s = serpentine
s-a = altered serpentine
Arrows indicate chromite boudinage.
Chromite-ore.
Kursumia, Vourinos, Greece.
Hand specimen, natural size.
(sample courtesy of Eng. G. Kanellopoulos).

Fig. 670b Leopard chromite-ore.
Arrow (a) shows almond-shaped chromite body due to the boudinage of the chromite mass.
Xerolivado, N. Greece.
Hand specimen, natural size.

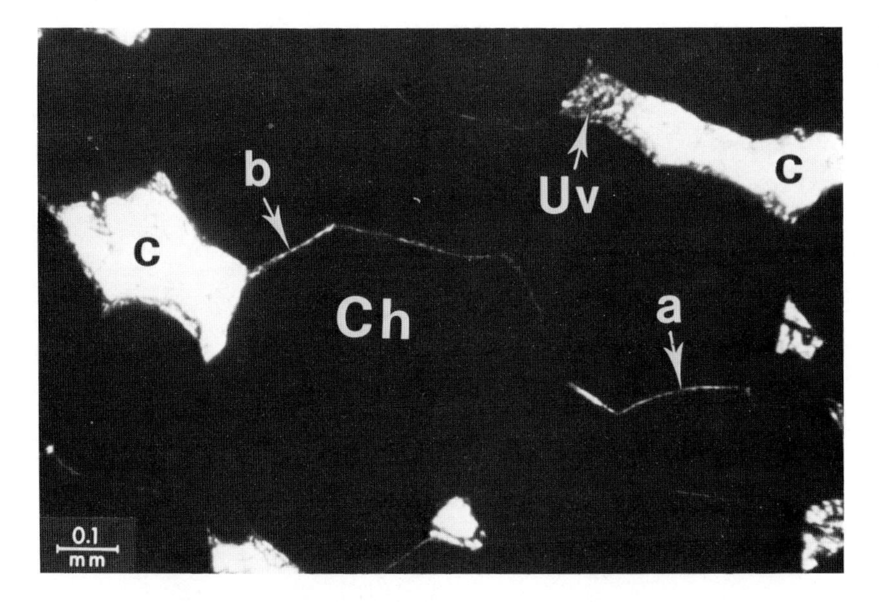

Fig. 671 In contrast to the rounded chromite forms shown in Figs. 672 and 673, Fig. 671 shows chromite grains exhibiting rounded outlines (see arrow "a") and simultaneously indicating crystal "faces" (see arrow "b").
Ch = chromite
Uv = uvarovite
c = chalcedony
Chromite, Bushveld Complex, Transvaal, South Africa.
Without crossed nicols.

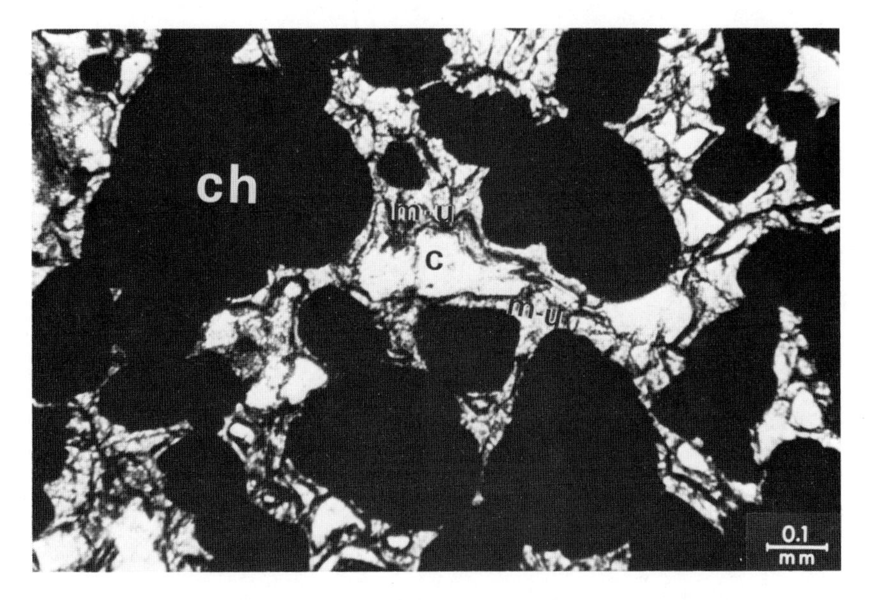

Figs. 672 and 673 Rounded granular chromite with a margin of blastically grown uvarovite and intergranular chalcedony
c = chalcedony
ch = chromite
m-u = uvarovite marginal to the granular chromite.
"Chromite rich band". Bushveld, Igneous Complex. Bushveld, Transvaal, South Africa.
Without crossed nicols.

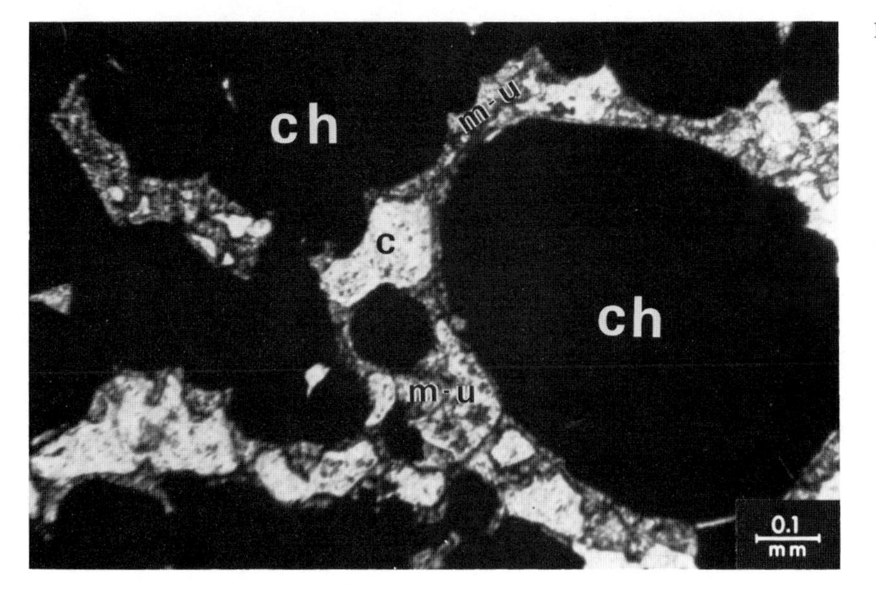

Fig. 673

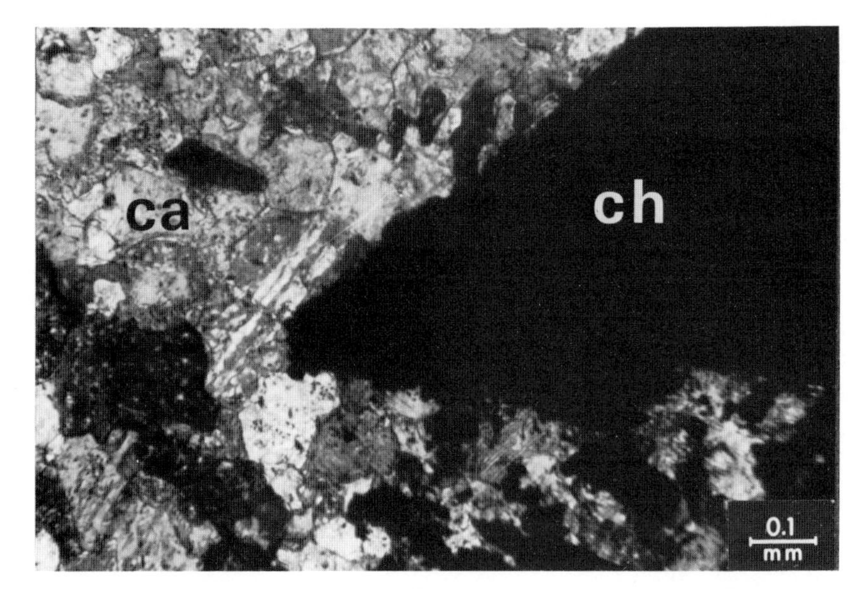

Fig. 674 and Fig. 675 Angular fragments of chromite in a recrystallised calcite mass.
ch = chromite
ca = recrystallised calcite (marble)
Brecciated chromite in marble.
Drepanon, Kozani, N. Greece.
With crossed nicols.

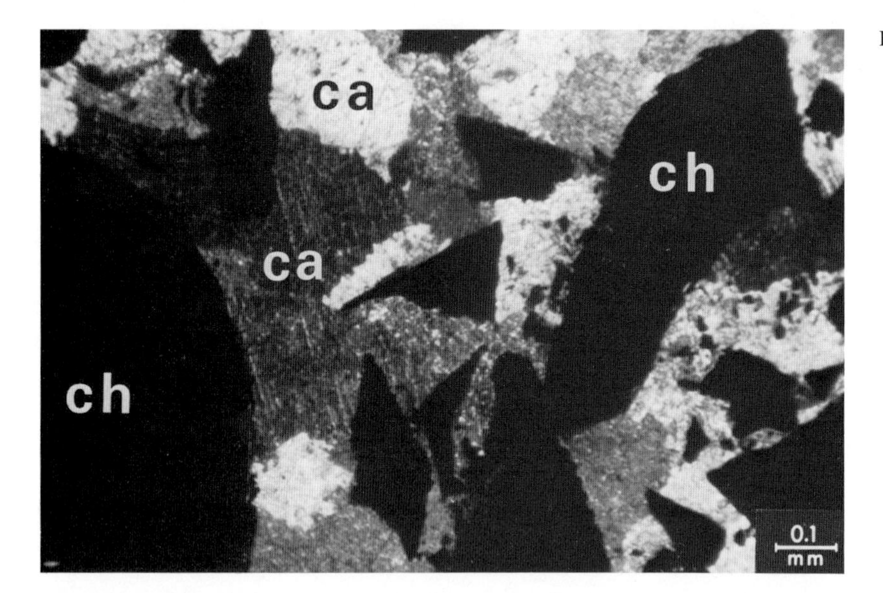

Fig. 675

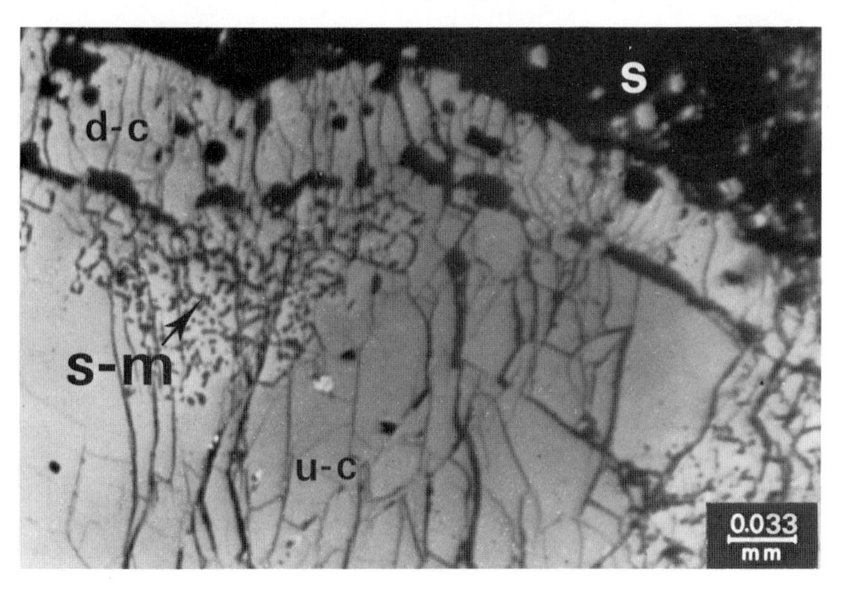

Fig. 676 Chromite with a decoloration margin (decoloration rim) in which serpentine is in myrmekitic intergrowth with chromite.
d-c = decolorised chromite
s-m = myrmekitic serpentine
u-c = unaffected chromite
s = serpentine
Chromite-ore. Rodiani, N. Greece.
Polished section (oil-immersion, with one nicol).

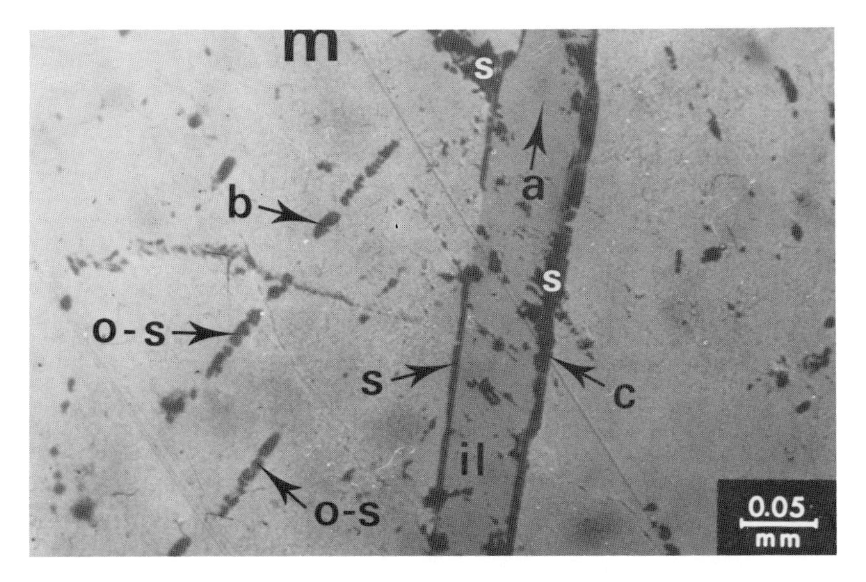

Fig. 677 Magnetite with lamellar body of ilmenite. Spinel bodies occur either parallel orientated (parallel to 100 of the magnetite) or occur as intergranular-interleptonic space fillings between the ilmenite and the magnetite.
m = magnetite
il = ilmenite lamellae orientated parallel to 111 of the magnetite (arrow "a").
o-s = orientated spinel bodies most probably parallel to the 100 of the magnetite (arrow "b").
s = spinel occupying the interleptonic space between the ilmenite lamella and the magnetite (arrow "c").
Lac de la Blache Saguenay comté, Prov. Quebec, Canada.
Polished section (oil-immersion with one nicol).

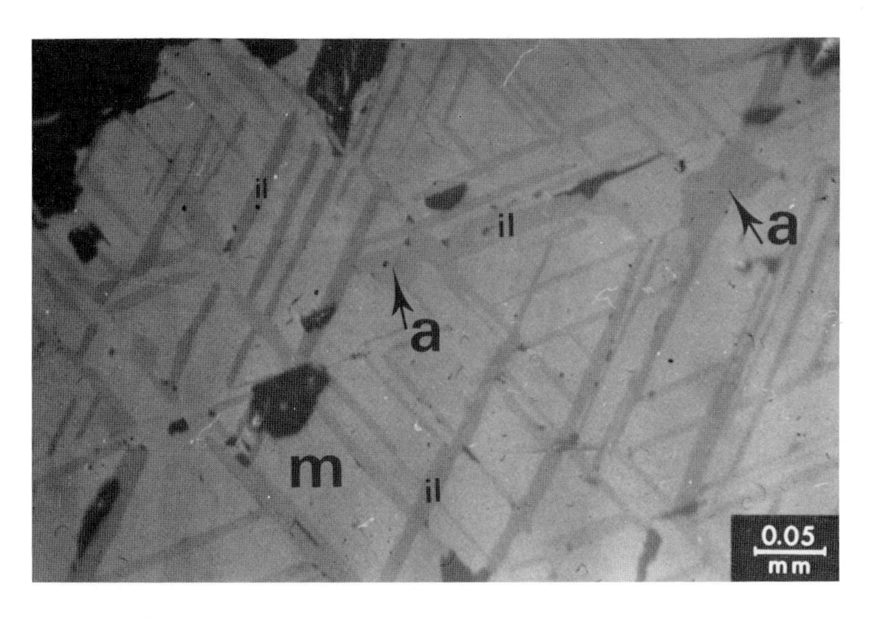

Fig. 678 Ilmenite lamellae orientated parallel to the octahedral magnetite face. However, as arrows "a" show the ilmenite are not delimited by a straight boundary with the ilmenite. It appears that ilmenite bodies defuse into the magnetite, most probably suggesting an ilmenite mobilization along the weakness planes 111 of the host magnetite.
m = magnetite
il = ilmenite (following the 111 of the magnetite).
Arrows "a" ilmenite not confined in the 111 planes of the magnetite.
Otanmäki, Finland.
Polished section (oil-immersion, with one nicol).

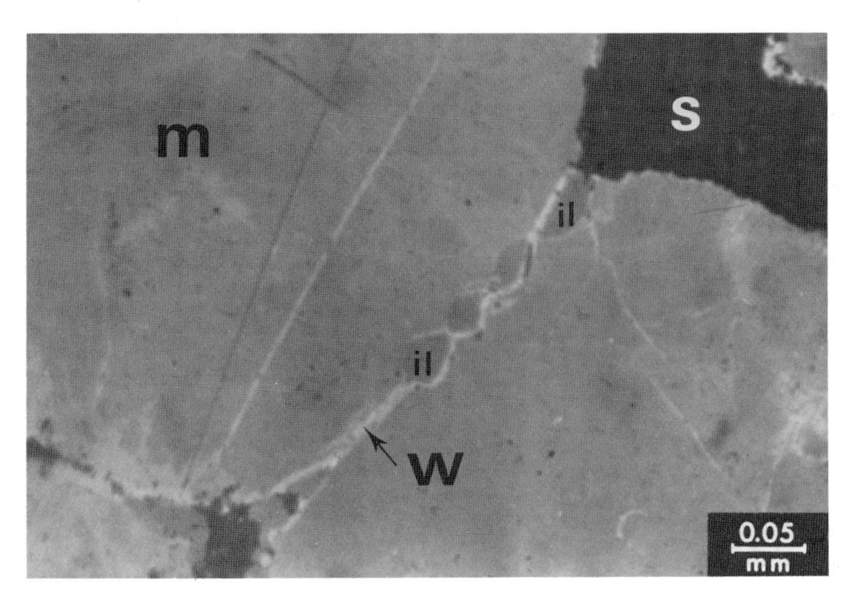

Fig. 679 Magnetite with lines (white in colour) and with lens shaped ilmenite bodies following the white lines.
m = magnetite
il = ilmenite
s = spinel
w = white lines in the magnetite
Bushveld Norite. Phinix, Pretoria, Transvaal, South Africa.
Polished section (oil-immersion, with one nicol).

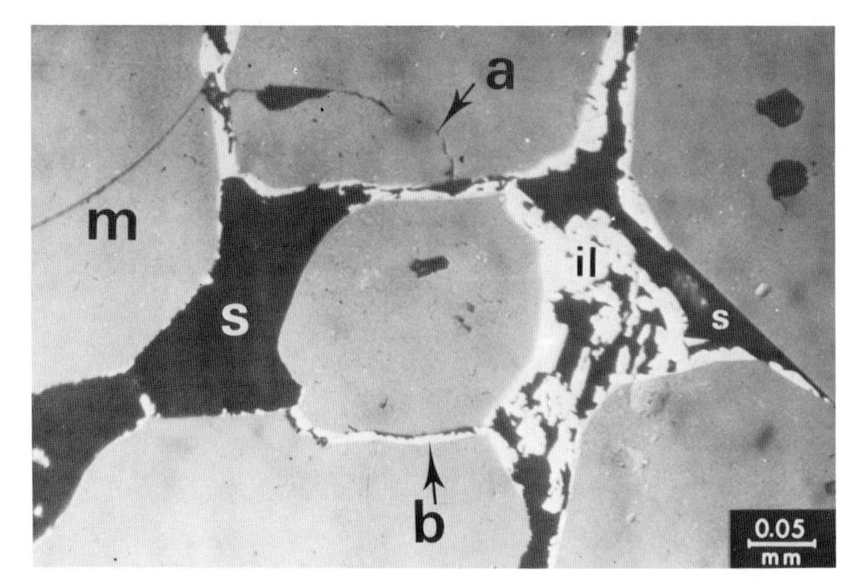

Fig. 680 Magnetite with a fracture system, some occupied by spinel and ilmenite.
m = magnetite
s = spinel
il = ilmenite
Arrows "a" show fractures not occupied by spinel or ilmenite.
Arrows "b" show fine fractures occupied by ilmenite.
Mooihoek, Lydenburg Dist. Transvaal, South Africa.
Polished section (oil-immersion, with one nicol).

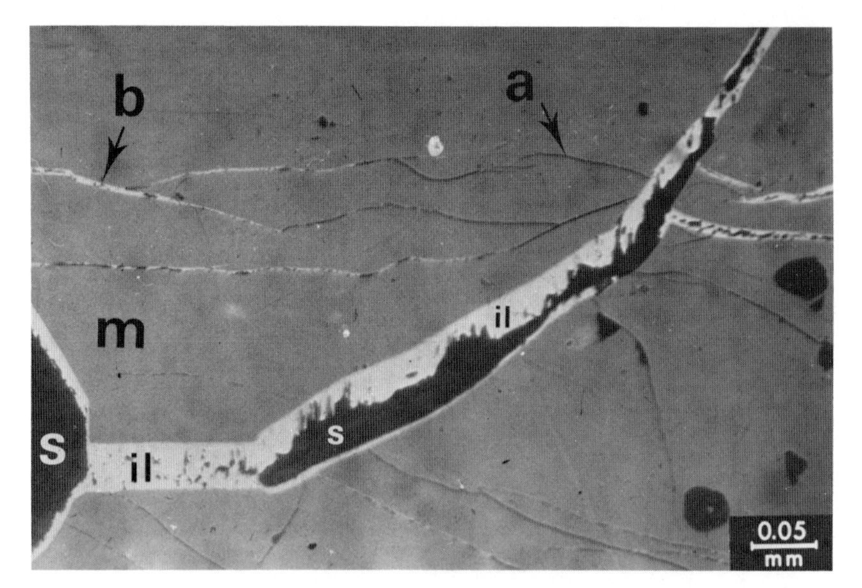

Fig. 681 Magnetite with a fracture system, some occupied by spinel and ilmenite.
m = magnetite
s = spinel
il = ilmenite
Arrows "a" show fractures not occupied by spinel or ilmenite.
Arrows "b" show fine fractures occupied by ilmenite.
Mooihoek, Lydenburg Dist. Transvaal, South Africa.
Polished section (oil-immersion, with one nicol).

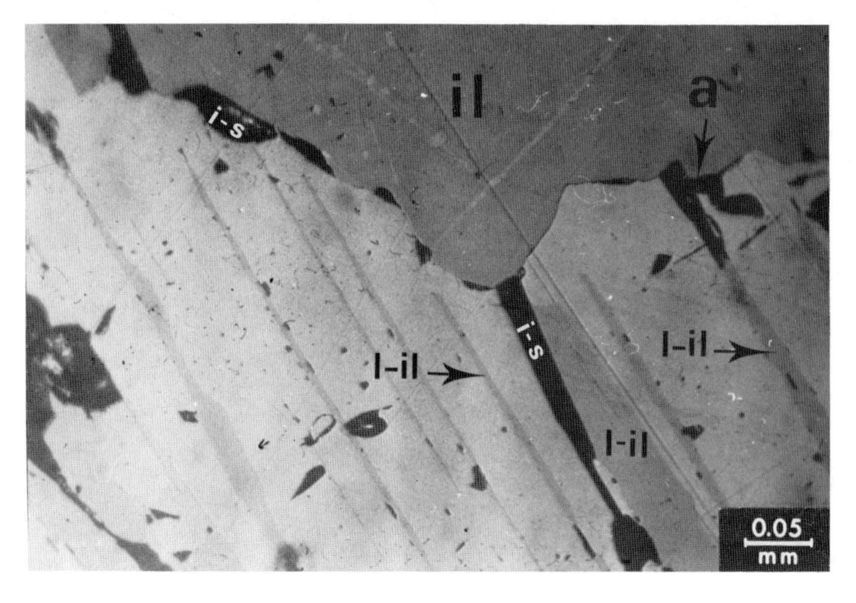

Fig. 682 Ilmenite in contact with magnetite. Ilmenite lamellae are parallely orientated to the (111) of host magnetite. Similarly, spinel occurs as lamellar bodies following the (111) of the host magnetite.
The spinel occurs also as intergranular bodies between the ilmenite and the magnetite and often sends extensions which extend as lamellar bodies following the (111) of the host magnetite.
m = magnetite
il = ilmenite
l-il = lamellar ilmenite following the 111 of the magnetite.
i-s = intergranular spinel between magnetite and ilmenite.
Arrow "a" shows intergranular spinel with extensions of it, as lamellar spinel into the adjacent magnetite.
Otanmäki, Finland. Polished section (oil-immersion, with one nicol).

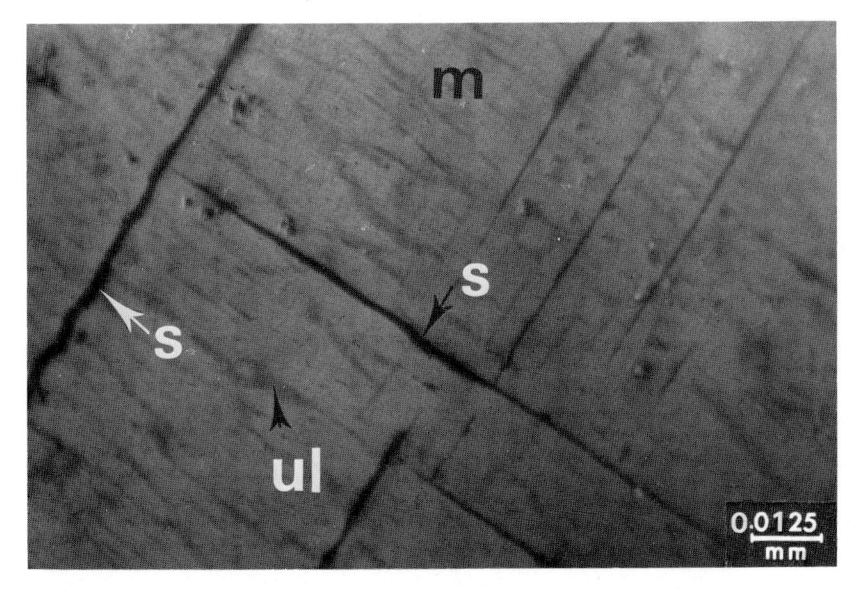

Fig. 683 Magnetite with a pattern of cleavage parallel to the 100 of the magnetite.
Spinel bodies are following and occupying the (100) cleavage pattern of the host magnetite. Also ulvite (despite a defused character) is following the 100 of the magnetite.
m = magnetite
s = spinel following the 100 of the magnetite.
ul = ulvospinel
Lac de la Blache Saguenay comté, Prov. Quebec, Canada.
Polished section (oil-immersion, with one nicol).

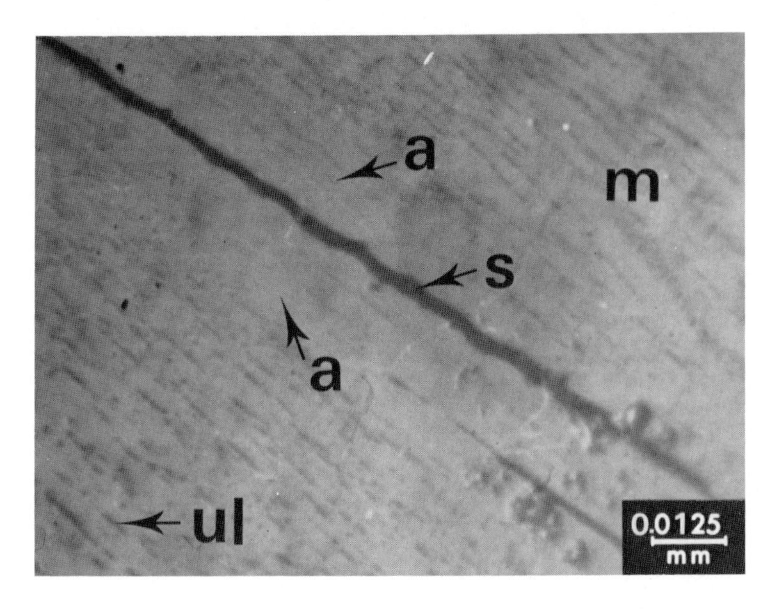

Fig. 684 A cleavage crack (100) of magnetite occupied by spinel.
ul = ulvite is following the (100) of the magnetite. It should be noticed that the magnetite adjacent to the cleavage crack which is occupied by spinel is free from ulvospinel bodies.
m = magnetite
s = spinel
Arrow "a" shows area of magnetite free of ulvite.
ul = ulvite
Lac de la Blache Saguenay comté, Prov. Quebec, Canada.
Polished section (oil-immersion, with one nicol).

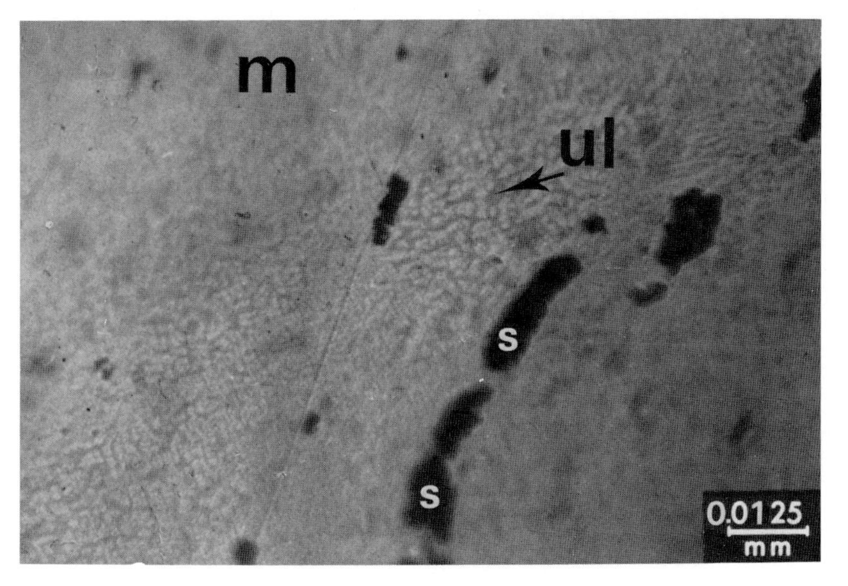

Fig. 685 Magnetite with spinel and table-cloth structure.
m = magnetite
ul = ulvite (table-cloth structure)
s = spinel
Lac de la Blache Saguenay comté, Prov. Quebec, Canada.
Polished section (oil-immersion, with one nicol).

Fig. 686 Rounded olivine, transversed by antigoritic veinlets and surrounded by pyroxene. The antigorite veinlets continue along cracks into the pyroxene as well.
ol = olivine
py = pyroxene
ant = antigorite veinlets
Picrite. Broadlands, Aberdeenshire, Scotland.
With crossed nicols.

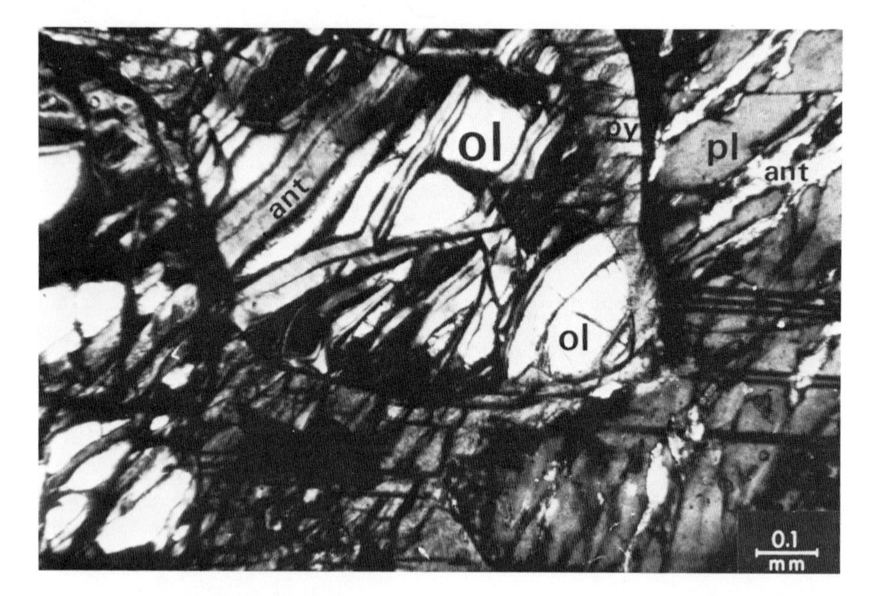

Fig. 687 Olivine rounded, transversed by a system of antigoritic veinlets and surrounded by pyroxene. A plagioclase is also shown transversed by veinlets of antigorite.
ol = olivine
ant = antigoritic veinlets
py = pyroxene
pl = plagioclase (transversed by veinlets of antigorite).
Picrite. Broadlands, Aberdeenshire, Scotland.
With crossed nicols.

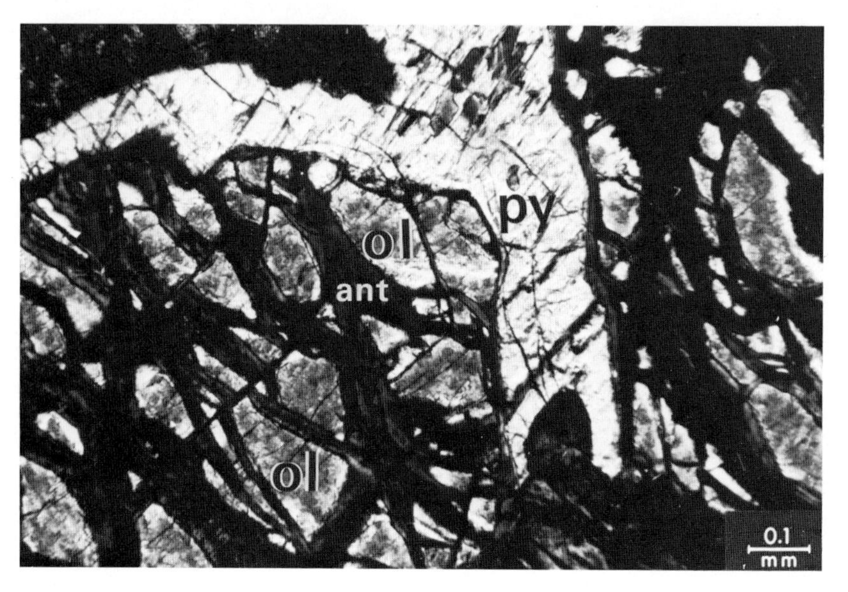

Fig. 688 Olivine, transversed by veinlets of antigorite and surrounded by later pyroxene.
ol = olivine
ant = antigoritic veinlets
py = pyroxene
Picrite. Broadlands, Aberdeenshire, Scotland.
With crossed nicols.

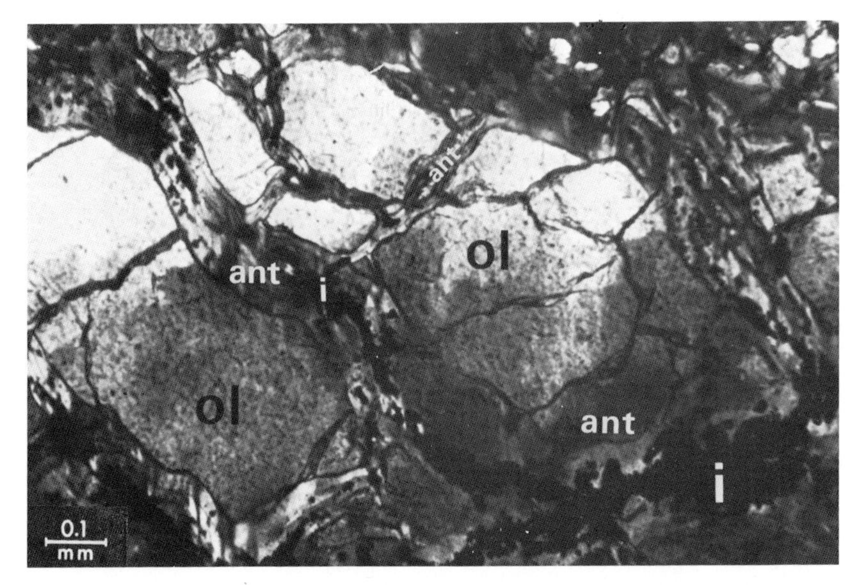

Fig. 689 Tectonically deformed olivine showing undulating extinction, with cracks of it followed by antigoritic veinlets. Also iron oxides are present often associated with the antigorite.
ol = tectonically deformed olivine
an = antigorite veinlets
i = iron oxides
Serpentinised dunite.
Xerolivado, Vourinos, North Greece.
With crossed nicols.

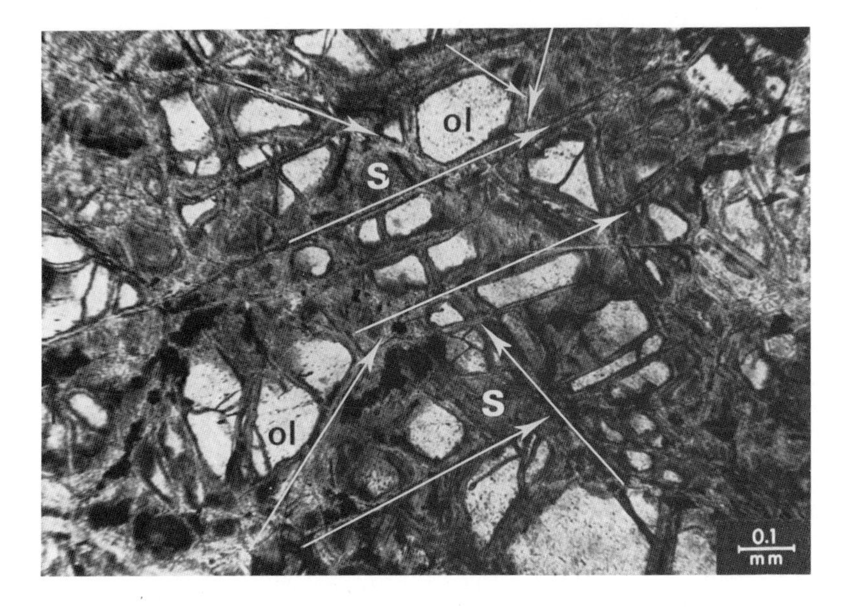

Fig. 690a Serpentinisation following a pattern of cracks transversing olivine grains.
ol = olivine relic in serpentine.
s = serpentine
Arrows indicate pattern of cracks which were followed by serpentinisation.
Serpentinised dunite. Xerolivado, Vourinos, North Greece.
Without crossed nicols.

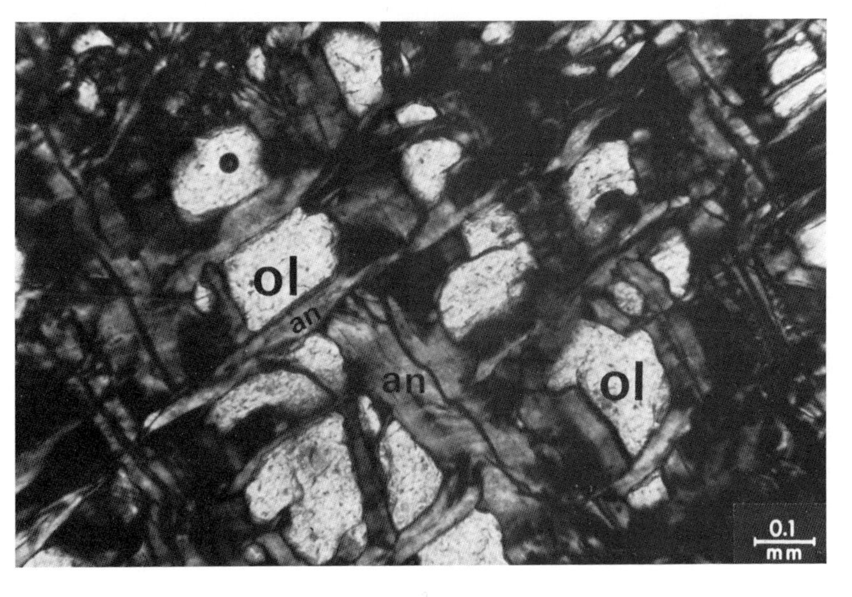

Fig. 690b Antigoritisation following a set of cracks "cleavage" perpendicularly intersected. A few olivine grains are left as relics.
ol = olivine
an = antigorite
Serpentinised dunite. Xerolivado, Vourinos, North Greece.
With crossed nicols.

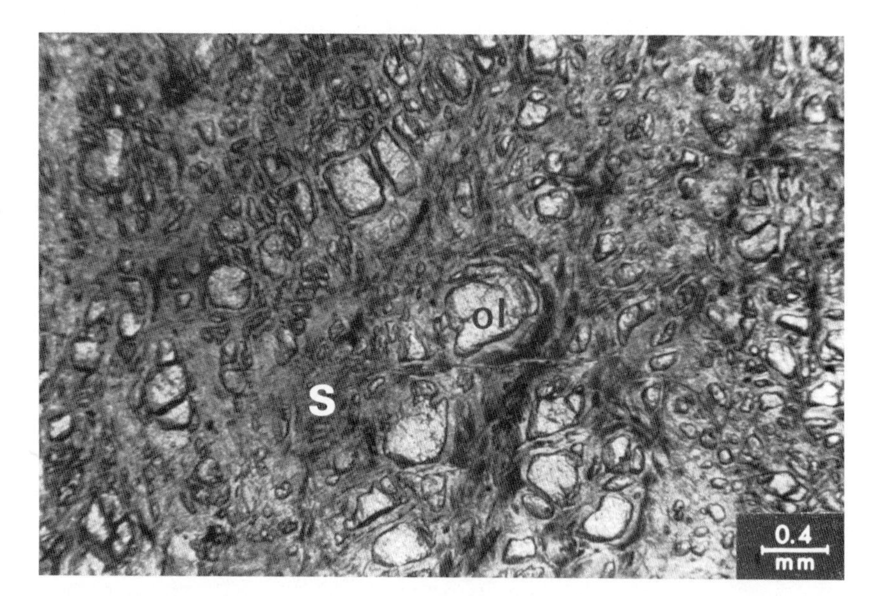

Fig. 691a Olivine remnants scattered at random in a predominant mass of serpentine, produced by the alteration of the olivine.
ol = olivine
s = serpentine
Serpentinised dunite.
Xerolivado, Vourinos, North Greece.
Without crossed nicols.

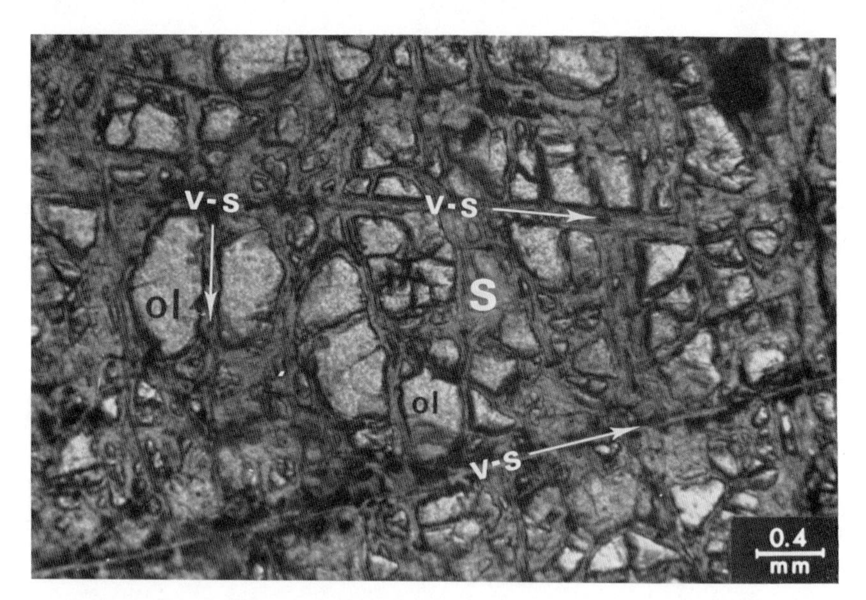

Fig. 691b Olivine grains transversed and altered by a pattern of serpentine veinlets.
ol = olivine
s = serpentine
v-s = veinlets of serpentine
Serpentinised dunite. Xerolivado, Vourinos, North Greece.
Without crossed nicols.

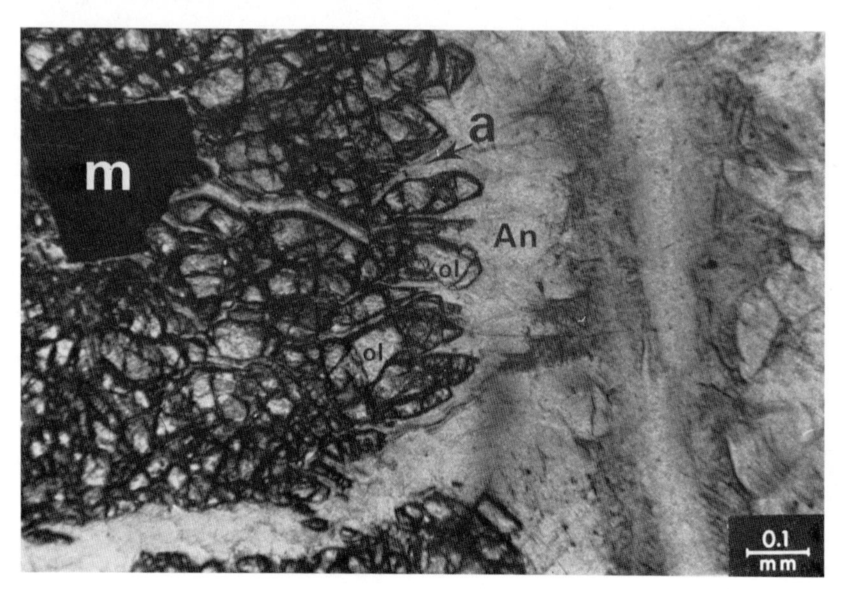

Fig. 692 Dunite (olivine predominant) with magnetite, crushed and invaded by mobilised antigorite "veinlets" which extend into the dunite following its cracks.
ol = olivine
m = magnetite
An = antigorite
Arrow "a" shows antigorite invading the dunite along cracks.
Serpentinised dunite.
Xerolivado, Vourinos, North Greece.
Without crossed nicols.

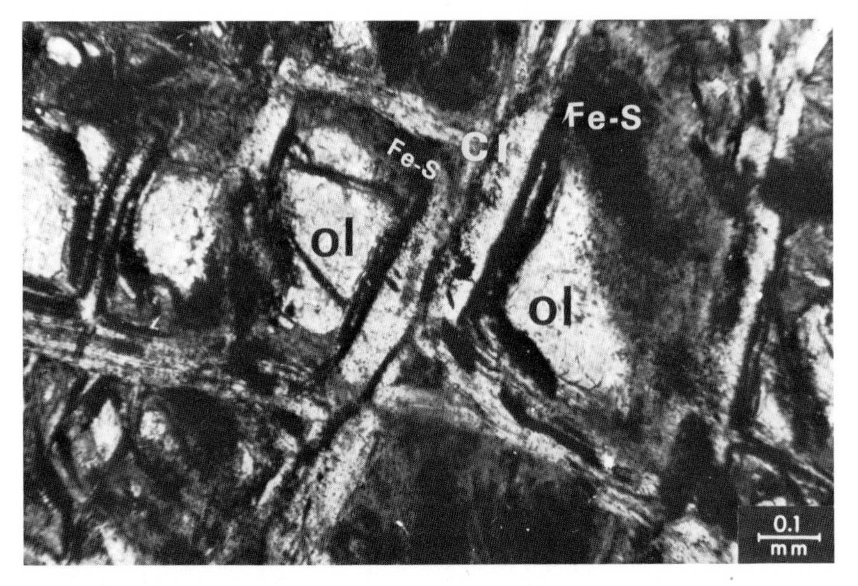

Fig. 693 Coarse grained olivines in which a system of cracks develops along which alteration of the olivine takes place. In addition to antigoritisation, differential leaching of Mg takes place and residual Fe shadows are formed within the olivine crack systems in the alteration zones and marginal to the olivine.
ol = olivine residual grains
cr = alteration cracks (with antigorite formation)
Fe-S = shadows of residual Fe formed as a result of differential leaching, due to olivine alteration.
Serpentinised dunite. Xerolivado, Vourinos, North Greece.
Without crossed nicols.

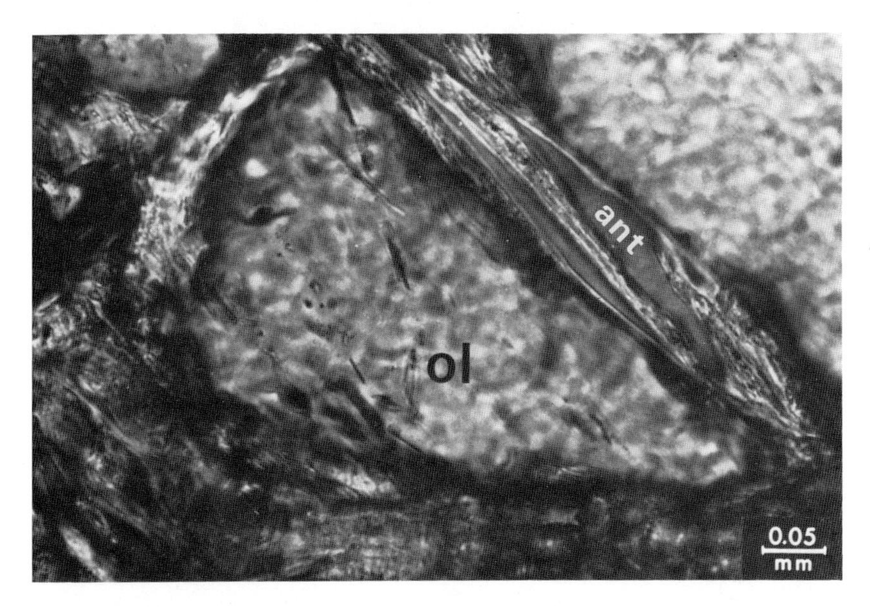

Fig. 694 Olivine grains marginally changed to antigorite.
ol = olivine grains
ant = antigorite
Dunite. Euboea, Greece.
With crossed nicols.

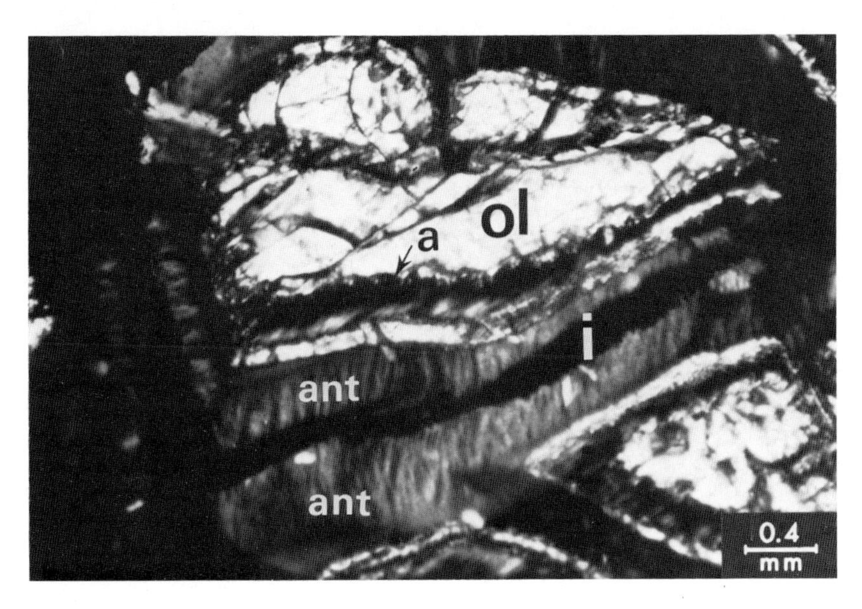

Fig. 695 Olivine with alteration to serpentine. Cross fibre-antigorite formation, with residual iron inter-granular between initial boundaries of olivines.
ol = olivine
ant = antigorite
i = iron concentration (also, see arrow "a")
Wehrlite. Frankenstein, Germany.
With crossed nicols.

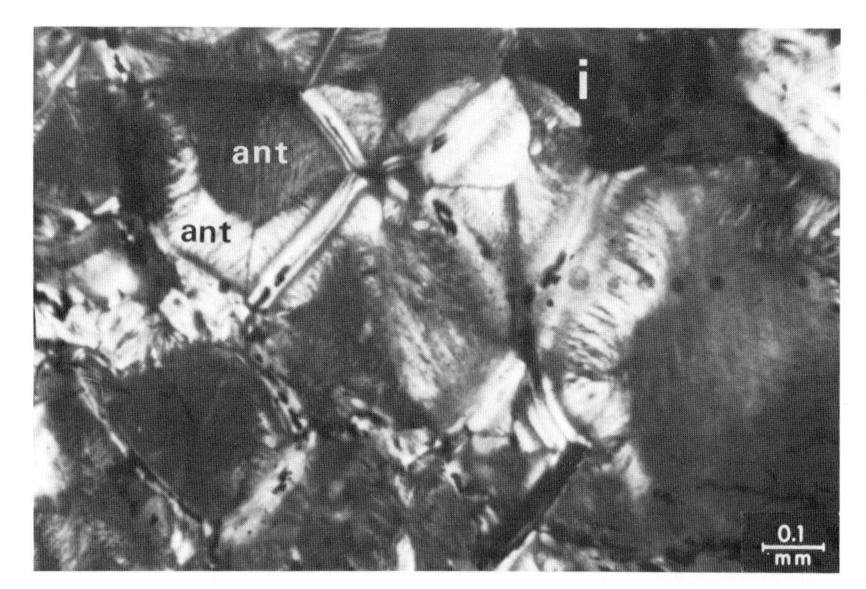

Fig. 696a Serpentinised wehrlite with development of a "cellular" structure approximately corresponding to the initial olivine granular texture of the initial dunite, actually wehrlite.
Within the serpentine cell structure, the fibrous antigorite is arranged perpendiculary to initial outlines of the initial olivine.
Residual iron concentrates as a less leached element along initial olivine intergranulars.
ant = antigorite (fibrous)
i = residual iron
Wehrlite. Frankenstein, Germany.
With crossed nicols.

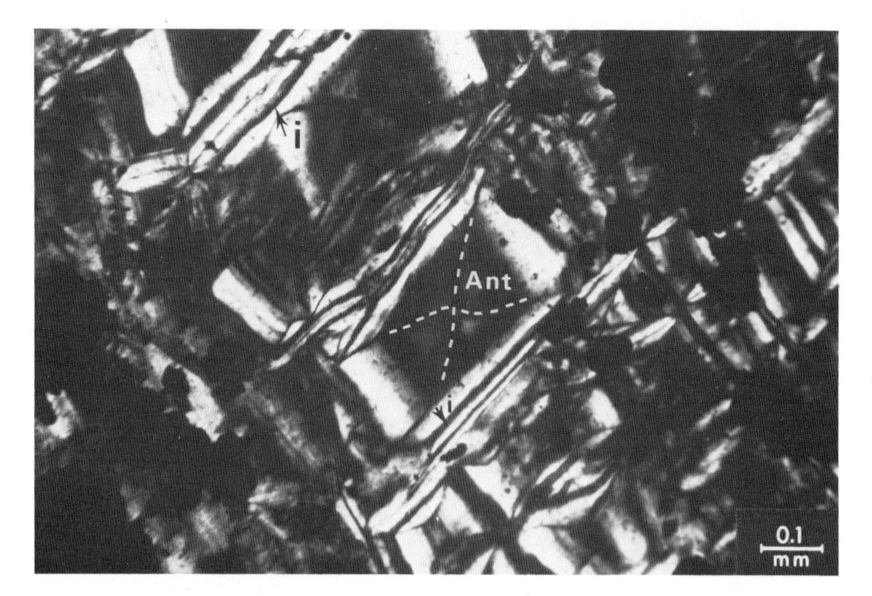

Fig. 696b Serpentinised dunite with development of a "cellular" structure approximately corresponding to the initial olivine granular texture of the initial dunite.
Within the serpentine 'cell' structure the fibrous antigorite is arranged perpendiculary to initial outlines of the initial olivine. Residual iron concentrates as a less leached element along initial olivine intergranulars.
Ant = Fibrous antigorite
i = residual iron.
Initially dunite, completely serpentinised.
Euboea, Greece.
With crossed nicols.

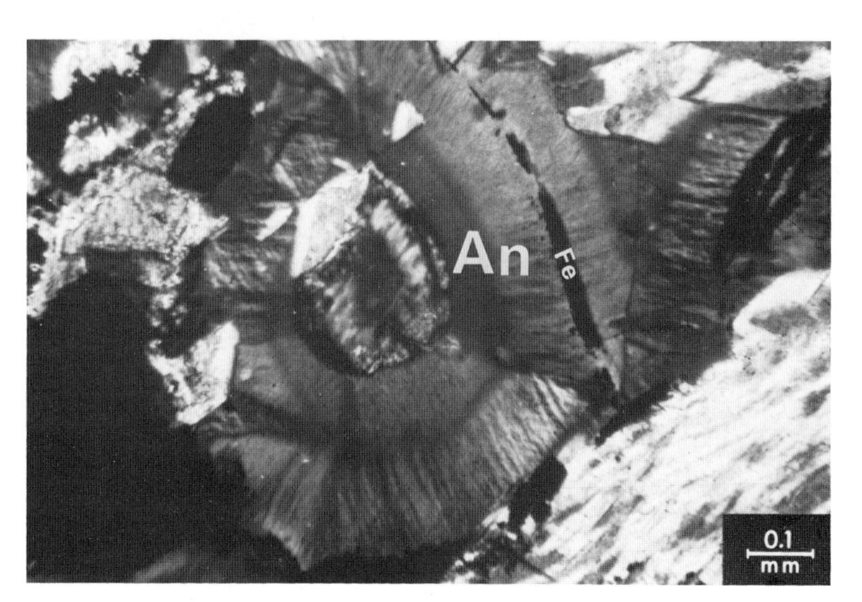

Fig. 697 Serpentinisation of olivine (wehrlite). The fibrous antigorite formed is approximately perpendicular to the initial olivine boundaries, residual iron (due to preferential Mg-leaching) concentrated along initial olivine boundaries.
An = fibrous antigorite
Fe = residual Fe
Wehrlite. Frankenstein, Germany.
With crossed nicols.

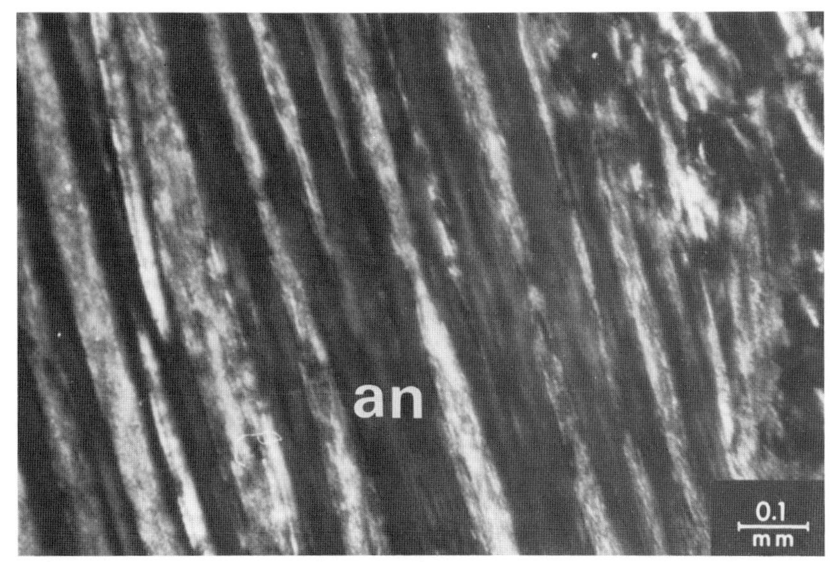

Fig. 698 Antigorite development in Dunite.
An = antigorite
Serpentinite. Lizard Peninsula, Cornwall, England.
With crossed nicols.

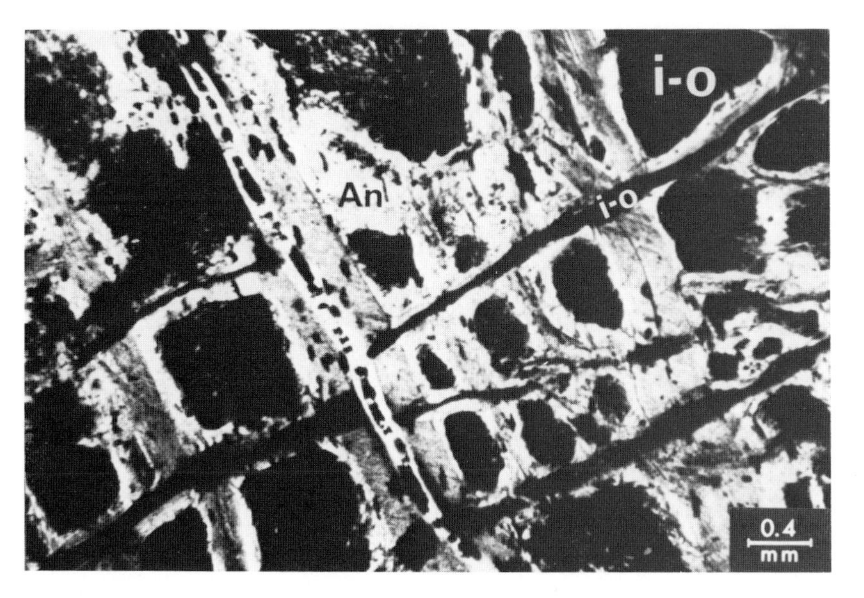

Fig. 699 Serpentinisation of dunite. The original cell-structure of dunite is preserved, with development of antigorite and iron-oxides.
An = antigorite
i-o = iron oxides due to the serpentinisation of the initial dunite.
Serpentinite. Lizard Peninsula, Cornwall, England.
With crossed nicols.

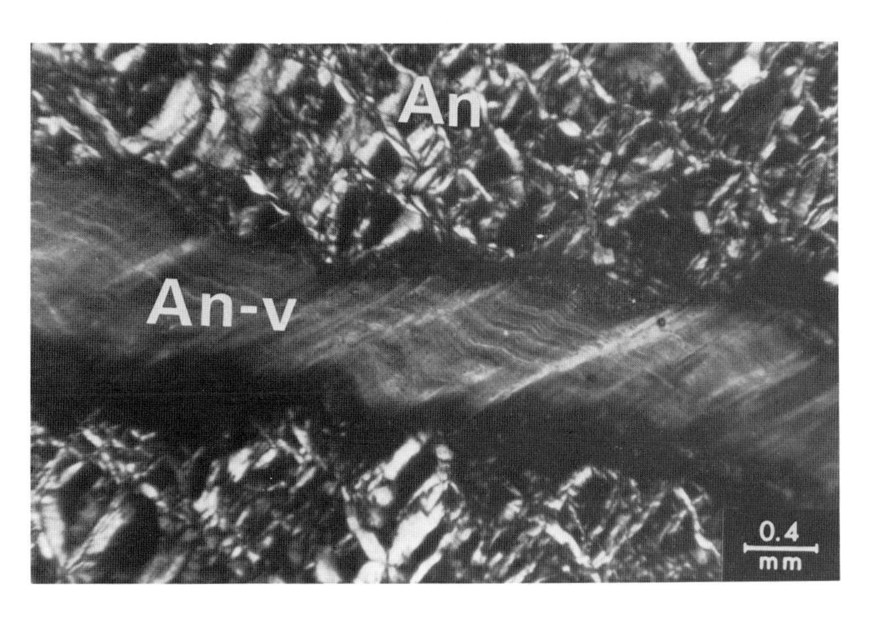

Fig. 700a A serpentinised dunite with typical cell-structure transversed by a banded veinlet of antigorite "showing a pattern of folding".
An = antigorite (indicating cell structure of initial dunite).
An-v = antigorite veinlet.
Serpentinised dunite.
Euboea, Greece.
With crossed nicols.

Fig. 700b A veinlet of banded antigorite transversed by a later veinlet.
b-an = banded antigorite
r-an = veinlet of later antigorite
Serpentinite. Lizard Peninsular, Cornwall, England.
With crossed nicols.

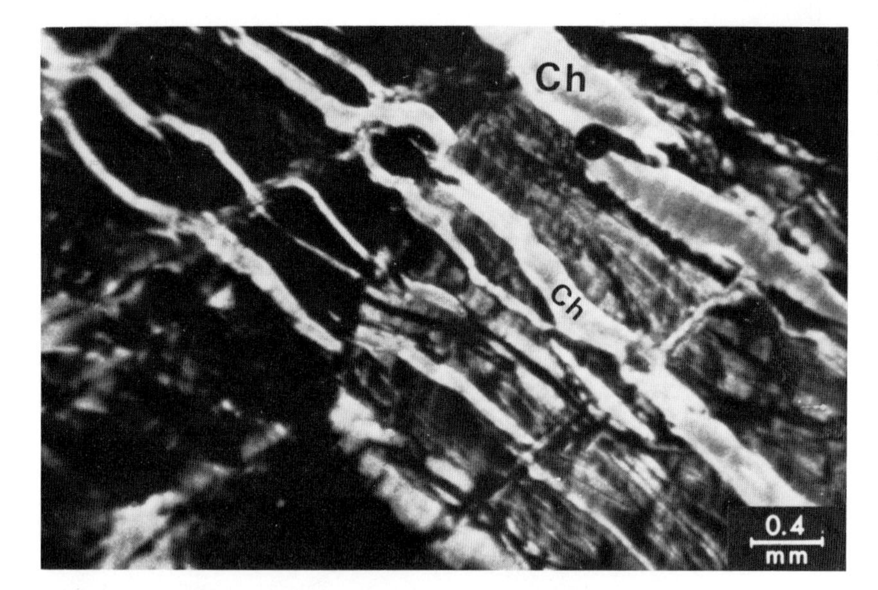

Fig. 701 Peridotite with veinlets of chrysotile asbestos.
ch = chrysotile asbestos
(Courtesy Gorni Institute, Leningrad, USSR.)
With crossed nicols.

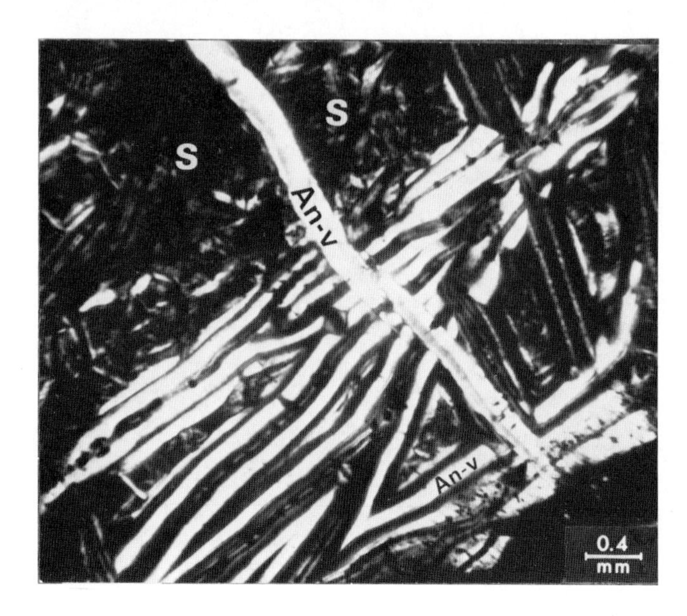

Fig. 702 A network of antigorite veinlets in serpentine.
An-v = antigorite veinlets
s = serpentine
Serpentine. Lizard Peninsular, Cornwall, England.
With crossed nicols.

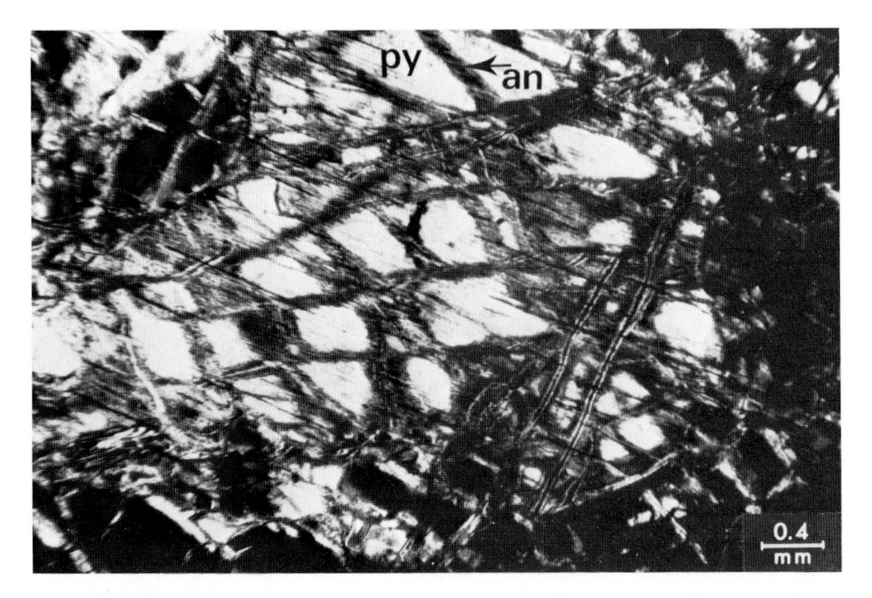

Fig. 703 Serpentinisation of pyroxene (orthopyroxene). The antigoritisation follows a pattern of cracks, and the cleavage of the pyroxene.
py = pyroxene
an = antigoritisation
Xerolivado. Vourinos, North Greece.
With crossed nicols.

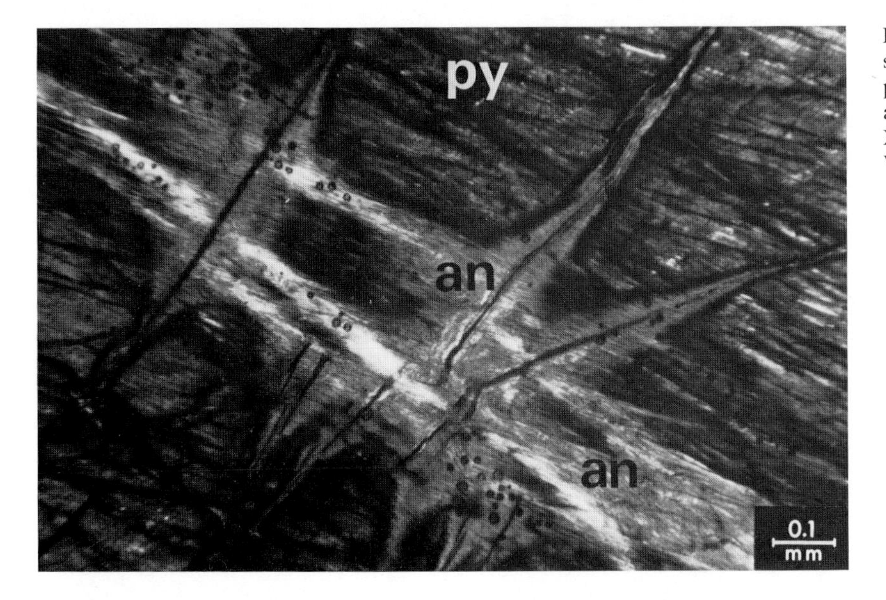

Fig. 704 Antigoritisation follows as a system of veinlets within the pyroxene
py = pyroxene
an = antigoritisation
Xerolivado. Vourinos, North Greece.
With crossed nicols.

Fig. 705 Antigoritisation of enstatite.
An = antigorite
en = enstatite
Enstatite-Serpentinite. Berg, Moravia.
With crossed nicols.

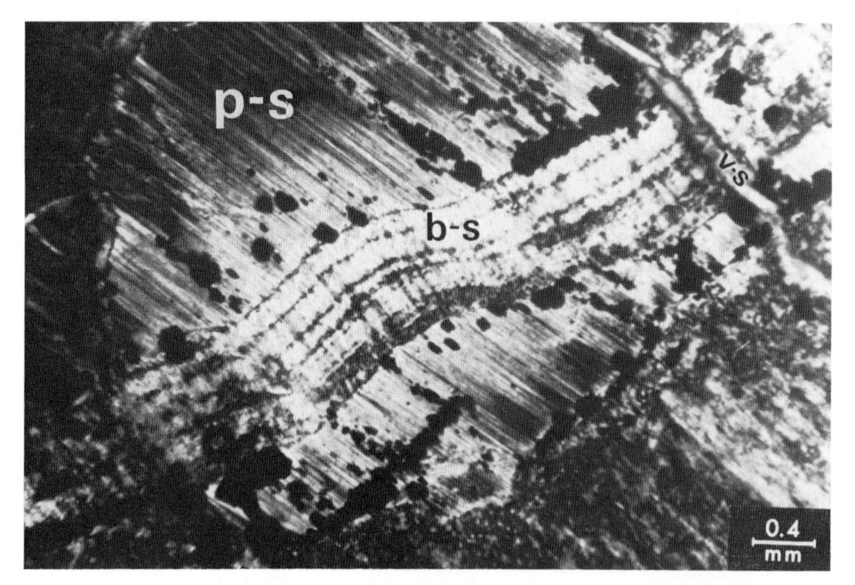

Fig. 706 An initial pyroxene serpentinised, transversed by banded veinlets of serpentine.
A still later veinlet of antigorite transversed the banded veinlet.
p-s = initial pyroxene serpentinised.
b-s = banded serpentine transversing the serpentinised pyroxene.
v-s = veinlet of serpentine.
Euboea, Greece.
With crossed nicols.

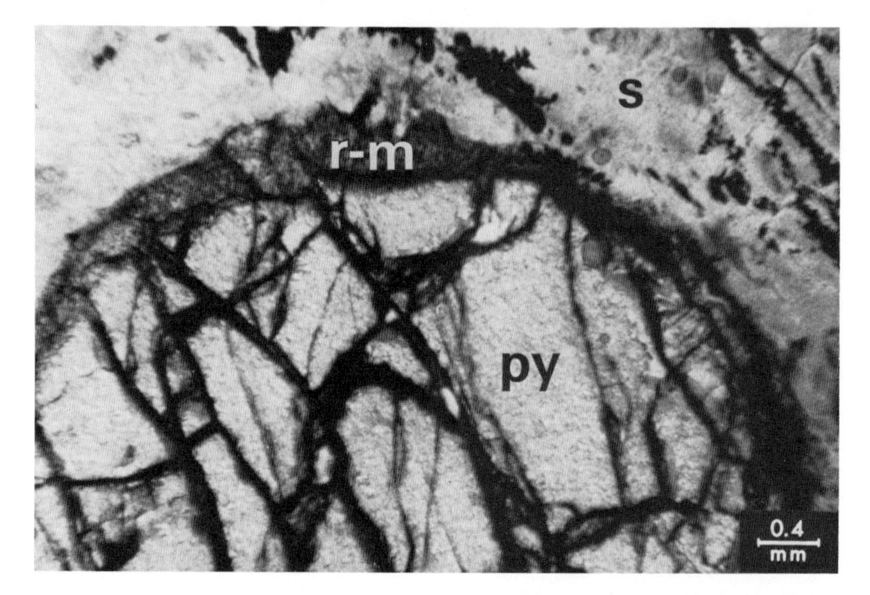

Fig. 707a Rounded pyrope crystal in serpentine. A reaction margin is indicated surrounding the pyrope.
py = pyrope
s = serpentine
r-m = reaction margin (calcite)
Serpentine with pyrope. Zoblitz, Saxony, Germany.
Without crossed nicols.

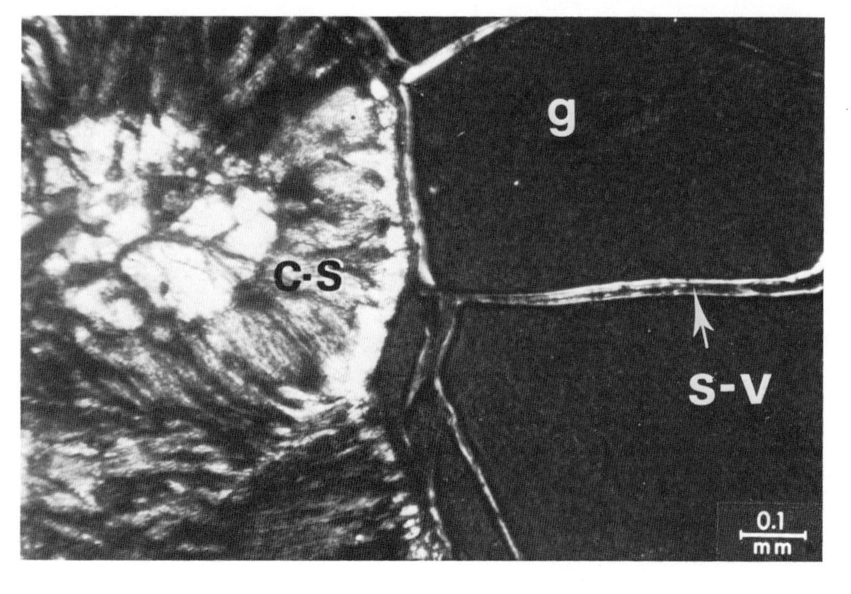

Fig. 707b Garnet with a corona reaction structure in serpentine. Veinlets of serpentine extend into the garnet.
g = garnet
c-s = corona reaction structure marginal to the serpentine.
s-v = serpentine veinlets extending into the garnet.
Magnetite-Serpentinite. Forbau near Hof, Saale River, Germany.
With crossed nicols.

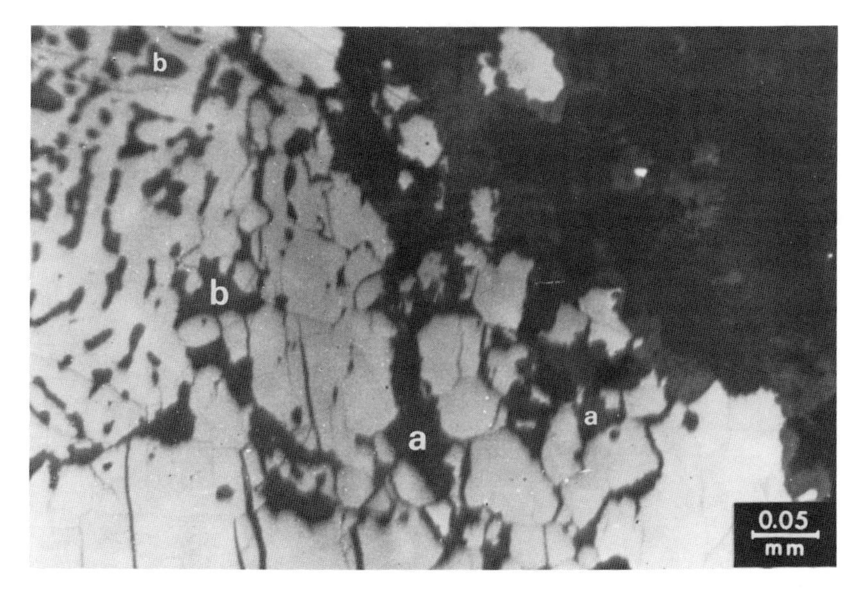

Fig. 708 Fine chromite grains are bound together by brown intergranular serpentine which extends as "myrmekitic" intergrowth in the adjacent larger chromite grain.
a = intergranular serpentine between fine chromite grains
b = serpentine attaining "myrmekitic intergrowths".
Chromite-ore.
Rodiani, Northern Greece.
Polished section (oil-immersion, with one nicol).

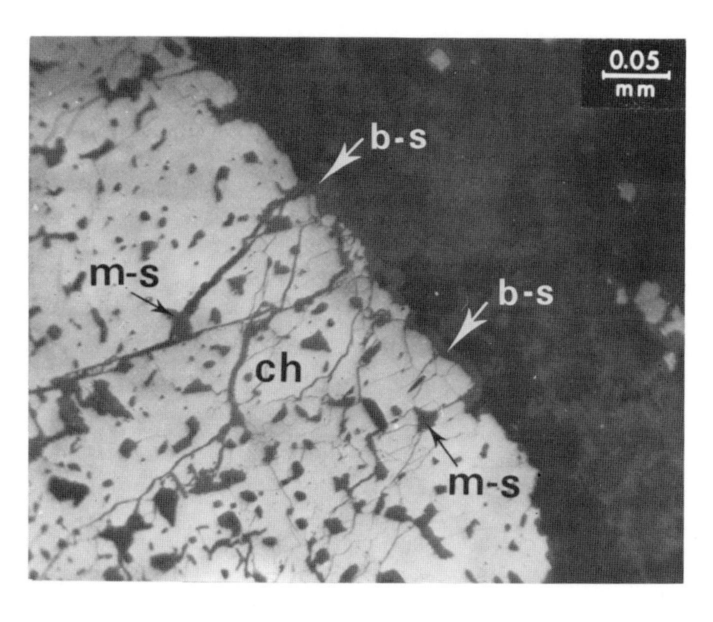

Fig. 709 Brown serpentine forms a margin outside the chromite (b-s), with extensions of it following cracks of the adjacent chromite and attaining a myrmekitic intergrowth with the chromite.
b-s = brown serpentine
ch = chromite
m-s = myrmekitic serpentine in intergrowth with the chromite.
Chromite-ore. Rodiani, North Greece.
Polished section (oil-immersion, with one nicol).

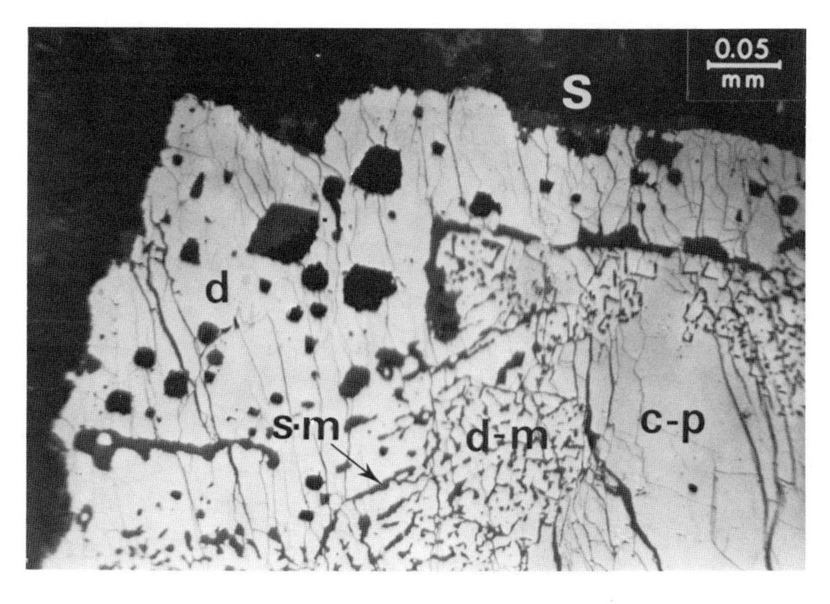

Fig. 710a Chromite grains in contact with serpentine. Marginally the chromite is indicating a decoloration (due to element leaching). Also myrmekitic serpentine is associated with a part of the decolorized margin.
d = decolorised margin without myrmekitic serpentine.
d-m = decolorised margin with serpentine in myrmekitic intergrowth with the chromite.
c-p = central part of the chromite.
s = serpentine
s-m = myrmekitic serpentine
Chromite-ore. Rodiani, North Greece.
Polished section (oil-immersion, with one nicol).

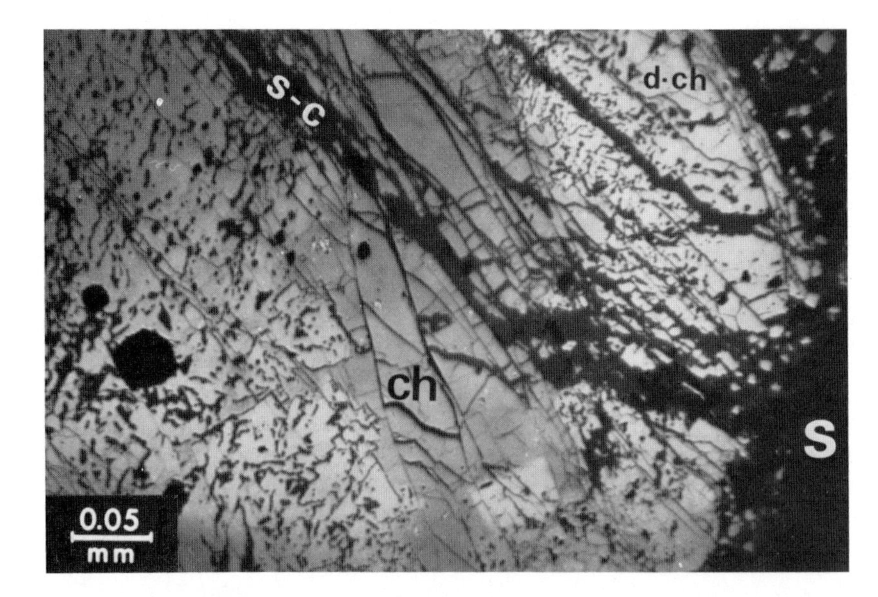

Fig. 710b Chromite in contact with serpentine. The chromite shows decoloration margins and patches with which serpentine is in myrmekitic intergrowth.
ch = chromite
d-ch = decolorised chromite with serpentine in myrmekitic intergrowth.
s-c = serpentine filling cataclastic cracks of the chromite.
s = serpentine
Chromite-ore. Rodiani, North Greece.
Polished section (oil-immersion, with one nicol).

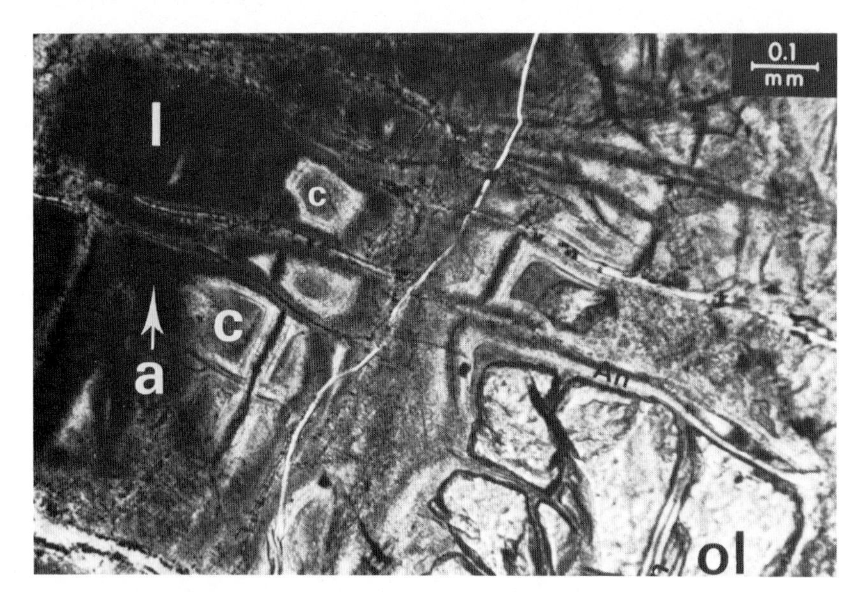

Fig. 711a All transition phases of the gradoseparation of the dunitic olivine to recrystallised silica are shown. Olivine of the initial dunite and olivine with transition phases to colloform banded SiO_2 with limonite are to be seen.
ol = olivine
c = colloform silica
l = limonitic concentrations (see arrow "a")
An = antigorite
Altered dunite, Mandoudi, Euboea, Greece.
With crossed nicols.

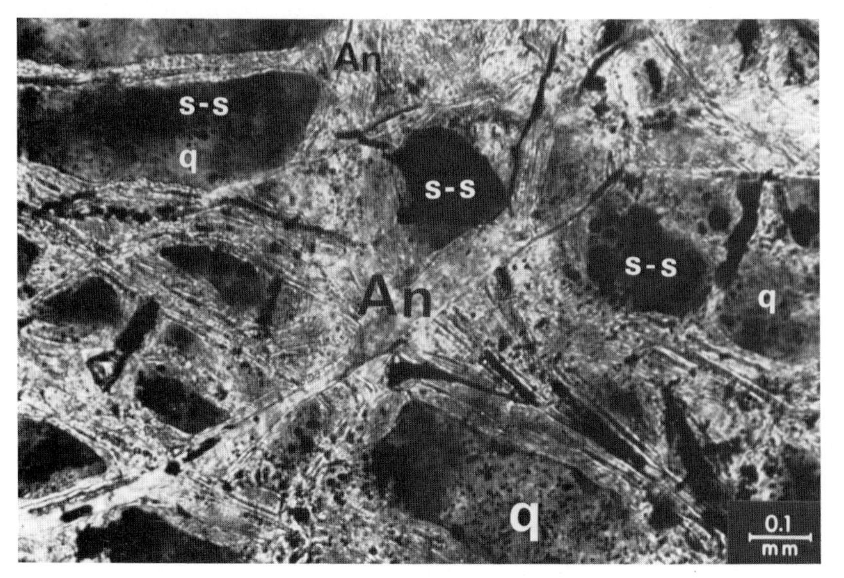

Fig. 711b Initial granular olivine birbiritised due to gradoseparation of the olivine. The remnants (SiO_2) show the transition phase from synisotropisation to quartz regeneration.
s-s = synisotropic SiO_2 with transitions to quartz.
q = quartz regeneration (from synisotropic SiO_2).
An = antigoritic serpentine
Altered dunite. Mandoudi, Euboea, Greece.
Without crossed nicols.

Fig. 712 Banded gel-structure consisting of small, radiating chalcedony crystals. In the central part between the gel-spheroids microcrystalline to well-crystalline quartz is present.
Birbirite. Yubdo, Western Ethiopia.
With crossed nicols.

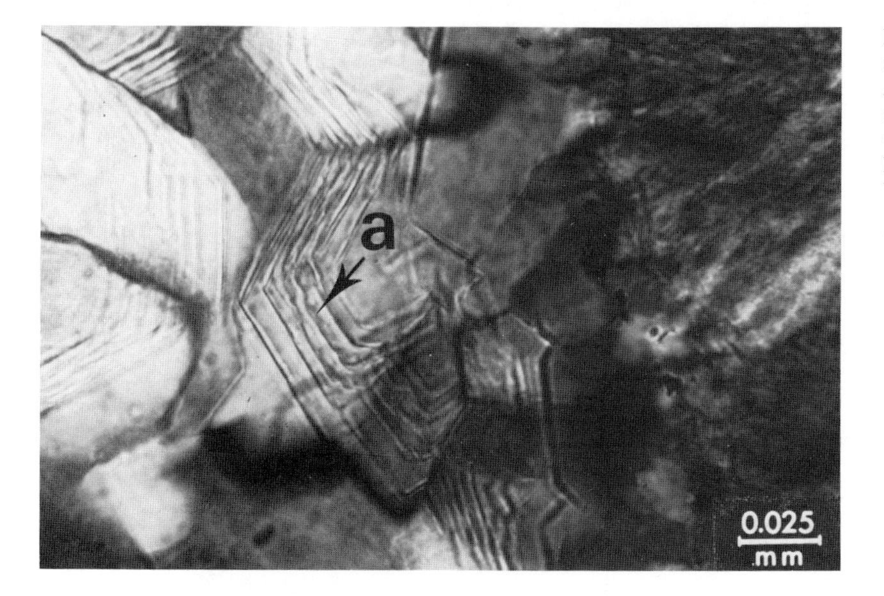

Fig. 713a Transitions from gel-silica to quartz.
Arrow (a) shows relic gel-structure resembling crystal-zoning.
Chalalaca/Chambe, Adola, South Ethiopia.
With crossed nicols.

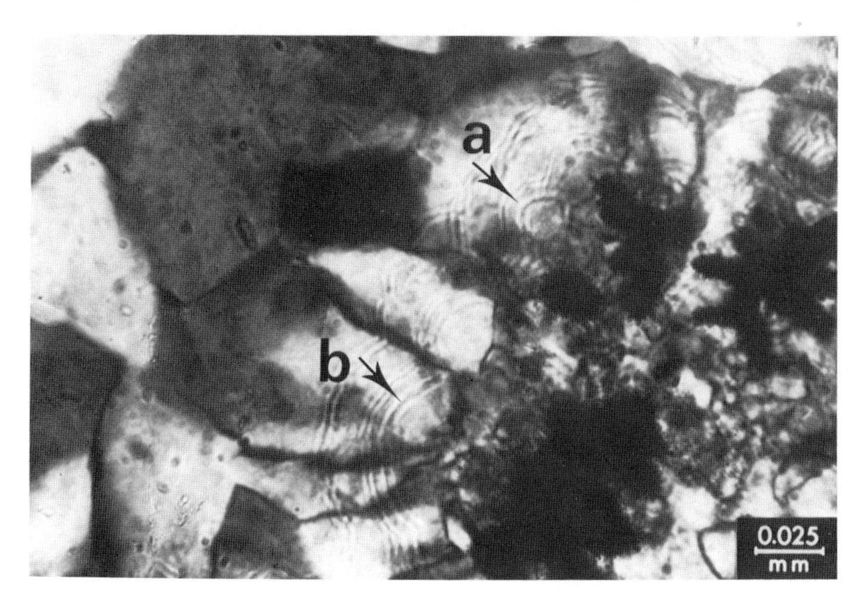

Fig. 713b Transitions from gel-silica to quartz.
Arrow (a) shows relic gel-structure resembling crystal-zoning. Arrow (b) shows relic gel-structure maintained in the different quartz crystals after the transformation gel-silica to quartz.
Chalalaca/Chambe, Adola, South Ethiopia.
With crossed nicols.

Fig. 714 Gel and crystalline structures co-existing in the same quartz individual.
Arrow 'a' shows a crystalline outline of the quartz nucleous; arrow 'b' indicates relic gel-structures in the same quartz individual.
Chalalaca/Chambe, Adola, South Ethiopia.
With crossed nicols.

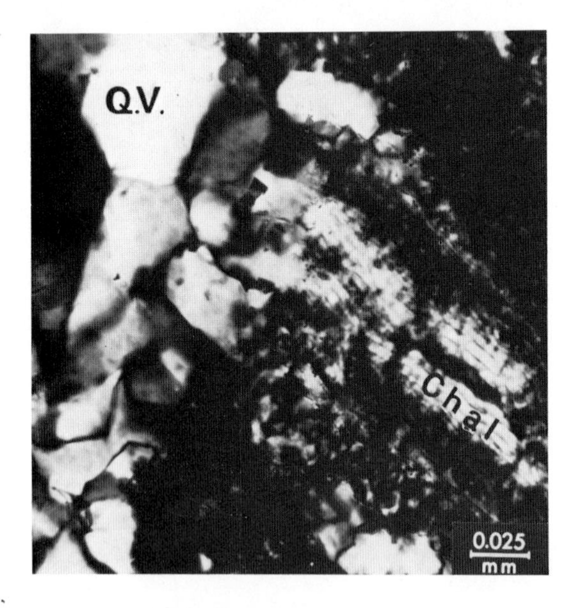

Fig. 715a Banded colloform silica (chalcedony) transversed by a veinlet of well crystallised quartz.
Chal = chalcedony
Q.V = quartz veinlet
Birbirite. Yubdo, Western Ethiopia.
With crossed nicols.

Fig. 715b Microgeode containing fine to coarse quartz crystals between limonitic masses.
Birbirite. Yubdo, Western Ethiopia.
With crossed nicols.

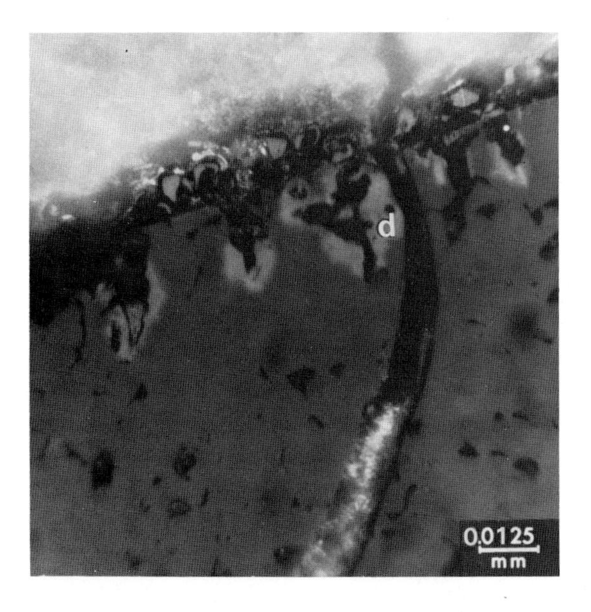

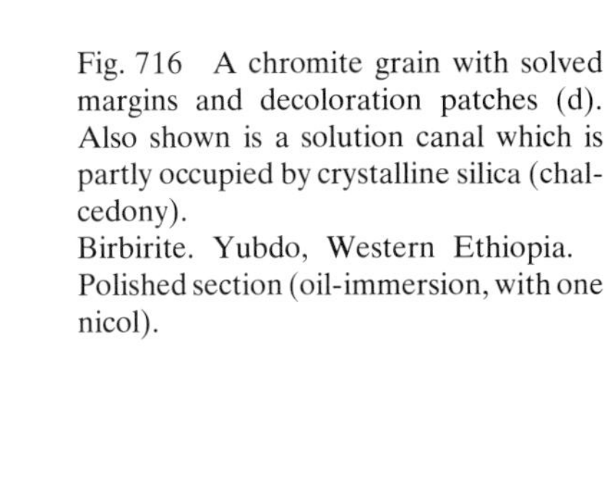

Fig. 716 A chromite grain with solved margins and decoloration patches (d). Also shown is a solution canal which is partly occupied by crystalline silica (chalcedony).
Birbirite. Yubdo, Western Ethiopia.
Polished section (oil-immersion, with one nicol).

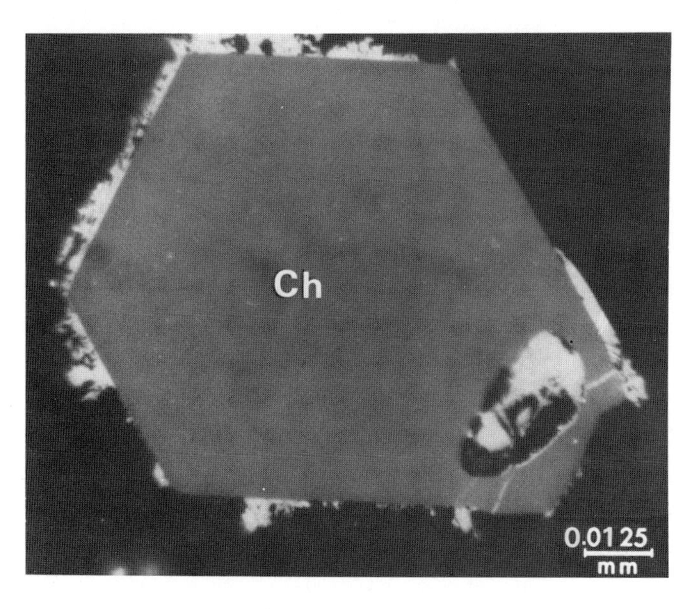

Fig. 717a Idiomorphic chromite (a well crystallised form is exhibited) with a margin of magnetite. Also, the magnetite occupies a cavity of the chromite.
Ch = chromite
M = magnetite
In the unaltered dunite of Yubdo, W.Ethiopia.
Dunite, Yubdo, Western Ethiopia.
Polished section (oil-immersion, with one nicol).

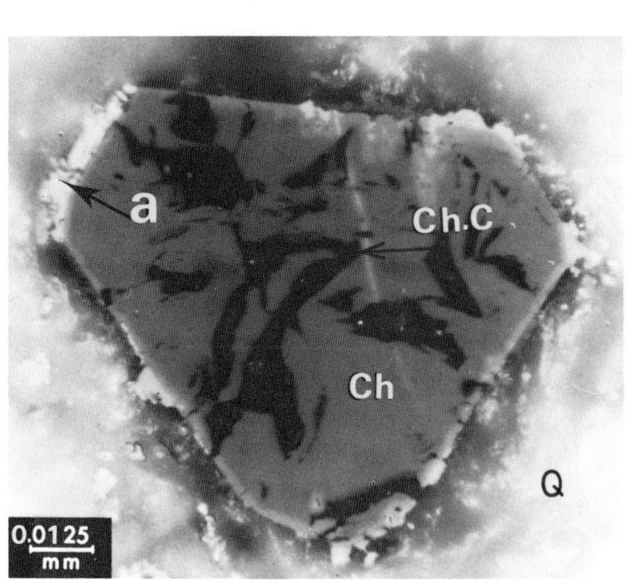

Fig. 717b An idiomorphic chromite comparable to Fig. 717a with solution canals due to solving of the chromium. In the birbiritic cover of the Yubdo/dunite.
Ch = chromite
Arrow "a" shows initial magnetitic margin now changed into limonite.
ch-c = solution canals
Birbirite. Yubdo, Western Ethiopia.
Polished section (oil-immersion, with one nicol).

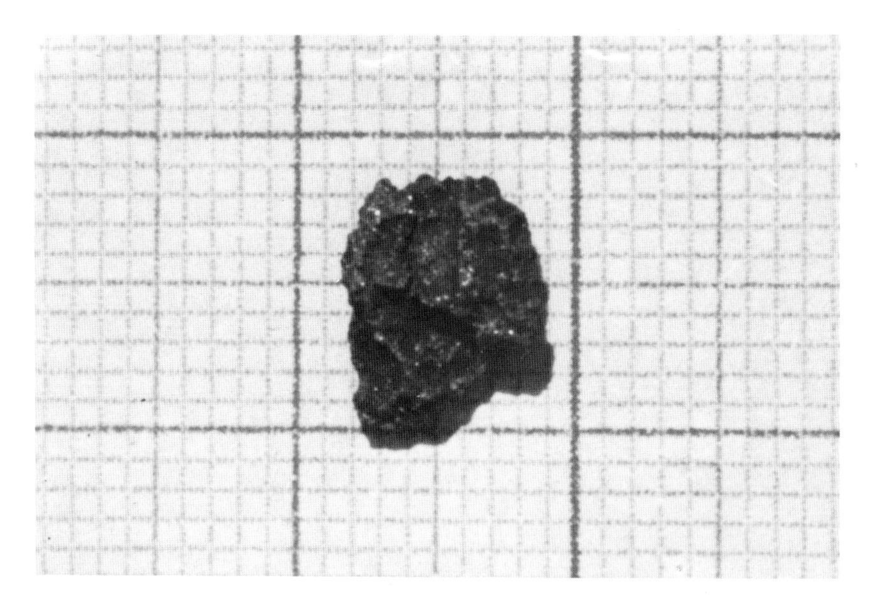

Fig. 718 A nugget of "ferroplatin" with a limonitic coating and with angular protuberances grown due to accretion in the lateritic eluvial cover of altered ultrabasics. Lateritic cover of Yubdo ultrabasic ring intrusion.
Yubdo, W. Ethiopia.

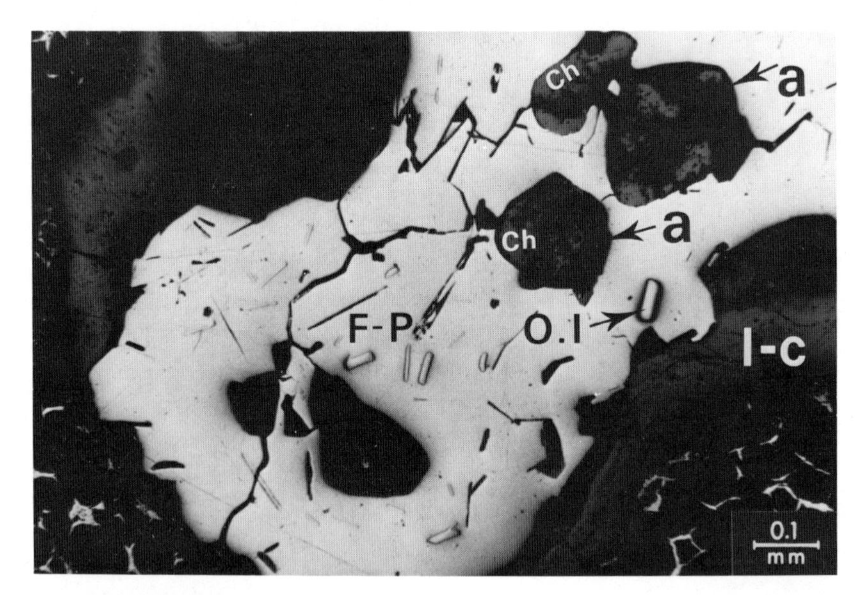

Fig. 719a A polished section view of a platinum "ferroplatin" nugget consisting essentially of "ferroplatin" and including corroded chromite grains. A limonitic coating of the ferroplatin is also indicated. Osmiridium is included in the "ferroplatin".
F-P = "ferroplatin"
Ch = chromite grains (arrow "a" shows corrosion prior to inclusion in the "ferroplatin")
l-c = limonitic coating
O.I = osmiridium
Eluvial lateritic cover of the Yubdo dunite.
Yubdo, W. Ethiopia.
Polished section (oil-immersion, with one nicol).

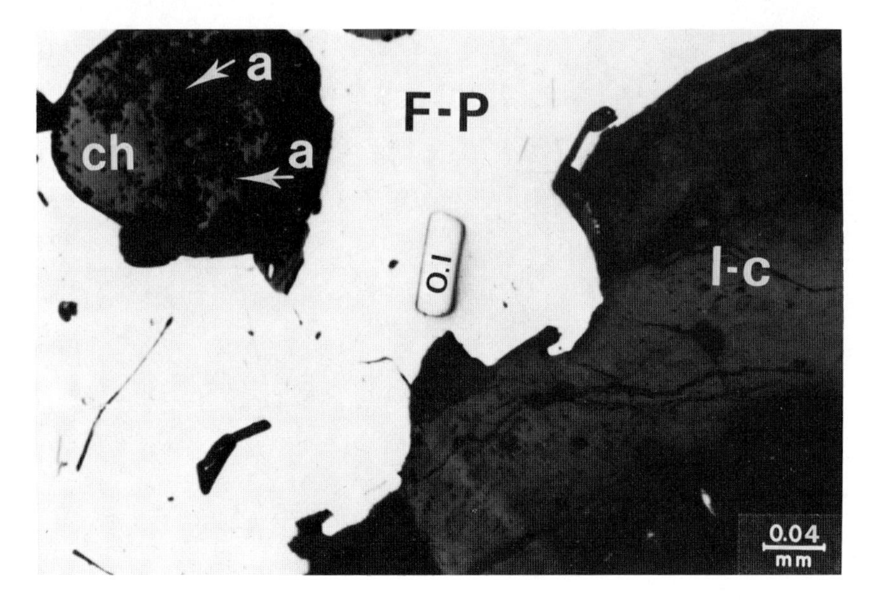

Fig. 719b detail of Fig. 719a The corrosion of the chromite grain (Ch) included in the ferroplatin (F-P) is indicated by arrow "a".
Yubdo, W. Ethiopia.
Polished section (oil-immersion, with one nicol).

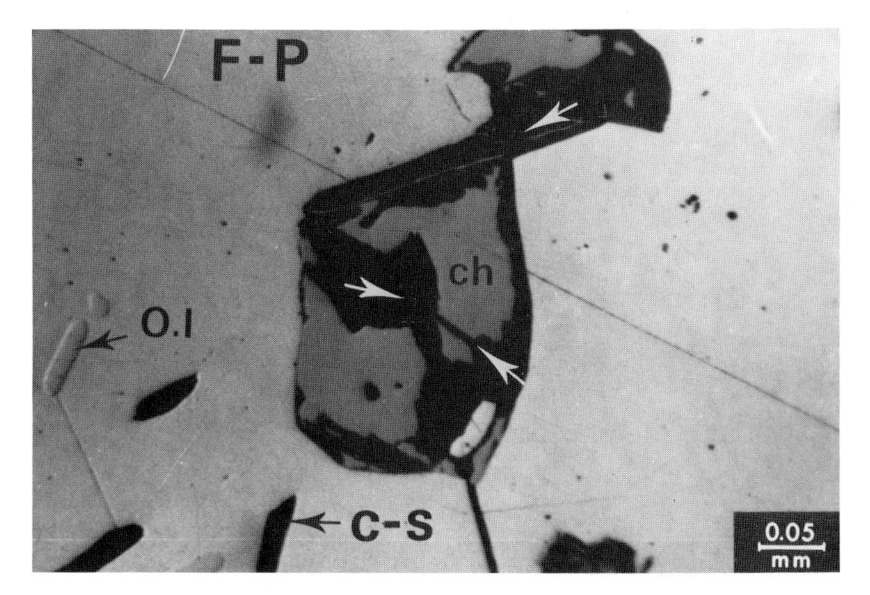

Fig. 719c "Ferroplatin" including corroded chromite grain and chain-silicate? (asbestos). Also in the "ferroplatin" occurs osmiridium.
F-P = "ferroplatin"
Ch = chromite grain (arrows show corrosion of the chromite grain).
O.I = osmiridium in "ferroplatin"
c-s = chain silicate?
Yubdo, W. Ethiopia.
Polished section (oil-immersion, with one nicol).

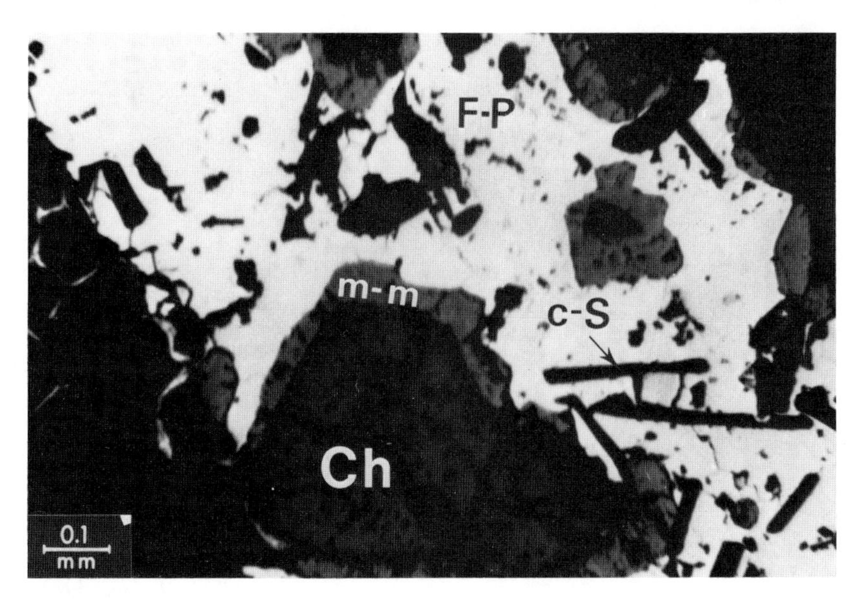

Fig. 720a and b Chromite with a martitised magnetite margin is included in a "ferroplatin" nugget.
Ch = chromite
m-m = martitised magnetite
F-P = "ferroplatin"
c-S = chain silicates (asbestos) included in the "ferroplatin" nugget.
Ferroplatin nugget in the eluvial cover of the ultrabasics.
Yubdo, W. Ethiopia.
Polished section (oil-immersion, with one nicol).

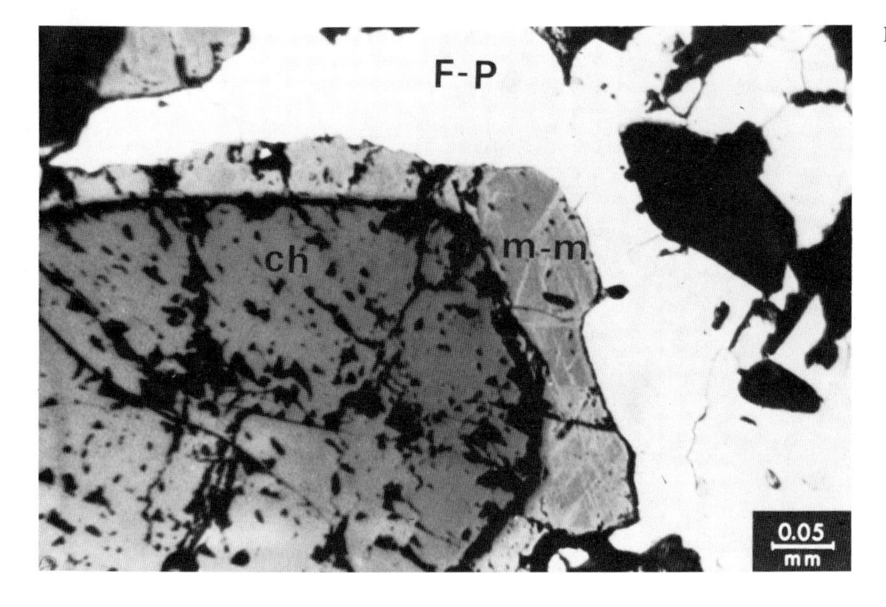

Fig. 720b

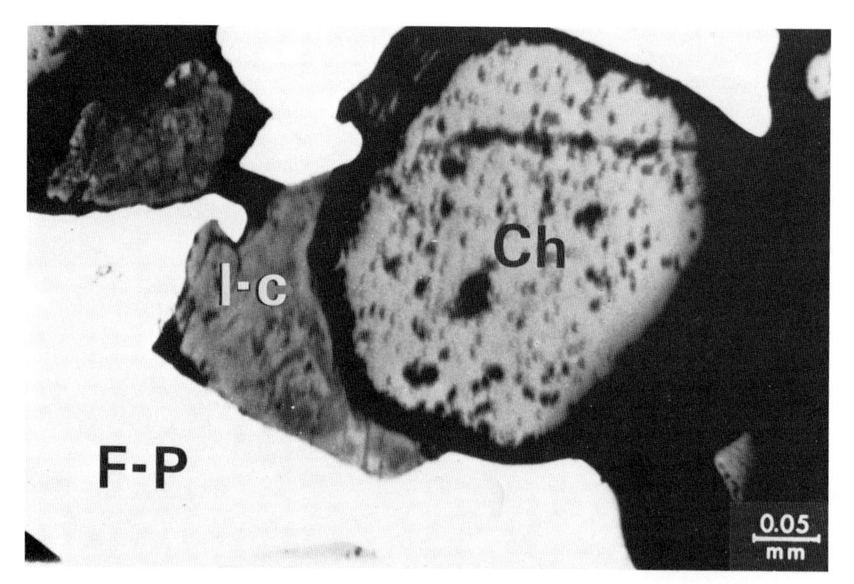

Fig. 721 Chromite grain with a limonitic coating included in "ferroplatin" nuggets.
Ch = chromite
l-c = limonitic cover
F-P = "ferroplatin"
Ferroplatin nugget, lateritic cover of Yubdo ultrabasics.
Yubdo ultrabasics.
Polished section (oil-immersion, with one nicol).

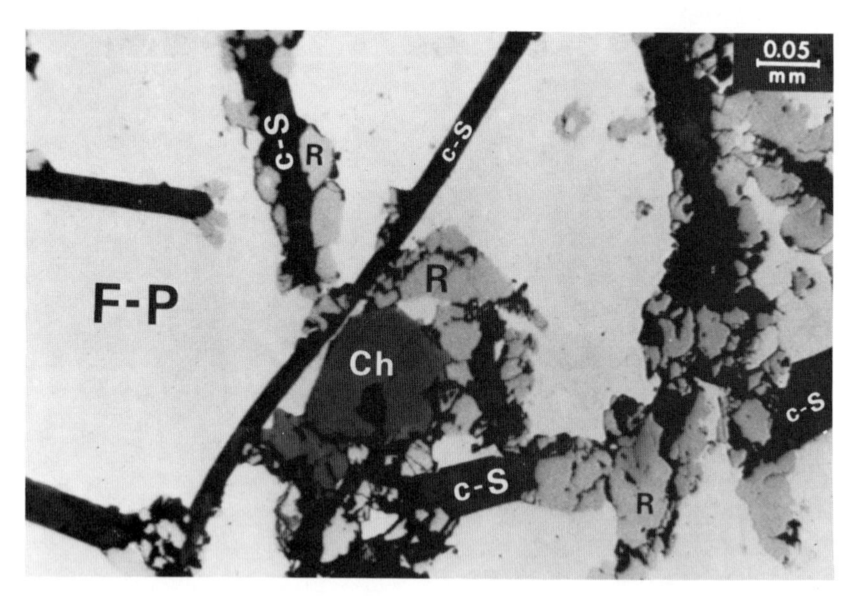

Fig. 722 Ferroplatin nugget including chain-silicates (asbestos) and associated with the silicates are "Roseite" (Os, Ir, S).
F-P = "ferroplatin"
Ch = chromite grain included in the "ferroplatin" nugget.
R = "Roseite"
c-S = chain silicates (asbestos) Fibrous silicates, asbestos, formed after the initial dunite-consolidation, probably within the dunite alteration processes.
Ferroplatin nugget, lateritic cover of the Yubdo ultrabasics.
Yubdo, W. Ethiopia.
Polished section (oil-immersion, with one nicol).

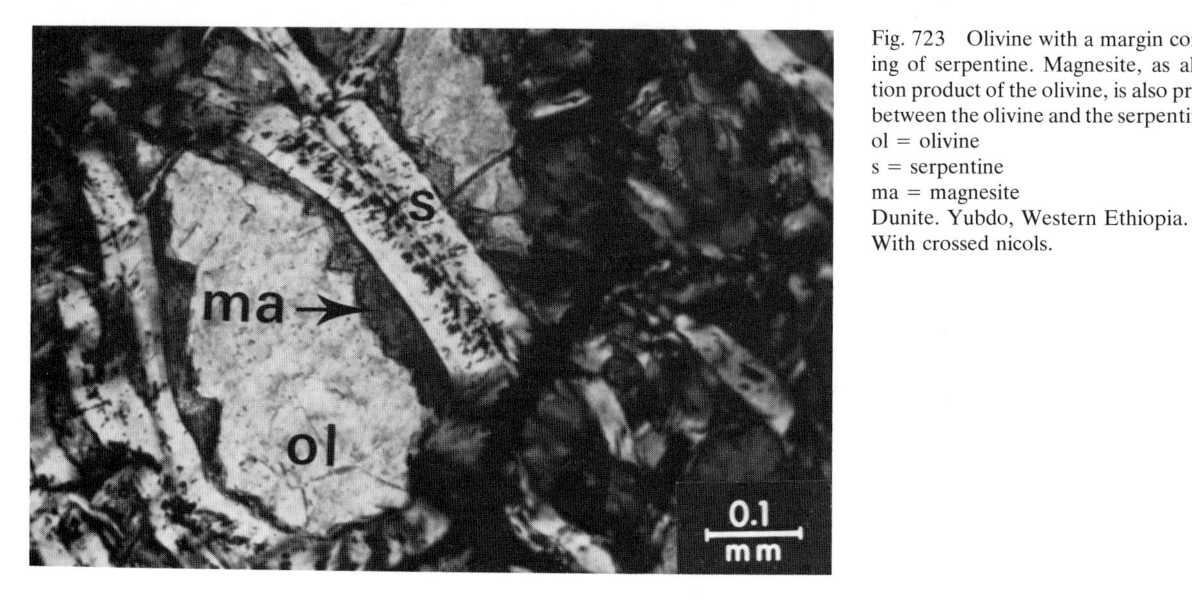

Fig. 723 Olivine with a margin consisting of serpentine. Magnesite, as alteration product of the olivine, is also present between the olivine and the serpentine.
ol = olivine
s = serpentine
ma = magnesite
Dunite. Yubdo, Western Ethiopia.
With crossed nicols.

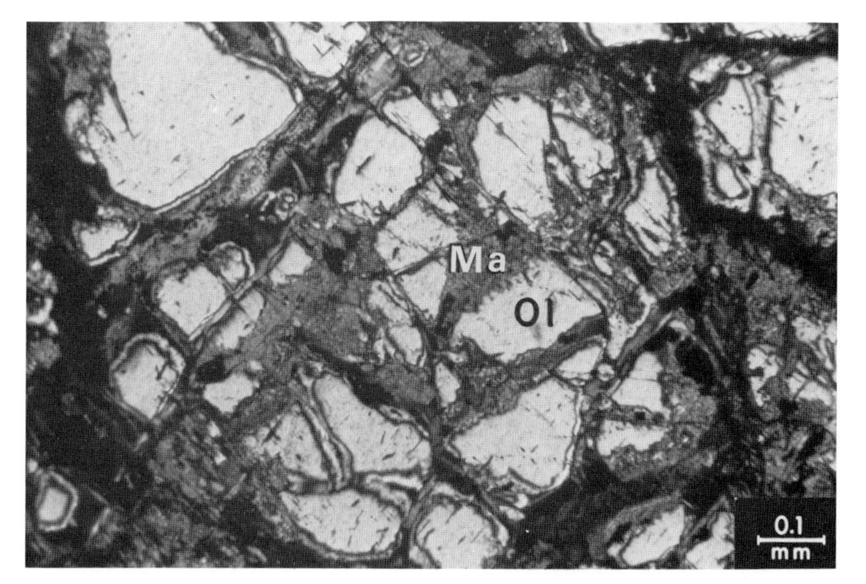

Fig. 724 Olivine grains (Ol) with marginal alteration consisting of magnesite (Ma).
This alteration product also follows cracks of the olivine.
Dunite. Yubdo, Western Ethiopia.
With crossed nicols.

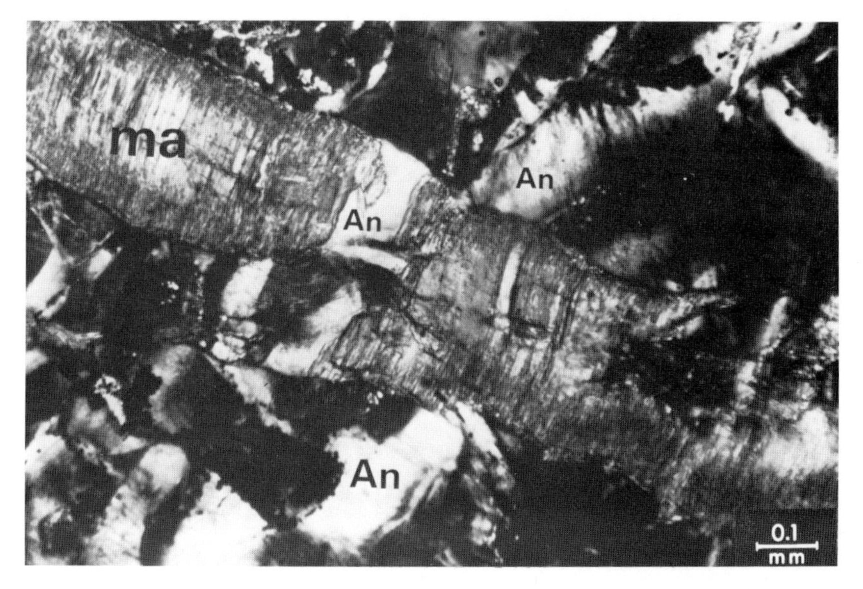

Fig. 725 An intimate association of antigorite with magnesite veinlets.
An = antigorite
ma = magnesite
Mandoudi, Euboea, Greece.
With crossed nicols.

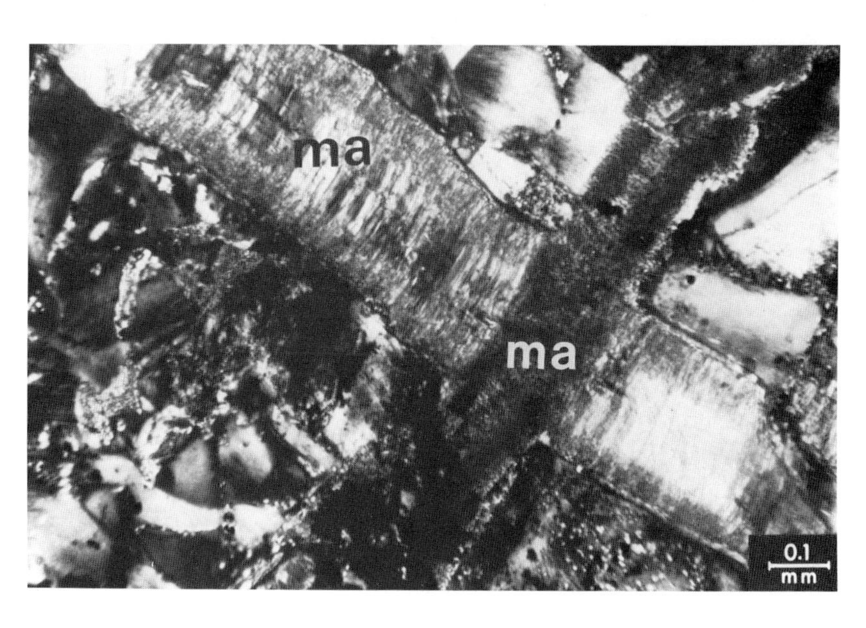

Fig. 726 Intersecting veinlets of magnesite.
ma = magnesite
Mandoudi, Euboea, Greece.
With crossed nicols.

Fig. 727a Birbirite (altered dunite due to Mg leaching) with a stock-work of magnesite veinlets.
Mandoudi, Euboea, Greece.
1/3 natural size.

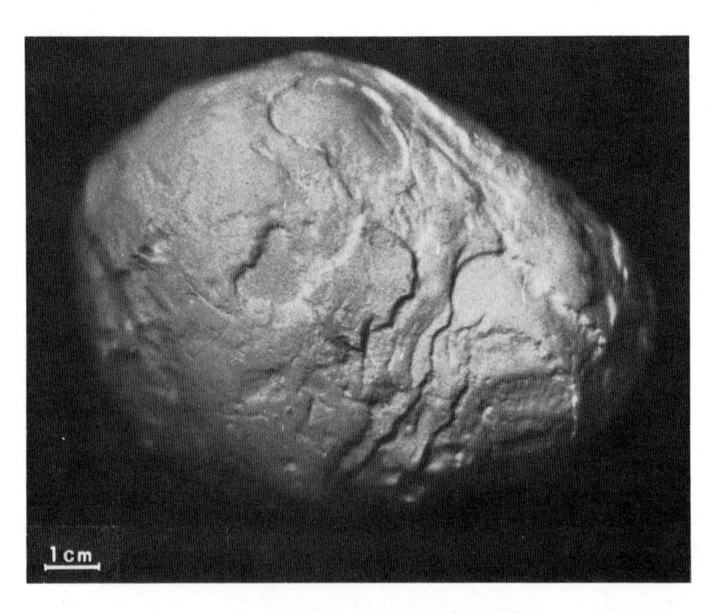

Fig. 727b Magnesite "potato".
Magnesite concretion found in birbiritised dunite.
Mandoudi, Euboea, Greece.
Sample courtesy of Skalistiris Mining Co. Greece.

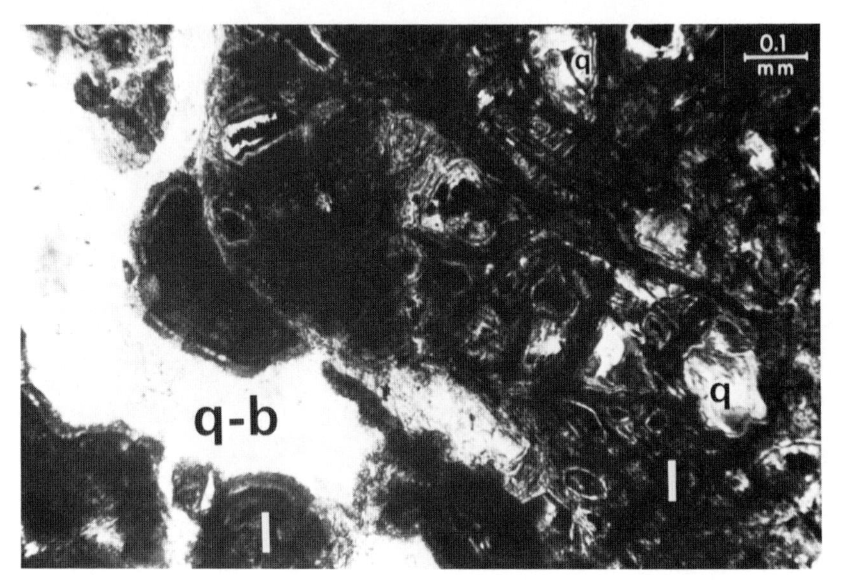

Fig. 728a The initial cellular olivine structure is preserved and maintained by limonite and silica colloform structures, relics of the gradoseparation of olivines after the leaching out of Mg. A band of silica is also present.
l = limonite colloform structures.
q = quartz recrystallization of a gel phase.
q-b = quartz band
Birbiritised dunite. Mandoudi, Euboea, Greece.
With crossed nicols.

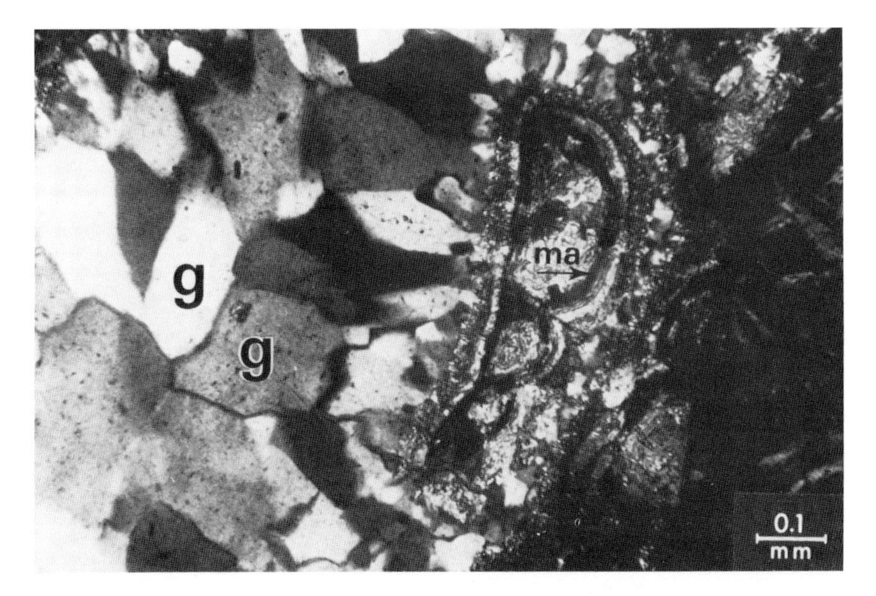

Fig. 728b Granular quartz recrystallised from a collomorph silica phase (SiO_2) remnant of initial olivine, out of which Mg has been leached due to gradoseparation. Also colloform banded magnesite is present.
g = granular quartz
ma = colloform banded magnesite
Birbiritised dunite. Mandoudi, Euboea, Greece.
With crossed nicols.

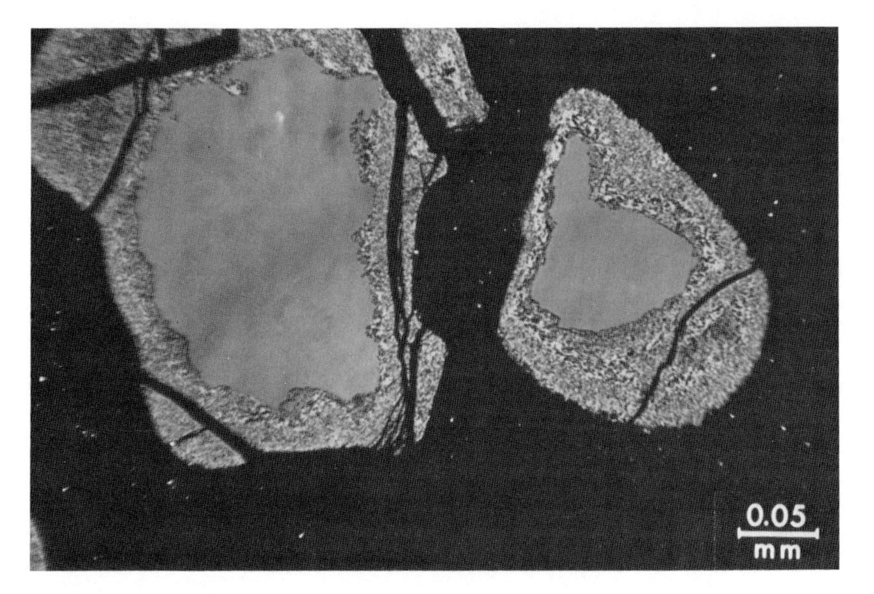

Fig. 729 Marginal alteration of chromite grains in talc (talc metasomatism).
Chromite-ore
Rodiani, N. Greece.
Polished section (oil-immersion, with one nicol).

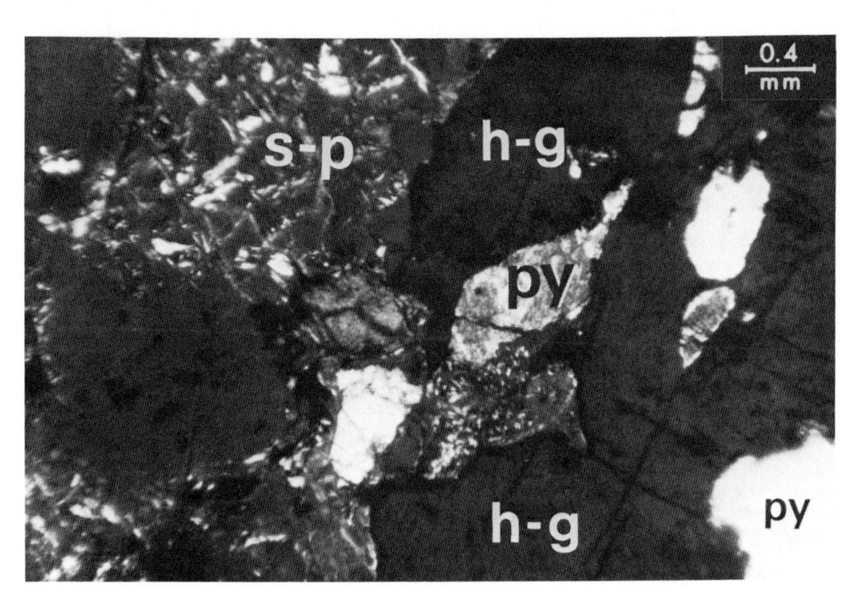

Fig. 730a Hydrogrossular in contact with pyroxenitic mass (partly serpentinised) and partly enclosing pyroxenes or serpentinised pyroxenes.
h-g = hydrogrossular
py = pyroxene
s-p = serpentinised pyroxene
Rodingite. Pagoda, Euboea, Greece.
With crossed nicols.

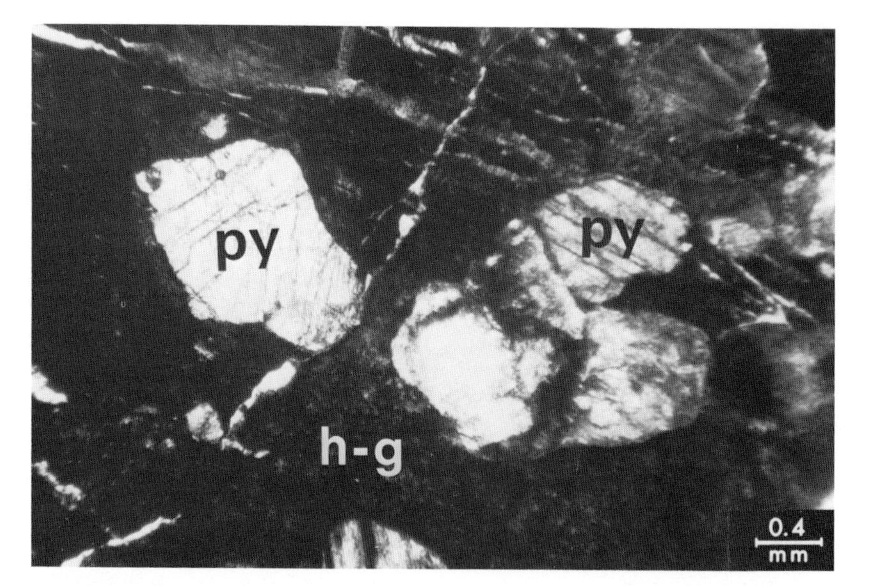

Fig. 730b Pyroxenes (diopside) enclosed by hydrogrossular.
py = pyroxene
h-g = hydrogrossular
Rodingite. Pagoda, Euboea, Greece.
With crossed nicols.

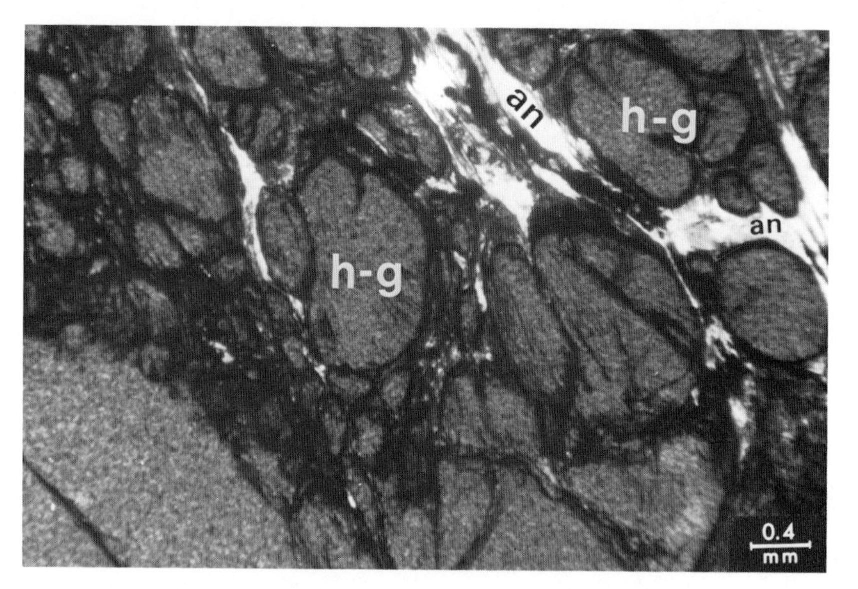

Figs. 732a and 732b Tectonically affected hydro-garnet with rounded forms and outlines (tectonic roundening is common in garnets) in which tectonically mobilised antigorite has been "intruded".
h-g = hydrogarnet
an = antigorite
Maroa, Tigre, Ethiopia.
With half crossed nicols.

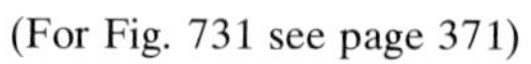
(For Fig. 731 see page 371)

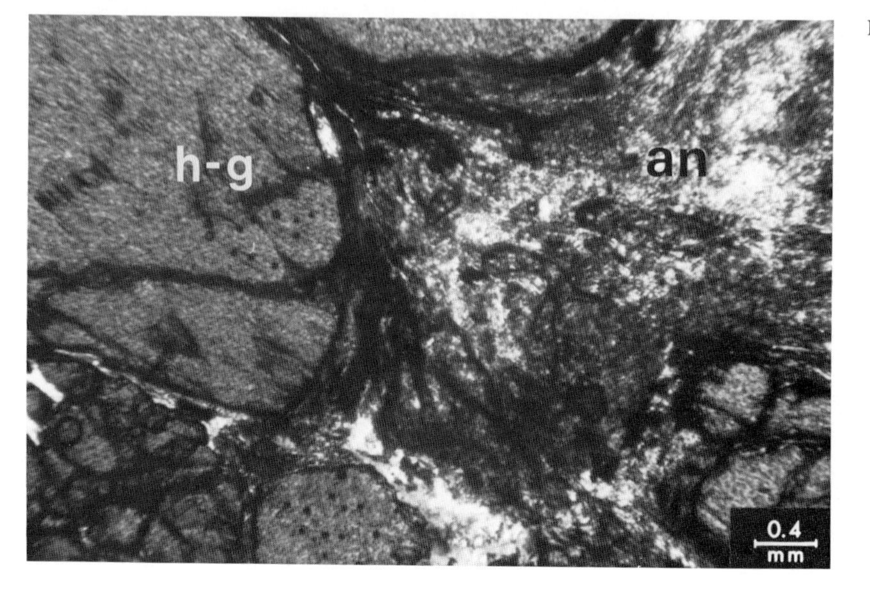

Fig. 732b

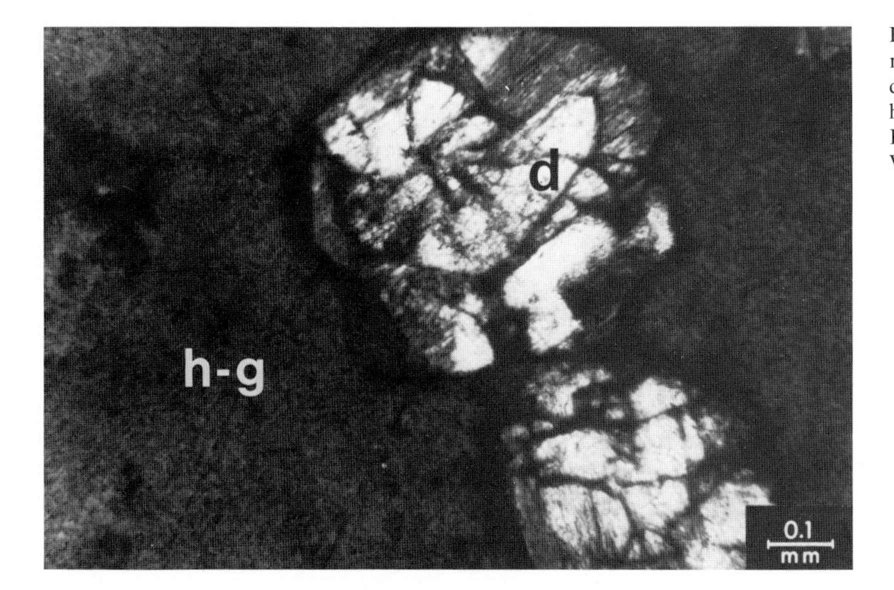

Fig. 731 Corroded diopside as a remnant phase in hydrogrossular.
d = diopside
h-g = hydrogrossular
Rodingite. Pagoda, Euboea, Greece.
With crossed nicols.

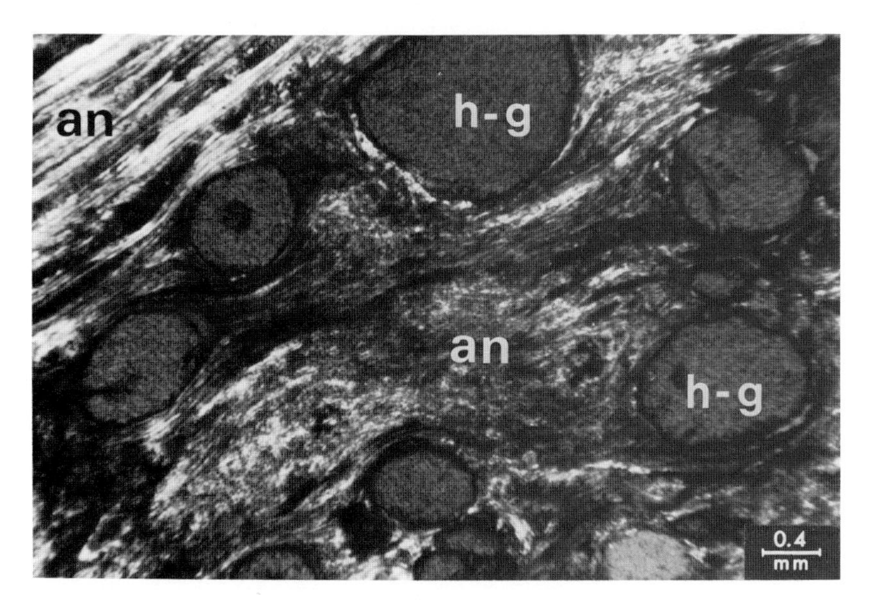

Figs. 733a and 733b The relatively harder hydro-garnet forms "augen" structures in tectonically mobilised antigorite.
h-g = hydro-garnet
an = antigorite tectonically mobilised
Maroa, Tigre, Ethiopia.
With crossed nicols.

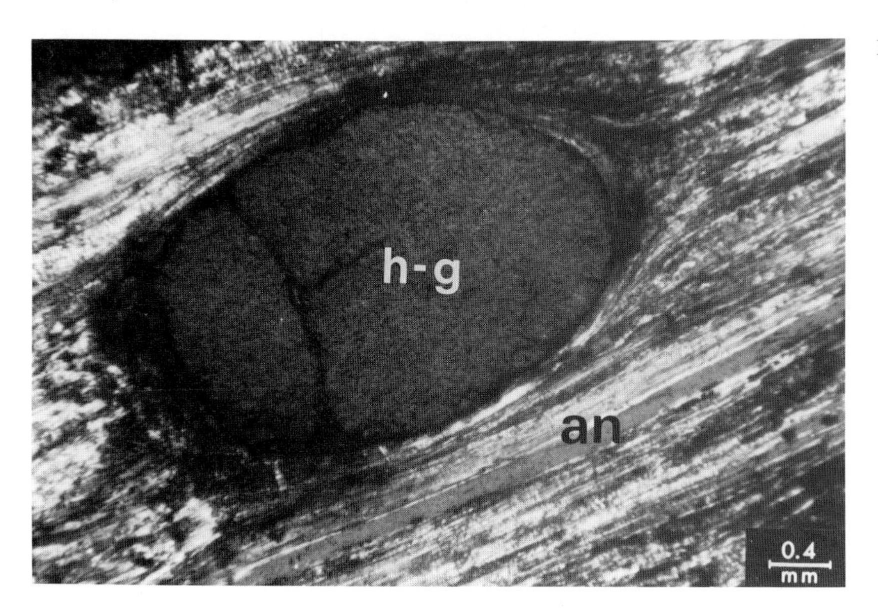

Fig. 733b

References

Abraham, E.M.: 1953. Geology of Sotham township Ont. .Dept.Mines.Ann.Rept. pt.6, pp. 1–36.

Agiorgitis, G.: 1973. Beitrag zur Kenntnis der Eisenooid-Bildung in Nickel- Eisen Vorkommen von Aigaleo, Attiki-Distrikt, Griechenland. Bull of the Geol. Soc. of Greece. Vol. X, No. 2 pp. 248–253.

Agiorgitis, G. and Grundlach, R.: Platinum content in manganese nodules. In Press.

Agiorgitis, G. and Wolf, R.: 1978. Aspects of osmium, ruthenium and iridium contents in some Greek chromites. Chemical Geology, 23, p 267–272.

Akimoto, S.: (1954). J. Geomagnet-Geoelect. 61.

Albandakis, N.: 1974. The nickeliferous iron deposits of Lokris, Euboea, Dr.Thesis, University of Athens. pp. 1–41 (in Greek).

Anhaeusser, C. R.: 1975. Book review "Atlas of the Textural Patterns of Granites, Gneisses and associated Rock types by S. S. Augustithis", Minerals Science and Engineering. South Africa. vol. 7 No. 4.

Arndt, N. T., Naldrett, A. J., and Pyke, D. R.: 1977. Komatiitic and Iron-rich Tholeiitic lavas of Munro Township, Northeast Ontario. Journal of Petrology, Vol. 18, Part 2, pp 319–369.

Aronis, G.: 1953. The bauxite of the Eleusis Mandra Area. Inst. for Geol. and Subsurface Research. *3*, Athens.

Augustithis, S.S.: 1960(a) Über Blastese in Gesteinen unterschiedlicher Genese, (Migmatit, Granit, Metamorphit. (Smirgel), Basalt). Hamburger Beitr. Angew. Mineral. Kristallphysik. Petrogen. 2. 40–68.

Augustithis, S. S.: 1960(b). Alterations of chromite ore – Microscopic observations on Chromite-ores from Rodiani, Greece. N.Jb.Miner. Abb. 94. pp. 890–904.

Augustithis, S. S.: 1962(a). Non-Eutectic, Graphic, Mircographic and Graphic-like "Myrmekitic" Structures and Textures. Beiträge zur Mineralogie und Petrographie. 8. 491–498.

Augustithis, S. S.: 1962(b). Mineralogical and Geochemical changes in the Diagenetic and Post-diagenetic phases of the Ni-Cr-Iron Oolitic Deposits of Larymna/Lokris, Greece. Chem. der Erde. Band 22. pp. 5–17.

Augustithis, S. S.: 1964(a). On the phenomenology of phenocrysts (tecoblasts, zoned phenocrysts, holophenocrystalline rock types) In S. S. Augustithis (Editor) Festband Prof. F. K. Drescher-Kaden. Petrogen. Geochem. Spec. Bull. p. 15–31.

Augustithis, S. S.: 1964(b). Blastic Magnetite with ilmenite Ex-Solutions in Epizonal chlorite, with Anthophyllite Blastic growths. Special Bull of Petrogenesis pub. by the Inst. of Petrog. and Geochemistry. pp. 32–48.

Augustithis, S. S.: 1964(c). Geochemical and ore-microscopic studies of hydrothermal and pegmatitic primary Uranium paragenes. Nova Acta Leopoldina. N.F. No. 170. Band 28. Leipzig, pp. 1–94.

Augustithis, S. S.: 1965. Mineralogical and Geochemical Studies of the Platiniferous Dunite-Birbirite Pyroxenite Complex of Yubdo/Birbir, W. Ethiopia. Chemie der Erde. Band 24. pp. 159–196.

Augustithis, S. S.: 1967(a). The accessory minerals as relics of a pre-granitic paragneissic phase. Chem. Erde 26 (1): 11–26.

Augustithis, S. S.: 1967(b). On the phenomenology and geochemistry of differential leaching and element agglutination processes. Chem. Geol. 2. 311–329.

Augustithis, S. S.: 1972. Mantle Fragments in Basalt. Bul. Geol. Society of Greece, Vol. VIII, No. 2, pp. 93–101.

Augustithis, S. S.: 1973. Atlas of the Textural Patterns of Granites, Gneisses and Associated Rock Types. Elsevier Scientific Publishing Company – Amsterdam–London–New York. p. 378.

Augustithis, S. S.: 1978. Atlas of the Textural Pattern of Basalts and their genetic Significance. Elsevier pp. 323

Bader, H.: 1961. Apatite und Zircone als sedimentäre Relikte in Metablastitgneisen der Oberpfalz. Neues Jahrb. Mineral., Monatsh. 1961. 169–179.

Barrass, P. F.: 1974. Platinoid Minerals in the gold reefs of the Witwatersrand basin. M. Sc. thesis Durham Univ. U.K.

Barth, T. F. W., 1962. Theoretical Petrology. A Textbook on the Origin and the Evolution of Rocks. John Wiley, New York, N. Y., 2nd ed., 416 pp.

Basta, E. Z.: 1960. Natural and Synthetic Titanomagnetites (the System Fe_3O_4–Fe_2TiO_4–$FeTiO_3$). N. Jb. Miner. Abh. 94. Festband Ramdohr. p. 1017–1048.

Becke, F.: 1908. Über Myrmekit. Tschermaks Mineral Petrogr. Mitt. 27: 377–390.

Berry, L. G.: 1940. Geology of the Languir-Sheraton area; Ont. Dept. Mines. Ann. Rept. vol. 49. Pt. 4. pp. 1–21.

Bilgrami, S.A.: 1963. Serpentine-Limestone contact at Teleri Mohammad Jan, Zhob Valley, West Pakistan. Am. Miner. Vol. 48. pp. 1176–1177.

Bilgrami, S. A. and Howie, R. A.: 1960. The Mineralogy and petrology of a rodingite dike, Hindubagh, Pakistan. Am. Miner. Vol. 45. pp. 781–801.

Boshoff, J. C.: 1942. Upper Zone of the Bushveld Complex at Tanteshoogte: D. Sc. dissertation. Univ. Pretoria.

Bowen, N. L.: 1928. The evolution of the igneous rocks: Princeton Univ. Press, Princeton, New Jersey. p. 334.

Bowen, N. L.: 1940. Progressive metamorphism of siliceous limestone and dolomite. Journ. Geol. Vol. 48. p. 225.

Bowen, N. L. and Schairer, J. F.: 1935. The system MgO–FeO–SiO_2. Amer. Journ. Sci., ser. 5, vol. 29: 197.

Brothers, R. N.: 1964. Petrofabric Analysis of Rhum and Skaergaard Layered Rocks. Journal of Petrology. Vol. 5. part 2, pp. 255–274.

Brown, G. M.: 1956. The layered ultrabasic rocks of Rhum, Inner Hebrides; Roy. Soc. London. Philos. Trans. Se. B.

Bruce, E. L.: 1926. Geology of McArthur, Bartlett, Douglas and Geikie Townships (Redstone River Area) Ont. Dept. Mines. vol. 35, pt. 6. pp. 37–56.

Brueckner, H. K.: 1977. A Crustal Origin of Eclogites and a Mantle Origin for Garnet Peridotites. Strontium Isotopic Evidence from Clinopyroxenes. Contrib. Mineral. Petrol. 60. 1–15.

Brynard, H.J., Villiers, J.P.R. and Viljoen, E.A.: 1976. A mineralogical Investigation of the Merensky Reef at the Western Platinum Mine, near Marikana, South Africa. Economic

Geology. Vol. 71. pp. 1299–1307.
Buddington, A. F.: 1936. Gravity stratification as a criterion in the interpretation of the structure of certain intrusives of the northwestern Adirondecks: 16th Internat. Geol. Cong. Rept. Washington (1933) 347–352.
Buddington, A. F.: 1943. Some petroligical concepts and the interior of the earth. Amer. Miner. 28, p. 119–140.
Buddington, A. F. and Lindsley, D. H.: 1964. Iron-titanium oxide minerals and synthetic equivalents. Journ. Petrolog. p. 310–356.
Cabri, L. J.: 1976. Glossary of Platinum-group minerals. Economic Geology. Vol. 71, No. 7. pp. 1475–1480.
Cabri, L. J. and Harris, D. C.: 1975. Zoning in Os-Ir Alloys and the relation of the Geological and tectonic environment of the source rocks to the bulk. Pt : Pt + Ir + Os ratio for placers. Canadian Mineralogist. Vol. 13. pp. 266–274.
Cameron, E. N.: 1963. Structure and rock sequence of the Critical Zone of Eastern Bushveld Complex: Mineralogical Society of America. Special Paper I. p. 93–107.
Cameron, E. N.: 1964. Chromite deposits of the eastern part of Bushveld Complex: The geology of Some Ore Deposits of Southern Africa, v. 2 pp. 131–168.
Carswell, D. A., Curtis, C. D. and Kanaris-Sotiriou, R.: 1974. Vein Metasomatism in Peridotite at Kalskaret near Tafjord, South Norway. Journal of Petrology. Vol. 15. part 2. pp. 383–402.
Chatterjee, N. D.: 1969. Aus welchen Erdtiefen stammen die diamantführenden Kimberlite? N. Jb. Miner. Mh. H. 7, P 289–305.
Challis, G. A.: 1965. The origin of New Zealand Ultramafic Intrusions – Journal of Petrology. Vol. 6. pp. 323–364.
Chesterman, C. W.: 1960. Intrusive ultrabasic rocks and their metamorphic relationships at Leech Lake Mountain. Mendocino County, California. Rept. XXIst Session Internat. Geol. Congr. Copenhagen, pt. xiii, 28, 259–83.
Chetelat, E.: 1947. La genèse et l'évolution des gisements de nickel de la Nouvelle-Calédonie. Bull. Soc. Geol. France. Ser. 5, 17. pp. 105–160.
Chyi, L. L. and Crocket, J. H.: 1976. Partition of Platinum, Palladium, Iridium and Gold among coexisting Minerals from the Deep Ore Zone, Strathcona Mine, Sudbury, Ontario. Economic Geology. Vol. 71. pp. 1196–1205.
Coertze, F. J.: 1958. Intrusive relationships and ore-deposits in the western part of the Bushveld Igneous Complex; Trans. geol. Soc. S. Afr. v. LXI. p. 387–392.
Cogulu, E. and Vuagnat, M.: 1965. Sur l'existance de rodingites dans les serpentinites des environs de Mihali ççik (Vilayet d'Eskisehir, Turquie). Schw. Miner. Petr. Mitt. Bd. 45. pp. 17. Zürich.
Coleman, P. J.: 1970. Geology of the Solomon and New Hebrides Islands, as part of the Melanesian re-entrant, South West Pacific. Pacific. Sci. 24. 289–314.
Coleman, R. G.: 1971. Plate tectonic emplacement of upper mantle peridotites along continental edges. J. Geophys. Res. 76. 1212–1222.
Coleman, R. G.: 1977. Ophiolites. Ancient Oceanic Lithosphere? Springer-Verlag. Berlin, Heidelberg, New-York. pp. 229.
Coleman, R. G., Irwin, W. P.: 1974. Ophiolites and ancient continental margins. In: The Geology of Continental-Margins. Burk, C. A., Drake, C. L. (eds.) New York: Springer. pp. 921–931.
Coleman, R. G. and Peterman, Z. E.: 1975. Oceanic Plagiogranite. J. Geophys. Res.gi". 1099–1108.
Collerson, K. D., Jesseau, C. W. and Bridgwater, D.: 1976. Contrasting types of bladed olivine in ultramafic rocks from the Archean of Labrador. Can. J. Earth Sci. Vol. 13. pp. 442–450.
Cooper, J. R.: 1936. Geology of the Southern half of the Bay of Islands Igneous Complex: Dept. Nat. Res. (Geol. Section) of Newfoundland and Bull. 4. 1–62.
Cousins, C. A.: 1973(a). Notes on the geochemistry of the platinum group elements. Trans. Geol. Soc. S. Africa. 76, 1, 77–81.
Cousins, C. A.: 1973(b). Platinoid in the Witwatersrand system. J. S. Afr. 76. 1. pp. 77–81.
Cousins, C. A. and Kinloch, E. D.: 1976. Some Observations on Textures and Inclusions in Alluvial Platinoids. Economic Geology. Vol. 71, pp. 1377–1393.
Craig, J. R. and Kullerud, G.: 1969. Phase Relations in the Cu-Fe-Ni-S System and their applications to Magmatic Ore Deposits. Magmatic Ore-Deposits Symposium. The Economic Geology Publishing Co. Lancaster, Penna. pp. 344–358.
Crncěvic, S., Grčev, K., Karamata, S. and Simić, I.: 1962. Rodingites from the Ljuboten Serpentine Massif. Referati V. Savetovanja geol. FNRJ. Vol. 2. pp. 87–92. Beograd.
Davies, J. L.: 1971. Peridotite-gabbro-basalt complex in eastern Papua: an over-thrust plate of oceanic mantle and crust. Australian Bur. Min. Resur. Bull. 128. p. 48.
Deer, W. A., Howie, R. A. and Zussmann, J.: 1962. Rock forming Minerals. Longmans. Vol. I. p. 333.
Demou, E. G.: 1972. On Rodingite occurrences in Greece. Bull. Geol. Soc. of Greece. Vol. VIII. pp. 142–162.
De Noske Salverker Av. Bergen (see Dittmann 1959/60)
Dewey, J. F.: 1974. Continental margins and ophiolite obduction: Appalachian Caledonian system: The Geology of Continental Margins. Burk, C. A., Drake, C. L. (eds.) New York: Springer, pp. 933–950.
Dewey, J. F., Bird, J. M.: 1971. Origin and emplacement of the ophiolite suite: Appalachian ophiolites in Newfoundland: J. Geophys. Res. 76, 3179–3206.
Dietz, R. S.: 1963. Alpine serpentinites as oceanic rind fragments. Geol. Soc. Am. Bull. 74. 947–952.
Dittmann, J.: 1959/60. Syntheseversuche zur Metasomatose von Magnesiumsilikaten. Hamb. Beitr. zur Angew. Min-Kristallphysik und Petrog. Band 2, pp. 86–98.
Drescher-Kaden, F. K.: 1948. Die Feldspat-Quartz-Reaktionsgefüge der Granite und Gneise. Springer-Verlag, Heidelberg: 259.
Drescher-Kaden, F. K.: 1961. Olivine-Metasomatose in Carbonatgesteinen aus der Umrandung des Bergeller Granitmassivs (Oberengadin) – Die Naturwissenschaften, 48. Jahrg. H. 8. p. 300.
Drescher-Kaden, F. K.: 1969. Granitprobleme Akademie-Verlag, Berlin. 586 pp.
Drever, H. I. and Johnston, R.: 1957. Crystal growth of forsteritic olivine in magmas and melts: Trans. Roy. Soc. Edin. Vol. LX-III, part 11. 1956–57 (No. 13) pp. 289–315.
Duchesne, J. C.: 1972. Iron-Titanium Oxide Minerals in the Bjerkrem Sogndal Massif, South Western Norway. Journal of Petrology, Vol. 13. Part. I. pp. 57–81.
Ducloz, Ch. and Vuagnat, M.: 1962. Apropos de l'âge des serpentinites de Cuba. Archiv. Sc. Genève, Vol. 15, Fasc. 2. pp. 309–332.
Duparc, L., Molly, E., et Borloz, A.: 1927. Sur la Birbirite, une roche nouvelle. Compt. Rend. Soc. physiq. de Genève 44, No. 3.
Eckstrand, O. R.: 1973: Spinifex, ultramafic flows and nickel deposits in the Abitibi Orogenic Belt. In Volcanism and Volcanic Rocks, G.F.C. Open File 164 pp. 111–128.
Edwards, A. B.: 1949. Natural exsolution intergrowth of magnetite and haematite. Amer. Min. 34. 759–761.
Elsdon, R.: 1971. Crystallization, History of the Upper Layered Series, Kap Edvard Holm, East Greenland. Journal of Petrology Vol. 12. Part 3. pp. 499–521.
Erdmannsdörffer, O. H.: 1950. Die Rolle der Endoblastese in Granit. Fortschr. Mineral. 28: 22–25.
Erhart, H.: 1973. Itinéraires Géochimiques et Cycle Géologique de l'Aluminium. Doin, Éditeurs, S.A. pp. 253.

Ernst, Th.: 1975. Erdmantelprobleme – Bericht und Theorie. Fortschr. Miner. 52(2) 106–140.

Eskola, P.: 1921. On the Eclogites of Norway. Norsk Vidensk, Selsk. Skr. Mat. Nat. Kl. No. 8. 118.

Ewart, A. and Bryan, W. B.: 1972. Petrography and geochemistry of the igneous rocks from Eua, Rongan Islands. Bull. Geol. Soc. 83, 3281–3298.

Ferguson, J. and Botha, E.: 1964. Some aspects of igneous layering in basic zones of the Bushveld Complex. Trans. Geol. Soc. S. Afr. v. LXVI, pp. 1–19.

Ferguson, J. and Currie, K. L.: 1971. Evidence of Liquid Immiscibility in Alkaline Ultrabasic Dikes at Callander Bay, Ontario. Journal of Petrology, Vol. 12, Part 3, pp. 561–85.

Fersman, A. E.: 1955. The Crystallography of diamonds (in Russian). Publication of the Academy of Science, USSR. p. 566.

Fleet, M. E.: 1975. The growth habits of Olivine – A structural Interpretation. Canadian Mineralogist. Vol. 13. pp. 293–297.

Foslie, S.: 1928. Fennia, 50. 1.

Frenzel, G.: 1953. Die Erzparagenese des Katzenbuckels im Odenwald. Heidelberger Beiträge zur Mineralogie und Petrographie. Bd. 3. pp. 409–444.

Frenzel, G.: 1954. Erzmikroskopische Beobachtungen an natürlich erhitzten, insbesondere pseudobookitführenden Vulkaniten. Heidelberger Beiträge zur Mineralogie und Petrographie. Bd. 4. pp. 343–376.

Frenzel, G.: 1954. Über einen ungewöhnlichen Hochtemperaturmagnetkies vom Katzenbuckel im Odenwald. Heidelberger Beiträge zur Mineralogie und Petrographie. Bd. 4. pp. 377–378.

Gass, I. G.: 1963. Is the Troodos Massif of Cyprus a fragment of Mesozoic Ocean Floor? Nature (London) 220. 39–42.

Gass, I. G. and Masson-Smith, D.: 1963. The Geology and Gravity Anomalies of the Troodos Massif, Cyprus. Philosophical Transactions of the Royal Society of London, A. Vol. 255. pp. 417–67.

Gass, I. G., Neary, C. R., Plant, J., Robertson, A. H. F., Simonian, K. O., Smewing, J. D., Spooner, E. T. C. and Wilson, R. A. M.: 1975. Comments on the "Troodos ophiolitic complex was probably formed in an island arc", by A. Miyashiro and subsequent correspondence by A. Hynes and A. Miyashiro, Earth Planet. Sci. Letters, 25. 236–238.

Gass, I. G. and Smewing, J. D.: 1973. Intrusion, extrusion and metamorphism at constructive margins: evidence from the Troodos Massif, Cyprus. Nature, 242, 26–29.

Geijer, P.: 1963. On the source of chromium in micas. Arkiv för Mineralogi och geologi. Utgivet av Kungl. Svenska vetenskapsakademicn Band 3 nr 23. pp. 415–422.

Geikie, A. and Teall, J. I. H.: 1894. On the banded structure of some Tertiary gabbros in the Isle of Skye. Quart. Jour. Geol. Soc. London, I. 645–659.

Genkin, A. D.: 1959. Composition of platinum group minerals in the Noril'sk deposit. Geol. Rudn. Mesterozhdenii, No. 6. pp. 74–84 (in Russian).

Gillery, F. H.: 1959. The x-ray study of synthetic Mg-Al serpentines and chlorites. Amer. Miner. 44, p. 143–152.

Giuseppetti, G., Tadini, C., Veniale, F.: 1962. Ulteriore ritrovamento della Lizardite in prodotti di alterazione di rocce serpentiniche. Estratto dai Rendiconti della Societá Mineralogica Italiana. Anno XIX, Pavia. pp. 1–19.

Glennie, K. W., Boeuf, M. G. A., Hughes-Clarke, M. W., Moody-Stuart, M., Pilaar, W. F. H., Reinhardt, B. M.: 1974. Geology of the Oman Mountains, Part One (Text), Part two (Tables and Illustrations), Part Three (Enclosures) Kon. Nederlands. Geol. Mijb. Gen. Ver. Verh. 31, 423 p.

Goldschmidt, V. M.: 1911. Die Kontaktmetamorphose in Kristianiagebiet. Videnskapsselsk. Skr., 1. Mat. Naturv. Kl. Kristiania. 1 : 226 pp.

Goldschmidt, V.: 1916. Atlas der Krystallenformen. Band III. Carl Winters Universitätbuchhandlung, Heidelberg. p. 244.

Goode, A. D. T.: 1976. Small Scale Primary Cumulus Igneous Layering in the Kalka Layered Intrusion, Giles Complex. Central Australia. Journal of Petrology. Vol. 17. Part 3, pp. 379–397.

Govett, G. J. S. and Pantazis, Th. M.: 1971. Distribution of Cu, Zn, Ni and Co in the Troodos pillow lava series, Cyprus. Inst. Min. Metall. Trans., 80. B 27–E 46.

Green, D. H.: 1959. The Geology of the Beaconsfield District, including the Anderson's Creek Ultrabasic Complex. Records of the Queen Victoria Museum, Launceston, Tasmania. New Series. No. 10, 1–26.

Green, D. H.: 1961. Ultramafic breccias from the Musa Valley, Eastern Papua. Geol. Mag. 98. pp. 4–26.

Green, D. H.: 1964. The Petrogenesis of the High Temperature Peridotite Intrusion in the Lizard Area, Cornwall. Journal of Petrology Vol. 5. Part. 1. pp. 134–188.

Greenwood, H. J.: 1961/62. Synthesis and Stability of Anthophyllite. Annual Report of the Director of the Geophysical Laboratory, 2801, Upton Street, Northwest, Washington. 8. D.C. pp. 85–86.

Greenwood, H. J.: 1963. The Synthesis and stability of Anthophyllite J. Petrol., 4. pp. 317–351.

Griffin, W. L.: 1971. Genesis of Coronas in Anorthosites of the Upper Jotun Nappe, Indre Sogn, Norway. Journal of Petrology, Vol. 12. Part 2. pp. 219–43.

Grout, F. F.: 1918. Internal structures of igneous rocks; their significance and origin; with special reference to the Duluth gabbro. Jour. Geology. 26. 439–457.

Gruenwald, G.: 1976. Sulfides in the Upper Zone of the Eastern Bushveld Complex. Economic Geology. Vol. 71. pp. 1324–1336.

Gübelin, E.: 1948. Gemstone Inclusions. The Journal of Gemmology. Vol. 1. No. 7. p. 7–31.

Gübelin, E.: 1952. Inclusions in Diamonds. The Journal of Gemmology. Vol. III. No. 1–13.

Gübelin, E.J., Meyer, H.O. and Hsiao-Ming Tsai,: 1978 Natur und Bedeutung der Mineral-Einschlüsse im natürlichen Diamanten. Eine Übersicht. Z.Dt. Gemmol. Ges. Jr.g.27, Nr.2 p.61–101.

Hall, A. L.: 1932. The Bushveld igneous complex of the Central Transvaal: Geol. Surv. S. Africa, Mem. 28. 560 p.

Harker, A.: 1904. The Tertiary igneous rocks of Skye.Geol. Survey United Kingdom. Mem. p. 481.

Harker, A.: 1908. The geology of the Small Islands of Invernessshire: Geol. Survey United Kingdom. Mem. (Sheet 60, Scotland) 210 p.

Harker, A.: 1950. Metamorphism. Methuen and Co. Ltd., London: 1–362.

Harris, J. W. and Vance, E. R.: 1974. Studies of the Reaction between Diamond and Heated Kimberlite. Contrib. Mineral. Petrol. 47. 237–244.

Hatch, F. H., Wells, A. K. and Wells, M. K.: 1949. The Petrology of the Igneous rocks. Thomas Murby and Co. London, 469 p.

Heckroodt, R. O.: 1958. Die platinumdraende dunietpyp op Driekop (Oos-Transvaal) en die samastelling van olivinen in die Bosveldstollingskompleks. M. Sc. thesis. Pretoria Univ.

Heflik, W. and Zabiński, W.: 1969. A chromian hydrogrossular from Jordanów, Lower Silesia, Poland. Miner. Mag. Vol. 37. No. 286, pp. 241–243.

Hess, H. H.: 1938. A primary peridotite magma. Am J. Sci. 35, 321–344.

Hess, H. H.: 1955. Serpentines, orogeny and epeirogeny. Spec. Paper Geol. Soc. Am. 62, 391–408.

Hess, H. H.: 1960. Stillwater Igneous Complex, Montana: a quantitative mineralogical study: Geol. Soc. America. Mem. 8. 225 p.

Hess, H. H.: 1960 Stillwater Igneous Complex, Montana, Mem. Geol. Soc. Am. 80. 230 pp.

Hess, H. H.: 1965. Mid-oceanic ridges and tectonics of the sea floor. In: Proc. 17th Symp. Colston Res. Society. Univ. Bristol. London: Butterworths, 317–333.

Hiemstra, S. A. and val Biljon, W. J.: 1959. The geology of the upper Magaliesberg stage and the lower Bushveld Complex in the vicinity of Steelpoort: Geol. Soc. of South Africa, Transactions, V. LXII. p. 239–225.

Himmelberg, G. R. and Ford, A. B.: 1976. Pyroxenes of the Dufek Intrusion, Antartica. Journal of Petrology, Vol. 17. Part. 2. pp. 219–243.

Hoppe, G.: 1962a. Petrogenetisch auswertbare morphologische Erscheinungen an akzessorischen Zirkonen. Neues Jahrb. Mineral. Abh. 98: 35–50.

Hoppe, G.: 1962b. Die akzessorischen Zirkone aus Gesteinen des Bergeller und des Adamello-Massivs. Chem. Erde. 22. 245–263.

Hoppe, G.: 1962c. Die Formen des akzessorischen Apatits. Ber. Geol. Ges. DDR. 7(2). 233–235.

Hoppe, G.: 1963. Die Verwertbarkeit morphologischer Erscheinungen an akzessorischen Zirkonen zu petrographischen Auswertungen. Abh. Dtsch. Akad. Wiss. Berl. Kl. Bergbau Hüttenw. Montangeol. 1. 1–130.

Hoppe, G.: 1964. Morphologische Untersuchungen als Beiträge zu einigen Zirkon-Alterbestimmungen. Neues Jahrb. Mineral. Abh. 102. 89–106.

Hoppe, G.: 1966. Zircone aus Granuliten. Ber. Dtsch. Ges. Geol. Wiss. Beit. Mineral. Lagerstättenf. 11(1). 47–81.

Hopson, C. A., Frano, S. J., Pessagno, E. A. Jr., Mattinson, J. M.: 1975. Preliminary report and geologic guide to the Jurassic ophiolite near Point Sal, Southern California coast. Prep. for 71st Ann. Meet. Cordilleran Section. GSA, Field Trip No. 5. 36 p.

Hotz, P. E.: 1964. Nickeliferous laterites in southwestern Oregon and northwestern California. Econ-Geol. 59. 355–396. 1964.

Hutton, C. O.: 1942. Fuchsite-bearing schists from Dead Horse Creek, Lake Wakatipu, Region, Western Otago. Trans. Royal Soc. New Zealand, 72 pp. 53–68.

Hynes, A.: 1975. Comment on "The Troodos ophiolitic complex was probably formed in an island arc", by A. Miyashiro, Earth Planet. Sci. Letters, 25. 213–216.

Irvine, T. N.: 1959. The ultrabasic complex and related rocks of Duke Island, South eastern Alaska. Ph. D. thesis, California Inst. of Technology.

Irvine, T. N.: 1963. Origin of the ultramafic complex at Duke Island, South eastern Alaska. Spec. Paper, Miner. Soc. Am. 1, 36–45.

Irvine, T. N.: 1965. Sedimentary Structures in igneous intrusions with particular reference to the Duke Island ultramafic complex. Soc. Econ. Palaeontologist and Mineralogist Spec. Publ. 12. 220–32.

Irvine, T. N.: 1967. The Duke Island ultramafic complex, Southeastern Alaska, in Wyllie, P. J. ed. Ultramafic and related rocks New York John Wiley and Sons. 84–96.

Jackson, E. D., 1961. Primary textures and mineral associations in the ultramafic zone of the Stillwater Complex, Montana. U.S. Geol. Survey. Prof. Paper 358. 106 p.

Jackson, E. D.: 1961. The ultramafic rocks by cumulus processes. Fortschr. Miner. 48, 111. 128–174.

Jackson, E. D.: 1967. Ultramafic cumulates in the Stillwater, Great Dyke and Bushveld intrusions, in Wyllie, P. J. ed. Ultramafic and related rocks, New York, John Wiley and Sons. 20–38.

Jackson, E. D., Green, H. W. II., Moores, E. M.: 1975. The Vourinos ophiolite, Greece: cyclic units of lineated cumulates overlying harzburgite tectonite. Geol. Soc. Am. Bull., 86. 390–398.

Jacobshagen, V.: 1977. Geodynamic evolution of the Hellenides. Collected Abstracts. VI. Colloquium on the Geology of the Aegean Region, Athens 1977. p. 173.

James, O. B.: 1971. Origin and Emplacement of the Ultrabasic Rocks of the Emigrant Gap area, California. Journal of Petrology. Vol. 12. Part. 3, pp. 523–60.

Kalinin, D. B., Stenina, N. G. and Denickina, N. D.: 1976. Natural and Syntectic Amphibole Asbestos. Geochemistry, Mineralogy, Petrology International Geological Congress (Report of soviet geologists). Published by The Academy of Science of the U.S.S.R. – Science 1976. pp. 210–222 (in Russian).

Keays, R. R. and Davison, R. M.: 1976. Palladium, Iridium, and gold in the Ores and Host Rocks of Nickel Sulfide Deposits in Western Australia. Economic Geology. Vol. 71. pp. 1214–1228.

Kemp, J. F.: 1916. The Mayari iron-ore deposits, Cuba. Am. Inst. Mining Eng. Trans. 51. 3–30.

Kern, H.: 1968. Zur Geochemie und Lagerstätten-Kunde des Chroms und zur Mikroskopie und Genese der Chromerze. Gebrüder Borntraeger – Berlin–Stuttgart. p. 236.

Kilburn, L. C., Wilson, H. D. B., Graham, A. R., Oyura, Y., Coats, C. J. A., and Scoates, R. F. J.: 1969. Nickel Sulfide Ores Related to ultrabasic Intrusions in Canada. Magmatic Ore deposits Symposium. The Economic Geology Publishing Co. Lancaster, Penna. pp. 276–293.

Koehnken, P. J.: 1976. Book Review: Atlas of the Textural Pattern of Granites, Gneisses and Associated Rock Types. Elsevier, 1973. Journal of Sedimentary Petrology. Vol. 46. No. 2. pp. 444–448.

Korzhinskii, D. S.: 1965. Abriss der metasomatischen Prozesse. pp. 117–118. Berlin.

Ktenas, K.: 1916. Research about the metallogenesis of the S. W. Aegeon region. Publication of the University of Athens, 13. pp. 85–129.

Kullerud, G., Yund, R. A., Moh, G. H.: 1969. Phase Relations in the Cu-Fe-S, Cu-Ni-S and Fe-Ni-S Systems. Magmatic Ore deposits. A symposium. The Economic Geology Publishing Company. Lancaster. Penna. pp. 323–343.

Kupferbürger, W., Lombaard, B. V., Wasserstein, B., and Schwellnus, C. M.: 1937. The chromite deposits of the Bushveld Igneous Complex, Transvaal. Geological Survey of S. Africa. Bull. 10. 48 p.

Kurzweil, H.: 1966. Zur Mineralogie Ni-führender Verwitterungs- und sedimentärer Fe-Erze in Lokris, Mittelgriechenland. Berg- und Hüttenmännische Monatshefte. 111 Jahrgang Heft 10 pp. 488–497.

Lappin, M. A.: 1974. Eclogites from the Sunndal-Grubse Ultramafic Mass. Almkovdalen, Norway and the T-P. History of the Almkovdalen Masses. Journal of Petrology, Vol. 15. Part. 3. pp. 567–601.

Lewis, J. D.: 1970. "Spinifex Texture" in a slag, as evidence for its origin in rocks: The Department of Mines, Western Australia, Ann. Rept. pp. 91–95.

Lindley (in Ramdohr and Strunz, 1967).

Lindsley, D. H.: 1961/62. Investigations in the system $FeO-Fe_2O_3 - TiO_2$. Annual Report of the Director of the Geophysical Laboratory, 2801 Upton Street, Northwest, Washington. 8. D.C. p. 100–101.

Lombaard, A. F., 1948. Die geologie van die Bosveldkompleks (augs Bloedrivier: Trans. Geol. Soc. S. Afr. v. LII, p. 343–376).

Lombaard, B. V.: 1934. On differentiation and relationships of the rocks of the Bushveld Igneous Complex. Trans. Geol. Soc. S. Afr. v. XXXVII, p. 5–52.

Lombaard, B. V.: 1956. Chromite and dunite of the Bushveld Igneous Complex. Trans. Geol. Soc. S. Afr. v. LIX, p. 59–76.

Loney, R. A., Himmelberg, G. R. and Coleman, R. G., 1971. Structure and Petrology of the Alpine-type Peridotite at Burro Mountain, California, U.S.A. Journal of Petrology Vol. 12, Part. 2. pp. 245–309.

Maaløe, S. and Aoki, K.: 1977. The Major Element Composition

of the Upper Mantle Estimated from Composition of Lherzolites. Contrib. Mineral. Petrol. 63. 161–173.
Majer, V.: 1960. Rodingit von Cap Bassit (Nordwestsyrien) N. Jb. Miner. Mh. 4. pp. 85–89.
Makris, J., Menzel, H., Zimmermann, H. and Gouin, P.: 1975. Gravity field and crustal structure of north Ethiopia. Afar Depression of Ethiopia: 135–144.
Makris, J.: 1977a "Plate tectonics" versus "Plumes" in the Aegean Tectonics. Collected Abstracts VI Colloquium on the Geology of the Aegean Region, Athens. p. 9.
Makris, J.: 1977b. Geophysical Investigations of the Hellenides. Hamburger Geophysikalische Einzelschriften. Reihe A: Wissenschaftliche Abhandlungen Heft 34. pp. 1–124.
Makšimovic, Z. and Crnković Božica: 1968. Halloysite and Kaolinite formed through the alteration of ultramafic rocks. 23 Intern. Geol. Congress. 14, 95–105 Prague.
Marinos, G.: 1951. The Geology and metallogenesis of the Island of Seriphos. Geol. and Geophysic. Studies of the Institute of Geology and Subsurface Research, Athens. pp. 95–127.
Maxwell, J. C.: 1970. The Mediterranean, ophiolites and continental drift. In: Megatectonics of continents and oceans. Johnson, H., Smith, B. L. (eds.). New Brunswick, New Jersey: Rutgers Univ. pp. 167–193.
Maxwell, J. C.: 1973. Ophiolites – old oceanic crust or internal diapirs? In: Symp. Ophiolites in the Earth's Crust. OVOSCOW: Acad. Sci: U.S.S.R. pp. 71–73.
Maxwell, J. C.: 1974. Early western margin of the United States. In: The Geology of Continental Margins. Burk, C. A., Drake, C. L. (eds.). New York: Springer. pp. 831–852.
McCallum, M. E. and Eggler, D. H.: 1971. Mineralogy of the Sloan diatreme, a kimberlite pipe in northern Larimer County, Colorado, Amer. Mineral. Vol. 50. Nos. 9–10. pp. 1735–49.
McCallum, M. E., Loucks, R. R., Carlson, R. R. and Cooley, E. F.: 1976. Platinum Metals Associated with Hydrothermal Copper Ores of the New Rambler Mine, Medicine Bow, Mountains, Wyoming. Economic Geology. Vol. 71. pp. 1429–1450.
Mehnert, K. R.: 1968. Migmatites. Elsevier Amsterdam, 393 pp.
Mercier, J. C. C. and Nicolas: 1975. Textures and Fabrics of Upper-Mantle Peridotites as Illustrated by Xenoliths from Basalts. Journal of Petrology, Vol. 16. No. 2. pp. 454–487.
Mesorian, H., Juteau, T., Lapierre, H., Nicolas, A., Parrot, J. F., Ricou, L. E., Rocci, G., Rollet, M.: 1973. Idées actuelles sur la constitution, l'origine et l'évolution des assemblages ophiolitiques mésogeens. Bull. Soc. Géol. France 15. 478–493.
Miyashiro, A.: 1973. The Troodos ophiolitic complex was probably formed in an island arc. Earth and Planetary Sci. Letters: Am. Jour, Sci., v. 272, p. 629–656.
Miyashiro, A.: 1975a. Origin of the Troodos and other ophiolites: a reply to Hynes. Earth Planet. Sci. Letters, 25. 217–222.
Miyashiro, A.: 1975b. Origin of the Troodos and other ophiolites; a reply to Moores. Earth Planet. Sci. Letters, 25. 227–235.
Miyashiro, A.: 1975c. Classification, Characteristics and Origin of Ophiolites. Jour. Geol., 83, p. 249–281.
Modesto Montoto San Miguel (1967) Dr. Thesis. Estudio petrológico y petrogenético de (as rocas graniticas de la cadena litoral catalana). Barcelona, p. 1–25.
Mogensen, E.: 1946. A Ferrotitanate ore from Södra Ulvöen. Geol. Foren. Förh. 68. 578–588.
Moore, A. C.: 1973. Studies of Igneous and Tectonic Textures and Layering in the Rocks of the Gosse Pilelutrusion Central Australia. Journal of Petrology, Vol. 14. Part. 1. pp. 49–79.
Moores, E. M.: 1975. Discussion of "Origin of Troodos and other ophiolites: a reply to Hynes". Earth Planet. Sci. Letters, 25, 223–226.
Moores, E. M. and Vine, E. J.: 1971. The Troodos massif, Cyprus and other ophiolites as oceanic crust: evaluation and implications. Royal Soc. (London) Philos. Trans. A., 268, 443–446.
Mposkos, E.: 1977. Scapolite in the Island of Seriphos; notes on the Composition of the granite. Announced in the Greek Geological Society. Nov. 1977.
Müller, P.: 1962. Kalksilikatfelse im Serpentin des Piz Lunghin bei Maloja. Ein Beitrag zur Genese des Rodingite. Chem. der Erde, Vol. 22. pp. 452–464.
Naldrett, A. J. and Cabri, L. J.: 1976. Ultramafic and related mafic rocks: Their Classification and genesis with special reference to the Concentration of Nickel Sulfides and Platinum-group Elements. Economic Geology, Vol. 71, pp. 1131–1158.
Naldrett, A. J. and Mason, G. C.: 1968. Contrasting Archean ultramafic igneous bodies in Dundonald and Clergue Townships. Ontario; Can. J. Earth. Sci. Vol. 5. pp. 111–143.
Nesbitt, R. W.: 1971. Skeletal Crystal Forms in the ultramafic rocks of the Yilyarn Block, Western Australia, Spec. Publ. 3. pp. 331–347.
Newhouse, W. H.: 1936. Opaque oxides and sulphides in common igneous rocks. Bull. Geol. Soc. Amer. 47. 1–52.
Niggli, P.: 1948. Gesteine und Minerallagerstätten. Verlag Birkhäuser Basel. I. Band. p. 539.
Noble, J. A. and Taylor, H. P. Jr.: 1960. Correlation of the Ultramafic Complexes of Southeastern Alaska with those of the other parts of North America and the World. Rep. 21st Int. Geol. Congr. pt. 13, 188–97.
Ödman, O.: 1932. Mineragraphic Study of the opaque minerals in the lavas from Mt. Elgon, British East Africa. Geol. Fören. Förh. Stockholm. 54. 285–304.
Oen, J. S.: 1968. Magnesium-metasomatism in basic hornfelses near Farminhão, viseu district (Northern Portugal), Chem. Geol. 3. 249–279.
Oliver, R. L. and Nesbitt, R. W.: 1972. Metamorphic Olivine in ultrabasic rocks from Western Australia. Contr. Mineral. and Petrol. 36, 335–342.
Oliver, R. L. and Ward, M.: 1971. A petrological study of serpentinous rocks associated with nickel sulphide mineralization at Pioneer, Western Australia. J. Geol. Soc. Australia. Spec. Publ. No. 3. 311–319.
Ottemann, J. and Augustithis, S. S.: 1967. Geochemistry and origin of "platinum-nuggets" in lateritic covers from ultrabasic rocks and birbirites of W. Ethiopia. Mineralium Deposita. Vol. I. p. 260–277.
Panagos, A. and Ottemann, J.: 1966. Chemical differentiation of chromite grains in nodular-chromite from Rodiani (Greece). Mineralium Deposita 1. 72–75.
Pantazis, Th. M.: 1973. Contribution to the petrology, metallogeny and geochemistry of the ophiolitic complex of Troodos, Cyprus. Dr. habil. Thesis, University of Athens. (In Greek).
Pantazis, Th. M.: 1977. Major oxide geochemistry and origin of the Troodos ophiolitic complex. (In print) Contrib. to Miner. and Petrol. Springer-Verlag, Germany.
Paraskevopoulos, G.: 1970. Rodingite in Serpentiniten von NW-Thessalien, Griechenland. Publication of the Univ. of Athens. Vol. 2. pp. 533–550.
Paulitch, P.: 1953. Olivinkornregelung und Genese des chromitführenden Dunits von Anghida auf der Chalkidike. Tschermaks min. und petr. Mitteilungen. Band 3. Heft 2. pp. 158–166.
Pavlicová, M.: 1963. Über die Kristallisation des Magnetites (Beitrag zur Kristallisation der Plutonite I) – Symposium Problems of postmagmatic ore-deposition, I. Prag. pp. 578–583.
Pavlicová, M.: 1964. Epidotbildung in tonalitischen und verwandten Gesteinen (Beitrag zur Kristallisation der Plutonite II) – Casopis pro mineralogi a geologi Roč 9. č 3, p. 317–21.
Peoples, J. W.: 1936. Gravity stratification as a criterion in the interpretation of the structure of the Stillwater Complex, Montana. 16th Internat. Geol. Cong. Rept. Washington (1933) 353–360.
Petrascheck, W. E.: 1939. Die Chromerzlagerstätten der öst-

lichen Rhodopen in Bulgarien. Zeitschrift für praktische Geologie. 47. 1939. Heft 4. pp. 61–67.
Petrascheck, W. E.: 1957. Die genetischen Typen der Chromerzlagerstätten und ihre Aufsuchung. Zeitschrift für Erzbergbau und Metallhüttenwesen, Band X (1957) Heft 6. pp. 1–10.
Petrascheck, W. E.: 1958. Zur Geologie der chromführenden Ophiolithe der Osttürkei. Bulletin of the Mineral Research and Exploration Institute of Turkey, No. 50. pp. 1–13.
Petrascheck, W. E.: 1959. Intrusiver und Extrusiver Peridotitmagmatismus im Alpinotypen Bereich. Geologische Rundschau Band 48. pp. 205–217.
Petrascheck, W. E.: 1966. Die metamorphe Chromit-Lagerstätte des Mont Ahito in Togo. Zeitschrift für Erzbergbau und Metallhüttenwesen. Band XIX (966) Heft 11. pp. 569–573.
Pieruccini, R.: 1962. Einige in der Theorie und Praxis gewonnene geochemische Erkenntnisse über das Verhalten des Wassers bei der Umwandlung von vulkanischen Tuffen und Sedimentgesteinen. Ber. Geol. Ges. DDR. Erfurt. Sonderh. I. 117–144.
Pirsson, L. V. and Weed, W. H.: 1895. Bull. Geol. Soc. Amer. vi (1895). p. 389.
Pitcher, W. S.: 1974. Review: Augustithis, S. S. Atlas of the textural pattern of Granites, gneisses and associated rock types. Elsevier, 1973. Published Sept. 1974. Mineralogical Magazine. London.
Poldervaart, A.: 1955, 1956. Zircon in rocks, 1. Sedimentary rocks: 2. Igneous rocks. Am. J. Sci. 253. 433–461. 254. 521–554.
Prest, V. K.: 1950. Geology of the Keith-Muskego townships area; Ont. Dept. Mines. Ann. Rept. Vol. 59. pt. 7 pp. 1–44.
Pyke, D. R.: 1970. Geology of Langmuir and Blackstock townships. Ont. Dept. Mines. Geol. Rept. 85. 65 p.
Pyke, D. R., Naldrett, A. J. and Eckstrand, O. R.: 1973. Archean ultramafic flows in Munro townships, Ontario. Geol. Soc. America. Bull. Vol. 84. pp. 955–978.
Qaiser, M. A., Akhter, S. M., Khan, A. H., 1970. Rodingite from Natanji Sar, Dargai ultramafic Complex, Malakand, West Pakistan. Miner. Mag. Vol. 37. No. 290. pp. 735–739.
Raal, F.: 1965. The transition between the Main and Upper Zone of the Bushveld Complex in the Western Transvaal. M. Sc. thesis. Univ. of Pretoria.
Ragan, D. M.: 1963. Emplacement of the Twin Sisters dunite, Washington Am. J. Sc. 261. 549–65.
Ragan, D. M.: 1967. The Twin Sisters dunite, Washington, in Wyllie, P. J. ed., ultramafic and related rocks, New York, John Wiley and Sons. 160–6.
Raleigh, C. B.: 1965. Structure and Petrology of an alpine peridotite on Cypress Island, Washington U.S.A. Beitr. Miner. Petrog. 11. 719. 41.
Ramdohr, P.: 1960. Die Erzmineralien und ihre Verwachsungen. Akademie Verlag. Berlin. p. 846.
Ramdohr, P., and Strunz, H.: 1967. Lehrbuch der Mineralogie. Ferdinand Enke Verlag Stuttgart. p. 820.
Remy, H.: 1961. Grundriß der Anorganischen Chemie, II Auflage. Akademische Verlagsgesellschaft. Geest und Portig K-G. Leipzig. pp. 328.
Reynolds, C. D., Havryluk, I., Saleh Bastaman, Soepomo Atmowidjojo: 1973. The exploration of the nickel laterite deposits in Irian Barat, Indonesia. Geol. Soc. Malaysia. Bull. 6. 309–323.
Reynolds, D. L.: 1947. The association of basic "fronts" with granitization Sci. Prog. 35. (138): 205–236.
Rickwood, P. C. and Mathias, M.: 1970. Diamondiferous Eclogite Xenoliths in Kimberlite – Lithos, 3 pp. 223–235.
Rosenbusch, H.: 1898 (1923). Elemente der Gesteinslehre. E. Schweizerbart'sche Verlagsbuchhandlung Stuttgart, p. 779.
Ross, C. S., Foster, M. D. and Myers, A. T.: 1954. Origin of dunites and of olivine rich inclusions in basaltic rocks. Amer. Min., Vol. 39. 693.
Routhier, P.: 1952. Les gisements de fer de la Nouvelle-Caledonie. 19th Intern. Geol. Cong. Symposium sur les gisements de fer du monde 11. 567–587.
Roy, D. M. and Roy, R. An experimental study of the formation and properties of synthetic serpentine and related layer silicate minerals. Amer. Miner. 32, p. 957–975.
Ruckmick, J. C. and Noble, J. A.: 1959. Origin of the ultramafic complex at Union Bay, South- eastern Alaska. Bull. geol. Soc. Am. 70. 981–1019.
San Miguel Arribas, A.: 1956. Caracteristicas estructurales del granite de la Costa Brava y su significación petrogenética. Publicado en la revista Estudios Geologicos. Números 29 y 30 (Tomo XII). C.S.I.C. Madrid. pp. 95–134.
Satterly, J.: 1951. Geology of Munro townships. Ont. Dept. Mines. Ann. Rept. Vol. 60. pt. 8. pp. 1–60.
Scharlau, T. A.: 1972. Petrographische und petrologische Untersuchungen in der Haupt-Zone des östlichen Bushveld-Komplexes, District Groblersdal/Transvaal, Südafrikanische Republik. Dr. Thesis. Univ. Frankfurt. pp. 124.
Schellmann, W.: 1964. Zur lateritischen Verwitterung von Serpentinit. Gel. Jb. 81. pp. 645–678.
Schneiderhöhn, H.: 1958. Die Erzlagerstätten der Erde. Stuttgart. Gustav Fischer, Vol. 1.
Schubert, W.: 1977. Reaktionen im alpinotypen Peridotitmassiv von Ronda (Spanien) und seinen parallelen Schmelzprodukten. Contrib. Mineral. 62. 205–220.
Schwellnus, J. S. I., 1956. The basal position of the Bushveld Igneous complex and the adjoining metamorphosed sediments in the north eastern Transvaal; D. Sc. dissertation. Univ. of Pretoria.
Sederholm, J. J.: 1910. Die regionale Umschmelzung (Anatexis) erläutert an typischen Beispielen. Compt. Rend. Int. Geol. Congr., 11 me, Stockholm. 1910. 573–586.
Sederholm, J. J.: 1926. On migmatites and associated Precambrian rocks of South western Finland. 2 Bull. Comm. Geol. Finl. 107.
Seeliger, E.: 1975. Was sind Lamprophyre? Sonderdruck aus "Der Aufschluss" Jahrgang 26, Heft 7/1975. p. 4–5.
Siegel, W.: 1954. Mineralogische Untersuchung der Eisenerze von Karditsa, Neo-Kokkinon und Skyros. Institute for Geology and subsurface Research. Athens.
Singewald, J. T.: 1913. Titaniferous iron ores of the U.S. U.S. Dep. of Mines. Bull. 64.
Skinner, B. J. and Peck, D. L.: 1969. An Immiscible Sulfide Melt from Hawaii. Magmatic ore-deposits Symposium. The Economic Geology Publishing Co. Lancaster, Penna. pp. 310–322.
Skunakis, D.: 1977. The presence and genesis of perwoskite in the chromite deposits of Rodiani (Kozani), N. Greece. Announced in the meetings of the Greek Geological Society.
Sobolev, N. V., Kužetsova and Zyuzin, N. I.: 1968. The Petrology of Grospydite Xenoliths from the Zagadochnaya Kimberlite Pipe in Yakutia. Journal of Petrology. Vol. 9. Part. 2. pp. 253–80.
Spathi, K.: 1966. Report: Microtectonic Studies of Chromites of Vourinon (Kozani), Greece. Pub. of the Geological Survey of Greece, pp. 1–6.
Stumpfl, E. F.: 1962. Some aspects of the genesis of platinum deposits. Econ. Geol. Vol. 57. pp. 619–623.
Stumpfl, E. F.: 1974. The genesis of Platinum deposits: Further thoughts. Minerals Science and Engineering. pp. 120–141.
Stumpfl, E. F. and Tarkian, M.: 1976. Platinum Genesis: New Mineralogical Evidence. Economic Geology. Vol. 71. pp. 1451–1460.
Suzuki, J.: 1954. On the Rodingitic Rocks within the serpentine Masses of Hokkaido. J. Fac. Sci. Hokkaido. Univ. Ser. IV. Geol. and Min. Vol. 8. pp. 419–430.
Taliaferro, N. L.: 1943. The Franciscan-Knoxville problem. Bull. Amer. Assoc. Petroleum Geologist, 27. 109–219.

Taubeneck, W. H. and Poldervaart, A.: 1960. Geology of the Elkorn Mountains, north eastern Oregon, Part. 2. Willow Lake intrusion: Geol. Soc. America. Bull., 71. pp. 1295–1322.
Taylor, H. P.: 1967. The zoned ultramafic complexes of South eastern Alaska, in Wyllie, P. J. ed. Ultramafic and related rocks, New York, John Wiley and Sons. 97–118.
Taylor, H. P. Jr. and Noble, J. A.: 1960. Origin of the Ultramafic complexes in South eastern Alaska. Rep. 21st Int. Geol. Congr. 13. 175–87.
Thayer, T. P.: 1946. Preliminary chemical correlation of chromite with the containing rocks. Econ. Geol. 41, 202–18.
Thayer, T. P.: 1960. Some critical differences between Alpine-type and Stratiform Peridotite-Gabbro Complexes. 21st Intern. Geol. Congress, Part XIII p. 247–259, Copenhagen.
Thayer, T. P.: 1963 Flow-layering in alpine peridotite-gabbro complexes. Miner. Soc. Am. Spec. Paper 1. 55–61.
Thayer, T. P.: 1967. Chemical and Structural relations of ultramafic and feldspathic rocks in Alpine intrusive complexes, in Wyllie, P. J. ed. Ultramafic and related rocks, New York, John Wiley and Sons, 222–38.
Thayer, T. P.: 1969. Peridotite-gabbro complexes as keys to petrology of mid-oceanic ridges. Geol. Soc. Am. Bull. 80. 1515–1522.
Thodoropulos, D.: 1971. The origin of the bauxites of Mandra Eleusis. Tech. Chronica. Vol. 9. (Sep. 1971) pp. 603–607.
Tilley, C. E.: 1947. The dunite-mylonites of St. Paul's Rocks (Atlantic) Amer. Journ. Sci., vol. 245. 483.
Tremblay, L. P.: 1950. Piedmont map-area. Abitibi Country, Quebec, Geol. Sur. Canada. Mem. 253.
Van der Merwe, M. J.: 1976. The Layered Sequence of the Potgietersrus Limb of the Bushveld Complex. Economic Geology. Vol. 71. pp. 1337–1351.
Van Zyl, J. P.: 1960. Die petrologie van die Merenskyrif en geassosieerde gesteentes in n aantal boorgate en mynprofiele op Swartklip 988, Rustenburg. M. Sc. thesis University of Potchefstroom.
Veniale, F.: 1962. Un minerale del gruppo del serpentino con caratteristiche della varietá lizardite e morfologia tubulare (S. Margherita Staffora, Appenino Pavese). Periodo di Mineralogia. Anno XXX, n, 2–3. Roma, pp. 307–332.
Verhoogen, J.: 1962. Distribution of titanium between silicates and oxides in igneous rocks. Amer. Journ. Sci., 260. 211–220.
Vermaak, C. F.: 1976. The Merensky Reef – Thoughts on its environment and Genesis. Economic Geology. Vol. 71. pp. 1270–1298.
Vgenopoulos, A.: 1977. Studium der Paragenese Hornblende-Granat-Biotit und den Pegmatiteinschlüssen Westthrakiens/Griechenland. Proceeding of the Academy of Athens. Vol. 52. pp. 199–211.
Viljoen, M. J. and Viljoen, R. P.: 1969. Evidence for the existence of a mobile extrusive peridotitic magma from the Komati formation of the Onverwacht group; Geol. Soc. S. Africa, Spec. Publ. 2. Upper Mantle Project. pp. 87–112.
Vincent, E. A. and Phillips, R.: 1954. Iron titanium minerals in layered gabbros of the Skaergaard intrusion. East Greenland. Geochemica Acta. 6, 1–26.
Vletter, R. de: 1955. How Cuban nickel ore was formed – a lesson in laterite genesis. Engin. Mining J. 156. 84–87.
Vuagnat, M.: 1965. Remarques sur une inclusion rodingitique de l'Alpe Champatsch (Basse-Engadine). Eclogae Geologicae Helvetiae. Vol. 58. No. 1. pp. 443–448.
Vuagnat, M., Pusztaszeri, I. L.: 1964. Ophisphérites et rodingites dans diverse serpentinites des Alpes. Schw. Min. Petr. Mitt. Bd. 44. pp. 12–15. Zürich.
Wager, L. R.: 1958. Beneath the Earth's Crust (Presidential address to Section C, British Association, Glasgow, 1958), The Advancement of Science, 58. 15 p.
Wager, L. R.: 1960. The major element variation of the layered series of the Skaergaard Intrusion and a re-estimation of the average composition of the hidden layered series and of the successive residual magmas. Jour. Petrology. 1. pt. 3. 364–398.
Wager, L. R.: 1963. The mechanism of adcumulus growth in the layered series of the Skaergaard intrusion. Mineral. Soc. America. Spec. Paper. 1, 1. 9.
Wager, L. R.: 1968. Rhythmic and Cryptic Layering in Mafic and ultramafic Plutons. Basalts, Vol. 2. Interscience Publishers. John Wiley & Sons. Inc. pp. 573–622.
Wager, L. R. and Brown, G. M.: 1951. A note on rhythmic layering in the ultrabasic rocks of Rhum: Geol. Mag. 88, 166–168.
Wager, L. R., Brown, G. M. and Wadsworth, W. J.: 1960. Types of Igneous Cumulates. Journ. Petrology, 1. 73–85.
Wager, L. R. and Brown, G. M.: 1968. Layered Igneous Rocks, Edinburgh. Oliver and Boyd. 588 pp.
Wager, L. R. and Deer, W. A.: 1939. The Petrology of the Skaergaard intrusion. Kangerdlugssuaq. East Greenland, 105. No. 4. Pt. 3. 35 p.
Wagner, P. A.: 1929. The platinum deposits and mines of South Africa. Edinburgh, Oliver and Boyd.
Warren, C. R.: 1918. On the microstructure of certain titanic iron ores. Econ. Geol. 13. 419–446.
Weeks, W. F.: 1956. A Thermochemical study of equilibrium relations during metamorphism of siliceous carbonate rocks. Journ. Geol. Vol. 64. p. 245.
Wegmann, C. E.: 1935. Zur Deutung der Migmatite. Geol. Rundsch. 26. 307.
Wegmann, C. E.: 1935. Über einige Fragen der Tiefentektonik. Geol. Rundsch. 26. 449.
Wells, A. K.: 1952. Textural features of some Bushveld norites. Miner. Mag. v. 29. p. 913–924.
Willemse, J.: 1964. "A brief outline of the geology of the Bushveld Igneous Complex" in "The Geology of some ore deposits of Southern Africa". II. pp. 91–128. Geol. Soc. S. Africa.
Willemse, J.: 1969. The Geology of the Bushveld Igneous Complex, the Largest Repository of Magmatic Ore Deposits in the World. Magmatic ore deposits (Symposium). The Economic Geology Publishing Co. Lancaster, Penna. 1969.
Willemse, J. and Bensch, J. J.: 1964. Inclusions of original carbonate rocks in gabbro and norite of the eastern part of the Bushveld Complex. Trans. Geol. Soc. S. Africa. v. LXVII. p. 1–87.
Wilson, H. D. B., Kilburn, L. C., Graham, A. R. and Ramlal, K.: 1969. Geochemistry of some Canadian Nickeliferous Ultrabasic Intrusions. Magmatic ore deposits Symposium. The Economic Geology Publishing Co. Lancaster, Penna. 1969. pp. 294–309.
Wilson, R. A. M.: 1959. The geology of the Xeros, Troodos area. Cyprus Geol. Survey. Dept. Mem. I. 1959. p. 184.
Wyatt, M.: 1954. Zircons as provenance indicators. Am. Mineral. 39. 983–990.
Yoder, H. S.: 1952. The MgO-Al_2O_3-SiO_2-H_2O system and the related metamorphic facies. Am. Jour. Sci. Bowen vol. p. 569–627.
Yoder, H. S. Jr. and Tilley, C. E.: 1962. Origin of basalt magmas, an experimental study of natural and synthetic rock systems. Jour. Petrology. 3. 342–532.
Zimmerman, J. Jr.: 1972. Emplacement of the Vourinos ophiolitic complex, northern Greece. Geol. Soc. Am. Mem. 132. 225–239.

Author Index

Subject Index to the Text Part

Subject Index to the Illustrations